D1219504

FIBONACCI AND LUCAS NUMBERS WITH APPLICATIONS

PURE AND APPLIED MATHEMATICS

A Wiley-Interscience Series of Texts, Monographs, and Tracts

A complete list of the titles in this series appears at the end of this volume.

FIBONACCI AND LUCAS NUMBERS WITH APPLICATIONS

THOMAS KOSHY
Framingham State College

A Wiley-Interscience Publication

JOHN WILEY & SONS, INC.

New York · Chichester · Weinheim · Brisbane · Singapore · Toronto

This book is printed on acid-free paper. ♾

Published simultaneously in Canada.

For ordering and customer service, call 1-800-CALL WILEY.

Library of Congress Cataloging-in-Publication Data:
Koshy, Thomas.
Fibonacci and Lucas numbers with applications/Thomas Koshy.
p. cm. — (Pure and applied mathematics: a Wiley-Interscience series of texts, monographs, and tracts)
"A Wiley-Interscience publication."
Includes bibliographical references and index.
ISBN 0-471-39969-8 (cloth : alk. paper)
1. Fibonacci numbers. 2. Lucas numbers. I. Title II. Pure and applied mathematics (John Wiley & Sons: Unnumbered)

QA246.5.K67 2001
512′.72—dc21 2001017506

Printed in the United States of America.

10 9 8 7 6 5 4 3 2

To

Suresh, Neethu, and Sheeba

CONTENTS

PREFACE

> Man has the faculty of becoming completely absorbed in one subject, no matter how trivial and no subject is so trivial that it will not assume infinite proportions if one's entire attention is devoted to it.
>
> —Tolstoy, *War and Peace*

The Twin Shining Stars

The Fibonacci sequence and the Lucas sequence are the two shining stars in the vast array of integer sequences. They have fascinated both amateurs and professional mathematicians for centuries, and they continue to charm us with their beauty, their abundant applications, and their ubiquitous habit of occurring in totally surprising and unrelated places. They continue to be a fertile ground for creative amateurs and mathematicians alike.

This book grew out of my fascination with the intriguing beauty and rich applications of the twin sequences. It has been my long-cherished dream to study and to assemble the myriad properties of both Fibonacci and Lucas numbers, developed over the centuries, and to catalog their applications to various disciplines in an orderly and enjoyable fashion.

An enormous amount of information is available in the mathematical literature on Fibonacci and Lucas numbers; but, unfortunately, most of it is widely scattered in numerous journals, so it is not easily accessible to many, especially to non-professionals. In this book, I have collected and presented materials from a wide range of sources, so that the finished volume represents, to the best of my knowledge, the largest comprehensive study of this area to date.

Although many Fibonacci enthusiasts know the basics of Fibonacci and Lucas numbers, there are a multitude of discoveries about properties and applications that may be less familiar. Fibonacci and Lucas numbers are also a source of great fun; teachers and professors often use them to generate excitement among students, who find that the sequences stimulate their intellectual curiosity and sharpen their mathematical skills, such as pattern recognition, conjecturing, proof techniques, and problem-solving.

Audience

This book is intended for a wide audience. College undergraduate and graduate students often opt to study Fibonacci and Lucas numbers because they find them challenging and exciting. Often many students propose new and interesting problems in periodicals. It is certainly delightful that students often pursue Fibonacci and Lucas numbers for their senior and master's theses.

High school students have enjoyed exploring this material for a number of years. Using Fibonacci and Lucas topics, students at Framingham High School in Massachusetts, for instance, have published many of their Fibonacci and Lucas discoveries in *Mathematics Teacher*.

I have also included a large amount of advanced material to challenge mathematically sophisticated enthusiasts and professionals in such diverse fields as art, biology, chemistry, electrical engineering, neurophysiology, physics, and music. It is my hope that this book will serve them as a valuable resource in exploring new applications and discoveries, and advance the frontiers of mathematical knowledge.

Organization

In the interest of manageability, the book is divided into forty-seven short chapters. Most conclude with numeric and theoretical exercises for Fibonacci enthusiasts to explore, conjecture, and confirm. I hope that the exercises are as exciting for readers as they are for me. Where the omission can be made without sacrificing the essence of development or focus, I have omitted some of the long, tedious proofs of theorems. The solutions to all odd-numbered exercises are given in the back of the book.

Salient Features

Salient features of this book include: a user-friendly, historical approach; a nonintimidating style; a wealth of identities, applications, and exercises of varying degrees of difficulty and sophistication; links to graph theory, matrices, geometry, and trigonometry; the stock market; and relationships to geometry and information from everyday life. For example, works of art are discussed vis-à-vis the *Golden Ratio*, one of the most intriguing irrational numbers.

Interdisciplinary Appeal

The book contains numerous and fascinating applications to a wide spectrum of disciplines and endeavors; These include art, architecture, biology, chemistry, chess, electrical engineering, geometry, graph theory, music, origami, poetry, physics, physiology, psychology, neurophysiology, sewage/water treatment, snow plowing, stock market trading, and trigonometry. Most of the applications are well within the reach of mathematically sophisticated amateurs, although they vary in difficulty and sophistication.

Historical Perspective

Throughout, I have tried to present historical background for the material, and to humanize the discourse by giving the name and affiliation of every contributor to the field, as well as the year of contribution. My apologies to any discoverers whose names or affiliations are missing; I would be pleased to hear of any such inadvertent omissions.

Puzzles

The book contains several numeric puzzles based on Fibonacci numbers. In addition, it contains several popular geometric paradoxes, again rooted in Fibonacci numbers, which are certainly a source of excitement and surprise.

List of Symbols

A glossary of symbols follows this preface. Readers can find a list of the fundamental properties from the theory of numbers and the theory of matrices in the Appendix. Those who are curious about their proofs will find them in my forthcoming book on number theory.

I would be delighted to hear from Fibonacci enthusiasts about any possible inadvertent errors. If any reader should have questions, or should discover any additional properties and applications, I would be more than happy to hear about them.

Acknowledgments

I am pleased to take this opportunity to thank a number of people who have helped to improve the manuscript with their constructive suggestions, comments, and support.

I am deeply indebted to the following reviewers for their boundless enthusiasm and input:

N. Gauthier	Royal Military College, Kingston, Canada
H. W. Gould	West Virginia University, Morgantown, West Virginia
Marjorie Johnson	Santa Clara, California
Thomas E. Moore	Bridgewater State College, Bridgewater, Massachusetts
Carl Pomerance	University of Georgia, Athens, Georgia
M. N. S. Swamy	Concordia University, Montreal, Canada
Monte J. Zerger	Adams State College, Alamosa, Colorado

Thanks also to Angelo DiDomenico of Framingham High School, who read an early version of the work and offered valuable suggestions; to Margarite Roumas for her superb editorial assistance; to Kevin Jackson-Mead for preparing the canonical prime factorizations of the Lucas numbers L_{51} through L_{100} and for proofreading the entire work; and to Madelyn Good at the Framingham State College Library, who tracked down copies of many articles cited in the references, and many books on loan from other institutions; and to the staff at Wiley, especially, Steve Quigley and Heather Haselkorn for their cooperation, support, and encouragement, and their confidence in the project.

THOMAS KOSHY
tkoshy@frc.mass.edu

Framingham, Massachusetts
June 2001

LIST OF SYMBOLS

SYMBOL	MEANING
$\mathbb{N}$	the set of positive integers 1, 2, 3, 4, ...
$\mathbf{W}$	the set of whole numbers 0, 1, 2, 3, ...
$\mathbb{R}$	the set of real numbers
$\{s_n\}_1^\infty = \{s_n\}$	sequence with general term s_n
$\sum_{i=k}^{i=m} a_i = \sum_{i=k}^{m} a_i = \sum_{k}^{m} a_i$	$a_k + a_{k+1} + \cdots + a_m$
$\sum_{i \in I} a_i$	sum of the values of a_i as i runs over the various values in I
$\sum_{P} a_i$	sum of the values of a_i, where i satisfies certain properties P
$\sum_{i} \sum_{j} a_{ij}$	$\sum_{i}(\sum_{j} a_{ij})$
$\prod_{i=k}^{i=m} a_i = \prod_{i=k}^{m} a_i = \prod_{k}^{m} a_i$	$a_k a_{k+1} \cdots a_m$
$n!$ (n factorial)	$n(n-1)\cdots 3 \cdot 2 \cdot 1$, where $0! = 1$
$\|x\|$	the absolute value of x
$\lfloor x \rfloor$ (the floor of x)	the greatest integer $\leq x$
$\lceil x \rceil$ (the ceiling of x)	the least integer $\geq x$
PMI	the principle of mathematical induction
a div b	the quotient when a is divided by b
a mod b	the remainder when a is divided by b
$a \mid b$	a is a factor of b
$a \nmid b$	a is not a factor of b
$\{x, y, z\}$	the set consisting of the elements x, y, and z

$\{x\|P(x)\}$	the set of elements with property $P(x)$
$\|A\|$	the number of elements in set A
$A \cup B$	the union of sets A and B
$A \cap B$	the intersection of sets A and B
(a, b)	the greatest common factor of a and b
$[a, b]$	the least common factor of a and b
$A = (a_{ij})_{m \times n}$	$m \times n$ matrix A whose *ij*th element is a_{ij}
$\|A\|$	the determinant of matrix A
$\in$	belongs to
$\approx$	is approximately equal to
$\equiv$	is congruent to
$\therefore$	therefore
∞	the infinity symbol
$\blacksquare$	the end of a proof, solution, or an example
$\overline{AB}$	the line segment with endpoints A and B
AB	the length of the line segment $\overline{AB}$
$\overleftrightarrow{AB}$	the line containing the points A and B
$\overrightarrow{AB}$	the ray AB
$\angle ABC$	the angle ABC
$\overleftrightarrow{AB} \parallel \overleftrightarrow{CD}$	the lines $\overleftrightarrow{AB}$ and $\overleftrightarrow{CD}$ are parallel
$\overleftrightarrow{AB} \perp \overleftrightarrow{CD}$	the lines $\overleftrightarrow{AB}$ and $\overleftrightarrow{CD}$ are perpendicular
RHS	right-hand side
LHS	the left-hand side

FIBONACCI AND LUCAS NUMBERS WITH APPLICATIONS

LEONARDO FIBONACCI

Leonardo Fibonacci, also called Leonardo Pisano or Leonard of Pisa, was the most outstanding mathematician of the European Middle Ages. Little is known about his life except for the few facts he gives in his mathematical writings. Ironically, none of his contemporaries mention him in any document that survives.

Fibonacci (Fig. 1.1) was born around 1170 into the Bonacci family of Pisa, a prosperous mercantile center. ("Fibonacci" is a contraction of "Filius Bonacci," son of Bonacci.) His father Guglielmo (William) was a successful merchant, who wanted his son to follow his trade.

Around 1190, when Guglielmo was appointed collector of customs in the Algerian city of Bugia (now Bougie), he brought Leonardo there to learn the art of computation. In Bougie, Fibonacci received his early education from a Muslim schoolmaster, who introduced him to the Indo-Arabic numeration system and Indo-Arabic computational techniques. He also introduced Fibonacci to a book on algebra, *Hisâb al-jabr w'al-muqabâlah*, written by the Persian mathematician, al-Khowarizmi (ca. 825). (The word *algebra* is derived from the title of this book.)

As an adult, Fibonacci made frequent business trips to Egypt, Syria, Greece, France, and Constantinople, where he studied the various systems of arithmetic then in use, and exchanged views with native scholars. He also lived for a time at the court of the Roman Emperor, Frederick II (1194–1250), and engaged in scientific debates with the Emperor and his philosophers.

Around 1200, at the age of about 30, Fibonacci returned home to Pisa. He was convinced of the elegance and practical superiority of the Indo-Arabic system over the Roman numeration system then in use in Italy. In 1202, Fibonacci published his pioneering work, *Liber Abaci* (*The Book of the Abacus*.) (The word *abaci* here does not refer to the hand calculator called an abacus, but to computation in general.) *Liber Abaci* was devoted to arithmetic and elementary algebra; it introduced the Indo-Arabic numeration system and arithmetic algorithms to Europe. In fact, Fibonacci

Figure 1.1. Fibonacci (*Source*: David Eugene Smith Collection, Rare Book and Manuscript Library, Columbia University.).

demonstrated in this book the power of the Indo-Arabic system more vigorously than in any mathematical work up to that time. *Liber Abaci*'s 15 chapters explain the major contributions to algebra by al-Khowarizmi and another Persian mathematician, Abu Kamil (ca. 900). Six years later, Fibonacci revised *Liber Abaci* and dedicated the second edition to Michael Scott, the most famous philosopher and astrologer at the court of Frederick II.

After *Liber Abaci*, Fibonacci wrote three other influential books. *Practica Geometriae* (*Practice of Geometry*), written in 1220, is divided into eight chapters and is dedicated to Master Domonique, about whom little is known. This book skillfully presents geometry and trigonometry with Euclidean rigor and some originality. Fibonacci employs algebra to solve geometric problems and geometry to solve algebraic problems, a radical approach for the Europe of his day.

His next two books, the *Flos* (*Blossom* or *Flower*) and the *Liber Quadratorum* (*The Book of Square Numbers*) were published in 1225. Although both deal with number theory, *Liber Quadratorum* earned Fibonacci his reputation as a major number theorist, ranked between the Greek mathematician Diophantus (ca. 250 A.D.) and the French mathematician Pierre de Fermat (1601–1665). *Flos* and *Liber Quadratorum* exemplify Fibonacci's brilliance and originality of thought, which outshine the abilities of most scholars of his time.

In 1225 Frederick II wanted to test Fibonacci's talents, so he invited him to his court for a mathematical tournament. The contest consisted of three problems. The first was to find a rational number x such that both $x^2 - 5$ and $x^2 + 5$ are squares of rational numbers. Fibonacci gave the correct answer 41/12: $(41/12)^2 - 5 = (31/12)^2$ and $(41/12)^2 + 5 = (49/12)^2$.

The second problem was to find a solution of the cubic equation $x^3 + 2x^2 + 10x - 20 = 0$. Fibonacci showed geometrically that it has no solutions of the form $\sqrt{a + \sqrt{b}}$, but gave an approximate solution, 1.3688081075, which is correct to nine decimal places. This answer appears in the *Flos* without any explanation.

The third problem, also recorded in the *Flos*, was to solve the following:

> Three people share 1/2, 1/3, and 1/6 of a pile of money. Each takes some money from the pile until nothing is left. The first person then returns one- half of what he took, the second one-third, and the third one-sixth. When the total thus returned is divided among them equally, each possesses his correct share. How much money was in the original pile? How much did each person take from the pile?

Fibonacci established that the problem was indeterminate and gave 47 as the smallest answer. In the contest, none of Fibonacci's competitors could solve any of these problems.

The Emperor recognized Fibonacci's contributions to the city of Pisa, both as a teacher and as a citizen. Today, a statue of Fibonacci stands in a garden across the Arno River, near the Leaning Tower of Pisa.

Not long after Fibonacci's death in about 1240, Italian merchants began to appreciate the power of the Indo-Arabic system and gradually adopted it for business transactions. By the end of the sixteenth century, most of Europe had accepted it. *Liber Abaci* remained the European standard for more than two centuries and played a significant role in displacing the unwieldy Roman numeration system.

2 THE RABBIT PROBLEM

Fibonacci's classic book, *Liber Abaci*, contains many elementary problems, including the following famous problem on rabbits:

> Suppose there are two newborn rabbits, one male and the other female. Find the number of rabbits produced in a year if:
>
> 1) each pair takes one month to become mature;
>
> 2) each pair produces a mixed pair every month, from the second month on; and
>
> 3) no rabbits die during the course of the year.

Suppose, for convenience, that the original pair of rabbits was born on January 1. They take a month to become mature, so there is still only one pair on February 1. On March 1, they are two months old and produce a new mixed pair, a total of two pairs. Continuing like this, there will be three pairs on April 1, five pairs on May 1, and so on. See the last row of Table 2.1.

TABLE 2.1.

Number of Pairs	Jan	Feb	Mar	Apr	May	Jun	Jul	Aug
Adults	0	1	1	2	3	5	8	13
Babies	1	0	1	1	2	3	5	8
Total	1	1	2	3	5	8	13	21

Figure 2.1. Lucas (*Source*: H. C. Williams, *Edouard Lucas and Primality Testing,* New York: Wiley, 1998. Copyright © 1998, reprinted with permission of John Wiley & Sons, Inc.).

FIBONACCI NUMBERS

The numbers in the bottom row are called *Fibonacci numbers*, and the number sequence 1, 1, 2, 3, 5, 8, . . . is the *Fibonacci sequence*. Table A.2 in the Appendix lists the first 100 Fibonacci numbers.

The sequence was given its name in May of 1876 by the outstanding French mathematician François-Edouard-Anatole-Lucas (Fig. 2.1),* who had originally called it "the series of Lamé," after the French mathematician Gabriel Lamé (1795–1870). It is a bit ironic that despite Fibonacci's numerous mathematical contributions, he is primarily remembered for this sequence that bears his name.

*François-Edouard-Anatole-Lucas was born in Amiens, France, in 1842. After completing his studies at the École Normale in Amiens, he worked as an assistant at the Paris Observatory. He served as an artillery officer in the Franco-Prussian war and then became professor of mathematics at the Lycee Saint-Louis and Lycee Charlemagne, both in Paris, and he was a gifted and entertaining teacher. Lucas died of a freak accident at a banquet; his cheek was gashed by a shard that flew from a plate that was accidently dropped; he died from infection within a few days, on October 3, 1891.

Lucas loved computing and developed plans for a computer, but it never materialized. Besides his contributions to number theory, he is known for his four-volume classic on recreational mathematics. Best known among the problems he developed is the *Tower of Brahma.*

The Fibonacci sequence is one of the most intriguing number sequences, and it continues to provide ample opportunities for professional and amateur mathematicians to make conjectures and to expand the mathematical horizon.

The sequence is so important that an organization of mathematicians, *The Fibonacci Association*, has been formed for the study of Fibonacci and related integer sequences. The association was founded in 1963 by Verner E. Hoggatt, Jr. (1921–1980) of San Jose State College (now San Jose State University), California, and Brother Alfred Brousseau (1907–1988) of St. Mary's College in California. The association publishes *The Fibonacci Quarterly*, devoted to articles related to integer sequences.

A close look at the Fibonacci sequence reveals that it has a fascinating property: every Fibonacci number, except the first two, is the sum of the two immediately preceding Fibonacci numbers. (At the given rate, there will be 144 pairs of rabbits on December 1. This can be verified by extending Table 2.1 through December.)

RECURSIVE DEFINITION

This observation yields the following recursive definition of the nth Fibonacci number, F_n:

$$\begin{aligned} &F_1 = F_2 = 1 & & \leftarrow \text{Initial conditions} \\ &F_n = F_{n-1} + F_{n-2} & n \geq 3 \quad & \leftarrow \text{Recurrence relation} \end{aligned} \tag{2.1}$$

We shall formally establish the validity of this recurrence relation shortly.

It is not known whether Fibonacci knew of this relation. If he did, no record exists to that effect. In fact, the first written confirmation of the recurrence relation appeared four centuries later, when the great German astronomer and mathematician Johannes Kepler (1571–1630) wrote that Fibonacci must have surely noticed this recursive relationship. In any case, it was first noticed by the Dutch mathematician Albert Girard (1595–1632).

However, according to P. Singh of Raj Narain College in Bihar, India, Fibonacci numbers and the recursive formulation were known in India several centuries before Fibonacci proposed the problem; they were given by Virahanka (between 600 and 800 A.D.), Gopala (prior to 1135 A.D.), and Hemacandra (about 1150 A.D.). In fact, Fibonacci numbers also occur as a special case of a formula established by Narayana Pandita (1356 A.D.).

The growth of the rabbit population can be displayed nicely in a tree diagram, as Figure 2.2 shows. Each new branch of the "dream-tree" becomes an adult branch in one month and each adult branch, including the trunk, produces a new branch every month.

Table 2.1 shows several interesting relationships among the numbers of adult pairs, baby pairs, and total pairs. To see these relationships, let A_n denote the number of adult pairs and B_n the number of baby pairs in month n, where $n \geq 1$. Clearly, $A_1 = 0$, and $A_2 = 1 = B_1$.

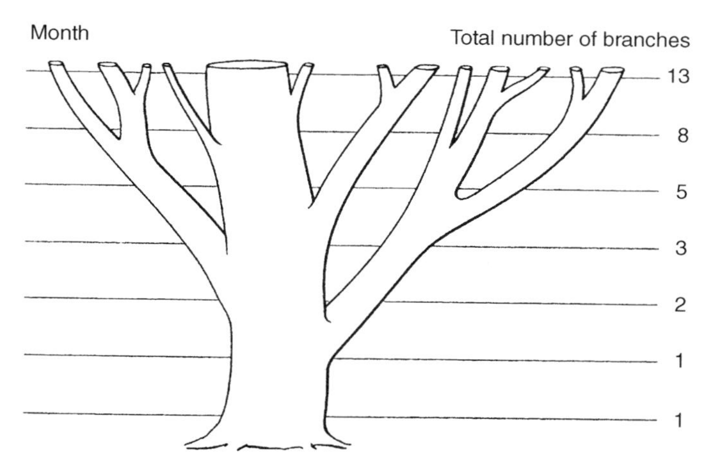

Figure 2.2. A Fibonacci tree.

Suppose $n \geq 3$. Since each adult pair produces a mixed baby pair in month n, the number of baby pairs in month n equals the number of adult pairs in the preceding month, that is, $B_n = A_{n-1}$. Then:

$$\begin{pmatrix}\text{Number of pairs}\\ \text{in month } n\end{pmatrix} = \begin{pmatrix}\text{Number of adult pairs}\\ \text{in month } n-1\end{pmatrix} + \begin{pmatrix}\text{Number of baby pairs}\\ \text{in month } n-1\end{pmatrix}$$

That is,

$$\begin{aligned} A_n &= A_{n-1} + B_{n-1} \\ &= A_{n-1} + A_{n-2} \qquad n \geq 3 \end{aligned}$$

Thus A_n satisfies the same recurrence relation as the Fibonacci recurrence relation (FRR), where $A_2 = 1 = A_3$. Consequently, $F_n = A_{n+1}$, $n \geq 1$.

Notice that:

$$\begin{pmatrix}\text{Total number of pairs}\\ \text{in month } n\end{pmatrix} = \begin{pmatrix}\text{Number of adult pairs}\\ \text{in month } n\end{pmatrix} + \begin{pmatrix}\text{Number of baby pairs}\\ \text{in month } n\end{pmatrix}$$

That is, $F_n = A_n + B_n = A_n + A_{n-1}$, where $n \geq 3$. Thus $F_n = F_{n-1} + F_{n-2}$, $n \geq 3$. This establishes the Fibonacci recurrence relation observed earlier.

Since $F_n = A_{n+1}$, where $n \geq 1$, every entry in row 1, beginning with the second element (February), is a Fibonacci number. In other words, the ith element in row 1 is F_{i-1}, where $i \geq 2$. Likewise, since $B_n = A_{n-1} = F_{n-2}$, where $n \geq 3$, the ith element in row 2 is F_{i-2}, where $i \geq 3$.

The recursive definition of F_n yields a straightforward method for computing it, as Algorithm 2.1 shows.

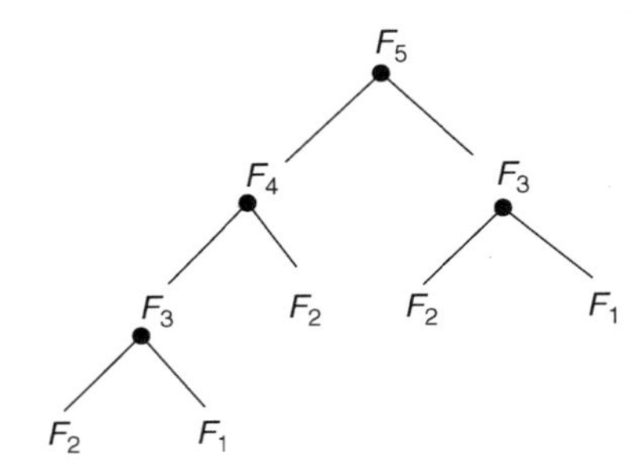

Figure 2.3. Tree diagram of recursive computing of F_5.

```
Algorithm Fibonacci(n)
(* This algorithm computes the nth Fibonacci number
using recursion. *)
Begin (* algorithm *)
  if n = 1 or n = 2 then (* base cases *)
    Fibonacci ← 1
  else                  (* general case *)
    Fibonacci ← Fibonacci(n - 1) + Fibonacci(n - 2)
End (* algorithm *)
```

Algorithm 2.1.

The tree diagram in Figure 2.3 illustrates the recursive computing of F_5, where each dot represents an addition.

Using the recurrence relation (Eq. 2.1), we can assign a meaningful value to F_0. When $n = 2$, Eq. (2.1) yields $F_2 = F_1 + F_0$, that is, $1 = 1 + F_0$, so $F_0 = 0$. This fact will come in handy in our later discussions.

In the case of a nontrivial triangle, it is well known that the sum of the lengths of any two sides is greater than the length of the third side. Accordingly, the FRR can be interpreted to mean that *no three consecutive Fibonacci numbers can be the lengths of the sides of a nontrivial triangle.*

LUCAS NUMBERS

Using the Fibonacci recurrence relation and different initial conditions, we can construct new number sequences. For instance, let L_n be the nth term of a sequence with $L_1 = 1$, $L_2 = 3$, and $L_n = L_{n-1} + L_{n-2}$, $n \geq 3$. The resulting sequence 1, 3, 4, 7, 11, ... is called the *Lucas sequence*, after Edouard Lucas; L_n is the nth term of the sequence. Table A.2 also lists the first 100 Lucas numbers.

We will see in later chapters that L_n and F_n are very closely related, and hence the title of this book. For instance, both L_n and F_n satisfy the same recurrence relation.

FIBONACCI AND LUCAS SQUARES AND CUBES

Of the infinitely many Fibonacci numbers, some have special characteristics. For example, only two distinct Fibonacci numbers are perfect squares, namely, 1 and 144. This was established in 1964 by J. H. E. Cohn of the University of London. In the same year, Cohn also established that 1 and 4 are the only Lucas squares (see Chapter 34).

In 1969, H. London of McGill University and R. Finkelstein of Bowling Green State University proved that there are exactly two distinct Fibonacci cubes, namely, 1 and 8, and that the only Lucas cube is 1.

A UBIQUITOUS FIBONACCI NUMBER AND ITS CONSTANT LUCAS COMPANION

Another Fibonacci number that appears to be ubiquitous is 89.

- Since 1/89 is a rational number, its decimal expansion is periodic:

$$\frac{1}{89} = 0.\overline{01123595505617977528\,0\,(\underset{\uparrow}{89})\,8876404494382022471 91}$$

  The period is 44, and a surprising number occurs in the middle of a repeating block.
- It is the eleventh Fibonacci number, and both 11 (the fifth Lucas number) and 89 are prime numbers. While 89 can be viewed as the $(8+3)$rd Fibonacci number, it can also be looked at as the $(8 \cdot 3)$rd prime.
- Concatenating 11 and 89 gives the number 1189. Since $1189^2 = (1 + 2 + 3 + \cdots + 1681)/2$, it is also a triangular number. Interestingly enough, there are 1189 chapters in the Bible, of which 89 are in the four gospels.
- Eighty-nine is the smallest number to stubbornly resist being transformed into a palindrome by the familiar "reverse the digits and then add" method. In this case, it takes 24 steps to produce a palindrome, namely, 8813200023188.
- $8+9$ is the sum of the four primes preceding 11, and $8 \cdot 9$ is the sum of the four primes succeeding it: $17 = 2+3+5+7$ and $72 = 13+17+19+23$.
- The most recent year divisible by 89 is 1958: $1958 = 2 \cdot 11 \cdot 89$. Notice the prominent appearance of 11 again.
- The next year divisible by 89 is $2047 = 2^{11} - 1$. Again, 11 makes a conspicuous appearance. It is, in fact, the smallest number of the form $2^p - 1$, which is not a prime, where p is of course a prime. Primes of the form $2^p - 1$ are called *Mersenne primes*, after the French Franciscan priest Marin Mersenne (1588–1648), so 2047 is the smallest Mersenne number that is *not* a prime.

- On the other hand, $2^{89} - 1$ is a Mersenne prime; in fact, it is the tenth Mersenne prime, discovered in 1911 by R. E. Powers. Its decimal value contains 27 digits and looks like this:

$$2^{89} - 1 = 6189700196 \cdots 11$$

  The first three digits are significant because that they are the first three decimal digits of an intriguing irrational number we shall encounter in Chapters 20–27. Once again, note the surprising appearance of 11 at the end.
- Multiply the two digits of 89; add its digits again; and their sum is again 89: $(8 \cdot 9) + (8 + 9) = 89$. (It would be interesting to check if there are other numbers that exhibit this remarkable behavior.) Also, $8/9 \approx 0.89$.
- There are only two consecutive positive integers, one of which is a square and the other a cube: $8 = 2^3$ and $9 = 3^2$.
- Square the digits of 89 and add them to obtain 145. Add the squares of its digits again. Continue like this. After eight iterations, we return to 89:

$$89 \to 145 \to 42 \to 20 \to 4 \to 16 \to 37 \to 58 \to 89$$

  In fact, if we apply this "sum the squares of the digits" method to any number, we will eventually attain 89 or 1.
- On 8/9 in 1974, an unfortunate and unprecedented event occurred in the history of the United States—the resignation of President Richard M. Nixon. Strangely enough, if we swap the digits of 89, we get the date on which Nixon was pardoned by his successor, President Gerald R. Ford.

All these fascinating observations about 11 and 89 were made in 1996 by M. J. Zerger of Adams State College, Colorado.

Soon after these Fibonacci curiosities appeared in *Mathematics Teacher*, G. J. Greenbury of England (private communication, 2000) contacted Zerger with two curiosities involving the decimal expansions of two primes:

$$\frac{1}{29} = 0.\overline{0344827586206(89)6551724137931}$$
$$\frac{1}{59} = 0.\overline{01694915254237288135593220220033(89)8305084745762711864406779661}$$

Curiously enough, 89 makes its remarkable appearance in the repeating block of each expansion.

R. K. Guy of the University of Calgary, Canada, in his fascinating book, *Unsolved Problems in Number Theory*, presents an interesting number sequence $\{x_n\}$. It has a quite remarkable and not immediately obvious relationship with 89. The sequence is defined recursively as follows:

$$x_0 = 1$$

$$x_n = \frac{1 + x_0^3 + x_1^3 + \cdots + x_{n-1}^3}{n}$$

For example, $x_0 = 1$, $x_1 = (1 + 1^3)/1 = 2$, and $x_2 = (1 + 1^3 + 2^3)/2 = 5$.

Surprisingly enough, x_n is integral for $0 \le n < 89$, but x_{89} is not.

FIBONACCI AND PRIMES

Zerger also observed that the product $F_6F_7F_8F_9$ is the product of the first seven prime numbers: $F_6F_7F_8F_9 = 13\cdot21\cdot34\cdot55 = 510,510 = 2\cdot3\cdot5\cdot7\cdot11\cdot13\cdot17$. Interestingly enough, 510 is the Dewey Decimal Classification Number for Mathematics.

FIBONACCI AND LUCAS PRIMES

Many Fibonacci and Lucas numbers are indeed primes. For example, the Fibonacci numbers 2, 3, 5, 13, 89, 233, and 1597 are primes, and so are the Lucas numbers 3, 7, 11, 29, 47, 199, and 521. Although it is widely believed that there are infinitely many Fibonacci and Lucas primes, their proofs still remain elusive.

The largest known Fibonacci prime is F_{9311}, and the largest known Lucas prime is L_{14449}. Discovered in 1999 by H. Dubner and W. Keller, they are 1946 and 3020 digits long, respectively. (Chapter 5 discusses a method for determining the number of digits in both F_n and L_n.)

Table A.3 lists the canonical prime factorizations of the first 100 Fibonacci numbers. Lucas had found the prime factorizations of the first 60 Fibonacci numbers before March 1877 and most likely even earlier. Boldface type in the table indicates the corresponding prime factor's first appearance in the list. For instance, the largest prime among the first 100 Fibonacci numbers is F_{83}.

Table A.4 gives the complete prime factorizations of the first 100 Lucas numbers.

CUNNINGHAM CHAINS

A *Cunningham chain*, named after Lt. Col. Allan J. C. Cunningham (1842–1928), an officer in the British Army, is a sequence of primes in which each element is one more than twice its predecessor. Interestingly enough, the smallest six-element chain begins with 89: 89, 179, 359, 719, 1439, 2879.

Are there Fibonacci and Lucas numbers that are one more than or one less than a square? A cube? We shall find the answers shortly.

FIBONACCI AND LUCAS NUMBERS $w^2 \pm 1, w \geq 0$

In 1973, R. P. Finkelstein of Bowling Green State University, Ohio, established yet another curiosity: The only Fibonacci numbers of the form $w^2 + 1$, where $w \geq 0$, are 1, 2, and 5: $1 = 0^2 + 1$, $2 = 1^2 + 1$, and $5 = 2^2 + 1$.

Two years later, Finkelstein proved that the only Lucas numbers of the same form are 2 and 1: $2 = 1^2 + 1$ and $1 = 0^2 + 1$.

In 1981, N. R. Robbins of Bernard M. Baruch College, New York, proved that the only Fibonacci numbers of the form $w^2 - 1$, where $w \geq 0$, are 3 and 8: $3 = 2^2 - 1$ and $8 = 3^2 - 1$. The only such Lucas number is 3.

FIBONACCI AND LUCAS NUMBERS $w^3 \pm 1, w \geq 0$

In the same year, Robbins also determined all Fibonacci and Lucas numbers of the form $w^3 \pm 1$, where $w \geq 0$. There are two Fibonacci numbers of the form $w^3 + 1$, namely, 1 and 2: $1 = 0^3 + 1$ and $2 = 1^3 + 1$. There are two Lucas numbers of the same form: 1 and 2.

There are no Fibonacci numbers of the form $w^3 - 1$, where $w \geq 0$. But there is exactly one such Lucas number, namely, 7: $7 = 2^3 - 1$.

FIBONACCI NUMBERS $(a^3 \pm b^3)/2$

Certain Fibonacci numbers can be expressed as one-half of the sum or difference of two cubes. For example, $1 = (1^3 + 1^3)/2$, $8 = (2^3 + 2^3)/2$, and $13 = (3^3 - 1^3)/2$. In fact, at the 1969 Summer Institute on Number Theory, held at Stony Brook, New York, H. M. Stark of the University of Michigan at Ann Arbor asked: Which Fibonacci numbers have this distinct property? This problem is linked to the finding of all complex quadratic fields with class 2. In 1983, J. A. Antoniadis tied such fields to solutions of certain diophantine equations.

FIBONACCI AND LUCAS TRIANGULAR NUMBERS

A *triangular number* is a positive integer of the form $n(n + 1)/2$. The first five triangular numbers are 1, 3, 6, 10, and 15; they can be represented geometrically, as Figure 2.4 shows.

In 1963, M. H. Tallman of Brooklyn, New York, observed that the Fibonacci numbers 1, 3, 21, and 55 are triangular numbers:

$$1 = \frac{1 \cdot 2}{2}, \qquad 3 = \frac{2 \cdot 3}{2}, \qquad 21 = \frac{6 \cdot 7}{2}, \qquad \text{and} \qquad 55 = \frac{10 \cdot 11}{2}$$

He asked if there were any other Fibonacci number that is also triangular.

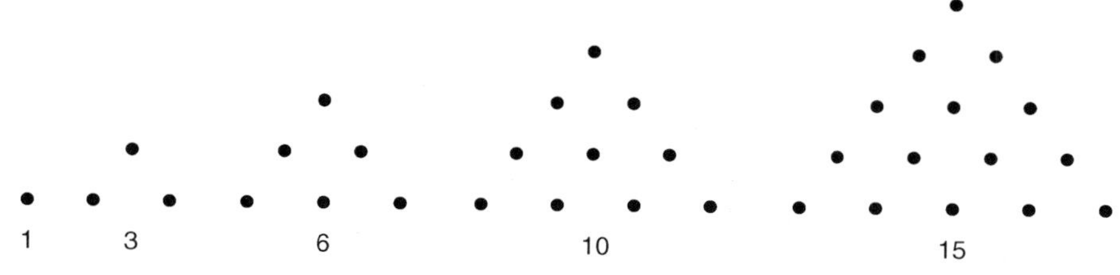

Figure 2.4. The first five triangular numbers.

Twenty-two years later, S. R. Wall of Trident Technical College, South Carolina, established that there are *no* other triangular numbers in the first one billion Fibonacci numbers. In fact, he conjectured that there are no other such Fibonacci numbers.

In 1976, Finkelstein proved that 1, 3, 21, and 55 are the only triangular Fibonacci numbers of the form F_{2n}.

In fact, eleven years later, L. Ming of Chongqing Teachers' College, China, proved conclusively that 1, 3, 21, and 55 are the only Fibonacci triangular numbers. This result is a byproduct of the two following results by Ming:

- $8F_n + 1$ is a perfect square if and only if $n = 0, \pm 1, 2, 4, 8, 10$.
- F_n is triangular if and only if $n = \pm 1, 2, 4, 8, 10$.

Are there Lucas numbers that are also triangular? Obviously, 1 and 3 are. In fact, in 1990, Ming also established that the only such Lucas numbers are 1, 3, and 5778:

$$1 = \frac{1 \cdot 2}{2}, \qquad 3 = \frac{2 \cdot 3}{2}, \qquad \text{and} \qquad 5778 = \frac{107 \cdot 108}{2}$$

FIBONACCI AND THE BEASTLY NUMBER

In 1989, C. Singh of St. Laurent's University in Quebec, Canada, discovered some mystical relationships between the infamous beastly number, 666, and Fibonacci numbers F_n:

- $666 = F_{15} + F_{11} - F_9 + F_1$, where $15 + 11 - 9 + 1 = 6 + 6 + 6$.
- $666 = F_1^3 + F_2^3 + F_4^3 + F_5^3 + F_6^3$, where the sum of the subscripts equals
$$1 + 2 + 4 + 5 + 6 = 6 + 6 + 6$$
- $666 = [F_1^3 + (F_2 + F_3 + F_4 + F_5)^3]/2$.

EXERCISES 2

1. Compute the first 20 Fibonacci numbers.
2. Compute the first 20 Lucas numbers.
3. Determine the value of L_0.
4. Using the FRR (Eq. 2.1), compute the value of F_{-n}, where $1 \leq n \leq 10$.
5. Using Exercise 4, predict the value of F_{-n} in terms of F_n.
6. Compute the value of L_{-n}, where $1 \leq n \leq 10$.
7. Using Exercise 6, predict the value of L_{-n} in terms of L_n.

To commemorate the publication of the maiden issue of the *Journal of Recreational Mathematics*, L. Bankoff of Los Angles published his discovery that $F_{20} - F_{19} - F_{15} - F_5 - F_1 = F_{17} + F_{13} + F_{11} + F_9 + F_7 + F_3$ and that each sum gives the year.

8. Find the year in which the journal was first published.
9. Verify that the sums of the subscripts of the Fibonacci numbers on either side are equal.

Compute the sum $\sum_1^n F_i$ for each value of n.

10. 3
11. 5
12. 7
13. 8
14. Using Exercises 10–13, predict a formula for $\sum_1^n F_i$.

15–18. Compute the sum $\sum_1^n L_i$ for each value of n in Exercises 10–13.

19. Using Exercises 15–18, predict a formula for $\sum_1^n L_i$.

20–23. Compute the sum $\sum_1^n F_i^2$ for each value of n in Exercises 10–13.

24. Using Exercises 20–23, predict a formula for the sum $\sum_1^n F_i^2$.

25–28. Compute the sum $\sum_1^n L_i^2$ for each value of n in Exercises 10–13.

29. Using Exercises 20–23, predict a formula for the sum $\sum_1^n L_i^2$.
30. Verify that $F_{2n} = F_n L_n$ for $n = 3$ and $n = 8$.
31. Verify that $L_n = F_{n-1} + F_{n+1}$ for $n = 4$ and $n = 7$.

Let a_n denote the number of additions needed to compute F_n by recursion:

32. Define a_n recursively.

33. Show that $a_n = F_n - 1, n \geq 1$.
34. Prove that $F_n < 1.75^n$ for every positive integer n (LeVeque, 1962).
35. Show that there are no four distinct Fibonacci numbers in arithmetic progression (Silverman, 1964).
36. Let $I_n = \int_0^1 x^{I_{n-1}} dx$, where $n \geq 2$ and $I_1 = \int_0^1 x\ dx$. Evaluate I_n (Lind, 1965).
37. If $F_n < x < F_{n+1} < y < F_{n+2}$, then $x + y$ cannot be a Fibonacci number (Hoggatt, 1982).

Suppose we introduce a mixed pair of 1-month-old rabbits into a large enclosure on the first day of a certain month. By the end of each month, the rabbits become mature and each pair produces $k-1$ mixed pairs of offspring at the beginning of the following month. (*Note:* $k \geq 2$.) For instance, at the beginning of the second month, there is one pair of 2-month-old rabbits and $k - 1$ pairs of 0-month-olds; at the beginning of the third month, there is one pair of 3-month-olds, $k - 1$ pairs of 1-month-olds, and $k(k - 1)$ pairs of 0-month-olds. Assume the rabbits are immortal. Let a_n denote the average age of the rabbit-pairs at the beginning of the nth month (Filipponi and Singmaster, 1990).

**38. Define a_n recursively.
**39. Predict an explicit formula for a_n.
**40. Prove the formula in Exercise 39.
41. (For those familiar with the concept of limits) Find $\lim_{n\to\infty} a_n$.

FIBONACCI NUMBERS IN NATURE

Come forth into the light of things,
let Nature be your teacher.
—William Wordsworth

Interestingly enough, the amazing Fibonacci numbers occur in quite unexpected places in nature.

FIBONACCI AND THE EARTH

Do Fibonacci numbers also appear elsewhere? Zerger observed that the equatorial diameter of the earth in miles is approximately the product of two alternate Fibonacci numbers, and that this in kilometers is approximately the product of three consecutive Fibonacci numbers:

$$55 \cdot 144 = 7920\,\text{miles} \quad \text{and} \quad 89 \cdot 144 = 12{,}816\,\text{kilometers}$$

For the curious-minded, the earth's diameter, according to *The 2000 World Almanac and Book of Facts*, is 7928 miles and 12,756 kilometers; the polar diameter is 7901 miles. The diameter of Jupiter, the largest planet, is 11 times that of the earth.

FIBONACCI AND ILLINOIS

In 1992, Zerger discovered some astonishing occurrences of Fibonacci numbers in relation to the state of Illinois:

- Illinois was admitted to the Union on the 3rd of December.
- Illinois is the fifth largest state, according to the 1990 census.
- Illinois' name consists of 8 letters.
- Illinois is the thirteenth state, when the states are arranged alphabetically.
- Illinois was the twenty-first state admitted to the Union. The postal abbreviation IL is formed with the ninth and twelfth letters: $9 + 12 = 21$.
- Interstate 55 begins in Chicago and roughly follows the 89th parallel to New Orleans.

FIBONACCI AND FLOWERS

The number of petals in many flowers is often a Fibonacci number. For instance, count the number of petals in the flowers pictured in Figure 3.1. Enchanter's nightshade has two petals, iris and trillium three, wild rose five, and delphinium and cosmos eight. Most daisies have 13, 21, or 34 petals; there are even daisies with 55 and 89 petals. Table 3.1 lists the Fibonacci number of petals in an assortment of flowers. Although some plants, such as buttercup and iris, always display the same number of petals, some do not. For example, delphinium blossoms sometimes have 5 petals and sometimes 8 petals, and some Michaelmas daisies have 55 petals, while some have 89 petals.

The cross section of an apple reveals a pentagonal shape with five pods. The starfish, with five limbs, also exhibits a Fibonacci number (see Fig. 3.2).

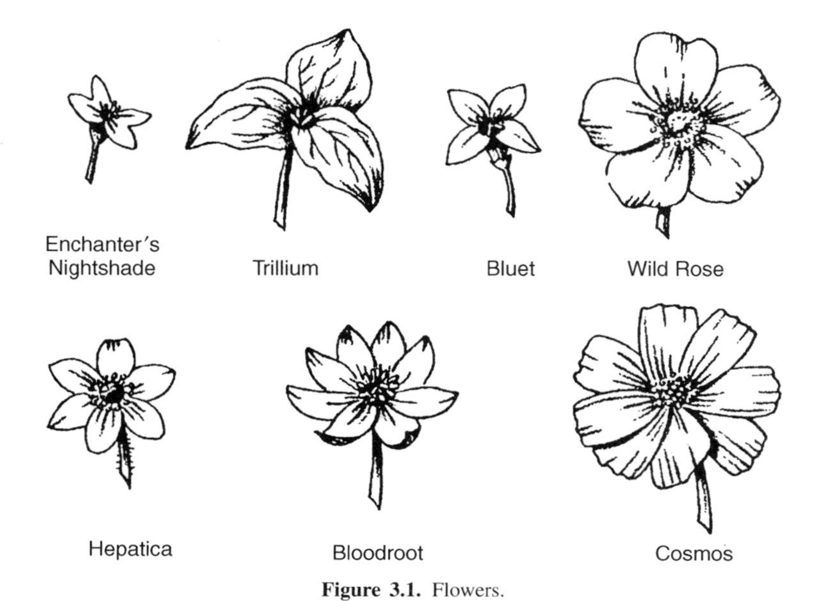

Figure 3.1. Flowers.

TABLE 3.1.

Plant	Number of Petals
Enchanter's nightshade	2
Iris, lilly	3
Buttercup, columbine, delphinium, larkspur, wall lettuce	5
Celandine, delphinium, field senecio, squalid senecio	8
Chamomile, cineraria, corn marigold, double delphinium, globeflower	13
Aster, black-eyed Susan, chicory, doronicum, helenium, hawkbit	21
Daisy, gailliardia, plantain, pyrethrum, hawkweed	34

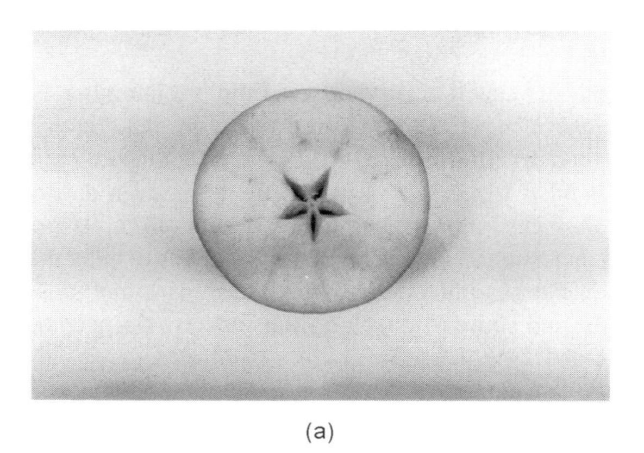

(a)

(b)

Figure 3.2. (*a*) Cross section of an apple; (*b*) Starfish.

FIBONACCI AND TREES

Fibonacci numbers are also found in some spiral arrangements of leaves on the twigs of plants and trees. From any leaf on a branch, count up the number of leaves until you reach the leaf directly above it; the number of leaves is often a Fibonacci number. On basswood and elm trees, this number is 2; on beech and hazel trees, it is 3; on apricot, cherry, and oak trees, it is 5; on pear and poplar trees, it is 8; and on almond and willow trees, it is 13 (see Fig. 3.3).

Here is another intriguing fact: The number of turns, clockwise or counterclockwise, we can take from the starting leaf to the terminal leaf is also usually a Fibonacci number. For example, on basswood and elm trees, it takes one turn; for beech and hazel trees, it is also 1; for apricot, cherry, and oak trees, it is 2; for pear and poplar trees, it is 3; and on almond and willow trees, it is 5.

The arrangement of leaves on the branches of *phyllotaxis*.* Accordingly, the ratio of the number of turns to the number of leaves is called the *phyllotactic ratio* of the tree. Thus, the phyllotactic ratio of basswood and elm is 1/2; for beech and hazel, it is 1/3; for apricot, cherry, and oak, it is 2/5; for pear and poplar, it is 3/8; and for almond and willow, it is 5/13. These data are summarized in Table 3.2. As an example, it takes 3/8 of a full turn to reach from one leaf to the next leaf on a pear tree.

FIBONACCI AND SUNFLOWERS

Mature sunflowers display Fibonacci numbers in a unique and remarkable way. The seeds of the flower are tightly packed in two distinct spirals, emanating from the center of the head to the outer edge (Figs. 3.4 and 3.5). One goes clockwise and the other counterclockwise. Studies have shown that although there are exceptions, the number of spirals, by and large, is adjacent Fibonacci numbers; usually, they are 34 and 55. Hoggatt reports a large sunflower with 89 spirals in the clockwise direction and 55 in the opposite direction, and a gigantic flower with 144 spirals clockwise and 89 counterclockwise.

It is interesting to note that Br. Alfred Brousseau once gave Hoggatt a sunflower with 123 clockwise spirals and 76 counterclockwise spirals, two adjacent Lucas spirals.

In 1951, John C. Pierce of Goddard College in Massachusetts reported in *The Scientific Monthly* that the Russians had grown a sunflower head with 89 and 144 spirals. After reading his article on Fibonacci numbers, Margaret K. O'Connell and Daniel T. O'Connell of South Londonderry, Vermont, examined their sunflowers, raised from seeds from Burpee's. They found heads with 55 and 89 spirals, some with 89 and 144 spirals, and one giant head with 144 and 233 spirals. The latter seems to be a world record.

*The word *phyllotaxis* is derived from the Greek words *phyllon*, meaning *leaf*, and *taxis*, meaning *arrangement*.

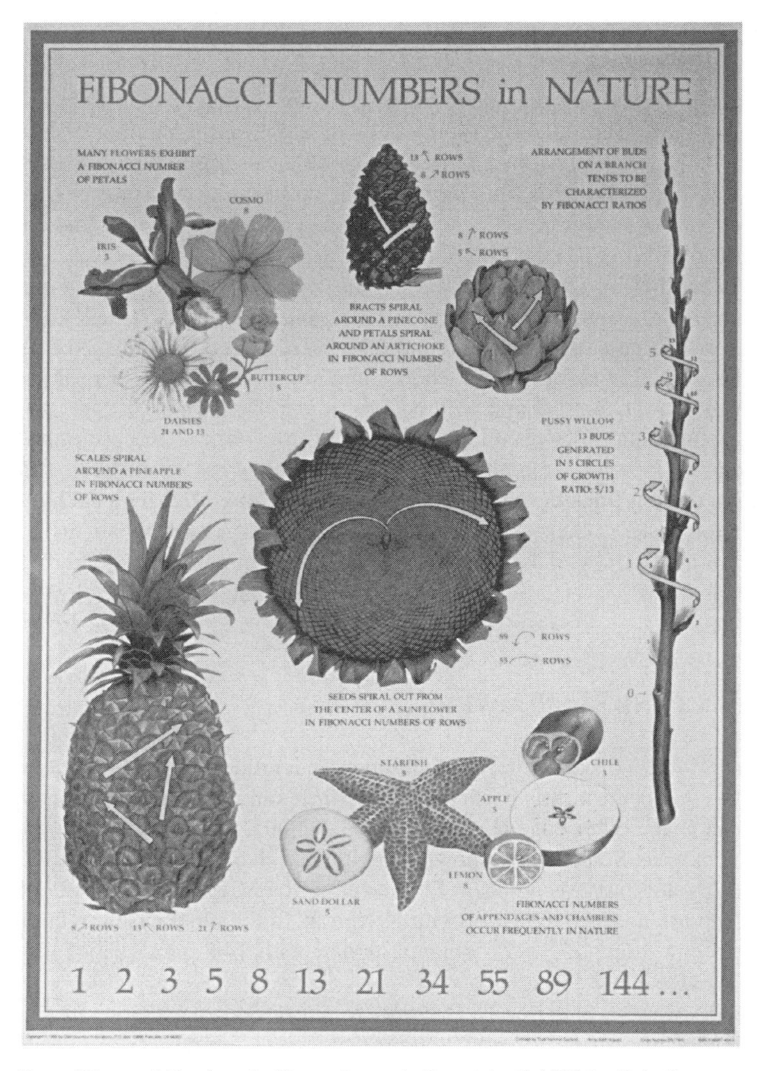

From *Fibonacci Numbers in Nature (poster)*. Copyright © 1988 by Dale Seymour Publications. Used with permission of Pearson Education.

FIBONACCI, PINECONES, ARTICHOKES, AND PINEAPPLES

The scale patterns on pinecones, artichokes, and pineapples provide excellent examples of Fibonacci numbers. The scales are in fact modified leaves closely packed on short stems, and they form two sets of spirals, called *parastichies*.* Some spirals

*The word *parastichies* is derived from the Greek words *para*, meaning *beside* and *stichos*, meaning *row*.

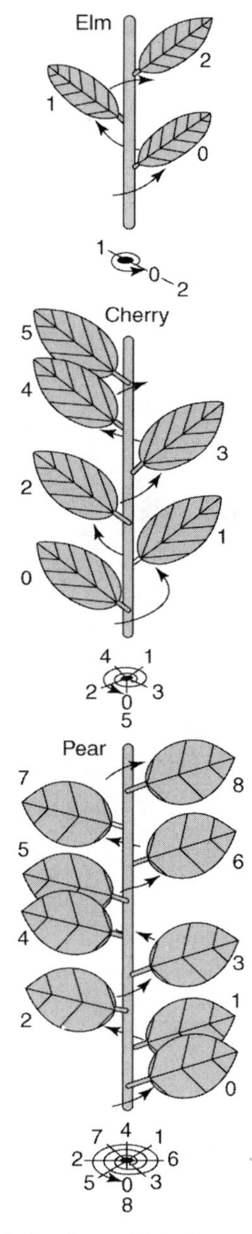

Figure 3.3. Elm, Cherry, and Pear limbs (*Source*: V. E. Hoggatt, Jr., *Fibonacci and Lucas Numbers*, Boston: Houghton Mifflin, 1968.).

TABLE 3.2.

Tree	Number of Turns	Number of Leaves	Phyllotactic Ratio
Basswood, elm	1	2	1/2
Beech, hazel	1	3	1/3
Apricot, cherry, oak	2	5	2/5
Pear, poplar	3	8	3/8
Almond, willow	5	13	5/13

Figure 3.4. Sunflower (*Source*: Runk/Schoenberger/Grant Heilman Photography, Inc.).

are clockwise and the rest are counterclockwise, as on a sunflower. Spiral numbers are often adjacent Fibonacci numbers. Some cones have 3 clockwise spirals and 5 counterclockwise spirals; some have 5 and 8; and some 8 and 13. Figure 3.6 and Figure 3.7 show the scale patterns on two pinecones.

Interestingly enough, some pinecones display three different spiral patterns. Their numbers, as you would expect, are also adjacent Fibonacci numbers.

Artichokes show a similar pattern, with the number of spirals in the two directions often adjacent Fibonacci numbers. Usually, there are 3 clockwise and 5 counterclockwise spirals, or 5 clockwise and 8 counterclockwise ones. Figure 3.8 shows two artichokes of the latter type.

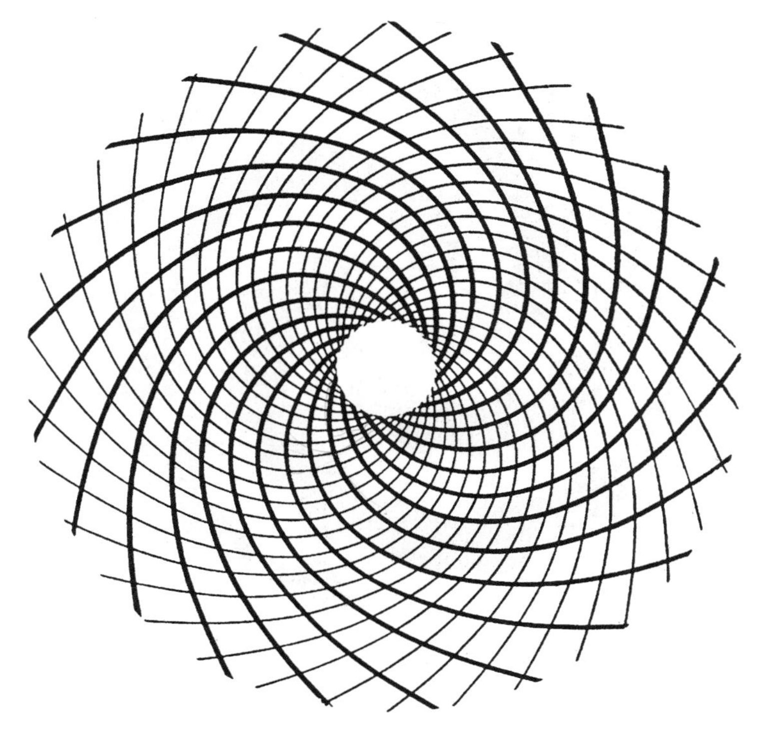

Figure 3.5. The spiral pattern in a Sunflower (*Source*: H. E. Huntley, *The Divine Proportion*, Mineola, NY: Dover, 1970. Reproduced with permission of Dover Publications.).

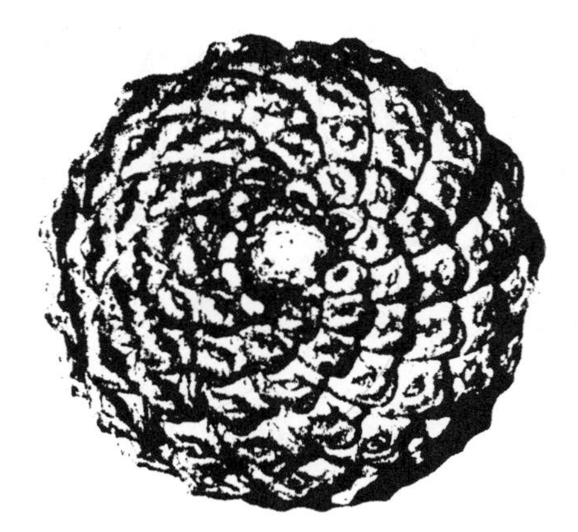

Figure 3.6. Pinecone (*Source*: Courtesy of American Museum of Natural History Library.).

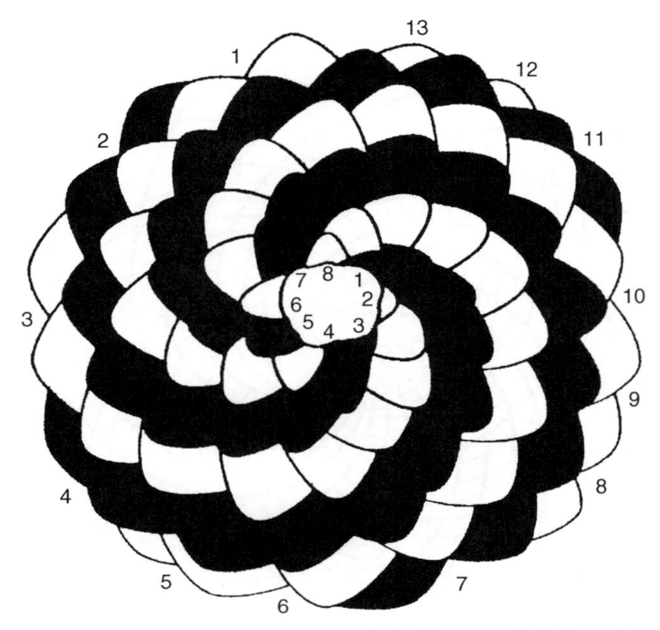

Figure 3.7. Scale patterns (8 clockwise spirals, 13 counterclockwise spirals).

Figure 3.8. Artichoke (*Source*: Trudi Hammel Garland, *Fascinating Fibonaccis: Mystery and Magic in Numbers*, Palo Alto, CA: Seymour, 1987. Copyright © 1987 by Dale Seymour Publications. Used by permission of Pearson Learning.).

The scales on pineapples are nearly hexagonal in shape (See Fig. 3.9). Since hexagons tessellate a plane perfectly and beautifully (see Fig. 3.10), the scales form three different spiral patterns. Once again, the number of spirals is adjacent Fibonacci numbers 8, 13, and 21. According to the *1977 Yearbook of Science and the Future*, a careful study of 2000 pineapples confirmed this most unusual Fibonacci pattern.

Figure 3.9. Pineapple.

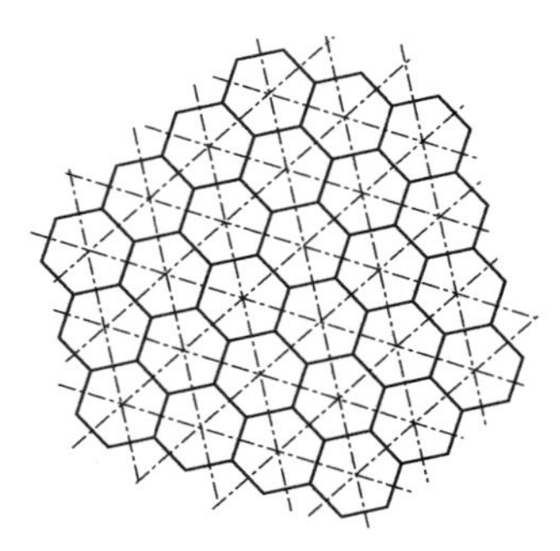

Figure 3.10. Hexagons (*Source*: V. E. Hoggatt, Jr., *Fibonacci and Lucas Numbers*, Santa Clara, CA: The Fibonacci Association, 1969.).

FIBONACCI AND MALE BEES

Male bees come from unfertilized eggs, so a male bee (M) has a mother but no father. A female bee (F), on the other hand, is developed from a fertilized egg, so it has both parents. Figure 3.11 shows the genealogical tree of a drone for seven generations. Count the total number of bees at each level, that is, in each generation. It is a Fibonacci number, as Table 3.3 demonstrates. Notice that it looks very much like Table 2.1.

Let a_n denote the number of female bees, b_n the number of male bees, and t_n the total number of bees, all in generation n, where $n \geq 1$. Clearly, $a_1 = 0$ and $b_1 = 1$. Since drones have exactly one parent, it follows that $b_n = a_{n-1}$, $a_n = a_{n-1} + b_{n-1}$, and $t_n = a_n + b_n$.

Since $a_n = a_{n-1} + a_{n-2}$, where $a_1 = 0$ and $a_2 = 1$, it follows that $a_n = F_{n-1}$. Now $t_n = a_n + b_n = a_n + a_{n-1} = a_{n+1}$, where $t_1 = a_2 = 1$ and $t_2 = a_3 = 1$, so $t_n = F_n$.

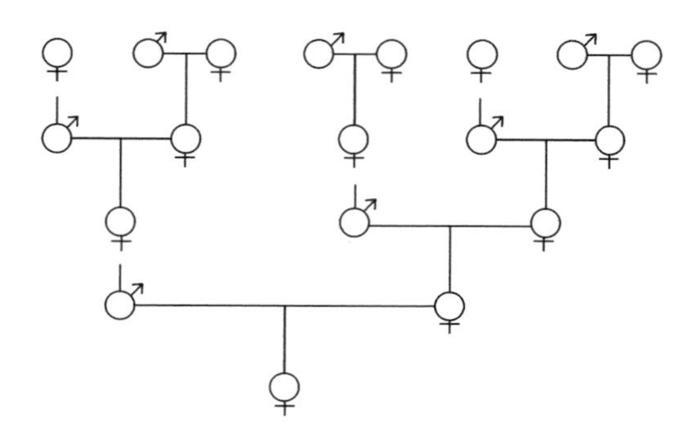

Figure 3.11. The family tree of a male bee.

TABLE 3.3. Number of Bees Per Generation is a Fibonacci Number

Generation	1	2	3	4	5	6	7	8
Number of female bees	0	1	1	2	3	5	8	13
Number of male bees	1	0	1	1	2	3	5	8
Total number of bees	1	1	2	3	5	8	13	21

Thus the number of ancestors of the drone in generation n is the Fibonacci number F_n. This fascinating relationship was originally presented in 1921 by W. Hope-Jones. It is examined further in Chapter 25.

FIBONACCI AND BEES

Consider two adjacent rows of cells in an infinite beehive, as pictured in Figure 3.12. We would like to find the number of paths the bee can take to crawl from one cell to an adjacent one. It can move in only one general direction, namely, to the right.

Let b_n denote the number of different paths to the nth cell. Since there is exactly one path to cell A (see Fig. 3.13), $b_1 = 1$. There are two distinct paths to cell B, as Figure 3.14 shows. So $b_2 = 2$. There are three different paths the bee can take to cell C (see Fig. 3.15), so $b_3 = 3$. There are five distinct paths the bee can take to cell D, as Figure 3.16 shows. Consequently, $b_4 = 5$, likewise, $b_5 = 8$.

Clearly, a pattern emerges, as shown in Table 3.4. It follows inductively that there are $b_n = F_{n+1}$ distinct paths for the bee to crawl to cell n (see Exercise 1 at the end of the chapter).

The next application was conceived in 1972 by L. Carlitz of Duke University.

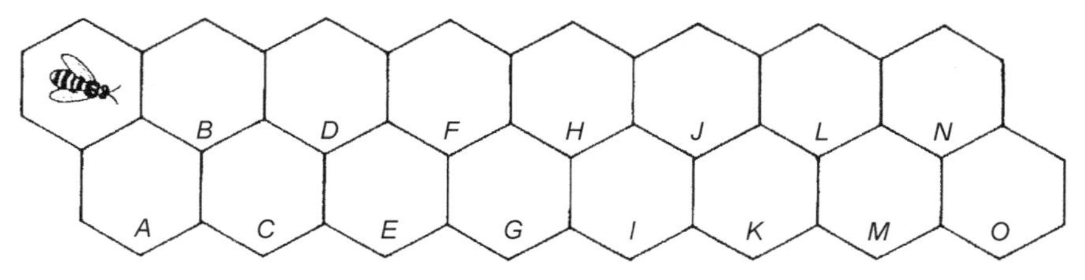

Figure 3.12.

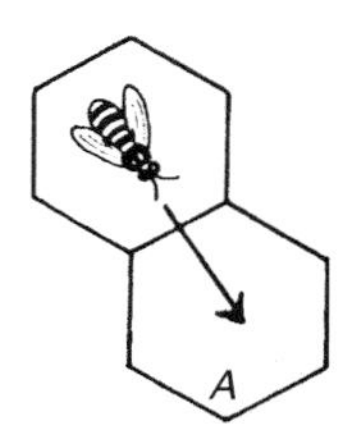

Figure 3.13.

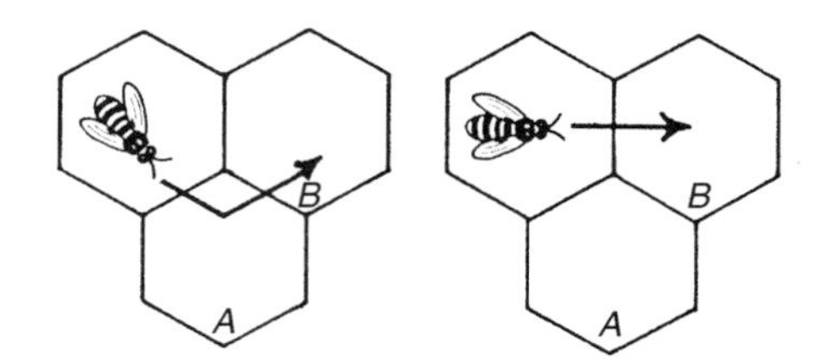

Figure 3.14.

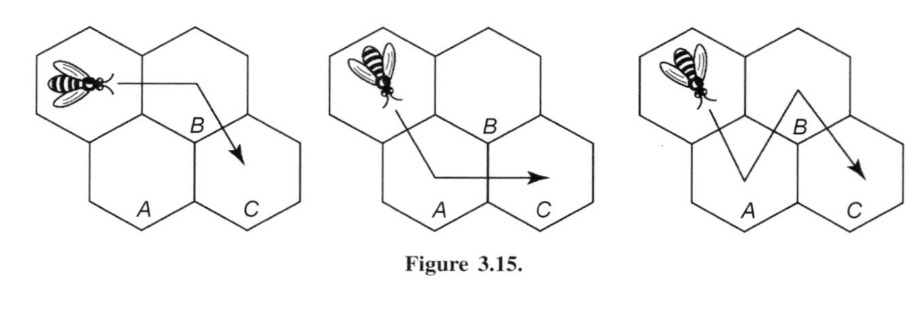

Figure 3.15.

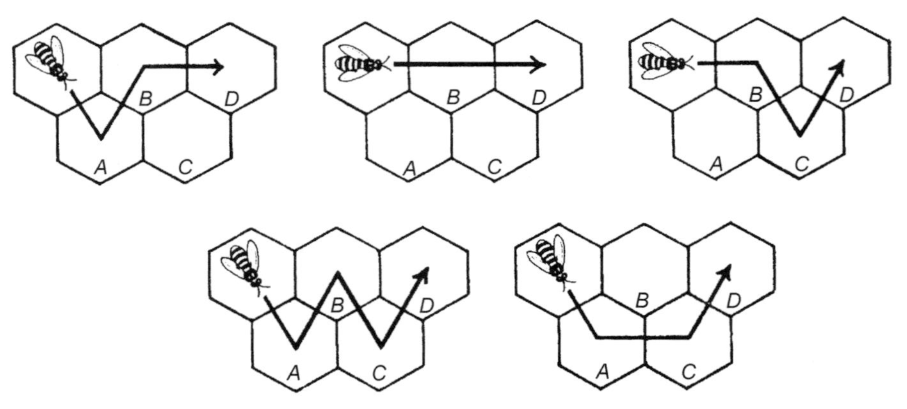

Figure 3.16.

TABLE 3.4.

n	1	2	3	4	5	...	n
b_n	1	2	3	5	8	...	?

FIBONACCI AND SUBSETS

Example 3.1. Find the number of subsets, including the null set, of a set of n points such that consecutive points are not allowed if the points lie on: (1) a line; and (2) a circle. ■

Solution.

1. Suppose the n points are linear. Let A_n denote the number of subsets. It follows from the following diagrams that $A_1 = 2$ and $A_2 = 3$.

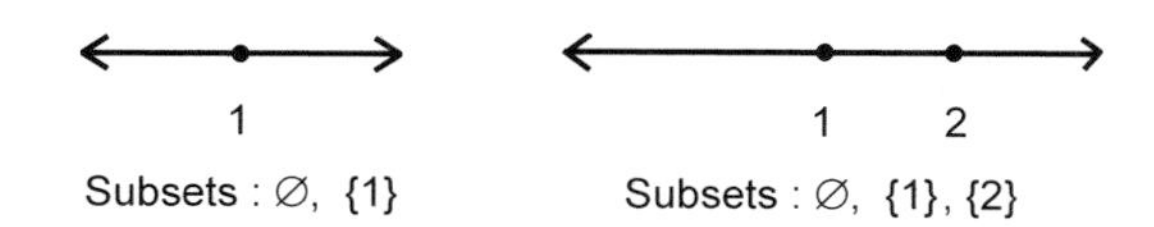

Let $n \geq 3$. Let n denote an extreme point, so it has just one neighbor. By definition, there are A_{n-1} subsets that do not contain n and A_{n-2} subsets that contain n as shown in the following diagram. Thus, by the addition principle, $A_n = A_{n-1} + A_{n-2}$:

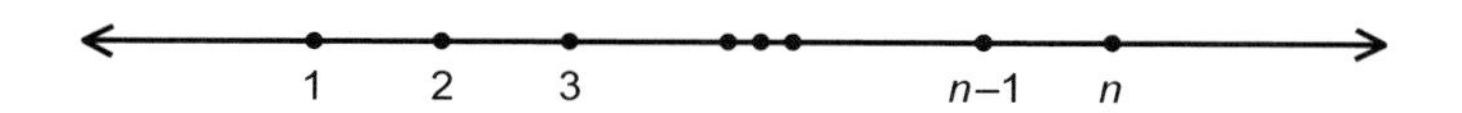

where $A_1 = 2$ and $A_2 = 3$. Therefore, $A_n = F_{n+2}$, $n \geq 1$. (Notice the similarity between this example and Example 4.1.)

2. Suppose the n points lie on a circle. Let B_n denote the number of subsets that do not contain consecutive points. If follows from the following diagrams that $B_1 = 2$, $B_2 = 3$, and $B_3 = 4$:

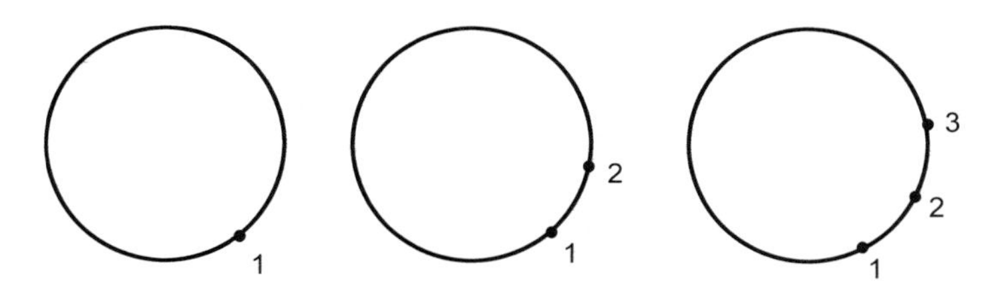

Let $n \geq 4$. Consider the point n (see the following diagram). There are A_{n-1} subsets that do not contain n and A_{n-3} subsets that do contain n. Therefore, $B_n = A_{n-1} + A_{n-3} = F_{n+1} + F_{n-1} = L_n$, $n \geq 2$.

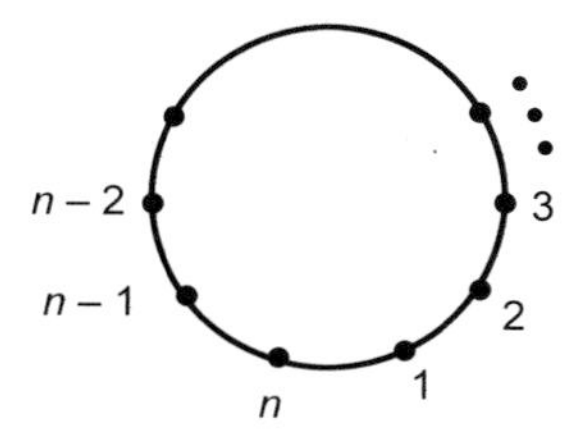

FIBONACCI AND SEWAGE TREATMENT

There are n towns on the bank of a river. They discharge their untreated sewage into the stream and pollute the water, so the towns would like to build treatment plants to control pollution. It is economically advantageous to build one or more central treatment plants along the main sewers and then send the wastewater from each city to another one. It is not economical to split the sewage of a town between two adjacent towns, since this would require the building of two sewers for the same town.

This problem was studied in 1972 by R. A. Deninger of the University of Michigan at Ann Arbor.

Let $f(n)$ denote the number of economic solutions. Clearly, $f(1) = 1$. Suppose $n = 2$. Then there are three possible solutions: Each town has its own plant, one plant at town 1, or one plant at town 2 (see Fig. 3.17). Thus $f(2) = 3$.

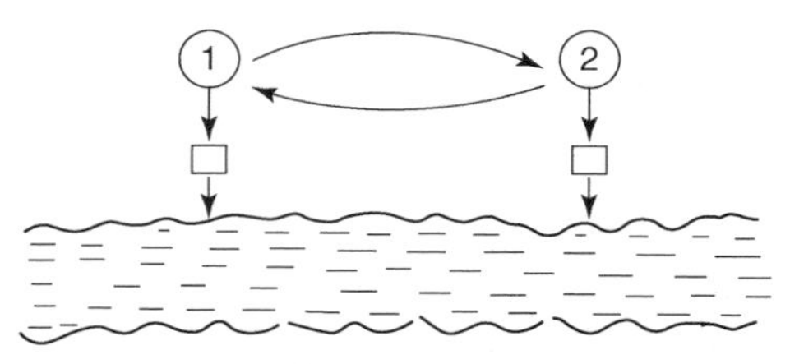

Figure 3.17.

Suppose there are three towns (see Fig. 3.18). Since there is no transfer of sewage between adjacent towns, each town can build its own treatment plant, send the sewage upstream ($\rightarrow$), or send it downstream ($\leftarrow$). Let 0 denote no transport between neighboring towns, 1 upstream transport and 2 downstream transport. Figure 3.19 shows the various economic solutions for three towns. They can be symbolically represented as follows:

00 01 02 10 11 12 20 ~~21~~ 22

where 21 is not a solution, since a town cannot simultaneously transfer waste both upstream and downstream. Thus $f(3) = 8 = 3f(2) - f(1)$. With $n = 4$ towns, there

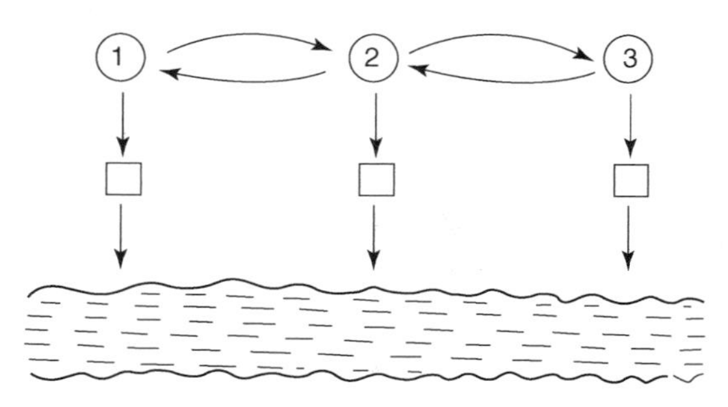

Figure 3.18.

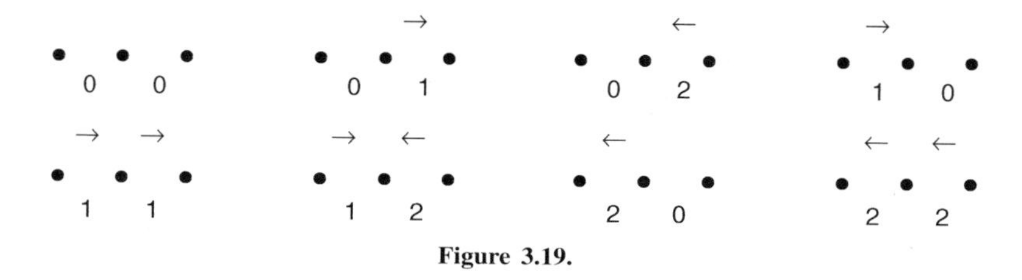

Figure 3.19.

are $f(4) = 21$ solutions:

000	001	002	010	011	012	~~021~~	022
100	101	102	110	111	112	~~121~~	122
200	201	202	210	211	212	~~221~~	222

Notice that $f(4) = 3f(3) - f(2)$, since there are three "words" that end in 21.

More generally, consider $n + 1$ towns with $f(n + 1)$ solutions. Adding one town increases the number of solutions to $3f(n)$. From this we must subtract the number of words ending in 21, namely, $f(n-1)$. Thus $f(n+1) = 3f(n) - f(n-1)$, where $n \geq 2$.

Using this recurrence relation, the value of $f(n)$ can be computed for various values of n. It appears from Table 3.5 that $f(n) = F_{2n}$.

TABLE 3.5.

n	1	2	3	4	5	6
$f(n)$	1	3	8	21	55	144

To confirm this formula, notice that $f(1) = 1 = F_2$ and $f(2) = 4 = F_4$. So it remains to show that F_{2n} satisfies the recurrence relation:

$$\begin{aligned} 3f(n) - f(n-1) &= 3F_{2n} - F_{2n-2} = 2F_{2n} + F_{2n-1} \\ &= F_{2n} + F_{2n+1} = F_{2n+2} \\ &= f(n+1) \end{aligned}$$

Thus $f(n) = F_{2n}$, $n \geq 1$.

FIBONACCI AND ATOMS

The atomic number Z of an atom is the number of protons in it. The periodic table shows an interesting relationship between the atomic numbers of inert gas and Fibonacci numbers (see Table 3.6).

There are six inert gases—helium, neon, argon, krypton, xenon, and radon—and they are exceptionally stable chemically. With the exception of helium, their atomic numbers are approximately the same as the Fibonacci numbers F_7 through F_{11}, as Table 3.6 shows. Suppose we compute $\lfloor Z/18 + 1/2 \rfloor$ for each gas; that is, divide each

TABLE 3.6.

Inert Gas	Atomic Number Z	Corresponding Fibonacci Numbers	$\lfloor Z/18 + 1/2 \rfloor$
Helium	2	8	0
Neon	10	13	1
Argon	18	21	1
Krypton	36	34	2
Xenon	54	55	3
Radon	86	89	5

TABLE 3.7.

N	2	8	14	20	28	50	82	126
$\lfloor N/10 + 1/2 \rfloor$	0	1	1	2	3	5	8	13

number by 18 and then find the nearest integer. It follows from column 4 that each is a Fibonacci number.

The nucleus of an atom consists of two kinds of particles: protons and neutrons. A proton has a charge equal but opposite to that of an electron, while a neutron is neutral. Let N denote the number of neutrons in the nucleus. Nuclei having the values 2, 8, 14, 20, 28, 50, 82, or 126 for N or Z are considered more stable than others. (The origin of these numbers is a mystery.) Let us compute $\lfloor N/10 + 1/2 \rfloor$ for each N, that is, compute $N/10$ rounded up to the nearest integer. Surprisingly enough, each is again a Fibonacci number! (see Table 3.7.)

FIBONACCI AND THE BALMER SERIES

In 1885, the Swiss schoolteacher Johann Jacob Balmer (1825–1898) discovered that the wavelengths (in angstroms) of four lines in the hydrogen spectrum (now known as the *Balmer series*) can be expressed as the product of the constant 364.5 (in nanometers) and a fraction:

$$656 = \frac{9}{5} \times 364.5$$

$$486 = \frac{4}{3} \times 364.5$$

$$434 = \frac{25}{21} \times 364.5$$

$$410 = \frac{9}{8} \times 364.5$$

Notice that the denominators of the fractions are Fibonacci numbers. This observation was made in 1973 by J. Wlodarski of Germany.

FIBONACCI AND REFLECTIONS

Optics, the branch of physics that deals with light and vision, has found yet another appearance of Fibonacci numbers in the real world. Consider two glass plates placed face-to-face. Such a stack has four reflective faces, as Figure 3.20 shows.

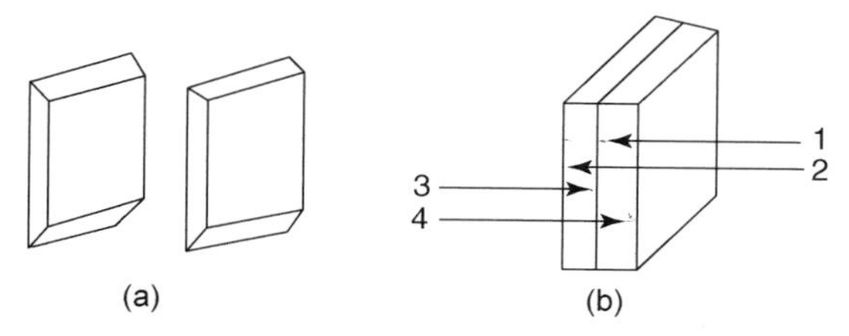

Figure 3.20. (a) Two separate glass plates (b) The stack has four reflective faces, labeled 1–4.

Suppose a ray of light falls on the stack. Let a_n denote the number of distinct reflective paths made with n reflections, where $n \geq 0$. We would like to determine the value of a_n.* To this end, let's first collect some data on a_n.

When $n = 0$, that is, when there are no reflections, the ray just passes through the glass plates, as Figure 3.21 shows, so $a_0 = 1$.

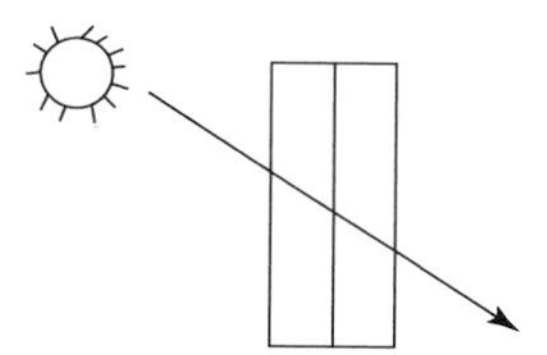

Figure 3.21. Stacked glass plates with no reflections.

Suppose the ray causes one reflection. Then there are two distinct possible paths, so $a_1 = 2$ (see Fig. 3.22).

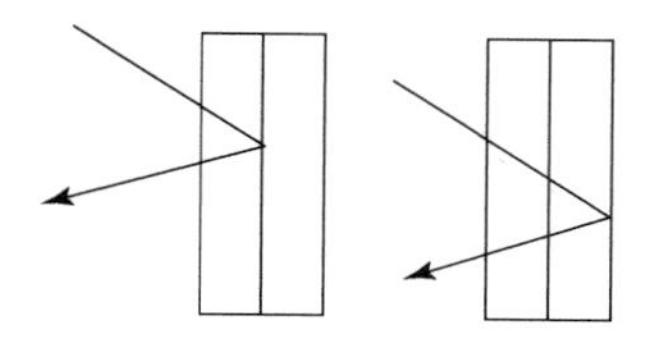

Figure 3.22. Stacked glass plates with one reflection.

If the ray is reflected twice, three possible paths can emerge, as Figure 3.23 illustrates, so $a_2 = 3$. If it is reflected thrice, there are five possible reflecting patterns, so $a_3 = 5$ (see Fig. 3.24). Likewise, $a_4 = 8$ (see Fig. 3.25).

*This problem was proposed in 1963 by L. Moser and M. Wyman, and solved by J. L. Brown.

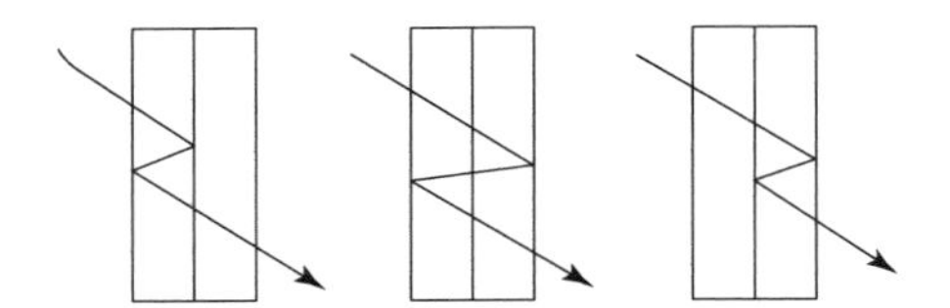

Figure 3.23. Stacked glass plates with two reflections.

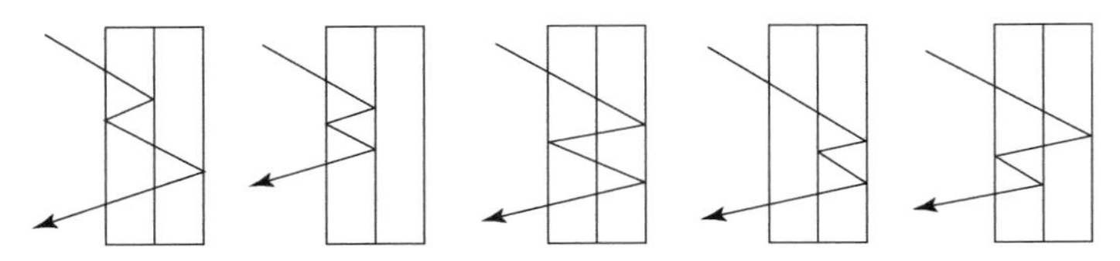

Figure 3.24. Stacked glass plates with three reflections.

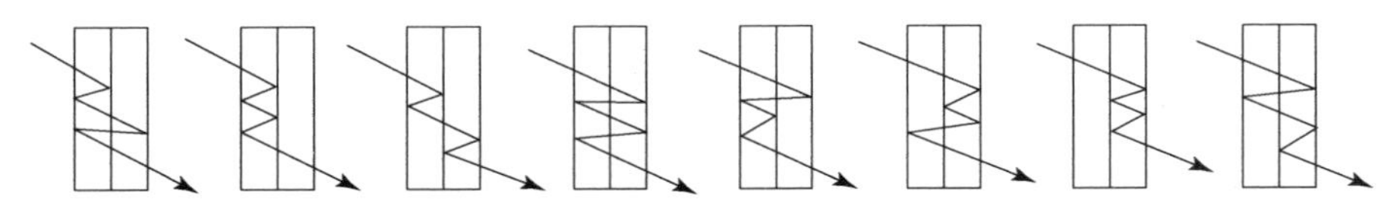

Figure 3.25. Stacked glass plates with four reflections.

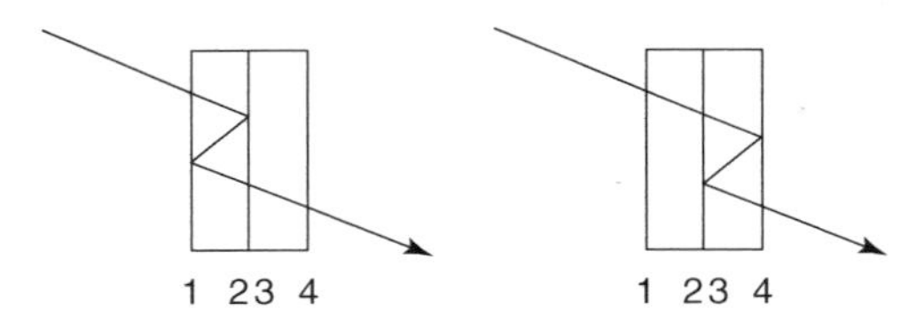

Figure 3.26.

More generally, suppose the ray is reflected n times, so the last reflection occurs at face 1 or 3. Then the previous reflection must have occurred on face 2 or 4, as Figure 3.26 shows. The number of paths with the nth reflection on face 1 equals the number of paths reaching 1 after $n - 1$ reflections, and there are a_{n-1} such paths.

Suppose the nth reflection takes place on face 3. The $(n - 1)$st reflection must have occurred on face 4. Such a ray must have already had $n - 2$ reflections before reaching face 4. By definition, the number of such paths is a_{n-2}.

Thus, by the addition principle, $a_n = a_{n-1} + a_{n-2}$, where $a_1 = 2$ and $a_2 = 3$, so $a_n = F_{n+2}$.

FIBONACCI, PARAFFINS, AND CYCLOPARAFFINS

Graph theory is a relatively new branch of mathematics. A *graph* is a finite, nonempty set of *vertices* and *edges* (arcs or line segments) joining them. Figures 3.27 and 3.28 are both graphs.

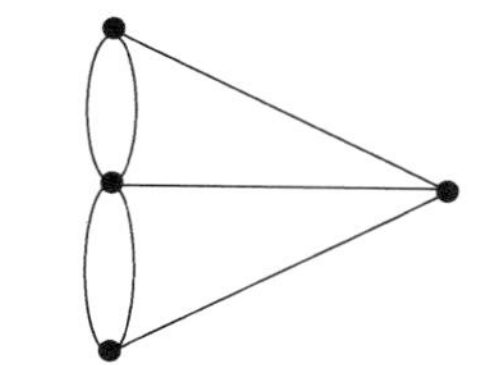

Figure 3.27.

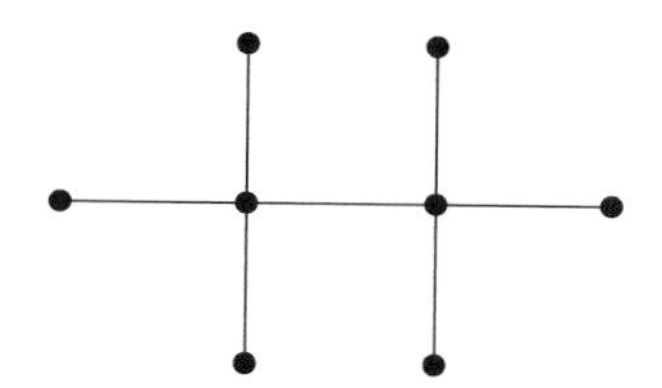

Figure 3.28. Ethane molecule C_2H_6.

Graphs are useful in the study of hydrocarbons. The English mathematician Arthur Cayley (1821–1895) was the first to employ graphs to examine hydrocarbon isomers.

A hydrocarbon molecule consists of hydrogen and carbon atoms. Each hydrogen atom (H) is bonded to a single carbon atom (C), whereas a carbon atom bonds to two, three, or four atoms, which can be carbon or hydrogen. But carbon atoms in saturated hydrocarbon molecules, such as ethane, contain only single bonds, as Figure 3.28 illustrates.

Deleting hydrogen atoms from the structural formulas of saturated hydrocarbons yields graphs consisting of carbon atoms and edges between two adjacent vertices. The *topological index* of such a graph G with n vertices is the total number of different ways the graph can be partitioned into disjoint subgraphs containing exactly k edges, where $k \geq 0$. For example, Figure 3.29 shows the carbon atom skeleton for the paraffin pentane, C_5H_{12}, and Figure 3.30 shows its various possible partitionings. Consequently, the topological index of pentane is $1 + 4 + 3 = 8$.

Figure 3.29.

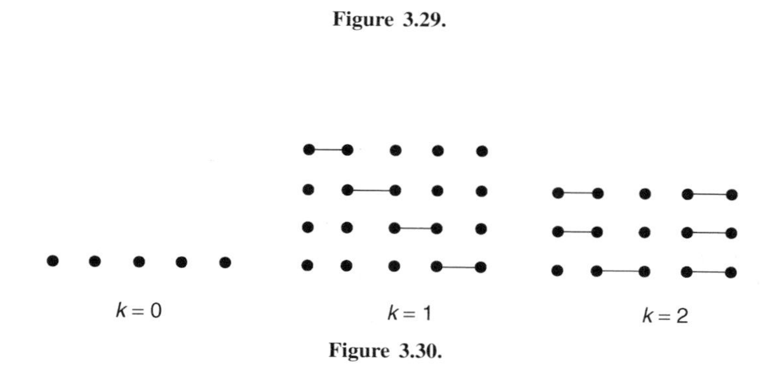

Figure 3.30.

TABLE 3.8. Topological Indices of Paraffins C_nH_{2n+2}

Paraffin	n	Graph	k = 0	1	2	3	4	5	Total
Methane	1		1						1
Ethane	2		1	1					2
Propane	3		1	2					3
Butane	4		1	3	1				5
Pentane	5		1	4	3				8
Hexane	6		1	5	6	1			13
Heptane	7		1	6	10	4			21
Octane	8		1	7	15	10	1		34
Nonane	9		1	8	21	20	5		55
Decane	10		1	9	28	35	15	1	89

↑ Fibonacci numbers

Table 3.8 shows the carbon atom graphs G_n and their topological indices of ten paraffins C_nH_{2n+2}, $n \geq 1$. For a graph consisting of a single vertex, the index is defined as one. It appears from the table that the index of G_n is F_{n+1}.

To confirm this observation, let t_n denote the topological index of the carbon atom graph G_n of a paraffin with n vertices, as Figure 3.31 shows.

Figure 3.31.

Case 1. Suppose the edge $v_{n-1} - v_n$ is not included. Then the edge $v_{n-2} - v_{n-1}$ may or may not be included. Consequently, the topological index of the remaining graph G_{n-1} is t_{n-1}.

Case 2. Suppose the edge $v_{n-1} - v_n$ is included. Then the edge $v_{n-2} - v_{n-1}$ is not included. This yields the graph G_{n-2}, and its index is t_{n-2}.

Thus, by the addition principle, $t_n = t_{n-1} + t_{n-2}$. But $t_1 = 1$ and $t_2 = 2$, so, $t_n = F_{n+1}$.

Table 3.9 shows the carbon atom skeleton C_n of ten cycloparaffins C_nH_{2n} and the corresponding indices. A similar argument shows that the index of C_n = index of G_n+ index of $G_{n-2} = F_{n+1} + F_{n-1}$, where $n \geq 3$. Notice that the index of C_n is in fact the Lucas number L_n. We shall confirm in Chapter 5 that $F_{n+1} + F_{n-1} = L_n$.

The triangular arrangements in Tables 3.8 and 3.9 are explored further in Chapter 13.

TABLE 3.9. Topological Indices of Cycloparaffins C_nH_{2n}

Cycloparaffin	n	Graph	k = 0	1	2	3	4	5	Total
·	1		1						1
·	2		1	2					3
Cyclopropane	3		1	3					4
Cyclobutane	4		1	4	2				7
Cyclopentane	5		1	5	5				11
Cyclohexane	6		1	6 ↘	9	2			18
Cycloheptane	7		1	7	14 ↓	7			29
Cyclooctane	8		1	8	20	16	2		47
Cyclononane	9		1	9	27	30	9		76
Cyclodecane	10		1	10	35	50	25	2	123

↑
Lucas numbers

FIBONACCI AND MUSIC

Fibonacci numbers occur in relation to music. They were also observed by Zerger:

- The word MUSIC begins with the thirteenth and twenty-first letters of the alphabet. With the eighth, thirteenth, and twenty-first letters, we can form the word HUM.
- The Library of Congress Classification Number for Music is M, the thirteenth letter.
- The Dewey Decimal Classification Number for Music is 780, where $780 = 2 \cdot 2 \cdot 3 \cdot 5 \cdot 13$, a product of Fibonacci numbers.
- Pianos are often tuned to a standard of 440 cycles per second, where $440 = 8 \cdot 55$.

The keyboard of a piano provides a fascinating visual illustration of the link between Fibonacci numbers and music. An octave on a keyboard represents a musical interval between two notes, one higher than the other. The frequency of the higher note is twice that of the lower. On the keyboard, the octave is divided into 5 black and 8 white keys, a total of 13 keys (see Fig. 3.32). The five black keys form two groups, one of two keys and the other of three keys.

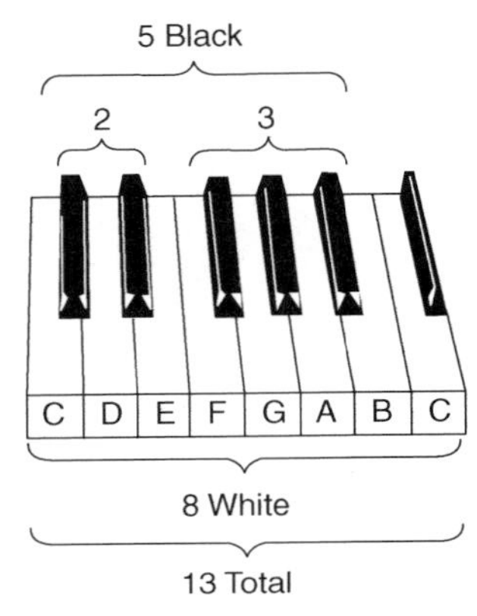

Figure 3.32. Fibonacci Numbers in the Octave of a Piano Keyboard (*Source*: Trudi Hammel Garland, *Fascinating Fibonaccis: Mystery and Magic in Numbers*, Palo Alto, CA: Seymour, 1987. Copyright © 1987 by Dale Seymour Publications. Used by permission of Pearson Learning.).

The 13 notes in an octave form the chromatic scale, the most popular scale in Western music. The chromatic scale was preceded by two other scales, the 5-note pentatonic scale and the 8-note diatonic scale. Popular tunes such as "Mary Had a Little Lamb," and "Amazing Grace" can be played using the pentatonic scale, while melodies such as "Row, Row, Row Your Boat" use the diatonic scale.

The major sixth and the minor sixth (six tones apart and $5\frac{1}{2}$ tones apart, respectively) are the two musical intervals most pleasing to the ear. A major sixth, for

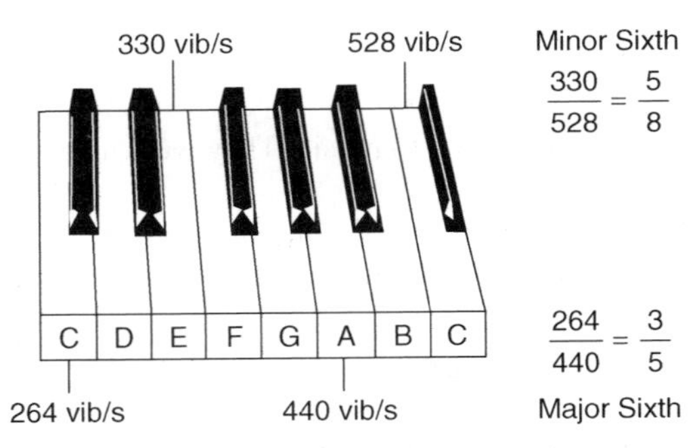

Figure 3.33. Fibonacci Ratios in Musical Intervals (*Source*: Trudi Hammel Garland, *Fascinating Fibonaccis: Mystery and Magic in Numbers*, Palo Alto, CA: Seymour, 1987. Copyright © 1987 by Dale Seymour Publications. Used by permission of Pearson Learning.).

example, consists of the notes C and A; they make 264 and 440 vibrations per second, respectively (see Fig. 3.33). Notice that $264/440 = 3/5$, a Fibonacci ratio.

A minor sixth interval, for instance, consists of the notes E and C, making 330 and 528 vibrations a second. Their ratio is also a Fibonacci ratio: $330/528 = 5/8$.

The ratios of consecutive Fibonacci numbers are discussed further in Chapter 20 on the Golden ratio.

FIBONACCI AND POETRY

Fibonacci numbers have found their way into the art of poetry also. A limerick, according to Webster's dictionary, is a nonsensical poem of 5 lines, of which the first, second, and fifth have 3 beats, and the other two have 2 beats, and rhyme. The following limerick*, for example, is made up of 5 lines; they contain 2 groups of 2 beats and 3 groups of 3 beats, for a total of 13 beats. Once again, all numbers involved are Fibonacci numbers:

A fly and a flea in a flue	3 beats
Were imprisoned, so what could they do?	3 beats
Said the fly, "Let us flee!"	2 beats
"Let us fly!" said the flea,	2 beats
So they fled through a flaw in the flue.	3 beats
Total	= 13 Beats

In the 1960s, G. E. Duckworth of Princeton University, New Jersey, analyzed the *Aeneid*, an epic poem written in Latin about 20 B.C. by Virgil (70–19 B.C.), the "greatest poet of ancient Rome and one of the outstanding poets of the world."† Duckworth discovered frequent occurrences of the Fibonacci numbers and several variations in this masterpiece:

$$1, 3, 4, 7, 11, \ldots \quad \leftarrow \text{Lucas sequence}$$

$$1, 4, 5, 9, 14, \ldots$$

$$1, 5, 6, 11, 17, \ldots$$

$$1, 6, 7, 13, 20, \ldots$$

$$2, 3, 5, 8, 13, \ldots$$

$$3, 7, 10, 17, 27, \ldots$$

$$4, 9, 13, 22, 35, \ldots$$

$$6, 13, 19, 32, 61, \ldots$$

*Based on T. H. Garland, *Fascinating Fibonaccis*, Dale Seymour Publications, Palo Alto, CA, 1987.
†*The World Book Encyclopedia*, Vol. 20, 1982.

The mathematical symmetry Virgil consciously employed in composing the *Aeneid* brings the harmony and aesthetic balance of music to the ear, since ancient poetry was written to be read out loud.

According to Duckworth's investigations into Virgil's structural patterns and proportions, there is evidence that even Virgil's contemporary poets, such as Catullus, Lucretius, Horace, and Lucan used the Fibonacci sequence in the structure of their poems. Duckworth's study lends credibility to the theory that the Fibonacci sequence and the Golden section (Chapter 20) were known to the ancient Greeks and Romans, although no such mention of it exists.

FIBONACCI AND COMPOSITIONS WITH 1s AND 2s

In the summer of 1974, Krishnaswami Alladi of Vivekananda College, India, and Hoggatt studied the compositions of positive integers, n, that is, expressing n as sums of 1s and 2s. For example, 3 has three such compositions and 4 has five, as Table 3.10 shows. Notice that $1 + 2$ and $2 + 1$ are considered distinct compositions, so order matters. Again, it appears from the table that the number of distinct compositions is a Fibonacci number. The next theorem confirms this conjecture.

TABLE 3.10.

n	Compositions of n	Number of Compositions
1	1	1
2	$1+1, 2$	2
3	$1+1+1, 1+2, 2+1$	3
4	$1+1+1+1, 1+1+2, 1+2+1,$ $2+1+1, 2+2$	5
5	$1+1+1+1+1, 1+1+1+2,$ $1+1+2+1, 1+2+1+1, 2+1+1+1,$ $1+2+2, 2+1+2, 2+2+1$	8

↑
Fibonacci Numbers

Theorem 3.1. (Alladi and Hoggatt, 1974). The number of distinct compositions C_n of a positive integer n in terms of 1s and 2s is F_{n+1}, where $n \geq 1$.

Proof. Let $C_n(1)$ and $C_n(2)$ denote the number of compositions of n that end in 1 and 2, respectively. Clearly, $C_1(1) = 1$ and $C_1(2) = 0$, so $C_1 = C_1(1) + C_1(2) = 1$. Likewise, $C_2 = C_2(1) + C_2(2) = 1 + 1 = 2$.

Now consider a composition of n, where $n \geq 3$.

Case 1. Suppose the composition ends in 1. Deleting the 1 at the end yields a composition of $n - 1$. On the other hand, adding a 1 at the end of a composition of $n - 1$ yields a composition of n that ends in 1. Thus $C_n(1) = C_{n-1}$.

Case 2. Suppose the composition ends in 2. Deleting the 2 at the end, we get a composition of $n-2$. On the other hand, by adding a 2 or two 1s, we get a composition of n. But the latter has already been counted in case 1, so $C_n(2) = C_{n-2}$.

Thus, by the addition principle, $C_n = C_n(1) + C_n(2) = C_{n-1} + C_{n-2}$, where $C_1 = 1$ and $C_2 = 2$. Therefore, $C_n = F_{n+1}$. ■

We shall re-prove this fact in Chapter 19 by an alternate method.

The next two results were also discovered in 1974 by Alladi and Hoggatt, where $f(n)$ denotes the total number of 1s in the various compositions of n and $g(n)$ denotes that of 2s. For example, $f(3) = 5$ and $g(3) = 2$.

Theorem 3.2.

1. $f(n) = f(n-1) + f(n-2) + F_n$
2. $g(n) = g(n-1) + g(n-2) + F_{n-1}$

where $n \geq 3$.

Proof. 1) As in the preceding proof, we have $C(n) = C_n(1) + C_n(2)$. Since $C_n(2) = C_{n-2}$, there are C_{n-2} compositions of n that end in 2. But C_{n-2} denotes the number of compositions of $n-2$. By definition, there is a total of $f(n-2)$ 1s in the various compositions of $n-2$.

Since $C_n(1) = C_{n-1}$, there are C_{n-1} compositions of n that end in a 1. Excluding this 1, they contain $f(n-1)$ 1s. Since each of the C_{n-1} compositions contains a 1 as the final addend, they contain a total of $f(n-1) + C_{n-1} = f(n-1) + F_n$ ones. Thus $f(n) = f(n-1) + f(n-2) + F_n$, where $n \geq 3$.

Similarly, $g(n) = g(n-1) + g(n-2) + F_{n-1}$, where $n \geq 3$. ■

For example,

$$f(5) = 20 = 10 + 5 + 5 = f(4) + f(3) + F_5$$

$$g(5) = 10 = 5 + 2 + 3 = g(4) + g(3) + F_3$$

Theorem 3.3. $f(n) = g(n+1), n \geq 1$.

Proof. [by the principle of mathematical induction (PMI)]. Since $f(1) = 1 = g(2)$ and $f(2) = 2 = g(3)$, the result is true when $n = 1$ and $n = 2$.

Assume it is true for all positive integers $< n$. By Theorem 3.2, we have

$$f(n) = f(n-1) + f(n-2) + F_n$$

$$g(n) = g(n-1) + g(n-2) + F_{n-1}$$

By the inductive hypothesis, $f(n-1) = g(n)$ and $f(n-2) = g(n-1)$, so $f(n) = g(n) + g(n-1) + F_n = g(n+1)$. Thus the given result is true for every $n \geq 1$. ■

For example, $f(3) = 5 = g(4)$ and $f(4) = 10 = g(5)$.

FIBONACCI AND NEUROPHYSIOLOGY

In 1976, Kurt Fischer of the University of Regensburg in Germany studied a model of the physiology of nerves and discovered yet another occurrence of Fibonacci numbers.

The impulses traveling along nerve fibers originate from sodium or potassium ions, and flow through identical transmembrane pores consisting of $n \geq 2$ cells. Tiny quantities of calcium ions, Ca^{2+}, can enter the pores and stop the flow of sodium ions, Na^{+}, in these pores. They can occupy one cell or two cells, except at the entrance of the pore. These two states are denoted by 1 and 2, respectively. Figure 3.34 shows a typical pore, where 0 indicates an empty cell.

2	0	1	2

Figure 3.34. A sample pore.

Suppose that sodium can enter or leave at either end of a pore, whereas calcium can do so only at the left side of the pore. Consequently, calcium ions within a pore impede the flow of sodium through this pore.

This Markovian stochastic process can be depicted by a tree structure; the vertices of the tree represent the possible states of a pore and its edges represent the possible transitions between states. Figure 3.35, for example, shows the various possible states of a pore with five nonempty cells.

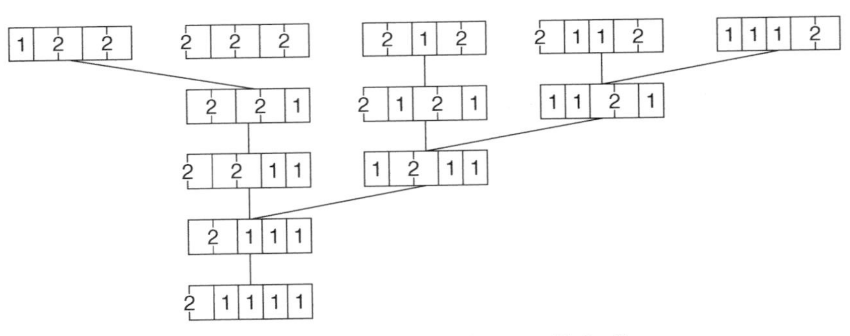

Figure 3.35. Tree of states of a pore with 5 cells.

Notice that the tree consists of two kinds of vertices, those with a 1 in the far right cell or a 2 in the middle of the two right cells. Every state in level five has the latter property, and shows that the translation of sodium ions to the right is no longer feasible because of the presence of calcium on the right side of each state.

Figure 3.36 depicts a tree-skeleton of Figure 3.35, which very much resembles the Fibonacci tree in Figure 2.1. It follows from either figure that a pore with five nonempty cells has $5 = F_5$ states at level 5.

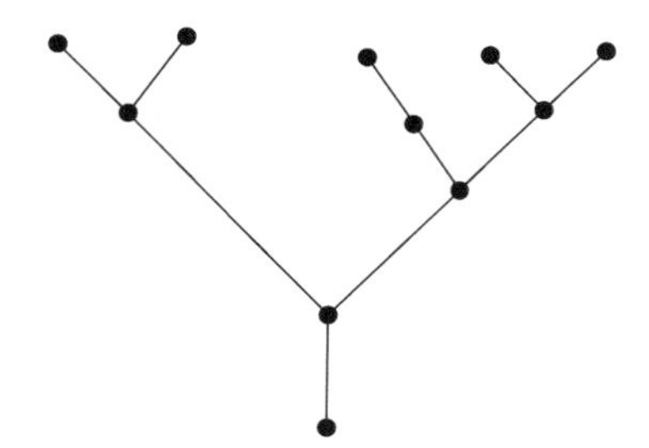

Figure 3.36. A tree diagram of Figure 3.35.

More generally, a pore with n nonempty cells has F_n states at level n. This follows from the fact that the number of states at level n satisfies the Fibonacci recurrence relation.

FIBONACCI AND ELECTRICAL NETWORKS

In 1963, S. L. Basin of then San Jose State College, California, wrote that "even those people interested in electrical networks cannot escape from our friend Fibonacci." And, in fact, Fibonacci numbers appear even in the study of electrical networks.

For example, consider a network of n resistors, arranged in the shape of a ladder, as Figure 3.37 illustrates. We shall show that the resistance $Z_o(n)$ (*output impedance*) across the output terminals C and D, the resistance $Z_i(n)$ (*input impedance*) between the input terminals A and B, and the *attenuation* $A(n) = Z_o/Z_i$ are all very closely related to Fibonacci numbers in an unusually special case.

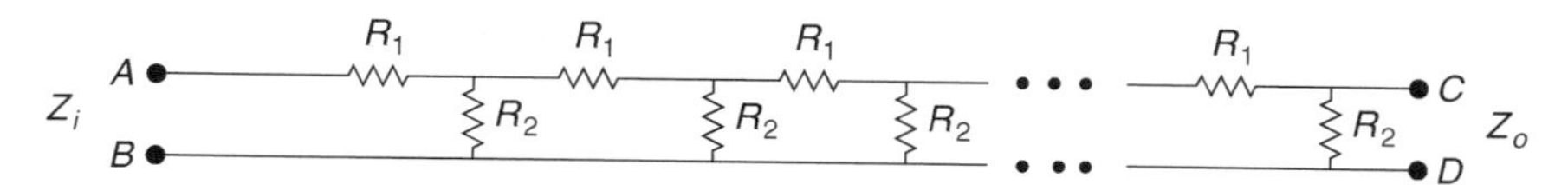

Figure 3.37. n ladder sections.

First, let us consider two resistors, R_1 and R_2, arranged in series. Let V denote the voltage drop across a resistor R due to current I (see Fig. 3.38). Then $V = IR = I(R_1 + R_2)$. So $R = R_1 + R_2$. On the other hand, suppose the resistors are connected in parallel, as Figure 3.39 shows. Then $V = I_1 R_1 = I_2 R_2 = (I_1 + I_2)R$, so

$$\frac{1}{R} = \frac{I_1 + I_2}{V} = \frac{I_1}{V} + \frac{I_2}{V} = \frac{1}{R_1} + \frac{1}{R_2}$$

Thus, if R_1 and R_2 are connected in parallel, then the resultant resistance R is given by

$$\frac{1}{R} = \frac{1}{R_1} + \frac{1}{R_2}$$

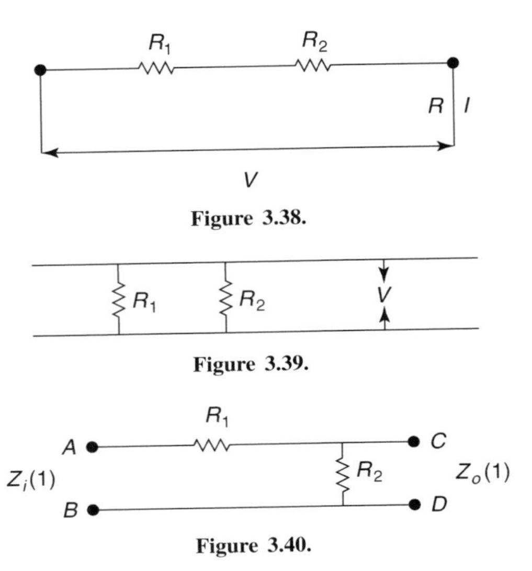

Figure 3.38.

Figure 3.39.

Figure 3.40.

We are now ready to tackle the ladder network problem step-by-step. Suppose $n = 1$; that is, the network consists of one section, as Figure 3.40 shows. Then $Z_o(1) = R_2$ and $Z_i(1) = R_1 + R_2$, so

$$A(1) = \frac{Z_i(1)}{Z_o(1)} = \frac{R_1}{R_2} + 1 \tag{3.1}$$

Suppose $n = 2$. The resulting circuit is obtained by adding a section to the one in Figure 3.40 (see Fig. 3.41). Since the resistors R_1 and R_2 in the extension are connected in series, they can be replaced by a resistor $R_3 = R_1 + R_2$; this yields the equivalent network in Figure 3.42. Now R_2 and R_3 are connected in parallel, so they can be replaced by a resistor R_4 (see Fig. 3.43). Then

$$\frac{1}{R_4} = \frac{1}{R_2} + \frac{1}{R_3} = \frac{1}{R_2} + \frac{1}{R_1 + R_2} = \frac{R_1 + 2R_2}{R_2(R_1 + R_2)}$$

$$\therefore R_4 = \frac{R_2(R_1 + R_2)}{R_1 + 2R_2}$$

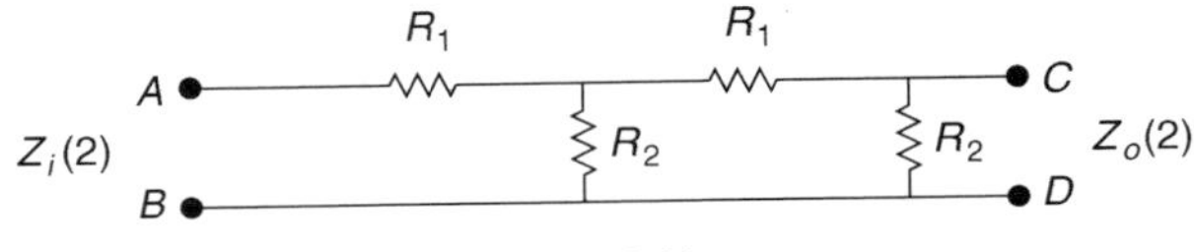

Figure 3.41.

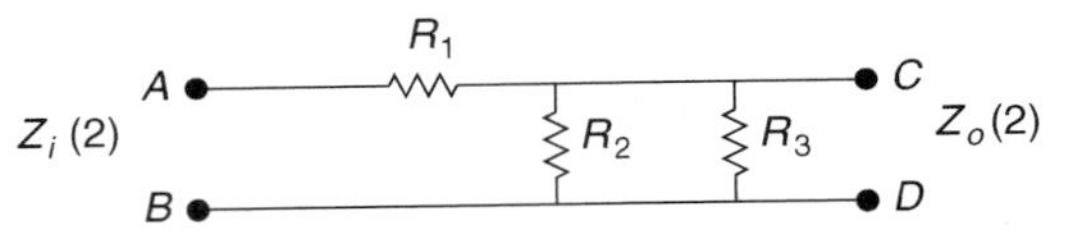

Figure 3.42.

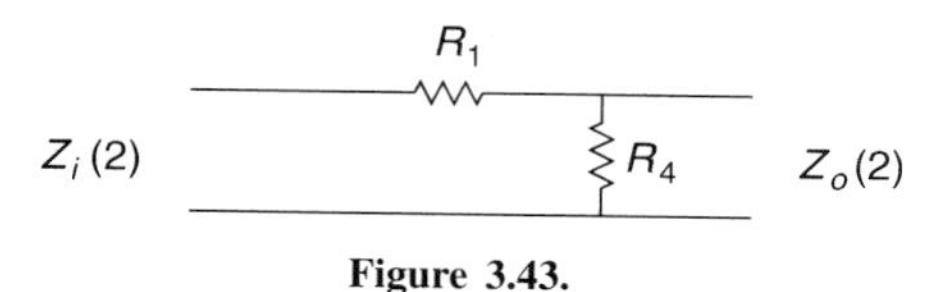

Figure 3.43.

Since the resistors R_1 and R_4 are connected in series,

$$\begin{aligned} Z_i(2) = R_1 + R_4 &= R_1 + \frac{R_2(R_1 + R_2)}{R_1 + 2R_2} \\ &= \frac{R_1(R_1 + 2R_2) + R_2(R_1 + R_2)}{R_1 + 2R_2} \end{aligned} \tag{3.2}$$

To compute the output impedance Z_o of the circuit in Figure 3.41, we traverse it in the opposite direction, that is, from left to right. The first resistor R_1 plays no role in its computation, so we simply ignore it (see Fig. 3.44). The resistors R_1 and R_2 are in series, so they can be replaced by a resistor $R_3 = R_1 + R_2$ (see Fig. 3.45). This yields a circuit with two parallel resistors R_3 and R_2, so

$$\frac{1}{Z_o(2)} = \frac{1}{R_3} + \frac{1}{R_2} = \frac{1}{R_1 + R_2} + \frac{1}{R_2} = \frac{R_1 + 2R_2}{R_2(R_1 + R_2)}$$

$$\therefore Z_o(2) = \frac{R_2(R_1 + R_2)}{R_1 + 2R_2} \tag{3.3}$$

(see Fig. 3.46).

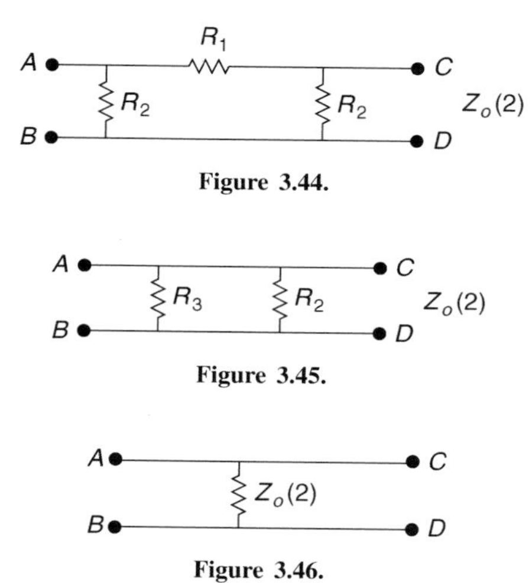

Figure 3.44.

Figure 3.45.

Figure 3.46.

Then

$$A(2) = \frac{Z_o(2)}{Z_i(2)}$$
$$= \frac{R_2(R_1 + R_2)}{R_1(R_1 + 2R_2) + R_2(R_1 + R_2)} \tag{3.4}$$

Now consider a ladder of $n = 3$ sections (see Fig. 3.47). Using Figures 3.48–3.51, it follows that

$$R_3 = R_1 + R_2 \qquad R_4 = \frac{R_2(R_1 + R_2)}{R_1 + 2R_2}$$

$$R_5 = R_1 + R_4 = R_1 + \frac{R_2(R_1 + R_2)}{R_1 + 2R_2}$$
$$= \frac{R_1(R_1 + 2R_2) + (R_1 + 2R_2)}{R_1 + 2R_2}$$

$$\frac{1}{R_6} = \frac{1}{R_2} + \frac{1}{R_5} = \frac{1}{R_5} + \frac{R_1 + 2R_2}{R_1(R_1 + 2R_2) + R_2(R_1 + R_2)}$$

$$\therefore R_6 = \frac{R_2(R_1^2 + 3R_1R_2 + R_2^2)}{R_1^2 + 4R_1R_2 + 3R_2^2}$$

$$Z_i(3) = R_1 + R_6 = R_1 + \frac{R_2(R_1^2 + 3R_1R_2 + R_2^2)}{R_1^2 + 4R_1R_2 + 3R_2^2}$$
$$= \frac{R_1^3 + 5R_1^2R_2 + 6R_1R_2^2 + R_2^3}{R_1^2 + 4R_1R_2 + 3R_2^2} \tag{3.5}$$

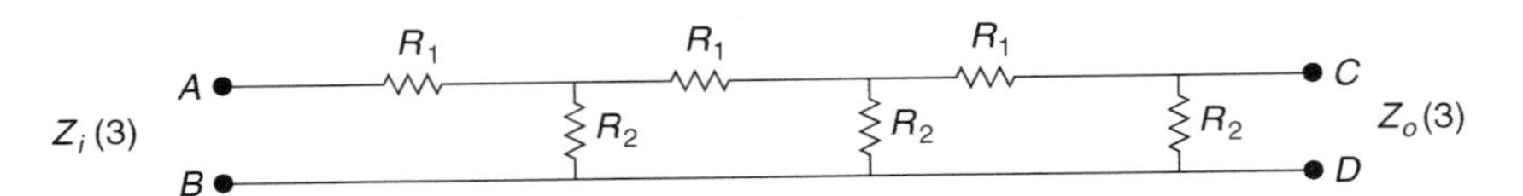

Figure 3.47.

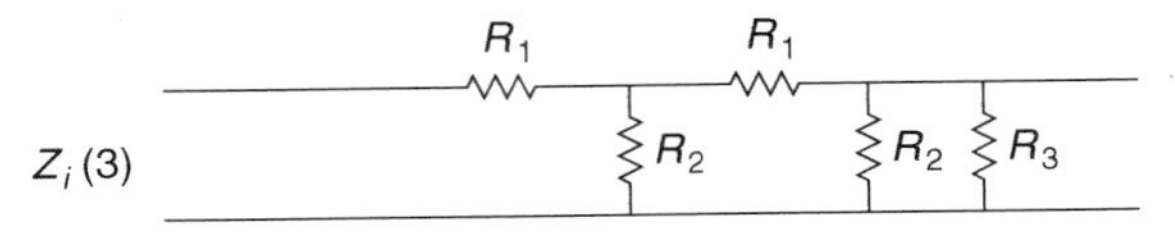

Figure 3.48.

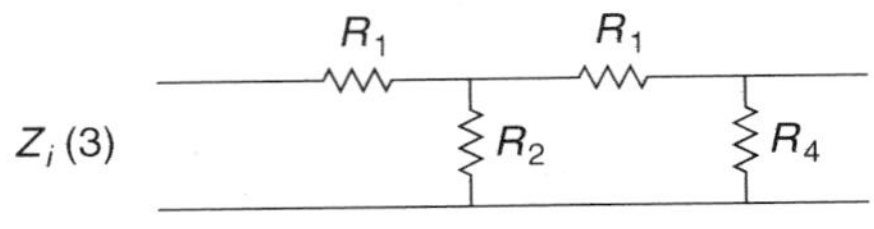

Figure 3.49.

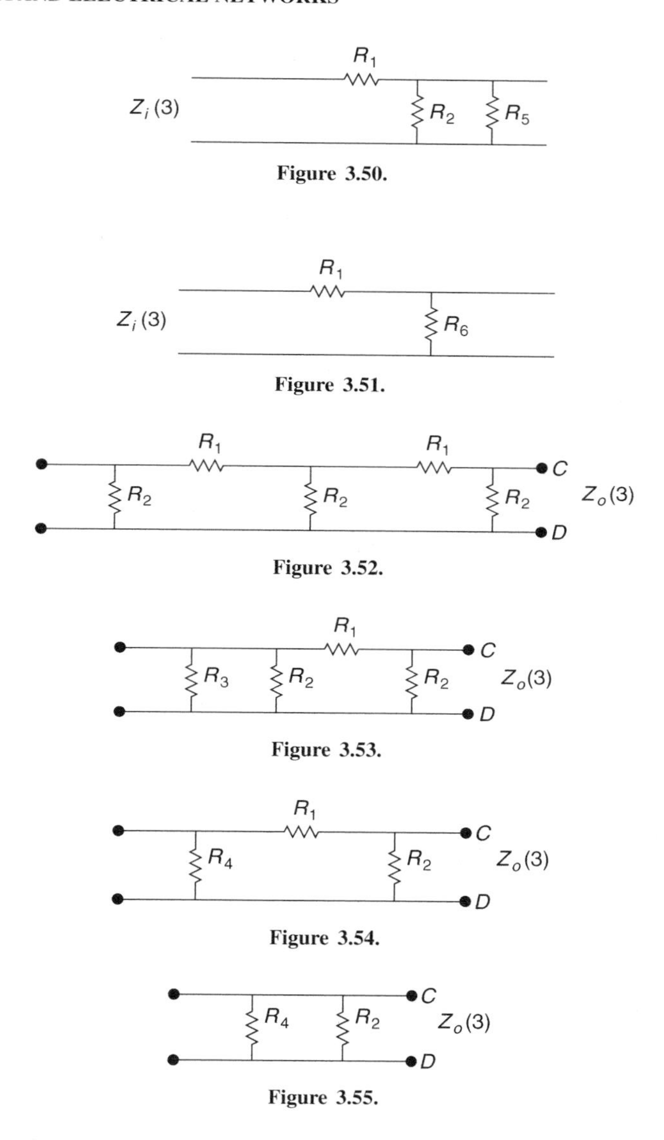

Using the same method employed for the case $n = 2$ and Figures 3.52–3.55, we have the following results:

$$R_3 = R_1 + R_2 \qquad \frac{1}{R_4} = \frac{1}{R_3} + \frac{1}{R_2} = \frac{1}{R_1 + R_2} + \frac{1}{R_2}$$

$$R_4 = \frac{R_2(R_1 + R_2)}{R_1 + 2R_2}$$

$$\begin{aligned}
R_5 &= R_1 + R_4 = R_1 + \frac{R_2(R_1 + R_2)}{R_1 + 2R_2} \\
&= \frac{R_1(R_1 + 2R_2) + R_2(R_1 + R_2)}{R_1(R_1 + 2R_2)} \\
\frac{1}{Z_o(3)} &= \frac{1}{R_5} + \frac{1}{R_2} = \frac{R_1(R_1 + 2R_2)}{R_1(R_1 + 2R_2) + R_2(R_1 + R_2)} + \frac{1}{R_2} \\
&= \frac{R_1^2 + 4R_1R_2 + 3R_1^2}{R_2(R_1^2 + 3R_1R_2 + 3R_1^2)} \\
\therefore Z_o(3) &= \frac{R_2(R_1^2 + 3R_1R_2 + R_2^2)}{R_1^2 + 4R_1R_2 + 3R_2^2} \qquad (3.6)
\end{aligned}$$

So

$$A(3) = \frac{R_2(R_1^2 + 3R_1R_2 + R_2^2)}{R_1^3 + 5R_1^2R_2 + 6R_1R_2^2 + R_2^3} \qquad (3.7)$$

In particular, let $R_1 = R_2 = 1$ ohm. Then Eqs. 3.1 through 3.7 yield the following result:

$$\begin{array}{lll}
Z_o(1) = \frac{1}{1} & Z_i(1) = \frac{2}{1} & A(1) = \frac{2}{1} \\
Z_o(2) = \frac{2}{3} & Z_i(2) = \frac{5}{3} & A(2) = \frac{5}{2} \\
Z_o(3) = \frac{5}{8} & Z_i(3) = \frac{13}{8} & A(3) = \frac{13}{5}
\end{array}$$

More generally, we predict that

$$Z_o(n) = \frac{F_{2n-1}}{F_{2n}}, \qquad Z_i(n) = \frac{F_{2n+1}}{F_{2n}}, \qquad \text{and} \qquad A(n) = \frac{F_{2n+1}}{F_{2n-1}}.$$

That these are true for a ladder network of $n \geq 1$ resistors can be established using PMI.

To prove that $Z_o(n) = F_{2n-1}/F_{2n}$, where $n \geq 1$: Since $Z_o(n) = 1/1 = F_1/F_1$, the statement is clearly true when $n = 1$. Assume it is true for an arbitrary number $k \geq 1$ resistors:

$$Z_o(k) = \frac{F_{2k-1}}{F_{2k}}$$

To show that the formula works for $n = k + 1$, consider a ladder network with $k + 1$ resistors, as in Figure 3.56. By the inductive hypothesis, the first k sections can be replaced by a resistor of resistance $Z_o(k)$. This yields the circuit in Figure 3.57.

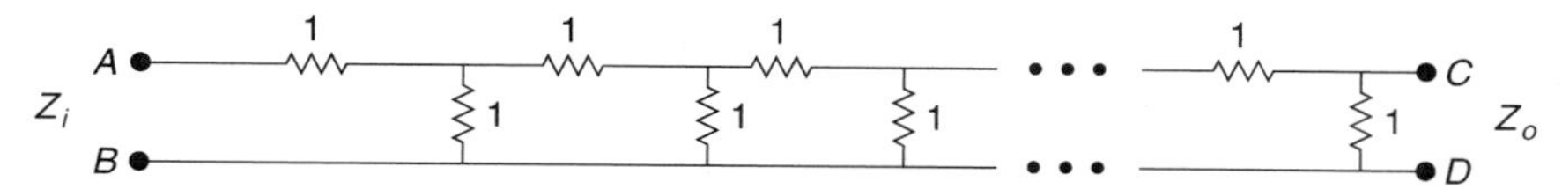

Figure 3.56. $k + 1$ ladder sections.

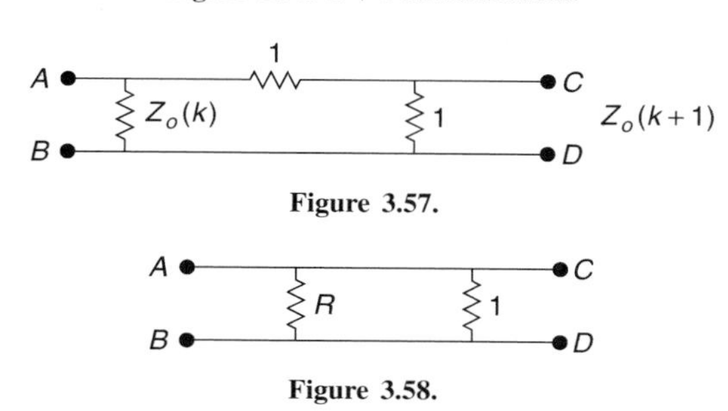

Figure 3.57.

Figure 3.58.

Using Figures 3.57 and 3.58, we have

$$R = Z_o(k) + 1 = \frac{F_{2k-1}}{F_{2k}} + 1 = \frac{F_{2k+1}}{F_{2k}}$$

$$\frac{1}{Z_o(k+1)} = \frac{1}{R} + \frac{1}{1} = \frac{F_{2k}}{F_{2k+1}} + 1 = \frac{F_{2k+2}}{F_{2k+1}}$$

$$\therefore \qquad Z_o(k+1) = \frac{F_{2k+1}}{F_{2k+2}}$$

So the formula works for $n = k + 1$. Thus, by PMI, $Z_o(n) = F_{2n-1}/F_{2n}$ for every ladder network of $n \geq 1$ resistors.

It can be similarly established that $Z_i(n) = F_{2n+1}/F_{2n}$, and hence $A(n) = F_{2n+1}/F_{2n-1}$ for all $n \geq 1$.

EXERCISES 3

1. Let b_n denote the number of distinct paths the bee in Figure 3.12 can take to crawl to cell n. Show that $b_n = F_{n+1}$, $n \geq 1$.

Exercises 2–10 require a knowledge of binary trees

The *Fibonacci tree* B_n, a binary tree, is defined recursively as follows: both B_1 and B_2 consist of a single vertex; when $n \geq 3$, B_n has a root, a left subtree B_{n-1}, and a right subtree T_{n-2}.

2. Draw the first five Fibonacci trees.
3. Is B_n a full binary tree?
4. Is B_6 a balanced binary tree?
5. Is B_5 a complete binary tree?
6. For what values of n is B_n a complete binary tree?
7. How many leaves l_n does B_n have?

Use B_n to find the following.

8. The number of internal vertices i_n.
9. The number of vertices v_n.
10. The height h_n.

Let $f(n, k)$ denote the element in row n and column k of the triangular array in Table 3.8, where $n \geq 1$ and $k \geq 0$.

11. Find $f(7, 2)$ and $f(10, 4)$.
12. Define $f(n, k)$ recursively.

Let $g(n, k)$ denote the element in row n and column k of the triangular array in Table 3.9, where $n \geq 1$ and $k \geq 0$.

13. Find $g(7, 2)$ and $g(10, 4)$.
14. Define $g(n, k)$ recursively.

FIBONACCI NUMBERS: ADDITIONAL OCCURRENCES

Fibonacci numbers appear in still many other unexpected places. For example, index cards are usually made in size 2×3 or 3×5; most oriental rugs come in five different sizes: 2×3, 3×5, 4×6, 6×9, or 9×12. In the first two cases, the dimensions are adjacent Fibonacci numbers; in the third and fourth cases, the ratio $4 : 6 = 6 : 9$ is the same as the ratio $2 : 3$; and in the last case, the ratio $9 : 12$ is the ratio $3 : 4$ of two adjacent Lucas numbers.

Before turning to our next example, we must make a formal definition of a *word*. A *word* is an ordered arrangement of symbols; it does not need to have a meaning. For example, *abc* is a word using the letters of the English alphabet, whereas 001101 is a binary word. A *bit* is a 0 (zero) or a 1 (one).

Example 4.1. Let a_n denote the number of n-bit words containing *no* two consecutive 1s. Define a_n recursively.

Solution. First, let us find the n-bit words containing no two consecutive 1s corresponding to $n = 1, 2, 3$, and 4 (see Table 4.1). It follows from the table that $a_1 = 2, a_2 = 3, a_3 = 5$, and $a_4 = 8$.

Now consider an arbitrary n-bit word. It may end in 0 or 1.

TABLE 4.1.

$n = 1$	$n = 2$	$n = 3$	$n = 4$
0	00	000	0000
1	01	010	0100
	10	100	1000
		001	0010
		101	1010
			0001
			0101
			1001

Case 1. Suppose the n-bit word ends in 0. Then the $(n - 1)$st bit can be a 0 or a 1, so there are no restrictions on the $(n - 1)$st bit:

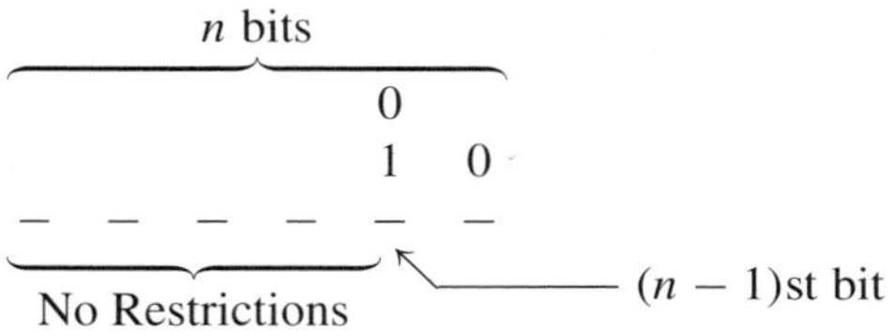

$$\overbrace{\underbrace{\_\ \_\ \_\ \_}_{\text{No Restrictions}}\ \ \begin{matrix}0\\1\\\_\end{matrix}\ \ \begin{matrix}\\0\\\_\end{matrix}}^{n\text{ bits}} \quad \nwarrow\text{——— } (n-1)\text{st bit}$$

Therefore, a_{n-1} n-bit words end in 0 and contain no two consecutive 1s.

Case 2. Suppose the n-bit word ends in 1. Then the $(n - 1)$st bit must be a zero. Further, there are no restrictions on the $(n - 2)$nd bit:

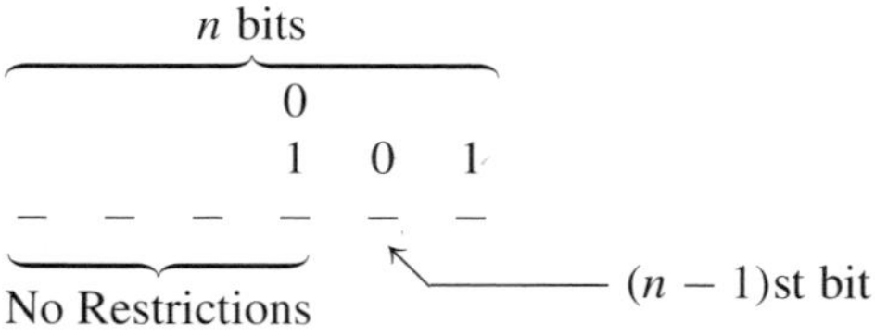

$$\overbrace{\underbrace{\_\ \_\ \_}_{\text{No Restrictions}}\ \ \begin{matrix}0\\1\\\_\end{matrix}\ \ \begin{matrix}\\0\\\_\end{matrix}\ \ \begin{matrix}\\1\\\_\end{matrix}}^{n\text{ bits}} \quad \nwarrow\text{——— } (n-1)\text{st bit}$$

Thus a_{n-2} n-bit words end in 1 and contain no two consecutive 1s.

Since the two cases are mutually exclusive, by the addition principle, we have:

$$a_1 = 2, \qquad a_2 = 3 \qquad \leftarrow \text{Initial conditions}$$

$$a_n = a_{n-1} + a_{n-2}, \qquad n \geq 3 \qquad \leftarrow \text{Recurrence relation}$$

Notice that this recurrence relation is exactly the same as the Fibonacci recurrence relation, but with different initial conditions. The resulting numbers are the Fibonacci numbers 2, 3, 5, 8, 13, ... Accordingly, $a_n = F_{n+2}$. ■

This example does not provide a constructive method for systematically listing all n-bit words with the desired property. That method is given in Exercise 4.1.

Example 4.1 can be interpreted as follows. Suppose n coins are flipped sequentially. The total number of outcomes so that no two consecutive coins fall heads is F_{n+2}. Consequently, the probability that no two adjacent coins fall heads is $F_{n+2}/2^n$.

Example 4.2. An n-storied apartment building needs to be painted green or yellow in such a way that no two adjacent floors can be painted yellow, where $n \geq 1$. Let b_n denote the number of ways of painting the building.

Figure 4.1 shows the various possible ways of painting the building when $n = 1$, 2, 3, and 4. It follows from the figure that $b_1 = 2$, $b_2 = 3$, $b_3 = 5$, and $b_4 = 8$.

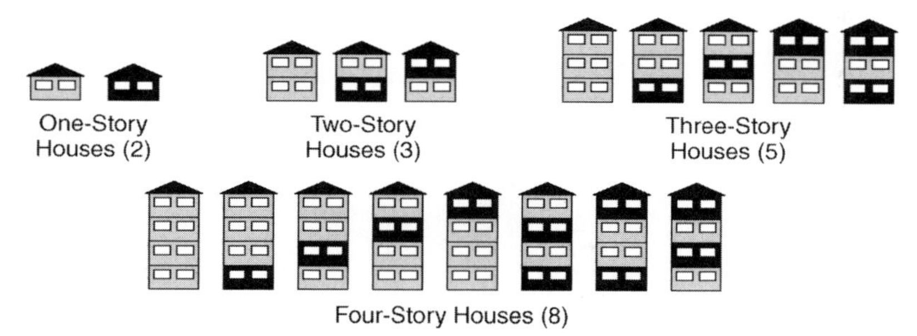

Figure 4.1. Possible ways of painting the building.

This problem is essentially the same as Example 4.1. With green = 0 and yellow = 1, every n-bit word that contains no consecutive 1s represents a possible way of painting, and vice versa. Thus $b_n = F_{n+2}$, where $n \geq 1$. ■

Example 4.3. Let a_n denote the number of $2 \times n$ rectangular grids that can be formed using n 1×2 dominoes. Find a formula for a_n.

Solution. When $n = 1$, $a_1 = 1$ (see Fig. 4.2). When $n = 2$, two 2×2 rectangular grids can be formed, so $a_2 = 2$ (see Fig. 4.3).

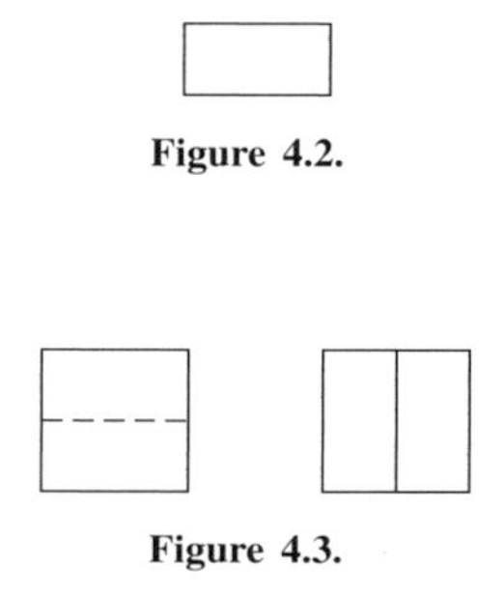

Figure 4.2.

Figure 4.3.

More generally, let $n \geq 3$. Since the pattern can begin with one horizontal domino or two vertical dominoes, it follows that $a_n = a_{n-1} + a_{n-2}$. This recurrence relation is exactly the same as the Fibonacci one with the initial conditions $a_0 = 1$, $a_1 = 2$. Thus it follows that $a_n = F_{n+1}$. ■

Example 4.4. Let a_n denote the number of subsets of the set $S_n = \{1, 2, \ldots, n\}$ that do not contain consecutive integers, where $n \geq 0$. We define $S_0 = \emptyset$. Find an explicit formula for a_n.*

Solution. To get an idea about a_n, let us find its value for $n = 0, 1, 2, 3$, and 4 by constructing a table, as in Table 4.2. It appears from the table that a_n is a Fibonacci number and $a_n = F_{n+2}$. We shall in fact prove that $a_n = F_{n+2}$ in two steps: first we shall define a_n recursively and then find an explicit formula.

TABLE 4.2.

n	Subsets of S That Do Not Contain Consecutive Integers	a_n
0	$\emptyset$	1
1	$\emptyset, \{1\}$	2
2	$\emptyset, \{1\}, \{2\}$	3
3	$\emptyset, \{1\}, \{2\}, \{3\}, \{1, 3\}$	5
4	$\emptyset, \{1\}, \{2\}, \{3\}, \{4\}, \{1, 3\}, \{1, 4\}, \{2, 4\}$	8

↑
F_{n+2}

To Define a_n Recursively. From Table 4.2, $a_0 = 1$ and $a_1 = 2$. So let $n \geq 2$. Let A be a subset of S_n that does not contain two consecutive integers. Then either $n \in A$ or $n \notin A$.

Case 1. Suppose $n \in A$. Then $n - 1 \notin A$. By definition, $S_{n-2} = \{1, 2, \ldots, n-2\}$ has a_{n-2} subsets not containing two consecutive integers. Add n to each of the subsets. The resulting sets are subsets of S_n satisfying the desired property, so S_n has a_{n-2} such subsets.

Case 2. Suppose $n \notin A$. By definition, there are a_{n-1} such subsets of S_n having the required property.

Since these two cases are mutually exclusive, by the addition principle, $a_n = a_{n-1} + a_{n-2}$, where $a_0 = 1$, $a_1 = 2$. It follows that $a_n = F_{n+2}$, $n \geq 0$. ■

Although Examples 4.1 and 4.4 look different, they are basically the same. For instance, the subsets of $S_3 = \{1, 2, 3\}$ that do not contain consecutive integers are $\emptyset$, $\{1\}$, $\{2\}$, $\{3\}$, and $\{1, 3\}$. Using the correspondence $\emptyset \leftrightarrow 000$, $\{1\} \leftrightarrow 100$, $\{2\} \leftrightarrow 010$, $\{3\} \leftrightarrow 001$, and $\{1, 3\} \leftrightarrow 101$, we can recover all 3-bit words that do not contain consecutive 1s.

*Proposed by Irving Kaplansky of The University of Chicago, Illinois.

More generally, let A be a subset of S_n that does not contain consecutive integers. The corresponding n-bit word has a 1 in position i if and only if $i \in A$, where $1 \leq i \leq n$.

Example 4.5. A subset of the set $S = \{1, 2, \ldots, n\}$ is said to be *alternating* if its elements, when arranged in increasing order, follow the pattern odd, even, odd, even, and so on. For example, $\{3\}$, $\{1, 2, 5\}$, and $\{3, 4\}$ are alternating subsets of $\{1, 2, 3, 4, 5\}$, whereas $\{1, 3, 4\}$ and $\{2, 3, 4, 5\}$ are not; $\oslash$ is considered alternating.* Let a_n denote the number of alternating subsets of S. Prove that $a_n = F_{n+2}$, where $n \geq 0$.

Solution. As in Example 4.4, let us collect some data on a_n to get a feel for it. It follows from Table 4.3 that $a_n = F_{n+2}$, where $n \geq 0$. To prove this, let A be an alternating subset of S.

TABLE 4.3.

n	Alternating Subsets of S	a_n
0	$\oslash$	1
1	$\oslash, \{1\}$	2
2	$\oslash, \{1\}, \{1, 2\}$	3
3	$\oslash, \{1\}, \{3\}, \{1, 2\}, \{1, 2, 3\}$	5
4	$\oslash, \{1\}, \{3\}, \{1, 2\}, \{1, 4\}, \{3, 4\}, \{1, 2, 3\}, \{1, 2, 3, 4\}$	8
		$\uparrow$ F_{n+2}

Case 1. Suppose $1 \notin A$, so $2 \notin A$. This leaves $n - 2$ elements in S. So, by definition, they can be used to form a_{n-2} alternating subsets of S.

Case 2. Suppose $1 \in A$. This leaves $n - 1$ elements in S. They can be used to form a_{n-1} alternating subsets of the set $S - \{1\}$. Now adding 1 to each of them yields a_{n-1} alternating subsets of S, each containing 1.

Thus, by the addition principle, $a_n = a_{n-1} + a_{n-2}$, where $a_0 = 1$ and $a_1 = 2$. So $a_n = F_{n+2}$, where $n \geq 0$. ■

Example 4.6. Let a_n denote the number of ways n can be written as an ordered sum of odd positive integers, where $n \geq 1$. Table 4.4 shows the various possibilities for integers 1–5. It appears from the table that $a_n = F_n$. In fact, $a_n = a_{n-1} + a_{n-2}$, so the conjecture is in fact true.

*Proposed by Olry Terquem (1782–1862).

TABLE 4.4.

n	Desired Ordered Sums	a_n
1	1	1
2	$1+1$	1
3	$3, 1+1+1$	2
4	$1+3, 1+1+1+1, 3+1$	3
5	$5, 1+1+3, 1+3+1,$ $3+1+1, 1+1+1+1+1$	5

↑
F_n

■

COMPOSITIONS WITH ODD SUMMANDS

Example 4.6 dealt with *compositions*, which are ordered sums of a positive integer n. Next, we explore compositions with summands greater than one. This problem was studied in 1901 by E. Netto of Germany.

Example 4.7. Let b_n denote the number of compositions of a positive integer n using summands greater than 1. Table 4.5 shows the various possibilities for integers 1–7. Once again, it appears from the table that $b_n = F_{n-1}$.

TABLE 4.5.

n	Compositions	b_n
1	—	0
2	2	1
3	3	1
4	$2+2, 4$	2
5	$2+3, 3+2, 5$	3
6	$2+2+2, 2+4, 3+3, 4+2, 6$	5
7	$2+2+3, 2+3+2, 2+5, 3+2+2,$ $3+4, 4+3, 5+2, 7$	8

↑
F_{n-1}

To confirm this, note that $b_1 = 0$ and $b_2 = 1$. So let $n \geq 3$. Notice, for example, that three compositions of 7 can be obtained by adding 2 as a summand to every composition of 5: $2+3+2$, $3+2+2$, and $5+2$; the other five compositions can be obtained by adding 1 to the last summand of every composition of 6: $2+2+3$, $2+5$, $3+4$, $4+3$, and 7.

More generally, every composition of n can be obtained from the compositions of $n-2$ and those of $n-1$ by inserting 2 as a summand to every composition of $n-2$ and by adding a 1 to the last summand of every composition of $n-1$. Since there is no overlapping between the two procedures, it follows that $b_n = b_{n-1} + b_{n-2}$, where $b_1 = 0$ and $b_2 = 1$. Thus $b_n = F_{n-1}$, where $n \geq 1$, as conjectured. ■

Example 4.8. Let b_n denote the number of n-bit words $x_1x_2x_3\cdots x_n$, where $x_1 \le x_2$, $x_2 \ge x_3$, $x_3 \le x_4$, $x_4 \ge x_5, \ldots$. When $n = 2$, there are two such binary words: 01 and 11. Table 4.6 shows such n-bit words for $1 \le n \le 4$. Again, it appears from the table that $b_n = F_{n+2}$, where $n \ge 1$. This is also, in fact, true.

TABLE 4.6.

n	Desired n-Bit Words	b_n
1	0, 1	2
2	00, 01, 11	3
3	000, 010, 011, 110, 111	5
4	0000, 0100, 0001, 0101, 0111, 1100, 1101, 1111	8
		↑ F_{n+2}

■

The next example combines Fibonacci numbers with permutations. A *permutation* on a set S is a function $f: S \to S$ that is both one-to-one and onto. In other words, a permutation is nothing but a rearrangement of the elements of S.

Example 4.9. Let p_n denote the number of permutations f of the set $S_n = \{1, 2, \ldots, n\}$ such that $|i - f(i)| \le 1$ for all $1 \le i \le n$, where $n \ge 1$. In other words, p_n counts the number of permutations that moves each element no more than one position from its natural position.

Figure 4.4 shows the various permutations for $n = 1$, 2, 3, and 4, and Table 4.7 summarizes these data. Once again, it appears from the table that $p_n = F_{n+1}$. We can, in fact, confirm this.

Case 1. Let $f(n) = n$. Then the remaining $n - 1$ elements can be used to form p_{n-1} permutations such that $|i - f(i)| \le 1$ for all i.

Case 2. Let $f(n) \ne n$. Then $f(n) = n - 1$ and $f(n - 1) = n$. The remaining $n - 2$ elements can be employed to form p_{n-2} permutations with the desired property.

Thus, by the addition principle, $p_n = p_{n-1} + p_{n-2}$, where $p_1 = 1$ and $p_2 = 2$. It now follows that $p_n = F_{n+1}$, where $n \ge 1$, as conjectured. ■

Since the total number of permutations of S_n is $n!$, it follows from this example that there are $n! - F_{n+1}$ permutations f of S_n such that $|i - f(i)| > 1$ for some integer i, where $1 \le i \le n$. In other words, there are $n! - F_{n+1}$ permutations of S_n that move at least one element of S_n by two spaces from its natural position.

In particular, there are $3! - F_4 = 3$ such permutations of the set $\{1, 2, 3\}$, as Figure 4.5 depicts.

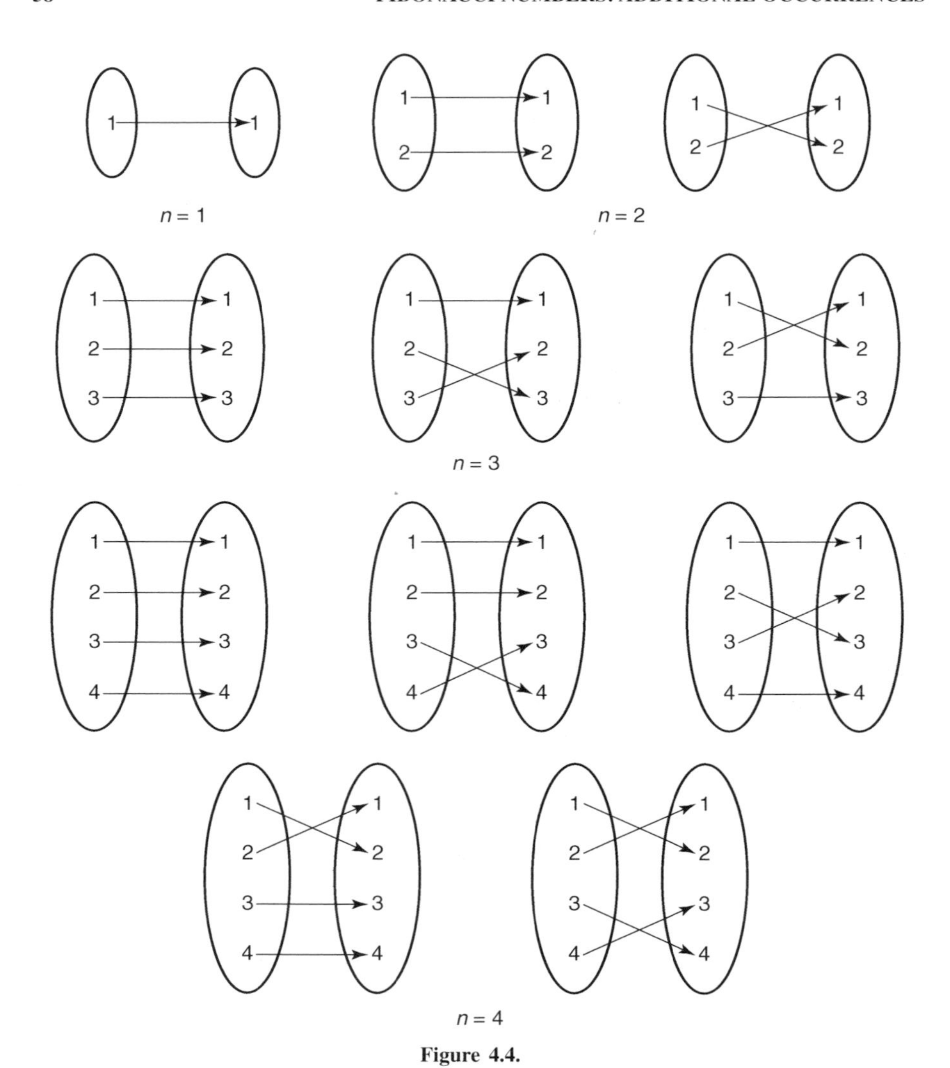

Figure 4.4.

TABLE 4.7.

n	1	2	3	4	$\cdots$	n
p_n	1	2	3	5	$\cdots$	?

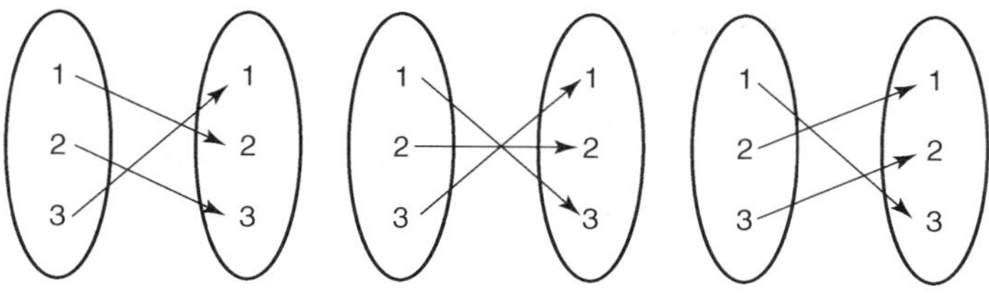

Figure 4.5.

GENERATING SETS AND FIBONACCI NUMBERS

The following example shows that Fibonacci numbers occur in the study of generating sets.

Example 4.10. Let $[n] = \{1, 2, 3, \ldots, n\}$, where $n \geq 1$. Let S be a nonempty subset of $[n]$. Let $S+1 = \{s+1 | s \in S\}$. For instance, let $S = \{1, 3, 6\}$, then $S+1 = \{2, 4, 6\}$.

A nonempty subset S of $[n]$ is said to *generate* $[n+1]$ if $S \cup (S+1) = [n+1]$. For example, let $n = 7$ and $S = \{1, 3, 5, 7\}$. Then $S+1 = \{2, 4, 6, 8\}$ and $S \cup (S+1) = \{1, 2, 3, 4, 5, 6, 7, 8\} = [8]$, so S generates the set $[8]$, as does the set $\{1, 3, 4, 5, 7\}$, but *not* $\{1, 3, 4, 7\}$.

Let s_n denote the number of subsets of $[n]$ that generate $[n+1]$. Clearly, $s_1 = 1 = s_2$. There are two subsets of $[3]$ that generate $[4]$: $\{1, 3\}$ and $\{1, 2, 3\}$: so $s_3 = 2$. There are three subsets of $[4]$ that generate $[5]$; they are $[1, 2, 4\}$, $\{1, 3, 4\}$, and $\{1, 2, 3, 4\}$. Thus $s_4 = 3$.

There are $s_5 = 5$ subsets of $[5]$ that generate the set $[6]$. Three of these subsets can be obtained by inserting the element 5 in each of the subsets $\{1, 2, 4\}$, $\{1, 3, 4\}$, and $\{1, 2, 3, 4\}$: $\{1, 2, 4, 5\}$, $\{1, 3, 4, 5\}$, and $\{1, 2, 3, 4, 5\}$. The remaining two can be obtained by inserting 5 in each of the subsets $\{1, 3\}$ and $\{1, 2, 3\}$: $\{1, 3, 5\}$ and $\{1, 2, 3, 5\}$. Thus s_5 equals the number of subsets of $[4]$ that generate $[5]$, plus that of $[3]$ that generate $[4]$.

More generally, the subsets of $[n]$ that generate $[n+1]$ can be obtained by inserting n in each of the subsets of $[n-1]$ that generate $[n]$, and by inserting n in each of the subsets of $[n-2]$ that generate $[n-1]$:

$$\begin{aligned} \therefore \quad s_n &= \text{Number of subsets of } [n-1] \text{ that generate}[n] \\ &\quad + \text{number of subsets of}[n-2]\text{that generate}[n-1] \\ &= s_{n-1} + s_{n-2} \end{aligned}$$

where $s_1 = 1 = s_2$. Thus $s_n = F_n$. ■

The next three examples link Fibonacci numbers with graphs, so it is helpful here to present additional basic terminology from graph theory.

BASIC GRAPH TERMINOLOGY

Recall that a *graph* $G = (V, E)$ consists of a set V of points, called *vertices*, and a set E of arcs or line segments, called *edges*, joining them. An edge connecting vertices v and w is denoted by v–w. A vertex v is *adjacent* to vertex w if there is an edge connecting them.

For example, the graph in Figure 4.6 has four vertices—A, B, C, and D—and seven edges. Vertex A is adjacent to B, but *not* to C.

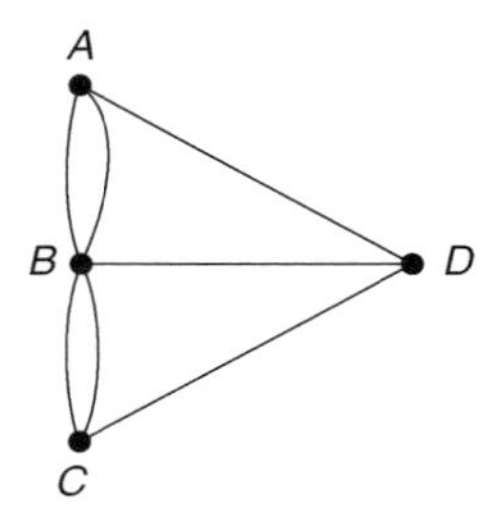

Figure 4.6.

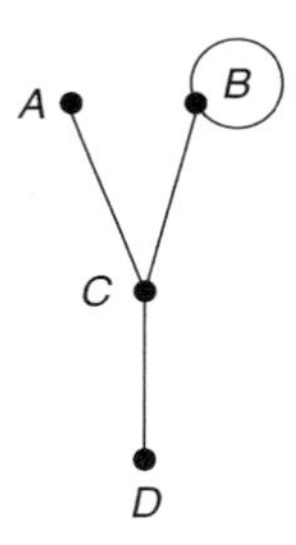

Figure 4.7.

An edge emanating from and terminating at the same vertex is a *loop*. *Parallel* edges have the same vertices. A loop-free graph that contains no parallel edges is a *simple graph*.

For example, the graph in Figure 4.7 has a loop at B, and the one in Figure 4.6 has parallel edges, while the graph in Figure 4.8 contains no loops or parallel edges, making it a simple graph.

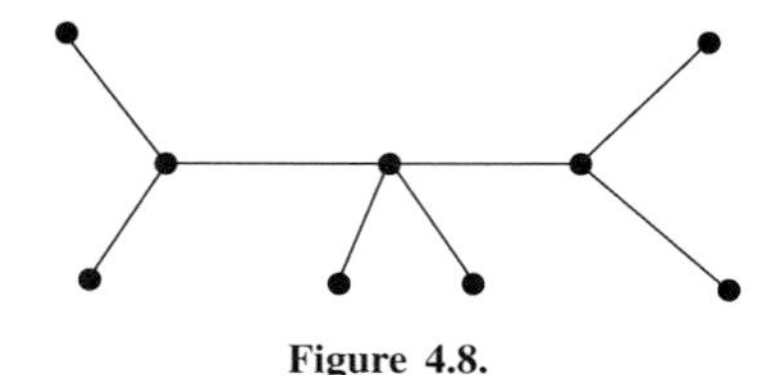

Figure 4.8.

A *subgraph* $H = (V', E')$ is a graph such that $V' \subseteq V$ and $E' \subseteq E$. A *path* between two vertices v_0 and v_n is a sequence $v_0 - v_1 - v_2 - \cdots - v_n$ of vertices and edges connecting them; its *length* is n. A graph is *connected* if there is a path between every two distinct vertices. For instance, consider the graph in Figure 4.6. The length of the path A–B is one and that of A–B–A–B–C is four.

Independent Subset of the Vertex Set

Let V denote the set of vertices of a graph. A subset I of V is *independent* if no two vertices in I are adjacent. In other words, if I is independent and $v, w \in I$, then the edge v–w does not exist. For example, consider the graph in Figure 4.9. Then $\{a, c, e\}$ and $\{b, d\}$ are independent, whereas $\{a, c, d, f\}$ is *not*.

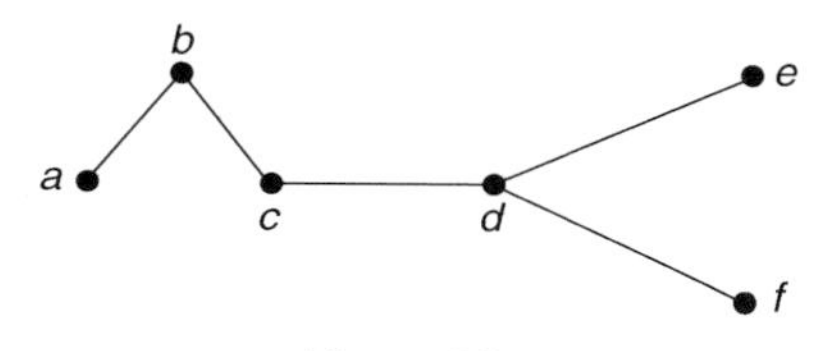

Figure 4.9.

We are now ready to examine the next graph-theoretic example.

Example 4.11. Let P_n denote the path $v_0 - v_1 - v_2 - \cdots - v_n$ of length n connecting the vertices $v_0, v_1, v_2, \ldots,$ and v_n in a simple graph, where $n \geq 0$. Let C_n denote the number of independent subsets of vertices of the path.

When $n = 0$, the path P_0 consists of a single point v_0, so there are two possible independent subsets: $\emptyset$, $\{v_0\}$.

When $n = 1$, the path is v_0–v_1. Then there are three independent subsets of $\{v_0, v_1\}$, namely, $\emptyset$, $\{v_0\}$, and $\{v_1\}$.

When $n = 2$, the path P_2 contains three vertices: v_0, v_1, and v_2. Accordingly, there are five independent subsets: $\emptyset$, $\{v_0\}$, $\{v_1\}$, $\{v_2\}$, and $\{v_0, v_2\}$.

These data are summarized in Table 4.8. Clearly, a pattern emerges. It seems safe to conjecture that $C_n = F_{n+3}$, where $n \geq 0$.

TABLE 4.8.

n	Path P_n	Independent Subsets	C_n
0	v_0	$\emptyset$, $\{v_0\}$	2
1	v_0 v_1	$\emptyset$, $\{v_0\}$, $\{v_1\}$	3
2	v_0 v_1 v_2	$\emptyset$, $\{v_0\}$, $\{v_1\}$, $\{v_2\}$, $\{v_0, v_2\}$	5
3	v_0 v_1 v_2 v_3	$\emptyset$, $\{v_0\}$, $\{v_1\}$, $\{v_2\}$, $\{v_3\}$, $\{v_0, v_2\}$, $\{v_0, v_3\}$, $\{v_1, v_3\}$	8

$\uparrow$
F_{n+3}

In fact, this example is basically the same as Example 4.4. It involves counting the number of subsets of the set $\{0, 1, 2, 3, \ldots, n\}$, where no subsets contain consecutive integers. Thus $C_n = F_{n+3}$, where $n \geq 0$. ■

We need a few more definitions before moving on to the next example.

A Few More Definitions

A *cycle* is a path with the same endpoints; it contains no repeated vertices. A graph is *acyclic* if it contains no cycles. A connected, acyclic graph is a *tree*. A tree with n vertices has exactly $n - 1$ edges.

For example, the graph in Figure 4.10 shows the family tree of the Bernoullis of Switzerland, the most distinguished family of mathematicians. The graph in Figure 4.11 is not a tree because it is cyclic. The tree in Figure 4.10 contains a specially designated vertex, called the *root*. Its root is it Nicolaus. The basic terminology of (rooted) trees reflects that of a family tree.

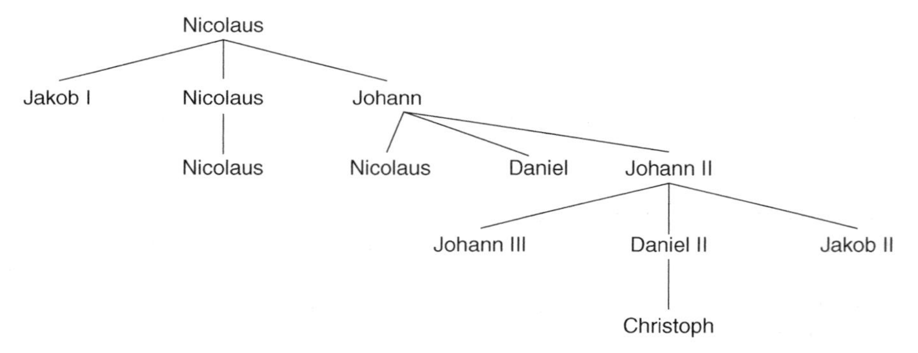

Figure 4.10. The Bernoulli family tree.

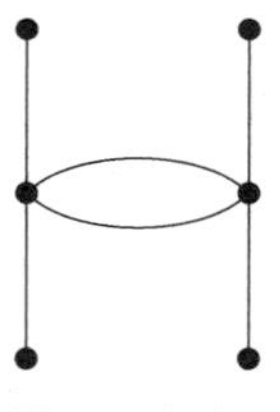

Figure 4.11.

Parent, Child, Sibling, Ancestor, Descendant, and Subtree. Let T be a tree with root v_0. Let $v_0 - v_1 - \cdots - v_{n-1} - v_n$ be the path from v_0 to v_n. Then:

- v_{i-1} is the *parent* of v_i.
- v_i is a *child* of v_{i-1}.

- The vertices $v_0, v_1, \ldots, v_{n-1}$ are *ancestors* of v_n.
- The *descendants* of a vertex v are those vertices for which v is an ancestor.
- A vertex with no children is a *leaf* or a *terminal vertex*.
- A vertex that is not a leaf is an *internal vertex*.
- The *subtree* rooted at v consists of v, its descendants, and all its edges.

For example, consider the tree in Figure 4.12. It is rooted at a. Vertex b is the parent of both e and f, so e and f are the children of b. Vertices a, b, and e are ancestors of i. Vertices b and e are descendants of a. Vertex f has no children, so it is a leaf. Vertices b and d have at least one child, so both are internal vertices. Figure 4.13 displays the subtree rooted at b.

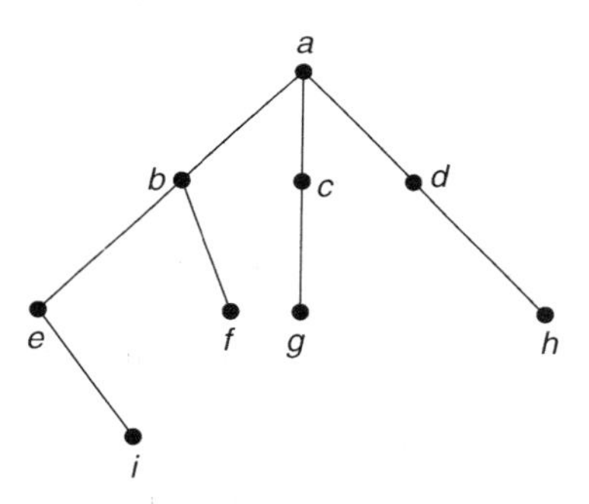

Figure 4.12.

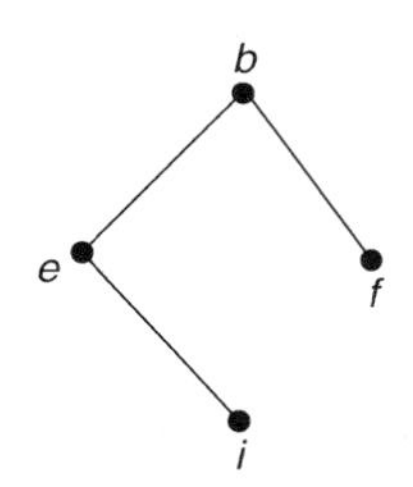

Figure 4.13.

Binary Tree. An *ordered rooted tree* is a rooted tree in which the vertices at each level are ordered as the first, second, third, and so on. Such a tree is a *binary tree* if every vertex has at most two children. For example, the tree in Figure 4.13 is a binary tree. Its *left subtree* is the binary tree rooted at e, and its *right subtree* is the binary tree rooted at f.

We are now ready to explore Fibonacci trees and their relationships with Fibonacci numbers.

Fibonacci Tree. The nth *Fibonacci tree* T_n is a binary tree, defined recursively as follows, where $n \geq 1$:

- Both T_1 and T_2 are binary trees with exactly one vertex each.
- Let $n \geq 3$. Then T_n is a binary tree whose left subtree is T_{n-1} and whose right subtree is T_{n-2}.

Figure 4.14 shows the Fibonacci trees T_1 through T_6.

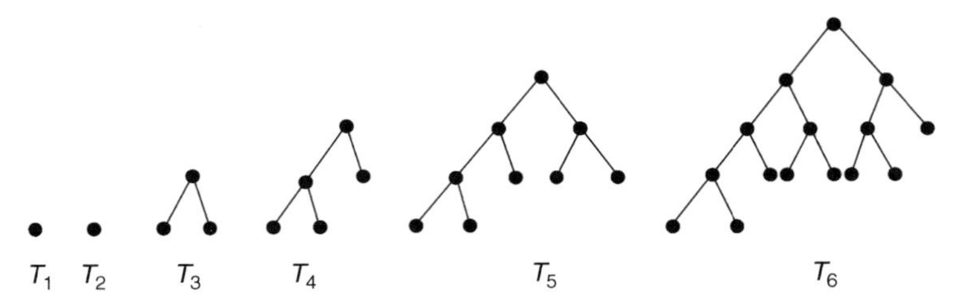

Figure 4.14. Fibonacci trees.

Next, we explore the number of vertices v_n, the number of leaves ℓ_n, the number of internal vertices i_n, and the number of edges e_n of a Fibonacci tree T_n. To facilitate our study, let us collect the needed data from Figure 4.14 and summarize them in tabular form, as in Table 4.9.

Using the table, we conjecture that $v_n = 2F_n - 1$, $\ell_n = F_n$, $i_n = \ell_n - 1 = F_n - 1$, and $e_n = v_n - 1 = 2F_n - 2$. Using the recursive definition of T_n, it is fairly easy to establish these results, as the next theorem shows.

TABLE 4.9.

n	1	2	3	4	5	6	$\cdots$	n
v_n	1	1	3	5	9	15	$\cdots$	?
ℓ_n	1	1	2	3	5	8	$\cdots$	?
i_n	0	0	1	2	4	7	$\cdots$	?
e_n	0	0	2	4	8	14	$\cdots$	?

Theorem 4.1. Let v_n, ℓ_n, i_n, and e_n denote the numbers of vertices, leaves, internal vertices, and edges of a Fibonacci tree T_n, where $n \geq 1$. Then $v_n = 2F_n - 1$, $\ell_n = F_n$, $i_n = F_n - 1$, and $e_n = 2F_n - 2$.

Proof.

1. Clearly, the formula works when $n = 1$ and 2. Suppose $n \geq 3$. Since T_n has T_{n-1} as its left subtree and T_{n-2} as its right subtree, it follows that $v_n = v_{n-1} + v_{n-2} + 1$, where $v_1 = 1 = v_2$. Let $b_n = v_n + 1$. Then $b_n = b_{n-1} + b_{n-2}$, where $b_1 = 2 = b_2$. So $b_n = 2F_n$, and hence $v_n = 2F_n - 1$, where $n \geq 3$. Thus the formula works for $n \geq 1$.

2. By the recursive definition of T_n, it follows that $\ell_n = \ell_{n-1} + \ell_{n-2}$, where $\ell_1 = 1 = \ell_2$. Thus $\ell_n = F_n$, where $n \geq 1$.

3. Clearly, $i_n = v_n - \ell_n = (2F_n - 1) - F_n = F_n - 1$.

4. Again, by the recursive definition of T_n, $e_n = e_{n-1} + e_{n-2} + 2$, where $e_1 = 0 = e_2$. Let $c_n = e_n + 2$. Then $c_n = c_{n-1} + c_{n-2}$, where $c_1 = 2 = c_2$. Thus $c_n = 2F_n$, so $e_n = 2F_n - 2$, where $n \geq 1$. ■

FIBONACCI NUMBERS AND THE STOCK MARKET

In the 1930s, Ralph Nelson Elliot, an engineer by training, made an extensive study of the fluctuations in the U.S. stock market. The Dow Jones Industrials Average (DJIA), an indicator based on stocks of 30 top companies, is often used as a measure of stock market activity and hence of the health of the economy. The DJIA varies according to human optimism and pessimism, as reflected by market conditions. Nevertheless, according to Prechter and Frost, Elliott discovered, based on his study and observations, "that the ever-changing stock market tended to reflect a basic harmony found in nature and from this discovery developed a rational system of stock market analysis."

In 1939, Elliott expressed his analysis as a theoretical principle, which has since been called the *Elliott Wave Principle*. In practice, the wave principle, corresponds to the performance of the DJIA.

Elliott observed that the stock market unfolds according to a fundamental pattern comprising a complete cycle of eight waves. Each cycle consists of two phases, the *numbered phase* and the *lettered phase*, as seen in Figure 4.15. The numbered phase consists of eight waves: five upward waves and three downward waves. The upward waves 1, 3, and 5 are *impulse waves*, and they reflect optimism in the stock market; the downward waves 2 and 4 are *corrective waves*, and they are corrections to those impulses, indicating pessimism. Wave 2 corrects wave 1 and wave 4 corrects wave 3.

The upward trend, depicted by the sequence 1–2–3–4–5, is then corrected by the downward trend, namely, the lettered phase *a–b–c*; the downward trend is, in fact, made up of two downward waves, *a* and *c*, and one upward wave *b*. The five-wave sequence 1–2–3–4–5 indicates a *bull market*, whereas the corrective sequence *a–b–c* indicates a *bear market*. Thus, one complete cycle consists of the sequence 1–2–3–4–5–*a*–*b*–*c*. According to the wave principle, this cycle of upward and downward turns continues.

Now the numbered phase can be considered a wave, say, wave ①, and the lettered phase wave ②. Thus there are waves within waves, as Figure 4.16 shows. Counting ① and ② separately, we get two waves. The pattern (1)–(2)–(3)–(4)–(5) consists of 21 smaller waves, and the downward trend (*a*)–(*b*)–(*c*) consists of 13 smaller waves; so the pattern (1)–(2)–(3)–(4)–(5)–(*a*)–(*b*)–(*c*) consists of 34 smaller waves. Thus we have the following pattern:

① – ② = 2 waves

$$(1) - (2) - (3) - (4) - (5) - (a) - (b) - (c) = 8 \text{ waves}$$

$$1 - 2 - 3 - 4 - 5 - \cdots - 1 - 2 - 3 - 4 - 5 = 34 \text{ waves}$$

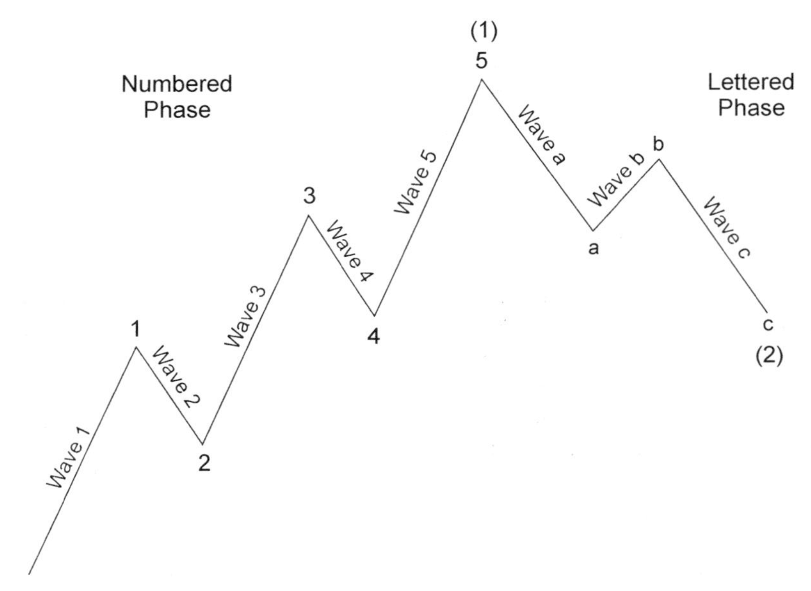

Figure 4.15. The Fundamental Behavior of the Elliott Wave Principle (*Source*: R. R. Prechter, Jr., and A. J. Frost, *Elliott Wave Principle—Key to Modern Behavior*, New Classics Library, Gainesville, GA, 1985. Copyright © 1978–2000, reproduced with permission from New Classics Library.).

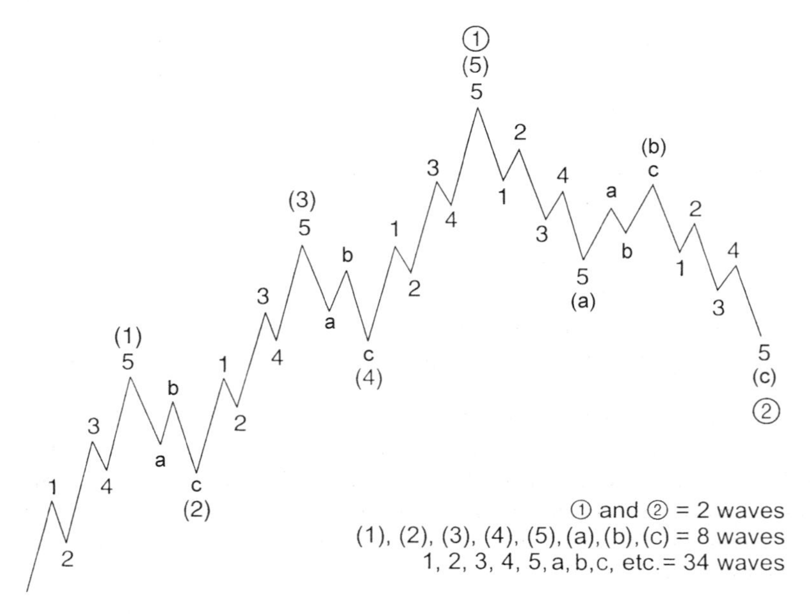

Figure 4.16. Waves Within Waves (*Source*: R. R. Prechter, Jr., and A. J. Frost, *Elliott Wave Principle—Key to Modern Behavior*, New Classics Library, Gainesville, GA, 1985. Copyright © 1978–2000, reproduced with permission from New Classics Library.).

That is, a wave of a large degree can be split into two waves of lower degree. These two waves can be divided into eight waves of next lower degree, and they in turn can be subdivided into 34 waves of even lower degree. This subdividing pattern also implies that waves can be combined to form waves of higher degrees. Whether waves are divided or combined, the underlying behavior remains invariant (see Fig. 4.17).

The complete cycle in Figure 4.17 comprises a bull market and a bear market. The bull market cycle consists of five primary waves, which can be subdivided into 21 intermediate waves, and they in turn can be resubdivided into 89 minor waves. The corresponding figures for the bear market are 1, 3, 13, and 55, respectively. This observation yields interesting dividends, as Table 4.10 shows. Odd as it may seem, according to the Elliott Wave Principle, this Fibonacci rhythmic pattern continues indefinitely.

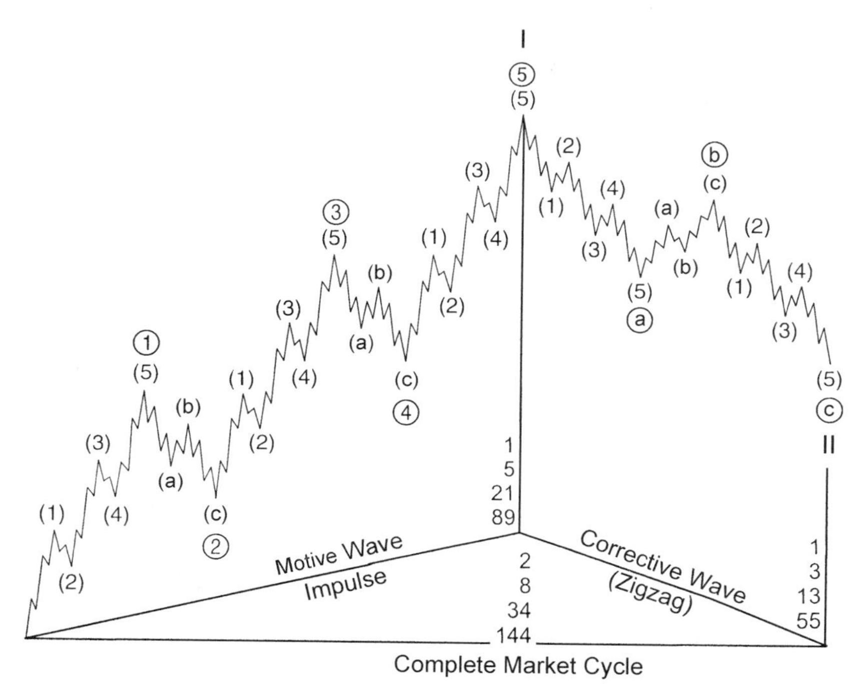

Figure 4.17. Forming Larger Waves (*Source*: R. R. Prechter, Jr., and A. J. Frost, *Elliott Wave Principle—Key to Modern Behavior*, New Classics Library, Gainesville, GA, 1985. Copyright © 1978–2000, reproduced with permission from New Classics Library.).

TABLE 4.10.

Waves	Bull Market Cycle	Bear Market Cycle	Total Waves
Cycle waves	1	1	2
Primary waves	5	3	8
Intermediate waves	21	13	34
Minor waves	89	55	144

EXERCISES 4

1. An n-bit word containing no two consecutive 1s can be constructed recursively as follows: Append a 0 to such $(n-1)$-bit words or append a 01 to such $(n-2)$-bit words. Using this procedure construct all 5-bit words containing no two consecutive 1s. There are 13 such words.
2. Let a_n denote the number of n-bit words that do not contain the pattern 111. Define a_n recursively.

Let a_n denote the number of ways a person can climb up a ladder with n rungs. At each step he can climb one or two rungs.*

3. Define a_n recursively.
4. Find an explicit formula for a_n.

Let b_n denote the number of ways of forming a sum of n (integral) dollars using only one- and two-dollar bills, taking order into consideration (Moser, 1963).

5. Define b_n recursively.
6. Find an explicit formula for b_n.

Let b_n denote the number of compositions of a positive integer n using 1, 2, and 3 as summands (Netto, 1901).

7. Find b_3 and b_4.
8. Define b_n recursively.

A set of integers A is *fat* if each of its elements is $\geq |A|$, where $|A|$ denotes the number of elements in A. For example, $\{5, 7, 91\}$ is a fat set, whereas $\{3, 7, 36, 41\}$ is not. $\emptyset$ is considered a fat set. Let a_n denote the number of fat subsets of the set $\{1, 2, 3, \ldots, n\}$ (Andrews).

9. Define a_n recursively.
10. Find an explicit formula for a_n.

An ordered pair of subsets $\{A, B\}$ of the set $S_n = \{1, 2, \ldots, n\}$ is *admissible* if $a > |B|$ for every $a \in A$ and $b > |A|$ for every $b \in B$, where $|X|$ denotes the number of elements of the set X. For example, $(\{2, 3\}, \{4\})$ is an admissible pair of subsets of S_4.

11. Find the various admissible ordered pairs of subsets of the sets S_0, S_1, and S_2.

*12. Predict the number of admissible ordered pairs of subsets of S_n.

**13. Let S_n denote the sum of the elements in the nth term of the sequence of sets of Fibonacci numbers $\{1\}$, $\{1, 2\}$, $\{3, 5, 8\}$, $\{13, 21, 34, 55\}$, ... Find a formula for S_n.

---

*Based on D. I. A. Cohen, *Basic Techniques of Combinatorial Theory*, Wiley, New York, 1978.

5 FIBONACCI AND LUCAS IDENTITIES

Both Fibonacci and Lucas numbers satisfy numerous identities that have been discovered over the centuries. In this chapter we explore several of these fundamental identities.

For example, Exercise 14 of Chapter 2 required that we conjecture a formula for the sum $\sum_{1}^{n} F_i$. In doing so, we notice the following interesting pattern:

$$\begin{aligned}
F_1 &= 1 = 2 - 1 = F_3 - 1 \\
F_1 + F_2 &= 2 = 3 - 1 = F_4 - 1 \\
F_1 + F_2 + F_3 &= 4 = 5 - 1 = F_5 - 1 \\
F_1 + F_2 + F_3 + F_4 &= 8 = 8 - 1 = F_6 - 1 \\
F_1 + F_2 + F_3 + F_4 + F_5 &= 12 = 13 - 1 = F_7 - 1
\end{aligned}$$

Following this pattern, we conjecture that $\sum_{1}^{n} F_i = F_{n+2} - 1$. We shall establish the validity of this formula in two ways, but we first state it as a theorem. See Exercise 23 for an alternate method.

Theorem 5.1. (Lucas, 1876)

$$\sum_{1}^{n} F_i = F_{n+2} - 1 \tag{5.1}$$

Proof. Using the Fibonacci recurrence relation, we have:

$$\begin{aligned} F_1 &= F_3 - F_2 \\ F_2 &= F_4 - F_3 \\ F_3 &= F_5 - F_4 \\ &\vdots \\ F_{n-1} &= F_{n+1} - F_n \\ F_n &= F_{n+2} - F_{n+1} \end{aligned}$$

Adding these equations, we get:

$$\sum_1^n F_i = F_{n+2} - F_2 = F_{n+2} - 1$$

■

AN ALTERNATE METHOD

An alternate method of proving Identity (5.1) is to apply the principle of mathematical induction (PMI). Since $F_1 = F_3 - 1$, the formula works for $n = 1$.

Now assume it is true for an arbitrary positive integer $k \geq 1$:

$$\sum_1^k F_i = F_{k+2} - 1$$

Then

$$\begin{aligned} \sum_1^{k+1} F_i &= \sum_1^k F_i + F_{k+1} \\ &= (F_{k+2} - 1) + F_{k+1}, \qquad \text{by the inductive hypothesis} \\ &= (F_{k+1} + F_{k+2}) - 1 \\ &= F_{k+3} - 1 \end{aligned}$$

Thus, by PMI, the formula is true for every positive integer n. ■

For example, $\sum_1^{20} F_i = F_{22} - 1 = 17,711 - 1 = 17,710$. You can verify this by direct computation.

This theorem is the basis of an interesting puzzle, conceived by W. H. Huff:

> Add up any finite number of consecutive Fibonacci numbers. Now add the second term to this sum. The resulting sum is a Fibonacci number.

The next example justifies the validity of this puzzle.

Example 5.1. Prove that $\sum_{j=0}^{k} F_{i+j} + F_{i+1} = F_{i+k+2}$. ■

Solution.

$$\begin{aligned}\sum_{j=0}^{k} F_{i+j} + F_{i+1} &= \sum_{1}^{i+k} F_r - \sum_{1}^{i-1} F_r + F_{i+1} \\ &= (F_{i+k+2} - 1) - (F_{i+1} - 1) + F_{i+1} \\ &= F_{i+k+2}\end{aligned}$$

This example, in fact, identifies the Fibonacci number that is the final sum in the puzzle. Obviously, this example and hence the puzzle can be extended to the generalized Fibonacci sequence (see Exercise 16 in Chapter 7).

Using the technique employed in Theorem 5.1, we can derive a formula for the sum of the first n Fibonacci numbers with odd subscripts.

Theorem 5.2. (Lucas, 1876)

$$\sum_{1}^{n} F_{2i-1} = F_{2n} \tag{5.2}$$

Proof. Using the Fibonacci recurrence relation, we have

$$\begin{aligned}F_1 &= F_2 - F_0 \\ F_3 &= F_4 - F_2 \\ F_5 &= F_6 - F_4 \\ &\vdots \\ F_{2n-3} &= F_{2n-2} - F_{2n-4} \\ F_{2n-1} &= F_{2n} - F_{2n-2}\end{aligned}$$

Adding these equations, we get

$$\sum_{1}^{n} F_{2i-1} = F_{2n} - F_0 = F_{2n}$$ ■

For example, $\sum_{1}^{10} F_{2i-1} = F_{20} = 6765$. Again, you can verify this by direct computation.

Corollary 5.1. (Lucas, 1876)

$$\sum_{1}^{n} F_{2i} = F_{2n+1} - 1 \tag{5.3}$$

Proof.

$$
\begin{aligned}
\sum_{1}^{n} F_{2i} &= \sum_{1}^{2n} F_i - \sum_{1}^{n} F_{2i-1} \\
&= (F_{2n+2} - 1) - F_{2n} \qquad \text{by Theorems 5.1 and 5.2.} \\
&= (F_{2n+2} - F_{2n}) - 1 \\
&= F_{2n+1} - 1 \qquad \text{by the Fibonacci recurrence relation (FRR).}
\end{aligned}
$$
■

This identity has a wonderful application to graph theory. But before we examine it, we need two definitions.

Spanning Tree of a Connected Graph

A *spanning tree* of a connected graph G is a subgraph of G that is a tree containing every vertex of G. The *complexity* $k(G)$ of a graph is the number of distinct spanning trees of the graph.

For example, the graph in Figure 5.1 has three distinct spanning trees (see Fig. 5.2), so its complexity is three.

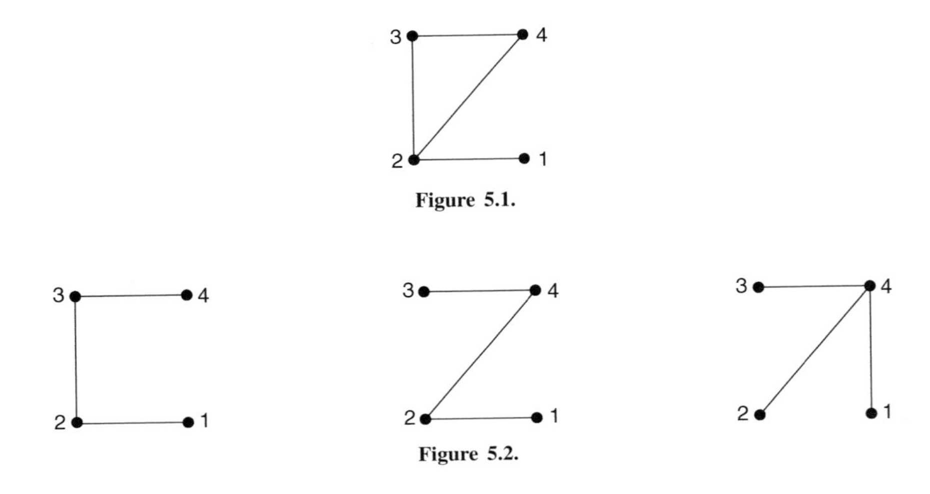

Figure 5.1.

Figure 5.2.

Fan Graph

A *fan graph* or simply a *fan*, G_1, of order 1 consists of two vertices, 0 and 1, and exactly one edge between them. A fan G_n of order n is obtained by adding a vertex n to a fan G_{n-1} and then connecting vertex n to vertices 0 and $n-1$, where $n \geq 2$. Figure 5.3 shows fans of orders 1 through 4.

Next, we look for the number of spanning trees s_n of a fan G_n. The fan G_1 has clearly one spanning tree, so $s_1 = 1$ (see Fig. 5.4). Fan G_2 has three spanning trees,

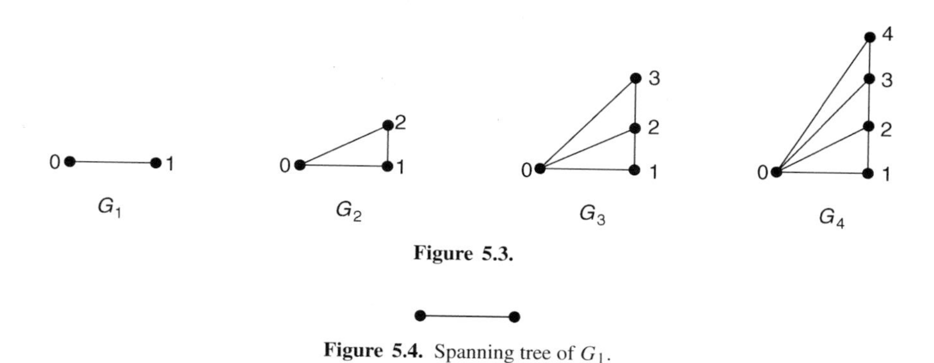

Figure 5.3.

Figure 5.4. Spanning tree of G_1.

so $s_2 = 3$ (see Fig. 5.5). G_3 has eight spanning trees, so $s_3 = 8$ (see Fig. 5.6). Thus $s_1 = 1 = F_2$, $s_2 = 3 = F_4$, and $s_3 = 8 = F_6$. So we predict that $s_4 = F_8$.

To confirm this, consider the possible ways of having vertex 4 in a spanning tree of G_4. It follows from Figure 5.7 that

$$\begin{aligned} s_4 &= s_3 + \sum_1^3 s_i + 1 \\ &= F_6 + (F_6 + F_4 + F_2) + 1 = F_6 + (F_7 - 1) \\ &= F_6 + F_7 = F_8 \end{aligned}$$

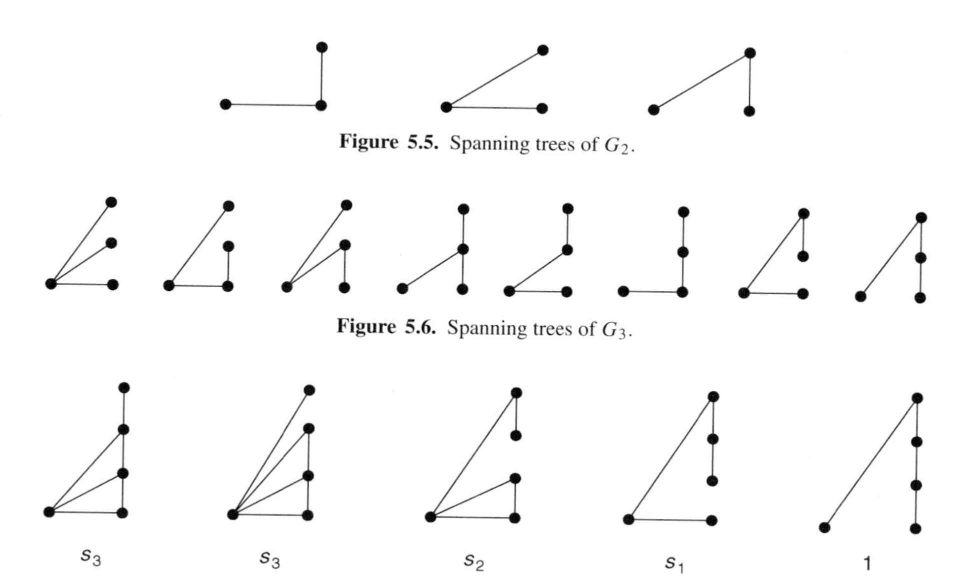

Figure 5.5. Spanning trees of G_2.

Figure 5.6. Spanning trees of G_3.

Figure 5.7. Possible ways of having vertex 4 in a spanning tree of G_4.

More generally,

$$\begin{aligned} s_n &= s_{n-1} + \sum_{1}^{n-1} s_i + 1 \\ &= F_{2n-2} + \sum_{1}^{n-1} F_{2i} + 1 = F_{2n-2} + (F_{2n-1} - 1) + 1 \\ &= F_{2n-2} + F_{2n-1} = F_{2n} \end{aligned}$$

For example, a fan of order 5 has $F_{10} = 55$ spanning trees.

Before we state the next stack exponents property, we need to study the following pattern:

$$\begin{aligned} F_1F_3 - F_2^2 &= 1 \cdot 2 - 1^2 = (-1)^2 \\ F_2F_4 - F_3^2 &= 1 \cdot 3 - 2^2 = (-1)^3 \\ F_3F_5 - F_4^2 &= 2 \cdot 5 - 3^2 = (-1)^4 \\ F_4F_6 - F_5^2 &= 3 \cdot 8 - 5^2 = (-1)^5 \\ &\vdots \end{aligned}$$

Clearly, a pattern emerges. We conjecture that $F_{n-1}F_{n+1} - F_n^2 = (-1)^n$, where $n \geq 1$. This leads to the next formula, which was discovered in 1680 by the Italian-born French astronomer and mathematician Giovanni Domenico Cassini (1625–1712), and discovered independently in 1753 by Robert Simson (1687–1768) of the University of Glasgow.

Theorem 5.3. (Cassini's Formula)

$$F_{n-1}F_{n+1} - F_n^2 = (-1)^n \tag{5.4}$$

where $n \geq 1$.

Proof. (by PMI) Since $F_0F_2 - F_1^2 = 0 \cdot 1 - 1 = -1 = (-1)^1$, the given statement is clearly true when $n = 1$.

Now we assume it is true for an arbitrary positive integer k: $F_{k-1}F_{k+1} - F_k^2 = (-1)^k$. Then

$$\begin{aligned} F_kF_{k+2} - F_{k+1}^2 &= (F_{k+1} - F_{k-1})(F_k + F_{k+1}) - F_{k+1}^2 \\ &= F_kF_{k+1} + F_{k+1}^2 - F_kF_{k-1} - F_{k-1}F_{k+1} - F_{k+1}^2 \\ &= F_kF_{k+1} - F_kF_{k-1} - F_k^2 - (-1)^k \qquad \text{by the IH} \\ &= F_kF_{k+1} - F_k(F_{k-1} + F_k) + (-1)^{k+1} \\ &= F_kF_{k+1} - F_kF_{k+1} + (-1)^{k+1} \\ &= (-1)^{k+1} \end{aligned}$$

Thus the formula works for $n = k + 1$. So, by PMI, the statement is true for every integer $n \geq 1$. ■

Cassini's formula yields the following fascinating by-product.

Corollary 5.2. Any two consecutive Fibonacci numbers are relatively prime; that is, $(F_{n+1}, F_n) = 1$ for every n.

Proof. Let p be a prime factor of both F_n and F_{n+1}. Then, by Cassini's formula, $p| \pm 1$, which is a contradiction. Thus $(F_{n+1}, F_n) = 1$. ■

Substituting for F_{n+1} in Cassini's formula yields $F_{n-1}^2 + F_nF_{n-1} - F_n^2 = (-1)^n$. This implies that the Diophantine equation $x^2 + xy - y^2 = \pm 1$ has infinitely many solutions, $x = F_{n-1}$ and $y = F_n$.

In 1972, Ira Gessel of Harvard University employed Cassini's formula to establish the following interesting result.

Theorem 5.4. A positive integer n is a Fibonacci number if and only if $5n^2 \pm 4$ is a perfect square.

Proof. We have

$$\begin{aligned}
(-1)^r + F_r^2 &= F_{r+1}F_{r-1} && \text{Cassini's formula}\\
L_r &= F_{r+1} + F_{r-1} && \text{by Exercise 32}\\
\therefore \quad L_r^2 - 4[(-1)^r + F_r^2] &= (F_{r+1} + F_{r-1})^2 - 4F_{r+1}F_{r-1}\\
&= (F_{r+1} - F_{r-1})^2\\
&= F_r^2\\
L_r^2 &= 5F_r^2 + 4(-1)^r
\end{aligned}$$

Thus if n is a Fibonacci number, then $5n^2 \pm 4$ is a perfect square.

Conversely, let $5n^2 \pm 4$ be a perfect square m^2. Then

$$m^2 - 5n^2 = \pm 4$$

$$\frac{m + n\sqrt{5}}{2} \cdot \frac{m - n\sqrt{5}}{2} = \pm 1$$

Since m and n have the same parity (both odd or both even), both $(m + n\sqrt{5})/2$ and $(m - n\sqrt{5})/2$ are integers in the extension field $Q(\sqrt{5}) = \{x + y\sqrt{5} | x, y \in Q\}$, where Q denotes the set of rational numbers. Since their product is ± 1, they must be units in the field. But the only integral units in $Q(\sqrt{5})$ are of the form $\pm\alpha^{\pm i}$. Then

$$\begin{aligned}\frac{m+n\sqrt{5}}{2} &= \alpha^i = \frac{1}{2}[(\alpha^i+\beta^i)+(\alpha^i-\beta^i)] \\ &= \frac{L_i+F_i\sqrt{5}}{2}\end{aligned}$$

Thus $n = F_i$, a Fibonacci number. ■

The identity $5F_n^2 + 4(-1)^n = L_n^2$ (see Exercise 39) was discovered in 1950 by P. Schub of the University of Pennsylvania. This result has an interesting application.

Let $n = 2m+1$. The resulting square, $L_{2m+1}^2 = 5F_{2m+1}^2 - 4$, is the discriminant of the quadratic equation $(F_{2m+1} \pm 1)x^2 - F_{2m+1}x - (F_{2m+1} \mp 1) = 0$. Consequently, its solutions are rational. For example,

$$\begin{aligned}1x^2 - 2x - 3 &= (1x+1)(1x-3) \\ 6x^2 - 5x - 4 &= (2x+1)(3x-4) \\ 12x^2 - 13x - 14 &= (3x+2)(4x-7) \\ 35x^2 - 34x - 33 &= (5x+3)(7x-11)\end{aligned}$$

More generally,

$$[F_{2m+1} + (-1)^m]x^2 - F_{2m+1}x - [F_{2m+1} - (-1)^m] = (F_{m+1}x + F_m)(L_mx - L_{m+1})$$

The truth of this rests on the following facts:

$$F_mF_{m-1} = F_m^2 - F_{m-1}^2 + (-1)^m, \qquad F_{m-1}F_{m+1} = F_m^2 + (-1)^m, \qquad \text{and}$$

$$F_m^2 + F_{m+1}^2 = F_{2m+1}$$

These observations were made in 1950 by A. Struyk.

What can we say about the sum of the squares of the first n Fibonacci numbers? Once again, let us look for a pattern:

$$F_1^2 + F_2^2 = 2 = F_2F_3$$

This result has a nice geometric interpretation: The sum of the areas of the squares of sizes $F_1 \times F_1$ and $F_2 \times F_2$ equals the area of the rectangle of size $F_2 \times F_3$, as Figure 5.8 demonstrates. Likewise, $F_1^2 + F_2^2 + F_3^2 = 1+1+4 = 6 = 2\cdot 3 = F_3F_4$ and $F_1^2 + F_2^2 + F_3^2 + F_4^2 = 1+1+4+9 = 15 = 3\cdot 5 = F_4F_5$. These results also can be interpreted geometrically in a similar manner, as Figure 5.9 shows.

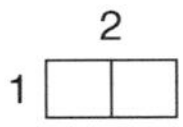

2

1

Figure 5.8.

More generally, we have the next result.

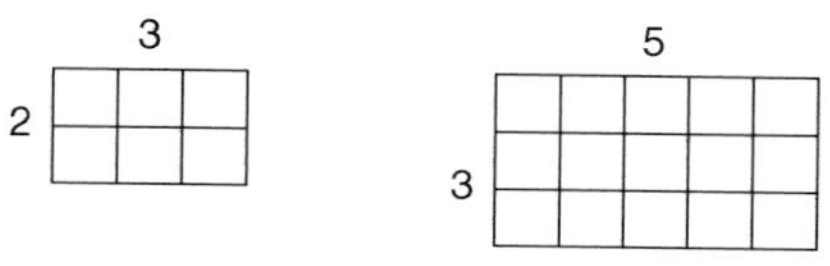

Figure 5.9.

Theorem 5.5. (Lucas, 1876)

$$\sum_{1}^{n} F_i^2 = F_n F_{n+1} \tag{5.5}$$

Proof. (by PMI) When $n = 1$, the left-hand side (LHS) $= \sum_{1}^{1} F_i^2 = F_1^2 = 1 = 1 \cdot 1 = F_1 \cdot F_2 =$ the right-hand side (RHS). So the result is true when $n = 1$.

Assume it is true for an arbitrary positive integer k : $\sum_{1}^{k} F_i^2 = F_k F_{k+1}$. Then

$$\begin{aligned}
\sum_{1}^{k+1} F_i^2 &= \sum_{1}^{k} F_i^2 + F_{k+1}^2 \\
&= F_k F_{k+1} + F_{k+1}^2 && \text{by the IH} \\
&= F_{k+1}(F_k + F_{k+1}) \\
&= F_{k+1} F_{k+2} && \text{by the Fibonacci recurrence relation (FRR)}
\end{aligned}$$

So the statement is true when $n = k + 1$. Thus it is true for every positive integer n.

■

For example,

$$\sum_{1}^{25} F_i^2 = F_{25} F_{26} = 75,025 \cdot 121,393 = 9,107,509,825$$

Interestingly enough, Identities 5.1 through 5.5 have analogous results for Lucas number also:

$$\sum_{1}^{n} L_i = L_{n+2} - 3 \tag{5.6}$$

$$\sum_{1}^{n} L_{2i-1} = L_{2n} - 2 \tag{5.7}$$

$$\sum_{1}^{n} L_{2i} = L_{2n+1} - 1 \tag{5.8}$$

$$L_{n-1}L_{n+1} - L_n^2 = 5(-1)^{n-1} \tag{5.9}$$

$$\sum_1^n L_i^2 = L_n L_{n+1} - 2 \tag{5.10}$$

These identities can be established using PMI (see Exercises 3–7). In addition, it is possible to establish Identity 5.10 using the Euclidean algorithm.

To derive new identities, we now present an explicit formula for F_n. To this end, let α and β be the roots of the quadratic equation $x^2 - x - 1 = 0$, so $\alpha = (1+\sqrt{5})/2$ and $\beta = (1-\sqrt{5})/2$. (The choice of the equation will become clear in Chapter 18.) Then $\alpha + \beta = 1$ and $\alpha\beta = -1$. Besides, $\alpha^2 = \alpha(1-\beta) = \alpha - \alpha\beta = \alpha + 1$, $\alpha^3 = \alpha(\alpha+1) = \alpha^2 + \alpha = 2\alpha + 1$, and $\alpha^4 = \alpha(2\alpha+1) = 2\alpha^2 + \alpha = 2(\alpha+1) + \alpha = 3\alpha + 2$. Thus we have:

$$\begin{aligned}
\alpha &= 1\alpha + 0 \\
\alpha^2 &= 1\alpha + 1 \\
\alpha^3 &= 2\alpha + 1 \\
\alpha^4 &= 3\alpha + 2
\end{aligned}$$

Notice an interesting pattern emerging: The constant term and the coefficient of α on the RHS appear to be adjacent Fibonacci numbers. Accordingly, we have the following result.

Lemma 5.1. $\alpha^n = \alpha F_n + F_{n-1}$, where $n \geq 0$. ■

This can be established easily using PMI (see Exercise 48).

Corollary 5.3. $\beta^n = \beta F_n + F_{n-1}$, where $n \geq 0$. ■

Let $u_n = (\alpha^n - \beta^n)/\sqrt{5}$, where $n \geq 1$. Then

$$u_1 = \frac{\alpha - \beta}{\sqrt{5}} = \frac{\sqrt{5}}{\sqrt{5}} = 1 \qquad \text{and} \qquad u_2 = \frac{\alpha^2 - \beta^2}{\sqrt{5}} = \frac{(\alpha+\beta)(\alpha-\beta)}{\sqrt{5}} = 1$$

Suppose $n \geq 3$. Then

$$\begin{aligned}
u_{n-1} + u_{n-2} &= \frac{\alpha^{n-1} - \beta^{n-1}}{\sqrt{5}} + \frac{\alpha^{n-2} - \beta^{n-2}}{\sqrt{5}} \\
&= \frac{\alpha^{n-2}(\alpha+1) - \beta^{n-2}(\beta+1)}{\sqrt{5}} = \frac{\alpha^{n-2}\cdot\alpha^2 - \beta^{n-2}\cdot\beta^2}{\sqrt{5}} \\
&= \frac{\alpha^n - \beta^n}{\sqrt{5}} = u_n
\end{aligned}$$

Thus, u_n satisfies the FRR (1) and the two initial conditions. This gives us an explicit formula for F_n : $F_n = u_n$.

Theorem 5.6. Let α be the positive root of the quadratic equation $x^2 - x - 1 = 0$ and β its negative root. Then

$$F_n = \frac{\alpha^n - \beta^n}{\alpha - \beta}$$

where $n \geq 1$. ■

This explicit formula for F_n is called *Binet's formula*, after the French mathematician Jacques-Phillipe-Marie Binet (1786–1856), who discovered it in 1843. In fact, it was first discovered in 1718 by the French mathematician Abraham De Moivre (1667–1754) using generating functions (see Chapter 18), and also arrived at independently in 1844 by the French engineer and mathematician Gabriel Lamé (1795–1870).

In any case, this formula, which we shall derive in two other ways in later chapters, can be employed to derive a myriad of Fibonacci identities.

Corollary 5.4. (Lucas, 1876)

$$F_{n+1}^2 + F_n^2 = F_{2n+1} \tag{5.11}$$

$$F_{n+1}^2 - F_{n-1}^2 = F_{2n} \tag{5.12}$$

■

For example, $F_8^2 + F_7^2 = 441 + 169 = 610 = F_{15}$ and $F_{11}^2 - F_9^2 = 7921 - 1156 = 6765 = F_{20}$.

In Chapter 34, we shall prove that there are only two distinct Fibonacci numbers that are perfect squares, namely, 1 and 144. Consequently, Identity (5.11) has a nice geometric interpretation: *No two consecutive Fibonacci numbers can be the lengths of the legs of a right triangle.*

The next theorem, however, provides a link between four consecutive Fibonacci numbers and the lengths of a Pythagorean triangle, as established in 1948 by C. W. Raine.

Theorem 5.7. Let ABC be a triangle with $AC = F_k F_{k+3}$, $BC = 2F_{k+1}F_{k+2}$, and $AB = F_{2k+3}$. Then ABC is a Pythagorean triangle, right-angled at C. ■

It suffices to verify that $AB^2 = AC^2 + BC^2$, so we leave its proof as an exercise.

For example, let $AC = F_7 F_{10} = 13 \cdot 55 = 715$, $BC = 2 \cdot F_8 F_9 = 2 \cdot 21 \cdot 34 = 1428$, and $AB = F_{17} = 1597$. Then $AC^2 + BC^2 = 715^2 + 1428^2 = 2,550,409 = 1597^2 = AB^2$, so ABC is a right triangle, right-angled at C.

Corresponding to Binet's formula for F_n, there is one for L_n also, as the next theorem shows. We invite you to confirm it (see Exercise 18).

Theorem 5.8. Let $n \geq 1$. Then

$$L_n = \alpha^n + \beta^n$$

■

The two Binet formulas can be used in tandem to derive an array of identities.

Corollary 5.5.

$$F_{2n} = F_n L_n \tag{5.13}$$

$$F_{n-1} + F_{n+1} = L_n \tag{5.14}$$

$$F_{n+2} - F_{n-2} = L_n \tag{5.15}$$

$$L_{n-1} + L_{n+1} = 5F_n \tag{5.16}$$

■

For example, $F_{20} = 6765 = 55 \cdot 123 = F_{10}L_{10}$, $F_{11} + F_{13} = 89 + 233 = 322 = L_{12}$, $F_{11} - F_7 = 89 - 13 = 76 = L_9$, and $L_{10} + L_{12} = 123 + 322 = 445 = 5 \cdot 89 = 5F_{11}$.

Identity 5.13 implies that when $n \geq 3$, every Fibonacci number F_{2n} with an even subscript has nontrivial factors. According to Identity 5.14, the sum of any two Fibonacci numbers that are two units away is a Lucas number. Likewise, by Identity 5.15, the difference of any two Fibonacci numbers that lie four units away is also a Lucas number.

Identity 5.13 has an interesting by-product. Let $2n = 2^m$, where $m \geq 1$. Then

$$\begin{aligned} F_{2^m} &= L_{2^{m-1}} F_{2^{m-1}} \\ &= L_{2^{m-1}}(L_{2^{m-2}} F_{2^{m-2}}) = L_{2^{m-1}} L_{2^{m-2}} F_{2^{m-2}} \\ &= L_{2^{m-1}} L_{2^{m-2}}(L_{2^{m-3}} F_{2^{m-3}}) = L_{2^{m-1}} L_{2^{m-2}} L_{2^{m-3}} F_{2^{m-3}} \end{aligned}$$

Continuing like this we get:

$$F_{2^m} = L_{2^{m-1}} L_{2^{m-2}} \cdots L_8 L_4 L_2 L_1$$

This can be established using PMI.

For example,

$$F_{32} = L_{16} L_8 L_4 L_2 L_1 = 2207 \cdot 47 \cdot 7 \cdot 3 \cdot 1 = 2,178,309$$

Identity 5.14 has interesting applications, as the next two examples demonstrate.

Example 5.2. Let us reconsider the same permutation problem as in Example 4.9, with the difference that the numbers are arranged around a circle (see Fig. 5.10). Let q_n denote the number of cyclic permutations g that move no element more than one position from its natural position on the circle, where $n \geq 3$.

Figure 5.11 shows the various such cyclic permutations for $n = 3$, so $q_3 = 4$. Notice that with 1 and 3 swapped, there is just one permutation; otherwise, there are three, for a total of four cyclic permutations.

More generally, let $g(n) = 1$. The remaining $n - 2$ elements can be rearranged in p_{n-2} ways such that no element is moved by more than one space from its natural position. On the other hand, let $g(n) \neq 1$. This case is precisely the same as the preceding problem, so there are p_n desired permutations. Thus, again by the addition principle, $q_n = p_{n-2} + p_n = F_{n-1} + F_{n+1} = L_n$.

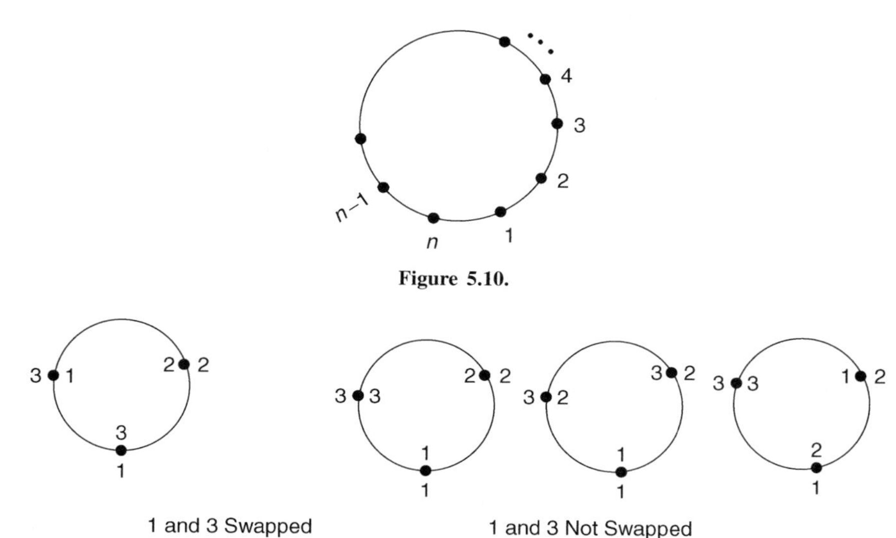

Figure 5.10.

Figure 5.11.

■

The next example is also a fine application of the identity $L_n = F_{n-1} + F_{n+1}$.

Example 5.3. In Example 4.1, we found that there are $a_n = F_{n+2}$ n-bit words that do not contain two consecutive 1s. Instead of arranging the bits linearly, suppose we arrange them around a circle in such a way that no two adjacent bits are 1s. Let b_n denote the number of such arrangements, where $n \geq 2$. Thus, b_n denotes the number of n-bit words such that:

1. No two adjacent bits are 1s;
2. If the word begins with a 1, then it cannot end in a 1.

Table 5.1 shows the possible such binary words for $n = 2$, 3, 4, and 5. It appears from the table that $b_n = L_n$.

TABLE 5.1.

n	n-Bit Words of the Desired Type	b_n
2	00, 01, 10	3
3	000, 001, 010, 100,	4
4	0000, 0001, 0010, 0100, 1000, 0101, 1010	7
5	00000, 00001, 00010, 00100, 01000, 10000, 01010, 01001, 10100, 10010, 00101	11
		↑ L_n

To establish this, suppose the word ends in a 0. There are $a_{n-1} = F_{n+1}$ binary words that meet the criteria. On the other hand, suppose the nth bit is 1. Then the $(n-1)$st bit, and the 1st bit cannot be 1:

$$\underset{1}{\overset{0}{-}} \quad \underbrace{- - - \cdots - -}_{n-3 \text{ bits}} \quad \underset{n-1}{\overset{0}{-}} \quad \underset{n}{\overset{1}{-}}$$

The remaining $n-3$ bits can be used to form $a_{n-3} = F_{n-1}$ words of the desired type. Thus $b_n = a_{n-1} + a_{n-3} = F_{n+1} + F_{n-1} = L_n$, where $n \geq 2$. ■

NUMBER OF DIGITS IN F_n AND L_n

Binet's formula can be successfully employed to predetermine the number of digits in F_n and L_n. We can show this by writing F_n as

$$F_n = \frac{\alpha^n}{\sqrt{5}}\left[1 - \left(\frac{\beta}{\alpha}\right)^n\right]$$

Since $|\beta| < |\alpha|$, $(\beta/\alpha)^n \to 0$ as $n \to \infty$. Therefore, when n is sufficiently large,

$$\begin{aligned} F_n &\approx \frac{\alpha^n}{\sqrt{5}} \\ \log F_n &\approx n \log \alpha - (\log 5)/2 \\ \text{Number of digits in } F_n &= 1 + \text{characteristic of } \log F_n = \lceil \log F_n \rceil \\ &= \lceil n \log \alpha - (\log 5)/2 \rceil \\ &= \lceil n[\log(1+\sqrt{5}) - \log 2] - (\log 5)/2 \rceil \end{aligned}$$

For example, the number of digits in F_{30} is given by

$$\lceil 30[\log(1+\sqrt{5}) - \log 2] - (\log 5)/2 \rceil = \lceil 5.920144205533 \rceil = 6$$

Notice that $F_{30} = 832,040$ does indeed contain six digits. Likewise, F_{45} consists of 10 digits.

Since $L_n = \alpha^n + \beta^n$, it follows that, when n is sufficiently large, $L_n \approx \alpha^n$, so that $\log L_n \approx n \log \alpha$. Thus the number of digits in L_n is given by

$$\lceil \log L_n \rceil = \lceil n \log \alpha \rceil = \lceil n[\log(1+\sqrt{5}) - \log 2] \rceil$$

For example, L_{39} contains $\lceil 39[\log(1+\sqrt{5}) - \log 2] \rceil = 9$ digits, whereas L_{50} contains $\lceil 50[\log(1+\sqrt{5}) - \log 2] \rceil = 11$ digits.

Using Binet's formula, we can generalize Cassini's formula, as the next theorem shows. It was established in 1879 by the Belgian mathematician Eugene Charles Catalan (1814–1894).

Theorem 5.9. (Catalan, 1879) Let k be a positive integer. Then

$$F_{n+k}F_{n-k} - F_n^2 = (-1)^{n+k+1}F_k^2 \tag{5.17}$$

where $n \geq k$.

Proof.

$$\begin{aligned}
\text{LHS} &= \frac{\alpha^{n+k} - \beta^{n+k}}{\sqrt{5}} \cdot \frac{\alpha^{n-k} - \beta^{n-k}}{\sqrt{5}} - \left(\frac{\alpha^n - \beta^n}{\sqrt{5}}\right)^2 \\
&= \frac{\alpha^{2n} - (\alpha^{n+k}\beta^{n-k} + \alpha^{n-k}\beta^{n+k}) + \beta^{2n}}{5} - \frac{\alpha^{2n} + \beta^{2n} - 2(-1)^n}{5} \\
&= \frac{-[(\alpha\beta)^n\alpha^{-k}\beta^k + (\alpha\beta)^n\alpha^k\beta^{-k}] + 2(-1)^n}{5} \\
&= \frac{2(-1)^n - (-1)^n(-1)^k\beta^{2k} - (-1)^n(-1)^k\alpha^{2k}}{5} \\
&= \frac{2(-1)^n - (-1)^{n+k}(\alpha^{2k} + \beta^{2k})}{5} \\
&= \frac{2(-1)^n + (-1)^{n+k+1}[5F_k^2 + 2(-1)^k]}{5} \qquad \text{by Exercise 42} \\
&= (-1)^{n+k+1}F_k^2 + \frac{2(-1)^n + 2(-1)^{n+2k+1}}{5} \\
&= (-1)^{n+k+1}F_k^2 + \frac{2(-1)^n + 2(-1)^{n+1}}{5} \\
&= (-1)^{n+k+1}F_k^2 \\
&= \text{RHS}
\end{aligned}$$

■

For example, let $n = 10$ and $k = 3$. Then $F_{13}F_7 - F_{10}^2 = 233 \cdot 13 - 55^2 = 4 = (-1)^{14}F_3^2$.

We can generalize Identity 5.17 even further (see Identity 19) on p. 88.

As with Theorem 5.4, we have the following result for Lucas numbers, developed by G. Wulczyn of Bucknell University in Pennsylvania in 1974.

Theorem 5.10. A positive integer n is a Lucas number if and only if $5n^2 \pm 20$ is a perfect square.

Proof. Let $n = L_{2m+1}$. Then

$$\begin{aligned}
5n^2 + 20 &= 5(\alpha^{2m+1} + \beta^{2m+1})^2 + 20 = 5[\alpha^{4m+2} + \beta^{4m+2} + 2(\alpha\beta)^{2m+1}] + 20 \\
&= 5[\alpha^{4m+2} + \beta^{4m+2} - 2(\alpha\beta)^{2m+1}] = 5(\alpha^{2m+1} - \beta^{2m+1})^2 \\
&= 5(\sqrt{5}F_{2m+1})^2 = 25F_{2m+1}^2
\end{aligned}$$

On the other hand, let $n = L_{2m}$. Then

$$\begin{aligned}5n^2 - 20 &= 5(\alpha^{2m} + \beta^{2m})^2 - 20 = 5[\alpha^{4m} + \beta^{4m} + 2(\alpha\beta)^{2m}] - 20 \\ &= 5[\alpha^{4m} + \beta^{4m} - 2(\alpha\beta)^{2m}] = 25F_{2m}^2\end{aligned}$$

Thus, if n is a Lucas number, then $5n^2 \pm 20$ is a perfect square.

Because the proof of the converse is a bit complicated, we omit it. ■

For example, let $n = 199 = L_{11}$. Then $5n^2 + 20 = 5 \cdot 199^2 + 20 = 198,025 = 445^2$, a perfect square. On the other hand, let $n = 843 = L_{14}$. Then $5n^2 - 20 = 5 \cdot 843^2 - 20 = 3,553,225 = 1,885^2$, again a perfect square.

With Binet's formulas at hand, we can extend the definitions of F_n and L_n to negative subscripts also. If we apply the FRR to the negative side, we get

...	F_{-4}	F_{-3}	F_{-2}	F_{-1}	F_0	F_1	F_2	F_3	F_4	...
...	-3	2	-1	1	0	1	1	2	3	...

So, it appears that $F_{-n} = (-1)^{n+1}F_n$, $n \geq 1$.

To prove this, assume Binet's formula holds for negative exponents

$$\begin{aligned}F_{-n} &= \frac{\alpha^{-n} - \beta^{-n}}{\sqrt{5}} = \frac{(-\beta)^n - (-\alpha)^n}{\sqrt{5}} \qquad \text{since } \alpha\beta = -1 \\ &= \frac{(-1)^n(\beta^n - \alpha^n)}{\sqrt{5}} = \frac{(-1)^{n+1}(\alpha^n - \beta^n)}{\sqrt{5}} \\ &= (-1)^{n+1}F_n \end{aligned} \tag{5.18}$$

Likewise,

$$L_{-n} = (-1)^n L_n. \tag{5.19}$$

Thus, $F_{-n} = F_n$ if and only if n is odd, and $L_{-n} = L_n$ if and only if n is even.

Since $F_{-1} = 1$, it is easy to verify that the identity $F_n^2 + F_{n+1}^2 = F_{2n+1}$ (5.11) follows from Catalan's formula 5.17.

A formula for α^{-n} can now be derived easily. Since $\alpha^n = \alpha F_n + F_{n-1}$ (Lemma 5.1), it follows that

$$\begin{aligned}\alpha^{-n} &= \alpha F_{-n} + F_{-n-1} \\ &= \alpha(-1)^{n+1}F_n + (-1)^{n+2}F_{n+1} \\ &= (-1)^{n+1}(\alpha F_n - F_{n+1}) \\ &= \begin{cases} \alpha F_n - F_{n+1} & \text{if } n \text{ is odd} \\ F_{n+1} - \alpha F_n & \text{otherwise} \end{cases}\end{aligned} \tag{5.20}$$

For example, $\alpha^{-12} = F_{11} - \alpha F_{12} = 89 - 144\alpha$.

Formula 5.20 can also be established by PMI or by showing that $\alpha^n(\alpha F_n - F_{n+1}) = (-1)^{n+1}$, using Binet's formula.

Likewise,

$$\beta^{-n} = \begin{cases} \beta F_n - F_{n+1} & \text{if } n \text{ is odd} \\ F_{n+1} - \beta F_n & \text{otherwise} \end{cases} \tag{5.21}$$

Notice two intriguing patterns that emerge from α^{-n}:

$$\alpha^{-1} = 1 \cdot \alpha - 1$$

$$\alpha^{-2} = 2 - 1 \cdot \alpha$$

$$\alpha^{-3} = 2 \cdot \alpha - 3$$

$$\alpha^{-4} = 5 - 3 \cdot \alpha$$

$$\alpha^{-5} = 5 \cdot \alpha - 8$$

$$\vdots$$

They are indicated by the two crisscrossing arrows: The absolute values of the coefficients of α are consecutive Fibonacci numbers, and so are the absolute values of the various constants.

The summation Formulas (5.1) through (5.3) are a special case of the generalized summation formula, given in the next theorem. In addition, the theorem yields an array of fascinating formulas as by-products. Its proof is a consequence of Binet's formulas and the geometric summation formula

$$\sum_{i=0}^{n-1} r^i = \frac{r^n - 1}{r - 1}$$

where $r \neq 1$.

Theorem 5.11. (Koshy, 1998) Let $k \geq 1$ and j any integer. Then

$$\sum_{i=0}^{n} F_{ki+j} = \begin{cases} \dfrac{F_{nk+k+j} - (-1)^k F_{nk+j} - F_j - (-1)^j F_{k-j}}{L_k - (-1)^k - 1} & \text{if } j < k \\[2ex] \dfrac{F_{nk+k+j} - (-1)^k F_{nk+j} - F_j + (-1)^k F_{j-k}}{L_k - (-1)^k - 1} & \text{otherwise} \end{cases} \tag{5.22}$$

Proof.

$$
\begin{aligned}
\sum_{i=0}^{n} F_{ki+j} &= \sum_{i=0}^{n} \frac{\alpha^{ki+j} - \beta^{ki+j}}{\sqrt{5}} \\
&= \frac{1}{\sqrt{5}}\left[\alpha^j \sum \alpha^{ki} - \beta^j \sum \beta^{ki}\right] \\
&= \frac{1}{\sqrt{5}}\left[\alpha^j \cdot \frac{\alpha^{nk+k}-1}{\alpha^k - 1} - \beta^j \cdot \frac{\beta^{nk+k}-1}{\beta^k-1}\right] \\
&= \frac{(\alpha^{nk+k+j} - \alpha^j)(\beta^k - 1) - (\beta^{nk+k+j} - \beta^j)(\alpha^k - 1)}{\sqrt{5}[(\alpha\beta)^k - (\alpha^k + \beta^k) + 1]} \\
&= \frac{-F_{nk+k+j} + (-1)^k F_{nk+j} + F_j + (\alpha^k\beta^j - \alpha^j\beta^k)/\sqrt{5}}{(-1)^k - L_k + 1}
\end{aligned}
$$

But

$$
\begin{aligned}
\alpha^k\beta^j - \alpha^j\beta^k &= \begin{cases} (\alpha\beta)^j(\alpha^{k-j} - \beta^{k-j}) & \text{if } j < k \\ (\alpha\beta)^k(\beta^{j-k} - \alpha^{j-k}) & \text{otherwise} \end{cases} \\
&= \begin{cases} (-1)^j\sqrt{5}F_{k-j} & \text{if } j < k \\ (-1)^{k+1}\sqrt{5}F_{j-k} & \text{otherwise} \end{cases}
\end{aligned}
$$

$$
\therefore \sum_{i=0}^{n} F_{ki+j} = \begin{cases} \dfrac{F_{nk+k+j} - (-1)^k F_{nk+j} - F_j - (-1)^j F_{k-j}}{L_k - (-1)^k - 1} & \text{if } j < k \\[2ex] \dfrac{F_{nk+k+j} - (-1)^k F_{nk+j} - F_j + (-1)^k F_{j-k}}{L_k - (-1)^k - 1} & \text{otherwise} \end{cases}
$$

■

Letting $j = 0$ in this formula yields the following result.

Corollary 5.6. (Koshy, 1998)

$$
\sum_{i=1}^{n} F_{ki} = \frac{F_{nk+k} - (-1)^k F_{nk} - F_k}{L_k - (-1)^k - 1} \tag{5.23}
$$

■

Corollary 5.7.

$$
\sum_{1}^{n} F_i = F_{n+2} - 1 \tag{5.1}
$$

$$
\sum_{1}^{n} F_{2i-1} = F_{2n} \tag{5.2}
$$

$$
\sum_{1}^{n} F_{2i} = F_{2n+1} - 1 \tag{5.3}
$$

Proof. When $k = 1$, Formula 5.23 yields

$$\sum_{i=1}^{n} F_i = \sum_{i=0}^{n} F_i = \frac{F_{n+1} + F_n - F_1}{L_1 + 1 - 1} = \frac{F_{n+2} - 1}{1} = F_{n+2} - 1$$

Formula 5.3 follows from formula 5.23 by letting $k = 2$, and Formula 5.2 from Formula 5.22 by letting $k = 2$ and $j = -1$. ■

It follows from Formula 5.23 that

$$\sum_{1}^{n} F_{3i} = \frac{F_{3n+3} + F_{3n} - F_3}{L_3 - 1 + 1} = \frac{F_{3n+3} + F_{3n} - 2}{4}$$

In particular,

$$\sum_{1}^{5} F_{3i} = \frac{F_{18} + F_{15} - 2}{4} = \frac{2584 + 610 - 2}{4} = 798$$

This may be verified by direct computation.

Corollary 5.8. (Koshy, 1998)

$$\sum_{i=0}^{n} F_{i+j} = \begin{cases} F_{n+j+2} - F_j - (-1)^j F_{1-j} & \text{if } j < 1 \\ F_{n+j+2} - F_{j+1} & \text{otherwise} \end{cases}$$

Proof. Since $L_1 - (-1)^1 - 1 = 1$, the corollary follows from Formula 5.22 when $k = 1$. ■

For example,

$$\sum_{0}^{5} F_{i+3} = F_{10} - F_4 = 55 - 3 = 52$$

and

$$\sum_{0}^{8} F_{i-5} = F_5 - F_{-5} - (-1)^{-5} F_6 = 5 - 5 + 8 = 8$$

Over the years, a vast array of Fibonacci and Lucas identities have been developed. The following list cites a substantial number of them. It would be a good exercise to establish the validity of each.

1. $F_{n+4}^3 - 3F_{n+3}^3 - 6F_{n+2}^3 + 3F_{n+1}^3 + F_n^3 = 0$ (Zeitlin and Parker, 1963)
2. $F_m F_n - F_{m+k} F_{n-k} = (-1)^{n-k} F_{m+k-n} F_k$
3. $5\sum_{1}^{n} F_{i-2} F_i = L_{2n-1} + 1 + 3\nu$, where $\nu = \begin{cases} 0 & \text{if } n \text{ is even} \\ 1 & \text{otherwise} \end{cases}$ (Koshy, 1998)

4. $\alpha^m F_{n-m+1} + \alpha^{m-1} F_{n-m} = \alpha^n$
5. $\beta^m F_{n-m+1} + \beta^{m-1} F_{n-m} = \beta^n$
6. $F_n = F_m F_{n-m+1} + F_{m-1} F_{n-m}$
7. $L_n = L_m F_{n-m+1} + L_{m-1} F_{n-m}$
8. $\left(\frac{L_n + \sqrt{5} F_n}{2}\right)^m = \frac{L_{mn} + \sqrt{5} F_{mn}}{2}$ (Fisk, 1963)
9. $\alpha^{-n} = (-1)^{n+1}(\alpha F_n - F_{n+1})$
10. $L_{2m} L_{2n} = L_{m+n}^2 + 5F_{m-n}^2$ (Wall, February 1964)
11. $L_{2m} L_{2n} = 5F_{m+n}^2 + L_{m-n}^2$ (Wall, February 1964)
12. $L_{2m} L_{2n} = L_{m+n}^2 + L_{m-n}^2 - 4(-1)^{m+n}$ (Lind and Hoggatt, Jr., 1964)
13. $L_{2m+2n} + L_{2m-2n} = L_{2m} L_{2n}$ (Koshy, 1998)
14. $L_{2m+2n} - L_{2m-2n} = 5F_{2m} F_{2n}$ (Koshy, 1998)
15. $L_{4n} = 5F_{2n}^2 + 2$ (Lucas, 1876)
16. $L_{4n+2} = 5F_{2n+1}^2 - 2$ (Lucas, 1876)
17. $2(F_n^4 + F_{n+1}^4 + F_{n+2}^4) = (F_n^2 + F_{n+1}^2 + F_{n+2}^2)^2$ (Candido, 1905)
18. $(F_n F_{n+3})^2 + (2F_{n+1} F_{n+2})^2 = F_{2n+3}^2$ (Raine, 1948)
19. $F_{n+h} F_{n+k} - F_n F_{n+h+k} = (-1)^n F_h F_k$ (Everman et al., 1960)
20. $\sum_1^n (-1)^{i-1} F_{i+1} = (-1)^{n-1} F_n$
21. $\sum_1^n F_{2i}^2 = (3F_{2n+1}^2 + 2F_{2n+2}^2 - 6F_{2n} F_{2n+2} - 2n - 5)/5$ (Rao, 1953)
22. $\sum_1^n F_{2i-1}^2 = (3F_{2n}^2 + 2F_{2n-1}^2 - 4F_{2n-2} F_{2n} + 2n - 2)/5$ (Rao, 1953)
23. $\sum_1^n F_{2i-1} F_{2i+1} = (2F_{2n+2}^2 - 3F_{2n+1}^2 + 3n + 1)/5$ (Rao, 1953)
24. $\sum_1^n F_{2i} F_{2i+2} = (3F_{2n+2}^2 - 2F_{2n+1}^2 - 3n - 1)/5$ (Rao, 1953)
25. $\sum_1^n F_i F_{i+2} = F_{2n+1} F_{2n+2} - 1$ (Rao, 1953)
26. $\sum_1^n F_{2i} F_{2i+1} = (F_{2n+2}^2 + F_{2n+1}^2 - n - 2)/5$ (Rao, 1953)
27. $\sum_1^n F_{2i-1} F_{2i} = (4F_{2n+1}^2 - F_{2n+2}^2 + n - 3)/5$ (Rao, 1953)

28. $\sum_{1}^{2n} F_i F_{i+1} = F_{2n+1}^2 - 1$ (Rao, 1953)

29. $\sum_{1}^{n} F_{2i-1} F_{2i+3} = (3F_{2n+2}^2 - 2F_{2n+1}^2 + 7n - 1)/5$ (Rao, 1953)

30. $\sum_{1}^{n} F_{2i} F_{2i+4} = (2F_{2n+4}^2 - 3F_{2n+3}^2 - 7n - 6)/5$ (Rao, 1953)

31. $\sum_{1}^{2n} F_i F_{i+4} = (F_{4n+6} - F_{2n+3}^2 - 4)/2$ (Rao, 1953)

32. $F_n F_{n+1} F_{n+2} = F_{n+1}^3 + (-1)^n F_{n+1}$

33. $F_{n+1}^3 = F_n^3 + F_{n-1}^3 + 3F_{n-1} F_n F_{n+1}$

34. $\sum_{1}^{n} F_{2i} F_{2i+1} F_{2i+2} = (F_{2n+2}^3 - F_{2n+2})/4$ (Rao, 1953)

35. $\sum_{1}^{n} F_{2i-1} F_{2i} F_{2i+1} = (F_{2n+1}^3 + F_{2n} - 2)/4$ (Rao, 1953)

36. $\sum_{1}^{n} F_{2i-1}^3 = (F_{2n}^3 + 3F_{2n})/4$ (Rao, 1953)

37. $\sum_{1}^{n} F_{2i}^3 = (F_{2n+1}^2 - 3F_{2n-1} + 2)/4$ (Rao, 1953)

38. $\sum_{1}^{n} F_i^3 = (F_{n+2}^3 - 3F_{n+1}^3 + 3(-1)^n F_n + 2)/4$ (Rao, 1953)

39. $\sum_{1}^{n} F_{2i-1} F_{2i+1} F_{2i+3} = (F_{2n+2}^3 + 7F_{2n+2} - 8)/4$ (Rao, 1953)

40. $\sum_{1}^{n} F_{2i} F_{2i+2} F_{2i+4} = (F_{2n+3}^3 - 7F_{2n+3} + 6)/4$ (Rao, 1953)

41. $\sum_{1}^{n} F_i F_{i+2} F_{i+4} = [F_{3n+8} - 16(-1)^n F_n - 5]/10$ (Rao, 1953)

42. $F_{3n} = 4F_{3n-3} + F_{3n-6}$

43. $F_{m+k} F_{m-k} - F_{m+s} F_{m-s} = (-1)^{m-s} F_s F_{k+s}$ (Halton, 1965)

44. $F_{m+n} = F_{m+1} F_{n+1} - F_{m-1} F_{n-1}$ (Mana, 1969)

45. $F_{r+s+t} = F_{r+1} F_{s+1} F_{t+1} + F_r F_s F_t - F_{r-1} F_{s-1} F_{t-1}$

46. $F_{3n} = 5F_n^3 + 3(-1)^n F_n$ (Halton, 1965)

47. $F_r F_{m+n} = F_{m+r} F_n - (-1)^r F_m F_{n-r}$ (Halton, 1965)

48. $[5F_m^2 + 2(-1)^m] F_{2r} = F_{m+r+1}^2 - F_{m+r-1}^2 - F_{m-r+1}^2 - F_{m-r-1}^2$ (Halton, 1965)

49. $[5F_m^2 + 2(-1)^m]F_r^2 + 2(-1)^r F_m^2 = F_{m+r}^2 + F_{m-r}^2$ (Halton, 1965)

50. $F_m F_{2m} F_{3r} = F_{m+r}^3 - (-1)^r F_{m-r}^3 - (-1)^m L_m F_r^3$ (Halton, 1965)

51. $F_{2m+1} F_{2n+1} = F_{m+n+1}^2 + F_{m-n}^2$ (Tadlock, 1965)

52. $L_{2m+1} L_{2n+1} = L_{m+n+1}^2 - F_{m-n}^2 + 4(-1)^{m-n}$ (Tadlock, 1965)

53. $F_n^2 + F_{n+2k}^2 = F_{n+2k-2} F_{n+2k+1} + F_{2k-1} F_{2n+2k-1}$ (Sharpe, 1965)

54. $F_n^2 + F_{n+2k+1}^2 = F_{2k+1} F_{2n+2k+1}$ (Sharpe, 1965)

55. $F_{n+2k}^2 - F_n^2 = F_{2k} F_{2n+2k}$ (Sharpe, 1965)

56. $F_{n+2k+1}^2 - F_n^2 = F_{n-1} F_{n+2} - F_{2k} F_{2n+2k+2}$ (Sharpe, 1965)

57. $\sum_1^n i F_i^2 = n F_n F_{n+1} - F_n^2 + [1 + (-1)^{n-1}]/2$ (Koshy, 1998)

58. $\sum_1^n (n - i + 1) F_i^2 = F_n F_{n+2} - [1 + (-1)^{n-1}]/2$ (Koshy, 1998)

59. $\sum_1^n (2n - i) F_i^2 = F_{2n}^2$ (Hoggatt, 1964)

60. $\sum_1^n F_i F_{i+1} = F_{n+1}^2 - [1 + (-1)^n]/2$ (Koshy, 1998)

61. $\sum_{i=0}^n \sum_{j=0}^i \sum_{k=0}^j \sum_{l=0}^k F_l^2 = F_{n+2}^2 - [2n^2 + 8n + 11 - 3(-1)^n]/8$ (Graham, 1965)

62. $5F_n = L_{n-1} L_{n+1} + (-1)^n$ (Koshy, 1998)

63. $L_n L_{n+1} = L_{2n+1} + (-1)^n$ (Hoggatt, 1965)

64. $F_n^2 + F_{n+4}^2 = F_{n+1}^2 + F_{n+3}^2 + 4F_{n+2}^2$ (Swamy, February 1966)

65. $F_{n+1}^5 - F_n^5 - F_{n-1}^5 = 5F_{n+1} F_n F_{n-1} [2F_n^2 + (-1)^n]$ (Carlitz, February 1967)

66. $L_{n+1}^5 - L_n^5 - L_{n-1}^5 = 5L_{n+1} L_n L_{n-1} [2L_n^2 - 5(-1)^n]$ (Carlitz, February 1967)

67. $F_{n+1}^7 - F_n^7 - F_{n-1}^7 = 7F_{n+1} F_n F_{n-1} [2F_n^2 + (-1)^n]^2$ (Carlitz, February 1967)

68. $L_{n+1}^7 - L_n^7 - L_{n-1}^7 = 7L_{n+1} L_n L_{n-1} [2L_n^2 - 5(-1)^n]^2$ (Carlitz, February 1967)

69. $L_{2n} = 1 + \lfloor \sqrt{5} F_{2n} \rfloor$ (Seamons, 1967)

70. $F_{4n+1} - 1 = L_{2n+1} F_{2n}$ (Hoggatt, 1967)

71. $F_{4n+3} - 1 = L_{2n+1} F_{2n+2}$ (Hoggatt, 1967)

72. $F_{n+1} L_{n+2} - F_{n+2} L_n = F_{2n+1}$ (Carlitz, 1967)

73. $F_n L_{n+r} - F_{n+r} L_{n-r} = (F_{2r} - F_r) F_{2n-r+1} + (F_{2r-1} - F_{r-1}) F_{2n-r} - (-1)^n [F_r + (-1)^r F_{2r}]$ (Koshy, 1998)

74. $F_n L_{n+r} - L_n L_{n-r} = F_{2n+r} - L_{2n-r} - (-1)^n [F_r + (-1)^r L_r]$ (Koshy, 1998)

75. $L_{n+r}^2 + L_{n-r}^2 = L_{2n}L_{2r} + 4(-1)^{n+r}$ (Koshy, 1998)
76. $L_{10n} = [(L_{4n} - 3)^2 + (5F_{2n})^2]L_{2n}$ (Jarden, 1967)
77. $5(F_{n+r}^2 + F_{n-r}^2) = L_{2n}L_{2r} - 4(-1)^{n+r}$ (Koshy, 1998)
78. $5(F_{m+r}F_{m+r+1} + F_{m-r}F_{m-r+1}) = L_{2m+1}L_{2r} - 2(-1)^{m+r}$ (Koshy, 1999)
79. $F_{m+r}F_{m+r-2} + F_{m-r}F_{m-r-2} = L_{2m-2}L_{2r} - 6(-1)^{m+r}$ (Koshy, 1999)
80. $L_{m+r}L_{m+r+1} + L_{m-r}L_{m-r+1} = L_{2m+2r+1} + L_{2m-2r+1} + 2(-1)^{m+r}$ (Koshy, 1999)
81. $L_{m+r}L_{m+r+1} + L_{m-r}L_{m-r+1} = L_{2m+1}L_{2r} + 2(-1)^{m+r}$ (Koshy, 1999)
82. $\frac{L_{m+n}+L_{m-n}}{F_{m+n}+F_{m-n}} = \begin{cases} 5F_m/L_m & \text{if } n \text{ is odd} \\ L_m/F_m & \text{otherwise} \end{cases}$ (Wall, 1967)
83. $2F_{m+n} = F_mL_n + F_nL_m$ (Ferns, 1967)
84. $2L_{m+n} = L_mL_n + 5F_nF_m$ (Ferns, 1967)
85. $L_{m+n} + L_{m-n} = \begin{cases} 5F_mF_n & \text{if } n \text{ is odd} \\ L_mL_n & \text{otherwise} \end{cases}$ (Koshy, 1998)
86. $L_{m+n} - L_{m-n} = \begin{cases} L_mL_n & \text{if } n \text{ is odd} \\ 5F_mF_n & \text{otherwise} \end{cases}$ (Koshy, 1998)
87. $L_{m+n}^2 - L_{m-n}^2 = 5L_{2m}F_{2n}$ (Koshy, 1998)
88. $(F_nF_{n+1} - F_{n+2}F_{n+3})^2 = (F_nF_{n+3})^2 + (2F_{n+1}F_{n+2})^2$ (Umansky and Tallman, 1968)
89. $(L_nL_{n+1} - L_{n+2}L_{n+3})^2 = (L_nL_{n+3})^2 + (2L_{n+1}L_{n+3})^2$ (Umansky and Tallman, 1968)
90. $F_{n+1}^3 - F_n^3 - F_{n-1}^3 = 3F_{n+1}F_nF_{n-1}$ (Carlitz, 1967)
91. $L_{n+1}^3 - L_n^3 - L_{n-1}^3 = 3L_{n+1}L_nL_{n-1}$ (Carlitz, 1967)
92. $L_n^2 - F_n^2 = 4F_{n-1}F_{n+1}$ (Hoggatt, 1969)
93. $L_nL_{n+2} + 4(-1)^n = 5F_{n-1}F_{n+3}$ (Hoggatt, 1969)
94. $\sum_1^n F_iF_{3i} = F_nF_{n+1}F_{2n+1}$ (Recke, 1969)
95. $F_{n-1}^4 + F_n^4 + F_{n+1}^4 = 2[2F_n^2 + (-1)^n]^2$ (Hunter, 1966)
96. $L_{n-1}^4 + L_n^4 + L_{n+1}^4 = 2[2L_n^2 - 5(-1)^n]^2$ (Carlitz and Hunter, 1969)
97. $F_{n-1}^6 + F_n^6 + F_{n+1}^6 = 2[2F_n^2 + (-1)^n]^3 + 3F_{n-1}^2F_n^2F_{n+1}^2$ (Koshy, 1999)
98. $F_{n-1}^8 + F_n^8 + F_{n+1}^8 = 2[2F_n^2 + (-1)^n]^4 + 8F_{n-1}^2F_n^2(F_{n-1}^4 + F_n^4 + 4F_{n-1}^2F_n^2 + 3F_{n-1}F_nF_{2n-1})$ (Koshy, 1999)
99. $F_{m+n}^2L_{m+n}^2 - F_m^2L_m^2 = F_{2n}F_{4m+2n}$ (Hunter, 1969)

100. $25\sum_{i=1}^{n-1}\sum_{j=1}^{i}\sum_{k=0}^{j} F_{2k-1}^2 = F_{4n} + n(5n^2 - 14)/3$ (Swamy, 1970)

101. $25\sum_{i=1}^{n}\sum_{j=1}^{i} F_{2j-1}^2 = L_{4n+2} + 5n(n+1) - 3$ (Peck, 1970)

102. $F_{n+2}^3 - F_{n-1}^3 - 3F_nF_{n+1}F_{n+2} = F_{3n}$ (Padilla, 1970)

103. $L_{n+1}^3 + L_n^3 - L_{n-1}^3 = 5L_{3n}$ (Koshy, 1999)

104. $(F_nF_{n+3})^2 + (2F_{n+1}F_{n+2})^2 = F_{2n+3}^2$ (Anglin, 1970)

105. $F_{m+n} = F_mL_n - (-1)^nF_{m-n}$ (Ruggles, 1963)

106. $L_{5n} = L_n[L_{2n}^2 - (-1)^nL_{2n} - 1]$ (Carlitz, 1970)

107. $L_{5n} = L_n\{[L_{2n} - 3(-1)^n]^2 + 25(-1)^nF_n^2\}$ (Carlitz, 1970)

108. $F_{n+3}^2 = 2F_{n+2}^2 + 2F_{n+1}^2 - F_n^2$

109. $F_{3n} = L_nF_{2n} - (-1)^nF_n$ (Cheves, 1970)

110. $F_{3n} = F_n[L_{2n} + (-1)^n]$ (Koshy, 1999)

111. $F_{3n} = [F_{2n} - (-1)^n]F_n$ (Koshy, 1999)

112. $F_n^2 + F_{n+3}^2 = 2(F_{n+1}^2 + F_{n+2}^2)$ (Thompson, 1929)

113. $(F_n + F_{n+6})F_k + (F_{n+2} + F_{n+4})F_{k+1} = L_{n+3}L_{k+1}$ (Blank, 1956)

114. $(F_n^2 + F_{n+1}^2 + F_{n+2}^2)^2 = 2(F_n^4 + F_{n+1}^4 + F_{n+2}^4)$ (Candido, 1905)

115. $\sum_{j=0}^{13} F_{i+j} = 29F_{i+8}$ (Heath, 1950)

116. $L_n^3 = 2F_{n-1}^3 + F_n^3 + 6F_{n-1}F_{n+1}^2$ (Barley, 1973)

117. $5F_{2n+3}F_{2n-3} = L_{4n} + 18$ (Blazej, 1975)

118. $1 + 4F_{2n+1}F_{2n+2}^2F_{2n+3} = (2F_{2n+2}^2 + 1)^2$ (Hoggatt and Bergum, 1977)

119. $1 + 4F_{2n+1}F_{2n+2}F_{2n+3}F_{2n+4} = (2F_{2n+2}F_{2n+3} + 1)^2$ (Hoggatt and Bergum, 1977)

120. $F_{8n} = L_{2n}\sum_{1}^{n} L_{2n+4k-2}$ (Higgins, 1976)

121. $F_nF_{n+3}^2 - F_{n+2}^3 = (-1)^{n+1}F_{n+1}$ (Hoggatt and Bergum, 1977)

122. $F_{n+3}F_n^2 - F_{n+1}^3 = (-1)^{n+1}F_{n+2}$ (Hoggatt and Bergum, 1977)

123. $F_nF_{n+3}^2 - F_{n+4}F_{n+1}^2 = (-1)^{n+1}L_{n+2}$ (Hoggatt and Bergum, 1977)

124. $F_nL_{n+3}^2 - F_{n+4}L_{n+1}^3 = (-1)^{n+1}L_{n+2}$ (Hoggatt and Bergum, 1977)

125. $7F_{n+2}^3 - F_{n+1}^3 - F_n^3 = 3L_{n+1}F_{n+2}F_{n+3}$ (Barley, 1973)

126. $F_{3^n} = \prod_0^{n-1} (L_{2\cdot 3^k} - 1)$ (Usiskin, 1974)

127. $L_{3^n} = \prod_0^{n-1} (L_{2\cdot 3^k} + 1)$ (Usiskin, 1974)

128. $F_{mn} = L_m F_{m(n-1)} + (-1)^{m+1} F_{m(n-2)}$ (Cheves, 1975)

129. $L_{(2m+1)(4n+1)} - L_{2m+1} = 5F_{2n(2m+1)} F_{(2m+1)(2n+1)}$ (Koshy, 1999)

130. $L_n^2 + L_{n+1}^2 = L_{2n} + L_{2n+2}$ (Koshy, 1999)

131. $F_m L_n + F_n L_m = 2F_{m+n}$ (Blazej, 1975)

132. $F_{n+k}^3 + (-1)^k F_{n-k}(F_{n-k}^2 + 3F_{n+k} F_n L_k) = L_k^3 F_n^3$ (Mana, 1978)

133. $F_{n+k}^3 - L_{3k} F_n^3 + (-1)^k F_{n-k}^3 = 3(-1)^n F_n F_k F_{2k})$ (Wulczyn, 1978)

134. $F_{n+10}^4 = 55(F_{n+8}^4 - F_{n+2}^4) - 385(F_{n+6}^4 - F_{n+4}^4) + F_n^4$ (Wulczyn, 1979)

135. $F_k F_{n+j} - F_j F_{n+k} = (-1)^j F_{k-j} F_n$ (Taylor, 1982a)

136. $F_k L_{n+j} - F_j L_{n+k} = (-1)^j F_{k-j} L_n$ (Taylor, 1982b)

Additional identities are presented in the exercises and in the following chapters.

FERMAT AND FIBONACCI

A judge by profession, the great French mathematician Pierre de Fermat (1601–1665) observed that the numbers 1, 3, 8, and 120 have a fascinating property. *One more than the product of any two of them is a perfect square*:

$$1 + 1\cdot 3 = 2^2 \qquad 1 + 1\cdot 8 = 3^2 \qquad 1 + 1\cdot 120 = 11^2$$
$$1 + 3\cdot 8 = 5^2 \qquad 1 + 3\cdot 120 = 19^2 \qquad 1 + 8\cdot 120 = 31^2$$

In 1969, Alan Baker and Harold Davenport of Trinity College, Cambridge proved that if 1, 3, 8, and x have this property, then x must be 120.

Intriguingly enough, notice that $1 = F_2$, $3 = F_4$, $8 = F_6$, and $120 = 4\cdot 2\cdot 3\cdot 5 = 4F_3F_4F_5$. Accordingly, eight years later, V. Hoggatt, Jr., and G. E. Bergum of South Dakota State University picked up on this observation and established the following generalization.

Theorem 5.12. The numbers F_{2n}, F_{2n+2}, F_{2n+4}, and $4F_{2n+1}F_{2n+2}F_{2n+3}$ have the property that one more than the product of any two of them is a perfect square.

Proof. It follows by Cassini's formula that $1 + F_{2n}F_{2n+2} = F_{2n+1}^2$. Similarly, $1 + F_{2n+1}F_{2n+3} = F_{2n+2}^2$ and $1 + F_{2n+2}F_{2n+4} = F_{2n+3}^2$. Next we have

$$
\begin{aligned}
1 + F_{2n}(4F_{2n+1}F_{2n+2}F_{2n+3}) \\
&= 1 + 4(F_{2n}F_{2n+2})(F_{2n+1}F_{2n+3}) \\
&= 1 + 4(F_{2n+1}^2 - 1)(F_{2n+2}^2 + 1) \qquad \text{by Cassini's formula} \\
&= 4F_{2n+1}^2F_{2n+2}^2 - 4(F_{2n+2}^2 - F_{2n+1}^2) - 3 \\
&= 4F_{2n+1}^2F_{2n+2}^2 - 4F_{2n+3}F_{2n} - 3 \\
&= 4F_{2n+1}^2F_{2n+2}^2 - 4F_{2n+3}(F_{2n+2} - F_{2n+1}) - 3 \\
&= 4F_{2n+1}^2F_{2n+2}^2 - 4F_{2n+3}F_{2n+2} + 4F_{2n+1}F_{2n+3} - 3 \\
&= 4F_{2n+1}^2F_{2n+2}^2 - 4F_{2n+3}F_{2n+2} + 4(F_{2n+2}^2 + 1) - 3 \\
&= 4F_{2n+1}^2F_{2n+2}^2 - 4F_{2n+2}(F_{2n+3} - F_{2n+2}) + 1 \\
&= 4F_{2n+1}^2F_{2n+2}^2 - 4F_{2n+1}F_{2n+2}) + 1 \\
&= (2F_{2n+1}F_{2n+2} - 1)^2
\end{aligned}
$$

Similarly, it can be shown that $1 + F_{2n+2}(4F_{2n+1}F_{2n+2}F_{2n+3}) = (2F_{2n+2}^2 + 1)^2$ and $1 + F_{2n+4}(4F_{2n+1}F_{2n+2}F_{2n+3}) = (2F_{2n+2}F_{2n+3} + 1)^2$. Thus, one more than the product of any two of the numbers is a square. ■

For example, $n = 1$ yields the Fermat's quadruple $(F_2, F_4, F_6, 4F_3F_4F_5) = (1, 3, 8, 120)$; $n = 2$ yields the quadruple $(F_4, F_6, F_8, 4F_5F_6F_7) = (3, 8, 21, 2080)$; and $n = 3$ yields the quadruple $(F_6, F_8, F_{10}, 4F_7F_8F_9) = (8, 21, 55, 37128)$.

Hoggatt and Bergum also proved the following theorem for Fibonacci numbers with consecutive subscripts.

Theorem 5.13. Let $x = 4F_{2n+2}F_{2n+3}F_{2n+4}$. Then,

$$
\begin{aligned}
1)\ 1 + xF_{2n+1} &= (2F_{2n+2}F_{2n+3} + 1)^2 \\
2)\ 1 + xF_{2n+3} &= (2F_{2n+3}^2 - 1)^2 \\
3)\ 1 + xF_{2n+5} &= (2F_{2n+3}F_{2n+4} - 1)^2
\end{aligned}
$$

■

For example, let $n = 3$. Then $x = 4F_8F_9F_{10} = 4 \cdot 21 \cdot 34 \cdot 55 = 157,080$. We have:

$$
\begin{aligned}
1 + xF_7 &= 1 + 157,080 \cdot 13 = 1429^2 = (2 \cdot 21 \cdot 34 + 1)^2 = (2F_8F_9 + 1)^2 \\
1 + xF_9 &= 1 + 157,080 \cdot 34 = 2311^2 = (2 \cdot 34^2 - 1)^2 = (2F_9^2 - 1)^2 \\
1 + xF_{11} &= 1 + 157,080 \cdot 89 = 3739^2 = (2 \cdot 34 \cdot 55 - 1)^2 = (2F_9F_{10} - 1)^2
\end{aligned}
$$

An unusual relationship exists between the geometric constant π and Fibonacci numbers.

FIBONACCI AND π

In 1985, Yuri V. Matiyasevich of St. Petersburg, Russia, developed a wonderful formula for π in terms of Fibonacci numbers:

$$\pi = \lim_{n\to\infty} \sqrt{\frac{6 \log F_1F_2\cdots F_n}{\log[F_1, F_2, \ldots, F_n]}}$$

where $[x, y]$ denotes the least common multiple (LCM) of the integers x and y. A proof of this formula, using some sophisticated number theory, appeared in the following year in *The American Mathematical Monthly*.

It is easy to verify that

$$\lim_{n\to\infty} \sqrt{\frac{6 \log F_1F_2\cdots F_{10}}{\log[F_1, F_2, \ldots, F_{10}]}} \approx 2.77322490387$$

and

$$\lim_{n\to\infty} \sqrt{\frac{6 \log F_1F_2\cdots F_{12}}{\log[F_1, F_2, \ldots, F_{12}]}} \approx 2.8454900617$$

So it is a valuable exercise to determine the value of n for which the formula yields a desired approximation of π. Additionally, does a corresponding formula exist for Lucas numbers?

We now turn to two simple but interesting Fibonacci puzzles.

1. Think of two positive integers. Add them to get a third number. Add the second number and the third number to get a fourth number. Continue like this until there are ten numbers. Add all ten numbers. The resulting sum is 11 times the seventh number. See Exercise 15 in Chapter 7. (This puzzle was discovered in 1950 by R. V. Heath.)

2. Write down four consecutive Fibonacci numbers. The (positive) difference of the squares of the two middle numbers equals the product of the other two. (See Exercise 35.)

EXERCISES 5

1. Prove Theorem 5.2 using PMI.
2. Prove Corollary 5.1 using PMI.

3–7. Prove Identities 5.6 through 5.10.

8. Prove Theorem 5.4 using PMI.

Verify each.

9. $F_{10} = F_5 L_5$
10. $L_{10} = F_9 + F_{11}$
11. $F_7^2 - F_5^2 = F_{12}$
12. $F_6^2 + F_5^2 = F_{11}$

Disprove each, where $n \geq 1$.

13. $L_{n+1}L_{n-1} - L_n^2 = (-1)^n$
14. $L_{n+1}(L_n + L_{n+2}) = L_{2n+2}$

Let $v_n = \alpha^n + \beta^n$, $n \geq 1$. Verify each.

15. $v_1 = 1$ and $v_2 = 3$.
16. $v_n = v_{n-1} + v_{n-2}$, where $n \geq 3$. (Exercises 15 and 16 prove that $v_n = L_n$.)

Prove each using PMI.

17. Binet's formula for F_n.
18. $L_n = \alpha^n + \beta^n$.
19. Prove that $F_n = L_n$ if and only if $n = 1$.

Find a quadratic equation with the given roots, where k is a real number.

20. α^n, β^n
21. $\alpha^n + k, \beta^n + k$
22. α^n, α^{-n}, where n is odd.

Using Lemma 5.1, prove each.

23. Identity 5.1
24. Identity 5.6
25. Prove that $F_{n+5} = 5F_{n+1} + 3F_n$, where $n \geq 0$.
26. Using Exercise 25, prove that $5|F_{5n}$ for every $n \geq 0$.
27. Establish Cassini's formula (5.4) using Binet's formula.
28. Solve the recurrence relation $D_{n+1} = D_n + L_{2n} - 1$, where $D_0 = 0$ (Hoggatt, 1972).

Prove each, where $m, n \geq 1$.

29. $F_{2n} = F_n L_n$

30. $F_{n+1}^2 + F_n^2 = F_{2n+1}$
31. $F_{n+1}^2 - F_{n-1}^2 = F_{2n}$
32. $F_{n-1} + F_{n+1} = L_n$
33. $F_{n+2} - F_{n-2} = L_n$
34. $L_{n-1} + L_{n+1} = 5F_n$
35. $F_{n+1}^2 - F_n^2 = F_{n-1}F_{n+2}$
36. $5(F_n^2 + F_{n-2}^2) = 3L_{2n-2} - 4(-1)^n$
37. $L_n^2 + L_{n+1}^2 = 5F_{2n+1}$
38. $L_{n-1}L_{n+1} - L_n^2 = 5(-1)^{n-1}$
39. $5F_n^2 = L_n^2 - 4(-1)^n$
40. $L_{n+1}^2 - L_n^2 = L_{n-1}L_{n+1}$
41. $L_n^2 = L_{2n} + 2(-1)^n$
42. $L_{2n} = 5F_n^2 + 2(-1)^n$
43. $L_{n+2} - L_{n-2} = 5F_n$
44. $L_{n+1}^2 - L_n^2 = 5F_{2n}$
45. $F_{2n-2} < F_n^2 < F_{2n-1}, \quad n \geq 2$ (Hoggatt, 1963)
46. $F_{2n-1} < L_{n-1}^2 < F_{2n}, \quad n \geq 2$ (Hoggatt, 1963)
47. $L_{-n} = (-1)^n L_n$
48. $\alpha^n = \alpha F_n + F_{n-1}$
49. $1 + \alpha^{2n} = \begin{cases} \sqrt{5}F_n\alpha^n & \text{if } n \text{ is odd} \\ L_n\alpha^n & \text{otherwise} \end{cases}$
50. $1 + \beta^{2n} = \begin{cases} -\sqrt{5}F_n\beta^n & \text{if } n \text{ is odd} \\ L_n\beta^n & \text{otherwise} \end{cases}$
51. $L_{2m+n} - (-1)^m L_n = 5F_m F_{m+n}$
52. $F_{2m+n} - (-1)^m F_n = F_m L_{m+n}$
53. $F_{2m+n} + (-1)^m F_n = F_{m+n} L_m$
54. $F_{2n} = F_n F_{n+1} + F_{n-1} F_n$
55. $L_{3n} = L_n[L_{2n} - (-1)^n]$
56. $F_{m+n} + F_{m-n} = \begin{cases} L_m F_n & \text{if } n \text{ is odd} \\ L_n F_m & \text{otherwise} \end{cases}$
57. $F_{m+n} - F_{m-n} = \begin{cases} F_m L_n & \text{if } n \text{ is odd} \\ L_m F_n & \text{otherwise} \end{cases}$
58. $F_{m+n}^2 - F_{m-n}^2 = F_{2m} L_{2n}$
59. $F_n F_{n+1} - F_{n-1} F_{n-2} = F_{2n-1}$ (Lucas, 1876)
60. $\sum_{i=1}^{n} F_{k+i} = F_{n+k+2} - F_{k+2}$
61. $F_{n+3}^2 = 2F_{n+2}^2 + 2F_{n+1}^2 - F_n^2$ (Gould, 1963)
62. $F_{n+1}^3 + F_n^3 - F_{n-1}^3 = F_{3n}$ (Lucas, 1876)
63. $F_{n+2}^3 - 3F_n^3 + F_{n-2}^3 = 3F_{3n}$ (Ginsburg, 1953)

64. $(\alpha F_n + F_{n-1})^{1/n} + (-1)^{n+1}(F_{n+1} - \alpha F_n)^{1/n} = 1$ (A. Sofo, 1999)

Find a solution of each equation.

65. $x^2 - 5y^2 = 4$
66. $x^2 - 5y^2 = -4$
67. Prove that $x^2 - x - 1$ is a factor of $x^{2n} - L_n x^n + (-1)^n$, where $n \geq 1$ (P. Mana, 1972).
68. Show that $[L_{2n} + 3(-1)^n]/5$ is the product of two Fibonacci numbers (Freitag, 1974).
69. Show that $L_{2n} - 3(-1)^n$ is the product of two Lucas numbers (Freitag, 1974).
70. Let $R_m = \sum_{0}^{m} F_{i+1} L_{m-i}$. Prove that $S_m = 10R_m/(m+2)$ is a sum of two Lucas numbers (H. T. Freitag, 1982).
71. Prove that $2L_{n-1}^3 + L_n^3 + 6L_{n+1}^2 L_{n-1}$ is a perfect cube (Wulczyn, 1977).
72. Show that the sum of any $2n$ consecutive Fibonacci numbers is divisible by F_n, where n is even (Lind, 1964).
73. Let $n \geq 1$ and $(1 + \sqrt{5})^n = a_n + b_n\sqrt{5}$, where a_n and b_n are positive integers. Prove that $2^{n-1}|a_n$ and $2^{n-1}|b_n$ (Mana, 1970).
74. Let $L(n) = L_n$ and $t_n = n(n+1)/2$. Prove that $L(n) = (-1)^{t_n}[L(t_{n-1})L(t_n) - L(n^2)$ (Freitag, 1982).
75. Prove that if $2F_{2n-1}F_{2n+1} - 1$ is a prime, then so are $2F_n^2 + 1$ and $F_{2n}^2 + F_{2n-1}F_{2n+1}$ (Guillotte, 1973).
76. Find a formula for $K_n = (K_1 + K_2 + \cdots + K_{n-1}) + F_{2n-1}$, where $K_1 = 1$ (V. Hoggatt, Jr., 1972).
77. Let $\{g_n\}$ be any number sequence. Show that $\sum_{1}^{n}(g_{k+2} + g_{k+1} - g_k)F_k = g_{n+2}F_n + g_{n+1}F_{n+1} - g_1$ (Recke, 1969).

Let f be a function defined by

$$f(n) = \begin{cases} f(n/2) & \text{if } n \text{ is even} \\ f((n+1)/2) + f((n-1)/2) & \text{otherwise} \end{cases}$$

where $f(1) = 1$. Prove each (D. Lind, 1970).

78. $f([2^{n+1} + (-1)^n]/3) = F_{n+1}$
79. $f([7 \cdot 2^{n-1} + (-1)^n]/3) = L_n$
80. Evaluate the sum $\sum_{1}^{n} F_i G_i$, where $G_{n+2} = 2G_{n+1} + G_n$, $G_1 = 1$ and $G_2 = 2$ (Mead, 1965).
81. Let S_n denote the sum of the numbers in row n of the triangular array of Fibonacci numbers in Figure 5.12. Derive a formula for S_n.

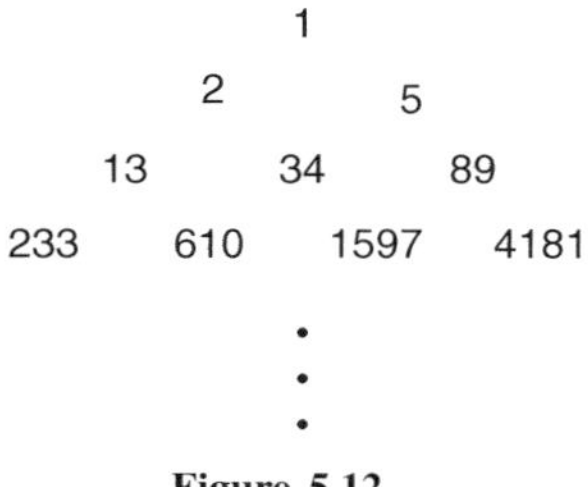

Figure 5.12.

82. Redo Exercise 81 using Figure 5.13.

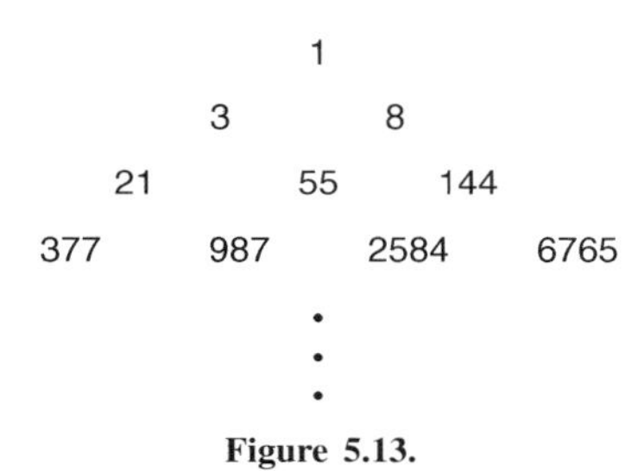

Figure 5.13.

83. Prove that the area of the trapeziod with bases F_{n+1} and F_{n-1}, and sides F_n is $\sqrt{3}F_{2n}/4$ (*Mathematics Teacher*, 1993).

GEOMETRIC PARADOXES

The preceding chapter established that $F_{n+1}F_{n-1} - F_n^2 = (-1)^n$. This identity is the cornerstone of two classes of fascinating geometric paradoxes. When n is even, say, $n = 2k$, the identity yields $F_{2k}^2 - F_{2k+1}F_{2k-1} = 1$; the first paradox is based on this result. When n is odd, say, $n = 2k - 1$, the identity yields $F_{2k-1}^2 - F_{2k+2}F_{2k} = -1$; the second paradox is based on this result.

The first paradox was a favorite of the famous English puzzlist, Charles Lutwidge Dodgson (1832–1898), better known by his pseudonym, Lewis Carroll. This puzzle, first proposed in 1774 by William Hooper in his *Rational Recreations*, reappeared in a mathematics periodical in Leipzig, Germany, in 1868, 666 years (watch for the beastly number) after Fibonacci published his rabbit-breeding problem.

W. W. Rouse Ball claims in his *Mathematical Recreations and Essays*, which is a jewel in recreational mathematics, that 1868 was the earliest date he could find for the first appearance of this puzzle. Although the origin of the puzzle is still a mystery, the elder Sam Loyd claimed that he had presented the puzzle to the American Chess Congress in 1858.

Consider an 8×8 square; cut it up into four pieces, A, B, C, and D, as in Figure 6.1. Now rearrange the pieces to form a 5×13 rectangle, as Figure 6.2 shows. While the area of the square is 64 units, that of the rectangle is 65 units. In other words, by reassembling the pieces of the original square, we have gained one unit. This is paradoxical.

How is that possible? In Figure 6.2, it appears that the "diagonal" *PQRS* of the rectangle is a line (segment). In fact, this appearance is deceptive. The points P, Q, R, and S are in fact the vertices of a very narrow parallelogram, as Figure 6.3 illustrates.

$$\begin{aligned} \therefore \quad \text{Area of the parallelogram} &= \text{area of the rectangle} - \text{area of the square} \\ &= 65 - 64 = 1 \end{aligned}$$

Thus, the area of the parallelogram equals $F_7F_5 - F_6^2 = 1$ unit.

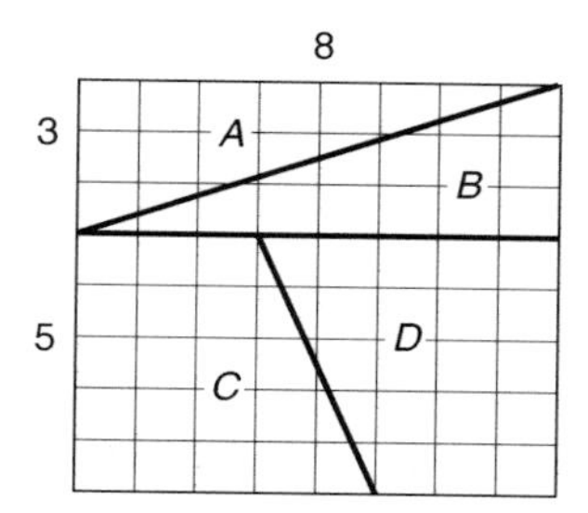

Figure 6.1.

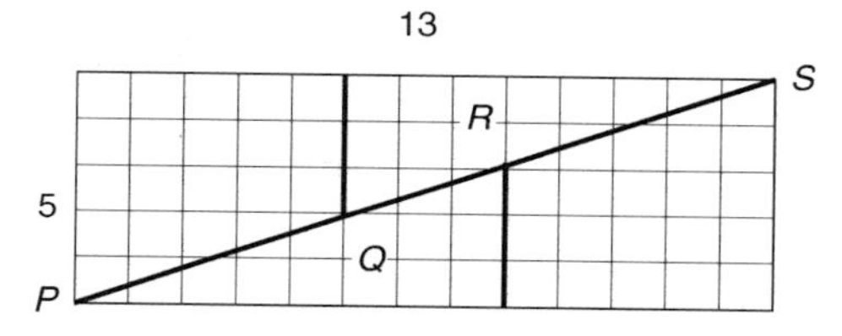

Figure 6.2.

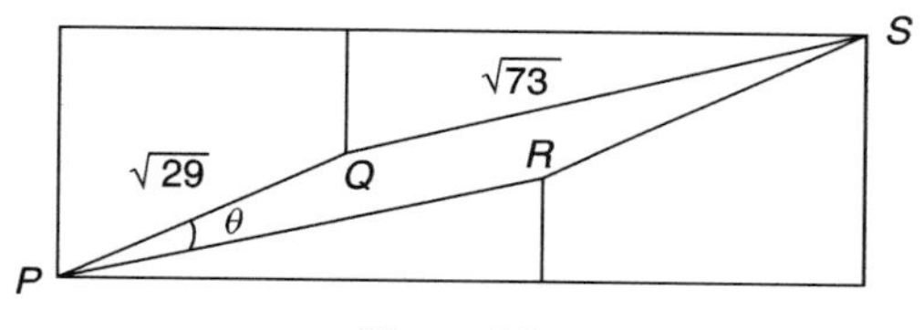

Figure 6.3.

Its sides are $\sqrt{29}$ and $\sqrt{73}$ units long, and the diagonal is $\sqrt{194}$ units long. Let θ be the acute angle between the adjacent sides of the parallelogram. Then, by the law of cosines in trigonometry,

$$\begin{aligned} \cos\,\theta/2 &= \frac{194 + 29 - 73}{2\sqrt{29 \cdot 194}} \\ &\approx 0.763898460833 \\ \theta &\approx 1^\circ 31' 40'' \end{aligned}$$

This explains why it is a very narrow parallelogram.

In fact, there is nothing sacred about the choice of the size of the square, except that $8 = F_6$ is a Fibonacci number with an even subscript and $F_7 = 13$ and $F_5 = 5$ are its adjacent neighbors.

Since $F_{n+1}F_{n-1} - F_n^2 = 1$, when n is even, the puzzle can be extended to any $F_n \times F_n$ square. Cut this square up into four squares, as in Figure 6.4, and these

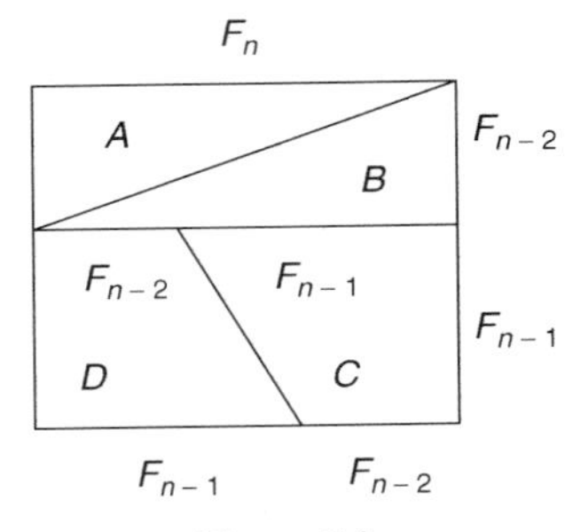

Figure 6.4.

squares can be rearranged to form the deceptive rectangle of size $F_{n-1} \times F_{n+1}$, as Figure 6.5 shows.

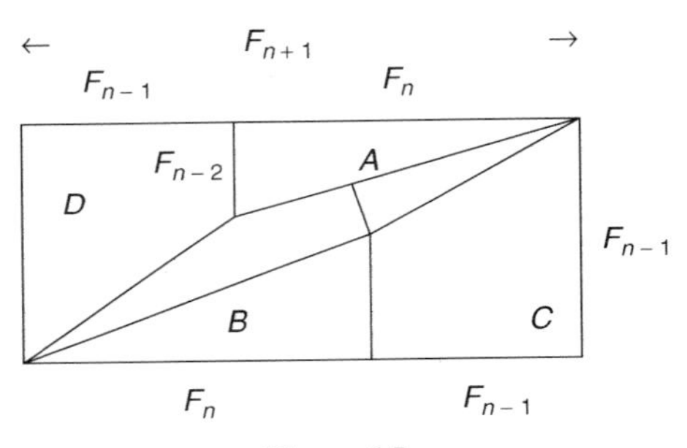

Figure 6.5.

The parallelogram, magnified in the figure, has an area of one unit. So we can now determine its height h:

$$\text{Area of the parallelogram} = \text{base} \times \text{height}$$

That is,

$$1 = h\sqrt{F_n^2 + F_{n-2}^2} \quad \text{by the Pythagorean theorem}$$

$$\therefore \quad h = \frac{1}{\sqrt{F_n^2 + F_{n-2}^2}}$$

Thus, as the size of the original square increases, the parallelogram becomes narrower and the gap becomes less and less noticeable.

Sam Loyd's son was the first person to discover that the four pieces in Figure 6.1 can be arranged to form an area of 63 square units, as Figure 6.6 shows. The son adopted his father's name and inherited his father's puzzle column in the Brooklyn *Daily Eagle*.

To illustrate a paradox of the second kind, consider a 5×5 square and cut it into four pieces, as Figure 6.7 shows. Now reassemble the pieces to form the 3×8

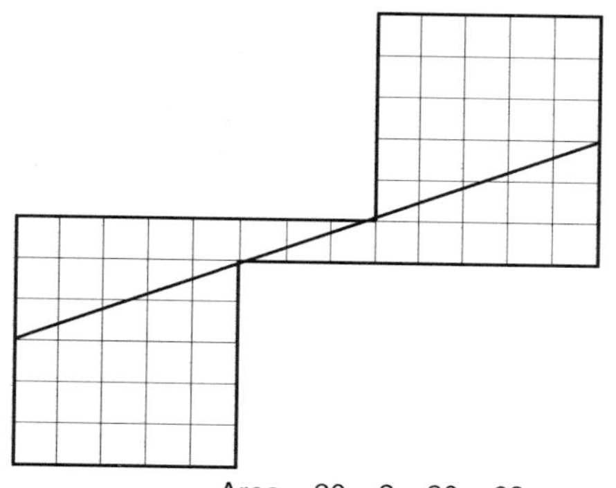

Figure 6.6.

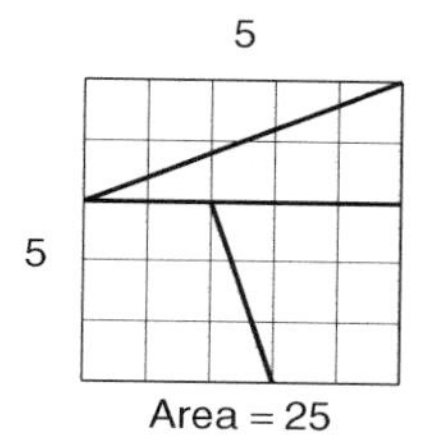

Figure 6.7.

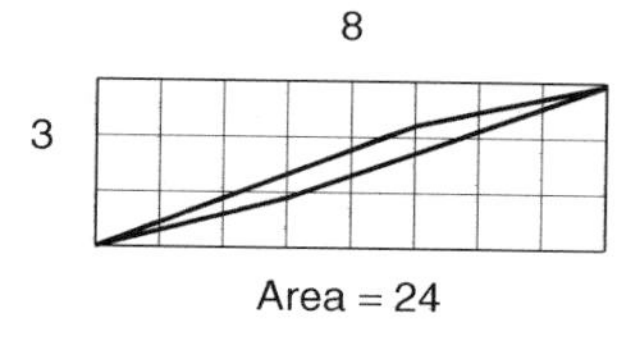

Figure 6.8.

"rectangle" (see Fig. 6.8). The area of the square is 25 units, whereas that of the rectangle is only 24 units, so we have lost one unit. The overlap along the diagonal accounts for the missing area. Notice that the area of the square $= F_5^2 = F_4F_6 + 1 =$ area of the rectangle + 1.

More generally, let n be odd; suppose an $F_n \times F_n$ square is cut into four pieces (see Fig. 6.9) and they are assembled into an $F_{n-1} \times F_{n+1}$ rectangle (see Fig. 6.10). Then we would be missing an area of one unit, because $F_{n+1}F_{n-1} - F_n^2 = -1$.

In 1962, A. F. Horadam of the University of New England, Australia, derived a formula for tan θ_n, where θ_n denotes the acute angle between the adjacent sides of the parallelogram. To derive the formula, we first consider the case n even, where $n \geq 4$.

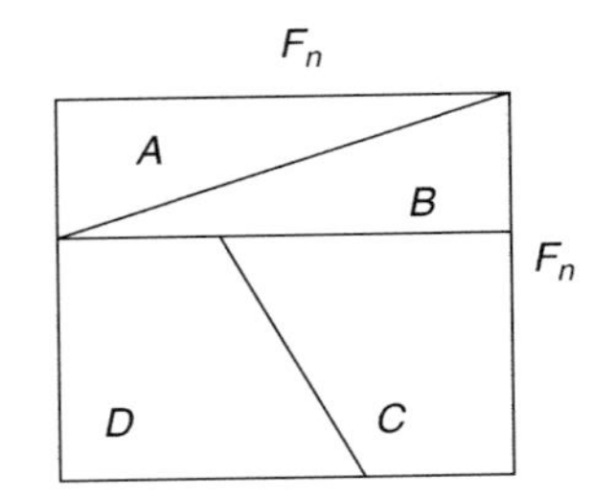

Figure 6.9.

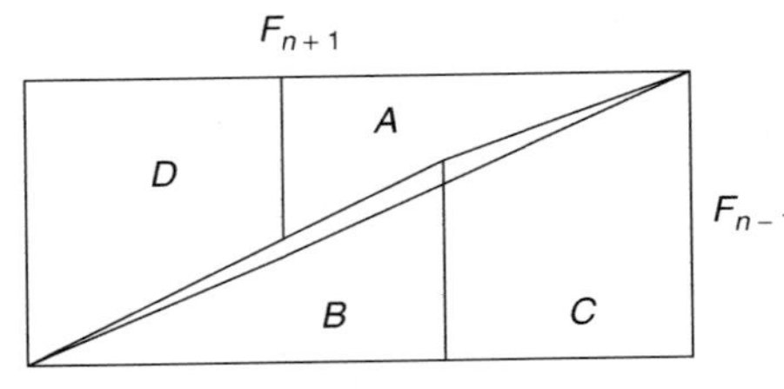

Figure 6.10.

Using Figure 6.11, we have:

$$
\begin{aligned}
\theta_n &= \frac{\pi}{2} - (\alpha_n + \beta_n) \\
&= \frac{\pi}{2} - \tan^{-1}\frac{F_{n-1}}{F_{n-3}} - \tan^{-1}\frac{F_{n-2}}{F_n} \\
&= \tan^{-1}\frac{F_{n-3}}{F_{n-1}} - \tan^{-1}\frac{F_{n-2}}{F_n} \qquad \text{since } \tan^{-1}x + \tan^{-1}1/x = \pi/2 \\
\tan\theta_n &= \frac{(F_{n-3}/F_{n-1}) - (F_{n-2}/F_n)}{1 + (F_{n-3}/F_{n-1})\cdot(F_{n-2}/F_n)} \\
&= \frac{F_{n-3}F_n - F_{n-1}F_{n-2}}{F_{n-1}F_n + F_{n-3}F_{n-2}} \\
&= \frac{F_{n-3}(F_{n-1} + F_{n-2}) - F_{n-2}(F_{n-2} + F_{n-3})}{F_{n-1}(F_{n-1} + F_{n-2}) + F_{n-3}F_{n-2}} \\
&= \frac{F_{n-1}F_{n-3} - F_{n-2}^2}{F_{n-1}^2 + F_{n-2}(F_{n-2} + F_{n-3}) + F_{n-3}F_{n-2}} \\
&= \frac{(-1)^{n-2}}{F_{n-1}^2 + F_{n-2}^2 + 2F_{n-3}F_{n-2}} \\
&= \frac{(-1)^n}{F_{2n-3} + 2F_{n-3}F_{n-2}} \qquad \text{by Identity 5.11} \\
&= \frac{1}{F_{2n-3} + 2F_{n-3}F_{n-2}} \qquad (6.1)
\end{aligned}
$$

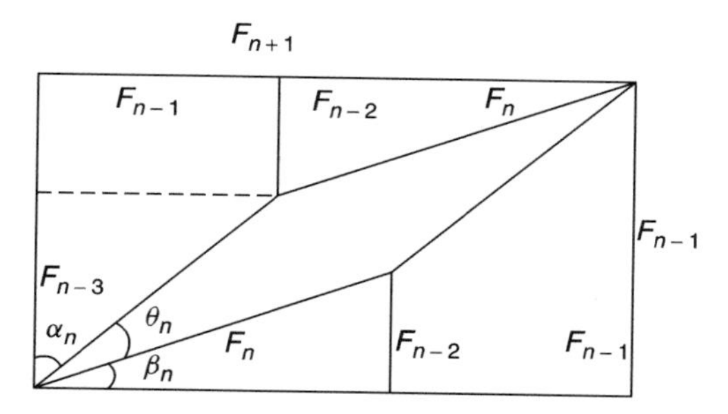

Figure 6.11.

Now, let n be odd. Then there is an overlap, as Figure 6.12 shows. It follows from the figure that:

$$\begin{aligned}\theta_n &= (\alpha_n + \beta_n) - \frac{\pi}{2} \\ &= \tan^{-1}\frac{F_{n-1}}{F_{n-3}} + \tan^{-1}\frac{F_{n-2}}{F_n} - \frac{\pi}{2} \\ &= \tan^{-1}\frac{F_{n-2}}{F_n} - \tan^{-1}\frac{F_{n-3}}{F_{n-1}}\end{aligned}$$

As before, this leads to the equation

$$\begin{aligned}\tan\theta_n &= \frac{(-1)^{n-1}}{F_{2n-3} + 2F_{n-3}F_{n-2}} \\ &= \frac{1}{F_{2n-3} + 2F_{n-3}F_{n-2}} \qquad \text{since } n \text{ is odd}\end{aligned}$$

Thus, in both cases,

$$\tan\theta_n = \frac{1}{F_{2n-3} + 2F_{n-3}F_{n-2}} \qquad n \geq 4$$

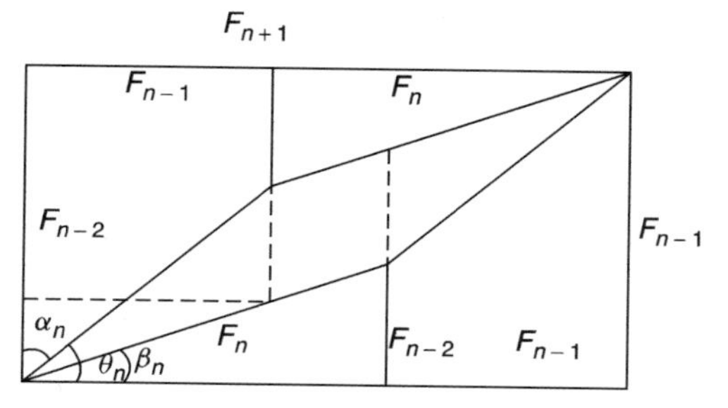

Figure 6.12.

TABLE 6.1.

n	Fibonacci Triplets F_{n-1}	F_n	F_{n+1}	θ_n
4	2	3	5	$\tan^{-1}\frac{1}{7} \approx 8°7'48''$
5	3	5	8	$\tan^{-1}\frac{1}{17} \approx 3°21'59''$
6	5	8	13	$\tan^{-1}\frac{1}{46} \approx 1°14'43''$
7	8	13	21	$\tan^{-1}\frac{1}{119} \approx 28'53''$
8	13	21	34	$\tan^{-1}\frac{1}{313} \approx 10'59''$
9	21	34	55	$\tan^{-1}\frac{1}{818} \approx 4'12''$
10	34	55	89	$\tan^{-1}\frac{1}{2,143} \approx 1'36''$
11	55	89	144	$\tan^{-1}\frac{1}{5,609} \approx 37''$
12	89	144	233	$\tan^{-1}\frac{1}{14,686} \approx 14''$

Table 6.1 shows the values of θ_n for the first few Fibonacci triplets F_{n-1}, F_n, and F_{n+1}. It follows from the table that as n increases slowly, $\theta_n \to 0$ rapidly, thus, $\theta_n \to 0$ as $n \to \infty$.

ADDITIONAL FIBONACCI-BASED PUZZLES

In fact, there are many delightful puzzles in which Fibonacci-based rectangles can be cut into several pieces, and the pieces arranged to form a rectangle of larger or smaller area. One such paradox is Langman's paradox, developed by H. Langman of New York City.

Langman's Paradox

Cut an 8×13 rectangle into four pieces, as Figure 6.13 shows. Now arrange the pieces to form a 5×21 rectangle, as in Figure 6.14. Thus we gain one unit square.

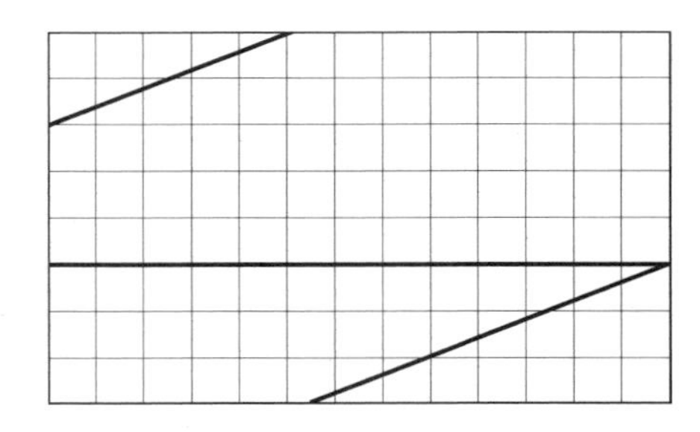

Area = 8 × 13 = 104

Figure 6.13.

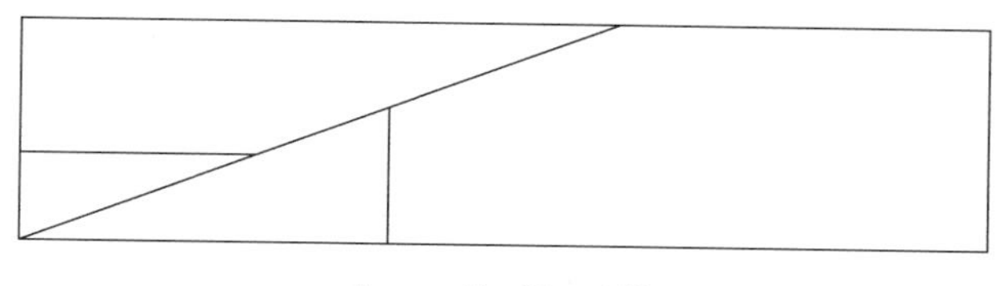

Figure 6.14.

Another Version of Langman's Paradox

Another version of Langman's paradox involves gaining two square units when the pieces of the 8×21 rectangle in Figure 6.15 are assembled. Cut out the shaded area and place it on top of the unshaded area in such a way that the diagonal cuts form one long diagonal; now switch pieces A and B. The resulting area is 170 square units.

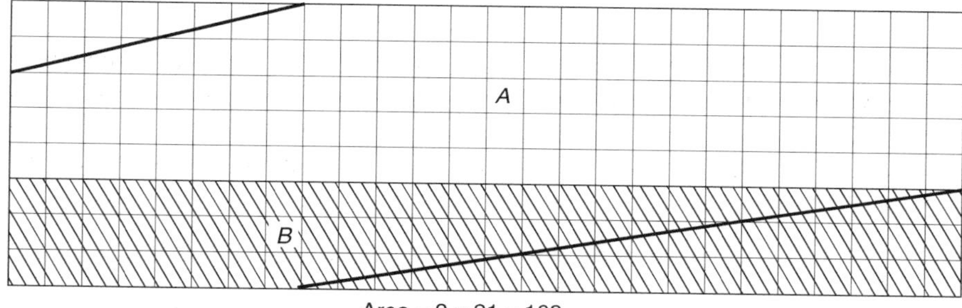

Figure 6.15.

The next paradox was developed in 1953 by Paul Curry, an amateur magician of New York City. It involves two alternate Fibonacci numbers.

Curry's Paradox

Swap the positions of the triangles B and C in Figure 6.16 to form the 5×13 rectangle in Figure 6.17. This results in an apparent loss of one square unit. In fact, the loss occurs in the shaded area. Figure 6.16 contains 15 shaded cells, whereas Figure 6.17 requires 16 cells to complete the 5×13 rectangle. In other words, we seem to lose one square area in the process.

AN INTRIGUING SEQUENCE

Finally, suppose we construct a number sequence beginning with two arbitrary real numbers a and b, and then use the Fibonacci recurrence relation to construct the remaining elements. All such sequences, except one, can be used to develop the

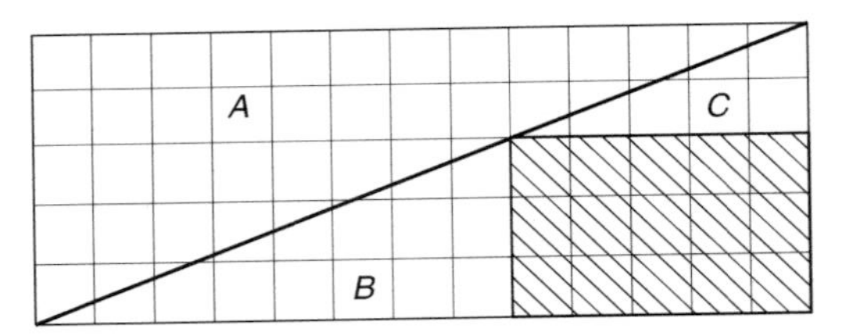

Figure 6.16.

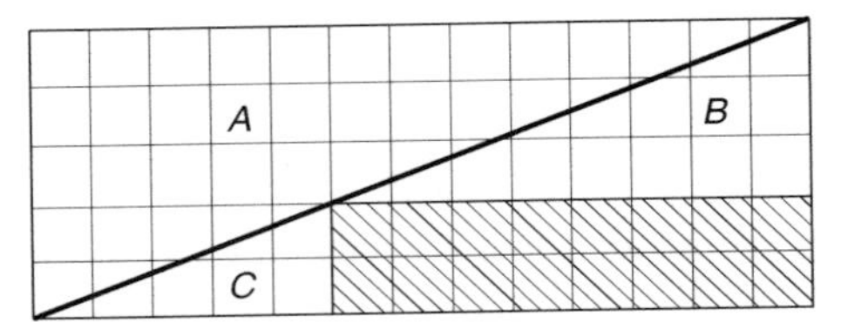

Figure 6.17.

preceding puzzles. So, the question is, which sequence will *not* produce the puzzle? In other words, under what conditions will the square and rectangle have exactly the same area?

To answer this, we must consider the following additive number sequence with $a = 1$ and $b = \alpha$:

$$1, \alpha, \alpha + 1, 2\alpha + 1, 3\alpha + 2, \ldots, s_n, \ldots$$

Suppose we pick any three consecutive terms: s_{n-2}, s_{n-1}, and s_n. Then $s_{n-1}s_{n+1} = s_n^2$, so the area of the square indeed equals that of the rectangle. This is so, because, by Lemma 5.1, $s_n = \alpha^n$ and $\alpha^{n-1} \cdot \alpha^{n+1} = \alpha^{2n}$.

Interestingly enough, $\{s_n\}$ is the only additive number sequence that has this striking behavior. The ratio of any two consecutive terms of the sequence is a constant: $s_{n+1}/s_n = \alpha$. Martin Gardner, who wrote a popular column called *Mathematical Games* in *Scientific American*, referred to this sequence as the "golden series," which all additive number sequences struggle to become.

GENERALIZED FIBONACCI NUMBERS

We can study properties common to Fibonacci and Lucas numbers by investigating a number sequence that satisfies the Fibonacci recurrence relation, but with arbitrary initial conditions.

GENERALIZED FIBONACCI NUMBERS

To this end, consider the sequence $\{G_n\}$, where $G_1 = a$, $G_2 = b$, and $G_n = G_{n-1} + G_{n-2}$, $n \geq 3$. The ensuing sequence

$$a, b, a+b, a+2b, 2a+3b, 3a+5b, \ldots \tag{7.1}$$

is called the *generalized Fibonacci sequence* (GFS).

Take a close look at the coefficients of a and b in the various terms of this sequence. They follow an interesting pattern: The coefficients of a and b are Fibonacci numbers. In fact, we can pinpoint these two Fibonacci coefficients, as the following theorem shows.

Theorem 7.1. Let G_n denote the nth term of the GFS. Then $G_n = aF_{n-2} + bF_{n-1}$, $n \geq 3$.

Proof. (by PMI). Since $G_3 = a + b = aF_1 + bF_2$, the statement is true when $n = 3$.

Let k be an arbitrary integer ≥ 3. Assume the given statement is true for all integers i, where $3 \leq i \leq k$: $G_i = aF_{i-2} + bF_{i-1}$. Then:

$$\begin{aligned} G_{k+1} &= G_k + G_{k-1} \\ &= (aF_{k-2} + bF_{k-1}) + (aF_{k-3} + bF_{k-2}) \\ &= a(F_{k-2} + F_{k-3}) + b(F_{k-1} + bF_{k-2}) \\ &= aF_{k-1} + bF_k \end{aligned}$$

Thus, by the principle of mathematical induction (PMI) the formula holds for every integer $n \geq 3$. ■

Notice that this theorem is in fact true for all $n \geq 1$.

GENERALIZED FIBONACCI NUMBERS AND BEES

The generalized Fibonacci numbers occur in the study of a bee colony. Suppose we start the colony with a male and b female bees. Table 7.1 shows their genealogical growth for five generations. It follows from the table that the drone has a total of $G_{n+2} = aF_n + bF_{n+1}$ descendants in generation n.

TABLE 7.1.

Generation	1	2	3	4	5
Number of female bees	b	$a+b$	$a+2b$	$2a+3b$	$3a+5b$
Number of male bees	a	b	$a+b$	$a+2b$	$2a+3b$
Total number of bees	$a+b$	$a+2b$	$2a+3b$	$3a+5b$	$5a+8b$

The Fibonacci identities of Chapter 5 can be extended to the GFS. We study a few in the following theorems.

Theorem 7.2.

$$\sum_{i=1}^{n} G_{k+i} = G_{n+k+2} - G_{k+2}$$

Proof. By Theorem 7.1,

$$\begin{aligned} \sum_{i=1}^{n} G_{k+i} &= a\sum_{i=1}^{n} F_{k+i-2} + b\sum_{i=1}^{n} F_{k+i-1} \\ &= a(F_{n+k} - F_k) + b(F_{n+k+1} - F_{k+1}) \\ &= (aF_{n+k} + bF_{n+k+1}) - (aF_k + bF_{k+1}) \\ &= G_{n+k+2} - G_{k+2} \end{aligned}$$

■

Notice that Formulas (5.1) and (5.6) follow from this theorem.

Theorem 7.1 also yields the next summation formula. Its proof is straightforward, so we leave it as an exercise (see Exercise 32).

Theorem 7.3. (Koshy, 1998)

$$\sum_{i=1}^{n} G_i G_{i+1} = a^2(F_{n-2}^2 - \nu) + b^2(F_{n-1}^2 - \nu + 1)$$
$$+ab(L_{2n-1} + 5F_{n-1}F_n + \nu + 1)/5,$$

where

$$\nu = \begin{cases} 1 & \text{if } n \text{ is odd} \\ 0 & \text{otherwise} \end{cases}$$

Theorem 7.1 can also be employed to find Binet's formula for G_n, as the next theorem shows.

Theorem 7.4. (Binet's formula). Let $c = a + (a - b)\beta$ and $d = a + (a - b)\alpha$. Then

$$G_n = \frac{c\alpha^n - d\beta^n}{\alpha - \beta}$$

Proof. By Theorem 7.1,

$$\begin{aligned}
G_n &= aF_{n-2} + bF_{n-1} \\
\sqrt{5}G_n &= a(\alpha^{n-2} - \beta^{n-2}) + b(\alpha^{n-1} - \beta^{n-1}) \\
&= \alpha^n\left(\frac{a}{\alpha^2} + \frac{b}{\alpha}\right) - \beta^n\left(\frac{a}{\beta^2} + \frac{b}{\beta}\right) \\
&= \alpha^n(a\beta^2 - b\beta) - \beta^n(a\alpha^2 - b\alpha) \\
&= \alpha^n[a + (a - b)\beta] - \beta^n[a + (a - b)\alpha] \\
\therefore \quad G_n &= \frac{c\alpha^n - d\beta^n}{\alpha - \beta}
\end{aligned}$$

as desired. ∎

Notice that

$$\begin{aligned}
cd &= [a + (a - b)\beta][a + (a - b)\alpha] \\
&= a^2 + (a - b)^2\alpha\beta + a(a - b)(\alpha + \beta) \\
&= a^2 - (a - b)^2 + a(a - b) \\
&= a^2 + ab - b^2
\end{aligned}$$

This constant occurs in many of the formulas for generalized Fibonacci numbers. It is called the *characteristic* of the GFS. We denote it by the Greek letter μ (mu):

$$\mu = a^2 + ab - b^2$$

The characteristic of the Fibonacci sequence is 1, and that of the Lucas sequence is -5.

Binet's formula for G_n opens the door for a myriad of formulas for the generalized Fibonacci numbers. The next theorem, for instance, is one such generalization of Cassini's formula. Additional formulas can be found in the exercises.

Theorem 7.5.

$$G_{n+1}G_{n-1} - G_n^2 = \mu(-1)^n$$

Proof.

$$\begin{aligned}
5(G_{n+1}G_{n-1} - G_n^2) &= (c\alpha^{n+1} - d\beta^{n+1})(c\alpha^{n-1} - d\beta^{n-1}) - (c\alpha^n - d\beta^n)^2 \\
&= -cd(\alpha^{n+1}\beta^{n-1} + \alpha^{n-1}\beta^{n+1}) + 2cd(\alpha\beta)^n \\
&= -\mu(\alpha\beta)^{n-1}(\alpha^2 + \beta^2) + 2\mu(\alpha\beta)^n \\
&= 5\mu(-1)^n
\end{aligned}$$

Therefore, $G_{n+1}G_{n-1} - G_n^2 = \mu(-1)^n$. ■

In particular, $L_{n+1}L_{n-1} - L_n^2 = 5(-1)^{n-1}$.

In 1956, H. L. Umansky of Emerson High School in Union City, New Jersey, extended Raine's result in Theorem 5.7, as the following theorem shows.

Theorem 7.6. Let ABC be a triangle with $AC = G_kG_{k+3}$, $BC = 2G_{k+1}G_{k+2}$, and $AB = G_{2k+3}$. Then ΔABC is a right triangle with hypotenuse AB. ■

EXERCISES 7

Find each generalized Fibonacci number.

1. G_8
2. G_{11}
3. G_0
4. G_{-3}
5. Let $\{A_n\}$ be a sequence such that $A_1 = 2$, $A_2 = 3$, and $A_n = A_{n-1} + A_{n-2}$, where $n \geq 3$. Find an explicit formula for A_n. (Jackson, 1969)

Let $c = a + (a - b)\beta$ and $d = a + (a - b)\alpha$. Evaluate each.

6. $c + d$

7. $c - d$

8. $\lim_{n\to\infty} \frac{G_n}{F_n}$

9. $\lim_{n\to\infty} \frac{G_n}{L_n}$

10. Solve the quadratic equation $G_{n-1}x^2 - G_nx - G_{n+1} = 0$. (Umansky, 1973)

Prove each, where $n \geq k \geq 0$.

11. $\sum_1^n G_i = G_{n+2} - b$

12. $\sum_1^n G_{2i-1} = G_{2n} + a - b$

13. $\sum_1^n G_{2i} = G_{2n+1} - a$

14. $\sum_1^n G_i^2 = G_nG_{n+1} + a(a - b)$

15. $\sum_1^{10} G_{k+i} = 11G_{k+7}$ (Hoggatt, 1963)

16. $\sum_{i=0}^n G_{k+i} = G_{n+k+2} - G_{k+1}$ (Huff)

17. $\sum_1^n iG_i = nG_{n+7} - G_{n+3} + a + b$ (Wall, 1964)

18. $\sum_1^n (n - i + 1)G_i = G_{n+4} - a - (n + 2)b$ (Wall, 1965)

19. $\sum_1^n F_iG_{3i} = F_nF_{n+1}G_{2n+1}$ (Krishna, 1972)

20. $\sum_1^n (-2)^i \binom{n}{i} G_i = 5^{(n-1)/2}[c(-1)^n - d]$ (Koshy, 1998)

21. $\sum_{i+j+k=n} \frac{(-1)^k G_{j+2k}}{i!j!k!} = 0$ (Brady, 1974)

22. $G_{-n} = a(-1)^{n+1}F_{n+2} + b(-1)^nF_{n+1}$ (Koshy, 1998)

23. $5G_{n+k}G_{n-k} = 5L_{2n} - (-1)^{n-k}\mu L_{2k}$ (Koshy, 1999)

24. $G_{n-k}G_{n+k} - G_n^2 = (-1)^{n+k-1}\mu F_k^2$ (Tagiuri, 1901)

25. $G_n^2 + G_{n-1}^2 = (3a - b)G_{2n-1} - \mu F_{2n-1}$ (Koshy, 1998)

26. $G_n = G_m F_{n-m+1} + G_{m-1} F_{n-m}$ (Ruggles, 1963)

27. $G_{m+n} = G_m F_{n+1} + G_{m-1} F_n$

28. $G_{m-n} = (-1)^n (G_m F_{n-1} - G_{m-1} F_n)$

29. $G_{m+n} + G_{m-n} = \begin{cases} G_m L_n & \text{if } n \text{ is even} \\ (G_{m+1} + G_{m-1}) F_n & \text{otherwise} \end{cases}$ (Koshy, 1998)

30. $G_{m+n} - G_{m-n} = \begin{cases} (G_{m+1} + G_{m-1}) F_n & \text{if } n \text{ is even} \\ G_m L_n & \text{otherwise} \end{cases}$ (Koshy, 1998)

31. $G_{m+k} G_{n-k} - G_m G_n = (-1)^{n-k+1} \mu F_k F_{m+k-n}$ (Tagiuri, 1901)

32. Theorem 7.3.

33. $G_n^2 = G_{n-3}^2 + 4G_{n-1} G_{n-2}$ (Umansky, 1956)

34. $G_n^4 + G_{n-1}^4 = 2G_n^2 G_{n-1}^2 + G_{n+1}^2 G_{n-2}^2$ (Umansky, 1956)

35. $G_n^2 + G_{n+3}^2 = 2(G_{n+1}^2 + G_{n+2}^2)$ (Horadam, 1971)

36. $G_{n+2}^2 - 3G_{n+1}^2 + G_n^2 = 2\mu(-1)^{n+1}$ (D. Zeitlin, 1965)

37. $(2G_m G_n)^2 + (G_m^2 - G_n^2)^2 = (G_m^2 + G_n^2)^2$

38. $(G_n^2 + G_{n+1}^2 + G_{n+2}^2)^2 = 2(G_n^4 + G_{n+1}^4 + G_{n+2}^4)$

39. $[G_m^2 + G_n^2 + (G_m + G_n)^2]^2 = 2[G_m^4 + G_n^4 + (G_m + G_n)^4]$

40. $5(G_{n+r}^2 + G_{n-r}^2) = (a^2 L_{2n-4} + 2ab L_{2n-3} + b^2 L_{2n-2}) L_{2r} - 4\mu(-1)^{n+r}$ (Koshy, 1998)

41. $5(G_{n+r} G_{n+r+1} + G_{n-r} G_{n-r+1}) = (a^2 L_{2n-3} + 2ab L_{2n-2} + b^2 L_{2n-1}) L_{2r} - 2\mu(-1)^{n+r}$ (Koshy, 1998)

42. $G_{n+r} G_{n+r+1} + G_{n-r} G_{n-r+1} = (a^2 F_{2n-3} + 2ab F_{2n-2} + b^2 F_{2n-1}) F_{2r}$ (Koshy, 1998)

43. $G_n^2 G_{n+3}^2 + 4G_{n+1}^2 G_{n+2}^2 = (G_{2n+1}^2 + G_{2n+2}^2)^2$ (Koshy, 1999)

44. $G_{n+1}^3 - G_n^3 - G_{n-1}^3 = 3G_{n+1} G_n G_{n-1}$ (Koshy, 1999)

45. $G_{n+1}^5 - G_n^5 - G_{n-1}^5 = 5G_{n+1} G_n G_{n-1} [2G_n^2 + \mu(-1)^n]$ (Koshy, 1999)

46. $G_{n+1}^7 - G_n^7 - G_{n-1}^7 = 7G_{n+1} G_n G_{n-1} [2G_n^2 + \mu(-1)^n]^2$ (Koshy, 1999)

47. $G_{n-1}^4 + G_n^4 + G_{n+1}^4 = 2[2G_n^2 + \mu(-1)^n]^2$ (Koshy, 1999)

48. $G_{n-1}^6 + G_n^6 + G_{n+1}^6 = 2[2G_n^2 + \mu(-1)^n]^3 + 3G_{n-1}^2 G_n^2 G_{n+1}^2$ (Koshy, 1999)

49. $G_{n-1}^8 + G_n^8 + G_{n+1}^8 = 2[2G_n^2 + \mu(-1)^n]^4 + G_{n-1}^4 + G_n^4 + 8G_{n-1}^2 G_n^2 \{G_{n-1}^4 + G_n^4 + 4G_{n-1}^2 G_n^2 + 3G_{n-1} G_n [(3a - b) G_{2n-1} - \mu F_{2n-1}\}$ (Koshy, 1999)

50. $G_{m-1} G_n - G_m G_{n-1} = \mu(-1)^{n-1} F_{m-n}$

Let $A_n = \left(\sum_1^n kF_k\right) \Big/ \left(\sum_1^n F_k\right)$. Verify each. (Ledin, 1966)

51. $\lim_{n\to\infty} (A_{n+1} - A_n) = 1$

52. $\lim_{n\to\infty} A_{n+1}/A_n = 1$

53. Let $\{H_n\}$ and $\{K_n\}$ be two GFSs with characteristics μ and ν respectively. Let $C_n = \sum_1^n H_i K_{n-i}$. Show that $C_{n+2} = C_{n+1} + C_n + A_n$, where $\{A_n\}$ is a GFS with characteristic $\mu\nu$. (Hoggatt, 1972)

54. Let p, q, r, and s be any four consecutive generalized Fibonacci numbers. Prove that $(pq - rs)^2 = (ps)^2 + (2qr)^2$. (Umansky and Tallman, 1968).

Deduce each from Exercise 54.

55. $(L_n L_{n+1} - L_{n+2} L_{n+3})^2 = (L_n L_{n+3})^2 + (2L_{n+1} L_{n+2})^2$. (Umansky and Tallman, 1968)

56. $(F_n F_{n+1} - F_{n+2} F_{n+3})^2 = (F_n F_{n+3})^2 + (2F_{n+1} F_{n+2})^2$

8 ADDITIONAL FIBONACCI AND LUCAS FORMULAS

Recall that Binet's formulas give explicit formulas for both F_n and L_n:

$$F_n = \frac{\alpha^n - \beta^n}{\sqrt{5}} \quad \text{and} \quad L_n = \alpha^n + \beta^n$$

where

$$\alpha = \frac{1+\sqrt{5}}{2}, \quad \beta = \frac{1-\sqrt{5}}{2}, \quad \text{and} \quad n \geq 1.$$

In this chapter, we derive additional explicit formulas for both.

To begin with, we can conjecture an explicit formula for F_n. To this end, recall that $|\beta| < 1$, so when n is large, $\beta^n \to 0$ and hence $F_n \approx \alpha^n/\sqrt{5}$. So we compute the value of $\alpha^n/\sqrt{5}$ for the first ten values of n and look for a pattern:

$$\frac{\alpha}{\sqrt{5}} \approx 0.72 \qquad \frac{\alpha^2}{\sqrt{5}} \approx 1.17 \qquad \frac{\alpha^3}{\sqrt{5}} \approx 1.89 \qquad \frac{\alpha^4}{\sqrt{5}} \approx 3.07$$

$$\frac{\alpha^5}{\sqrt{5}} \approx 4.96 \qquad \frac{\alpha^6}{\sqrt{5}} \approx 8.02 \qquad \frac{\alpha^7}{\sqrt{5}} \approx 12.98 \qquad \frac{\alpha^8}{\sqrt{5}} \approx 21.00$$

$$\frac{\alpha^9}{\sqrt{5}} \approx 33.99 \qquad \frac{\alpha^{10}}{\sqrt{5}} \approx 55.00$$

Since the pattern is not yet quite obvious, we go one step further. Add $\frac{1}{2}$ to each and see if a pattern emerges:

$$\frac{\alpha}{\sqrt{5}}+\frac{1}{2}\approx 1.22 \quad \frac{\alpha^2}{\sqrt{5}}+\frac{1}{2}\approx 1.67 \quad \frac{\alpha^3}{\sqrt{5}}+\frac{1}{2}\approx 2.39 \quad \frac{\alpha^4}{\sqrt{5}}+\frac{1}{2}\approx 3.57$$

$$\frac{\alpha^5}{\sqrt{5}}+\frac{1}{2}\approx 5.46 \quad \frac{\alpha^6}{\sqrt{5}}+\frac{1}{2}\approx 8.52 \quad \frac{\alpha^7}{\sqrt{5}}+\frac{1}{2}\approx 13.48 \quad \frac{\alpha^8}{\sqrt{5}}+\frac{1}{2}\approx 21.51$$

$$\frac{\alpha^9}{\sqrt{5}}+\frac{1}{2}\approx 34.49 \quad \frac{\alpha^{10}}{\sqrt{5}}+\frac{1}{2}\approx 55.50$$

A pattern, surprisingly enough, does emerge:

$$\left\lfloor\frac{\alpha}{\sqrt{5}}+\frac{1}{2}\right\rfloor=1 \quad \left\lfloor\frac{\alpha^2}{\sqrt{5}}+\frac{1}{2}\right\rfloor=1 \quad \left\lfloor\frac{\alpha^3}{\sqrt{5}}+\frac{1}{2}\right\rfloor=2 \quad \left\lfloor\frac{\alpha^4}{\sqrt{5}}+\frac{1}{2}\right\rfloor=3$$

$$\left\lfloor\frac{\alpha^5}{\sqrt{5}}+\frac{1}{2}\right\rfloor=5 \quad \left\lfloor\frac{\alpha^6}{\sqrt{5}}+\frac{1}{2}\right\rfloor=8 \quad \left\lfloor\frac{\alpha^7}{\sqrt{5}}+\frac{1}{2}\right\rfloor=13 \quad \left\lfloor\frac{\alpha^8}{\sqrt{5}}+\frac{1}{2}\right\rfloor=21$$

$$\left\lfloor\frac{\alpha^9}{\sqrt{5}}+\frac{1}{2}\right\rfloor=34 \quad \left\lfloor\frac{\alpha^{10}}{\sqrt{5}}+\frac{1}{2}\right\rfloor=55$$

Thus we conjecture that

$$\left\lfloor\frac{\alpha^n}{\sqrt{5}}+\frac{1}{2}\right\rfloor=F_n.$$

Fortunately, the next theorem confirms this result. To establish it, we need the following lemma.

Lemma 8.1.

$$0<\frac{\beta^n}{\sqrt{5}}+\frac{1}{2}<1$$

Proof. Since $\beta<0$, $|\beta|=-\beta$. Also, since $0<|\beta|<1$, $0<|\beta|^n<1$. So

$$0<|\beta|^n<\frac{\sqrt{5}}{2}$$

that is,

$$0<\frac{|\beta|^n}{\sqrt{5}}<\frac{1}{2}.$$

Case 1. Let n be even. Then $|\beta|^n=\beta^n$, so $0<(\beta^n/\sqrt{5})<\frac{1}{2}$, and hence $\frac{1}{2}<(\beta^n/\sqrt{5})<1$.

Case 2. Let n be odd. Then $|\beta|^n = -\beta^n$, so $0 < -(\beta^n/\sqrt{5}) < \frac{1}{2}$, and hence $-\frac{1}{2} < (\beta^n/\sqrt{5}) < 0$.

Therefore,

$$0 < \frac{\beta^n}{\sqrt{5}} + \frac{1}{2} < \frac{1}{2}$$

Thus, in both cases, $0 < (\beta^n/\sqrt{5}) + \frac{1}{2} < 1$. This establishes the lemma. ■

We are now ready to state and prove the conjecture.

Theorem 8.1.

$$F_n = \left\lfloor \frac{\alpha^n}{\sqrt{5}} + \frac{1}{2} \right\rfloor$$

Proof. By Binet's formula,

$$\begin{aligned} F_n &= \frac{\alpha^n - \beta^n}{\sqrt{5}} \\ &= \left(\frac{\alpha^n}{\sqrt{5}} + \frac{1}{2}\right) - \left(\frac{\beta^n}{\sqrt{5}} + \frac{1}{2}\right) \qquad (8.1) \\ \therefore \quad \frac{\alpha^n}{\sqrt{5}} + \frac{1}{2} &= F_n + \left(\frac{\beta^n}{\sqrt{5}} + \frac{1}{2}\right) \\ &< F_n + 1, \qquad \text{by Lemma 8.1} \end{aligned}$$

Since $(\beta^n/\sqrt{5}) + \frac{1}{2} > 0$, it follows from Eq. (8.1) that $F_n < (\alpha^n/\sqrt{5}) + \frac{1}{2}$.
Thus $F_n < (\alpha^n/\sqrt{5}) + \frac{1}{2} < F_{n+1}$. Consequently,

$$F_n = \left\lfloor \frac{\alpha^n}{\sqrt{5}} + \frac{1}{2} \right\rfloor.$$

■

For example, $(\alpha^{20}/\sqrt{5}) + \frac{1}{2} \approx 6765.5$,
so

$$\left\lfloor \frac{\alpha^{20}}{\sqrt{5}} + \frac{1}{2} \right\rfloor = 6765 = F_{20}$$

as expected.

Since $\lfloor x \rfloor = \lceil x \rceil - 1$ for nonintegral real numbers x, it follows that

$$F_n = \left\lceil \frac{\alpha^n}{\sqrt{5}} + \frac{1}{2} \right\rceil - 1.$$

But $\lceil x + n \rceil = \lceil x \rceil + n$ for integer n:

$$\therefore F_n = \left\lceil \frac{\alpha^n}{\sqrt{5}} + \frac{1}{2} - 1 \right\rceil = \left\lceil \frac{\alpha^n}{\sqrt{5}} - \frac{1}{2} \right\rceil$$

Accordingly, we have the following result.

Corollary 8.1.

$$F_n = \left\lceil \frac{\alpha^n}{\sqrt{5}} - \frac{1}{2} \right\rceil \qquad \blacksquare$$

For example,

$$\frac{\alpha^{15}}{\sqrt{5}} - \frac{1}{2} \approx 609.4997$$

$$\therefore \quad \left\lceil \frac{\alpha^{15}}{\sqrt{5}} - \frac{1}{2} \right\rceil = 610 = F_{15}$$

Likewise,

$$\left\lceil \frac{\alpha^{20}}{\sqrt{5}} - \frac{1}{2} \right\rceil = 6765 = F_{20}$$

Here is yet another interesting observation:

$$\left\lceil \frac{\alpha}{\sqrt{5}} \right\rceil = F_1 \quad \left\lceil \frac{\alpha^3}{\sqrt{5}} \right\rceil = F_3 \quad \left\lceil \frac{\alpha^5}{\sqrt{5}} \right\rceil = F_5 \quad \left\lceil \frac{\alpha^7}{\sqrt{5}} \right\rceil = F_7 \quad \left\lceil \frac{\alpha^9}{\sqrt{5}} \right\rceil = F_9$$

$$\left\lfloor \frac{\alpha^2}{\sqrt{5}} \right\rfloor = F_2 \quad \left\lfloor \frac{\alpha^4}{\sqrt{5}} \right\rfloor = F_4 \quad \left\lfloor \frac{\alpha^6}{\sqrt{5}} \right\rfloor = F_6 \quad \left\lfloor \frac{\alpha^8}{\sqrt{5}} \right\rfloor = F_8 \quad \left\lfloor \frac{\alpha^{10}}{\sqrt{5}} \right\rfloor = F_{10}$$

Thus, we make another conjecture:

$$\left\lfloor \frac{\alpha^{2n}}{\sqrt{5}} \right\rfloor = F_{2n} \qquad \text{and} \qquad \left\lceil \frac{\alpha^{2n+1}}{\sqrt{5}} \right\rceil = F_{2n+1}$$

The following corollary confirms these two observations.

Corollary 8.2.

$$\left\lfloor \frac{\alpha^{2n}}{\sqrt{5}} \right\rfloor = F_{2n} \qquad \text{and} \qquad \left\lceil \frac{\alpha^{2n+1}}{\sqrt{5}} \right\rceil = F_{2n+1}$$

Proof. Let n be even. Then, from the proof of Lemma 8.1, we have

$$\frac{1}{2} < \frac{\beta^n}{\sqrt{5}} < 1, \qquad \text{so} \qquad -\frac{1}{2} > -\frac{\beta^n}{\sqrt{5}} > -1$$

Then

$$\frac{\alpha^n}{\sqrt{5}} - \frac{1}{2} > F_n > \frac{\alpha^n}{\sqrt{5}} - 1$$

That is,

$$\frac{\alpha^n}{\sqrt{5}} - 1 < F_n < \frac{\alpha^n}{\sqrt{5}} - \frac{1}{2}$$

But $\lfloor x \rfloor \le x$ and $\lfloor x + n \rfloor = \lfloor x \rfloor + n$:

$$\therefore \quad \left\lfloor \frac{\alpha^n}{\sqrt{5}} \right\rfloor - 1 < F_n < \frac{\alpha^n}{\sqrt{5}} - \frac{1}{2}$$

That is,

$$\left\lfloor \frac{\alpha^n}{\sqrt{5}} \right\rfloor - 1 < F_n < \frac{\alpha^n}{\sqrt{5}}$$

Thus

$$F_n = \left\lfloor \frac{\alpha^n}{\sqrt{5}} \right\rfloor$$

We can establish the case when n is odd in a similar fashion. ■

Theorem 8.1 has an analogous result for Lucas numbers also. We leave its proof as an exercise.

Theorem 8.2. (Hoggatt)

$$L_n = \left\lfloor \alpha^n + \frac{1}{2} \right\rfloor$$ ■

For example,

$$\alpha^{13} + \frac{1}{2} \approx 521.5019$$

$$\therefore \quad \left\lfloor \alpha^{13} + \frac{1}{2} \right\rfloor = 521 = L_{13}$$

Corollaries 8.1 and 8.2 also have their counterparts in Lucas numbers, as the next corollary reveals.

Corollary 8.3.

$$(1) \qquad L_n = \left\lceil \alpha^n - \frac{1}{2} \right\rceil$$

$$(2) \qquad L_{2n} = \lceil \alpha^{2n} \rceil \qquad \text{and} \qquad L_{2n+1} = \lfloor \alpha^{2n+1} \rfloor$$ ■

For example,

$$\alpha^7 - \frac{1}{2} \approx 28.5344$$

$$\therefore \quad \left\lceil \alpha^7 - \frac{1}{2} \right\rceil = 29 = L_7$$

$$\lceil \alpha^8 \rceil = \lceil 46.9787\cdots \rceil = 47 = L_8$$

$$\lfloor \alpha^{11} \rfloor = \lfloor 199.0050\cdots \rfloor = 199 = L_{11}$$

In every explicit formula we have developed thus far, we needed to know the value of n in order to compute F_n. Surprisingly enough that is no longer the case: knowing a Fibonacci number, we can easily compute its successor. The next theorem provides such a formula, but first we need to lay some groundwork in the form of a lemma, similar to Lemma 8.1.

Lemma 8.2. If $n \geq 2$, then $0 < \frac{1}{2} - \beta^n < 1$.

Proof. We have $|\beta| < 0.62$, $|\beta|^2 < 1/2$, so $|\beta|^n < 1/2$, when $n \geq 2$. Since $|\beta|^n = |\beta^n|$, this yields $-\frac{1}{2} < \beta^n < 1/2$. Then $-1 < \beta^n - 1/2 < 0$; that is, $0 < 1/2 - \beta^n < 1$. ■

We are now ready to state and prove the recursive formula.

Theorem 8.3.

$$F_{n+1} = \lfloor \alpha F_n + 1/2 \rfloor \qquad n \geq 2$$

Proof. By Binet's formula, we have

$$F_n = \frac{\alpha^n - \beta^n}{\sqrt{5}}$$

$$\alpha F_n = \frac{\alpha^{n+1} - \alpha\beta^n}{\sqrt{5}} = \frac{\alpha^{n+1} - \alpha\beta(\beta^{n-1}) + \beta^{n+1} - \beta^{n+1}}{\sqrt{5}}$$

$$= \frac{(\alpha^{n+1} - \beta^{n+1}) + \beta^{n-1} + \beta^{n+1}}{\sqrt{5}}$$

$$= F_{n+1} + \frac{\beta^{n-1}(\beta^2+1)}{\sqrt{5}}$$

$$= F_{n+1} + \frac{\beta^{n-1}(-\sqrt{5}\beta)}{\sqrt{5}} = F_{n+1} - \beta^n$$

$$\therefore \quad \alpha F_n + 1/2 = F_{n+1} + (1/2 - \beta^n) \tag{8.2}$$

Since $1/2-\beta^n > 0$, this implies $F_{n+1} < \alpha F_n + 1/2$. Besides, since $1/2-\beta^n < 1$, Eq. (8.2) yields $\alpha F_n + 1/2 < F_{n+1} + 1$. Thus $F_{n+1} < \alpha F_n + 1/2 < F_{n+1} + 1$, so $F_{n+1} = \lfloor \alpha F_n + \frac{1}{2} \rfloor$, as desired. ■

For instance, let $F_n = 4181$. Its successor is given by $\lfloor 4181\alpha + 1/2 \rfloor = \lfloor 6765.500 \cdots \rfloor = 6765$, as expected.

Substituting for α in the formula for F_n yields the following result due to Hoggatt.

Corollary 8.4.

$$F_{n+1} = \left\lfloor \frac{F_n + \sqrt{5}F_n + 1}{2} \right\rfloor \qquad n \geq 2$$ ■

We can use the recursive formula in Theorem 8.3 (or Corollary 8.4) to compute the ratio F_{n+1}/F_n as $n \to \infty$, as the following corollary demonstrates. Its proof employs the following fact: If $\lfloor x \rfloor = k$, then $x = k + \theta$, where $0 \leq \theta < 1$.

Corollary 8.5.

$$\lim_{n\to\infty} \frac{F_{n+1}}{F_n} = \alpha$$

Proof. By Theorem 8.3,

$$F_{n+1} = \alpha F_n + \frac{1}{2} + \theta \qquad \text{where} \quad 0 \leq \theta < 1$$

$$\frac{F_{n+1}}{F_n} = \alpha + \frac{1}{2F_n} + \frac{\theta}{F_n}$$

$$\lim_{n\to\infty} \frac{F_{n+1}}{F_n} = \alpha + 0 + 0 = \alpha$$ ■

Since $\lfloor x \rfloor = \lceil x \rceil - 1$, for any nonintegral real number x, we can express these two formulas in terms of the ceiling function, as the next corollary shows.

Corollary 8.6.

$$(1) \qquad F_{n+1} = \lceil \alpha F_n - 1/2 \rceil \qquad n \geq 2$$

$$(2) \qquad F_{n+1} = \left\lceil \frac{F_n + \sqrt{5}F_n - 1}{2} \right\rceil \qquad n \geq 2 \qquad ■$$

For instance, the successor of the Fibonacci number 1597 is given by $\lceil 1597\alpha - 1/2 \rceil = \lceil 2583.5002 \cdots \rceil = 2584$.

Using Corollary 8.5, we can evaluate the limit of $\tan\theta_n/\tan\theta_{n+1}$ as $n \to \infty$, where θ_n denotes the acute angle between the adjacent sides of the parallelogram in Figure 6.11. To this end, let $s_n = F_{2n-3} + 2F_{n-3}F_{n-2} = F_{n-1}F_n + F_{n-3}F_{n-2}$. Then

$$\begin{aligned}
\frac{s_{n+1}}{s_n} &= \frac{F_nF_{n+1} + F_{n-2}F_{n-1}}{F_{n-1}F_n + F_{n-3}F_{n-2}} \\
&= \frac{(F_{n+1}/F_{n-1}) + (F_{n-2}/F_n)}{1 + (F_{n-3}F_{n-2})/(F_{n-1}F_n)} \\
&= \frac{(F_{n+1}/F_n)\cdot(F_n/F_{n-1}) + (F_{n-2}/F_{n-1})\cdot(F_{n-1}/F_n)}{1 + (F_{n-3}/F_{n-2})\cdot(F_{n-2}/F_{n-1})\cdot(F_{n-2}/F_{n-1})\cdot(F_{n-1}/F_n)} \\
\therefore \quad \lim_{n\to\infty}\frac{s_{n+1}}{s_n} &= \frac{\alpha\cdot\alpha + (1/\alpha)\cdot(1/\alpha)}{1 + (1/\alpha)\cdot(1/\alpha)\cdot(1/\alpha)\cdot(1/\alpha)} = \frac{\alpha^2 + (1/\alpha^2)}{1 + (1/\alpha^4)} = \alpha^2
\end{aligned}$$

That is,

$$\lim_{n\to\infty}\frac{\tan\theta_n}{\tan\theta_{n+1}} = \alpha^2$$

Let u_n and v_n denote the lengths of the sides of the parallelogram in Figure 6.11, or 6.12, where $u_n > v_n$. Then $u_n = \sqrt{F_n^2 + F_{n-2}^2}$ and $v_n = \sqrt{F_{n-1}^2 + F_{n-3}^2}$, so

$$\lim_{n\to\infty}\frac{u_n}{v_n} = \alpha$$

Returning to Theorem 8.3 and its corollaries, we find that they have analogous recursive results for Lucas numbers as well. We leave their proofs as routine exercises.

Theorem 8.4. (Hoggatt)

$$L_{n+1} = \lfloor \alpha L_n + 1/2 \rfloor \qquad n \geq 2 \qquad ■$$

For example, the successor of the Lucas number 1364 is given by $\lfloor 1364\alpha + 1/2 \rfloor = \lfloor 2207.4983\cdots \rfloor = 2207$. Notice that $1364 = L_{15}$ and $2207 = L_{16}$.

We can use this theorem to compute the value of L_{n+1}/L_n as $n \to \infty$, as the next corollary shows. Again we leave its proof as a routine exercise, (see Exercise 33).

Corollary 8.7.

$$\lim_{n\to\infty} \frac{L_{n+1}}{L_n} = \alpha \qquad \blacksquare$$

Corollary 8.6 also has corresponding results to Lucas numbers.

Corollary 8.8.

$$(1) \qquad L_{n+1} = \left\lfloor \frac{L_n + \sqrt{5}L_n + 1}{2} \right\rfloor \qquad n \geq 2 \qquad \text{(Hoggatt)}$$

$$(2) \qquad L_{n+1} = \lceil \alpha L_n - 1/2 \rceil \qquad n \geq 2$$

$$(3) \qquad L_{n+1} = \left\lceil \frac{L_n + \sqrt{5}L_n - 1}{2} \right\rceil \qquad n \geq 2 \qquad \blacksquare$$

For example, the successor of the Lucas number 521 is given by $\lceil 521\alpha - \frac{1}{2} \rceil = \lceil 842.4957\cdots \rceil = 843$. For the curious-minded, $521 = L_{13}$ and $843 = L_{14}$.

There is yet another recursive formula that expresses each Fibonacci number in terms of its predecessor and one that expresses each Lucas number in terms of its predecessor. We find both in the following theorem.

Theorem 8.5.

$$(1) \qquad F_{n+1} = \frac{F_n + \sqrt{5F_n^2 + 4(-1)^n}}{2}$$

$$(2) \qquad L_{n+1} = \frac{L_n + \sqrt{5[L_n^2 + 4(-1)^n]}}{2} \qquad \blacksquare$$

These formulas, discovered by Basin of Sylvania Electronics Systems, Mountain View, California, can be derived using the following identities:

$$2F_{n+1} = F_n + L_n \tag{8.3}$$

$$2L_{n+1} = 5F_n + L_n \tag{8.4}$$

$$L_n^2 - 5F_n^2 = 4(-1)^n \tag{8.5}$$

Formulas 8.3 and 8.4 are consequences of Identities 5.14 and 5.16, and Formula (8.5) follows by Exercise 39 in Chapter 5.

There is still another formula that expresses a Fibonacci number in terms of its predecessor, discovered by Hoggatt and Lind in 1967.

Theorem 8.6.

$$F_{n+1} = \left\lfloor \frac{F_n + 1 + \sqrt{5F_n^2 - 2F_n + 1}}{2} \right\rfloor \qquad n \ge 2$$

Proof. Notice that $L_n - F_n = (F_{n-1} + F_{n+1}) - F_n = 2F_{n-1}$. By Exercise 39 in Chapter 5, $L_n^2 - 5F_n^2 = 4(-1)^n$, where $n \ge 1$. When $n \ge 2$, $4(-1)^n \le 4F_{n-1}$. Therefore, when $n \ge 2$, we have

$$L_n^2 - 5F_n^2 \le 4F_{n-1}$$

That is,

$$\begin{aligned} L_n^2 - 5F_n^2 &\le 2(L_n - F_n) \\ (L_n - 1)^2 &\le 5F_n^2 - 2F_n + 1 \end{aligned}$$

But

$$\begin{aligned} L_n = F_{n-1} + F_{n+1} = (F_{n+1} - F_n) + F_{n+1} = 2F_{n+1} - F_n \\ \therefore \quad (2F_{n+1} - F_n - 1)^2 \le 5F_n^2 - 2F_n + 1 \end{aligned}$$

Thus

$$\begin{aligned} 2F_{n+1} - F_n - 1 &\le \sqrt{5F_n^2 - 2F_n + 1} \\ F_{n+1} &\le \frac{F_n + 1 + \sqrt{5F_n^2 - 2F_n + 1}}{2} \end{aligned} \tag{8.6}$$

Notice also that $L_n + F_n = (F_{n-1} + F_{n+1}) + F_n = 2F_{n+1}$. So, when $n \ge 2$,

$$\begin{aligned} 4(-1)^n &> -4F_{n+1} \\ L_n^2 - 5F_n^2 &> -2(L_n + F_n) \end{aligned}$$

That is,

$$\begin{aligned} L_n^2 + 2L_n &> 5F_n^2 - 2F_n \\ (L_n + 1)^2 &> 5F_n^2 - 2F_n + 1 \\ \therefore \quad (2F_{n+1} - F_n + 1)^2 &> 5F_n^2 - 2F_n + 1 \end{aligned}$$

Thus

$$2F_{n+1} - F_n + 1 > \sqrt{5F_n^2 - 2F_n + 1}$$

$$F_{n+1} > \frac{F_n - 1 + \sqrt{5F_n^2 - 2F_n + 1}}{2} \tag{8.7}$$

$$F_{n+1} > \left\lfloor \frac{F_n - 1 + \sqrt{5F_n^2 - 2F_n + 1}}{2} \right\rfloor \tag{8.8}$$

From Eqs. (8.6) and (8.7), we have

$$\left\lfloor \frac{F_n - 1 + \sqrt{5F_n^2 - 2F_n + 1}}{2} \right\rfloor < F_{n+1} \leq \frac{F_n + 1 + \sqrt{5F_n^2 - 2F_n + 1}}{2}$$

Since F_{n+1} is an integer, it follows that

$$F_{n+1} = \left\lfloor \frac{F_n + 1 + \sqrt{5F_n^2 - 2F_n + 1}}{2} \right\rfloor, \qquad n \geq 2$$

■

For example, the successor of the Fibonacci number 987 is given by

$$\left\lfloor \frac{987 + 1 + \sqrt{5 \cdot 987^2 - 2 \cdot 987 + 1}}{2} \right\rfloor = \lfloor 1597.2760 \cdots \rfloor = 1597$$

Analogously, we have the following result for Lucas numbers. It was also developed by Hoggatt and Lind in 1967. Its proof is quite similar, so we leave it as an exercise.

Theorem 8.7.

$$L_{n+1} = \left\lfloor \frac{L_n + 1 + \sqrt{5L_n^2 - 2L_n + 1}}{2} \right\rfloor, \qquad n \geq 4$$

■

For instance, the successor of the Lucas number 1364 is given by

$$\left\lfloor \frac{1364 + 1 + \sqrt{5 \cdot 1364^2 - 2 \cdot 1364 + 1}}{2} \right\rfloor = \lfloor 2207.2748 \cdots \rfloor = 2207$$

Interestingly enough, we can use Theorem 8.3 in the reverse direction also. It can be employed to compute the predecessor of a given Fibonacci number, as the next theorem shows.

Theorem 8.8.

$$F_n = \left\lfloor \frac{1}{\alpha}\left(F_{n+1} + \frac{1}{2}\right) \right\rfloor \qquad n \ge 2$$

Proof. Since $x - 1 < \lfloor x \rfloor \le x$, Theorem 8.3 yields the double inequality

$$\alpha F_n - \frac{1}{2} < F_{n+1} \le \alpha F_n + \frac{1}{2}$$

$$F_n - \frac{1}{2\alpha} < \frac{F_{n+1}}{\alpha} < F_n + \frac{1}{2\alpha}$$

Then

$$F_n < \frac{1}{\alpha}\left(F_{n+1} + \frac{1}{2}\right) \qquad \text{and} \qquad F_n \ge \frac{1}{\alpha}\left(F_{n+1} - \frac{1}{2}\right)$$

$$\frac{1}{\alpha}\left(F_{n+1} - \frac{1}{2}\right) < F_n \le \frac{1}{\alpha}\left(F_{n+1} + \frac{1}{2}\right)$$

Since $(1/\alpha)(F_{n+1} + \frac{1}{2}) - (1/\alpha)(F_{n+1} - \frac{1}{2}) = \frac{1}{\alpha} \approx 0.618$ and F_n is an integer, it follows that

$$F_n = \left\lfloor \frac{1}{\alpha}\left(F_{n+1} + \frac{1}{2}\right) \right\rfloor, \qquad n \ge 2 \qquad \blacksquare$$

For example, the predecessor of the Fibonacci number 4181 is given by $\lfloor 4181.5/\alpha \rfloor = \lfloor 2584.3091 \cdots \rfloor = 2584$. For the curious-minded, $4181 = F_{19}$ and $2584 = F_{18}$.

Analogously, we have the following result for Lucas numbers. We shall leave its proof as an exercise.

Theorem 8.9.

$$L_n = \left\lfloor \frac{1}{\alpha}\left(L_{n+1} + \frac{1}{2}\right) \right\rfloor, \qquad n \ge 2 \qquad \blacksquare$$

For example, the predecessor of the Lucas number $L_{20} = 15{,}127$ is given by $\lfloor 15{,}127.5/\alpha \rfloor = \lfloor 9349.3091 \cdots \rfloor = 9349 = L_{19}$.

In 1972, Anaya and Crump of then San Jose State College, California, established the following generalization of Theorem 8.3.

Theorem 8.10.

$$\lfloor \alpha^k F_n + 1/2 \rfloor = F_{n+k} \qquad n \ge k \ge 1$$

Proof. Since the theorem is true for $k = 1$, assume that $n \geq k \geq 2$. By Binet's formula,

$$\begin{aligned}
\alpha^k F_n &= \frac{\alpha^{n+k} - \alpha^k \beta^n}{\sqrt{5}} = \frac{\alpha^{n+k} - \beta^{n+k}}{\sqrt{5}} + \frac{\beta^{n+k} - \alpha^k \beta^n}{\sqrt{5}} \\
&= F_{n+k} - \beta^n F_k \\
\alpha^k F_n + \frac{1}{2} &= F_{n+k} + \left(\frac{1}{2} - \beta^n\right) \qquad (8.9)
\end{aligned}$$

Next we shall prove that $0 < 1/2 - \beta^n F_k < 1$. When $n = k$, $|\beta^n F_k|$ has its largest value. Notice that $|\beta^n| \to 0$ as $n \to \infty$. Also,

$$|\beta^k F_k| = \left|\frac{\beta^k(\alpha^k - \beta^k)}{\sqrt{5}}\right| = \left|\frac{(-1)^k - \beta^{2k}}{\sqrt{5}}\right|$$

Case 1. Let k be even. Then

$$\begin{aligned}
|\beta^k F_k| &= \left|\frac{1 - \beta^{2k}}{\sqrt{5}}\right| \\
\lim_{k\to\infty} |\beta^k F_k| &= \left|\frac{1 - 0}{\sqrt{5}}\right| = \frac{1}{\sqrt{5}} < \frac{1}{2}
\end{aligned}$$

Since $|\beta^n| < |\beta^k|$, it follows that $0 < |\beta^n F_k| < \frac{1}{2}$.

Case 2. Let k be odd. Then

$$|\beta^k F_k| = \left|\frac{-1 - \beta^{2k}}{\sqrt{5}}\right| = \left|\frac{1 + \beta^{2k}}{\sqrt{5}}\right|$$

When $k = 3$, $\beta^{2k} \approx 0.055726$, so

$$|\beta^k F_k| \approx \frac{1.055726}{\sqrt{5}} \approx 0.472135 < \frac{1}{2}$$

As k increases, β^{2k} gets smaller and smaller. So $|\beta^k F_k| < \frac{1}{2}$ for $k > 3$ also. Thus $0 < |\beta^n F_k| < \frac{1}{2}$, since $|\beta^n| < |\beta^k|$.

Consequently, $0 < |\beta^n F_k| < \frac{1}{2}$ for all $n \geq k \geq 2$; that is,

$$\begin{aligned}
-\frac{1}{2} < \beta^n F_k < \frac{1}{2} \\
0 < \frac{1}{2} - \beta^n F_k < 1
\end{aligned}$$

Therefore, by Eq. (8.9), $F_{n+k} < \alpha^k F_n + \frac{1}{2} < F_{n+k} + 1$. Thus

$$\left\lfloor \alpha^k F_n + \frac{1}{2} \right\rfloor = F_{n+k}$$

■

For example, $\lfloor \alpha^7 F_8 + 1/2 \rfloor = \lfloor 21\alpha^7 + 1/2 \rfloor = \lfloor 610.223\cdots \rfloor = 610 = F_{15} = F_{8+7}$. Notice that $\lfloor \alpha^8 F_7 + 1/2 \rfloor = 611 \neq F_{15}$.

Corollary 8.9. $\lceil \alpha^k F_n - 1/2 \rceil = F_{n+k}$, where $n \geq k \geq 1$. ■

For example, $\lceil \alpha^9 F_{11} - 1/2 \rceil = \lceil 89\alpha^9 - 1/2 \rceil = \lceil 6764.6708\cdots \rceil = 6765 = F_{20} = F_{11+9}$.

In 1972, Anaya and Crump conjectured a similar formula for L_{n+k}. It was proved in the same year by Carlitz of Duke University.

Theorem 8.11. $\lfloor \alpha^k L_n + \frac{1}{2} \rfloor = L_{n+k}$, where $n \geq 4$ and $k \geq 1$.

Proof.

$$\begin{aligned} \alpha L_n - L_{n+1} &= \alpha(\alpha^n + \beta^n) - (\alpha^{n+1} + \beta^{n+1}) \\ &= \beta^n(\alpha - \beta) = \sqrt{5}\beta^n \end{aligned}$$

When $n \geq 4$,

$$\begin{aligned} |\sqrt{5}\beta^n| &\leq \sqrt{5}\beta^4 \\ &= \sqrt{5}(7 - 3\sqrt{5})/2 \\ &< 1/2 \\ \therefore \quad |\alpha L_n - L_{n+1}| &< 1/2 \end{aligned}$$

that is, $0 < \alpha L_n - L_{n+1} + 1/2 < 1$, so $\lfloor \alpha L_n + 1/2 \rfloor = L_{n+1}$. Thus, the theorem is true for $k = 1$.

Now, assume $n \geq k + 2$, where $k \geq 2$. Notice that

$$\begin{aligned} \alpha^{-2} + \alpha^{-6} &= \beta^2 + \beta^6 = \frac{3 - \sqrt{5}}{2} + 9 - 4\sqrt{5} \\ &= \frac{(21 - 9\sqrt{5})}{2} \end{aligned}$$

Since $k \geq 2$, this implies $\alpha^{-2} + \alpha^{-2k-2} < 1/2$; that is, $\alpha^{-k-2}(\alpha^k + \alpha^{-k}) < 1/2$. Since $n \geq k + 2$, this means $\alpha^{-n}(\alpha^k + \alpha^{-k}) < 1/2$:

$$\therefore \quad |\beta^n(\alpha^k - \beta^k)| < 1/2$$

That is,

$$\begin{aligned} |\alpha^k(\alpha^n + \beta^n) - (\alpha^{n+k} + \beta^{n+k})| &< 1/2 \\ |\alpha^k L_n - L_{n+k}| &< 1/2 \end{aligned}$$

As before, this implies that $\lfloor \alpha^k L_n + 1/2 \rfloor = L_{n+k}$. ■

For example, let $n = 11$ and $k = 3$. Then $\lfloor \alpha^3 L_{11} + 1/2 \rfloor = \lfloor \alpha^3 L_{11} + 1/2 \rfloor = \lfloor 199\alpha^3 + 1/2 \rfloor = \lfloor 843.4775 \cdots \rfloor = 843 = L_{14} = L_{11+3}$.

Corollary 8.10. $\lceil \alpha^k L_n - 1/2 \rfloor = L_{n+k}$, where $n \geq 4$ and $k \geq 1$. ■

For example, $\lceil \alpha^4 L_{10} - 1/2 \rceil = \lceil 123\alpha^4 - 1/2 \rceil = \lceil 842.5545 \cdots \rceil = 843 = L_{14} = L_{10+4}$.

EXERCISES 8

Using Theorem 8.1, compute F_n for the given value of n.

1. 15
2. 19
3. 23
4. 25

5–8. Compute F_n for each value in Exercises 1–4 using Corollary 8.1.

Verify that $\lfloor \alpha^n / \sqrt{5} \rfloor = F_n$ for each value of n.

9. 12
10. 20

Verify that $\lceil \alpha^n / \sqrt{5} \rceil = F_n$ for each value of n.

11. 15
12. 23

Using Theorem 8.2, compute L_n for each given value of n.

13. 8
14. 10
15. 15
16. 20

17–20. Compute L_n for each value in Exercises 13–16 using the formula $L_n = \lceil \alpha^n - \frac{1}{2} \rceil$.

Verify that $L_n = \lceil \alpha^n \rceil$ for each value of n.

21. 10
22. 16

Verify that $L_n = \lfloor \alpha^n \rfloor$ for each value of n.

23. 13
24. 19

Compute the successor of each Fibonacci number using Theorem 8.3.

25. 2584
26. 6765

27–28. Redo Exercises 25 and 26 using Corollary 8.5.

Compute the successor of each Lucas number using Theorem 8.4.

29. 843

30. 9349

31–32. Redo Exercises 29 and 30 using Corollary 8.6.

33. Using Theorem 8.4, evaluate $\lim_{n\to\infty} L_{n+1}/L_n$.

34. Using Theorem 8.5, evaluate $\lim_{n\to\infty} F_{n+1}/F_n$.

35. Using Theorem 8.7, evaluate $\lim_{n\to\infty} L_{n+1}/L_n$.

Compute the predecessor of each Fibonacci number.

36. 610

37. 17,711

Compute the predecessor of each Lucas number.

38. 1364

39. 39,603

Let u_n and v_n denote the lengths of the sides of the parallelogram in Figure 6.11, where $u_n > v_n$. Verify each (Horadam, 1962).

40. $\lim_{n\to\infty} u_n/v_n = \alpha$

41. $\lim_{n\to\infty} u_n/F_{n+1} = -\sqrt{3}\beta$

42. $\lim_{n\to\infty} v_n/F_n = -\sqrt{3}\beta$

Suppose every Fibonacci number F_n in Figure 6.11 is replaced by the corresponding generalized Fibonacci number G_n. Let θ_n denote the acute angle between the adjacent sides of the parallelogram and let $t_n = (3a-b)G_{2n-1} - \mu F_{2n-1} + 2G_{n-2}G_{n-1}$. Prove each (Horadam, 1962).

43. The lengths of the sides of the parallelogram are $x_n = \sqrt{G_n^2 + G_{n-2}^2}$ and $y_n = \sqrt{G_{n-1}^2 + G_{n-3}^2}$, where $x_n > y_n$.

44. $\lim_{n\to\infty} t_{n+1}/t_n = \alpha^2$

45. $\lim_{n\to\infty} x_n/y_n = \alpha$

46. $\tan\theta_n = \mu/t_n$

THE EUCLIDEAN ALGORITHM

This chapter continues our investigation of the properties of Fibonacci numbers. We reconfirm, using the Euclidean algorithm, that any two consecutive Fibonacci numbers are relatively prime. To this end, we first lay the necessary foundation for justifying the algorithm.

Among the several procedures for finding the greatest common divisor (gcd) of two positive integers, one efficient algorithm is the *Euclidean algorithm*. Although it seems to have been known before him, it is named after the great Greek mathematician Euclid*, who published it in Book VII of his extraordinary work, *The Elements*.

This next theorem paves the way for the Euclidean algorithm.

Theorem 9.1. Let a and b be any positive integers, and r the remainder, when a is divided by b. Then $(a, b) = (b, r)$.

Proof. Let $d = (a, b)$ and $d' = (b, r)$. To prove that $d = d'$, it suffices to show that $d|d'$ and $d'|d$. By the division algorithm, a unique quotient q exists such that

$$a = bq + r \tag{9.1}$$

To show that d/d':
Since $d = (a, b)$, $d|a$ and $d|b$, so $d|bq$, by Theorem A.10. Then $d|(a - bq)$, again by Theorem A.10. In other words, $d|r$, by Eq. 9.1. Thus, $d|b$ and $d|r$, so $d|(b, r)$; that is, $d|d'$.

*Little is known about Euclid's life. He was on the faculty at the University of Alexandria and founded the Alexandrian School of Mathematics. When the Egyptian ruler, King Ptolemy I, asked Euclid if there were an easier way to learn geometry than by studying *The Elements*, Euclid replied, "There is no royal road to geometry." Euclid is called the father of geometry.

Similarly, it can be shown that $d'|d$ (see Exercise 17). Thus, by Theorem A.9, $d = d'$, that is, $(a, b) = (b, r)$. ■

The following example elucidates this theorem.

Example 9.1. llustrate Theorem 9.1 with $a = 120$ and $b = 28$.

Solution. First, you can verify that $(120, 28) = 4$. Now, by the division algorithm, $120 = 4 \cdot 28 + 8$, so, by Theorem 9.1, $(120, 28) = (28, 8)$. But $(28, 8) = 4$; $\therefore (120, 28) = 4$. ■

Before formally presenting the Euclidean algorithm, we illustrate it in the next example.

Example 9.2. Illustrate the Euclidean algorithm by evaluating (2076, 1776).

Solution. Apply the division algorithm with 2076 (the larger of the two numbers) as the dividend and 1776 as the divisor:

$$2076 = 1 \cdot 1776 + 300$$

Apply the division algorithm with 1776 as the dividend and 300 as the divisor:

$$1776 = 5 \cdot 300 + 276$$

Continue this procedure until a zero remainder is reached:

$$\begin{aligned} 2076 &= 1 \cdot 1776 + 300 \\ 1776 &= 5 \cdot 300 + 276 \\ 300 &= 1 \cdot 276 + 24 \\ 276 &= 11 \cdot 24 + \boxed{12} \quad \leftarrow \text{last nonzero remainder} \\ 24 &= 2 \cdot 12 + 0 \end{aligned}$$

The last nonzero remainder in this procedure is the gcd, so $(2076, 1776) = 12$. ■

Take a close look at the preceding steps to see why the gcd is 12. By the repeated application of Theorem 9.1, we have

$$\begin{aligned} (2076, 1776) &= (1776, 300) = (300, 276) \\ &= (276, 24) = (24, 12) \\ &= 12 \end{aligned}$$

We now turn our attention to a justification of this algorithm, although it is somewhat obvious.

THE EUCLIDEAN ALGORITHM

Let a and b be any two positive integers, with $a \geq b$. If $a = b$, then $(a, b) = a$, so assume $a > b$. (If this is not true, simply switch them.) Let $r_0 = b$. Then by successive application of the division algorithm, we get a sequence of equations:

$$\begin{aligned} a &= q_0 r_0 + r_1 \quad 0 \leq r_1 < r_0 \\ r_0 &= q_1 r_1 + r_2 \quad 0 \leq r_2 < r_1 \\ r_1 &= q_2 r_2 + r_3 \quad 0 \leq r_3 < r_2 \\ &\vdots \end{aligned}$$

Continuing like this, we get the following sequence of remainders:

$$b = r_0 > r_1 > r_2 > r_3 > \cdots \geq 0$$

Since the remainders are nonnegative, and getting smaller and smaller, this sequence should eventually terminate with remainder $r_n = 0$. Thus, the last two equations in the preceding procedure are

$$r_{n-2} = q_{n-1} r_{n-1} + r_n \quad 0 \leq r_n < r_{n-1}$$

and

$$r_{n-1} = q_n r_n$$

It follows by the principle of induction (PMI) that $(a, b) = (a, r_0) = (r_0, r_1) = (r_1, r_2) = \cdots = (r_{n-1}, r_n) = r_n$, the last nonzero remainder (see Exercise 18).

The following example also demonstrates the Euclidean algorithm.

Example 9.3. Apply the Euclidean algorithm to find (4076,1024).

Solution. By the successive application of the division algorithm, we get:

$$\begin{aligned} 4076 &= 3 \cdot 1024 + 1004 \\ 1024 &= 1 \cdot 1004 + 20 \\ 1004 &= 50 \cdot 20 + \boxed{4} \quad \leftarrow \text{last nonzero remainder} \\ 20 &= 5 \cdot 4 + 0 \end{aligned}$$

Since the last nonzero remainder is 4, (4076, 1024) = 4. ■

The Euclidean algorithm is purely mechanical. All we need is to make our divisor the new dividend and the remainder the new divisor. That is, we just follow the southwest arrows in the solution.

The Euclidean algorithm is formally presented in Algorithm 9.1.

```
Algorithm Euclid (x,y, divisor)
(*  This algorithm returns the gcd (x,y) in divisor,
    where x ≥ y > 0.   *)
    Begin (* algorithm *)
       dividend ← x
       divisor ← y
       remainder ← dividend mod divisor
       while remainder > 0 do (* update dividend,
                                 divisor, and remainder *)
       begin (* while *)
         dividend ← divisor
         divisor ← remainder
         remainder ← dividend mod divisor
       endwhile
     End (* algorithm *)
```

Algorithm 9.1.

The Euclidean algorithm provides a procedure for expressing the gcd (a, b) as a linear combination of a and b, as the next example shows.

Example 9.4. Use the Euclidean algorithm to express (4076, 1024) as a linear combination of 4076 and 1024. ■

Solution. All we need is to use the equations in Example 9.3 in the reverse order, each time substituting for the remainder from the previous equation:

$$
\begin{aligned}
(4076, 1024) &= 4 = \text{last nonzero remainder}\\
&= 1004 - 50 \cdot 20\\
&= 1004 - 50(1024 - 1 \cdot 1004) \qquad \text{(substitute for 20)}\\
&= 51 \cdot 1004 - 50 \cdot 1024\\
&= 51(3076 - 3 \cdot 1024) - 50 \cdot 1024 \qquad \text{(substitute for 1004)}\\
&= 51 \cdot 4076 + (-203) \cdot 1024
\end{aligned}
$$

(We can confirm this by direct computation.)

In Chapter 5, we proved that any two consecutive Fibonacci numbers are relatively prime. We now establish it using the Euclidean algorithm.

Example 9.5. Prove that any two consecutive Fibonacci numbers are relatively prime.

Proof. Using the Euclidean algorithm with F_n as the original dividend and F_{n-1} as the original divisor. This yields the following system of linear equations:

$$\begin{aligned} F_n &= 1 \cdot F_{n-1} + F_{n-2} \\ F_{n-1} &= 1 \cdot F_{n-2} + F_{n-3} \\ F_{n-2} &= 1 \cdot F_{n-3} + F_{n-4} \\ &\vdots \\ F_4 &= 1 \cdot F_3 + F_2 \\ F_3 &= 2 \cdot F_2 + 0 \end{aligned}$$

Thus, it follows by the Euclidean algorithm that $(F_n, F_{n-1}) = F_2 = 1$. ■

THE EUCLIDEAN ALGORITHM AND THE LUCAS FORMULA (5.5)

In 1990, Ian Cook of the University of Essex, United Kingdom, developed Identity (5.5) as an application of the Euclidean algorithm.

By the Euclidean algorithm, we have:

$$\begin{aligned} 1976 &= 1 \cdot 1776 + 200 \\ 1776 &= 8 \cdot 200 + 176 \\ 200 &= 1 \cdot 176 + 24 \\ 176 &= 7 \cdot 24 + \boxed{8} \quad \leftarrow \text{gcd} \\ 24 &= 3 \cdot 8 + 0 \\ &\quad\uparrow \\ &\quad\text{quotients} \end{aligned}$$

It follows from these equations by successive substitutions that

$$1976 \cdot 1776 = 1 \cdot 1776^2 + 8 \cdot 200^2 + 1.176^2 + 7 \cdot 24^2 + 3 \cdot 8^2$$

Notice that the coefficients on the right-hand side (RHS) are the various quotients in the algorithm and the corresponding factors are the squares of the corresponding divisors.

More generally, the equations

$$
\begin{aligned}
a &= q_0 r_0 + r_1 \\
r_0 &= q_1 r_1 + r_2 \\
&\vdots \\
r_{i-1} &= q_i r_i + r_{i+1} \\
&\vdots \\
r_{n-2} &= q_{n-1} r_{n-1} + \boxed{r_n} \quad \leftarrow \text{gcd of } a \text{ and } b \\
r_{n-1} &= q_n r_n + 0
\end{aligned}
$$

imply that

$$ab = \sum_{i=0}^{n} q_i r_i^2 \tag{9.2}$$

We can confirm this using PMI.

In particular, let $a = F_{n+1}$ and $b = F_n$. By the Euclidean algorithm, we have

$$
\begin{aligned}
F_{n+1} &= 1 \cdot F_n + F_{n-1} \\
F_n &= 1 \cdot F_{n-1} + F_{n-2} \\
&\vdots \\
3 &= 1 \cdot 2 + 1 + \boxed{1} \quad \leftarrow \text{gcd } (F_{n+1}, F_n) \\
2 &= 2 \cdot 1 + 0
\end{aligned}
$$

With $q_n = 2$ and $q_i = 1$ for $0 \leq i < n$, Formula (9.1) yields

$$F_{n+1} F_n = \sum_{3}^{n} 1 \cdot F_i^2 + 2 \cdot 1^2$$

Since $2 \cdot 1^2 = 1^2 + 1^2$, this sum can be rewritten as

$$\sum_{1}^{n} F_i^2 = F_n F_{n+1} \tag{5.5}$$

which is Identity (5.5).

Next we estimate the number of divisions in the Euclidean algorithm, for which we need the following result.

Lemma 9.1. Let F_n denote the nth Fibonacci number and $\alpha = (1 + \sqrt{5})/2$. Prove that $\alpha^{n-2} < F_n < \alpha^{n-1}, n \geq 3$.

Proof by Strong Induction. (We shall prove that $\alpha^{n-2} < F_n$ and leave the other half as an exercise.) Let $P(n)$: $\alpha^{n-2} < F_n$, where $n \geq 3$.

Basis Step. Since the induction step below uses the recurrence relation $F_{k+1} = F_k + F_{k-1}$, the basis step involves verifying that both $P(3)$ and $P(4)$ are true.

1. To show that $P(3)$ is true: When $n = 3$,

$$\alpha^{n-2} = \alpha = \frac{1 + \sqrt{5}}{2} < \frac{1 + 3}{2} = 2 = F_3$$

$\therefore$ $P(3)$ is true.

2. To show that $P(4)$ is true:

$$\alpha^2 = \left(\frac{1 + \sqrt{5}}{2}\right)^2 = \frac{3 + \sqrt{5}}{2}$$

$$< \frac{3 + 3}{2} = 3 = F_4$$

$\therefore$ $P(4)$ is also true.

Induction Step. Assume $P(3), P(4), \ldots, P(k)$ are true; that is, assume $\alpha^{i-2} < F_i$ for $5 \leq i \leq k$. We must show that $P(k+1)$ is true; that is, $\alpha^{k-1} < F_{k+1}$. We have

$$\alpha^2 = \alpha + 1$$

Multiplying both sides by α^{k-3},

$$\begin{aligned} \alpha^{k-1} &= \alpha^{k-2} + \alpha^{k-3} && (Note : k - 3 \geq 2) \\ &< F_k + F_{k-1} && \text{by the IH} \\ &= F_{k+1} && \text{by the Fibonacci recurrence relation} \end{aligned}$$

Thus $P(k+1)$ is true.

Therefore, by the strong version of induction, $P(n)$ is true for every $n \geq 3$; that is, $\alpha^{n-2} < F_n$ for every $n \geq 3$. ■

The following theorem estimates the number of divisions required by the Euclidean algorithm to compute the gcd (a, b). It was established in 1844 by the French mathematician Gabriel Lamé (1795–1870).

Theorem 9.2. (Lamé's Theorem) The number of divisions needed to compute (a, b) by the Euclidean algorithm is no more than five times the number of decimal digits in b, where $a \geq b \geq 2$.

Proof. Let F_n denote the nth Fibonacci number, $a = r_0$ and $b = r_1$. By the repeated application of the division algorithm, we have:

$$\begin{aligned} r_0 &= r_1 q_1 + r_2 && 0 \le r_2 < r_1 \\ r_1 &= r_2 q_2 + r_3 && 0 \le r_3 < r_2 \\ &\vdots \\ r_{n-2} &= r_{n-1} q_{n-1} + r_n && 0 \le r_n < r_{n-1} \\ r_{n-1} &= r_n q_n \end{aligned}$$

Clearly, it takes n divisions to evaluate $(a, b) = r_n$. Since $r_i < r_{i-1}$, $q_i \ge 1$ for $1 \le i \le n$. In particular, since $r_n < r_{n-1}$, $q_n \ge 2$, so $r_n \ge 1$ and $r_{n-1} \ge 2 = F_3$. Consequently, we have

$$\begin{aligned} r_{n-2} &= r_{n-1} q_{n-1} + r_n \\ &\ge r_{n-1} + r_n \\ &\ge F_3 + 1 \\ &= F_3 + F_2 = F_4 \\ r_{n-3} &= r_{n-2} q_{n-2} + r_{n-1} \\ &\ge r_{n-2} + r_{n-1} \\ &\ge F_4 + F_3 = F_5 \end{aligned}$$

Continuing like this,

$$\begin{aligned} r_1 &= r_2 q_2 + r_3 \\ &\ge r_2 + r_3 \\ &\ge F_n + F_{n-1} = F_{n+1} \end{aligned}$$

That is,

$$b \ge F_{n+1}$$

By Lemma 9.1, $F_{n+1} > \alpha^{n-1}$, where $\alpha = (1 + \sqrt{5})/2$ and $n \ge 3$.

$$\begin{aligned} \therefore \quad b &> \alpha^{n-1} \\ \log b &> (n-1) \log \alpha \end{aligned}$$

Since $\alpha = (1+\sqrt{5})/2 \approx 1.618033989$, $\log\alpha \approx 0.2089876403 > 1/5$.

$$\therefore \quad \log b > \frac{n-1}{5}$$

Suppose b contains k decimal digits. Then $b < 10^k$. Therefore, $\log\ b < k$, and hence $k > (n-1)/5$. Thus $n < 5k+1$ or $n \le 5k$. That is, the number of divisions needed by the algorithm is no more than five times the number of decimal digits in n. ∎

Let us pursue this example a bit further. Since $\log b > (n-1)/5, n < 1+5\ \log\ b$. Also, since $b \ge 2$,

$$\begin{aligned} 5\ \log\ b &\ge 5\ \log 2 \\ &> 1 \\ \therefore \quad n &< 1+5\ \log b \\ &< 5\ \log b + 5\ \log b \\ &= 10\ \log b \\ &= O(\log b) \end{aligned}$$

Thus it takes $O(\log b)$ divisions to compute (a, b) by the Euclidean algorithm.*

EXERCISES 9

Use the Euclidean algorithm to find the gcd of the given integers.

1. 1024, 1000
2. 2024, 1024
3. 2076, 1076
4. 2076, 1776
5. 1976, 1776
6. 3076, 1776
7. 3076, 1976
8. 4076, 2076

9–16. Use the Euclidean algorithm to express the gcd of each pair in Exercises 1–8 as a linear combination of the given numbers.

17. Let a and b any two positive integers and r the remainder when a is divided by b. Let $d = (a, b)$ and $d' = (b, r)$. Prove that $d'|d$.

*Let $f, g\colon \mathbb{N} \to \mathbb{R}$. Then $f(n)$ is said to be of *order at most* $g(n)$ if there exist a positive constant C and a positive integer n such that $|f(n)| \le C|g(n)|$ for every $n \ge n$. In symbols, we write $f(n) = O(g(n))$. (Read this as *$f(n)$ is big-oh of $g(n)$*.)

18. Let a and b be any two positive integers with $a \geq b$. Using the sequence of equations in the Euclidean algorithm, prove that $(a, b) = (r_{n-1}, r_n)$, where $n \geq 1$.
19. Prove Identity (5.10) using the Euclidean algorithm.

SOLVING RECURRENCE RELATIONS

We shall now develop a method for solving a large and important class of recurrence relations, which are defined below. We will use this method to confirm Binet's explicit formula for F_n and L_n.

LINEAR HOMOGENEOUS RECURRENCE RELATIONS WITH CONSTANT COEFFICIENTS

A kth-*order linear homogeneous recurrence relation with constant coefficients* (LHRRWCCs) is a recurrence relation of the form

$$a_n = c_1 a_{n-1} + c_2 a_{n-2} + \cdots + c_k a_{n-k} \tag{10.1}$$

(where $c_1, c_2, \ldots, c_k \in \mathbb{R}$ and $c_k \neq 0$.)

We need a few words of explanation about the definitional terms. The term *linear* means every term on the right-hand side (RHS) of Eq. (10.1) contains at most the first power of any predecessor a_i. A recurrence relation is *homogeneous* if every term on the RHS is a multiple of some a_i; in other words, the relation is satisfied by the sequence $\{0\}$, that is, $a_n = 0$ for every n. All coefficients c_i are constants. Since a_n depends on its k immediate predecessors, the *order* of the recurrence relation is k. Accordingly, to solve a kth order LHRRWCC, we will need k initial conditions, say, $a_0 = C_0, a_1 = C_1, \ldots, a_{k-1} = C_k$.

The following example illustrates in detail the various terms in this definition.

Example 10.1.

1. The recurrence relation $s_n = 2s_{n-1}$ is a LHRRWCC. Its order is one.
2. The recurrence relation $a_n = na_{n-1}$ is linear and homogeneous. But the coefficient on the RHS is not a constant. Therefore, it is not a LHRRWCC.
3. $h_n = h_{n-1} + (n-1)$ is a linear recurrence relation. But, because of the term $n-1$, it is not homogeneous.
4. The recurrence relation $a_n = a_{n-1}^2 + 3a_{n-2}$ is homogeneous. But it is not linear, since the power of a_{n-1} is 2.
5. $a_n = a_{n-1} + 2a_{n-2} + 3a_{n-6}$ is a LHRRWCC of order six. ■

Before we discuss solving second-order LHRRWCCs, notice that the solution of the recurrence relation $s_n = 2s_{n-1}$, where $s_0 = 1$, is $s_n = 2^n$, $n \geq 0$. More generally, the solution of the recurrence relation $a_n = \gamma a_{n-1}$ where $a_0 = c$, is $a_n = c\gamma^n$, $n \geq 0$.

Let us now turn our attention to the second-order LHRRWCC

$$a_n = aa_{n-1} + ba_{n-2} \tag{10.2}$$

where a and b are nonzero constants. If it has a nonzero solution of the form $c\gamma^n$, then $c\gamma^n = ac\gamma^{n-1} + bc\gamma^{n-2}$. Since $c\gamma \neq 0$, this yields $\gamma^2 = a\gamma + b$, that is, $\gamma^2 - a\gamma - b = 0$, so γ must be a solution of the *characteristic equation*

$$x^2 - ax - b = 0 \tag{10.3}$$

of the recurrence relation Eq. (10.2). The roots of Eq. (10.3) are called the *characteristic roots* of Recurrence Relation (10.2).

Theorem 10.1 shows how characteristic roots help solve LHRRWCCs.

Theorem 10.1. Let γ and δ be the distinct solutions of the equation $x^2 - ax - b = 0$, where $a, b \in \mathbb{R}$ and $b \neq 0$. Then every solution of the LHRRWCC $a_n = aa_{n-1} + ba_{n-2}$, where $a_0 = C_0$ and $a_1 = C_1$, is of the form $a_n = A\gamma^n + B\delta^n$ for some constants A and B.

Proof. The proof consists of two parts: (1) We will show that $a_n = A\gamma^n + B\delta^n$ is a solution of the recurrence relation for any constants A and B, (2) we will find the values of A and B satisfying the given initial conditions.

First, notice that since γ and δ are solutions of Eq. (10.3), $\gamma^2 = a\gamma + b$ and $\delta^2 = a\delta + b$.

1. *To show that* $\mathrm{a_n} = \mathrm{A\gamma^n} + \mathrm{B\delta^n}$ *is a solution of the recurrence relation*:

$$\begin{aligned} aa_{n-1} + ba_{n-2} &= a(A\gamma^{n-1} + B\delta^{n-1}) + b(A\gamma^{n-2} + B\delta^{n-2}) \\ &= A\gamma^{n-2}(a\gamma + b) + B\delta^{n-2}(a\delta + b) \end{aligned}$$

$$= A\gamma^{n-2} \cdot \gamma^2 + B\delta^{n-2} \cdot \delta^2$$
$$= A\gamma^n + B\delta^n$$
$$= a_n$$

Thus $a_n = A\gamma^n + B\delta^n$ is a solution of Recurrence Relation (10.2).

2. Secondly, let $a_n = A\gamma^n + B\delta^n$ be a solution of Eq. (10.2). To find the values of A and B, notice that the conditions $a_0 = C_0$ and $a_1 = C_1$ yield the following linear system:

$$C_0 = A + B$$
$$C_1 = A\gamma + B\delta$$

Solving this system we get

$$A = \frac{C_1 - C_0\delta}{\gamma - \delta} \quad \text{and} \quad B = \frac{C_0\gamma - C_1}{\gamma - \delta} \quad (\text{remember, } \gamma \neq \delta)$$

With these values for A and B, a_n satisfies the initial conditions and the recurrence relation. Since the recurrence relation and the initial conditions determine a unique sequence $\{a_n\}$, $a_n = A\gamma^n + B\delta^n$ is indeed the unique solution of the recurrence relation. ■

Note

(1) The solutions γ and δ are nonzero, since $\gamma = 0$, for instance, would imply that $b = 0$.

(2) Theorem 10.1 *cannot* be applied if $\gamma = \delta$. However, it works even if γ and δ are complex numbers.

(3) The solutions γ^n and δ^n are the *basic solutions* of the recurrence relation. In general, the number of basic solutions equals the order of the recurrence relation. The *general solution* $a_n = A\gamma^n + B\delta^n$ is a *linear combination* of the basic solutions. The particular solution is obtained by selecting A and B in such a way that the initial conditions are satisfied, as in Theorem 10.1.

The next two examples illustrate how to solve second-order LHRRWCCs using their characteristic equations.

Example 10.2. Solve the recurrence relation $a_n = 5a_{n-1} - 6a_{n-2}$, where $a_0 = 4$ and $a_1 = 7$.

Solution.

1. *To find the general solution of the recurrence relation*:

   The characteristic equation of the recurrence relation is $x^2 - 5x + 6 = 0$; the characteristic roots are 2 and 3. Therefore, by Theorem 10.1, the general solution of the recurrence relation is $a_n = A2^n + B3^n$.

2. *To find the values of* A *and* B:

   Using the initial conditions we find:

$$a_0 = A + B = 4$$
$$a_1 = 2A + 3B = 7$$

Solving this linear system yields $A = 5$ and $B = -1$.

Thus the solution of the recurrence relation satisfying the given conditions is $a_n = 5 \cdot 2^n - 3^n, n \geq 0$. ■

The next example finds an explicit formula for the nth Fibonacci number, which we have been waiting for.

Example 10.3. Solve the Fibonacci recurrence relation $F_n = F_{n-1} + F_{n-2}$, where $F_1 = 1 = F_2$.

Solution. The characteristic equation of the recurrence relation is $x^2 - x - 1 = 0$ and its solutions are $\alpha = (1 + \sqrt{5})/2$ and $\beta = (1 - \sqrt{5})/2$. Recall that $\alpha + \beta = 1$ and $\alpha\beta = -1$.

The general solution is $F_n = A\alpha^n + B\beta^n$. To find A and B, we have

$$F_1 = A\alpha + B\beta = 1$$
$$F_2 = A\alpha^2 + B\beta^2 = 1$$

Solving these two equations, we get

$$A = \frac{\alpha}{1 + \alpha^2} = \frac{(1 + \sqrt{5})/2}{(5 + \sqrt{5})/2} = \frac{1 + \sqrt{5}}{5 + \sqrt{5}}$$
$$= \frac{1 + \sqrt{5}}{\sqrt{5}(1 + \sqrt{5})} = \frac{1}{\sqrt{5}}$$

and similarly $B = \beta/(1 + \beta^2) = -1/\sqrt{5}$.

Thus the solution of the recurrence relation satisfying the given conditions is

$$F_n = \frac{\alpha^n - \beta^n}{\sqrt{5}} = \frac{\alpha^n - \beta^n}{\alpha - \beta}$$

which is the Binet form for the Fibonacci number F_n. ■

The same method can be employed to derive Binet's formula for L_n (see Exercise 15).

EXERCISES 10

Determine if each recurrence relation is an LHRRWCC.

1. $L_n = L_{n-1} + L_{n-2}$
2. $D_n = nD_{n-1} + (-1)^n$
3. $a_n = 1.08a_{n-1}$
4. $b_n = 2b_{n-1} + 1$
5. $a_n = a_{n-1} + n$
6. $a_n = 2a_{n-1} + (2^n - 1)$
7. $a_n = a_{n-1} + 2a_{n-2} + 3a_{n-5}$
8. $a_n = a_{n-1} + 2a_{n-3} + n^2$

Solve the following LHRRWCCs.

9. $a_n = a_{n-1} + 2a_{n-2}, a_0 = 3, a_1 = 0$
10. $a_n = 5a_{n-1} - 6a_{n-2}, a_0 = 4, a_1 = 7$
11. $a_n = a_{n-1} + 6a_{n-2}, a_0 = 5, a_1 = 0$
12. $a_n = 4a_{n-2}, a_0 = 2, a_1 = -8$
13. $a_n = a_{n-1} + a_{n-2}, a_0 = 1, a_1 = 2$
14. $a_n = a_{n-1} + a_{n-2}, a_0 = 2, a_1 = 3$
15. $L_n = L_{n-1} + L_{n-2}, L_1 = 1, L_2 = 3$
16. $L_n = L_{n-1} + L_{n-2}, L_1 = 2, L_2 = 3$

11 COMPLETENESS THEOREMS

This chapter, like the preceding ones, provides numerous opportunities for studying patterns and making conjectures.

We begin with yet another interesting pattern:

$$\begin{array}{ll} 1 = 1 & 2 = 2 \\ 3 = 2 + 1 & 4 = 3 + 1 \\ 5 = 3 + 2 & 6 = 5 + 1 \\ 7 = 5 + 2 & 8 = 5 + 3 \\ 9 = 8 + 1 & 10 = 8 + 2 \end{array}$$

Every integer on the left-hand side (LHS) of each equation is a positive integer; and each number on the right-hand side (RHS) is a Fibonacci number, and each occurs exactly once.

More generally, we have the following result.

Theorem 11.1. (Completeness Theorem) Every positive integer n can be expressed as a finite sum of distinct Fibonacci numbers.

Proof. Let F_m be the largest Fibonacci number $\leq n$. Then $n = F_m + n_1$, where $n_1 \leq F_m$. Let F_{m_1} be the largest Fibonacci number $\leq n_1$. Then $n = F_m + F_{m_1} + n_2$, $n \geq F_m > F_{m_1}$. Continuing like this, we get $n = F_m + F_{m_1} + F_{m_2} + \cdots$, where $n \geq F_m > F_{m_1} > F_{m_2} \cdots$. Since this sequence of decreasing positive integers must terminate, the result follows. ■

We must emphasize that the representation of an integer n in terms of Fibonacci numbers is not unique. For example, $25 = 21 + 3 + 1 = 13 + 8 + 3 + 1$.

THE EGYPTIAN ALGORITHM FOR MULTIPLICATION

Every positive integer can be expressed as a sum of distinct powers of 2. This fact is the basis of the well-known *Egyptian algorithm for multiplication*. For example, let $b = \sum_{i=0}^{n} b_i 2^i$, where $b_i = 0$ or 1. Then $ab = \sum_{i=0}^{n} (ab_i)2^i$. Thus, to compute ab, we need only keep doubling a until the product gets larger than 2^n and then add the products corresponding to the ones in the binary representation of b. This algorithm is illustrated in the following example.

Example 11.1. Use the Egyptian method of multiplication to compute $47 \cdot 73$.

Solution. First, express 47 as a sum of powers of 2:

$$47 = 1 + 2 + 4 + 8 + 32$$

$$\therefore \quad 47 \cdot 73 = 1 \cdot 73 + 2 \cdot 73 + 4 \cdot 73 + 8 \cdot 73 + 32 \cdot 73$$

Next, construct a table (see Table 11.1) consisting of two rows, one headed by 1 and the other by 73; each successive column is obtained by doubling the preceding column. Identify the numbers in the second row that correspond to the powers of 2 used in the representation of 47 by asterisks; they correspond to the terms in the binary expansion of 47.

TABLE 11.1.

1	2	4	8	16	32
73*	146*	292*	584*	1168	2336*

To find the desired product, we add the starred numbers:

$$\begin{aligned} 47 \cdot 73 &= 73 + 146 + 292 + 584 + 2336 \\ &= 3431 \end{aligned}$$

■

By virtue of Theorem 11.1, we can also use Fibonacci numbers to effect multiplication of positive integers, as the next example demonstrates.

Example 11.2. Use Fibonacci numbers to compute $47 \cdot 73$.

Solution. First, we express 47 as a sum of distinct Fibonacci numbers:

$$47 = 2 + 3 + 8 + 13 + 21$$

$$\therefore \quad 47 \cdot 73 = 2 \cdot 73 + 3 \cdot 73 + 8 \cdot 73 + 13 \cdot 73 + 21 \cdot 73$$

Now construct a table as before (see Table 11.2). It follows from the table that $47 \cdot 73 = 146 + 219 + 584 + 949 + 1533 = 3,431$, as expected.

TABLE 11.2.

1	2	3	5	8	13	21
73	146*	219*	365	584*	949*	1533*

■

Although the Fibonacci numbers F_n, where $n \geq 1$, are complete, Lucas numbers L_n are not. For example, with the numbers $1, 3, 4, \ldots,$ we cannot represent 2. But all is not lost. If we add 2 to the list, the resulting set is complete.

But, as before, the representation need not be unique. For instance, $43 = 29 + 11 + 3 = 29 + 7 + 4 + 3 = 29 + 11 + 2 + 1$.

Accordingly, we have the following result.

Theorem 11.2. (Completeness Theorem) The set of Lucas numbers L_n is complete, where $n \geq 0$. ■

Consequently, we can also employ Lucas numbers to perform integer multiplication, as the next example illustrates.

Example 11.3. Use Theorem 11.2 to compute $47 \cdot 73$.

Solution. We have $47 = 3 + 4 + 11 + 29$. So it follows from Table 11.3 that $47 \cdot 73 = 219 + 292 + 803 + 2117 = 3431$.

TABLE 11.3.

2	1	3	4	7	11	18	29
146	73	219*	292*	511	803*	1314	2117*

■

EXERCISES 11

Express each number as a sum of distinct Fibonacci numbers.

1. 43
2. 99
3. 137
4. 343

5–8. Express each number in Exercises 1–4 as a sum of distinct Lucas numbers.

Use Fibonacci numbers to compute each.

9. $43 \cdot 49$
10. $99 \cdot 101$

11. $111 \cdot 121$
12. $243 \cdot 342$

13–16. Use Lucas numbers to compute each product in Exercises 9–12.

17. Prove Theorem 11.2.

PASCAL'S TRIANGLE

We shall see how Fibonacci numbers can be computed in a systematic way from the well-known Pascal's triangle. In addition, we will be able to derive a host of new Fibonacci and Lucas identities.

We begin with a discussion of binomial coefficients, which are coefficients occurring in the binomial expansion of an expression of the form $(x + y)^n$.

BINOMIAL COEFFICIENTS

Let n and k be nonnegative integers. The *binomial coefficient* $\binom{n}{k}$ is defined by

$$\binom{n}{k} = \frac{n!}{k!(n-k)!}$$

if $k \leq n$, and is 0 otherwise. It is also denoted by $C(n, r)$ and nCr.

For example,

$$\binom{5}{3} = \frac{5!}{3!(5-3)!} = \frac{5 \cdot 4 \cdot 3 \cdot 2 \cdot 1}{3 \cdot 2 \cdot 1 \cdot 2 \cdot 1} = 10$$

Using a TI-86, however, a large number such as $\binom{45}{22}$ can be found in seconds. Press the keys 2nd, MATH, and PROB; enter 45; press nCr; enter 22; then press the ENTER key. The answer is 4,116,715,363,800.

Suppose we let $k = 0$ in the definition. Then

$$\binom{n}{0} = \frac{n!}{0!(n-0)!} = \frac{n!}{1 \cdot n!} = 1$$

Besides, if $k = n$, then

$$\binom{n}{n} = \frac{n!}{n!(n-n)!} = \frac{n!}{n!0!} = 1$$

Thus we have two useful results:

$$\binom{n}{0} = 1 = \binom{n}{n}$$

There are many instances when we need to compute the binomial coefficients $\binom{n}{k}$ and $\binom{n}{n-k}$. The next theorem shows there is no need to evaluate both, since they are equal; this will certainly reduce our workload.

Theorem 12.1. Let n and k be nonnegative integers such that $k \le n$. Then $\binom{n}{k} = \binom{n}{n-k}$.

Proof.

$$\binom{n}{n-k} = \frac{n!}{(n-k)![n-(n-k)]!} = \frac{n!}{(n-k)!k!} = \frac{n!}{k!(n-k)!}$$
$$= \binom{n}{k} \qquad \blacksquare$$

For example, $\binom{25}{20} = \binom{25}{25-20} = \binom{25}{5} = 53{,}130$ by our earlier discussion. (See how useful the theorem is.)

The following theorem shows an important recurrence relation satisfied by binomial coefficients. It is called *Pascal's identity*, after the outstanding French mathematician and physicist, Blaise Pascal (1623–1662).

Theorem 12.2. (Pascal's identity) Let n and k be positive integers, where $k \le n$. Then $\binom{n}{k} = \binom{n-1}{k-1} + \binom{n-1}{k}$.

Proof. We shall simplify the right-hand side (RHS) and show that it is equal to the left-hand side (LHS).

$$\binom{n-1}{k-1} + \binom{n-1}{k} = \frac{(n-1)!}{(k-1)!(n-k)!} + \frac{(n-1)!}{k!(n-k-1)!}$$
$$= \frac{k(n-1)!}{k(k-1)!(n-k)!} + \frac{(n-k)(n-1)!}{k!(n-k)(n-k-1)!}$$
$$= \frac{k(n-1)!}{k!(n-k)!} + \frac{(n-k)(n-1)!}{k!(n-k)!} = \frac{(n-1)![k+(n-k)]}{k!(n-k)!}$$
$$= \frac{(n-1)!n}{k!(n-k)!} = \frac{n!}{k!(n-k)!}$$
$$= \binom{n}{k} \qquad \blacksquare$$

PASCAL'S TRIANGLE

The various binomial coefficients $\binom{n}{k}$, where $0 \leq k \leq n$, can be arranged in the form of a triangle, called *Pascal's triangle*,* as shown in Figures 12.1 and 12.2.

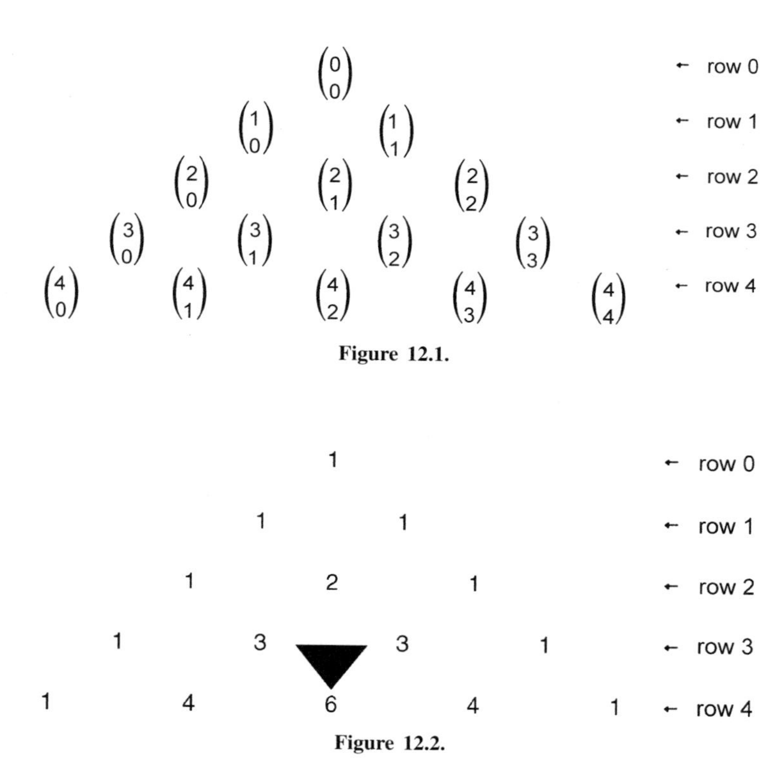

Figure 12.1.

Figure 12.2.

Pascal's triangle has many intriguing properties:

- Every row begins with and ends in 1.
- Pascal's triangle is symmetric about a vertical line through the middle. This is so by Theorem 12.1.
- Any interior number in each row is the sum of the numbers immediately to its left and to its right in the preceding row. This is so by virtue of Pascal's identity.
- The sum of the numbers in any row is a power of 2.

*Although Pascal's triangle is named after Pascal in the West, the array appeared in a 1303 work by the Chinese mathematician, Chu Shi-Kie.

The next theorem shows how the binomial coefficients can be used to find the *binomial expansion* of $(x + y)^n$.

Theorem 12.3. (The Binomial Theorem)* Let x and y be any real numbers, and n any nonnegative integer. Then $(x + y)^n = \sum_{r=0}^{n} \binom{n}{r} x^{n-r} y^r$.

Proof. (by Weak Induction) When $n = 0$, LHS $= (x + y)^0 = 1$ and RHS $= \sum_{r=0}^{0} \binom{0}{r} x^{0-r} y^r = x^0 y^0 = 1$, so LHS = RHS.

Assume $P(k)$ is true for some $k \geq 0$:

$$(x + y)^k = \sum_{r=0}^{k} \binom{k}{r} x^{k-r} y^r$$

Then

$$\begin{aligned}
(x + y)^{k+1} &= (x + y)^k (x + y) \\
&= \left[\sum_{0}^{k} \binom{k}{r} x^{k-r} y^r \right] (x + y) \quad \text{by the IH} \\
&= \sum_{0}^{k} \binom{k}{r} x^{k+1-r} y^r + \sum_{0}^{k} \binom{k}{r} x^{k-r} y^{r+1} \\
&= \left[\binom{k}{0} x^{k+1} + \sum_{1}^{k} \binom{k}{r} x^{k+1-r} y^r \right] \\
&\quad + \left[\sum_{0}^{k} \binom{k}{r} x^{k-r} y^{r+1} + \binom{k}{k} y^{k+1} \right] \\
&= \binom{k+1}{0} x^{k+1} + \sum_{1}^{k} \binom{k}{r} x^{k+1-r} y^r \\
&\quad + \sum_{1}^{k} \binom{k}{r-1} x^{k+1-r} y^r + \binom{k+1}{k+1} y^{k+1} \\
&= \binom{k+1}{0} x^{k+1} + \sum_{1}^{k} \left[\binom{k}{r} + \binom{k}{r-1} \right] x^{k+1-r} y^r + \binom{k+1}{k+1} y^{k+1}
\end{aligned}$$

*The binomial theorem for $n = 2$ can be found in Euclid's work (ca. 300 B.C.).

$$= \binom{k+1}{0} x^{k+1} + \sum_{1}^{k} \binom{k+1}{1} x^{k+1-r} y^r$$

$$+ \binom{k+1}{k+1} y^{k+1} \qquad \text{by Theorem 12.2}$$

$$= \sum_{0}^{k+1} \binom{k+1}{r} x^{k+1-r} y^r$$

Thus, by the principle of mathematical induction (PMI), the formula is true for every integer $n \geq 0$. ■

It follows from the binomial theorem that the binomial coefficients in the expansion of $(x + y)^n$ are the various numbers in row n of Pascal's triangle.

Corollary 12.1.

$$(1 + x)^n = \sum_{0}^{n} \binom{n}{i} x^n \qquad \text{and} \qquad (1 - x)^n = \sum_{0}^{n} \binom{n}{i} (-1)^n x^n$$ ■

Corollary 12.2.

$$\sum \binom{n}{2i} = \sum \binom{n}{2i-1}$$

That is, the sum of the 'even' binomial coefficients equals that of the 'odd' binomial coefficients. ■

The proof of this corollary follows by letting $x = -1$ in the first result in Corollary 12.1.

FIBONACCI NUMBERS

But how are Fibonacci numbers related to Pascal's triangle? To see this, we return to the triangular arrangement (see Fig. 12.3, for example). Now add the numbers along the northeast diagonals. The sums are 1, 1, 2, 3, 5, 8, . . . , and they seem to be the Fibonacci numbers. Indeed, they are, as the next theorem, discovered by E. Lucas in 1876, confirms.

Theorem 12.4. (Lucas Formula, 1876)

$$F_{n+1} = \sum_{i=0}^{\lfloor n/2 \rfloor} \binom{n-i}{i} \qquad n \geq 0 \qquad (12.1)$$

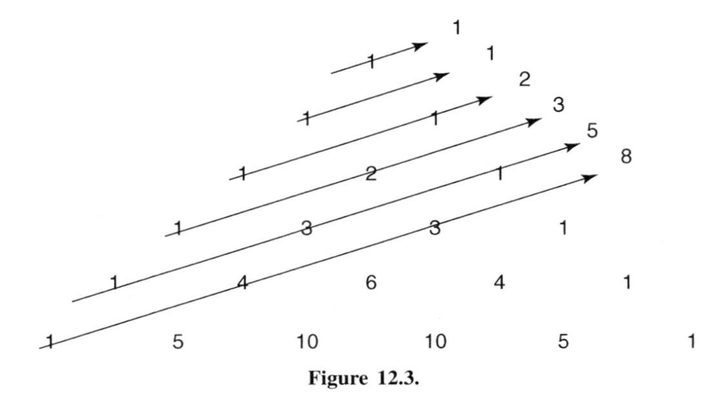

Figure 12.3.

Proof. (by the strong version of PMI) When $n = 0$, RHS $= \sum_{i=0}^{0} \binom{0-i}{i} = \binom{0}{0} = 1 = F_1 =$ LHS. So the statement is true when $n = 0$.

Now assume it is true for all nonnegative integers $\leq k$, where k is an arbitrary integer ≥ 0:

$$F_{k+1} = \sum_{i=0}^{\lfloor k/2 \rfloor} \binom{k-i}{i}$$

By Pascal's identity,

$$\sum_{i=0}^{\lfloor (k+2)/2 \rfloor} \binom{k+2-i}{i} = \sum_{i=0}^{\lfloor (k+2)/2 \rfloor} \binom{k+1-i}{i-1} + \sum_{i=0}^{\lfloor (k+2)/2 \rfloor} \binom{k+1-i}{i}$$

Suppose k is even. Then

$$\begin{aligned}
\sum_{i=0}^{\lfloor (k+2)/2 \rfloor} \binom{k+2-i}{i} &= \sum_{j=0}^{k/2} \binom{k-j}{j} + \sum_{i=0}^{k/2} \binom{k+1-i}{i} + \binom{k/2}{k/2+1} \\
&= \sum_{j=0}^{\lfloor (k+1)/2 \rfloor} \binom{k-j}{j} + \sum_{i=0}^{\lfloor (k+1)/2 \rfloor} \binom{k+1-i}{i} + 0 \\
&= F_k + F_{k+1} \\
&= F_{k+2}
\end{aligned}$$

It can be shown similarly that, when k is odd,

$$\sum_{i=0}^{\lfloor (k+2)/2 \rfloor} \binom{k+2-i}{i} = F_{k+2}$$

Thus, by the strong verision of PMI, the formula holds for all integers $n \geq 0$. ■

For example,

$$F_6 = \sum_0^2 \binom{5-i}{i} = \binom{5}{0} + \binom{4}{1} + \binom{3}{2} = 1 + 4 + 3 = 8$$

$$F_7 = \sum_0^3 \binom{6-i}{i} = \binom{6}{0} + \binom{5}{1} + \binom{4}{2} + \binom{3}{3} = 1 + 5 + 6 + 1 = 13$$

We can use Binet's and Lucas' formulas, and the binomial theorem in tandem to derive an array of Fibonacci and Lucas identities. For example, notice that

$$\begin{aligned}\sum_0^5 \binom{5}{i} &= \binom{5}{0}F_0 + \binom{5}{1}F_1 + \binom{5}{2}F_2 + \binom{5}{3}F_3 + \binom{5}{4}F_4 + \binom{5}{5}F_5 \\ &= 0 + 5 + 10 + 20 + 15 + 5 = 55 = F_{10}\end{aligned}$$

More generally, we have the following identity.

Theorem 12.5. (Lucas)

$$\sum_{i=0}^{n} \binom{n}{i} F_i = F_{2n} \qquad n \ge 0 \tag{12.2}$$

Proof. By Binet's formula,

$$\begin{aligned}\sum_{i=0}^{n} \binom{n}{i} F_i &= \sum_{i=0}^{n} \binom{n}{i} \left(\frac{\alpha^i - \beta^i}{\alpha - \beta}\right) \\ &= \frac{1}{\alpha - \beta}\left[\sum_0^n \binom{n}{i}\alpha^i - \sum_0^n \binom{n}{i}\beta^i\right] \\ &= \frac{(1+\alpha)^n - (1+\beta)^n}{\alpha - \beta} \qquad \text{by Corollary 12.1} \\ &= \frac{\alpha^{2n} - \beta^{2n}}{\alpha - \beta} \qquad \text{since } \alpha^2 = \alpha + 1 \text{ and } \beta^2 = \beta + 1 \\ &= F_{2n}\end{aligned}$$

■

A similar argument yields yet another identity by Lucas:

$$\sum_{i=0}^{n}\binom{n}{i}L_i = L_{2n} \qquad n \geq 0 \tag{12.3}$$

For example,

$$\sum_{0}^{4}\binom{4}{i} = \binom{4}{0}L_0 + \binom{4}{1}L_1 + \binom{4}{2}L_2 + \binom{4}{3}L_3 + \binom{4}{4}L_4$$
$$= 2 + 4 + 18 + 16 + 7 = 47 = L_8$$

Theorem 12.6.

$$\sum_{i=0}^{n}\binom{n}{i}(-1)^i F_i = (-1)^{n-1}F_n, \qquad n \geq 0 \tag{12.4}$$

Proof. By Binet's formula,

$$\begin{aligned}
\sum_{i=0}^{n}\binom{n}{i}(-1)^i F_i &= \sum_{i=0}^{n}\binom{n}{i}(-1)^i\left(\frac{\alpha^i - \beta^i}{\alpha - \beta}\right) \\
&= \frac{1}{\alpha - \beta}\left[\sum_{0}^{n}\binom{n}{i}(-\alpha)^i - \sum_{0}^{n}\binom{n}{i}(-\beta)^i\right] \\
&= \frac{(1-\alpha)^n - (1-\beta)^n}{\alpha - \beta} \qquad \text{by Corollary 12.1} \\
&= \frac{(-\beta)^n - (-\alpha^n)}{\alpha - \beta} = (-1)^{n-1}\frac{\alpha^n - \beta^n}{\alpha - \beta} \\
&= (-1)^{n-1}F_n
\end{aligned}$$

■

For example,

$$\sum_{0}^{5}\binom{5}{i}(-1)^i = \binom{5}{0}F_0 - \binom{5}{1}F_1 + \binom{5}{2}F_2 - \binom{5}{3}F_3 + \binom{5}{4}F_4 - \binom{5}{5}F_5$$
$$= 0 - 5 + 10 - 20 + 15 - 5 = -5 = (-1)^5 F_5$$

It similarly can be shown that

$$\sum_{i=0}^{n}\binom{n}{i}(-1)^i L_i = (-1)^n L_n \qquad n \geq 0 \tag{12.5}$$

For example,

$$\sum_{0}^{4} \binom{4}{i} (-1)^i L_i = \binom{4}{0} L_0 - \binom{4}{1} L_1 + \binom{4}{2} L_2 - \binom{4}{3} L_3 + \binom{4}{4} L_4$$

$$= 2 - 4 + 18 - 16 + 7 = 7 = (-1)^4 L_4$$

FIBONACCI PATHS OF A ROOK ON A CHESSBOARD

In 1970, Edward T. Frankel of New York showed that Fibonacci numbers can be derived by enumerating the number of different paths open to a rook on an empty chessboard from one corner to the opposite corner where its moves are restricted by a pattern of horizontal and vertical fences.

To see this, consider Pascal's triangle with the top nine rows, rows 0 through 8. This time, left-justify the elements in every row and then move up each column j-by-j elements, where $j \geq 0$. In other words, rotate Pascal's triangle to its left by 45°. Figure 12.4 shows the resulting 8×8 square array. Every rising diagonal of this array is a row of the Pascal's triangle in Figure 12.1. Every element $A(i, j)$ of this array can be realized by adding the element immediately to its left and the element immediately above it:

$$A(i, j) = A(i, j-1) + A(i-1, j)$$

where $i, j \geq 1$. Clearly, $A(0, j) = 1 = A(i, 0)$.

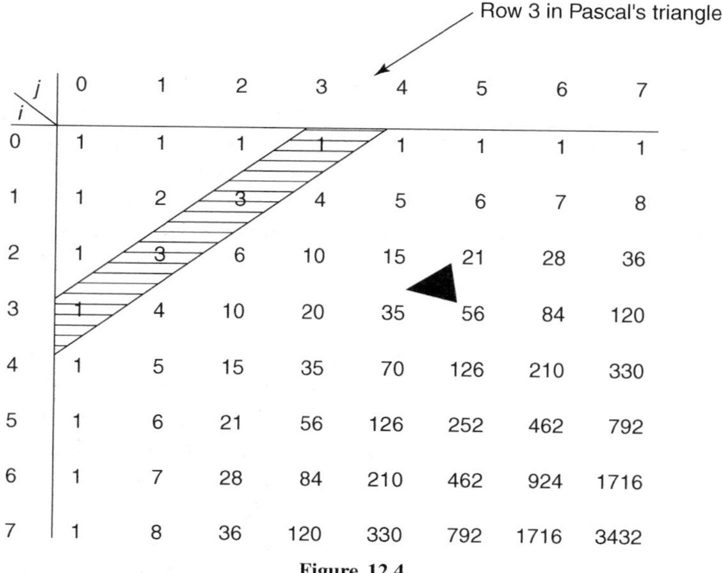

i \ j	0	1	2	3	4	5	6	7
0	1	1	1	1	1	1	1	1
1	1	2	3	4	5	6	7	8
2	1	3	6	10	15	21	28	36
3	1	4	10	20	35	56	84	120
4	1	5	15	35	70	126	210	330
5	1	6	21	56	126	252	462	792
6	1	7	28	84	210	462	924	1716
7	1	8	36	120	330	792	1716	3432

Figure 12.4.

$i \backslash j$	0	1	2	3	4	5	6	7
0	$\binom{0}{0}$	$\binom{1}{0}$	$\binom{2}{0}$	$\binom{3}{0}$	$\binom{4}{0}$	$\binom{5}{0}$	$\binom{6}{0}$	$\binom{7}{0}$
1	$\binom{1}{1}$	$\binom{1}{1}$	$\binom{2}{1}$	$\binom{3}{1}$	$\binom{4}{1}$	$\binom{5}{1}$	$\binom{6}{1}$	$\binom{7}{1}$
2	$\binom{2}{2}$	$\binom{3}{2}$	$\binom{4}{2}$	$\binom{5}{2}$	$\binom{6}{2}$	$\binom{7}{2}$	$\binom{8}{2}$	$\binom{9}{2}$
3	$\binom{3}{3}$	$\binom{4}{3}$	$\binom{5}{3}$	$\binom{6}{3}$	$\binom{7}{3}$	$\binom{8}{3}$	$\binom{9}{3}$	$\binom{10}{3}$
4	$\binom{4}{4}$	$\binom{5}{4}$	$\binom{6}{4}$	$\binom{7}{4}$	$\binom{8}{4}$	$\binom{9}{4}$	$\binom{10}{4}$	$\binom{11}{4}$
5	$\binom{5}{5}$	$\binom{6}{5}$	$\binom{7}{5}$	$\binom{8}{5}$	$\binom{9}{5}$	$\binom{10}{5}$	$\binom{11}{5}$	$\binom{12}{5}$
6	$\binom{6}{6}$	$\binom{7}{6}$	$\binom{8}{6}$	$\binom{9}{6}$	$\binom{10}{6}$	$\binom{11}{6}$	$\binom{12}{6}$	$\binom{13}{6}$
7	$\binom{7}{7}$	$\binom{8}{7}$	$\binom{9}{7}$	$\binom{10}{7}$	$\binom{11}{7}$	$\binom{12}{7}$	$\binom{13}{7}$	$\binom{14}{7}$

Figure 12.5.

A rook on a chessboard moves any number of cells either horizontally or vertically, but not in both directions in the same move. Suppose it moves horizontally to the right(R) or vertically down(D). It is well known that each entry in Figure 12.4 indicates the number of moves of a rook from the upper left-hand corner to that cell. For example, $A(2, 1) = 3$ implies that there are three different ways the rook can move from position (0,0) to position (2,1); they are 1R, 2D; 1D, 1R, 1D; and 2D, 1R. The rook has 3432 possible moves from the upper left corner to the lower right-hand corner (7,7). Using the combinatorial notation, we can rewrite the array in Figure 12.4, as Figure 12.5 shows.

Figure 12.6 shows the number of moves of the rook from the upper left-hand corner, where its moves are restricted by staggered horizontal and vertical fences. Oddly enough, all entries in this band array are Fibonacci numbers. The band is made up of four strands of Fibonacci numbers. The array begins with a 1 in the top left corner. Every other entry is the sum of the entries immediately to its left and immediately above it, assuming the entries outside the square are zeros.

Figure 12.7 displays the same chessboard array using the Fibonacci notation. Notice that the subscripts of any two adjacent Fibonacci numbers on each strand differ

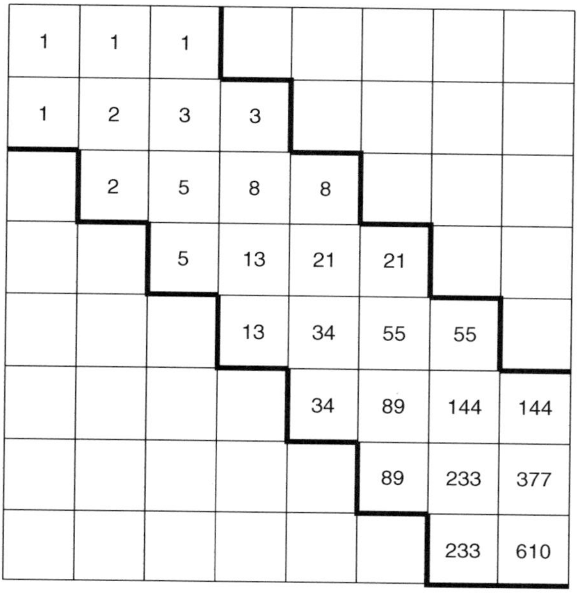

1	1	1					
1	2	3	3				
	2	5	8	8			
		5	13	21	21		
			13	34	55	55	
				34	89	144	144
					89	233	377
						233	610

Figure 12.6.

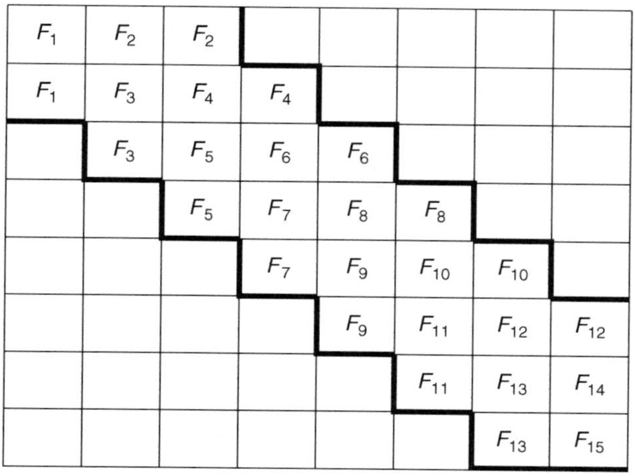

F_1	F_2	F_2					
F_1	F_3	F_4	F_4				
	F_3	F_5	F_6	F_6			
		F_5	F_7	F_8	F_8		
			F_7	F_9	F_{10}	F_{10}	
				F_9	F_{11}	F_{12}	F_{12}
					F_{11}	F_{13}	F_{14}
						F_{13}	F_{15}

Figure 12.7.

by two. Also, the Fibonacci numbers on the two upper strands have odd subscripts, whereas those on the two lower ones have even subscripts.

It follows from Figures 12.5–12.7 that the rook has $F_{15} = 610$ possible moves from position (0,0) to position (7,7). More generally, on an $n \times n$ chessboard, the rook has F_{2n-1} restricted moves from the top left corner to the bottom right corner (see Exercise 32).

EXERCISES 12

Using Lucas' formula, compute each.

1. F_5
2. F_9
3. F_{11}
4. F_{12}

Prove each.

5. $\sum_{i=0}^{n} \binom{n}{i} L_i = L_{2n}$
6. $\sum_{i=0}^{n} \binom{n}{i} (-1)^i L_i = (-1)^n L_n$
7. $\sum_{i=0}^{n} \binom{n}{i} F_{i+j} = F_{2n+j}$
8. $\sum_{i=0}^{n} \binom{n}{i} L_{i+j} = L_{2n+j}$
9. $\sum_{i=0}^{n} \binom{n}{i} (-1)^i F_{i+j} = (-1)^{j+1} F_{n-j}$
10. $\sum_{i=0}^{n} \binom{n}{i} (-1)^i L_{i+j} = (-1)^j L_{n-j}$
11. Verify that $5 \sum_{i=0}^{n} \binom{n}{i} F_i^2 = \sum_{i=0}^{n} \binom{n}{i} L_{2i}$ for $n = 4$ and $n = 5$.

Establish the formula in Exercise 11 using:

12. The binomial theorem.
13. Exercise 42 in Chapter 5.
14. Verify that $\sum_{i=0}^{n} \binom{n}{i} L_i^2 = \sum_{i=0}^{n} \binom{n}{i} L_{2i}$ for $n = 4$ and $n = 5$.
15. Establish the formula in Exercise 14.

Prove each.

16. $2^{n-1} F_n = \sum_{0}^{\lfloor (n-1)/2 \rfloor} \binom{n}{2i+1} 5^i$ (Catalan, 1846)
17. $2^{n-1} L_n = \sum_{0}^{\lfloor n/2 \rfloor} \binom{n}{2i} 5^i$ (Catalan)
18. $\sum_{0}^{n-1} (-2)^i \binom{n}{i} F_i = \begin{cases} -2^i F_i & \text{if } n \text{ is even} \\ 2^i F_i - 2 \cdot 5^{(n-1)/2} & \text{otherwise} \end{cases}$ (Ferns, 1964)
19. $\sum_{0}^{n-1} (-2)^i \binom{n}{i} G_i = 5^{(n-1)/2} [c(-1)^n - d]$ (Koshy, 1998)
20. $\sum_{0}^{n} (-)^i \binom{n}{i} L_{2i} = (-)^n L_n$ (Gould, 1963)

21. $\sum_0^n (-)^i \binom{n}{i} F_{2i} = (-)^n F_n$ (Gould, 1963)

22. $\sum_0^n \binom{n}{i} G_i = G_{2n}$ (Ruggles, 1963)

23. $\sum_0^n (-)^i \binom{n}{i} G_i = (-)^n G_{-n}$ (Koshy, 1998)

24. $\sum_0^n \binom{n}{i} G_{i+j} = G_{2n+j}$ (Koshy, 1998)

25. $\sum_0^n (-)^i \binom{n}{i} G_{i+j} = (-)^n G_{j-n}$ (Koshy, 1998)

26. $\sum_0^n \binom{n}{i} F_m^i F_{m-1}^{n-i} F_{r+i} = F_{mn+r}$ (Vinson, 1963)

27. $\sum_0^n \binom{n}{i} F_{k+2i} = \begin{cases} 5^{(n-1)/2} L_{n+k} & \text{if } n \text{ is odd} \\ 5^{n/2} F_{n+k} & \text{otherwise} \end{cases}$ (Carlitz, 1967)

28. $\sum_0^n \binom{n}{i} L_{k+2i} = \begin{cases} 5^{(n+1)/2} F_{n+k} & \text{if } n \text{ is odd} \\ 5^{n/2} L_{n+k} & \text{otherwise} \end{cases}$ (Carlitz, 1967)

29. $\sum_0^{2n} (-)^i \binom{2n}{i} 2^{i-1} L_i = 5^n$ (Brown, 1967)

30. $\sum_0^{2n} (-)^i \binom{2n}{i} 2^{i-1} F_i = 0$ (Brown, 1967)

31. $\sum_0^n \binom{n}{k} F_{4mk} = L_{2m}^n F_{2mn}$ (Hoggatt, 1968)

32. A rook on an $n \times n$ chessboard has F_{2n-1} restricted moves from the top left corner to the bottom right corner.

13 PASCAL-LIKE TRIANGLES

We have seen how Fibonacci numbers can be generated from Pascal's triangle. We now turn to how Fibonacci and Lucas numbers can be constructed from similar triangular arrays that have Pascal-like properties.

In 1966, N. A. Draim of Ventura, California, and M. Bicknell of A. C. Wilcox High School, Santa Clara, California, studied the sums and differences of like powers of the solutions of an arbitrary quadratic equation $x^2 - px - q = 0$. These sums and differences were also studied in 1997 by J. E. Woko of Abia State Polytechnic, Aba, Nigeria. As we will see shortly, an intriguing relationship exists between these expressions, and Fibonacci and Lucas numbers.

SUMS OF THE nth POWERS

Let r and s be the solutions of the quadratic equation $x^2 - px - q = 0$. Then

$$r = \frac{p + \sqrt{p^2 + 4q}}{2} \qquad \text{and} \qquad s = \frac{p - \sqrt{p^2 + 4q}}{2}$$

so $r + s = p$ and $rs = -q$. Consequently,

$$r^2 + s^2 = (r + s)^2 - 2rs = p^2 + 2q$$

and

$$r^3 + s^3 = (r + s)^3 - 3rs(r + s) = p^3 + 2pq$$

Continuing like this, we can compute the values of the various sums $r^n + s^n$:

$$
\begin{aligned}
r + s &= p \\
r^2 + s^2 &= p^2 + 2q \\
r^3 + s^3 &= p^3 + 3pq \\
r^4 + s^4 &= p^4 + 4p^2q + 2q^2 \\
r^5 + s^5 &= p^5 + 5p^3q + 5pq^2 \\
r^6 + s^6 &= p^6 + 6p^4q + 9p^2q^2 + 2q^3 \\
r^7 + s^7 &= p^7 + 7p^5q + 14p^3q^2 + 7pq^3
\end{aligned}
$$

More generally, using the principle of mathematical induction (PMI), Draim and Bicknell showed that

$$r^n + s^n = \sum_{0}^{\lfloor n/2 \rfloor} A(n, i) p^{n-2i} q^i \tag{13.1}$$

where $A(n, i) = 2\binom{n-i}{i} - \binom{n-i-1}{i}$.

Using Pascal's identity, we can simplify the formula for $A(n, i)$:

$$
\begin{aligned}
A(n, i) &= \binom{n-i}{i} + \left[\binom{n-i}{i} - \binom{n-i-1}{i}\right] \\
&= \binom{n-i}{i} + \binom{n-i-1}{i-1} \\
&= \binom{n-i}{i} + \frac{i}{n-i}\binom{n-i-1}{i-1} \\
&= \binom{n-i}{i}\left(1 + \frac{i}{n-i}\right) \\
&= \frac{n}{n-i}\binom{n-i}{i} \qquad (13.2)
\end{aligned}
$$

Thus we can rewrite Formula (13.1) as

$$r^n + s^n = \sum_{0}^{\lfloor n/2 \rfloor} A(n, i) p^{n-2i} q^i \tag{13.3}$$

where

$$A(n, i) = \frac{n}{n-i}\binom{n-i}{i}, \qquad 0 \le i \le \lfloor n/2 \rfloor$$

We can arrange the various values of $A(n, i)$ in a Pascal-like triangle, as Table 13.1 shows.

TABLE 13.1.

n \ i	0	1	2	3	4	5	...	Row sum
1	1							1
2	1	2						3
3	1	3						4
4	1	4	2					7
5	1	5	5					11
6	1	6 ↘	9	2				18
7	1	7	14 ↓	7				29
8	1	8	20	16	2			47
9	1	9	27	30	9			76
10	1	10	35	50	25	2		123
								↑ Lucas numbers

In particular, let r and s be the solutions of the equation $x^2 = x + 1$. Then $r = \alpha$, $s = \beta$, so Eq. (13.3) yields the formula

$$L_n = \sum_{r=0}^{\lfloor n/2 \rfloor} A(n, i) \tag{13.4}$$

This should not come as a surprise, since Table 13.1 is the same triangular arrangement we obtained by computing the topological indices of cycloparaffins C_nH_{2n} in Chapter 3; see Table 3.9 on p. 37.

Using Eq. (13.2), we can rewrite this formula as

$$L_n = \sum_{i=0}^{\lfloor n/2 \rfloor} \frac{n}{n-i} \binom{n-i}{i} \tag{13.5}$$

For example,

$$\sum_{0}^{2} \frac{5}{5-i} \binom{5-i}{i} = \frac{5}{5}\binom{5}{0} + \frac{5}{4}\binom{4}{1} + \frac{5}{3}\binom{3}{2} = 1 + 5 + 5 = 11 = L_5$$

See row 5 in the table.

The triangular array in Table 13.1 satisfies the following interesting properties:

- Since $A(n, 0) = 1$, every row begins with a 1.
- $A(n, i)$ satisfies the recurrence relation

$$A(n, i) = A(n-1, i) + A(n-2, i-1) \tag{13.6}$$

  If $n \geq 3$, every entry $A(n, i)$ can be obtained by adding the entry just above it in the previous row and the entry to its left in the row above it. See, for example, the arrows in the table.

- If $i > \lfloor n/2 \rfloor$, then $A(n, i) = 0$.
- If n is even:

  1. Row n ends in 2;
  2. Row n and row $n + 1$ contain the same number of entries.

- If n is odd, then:

  1. Row n ends in n;
  2. Row $n + 1$ contains one more entry than row n.

These results can be proved fairly easily. For example, suppose n is odd. Then

$$A(n, \lfloor n/2 \rfloor) = A(n, (n-1)/2) = \frac{2n}{n+1}\binom{(n+1)/2}{(n-1)/2}$$

$$= \frac{2n}{n+1} \cdot \frac{((n+1)/2)!}{((n-1)/2)!} = \frac{2n}{n+1} \cdot \frac{n+1}{2} = n$$

When n is odd,

$$\lfloor (n+1)/2 \rfloor = (n+1)/2 = (n-1)/2 + 1 = \lfloor n/2 \rfloor + 1$$

so row $n + 1$ contains one entry more than row n.

AN ALTERNATE FORMULA FOR L_n

The preceding discussion yields a wonderful dividend in the form of an alternate formula for L_n.

Let $\Delta = \sqrt{p^2 + 4q}$. Then $2r = p + \Delta$ and $2s = p - \Delta$, so

$$(2r)^n = (p + \Delta)^n = \sum_0^n \binom{n}{i} p^{n-i} \Delta^i$$

$$(2s)^n = (p - \Delta)^n = \sum_0^n \binom{n}{i} p^{n-i} (-\Delta)^i$$

$$(2r)^n + (2s)^n = 2 \sum_{i \text{ even}} \binom{n}{i} p^{n-i} \Delta^i$$

$$2^n(r+s)^n = 2\sum_{0}^{\lfloor n/2 \rfloor} \binom{n}{2i} p^{n-2i}\Delta^{2i}$$

$$2^{n-1}(r+s)^n = \sum_{0}^{\lfloor n/2 \rfloor} \binom{n}{2i} p^{n-2i}\Delta^{2i}$$

In particular, let $p = 1 = q$. Then $\Delta = \sqrt{5}$, so this yields the formula

$$L_n = \frac{1}{2^{n-1}} \sum_{0}^{\lfloor n/2 \rfloor} \binom{n}{2i} 5^i \tag{13.7}$$

For example,

$$\begin{aligned} L_5 &= \frac{1}{2^4}\sum_{0}^{2}\binom{5}{2i}5^i \\ &= \frac{1}{2^4}\left[\binom{5}{0} + \binom{5}{2}5 + \binom{5}{4}5^5\right] = \frac{1}{16}(1+50+125) = 11 \end{aligned}$$

DIFFERENCE OF THE nth POWERS

Let us now turn our attention to the difference of the nth powers of r and s:

$$\begin{aligned} r - s &= (p+\Delta)/2 - (p-\Delta)/2 = \Delta \\ r^2 - s^2 &= [(p+\Delta)/2]^2 - [(p-\Delta)/2]^2 = p\Delta \end{aligned}$$

Continuing on like this, we get

$$\begin{aligned} r - s &= \Delta \\ r^2 - s^2 &= p\Delta \\ r^3 - s^3 &= (p^2 + q)\Delta \\ r^4 - s^4 &= (p^3 + 2pq)\Delta \\ r^5 - s^5 &= (p^4 + 3p^2q + q^2)\Delta \\ r^6 - s^6 &= (p^5 + 4p^3q + 3pq^2)\Delta \\ r^7 - s^7 &= (p^6 + 5p^4q + 6p^2q^2 + q^3)\Delta \end{aligned}$$

More generally,

$$r^n - s^n = \Delta \sum_{0}^{\lfloor n/2 \rfloor} \binom{n-i-1}{i} p^{n-2i} q^i \tag{13.8}$$

We can establish this also by PMI (see Exercise 13).

As before, the various coefficients

$$B(n, i) = \binom{n-i-1}{i} \qquad 0 \le r \le \lfloor (n-1)/2 \rfloor$$

can be arranged in a triangular array, as Table 13.2 shows. We saw this triangular arrangement in Table 3.8 on page 36 when we computed the topological indices of paraffins C_nH_{2n+2}.

Table 13.2 satisfies several important properties:

- $B(n, i)$ satisfies the recurrence relation

$$B(n, i) = B(n-1, i) + B(n-2, i-1)$$

This is so since

$$\begin{aligned} B(n-1, i) + B(n-2, i-1) &= \binom{n-i-2}{i} + \binom{n-i-2}{i} \\ &= \binom{n-i-1}{i} = B(n, i) \end{aligned}$$

- Since $B(n, 0) = 1$, every row begins with a 1.

TABLE 13.2.

n \ i	0	1	2	3	4	5	...	Row Sum
1	1							1
2	1							1
		↓ positive integers						
3	1	1						2
4	1	2						3
			↓ triangular numbers					
5	1	3	1					5
6	1	4	3					8
				↓ tetrahedral numbers				
7	1	5	6	1				13
8	1	6	10	4				21
9	1	7	15	10	1			34
10	1	8	21	20	5			55
								↑ F_n

- If n is odd, row n ends in a 1. This is so since

$$B(n, \lfloor (n-1)/2 \rfloor) = B(n, (n-1)/2) = \binom{(n-1)/2}{(n-1)/2} = 1$$

- Suppose n is odd. Since $\lfloor (n-1)/2 \rfloor = (n-1)/2 = \lfloor n/2 \rfloor$, row n and row $n+1$ contain the same number of entries.
- Suppose n is even. Then row n ends in $n/2$ and row $n+1$ contains one entry more than row n.
- Since $B(n, 3) = \binom{n-3}{r}$, column 2 yields the various triangular numbers, where $n \geq 5$.
- $B(n, 4) = \binom{n-4}{r}$, so column 3 yields the tetrahedral numbers. Similarly, the remaining columns give higher dimensional figurate numbers.

Suppose we let $p = 1 = q$. Then $r = \alpha$, $s = \beta$, and $\Delta = \sqrt{5}$, so Eq. (13.8) yields the combinatorial formula for F_n:

$$\begin{aligned} F_n &= \sum_{0}^{\lfloor n/2 \rfloor} \binom{n-i-1}{i} \qquad (13.9) \\ &= \text{sum of the elements in row } n \text{ in Table 13.2} \end{aligned}$$

We saw this formula Eq. (12.1) in the preceding chapter.

We are now ready to present an alternate formula for F_n.

AN ALTERNATE FORMULA FOR F_n

From Eqs. (13.1) and (13.8), it follows that

$$\begin{aligned} (2r)^n - (2s)^n &= 2 \sum_{i \text{ odd}} \binom{n}{i} p^{n-i} \Delta^i \\ 2^n (r-s)^n &= 2 \sum_{0}^{\lfloor (n-1)/2 \rfloor} \binom{n}{2i+1} p^{n-2i-1} \Delta^{2i+1} \\ (r-s)^n &= \frac{1}{2^{n-1}} \sum_{0}^{\lfloor (n-1)/2 \rfloor} \binom{n}{2i+1} p^{n-2i-1} \Delta^{2i} \end{aligned}$$

In particular, let $p = 1 = q$. Since $\Delta = \sqrt{5}$, this yields yet another combinatorial formula for F_n:

$$F_n = \frac{1}{2^{n-1}} \sum_{0}^{\lfloor (n-1)/2 \rfloor} \binom{n}{2i+1} 5^i \qquad (13.10)$$

For example,

$$F_6 = \frac{1}{2^5}\sum_{0}^{2}\binom{6}{2i+1}5^i$$

$$= \frac{1}{2^5}\left[\binom{6}{1}+\binom{6}{3}5+\binom{6}{5}5^2\right] = \frac{1}{32}(6+100+150) = 8$$

A LUCAS TRIANGLE

In 1967, M. Feinberg, then a student at the University of Pennsylvania, studied the coefficients in the expansion of the polynomial $f_n(x, y) = (x+y)^{n-1}(x+2y)$, where $n \geq 1$, and discovered an invaluable treasure.

The first six expansions are

$$f_1(x, y) = x + 2y$$
$$f_2(x, y) = x^2 + 3xy + 2y^2$$
$$f_3(x, y) = x^3 + 4x^2y + 5xy^2 + 2y^3$$
$$f_4(x, y) = x^4 + 5x^3y + 9x^2y^2 + 7xy^3 + 2y^4$$
$$f_5(x, y) = x^5 + 6x^4y + 14x^3y^2 + 16x^2y^3 + 9xy^4 + 2y^5$$
$$f_6(x, y) = x^6 + 7x^5y + 20x^4y^2 + 30x^3y^3 + 25x^2y^4 + 11xy^5 + 2y^5$$

We can verify these. Arranging the various coefficients in these polynomials in a triangular array, we get the truncated arrangement in Figure 13.1. Feinberg called it a *Lucas triangle*. Every row begins with a 1 and ends in a 2; this is so because the

i \ j	0	1	2	3	4	5	6
1	1	2					
2	1	3	2				
3	1	4 →	5 ↓	2			
4	1	5	9	7	2		
5	1	6	14	16	9	2	
6	1	7	20	30	25	11	2

Figure 13.1. A Lucas triangle

coefficient of x^n in $f_n(x, y)$ is 1 and that of y^n is 2. Since $f_n(1, 1) = 3 \cdot 2^{n-1}$, the sum of the numbers in row n is $3 \cdot 2^{n-1}$, where $n \geq 1$.

Let $C(n, j)$ denote the entry in row n and column j, where $n \geq 1$ and $j \geq 0$. We can find an explicit formula for $C(n, j)$ as follows:

$$\begin{aligned}
(x+y)^{i-1} &= \sum_0^{i-1} \binom{i-1}{j} x^{i-j-1} y^j \\
f_i(x, y) &= (x+y)^{i-1}(x+2y) = \left[\sum_0^{i-1} \binom{i-j-1}{j} x^{i-j-1} y^j\right](x+2y) \\
C(n, j) &= \text{Coefficient of } x^{n-j} y^j \\
&= \binom{n-1}{j} + 2\binom{n-1}{j-1} \\
&= \binom{n}{j} + \binom{n-1}{j-1} \qquad (13.11)
\end{aligned}$$

by Pascal's identity.

For example,

$$C(6, 3) = \binom{6}{3} + \binom{5}{2} = 20 + 10 = 30$$

The triangular array in Figure 13.1 satisfies three additional properties:

- $\sum_{k=1}^{n} C(k, j) = C(n+1, j+1)$
- $\sum_{k=1}^{n} C(k, 1) = C(n+1, 2)$
- $C(n, n-2) = (n-1)^2 \qquad n \geq 2$

See Exercises 17–19.

A RECURSIVE DEFINITION FOR $C(n, j)$

Using Formula 13.11, we can easily verify that $C(n, j)$ satisfies the following recursive definition:

$$C(n, 0) = 1 \qquad C(1, 1) = 2$$
$$C(n, j) = C(n-1, j-1) + C(n-1, j) \qquad n, j \geq 1$$

This recurrence relation is the same as Pascal's identity. Thus we obtain $C(n, j)$ by adding the element $C(n-1, j)$ just above it and the element $C(n-1, j-1)$ to its left. Notice that $C(n, n) = 1$. For example,

$$C(4, 2) = 14 = 5 + 9 = C(3, 1) + C(3, 2)$$

Formula (13.11) contains a hidden secret: We can obtain every term $C(n, j)$ from rows n and $n-1$ of Pascal's triangle. Shift row $n-1$ by one place to the right and place the resulting row just above row n; then add the corresponding elements to yield the various elements $C(n, j)$ in row n. This algorithm for $n = 4$ is illustrated below:

$$\begin{array}{rccccc}
 & & 1 & 3 & 3 & 1 \\
+ & 1 & 4 & 6 & 4 & 1 \\
\hline
 & 1 & 5 & 9 & 7 & 2
\end{array}
\qquad
\begin{array}{l}
\leftarrow \text{Row 3 in Pascal's triangle} \\
\leftarrow \text{Row 4 in Pascal's triangle} \\
\leftarrow \text{Row 4 in Table 13.3}
\end{array}$$

Suppose we add the elements on the rising diagonals in Figure 13.1. It appears from Figure 13.2 that the sums are Lucas numbers. This result is not a fluke. To see why this is true, the sum of the elements on the nth rising diagonal is given by

$$\begin{aligned}
\sum_{0}^{\lfloor n/2 \rfloor} C(n-j, j) &= \sum_{0}^{\lfloor n/2 \rfloor} \binom{n-j}{j} + \sum_{0}^{\lfloor n/2 \rfloor} \binom{n-j-1}{j-1} \\
&= \sum_{0}^{\lfloor n/2 \rfloor} \binom{n-j}{j} + \sum_{0}^{\lfloor (n-2)/2 \rfloor} \binom{n-j-2}{j} \\
&= F_{n+1} + F_{n-1} = L_n
\end{aligned}$$

For example, the sum of the elements on the sixth rising diagonal is $18 = L_6$.

Suppose we flip the Lucas triangle in Figure 13.1 about a vertical line on the left and left-justify the elements. Figure 13.3 shows the resulting triangular arrangement.

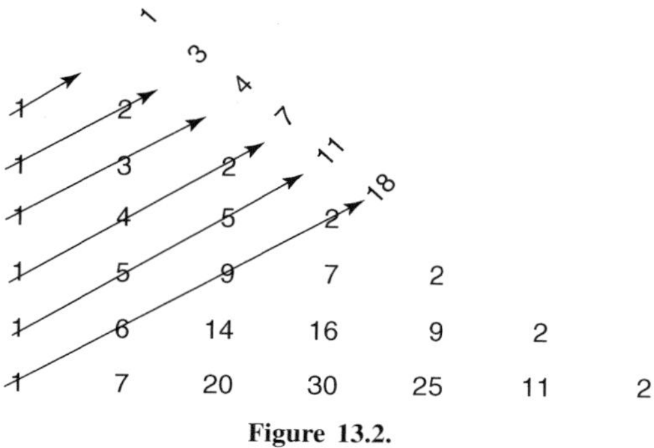

Figure 13.2.

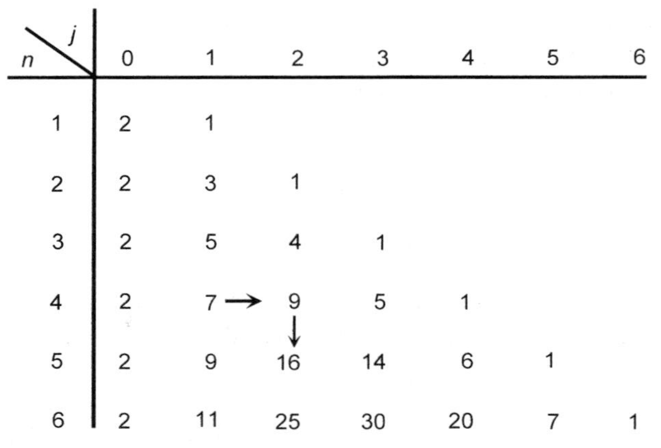

n \ j	0	1	2	3	4	5	6
1	2	1					
2	2	3	1				
3	2	5	4	1			
4	2	7 →	9 ↓	5	1		
5	2	9	16	14	6	1	
6	2	11	25	30	20	7	1

Figure 13.3. A reflection of the Lucas triangle

Let $D(n, j)$ denote the element of this array that lies in row n and column j. Then

$$D(1, 0) = 2 \qquad D(1, 1) = 1$$

and

$$D(n, j) = D(n-1, j-1) + D(n-1, j) \qquad n \geq 2$$

This is the same as the Fibonacci recurrence relation satisfied by $C(n, j)$.

Since Figure 13.3 is a reflection of the Lucas triangle, it follows that:

$$\begin{aligned} D(n, j) &= C(n, n-j) \\ &= \binom{n}{n-j} + \binom{n-1}{n-j-1} \\ &= \binom{n}{j} + \binom{n-1}{j} \end{aligned}$$

Consequently, we can obtain every row of the array in Figure 13.3 by adding rows $n-1$ and n (both left-justified) of Pascal's triangle, as the following algorithm demonstrates:

$$\begin{array}{rccccc l} & 1 & 3 & 3 & 1 & & \leftarrow \text{Row 3 of Pascal's triangle} \\ + & 1 & 4 & 6 & 4 & 1 & \leftarrow \text{Row 4 of Pascal's triangle} \\ \hline & 2 & 7 & 9 & 5 & 1 & \leftarrow \text{Row 4 in Table 13.6} \end{array}$$

The array in Figure 13.3 also provides a fascinating bonus. Add the elements on each rising diagonal; every sum is a Fibonacci number (see Fig. 13.4).

This is so because

$$\begin{aligned} \sum_{0}^{\lfloor n/2 \rfloor} D(n-j, j) &= \sum_{0}^{\lfloor n/2 \rfloor} \binom{n-j}{j} + \sum_{0}^{\lfloor n/2 \rfloor} \binom{n-j-1}{j} \\ &= F_{n+1} + F_n = F_{n+2} \end{aligned}$$

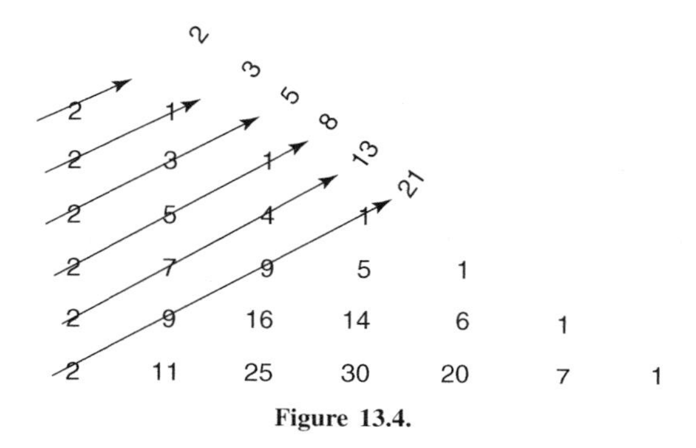

Figure 13.4.

The array in Figure 13.3 satisfies three additional properties:

- $$\sum_{k=1}^{n} D(k, j) = D(n+1, j+1)$$
- $$D(n, 2) = (n-1)^2 \qquad n \geq 2$$
- $$\sum_{k=1}^{n} D(k, 1) = n^2$$

See Exercises 22–24.

In 1970, V. E. Hoggatt described an interesting relationship between the Lucas triangle in Figure 13.1 and the triangular array of coefficients in the expansion of L_n^m, as we will see a bit later.

To establish such a link, shift down column $j(\geq 0)$ of the Lucas triangle by j elements. Figure 13.5 shows the resulting array. Let $E(n, j)$ denote the element in row n and column j of this array. Then $E(n, j) = A(n-j, j)$, where $0 \leq j \leq \lfloor n/2 \rfloor$. $E(n, j)$ satisfies the recurrence relation $E(n, j) = E(n-1, j) + E(n-2, j-1)$, where $E(n, 0) = 1$, $E(2j, j) = 2$, $E(2j+1, j) = 2j+1$, $E(2j+1, j+1) = 0$, and $1 \leq j \leq \lfloor n/2 \rfloor$.

POWERS OF LUCAS NUMBERS

In 1970, Harlan L. Umansky of Emerson High School in Union City, New Jersey, developed the following formulas for powers of Lucas numbers:

$$L_n^1 = L_n$$
$$L_n^2 = L_{2n} + 2(-1)^n$$
$$L_n^3 = L_{3n} + 3(-1)^n L_n$$
$$L_n^4 = L_{4n} + 4(-1)^n L_n^2 - 2$$
$$L_n^5 = L_{5n} + 5(-1)^n L_n^3 - 5L_n$$

$$\begin{aligned}
L_n^6 &= L_{6n} + 6(-1)^n L_n^4 - 9L_n^2 + 2(-1)^n \\
L_n^7 &= L_{7n} + 7(-1)^n L_n^5 - 14L_n^3 + 7(-1)^n L_n \\
L_n^8 &= L_{8n} + 8(-1)^n L_n^6 - 20L_n^4 + 16(-1)^n L_n^2 - 2
\end{aligned}$$

A few months later, Hoggatt observed that the absolute values of the coefficients of the various Lucas numbers and their powers on the right-hand side are the same as the entries in the triangular array in Figure 13.5. Accordingly, he established the following result.

Theorem 13.1. (Hoggatt, 1970).

$$L_n^m = L_{mn} + \sum_{j=1}^{\lfloor m/2 \rfloor} E(m, j)(-1)^{nj+j-1} L_n^{m-2j}$$

Proof. (by PMI). The formula is clearly true when $m = 1$, since the sum is zero. Assume it is true for all positive integers k, where $k \leq m$:

$$L_n^k = L_{kn} + \sum_{j=1}^{\lfloor k/2 \rfloor} E(k, j)(-1)^{nj+j-1} L_n^{k-2j}$$

Then

$$L_n^{m+1} = L_n L_{mn} + \sum_{j=1}^{\lfloor m/2 \rfloor} E(m, j)(-1)^{nj+j-1} L_n^{m-2j+1}$$

n \ j	0	1	2	3	4	5	6
1	1						
2	1	2					
3	1	3					
4	1	4	2				
5	1	5	5				
6	1	6	9	2			
7	1	7	14	7			
8	1	8	20	16	2		

Figure 13.5.

But $L_n L_{mn} = L_{(m+1)n} + (-1)^n L_{(m-1)n}$, so

$$L_n^{m+1} = L_{(m+1)n} + (-1)^n L_{(m-1)n} + \sum_{j=1}^{\lfloor m/2 \rfloor} E(m, j)(-1)^{nj+j-1} L_n^{m-2j+1} \tag{13.12}$$

By the inductive hypothesis,

$$L_{(m-1)n} = L_n^{m-1} - \sum_{j=1}^{\lfloor (m-1)/2 \rfloor} E(m-1, j)(-1)^{nj+j-1} L_n^{m-2j-1}$$

$$(-1)^n L_{(m-1)n} = (-1)^n L_n^{m-1} - \sum_{j=1}^{\lfloor (m-1)/2 \rfloor} E(m-1, j)(-1)^{n(j+1)+(j+1)-1} L_n^{m-2j-1}$$

Let $j + 1 = r$. Then this becomes

$$(-1)^n L_{(m-1)n} = (-1)^n L_n^{m-1} - \sum_{r=2}^{\lfloor (m+1)/2 \rfloor} E(m-1, r-1)(-1)^{nr+r-1} L_n^{m-2r+1}$$

Then Eq. (13.12) becomes

$$L_n^{m+1} = L_{(m+1)n} + (-1)^n L_n^{m-1} + \sum_{r=2}^{\lfloor (m+1)/2 \rfloor} E(m-1, r-1)(-1)^{nr+r-1} L_n^{m-2r+1} + \sum_{r=2}^{\lfloor m/2 \rfloor} E(m, r)(-1)^{nr+r-1} L_n^{m-2r+1} \tag{13.13}$$

Suppose $m = 2t$. Then $\lfloor m/2 \rfloor = \lfloor (m+1)/2 \rfloor$, $E(2t, t) = 2$, $E(2t-1, t-1) = 2t - 1$; so $E(2t+1, t) = 2t + 1$. On the other hand, let $m = 2t + 1$. Then $\lfloor m/2 \rfloor + 1 = \lfloor (m+1)/2 \rfloor = t + 1$, $E(2t+1, t+1) = 0$ and $E(2t, t) = 2$; thus $E(2t+2, t+1) = 2$.

Thus, Eq. (13.13) becomes

$$L_n^{m+1} = L_{(m+1)n} + \sum_{r=1}^{\lfloor (m+1)/2 \rfloor} E(m+1, r)(-1)^{nr+r-1} L_n^{m-2r+1}$$

Consequently, by the strong version of PMI, the formula works for every $n \geq 1$. ■

EXERCISES 13

1. Find a quadratic equation whose roots are α^n and β^n.
2. Find a quadratic equation whose roots are α^n and $-\beta^n$.

Compute L_8 using each formula.

3. Formula 13.5.
4. Formula 13.7.

Verify each, where r and s are the roots of equation $x^2 - px - q = 0$.

5. $r^5 + s^5 = p^5 + 5p^3q + 5pq^2$
6. $r^6 + s^6 = p^6 + 6p^4q + 9p^2q^2 + 2q^3$
7. Establish Formula 13.3 using PMI.
8. Prove that $\sum_{0}^{n}(-1)^i A(n, i) = F_{n-1}$.

Compute F_{10} using each formula.

9. Formula 13.9.
10. Formula 13.10.

Verify each, where r and s are the roots of equation $x^2 - px - q = 0$.

11. $r^5 - s^5 = (p^4 + 3p^2q + q^2)\Delta$
12. $r^6 - s^6 = (p^5 + 4p^3q + 3pq^2)\Delta$
13. Establish Formula 13.8 using PMI.
14. Verify that $C(n, j)$ satisfies the recursive definition:

$$C(n, 0) = 1 \qquad C(1, 1) = 2,$$
$$C(n, j) = C(n-1, j-1) + C(n-1, j) \qquad n, j \geq 1$$

Compute the sum of the elements on the given rising diagonal in Figure 13.2.

15. Diagonal 6.
16. Diagonal 7.

Prove each.

17. $\sum_{k=1}^{n} C(k, j) = C(n+1, j+1)$
18. $\sum_{k=1}^{n} C(k, 1) = C(n+1, 2)$
19. $C(n, n-2) = (n-1)^2$

Compute the sum of the elements on the given rising diagonal in Figure 13.4.

20. Diagonal 5.
21. Diagonal 6.

Prove each.

22. $\sum_{k=1}^{n} D(k, j) = D(n + 1, j + 1)$
23. $D(n, 2) = (n - 1)^2$
24. $\sum_{k=1}^{n} D(k, 1) = n^2$

ADDITIONAL PASCAL-LIKE TRIANGLES

We now turn to some more Pascal-like triangles that contain Fibonacci and Lucas numbers as hidden treasures.

The following three variants of Pascal's triangle were studied extensively by H. W. Gould of West Virginia University.

At the 1963 Joint Automatic Control Conference held at the University of Minnesota, P. C. Parks presented the variant of Pascal's triangle in Figure 14.1. The first few row sums are Fibonacci numbers, so we conjecture that the row sum in row n is F_{n+1}, where $n \geq 0$.

To establish this, let $f(i, j)$ denote the entry in row i and column j, where $i \geq j \geq 0$; $f(i, j) = 0$ if $j > i$; $f(i, 0) = 1$; and $f(i, i) = 1$ for every i. The inner elements are defined by the recurrence relations

$$f(i+1, 2j+1) = f(i, 2j) \qquad \text{and} \qquad f(i+1, 2j) = f(i, 2j-1) + f(i, 2j)$$

See the arrows in Figure 14.1. These two conditions can in fact be combined into a single recurrence relation:

$$f(i+1, j) = f(i, j-1) + \frac{1+(-1)^j}{2} f(i, j)$$

Using the principle of mathematical induction (PMI), it can be shown that

$$f(n, 2k) = \binom{n-k}{k} \qquad \text{and} \qquad f(n, 2k+1) = \binom{n-k-1}{k}$$

Consequently, they can be employed to produce a single formula for $f(n, r)$:

$$f(n, r) = \binom{n - \lfloor (r+1)/2 \rfloor}{\lfloor r/2 \rfloor}$$

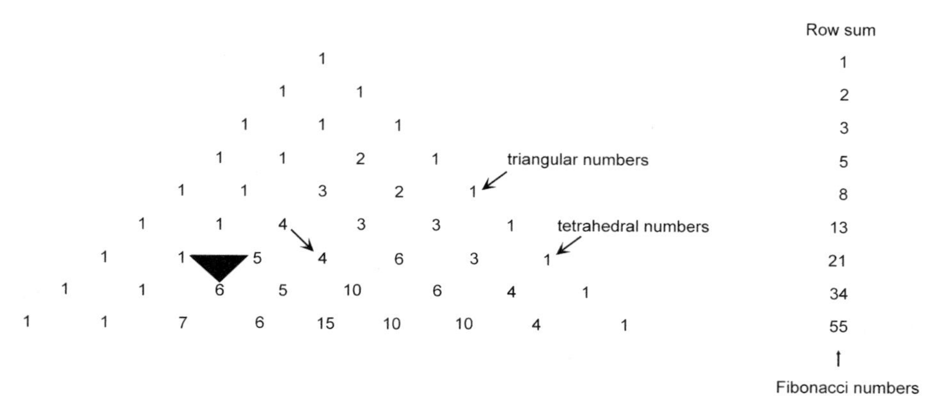

Figure 14.1.

We can verify this.

We are now ready to prove that every row sum in Figure 14.1 is a Fibonacci number. The proof is essentially the same as the one given by Gould in 1965.

Theorem 14.1.

$$\sum_{r=0}^{n} f(n,r) = F_{n+2} \qquad n \geq 0$$

Proof. (by PMI) Since $\sum_{0}^{0} f(n,r) = f(0,0) = 1 = F_2$, the statement is true when $n = 0$.

Assume it is true for all integers $i \leq k$, where $i \geq 0$ and k is arbitrary. Then:

$$\begin{aligned}
\sum_{r=0}^{k+1} f(k+1,r) &= \sum_{r \text{ even}} f(k+1,r) + \sum_{r \text{ odd}} f(k+1,r) \\
&= \sum_{r=0}^{\lfloor (k+1)/2 \rfloor} f(k+1,r) + \sum_{r=0}^{\lfloor k/2 \rfloor} f(k+1,r) \\
&= F_{k+2} + F_{k+1} \qquad \text{by the IH} \\
&= F_{k+3}
\end{aligned}$$

Thus, by the strong version of PMI, the formula is true for every $n \geq 0$. ■

The next theorem provides another fascinating property of the triangular array.

Theorem 14.2.

$$\sum_{r=0}^{n} (-1)^r f(n,r) = F_{n-1} \qquad n \geq 0$$ ■

For example,

$$\sum_{r=0}^{8}(-1)^r f(8, r) = 1 - 1 + 7 - 6 + 15 - 10 + 10 - 4 + 1$$
$$= 13 = F_7$$

Using the same rules of definition as in the previous triangular array, we can construct a new variant of Pascal's triangle by simply changing $f(1, 1)$ to 2, as Figure 14.2 demonstrates. This time, the row sums yield Lucas numbers. Let $g(i, j)$ denote the element in row i and column j, where $i \geq j \geq 0$; $g(i, j) = 0$ if $j > i$; $g(i, 0) = 1$; $g(1, 1) = 2$; $g(i + 1, 2j + 1) = g(i, 2j)$; and $g(i + 1, 2j) = g(i, 2j-1) + g(i, 2j)$. We can combine the last two conditions into a single recurrence relation:

$$g(i + 1, j) = g(i, j - 1) + \frac{1 + (-1)^j}{2} g(i, j)$$

Using PMI, we can show that

$$g(n, 2r) = \frac{n}{n - r}\binom{n - r}{r} \qquad \text{and} \qquad g(n, 2r + 1) = \frac{n - 1}{n - r - 1}\binom{n - r - 1}{r}$$

where $g(1, 1) = 2$. Consequently,

$$g(n, r) = \frac{n}{n - r}\binom{n - r}{r}$$

									Row sums
1									1
1	2								3
1	1	2							4
1	1	3	2						7
1	1	4	3	2					11
1	1	5	4	5	2				18
1	1	6	5	9	5	2			29
1	1	7	6	14	9	7	2		47
1	1	8	7	20	14	14	9	2	76

↑ Lucas numbers

Figure 14.2.

Since

$$L_n = \sum_{r=0}^{\lfloor n/2 \rfloor} \frac{n}{n-r} \binom{n-r}{r},$$

this triangular array satisfies two properties corresponding to Theorems 14.1 and 14.2.

Theorem 14.3.

$$(1) \quad \sum_{r=0}^{n} g(n,r) = L_{n+1} \qquad n \geq 0$$

$$(2) \quad \sum_{r=0}^{n} (-1)^r g(n,r) = L_{n-2} \qquad n \geq 0$$

■

For example,

$$\sum_{r=0}^{5} g(5,r) = 1+1+5+4+5+2 = 18 = L_6$$

$$\sum_{r=0}^{7} (-1)^r g(n,r) = 1-1+7-6+14-9+7-2 = 11 = L_5$$

Interestingly enough, the triangular arrays in Figures 14.1 and 14.2 can be generalized, as Figure 14.3 shows. Let $h(i, j)$ denote the element in row i and column j, where $i \geq j \geq 0$; $h(i, j) = 0$ if $j > i$; $h(i, 0) = a$; $h(1, 1) = b$; and

$$h(i+1, j) = h(i, j-1) + \frac{1+(-1)^j}{2} h(i, j) \qquad i \geq 1$$

a

$a \quad b$

$a \quad a \quad b$

$a \quad a \quad a+b \quad b$

$a \quad a \quad 2a+b \quad a+b \quad b$

$a \quad a \quad 3a+b \quad 2a+b \quad a+2b \quad b$

$a \quad a \quad 4a+b \quad 3a+b \quad 3a+3b \quad a+2b \quad b$

$a \quad a \quad 5a+b \quad 4a+b \quad 6a+4b \quad 3a+3b \quad a+2b \quad a+3b \quad b$

Figure 14.3.

In this recurrence relation, we have imposed $i \geq 1$ to avoid an awkward situation. To see this, if we let $i = 0$ and $j = 1$, then we get $h(1, 1) = h(0, 0) + 0 = a$, but $h(1, 1) = b$.

What can we say about the row sums $S_n(a, b)$ in row n? First, notice that:

$$\begin{aligned}
S_0(a, b) &= a \\
S_1(a, b) &= a + b \\
S_2(a, b) &= 2a + b \\
S_3(a, b) &= 3a + 2b \\
S_4(a, b) &= 5a + 3b \\
&\vdots
\end{aligned}$$

Clearly, a pattern emerges: $S_n(a, b) = aF_{n+1} + bF_n$, $n \geq 0$.

Likewise, let us check if the alternating row sums $T_n(a, b)$ follow any pattern:

$$\begin{aligned}
T_0(a, b) &= a \\
T_1(a, b) &= a - b \\
T_2(a, b) &= b \\
T_3(a, b) &= a \\
T_4(a, b) &= a + b \\
T_5(a, b) &= 2a + b \\
T_6(a, b) &= 3a + 2b \\
&\vdots
\end{aligned}$$

More generally, $T_n(a, b) = aF_{n-2} + bF_{n-3}$, $n \geq 1$.

These discussions lead to the following theorem.

Theorem 14.4. Let $S_n(a, b)$ denote the row sum of the entries in row n in Figure 14.3 and $T_n(a, b)$ their alternating row sum. Then

$$S_n(a, b) = aF_{n+1} + bF_n \qquad n \geq 0$$

and

$$T_n(a, b) = aF_{n-2} + bF_{n-3} \qquad n \geq 1$$

■

In particular, $S_n(1, 1) = F_{n+1} + F_n = F_{n+2}$ and $T_n(1, 1) = F_{n-2} + F_{n-3} = F_{n-1}$; this is consistent with Theorems 14.1 and 14.2. Likewise, $S_n(1, 2) = F_{n+1} + 2F_n =$

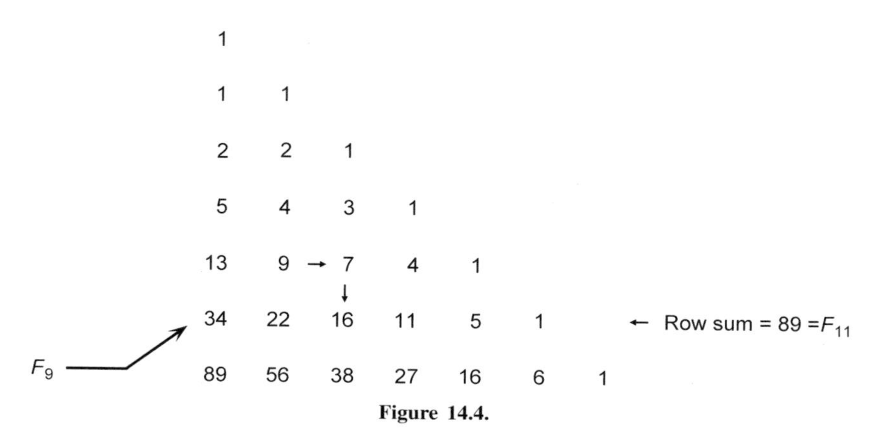

Figure 14.4.

L_{n+1} and $T_n(1, 2) = F_{n-2} + 2F_{n-3} = L_{n-2}$; this is consistent with Theorem 14.3.

Figure 14.4 shows yet another triangular array developed in 1971 by Hoggatt. This array possesses several interesting properties, in addition to the obvious ones:

- The first entry in row n is F_{2n-1}.
- Every internal entry is obtained by adding the number immediately above and the number to its left in the same row. For instance, $16 = 7 + 9$.
- The sum of the elements in row n is F_{2n+1} (see Exercise 2).

EXERCISES 14

1. Let $A(n, j)$ denote the element in row n and column j of the array in Figure 14.4, where $n, j \geq 0$. Define $A(n, j)$ recursively.
2. Prove that the sum of the elements in row n of the triangular array in Figure 14.4 is F_{2n+1}, where $n \geq 0$.

Use the array in Figure 14.5, developed in 1972 by Hoggatt, to prove the statements in Exercises 3–5, where $n \geq 0$.

1				
1	1			
3	1	1		
8	3	1	1	
21	8	3	1	1
55	21	8	3	1

Figure 14.5.

3. The nth row sum is F_{2n+1}.
4. If the columns are multiplied by 1, 2, 3, ... to the right, then the nth row sum is F_{2n+1}.
5. The sum of the elements in the nth rising diagonal is F_{n+1}^2.

Use the array in Figure 14.6, developed in 1977 by Hoggatt, to prove Exercises 6–8, where the Fibonacci numbers F_{2n+1} are written in staggered columns, where $n \geq 0$.

Figure 14.6.

6. Every row sum is F_{2n+2}.
7. The sum of every rising diagonal sum is $F_{n+1}F_{n+2}$.
8. Multiply the columns by 1, 2, 3, ... to the right. Then the nth row sum is $F_{2n+3} - 1$.

Let $S_n(a, b)$ denote the sum of the elements in row n in Figure 14.3.

9. Define $S_n(a, b)$ recursively.
10. Show that $S_n(a, b) = S_{n-1}(a, b) + S_{n-2}(a, b)$, where $n \geq 2$.
11. Prove that $S_n(a, b) = aF_{n+1} + bF_n$, where $n \geq 0$.

Let $T_n(a, b)$ denote the alternating sum of the elements in row n in Figure 14.3.

12. Define $T_n(a, b)$ recursively.
13. Prove that $T_n(a, b) = aF_{n-2} + bF_{n-3}$, where $n \geq 0$.

HOSOYA'S TRIANGLE

In 1976, H. Hosoya of Ochanomizu University in Tokyo introduced the triangular array in Figure 15.1, which is closely linked to Fibonacci numbers. We call it *Hosoya's triangle*. Besides the array being symmetric about the vertical line through the middle, the top two northeast and southeast diagonals consist of Fibonacci numbers. Every interior number can be obtained by adding the two previous numbers, on its diagonal; for example, $16 = 8 + 8 = 10 + 6$.

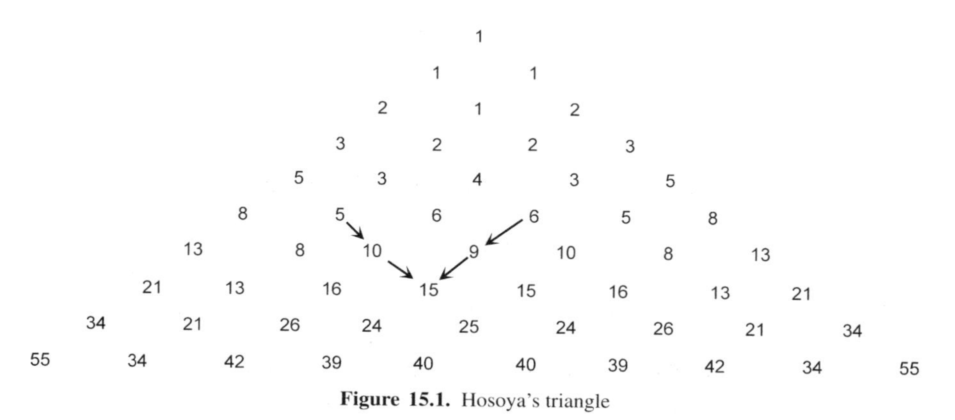

Figure 15.1. Hosoya's triangle

A RECURSIVE DEFINITION

In fact, we can define recursively every entry $H(n, j)$ of the array:

$$H(0, 0) = H(1, 0) = H(1, 1) = H(2, 1) = 1$$

$$H(n, j) = H(n-1, j) + H(n-2, j) \tag{15.1}$$
$$= H(n-1, j-1) + H(n-2, j-2) \tag{15.2}$$

where $n \geq j \geq 0$ and $n \geq 2$.

Since $H(n, 0) = H(n-1, 0) + H(n-2, 0)$, where $H(0, 0) = 1 = F_1$ and $H(1, 0) = 1 = F_2$, it follows that $H(n, 0) = F_{n+1}$; likewise, since $H(n, n) = H(n-1, n) + H(n-2, n)$, it follows that $H(n, n) = F_{n+1}$. Similarly, we can show that $H(n, 1) = H(n, n-1) = F_n$ (see Exercises 1–3).

Successive application of the recurrence relation (Eq. 15.1) yields:

$$\begin{aligned} H(n, j) &= H(n-1, j) + H(n-2, j) \\ &= [H(n-2, j) + H(n-3, j)] + H(n-2, j) \\ &= 2H(n-2, j) + H(n-3, j) \\ &= 2[H(n-3, j) + H(n-4, j)] + H(n-3, j) \\ &= 3H(n-3, j) + 2H(n-4, j) \end{aligned}$$

Continuing like this, we get a close link between $H(n, j)$ and Fibonacci numbers:

$$H(n, j) = F_{k+1}H(n-k, j) + F_k H(n-k-1, j) \tag{15.3}$$

where $1 \leq k \leq n-j-1$ (see Exercise 4). In particular, let $k = n-j-1$. Then

$$\begin{aligned} H(n, j) &= F_{n-j}H(j+1, j) + F_{n-j-1}H(j, j) \\ &= F_{n-j}F_{j+1} + F_{n-j-1}F_{j+1} \qquad \text{by Exercise 3} \\ &= F_{j+1}(F_{n-j} + F_{n-j-1}) \\ &= F_{j+1}F_{n-j+1} \end{aligned} \tag{15.4}$$

Thus every entry in the array is the product of two Fibonacci numbers.

For example, $H(7, 3) = 15 = 3 \cdot 5 = F_4 F_5$ and $H(9, 6) = 39 = 3 \cdot 13 = F_4 F_7$.

Since $H(n, j) = H(n, n-j)$, it follows from Eq. (15.4) that $H(n, j) = H(n, n-j) = F_{j+1}F_{n-j+1}$.

Let $n = 2m$ and $j = m$. Then Eq. 15.4 yields $H(2m, m) = F_{m+1}F_{m+1} = F_{m+1}^2$. Thus $H(2m, m)$ is the square of a Fibonacci number. In other words, the numbers along the vertical line through the middle are Fibonacci squares.

For example, $H(8, 4) = 25 = F_5^2$ and $H(10, 5) = 64 = F_6^2$.

A LINK BETWEEN $H(n, j)$ AND L_m

Using Eq. 15.4, we can compute $H(n, j)$ using Lucas numbers:

$$\begin{aligned} 5H(n, j) &= (\alpha^{j+1} - \beta^{j+1})(\alpha^{n-j+1} - \beta^{n-j+1}) \\ &= \alpha^{n+2} + \beta^{n+2} - \alpha^{j+1}\beta^{n-j+1} - \alpha^{n-j+1}\beta^{j+1} \end{aligned}$$

$$= (\alpha^{n+2} + \beta^{n+2}) + (\alpha\beta)^j(\alpha^{n-2j} + \beta^{n-2j})$$

$$H(n, j) = \frac{L_{n+2} + (-1)^j L_{n-2j}}{5} \tag{15.5}$$

For example, let $n = 10$ and $j = 3$. Then

$$\frac{L_{12} + (-1)^3 L_4}{5} = \frac{322 - 7}{5} = 63 = H(10, 3)$$

As a bonus, it follows from Eq. (15.5) that $L_{n+2} \equiv (-1)^{j-1} L_{n-2j} \pmod 5$. In particular, $L_{2m} \equiv 2(-1)^m \pmod 5$ and $L_{2m+1} \equiv (-1)^m \pmod 5$.

For example, $L_{12} = 322 \equiv 2 \equiv 2(-1)^6 \pmod 5$ and $L_{15} \equiv 1364 \equiv -1 \equiv (-1)^7 \pmod 5$.

A MAGIC RHOMBUS

Notice that Hosoya's triangle was constructed using four initial conditions, that is, four 1s, and they form a rhombus. In fact, we can employ any rhombus with vertices $H(i, j)$, $H(i-1, j-1)$, $H(i-2, j-1)$, and $H(i-1, j)$ to generate their nearest neighbors.

For example, consider the rhombus in Figure 15.2, where the letters A through H represent the numbers 4, 6, 9, 6, 5, 25, 5, and 1, respectively. Then $F = A + B + C + D$, $H = A + D - B - C$, $E = C + D - A - B$, and $G = B + D - A - C$ (see Exercise 7). We can represent these facts pictorially, as Figure 15.3 shows.

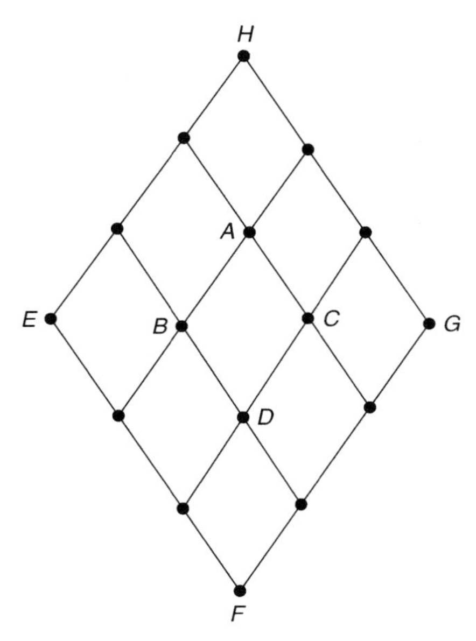

Figure 15.2.

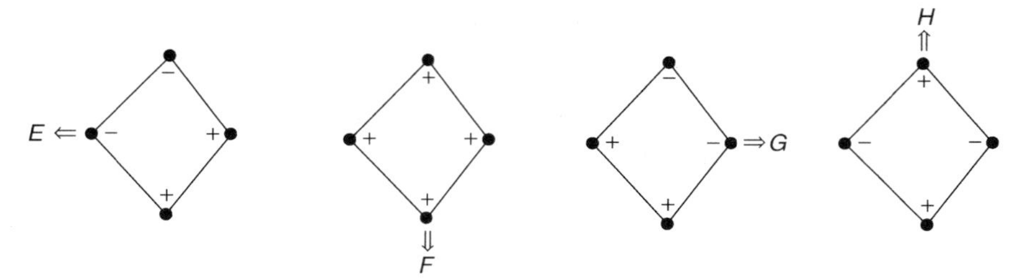

Figure 15.3.

ADDITIONAL FORMULAS

More generally, we have the following additional formulas:

$$H(n-1, j-2) = H(n, j) + H(n-1, j) - H(n-1, j-1) - H(n-2, j-1)$$
$$H(n+2, j+1) = H(n, j) + H(n-1, j) + H(n-1, j-1) + H(n-2, j-1)$$
$$H(n-1, j+1) = H(n-1, j-1) + H(n-1, j) - H(n-2, j-1) - H(n, j)$$

and

$$H(n-4, j-2) = H(n-2, j-1) + H(n-1, j) - H(n-1, j-1) - H(n, j)$$

See Figure 15.4.

Since $(F_{j+1}F_{n-j+1})(F_jF_{n-j}) = (F_jF_{n-j+1})(F_{j+1}F_{n-j})$, it follows by Eq. (15.4) that

$$H(n, j) \cdot H(n-2, j-1) = H(n-1, j-1) \cdot H(n-1, j) \tag{15.6}$$

that is, the product of the opposite vertices A and D in the rhombus $ABCD$ equals that of the remaining two opposite vertices B and C.

For example, consider the rhombus formed by 15, 24, 40, and 25. Clearly, $15 \cdot 40 = 24 \cdot 25$. Likewise, $8 \cdot 26 = 13 \cdot 16$.

We can write Eq. (15.6) as

$$\{[H(n, j) \div H(n-1, j)] * H(n-2, j-1)\} \div H(n-1, j-1) = 1 \tag{15.7}$$

See Figure 15.5.

Interestingly enough, we can extend Eq. (15.6) and hence Eq. (15.7) to the corners of any parallelogram:

$$H(n, j) \cdot H(n-k-l, j-k) = H(n-k, j-k) \cdot H(n-l, j)$$

See Figure 15.6.

For example, consider the array of parallelograms in Figure 15.7. Notice that $40 \cdot 4 = 16 \cdot 10$ and $15 \cdot 10 = 25 \cdot 6$. The other products can be verified similarly.

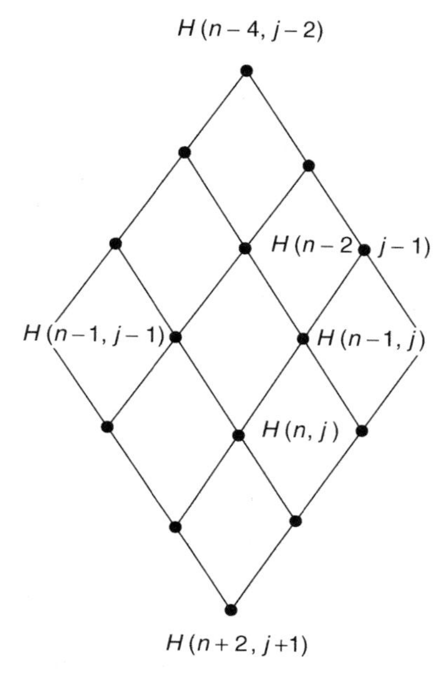

Figure 15.4.

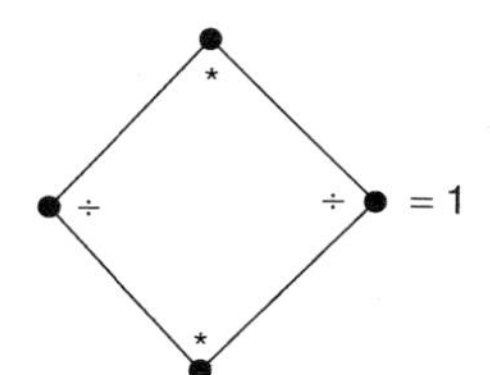

Figure 15.5.

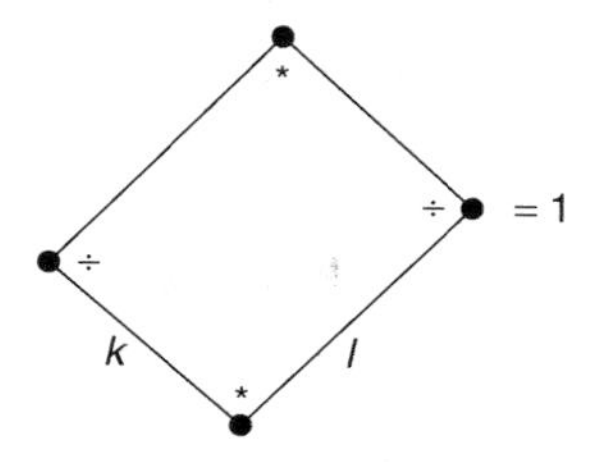

Figure 15.6.

Consider the downward pointing triangles with vertices belonging to two adjacent rows. For example, consider the adjacent rows in Figure 15.8. The sum of their vertices is a constant, namely, 34.

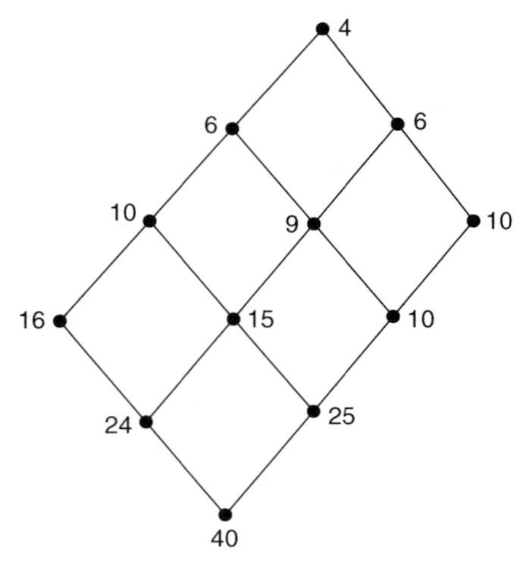

Figure 15.7.

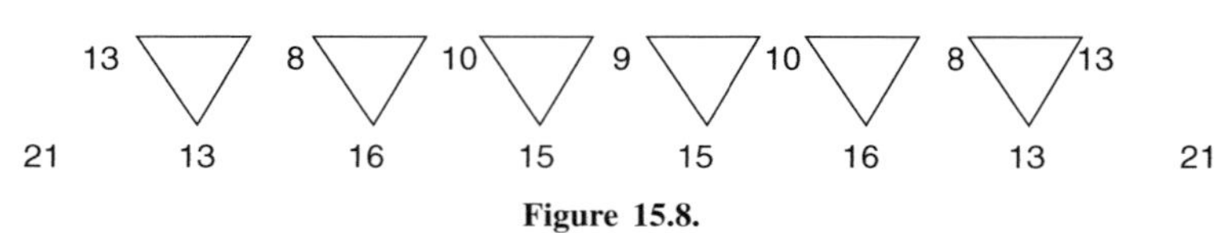

Figure 15.8.

More generally, $H(n, j) + H(n-1, j) + H(n-1, j-1)$ is a constant for every n (see Figs. 15.9 and 15.10, and see Exercise 8).

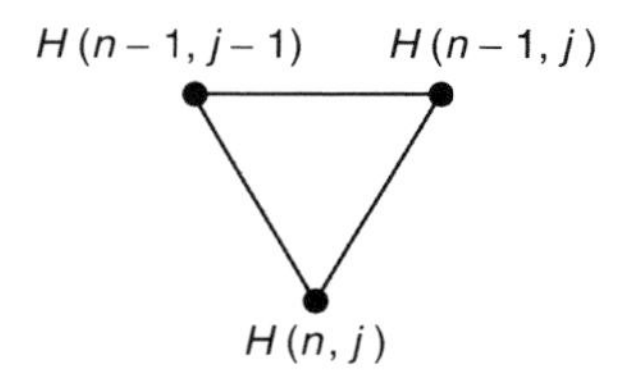

Figure 15.9.

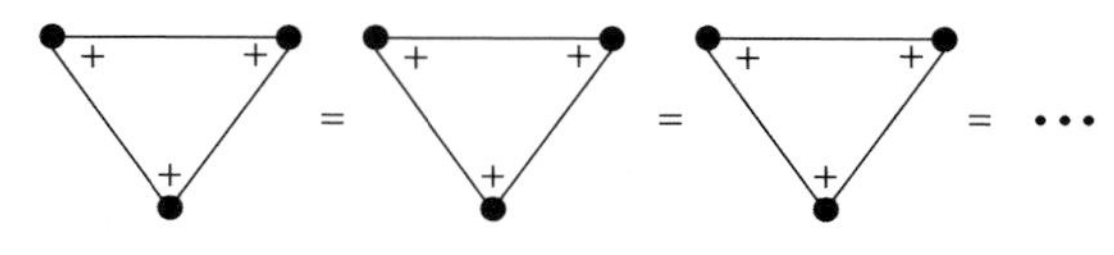

Figure 15.10.

In particular,

$$\begin{aligned} H(n, 0) + H(n-1, 0) + H(n-1, -1) &= F_{n+1} + F_n + 0 \\ &= F_{n+2} \end{aligned}$$

Thus

$$H(n, j) + H(n-1, j) + H(n-1, j-1) = F_{n+2} \tag{15.8}$$

In words, the magic constant for the downward-pointing triangle with lowest vertex on row n is F_{n+2}.

For instance, the constant for the triangles in Figure 15.8 is $34 = F_9$, as observed earlier.

Using the recurrence relation (Eq. 15.1), we can write Eq. (15.8) as

$$H(n, j) + H(n-2, j-1) = F_{n+1} \tag{15.9}$$

Thus the sum of any two vertical neighbors is a constant for a horizontal slide. That is, the sum of the north and south vertices in a magic rhombus is a Fibonacci number, as Figure 15.11 shows.

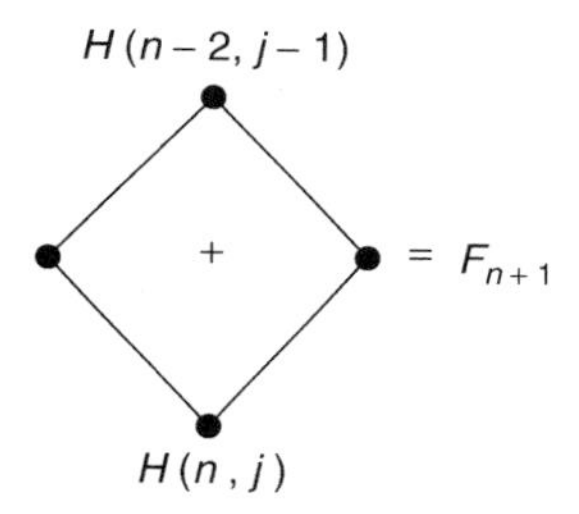

Figure 15.11.

For instance, the sum of the north and south vertices in the rhombus in Figure 15.12 is $25 + 64 = 89 = F_{11}$.

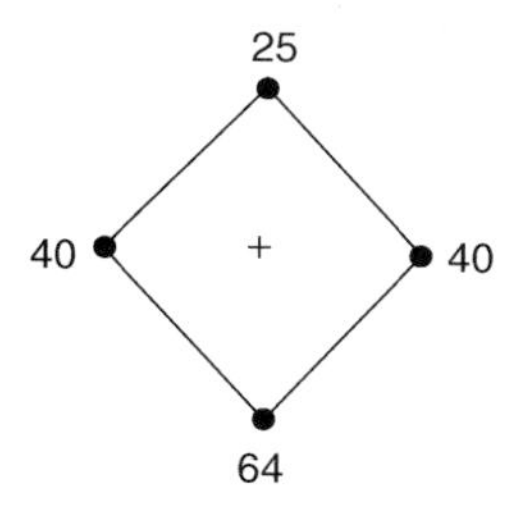

Figure 15.12.

It follows from Eqs. (15.8) and (15.9) that

$$H(n, j) + H(n, j-1) - H(n-1, j-1) = F_{n+1} \tag{15.10}$$

(see Exercise 9). That is, the sum of the two lower vertices of an upward-pointing triangle minus the vertex in row $n-1$ is F_{n+1} (See Figs. 15.13–15.15).

Using Eq. (15.9), we can show that

$$H(n, j) + H(n-6, j-3) = 2F_{n-1} \tag{15.11}$$

See Exercise 10 and Figure 15.16. For example, $H(10, 4) + H(4, 1) = 65 + 3 = 68 = 2F_9$.

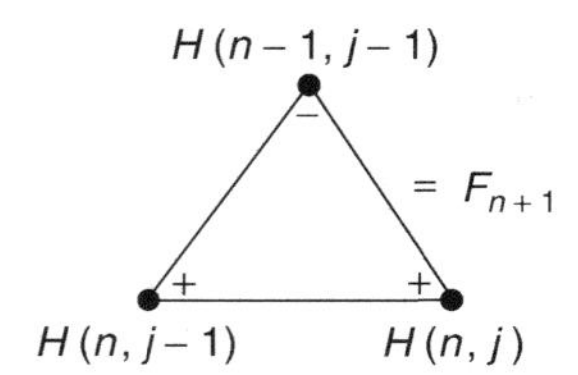

Figure 15.13.

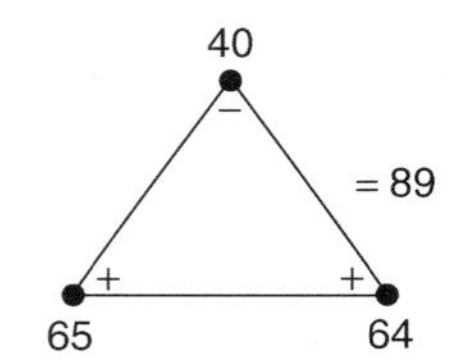

Figure 15.14.

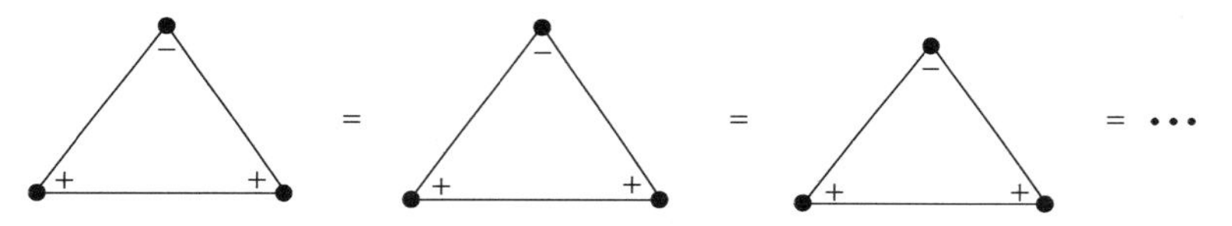

Figure 15.15.

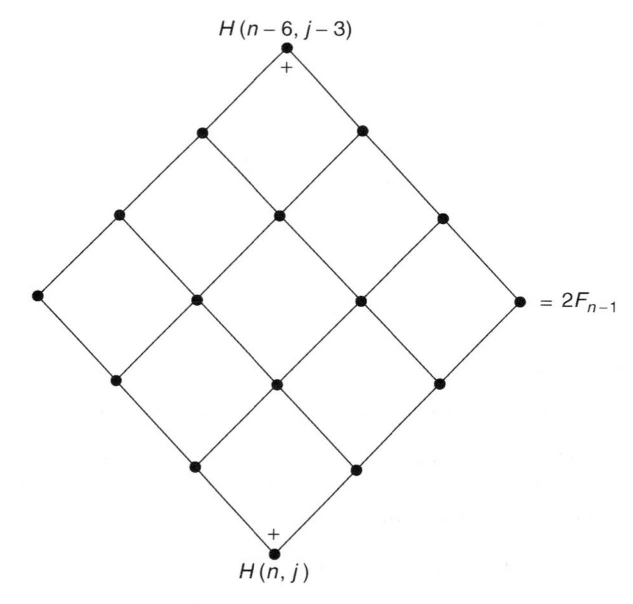

Figure 15.16.

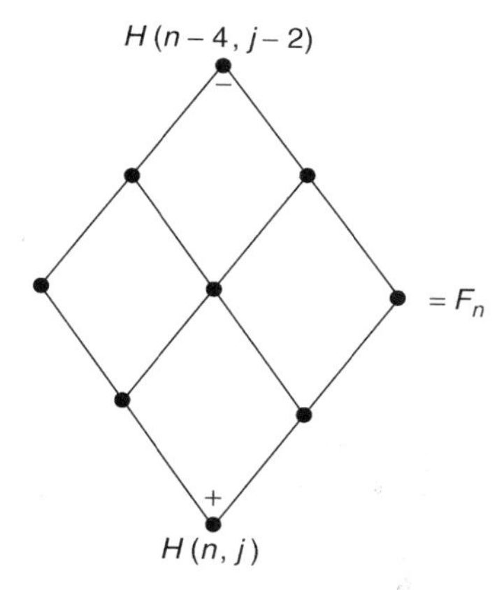

Figure 15.17.

Equation (15.10) yields yet another one:

$$H(n, j) - H(n-4, j-2) = F_n \tag{15.12}$$

See Exercise 11 and Figure 15.17.

For instance, $H(10, 4) - H(6, 2) = 65 - 10 = 55 = F_{10}$.

EXERCISES 15

Use Hosoya's triangle to answer each exercise.
Prove each, where $n \geq 1$.

1. $H(n, 1) = F_n$
2. $H(n, j) = H(n, n - j)$
3. $H(n, n - 1) = F_n$
4. $H(n, j) = F_{k+1}H(n - k, j) + F_k H(n - k - 1, j), 1 \leq k \leq n - j - 1$
5. $L_{2m} \equiv 2(-1)^m \pmod 5$
6. $L_{2m+1} \equiv (-1)^m \pmod 5$
7. Using Figure 15.2 show that $F = A + B + C + D$, $H = A + D - B - C$, $E = C + D - A - B$, and $G = B + D - A - C$.

Prove each.

8. $H(n, j) + H(n - 1, j) + H(n - 1, j - 1)$ is a constant for every row n.
9. $H(n, j) + H(n, j - 1) - H(n - 1, j - 1) = F_{n+1}$
10. $H(n, j) + H(n - 6, j - 3) = 2F_{n-1}$
11. $H(n, j) - H(n - 4, j - 2) = F_n$

16

DIVISIBILITY PROPERTIES

In Chapter 5 we found that $F_{2n} = F_n L_n$, so $F_n | F_{2n}$. Can we generalize this? In other words, under what conditions does $F_i | F_j$? The next theorem shows that if $i | j$, then $F_i | F_j$.

Theorem 16.1. $F_m | F_{mn}$.

Proof. (by PMI) The given statement is clearly true when $n = 1$. Now assume it is true for all integers 1 through k, where $k \geq 1$: $F_m | F_{mi}$ for every i, where $1 \leq i \leq k$.

To show that $F_m | F_{m(k+1)}$, we invoke Identity (32.3):

$$F_{r+s} = F_{r-1} F_s + F_r F_{s+1}$$

$$\therefore \quad F_{m(k+1)} = F_{mk+m} = F_{mk-1} F_m + F_{mk} F_{m+1}$$

Since $F_m | F_{mk}$, by the induction hypothesis, it follows that $F_m | F_m(k+1)$.

Thus, by the strong version of the PMI, the result is true for all integers $n \geq 1$. ■

For example, $F_6 = 8$, and $F_{24} = 46,368$. Since $6|24$, it follows by the theorem that $8|46368$, which can be verified.

Corollary 16.1. Every mth Fibonacci number is divisible by F_m. ■

For example, every third Fibonacci number is even and every fifth is divisible by $F_5 = 5$, that is, F_5, F_{10}, F_{15}, F_{20}, ... are all divisible by 5. Likewise, F_6, F_{12}, F_{18}, F_{24}, ... are all divisible by $F_6 = 8$.

In 1964, L. Carlitz of Duke University established the converse of this theorem using the identity

$$F_n = F_{n-m+1} F_m + F_{n-m} F_{m-1} \tag{16.1}$$

where $r \geq s \geq 1$.

Theorem 16.2. If $F_m | F_n$, then $m | n$.

Proof. By the division algorithm, $n = qm + r$, where $0 \leq r < m$. Suppose $F_m | F_n$. Then, by Theorem A.10 and Identity (16.1), $F_m | F_{n-m} F_{m-1}$. But $(F_m, F_{m-1}) = 1$, so $F_m | F_{n-m}$.

Similarly, $F_m | F_{n-2m}$. Continuing like this, $F_m | F_{n-qm}$, that is, $F_m | F_r$. This is impossible unless $r = 0$. $\therefore n = qm$. Thus $F_m | F_n$ implies $m | n$. ■

Corollary 16.2. $F_m | F_n$ if and only if $m | n$. ■

This follows from Theorems 16.1 and 16.2.

Corollary 16.3. If $(m, n) = 1$, then $F_m F_n | F_{mn}$.

Proof. By Theorem 16.1, $F_m | F_{mn}$ and $F_n | F_{mn}$. Therefore, $[F_m, F_n] | F_{mn}$. But $(F_m, F_n) = F_{(m,n)} = F_1 = 1$, so $[F_m, F_n] = F_m F_n$. Thus $F_m F_n | F_{mn}$. ■

For example, $(4, 7) = 1$, $F_4 = 3$, $F_7 = 13$, and $F_{28} = 317,811$. We can verify that $3 \cdot 13 | 317811$; that is, $F_4 F_7 | F_{28}$.

What are the chances that F_n is divisible by F_k, where $k \geq 3$? This problem, studied in 1964 by F. D. Parker of the University of Alaska, is pursued in the following example.

Example 16.1. Find the probability that a Fibonacci number F_n is divisible by another Fibonacci number F_k, where $k \geq 3$.

Solution. By Corollary 16.3, $F_3 | F_{3m}$, that is, every third Fibonacci number is divisible by 3. So the probability that F_n is divisible by 2 is 1/3. The probability that F_n is divisible by F_4 is 1/4; so the probability that F_n is divisible by 3, but not by 2, is $1/4 \cdot 2/3 = 2/(3 \cdot 4)$. Likewise, the probability that F_n is divisible by 5, but not by 2 or 3, is $2/(4 \cdot 5)$. In general, the probability that F_n is divisible by F_k and not by F_j, is $2/(k-1)k$, where $3 \leq j < k$.

Thus, by the addition principle, the probability that F_n is divisible by F_k, where $k \geq 3$, is

$$\begin{aligned}\sum_{2}^{k-1} \frac{2}{i(i+1)} &= 2 \sum_{2}^{k-1} \left(\frac{1}{i} - \frac{1}{i+1} \right) \\ &= 2 \left(\frac{1}{2} - \frac{1}{k} \right) = \frac{k-2}{k}\end{aligned}$$

As $k \to \infty$, this probability approaches unity. ■

In Corollary 5.2 and Example 9.5, we found that $(F_{n-1}, F_n) = 1$ for every $n \geq 1$. The next lemma generalizes this result in the light of Theorem 16.1.

Lemma 16.1. $(F_{qn-1}, F_n) = 1$.

Proof. Let $d = (F_{qn-1}, F_n)$. Then $d|F_{qn-1}$ and $d|F_n$. Since $F_n|F_{qn}$ by Theorem 16.1, $d|F_{qn}$. Thus $d|F_{qn-1}$ and $d|F_{qn}$. But $(F_{qn-1}, F_{qn}) = 1$, by Corollary 5.2. Therefore $d|1$, so $d = 1$. Thus $(F_{qn-1}, F_n) = 1$. ■

We are now ready for the next lemma. Its proof employs Identity (34.9).

Lemma 16.2. Let $m = qn + r$. Then $(F_m, F_n) = (F_n, F_r)$.

Proof.

$$\begin{aligned}
(F_m, F_n) &= (F_{qn+r}, F_n) \\
&= (F_{qn-1}F_r + F_{qn}F_{r+1}F_n, F_n) \qquad \text{by Identity (34.9)} \\
&= (F_{qn-1}F_r, F_n) \\
&= (F_r, F_n) \qquad \text{by Lemma 16.1} \\
&= (F_n, F_r)
\end{aligned}$$

■

The next theorem shows that the greatest common divisor (gcd) of two Fibonacci numbers is always a Fibonacci number. Its proof uses the Euclidean algorithm, Theorem 16.1, and this lemma.

Theorem 16.3. $(F_m, F_n) = F_{(m,n)}$.

Proof. Suppose $m \geq n$. Applying the Euclidean algorithm with m as the dividend and n as the divisor, we get the following sequence of equations:

$$\begin{aligned}
m &= q_0 n + r_1 && 0 \leq r_1 < n \\
n &= q_1 r_1 + r_2 && 0 \leq r_2 < r_1 \\
r_1 &= q_2 r_2 + r_3 && 0 \leq r_3 < r_2 \\
&\vdots \\
r_{n-2} &= q_{n-1} r_{n-1} + r_n && 0 \leq r_n < r_{n-1} \\
r_{n-1} &= q_n r_n + 0
\end{aligned}$$

By Lemma 16.2, $(F_m, F_n) = (F_n, F_{r_1}) = (F_{r_1}, F_{r_2}) = \cdots = (F_{r_{n-1}}, F_{r_n})$. But $r_n|r_{n-1}$, so $F_{r_n}|F_{r_{n-1}}$, by Theorem 16.1. Therefore, $(F_{r_{n-1}}, F_{r_n}) = F_{r_n}$. Thus $(F_m, F_n) = F_{r_n}$. But, by Euclidean algorithm, $r_n = (m, n)$; therefore, $(F_m, F_n) = F_{(m,n)}$. ■

For example, $(F_{12}, F_{18}) = F_{(12,18)} = F_6 = 8$. That is, $(144, 2584) = 8$. We can verify this using any of the traditional methods or by using the *gcd* function in the MATH menu in a TI-86 calculator.

Now we turn to an alternate proof of this theorem, given by G. Michael of Washington State University in 1964.

AN ALTERNATE PROOF

Let $d = (m, n)$ and $d' = (F_m, F_n)$. By Theorem 16.1, $F_d | F_m$ and $F_d | F_n$, so $F_d | d'$. Since $d = (m, n)$, there exist integers a and b such that $d = am + bn$. Since $d, m, n > 0$, either $a \leq 0$ or $b \leq 0$. Suppose $a \leq 0$. Let $a = -k$, where $k \geq 0$. Then $bn = d + km$.

By Identity (34.9),

$$F_{bn} = F_{d+km} = F_{d-1} F_{km} + F_d F_{km+1} \tag{16.2}$$

Since $d' | F_m$, $d' | F_{km}$ by Theorem 16.1. Now $d' | F_n$ and $F_n | F_{bn}$, so $d' | F_{bn}$. Thus $d' | F_{km}$ and $d' | F_{bn}$. Therefore, by Eq. (16.2), $d' | F_d F_{km+1}$. But $(d', F_{km+1}) = 1$, since $d' | F_{km}$ and $(F_{km}, F_{km+1})) = 1$; therefore, $d' | F_d$.

Thus, $F_d | d'$ and $d' | F_d$, so $d' = F_d$. In other words, $(F_m, F_n) = F_{(m,n)}$.

Corollary 16.4. If m and n are relatively prime, then so are F_m and F_n. ■

For instance, $(12, 25) = 1$, so $(F_{12}, F_{25}) = (144, 75025) = 1$.

Theorem 16.2 follows easily from Theorem 16.3.

Corollary 16.5. If $F_m | F_n$, then $m | n$.

Proof. Suppose $F_m | F_n$. Then $(F_m, F_n) = F_m = F_{(m,n)}$, by Theorem 16.3; $\therefore m = (m, n)$. Thus $m | n$. ■

Corollary 16.5 coupled with Theorem 16.1 provides an alternate proof of Corollary 16.2.

Theorem 16.5 has an intriguing by-product. In 1965, M. Wunderlich of the University of Colorado employed the theorem to provide a beautiful proof that there are infinitely primes, a fact that is universally known. The next corollary leads us to that proof.

Corollary 16.6. There are infinitely many primes.

Proof. Suppose there is only a finite number of primes, $p_1, p_2, \ldots,$ and p_k. Then consider the Fibonacci numbers $F_{p_1}, F_{p_2}, \ldots,$ and F_{p_k}. Clearly, they are pairwise relatively prime. Since there are only k primes, each of these Fibonacci numbers has exactly one prime factor, that is, each is a prime. This is a contradiction, since $F_{19} = 4181 = 37 \cdot 113$. Thus our assumption that there are only finitely many primes is false. In other words, thus there are infinitely many primes. ■

In 1966, L. Weinstein of the Massachusetts Institute of Technology established the following divisibility property, which is a direct consequence of Erdös's theorem in the Appendix and Theorem 16.3.

Theorem 16.4. (Weinstein, 1966). Every set S of $n+1$ Fibonacci numbers, selected from $F_1, F_2, \ldots, F_{2n}$, contains two elements such that one divides the other.

Proof. Let $S = \{F_{a_1}, F_{a_2}, \ldots, F_{a_n}, F_{a_{n+1}}\}$, where $1 \leq a_i \leq 2n$ and $1 \leq i \leq n+1$. Since $A = \{a_1, a_2, \ldots, a_n, a_{n+1}\} \subseteq \{1, 2, \ldots, 2n\}$, by Erdös' theorem, A contains two elements a_i and a_j such that $a_i | a_j$. Then $(a_i, a_j) = a_i$, so $(F_{a_i}, F_{a_j}) = F_{(a_i, a_j)} = F_{a_i}$, by Theorem 16.3. Thus $F_{a_i} | F_{a_j}$, as desired. ■

A quick look at Lucas numbers shows that every third Lucas number is even, that is, $2|L_{3n}$. This is, in fact, always true (see Exercise 40).

The next two divisibility properties were discovered by L. Carlitz in 1964.

Theorem 16.5. $L_m | F_n$ if and only if $2m|n$, where $m \geq 2$. ■

For example, 10|20, so $L_5 | F_{20}$; that is, $11|6765$.

Theorem 16.6. $L_m | L_n$ if and only if $n = (2k-1)m$, where $m \geq 2$ and $k \geq 1$. ■

For example, let $m = 4$, and $n = 3 \cdot 4 = 12$. We have $L_4 = 7$ and $L_{12} = 322$. Clearly, $L_4 | L_{12}$.

In 1965, George C. Cross and Helen G. Renzi of Williamtown Public Schools in Massachusetts proved that if the ratio $a{:}b = 2{:}3$, then $[a, b] - (a, b) = a + b$. For example, let $a = 12$ and $b = 18$. Then $[a, b] - (a, b) = 36 - 6 = 30 = 12 + 18$. Cross and Renzi also proved that if $a{:}b = 3{:}5$, then $[a, b] + (a, b) = 2(a + b)$. For instance, let $a = 45$ and $b = 75$. Then $[a, b] + (a, b) = 225 + 15 = 240 = 2(45 + 75)$.

More generally, suppose $a{:}b = F_n{:}F_{n+1}$ or $a{:}b = L_n{:}L_{n+1}$. How are $[a, b]$, (a, b), and $a + b$ related? These two questions were investigated two years later by G. F. Freeman of Williams College. The next two theorems were discovered by him.

Theorem 16.7. (Freeman, 1967)

1. Let $a{:}b = F_n{:}F_{n+1}$. Then $(a + b)F_{n-1} = [a, b] + (-1)^n (a, b)$, where $n \geq 2$.

2. Let $(c, d) = 1$ such that $a{:}b = c{:}d$. Let $(a + b)F_{n-1} = [a, b] + (-1)^n (a, b)$, where $n \geq 3$. Then the number of solutions of the ratio $c{:}d$ equals one-half the number of positive factors of $F_n F_{n-2}$, one of them being $F_n F_{n+1}$.

Proof.

1. Let $a{:}b = F_n{:}F_{n+1}$. Then, since $(F_n, F_{n+1}) = 1, a = F_n k, b = F_{n+1}k, (a, b) = k, [a, b] = F_n F_{n+1} k$ for some positive integer k.

$$\begin{aligned}\therefore \quad (a+b)F_{n-1} &= F_{n-1}(F_n + F_{n+1})k = F_{n-1}F_{n+2}k \\ &= (F_{n+1} - F_n)F_{n+2}k = F_{n+1}(F_n + F_{n+1})k - F_n F_{n+2}k \\ &= F_n F_{n+1}k + (F_{n+1}^2 - F_n F_{n+2})k \\ &= [a, b] + (-1)^n(a, b) \qquad \text{by Cassini's rule}\end{aligned}$$

2. Let $a{:}b = c{:}d$, where $(c, d) = 1$. Then $a = ck$; $B = dk, (a, b) = k$; and $[a, b] = cdk$ for some positive integer k. Since $(a + b)F_{n-1} = [a, b] + (-1)^n(a, b)$, we have

$$(c + d) = cd + (-1)^n$$

This yields

$$\begin{aligned} c &= \frac{dF_{n-1} - (-1)^n}{d - F_{n-1}} \\ &= F_{n-1} + \frac{F_{n-1}^2 - (-1)^n}{d - F_{n-1}} \\ &= F_{n-1} + \frac{F_n F_{n-2}}{d - F_{n-1}} \end{aligned} \tag{16.3}$$

If $0 < d < F_{n-1}$, then $c < 0$; so $d > F_{n-1}$. Since c is an integer, $d - F_{n-1} | F_n F_{n-2}$. Thus Eq. (16.3) yields a value of c for every positive factor of $F_n F_{n-2}$. But, if $c = A, d = B$ is a solution of the ratio $c{:}d$, then so is $c = B, d = A$. Thus the number of distinct values of the ratio $c{:}d$ equals the number of positive factors of $F_n F_{n-2}$.

In particular, let $d = F_{n+1}$. Then

$$\begin{aligned} c &= F_{n-1} + \frac{F_n F_{n-2}}{F_{n+1} - F_{n-1}} \\ &= F_{n-1} + \frac{F_n F_{n-2}}{F_n} = F_n \end{aligned}$$

Thus $c{:}d = F_n{:}F_{n+1}$ is also a value of the ratio. ■

The following example demonstrates this theorem.

Example 16.2.

1. Let $a{:}b = F_9{:}F_{10} = 34{:}55$, so $n = 9$. Let $a = 238$, $b = 385$, so $a{:}b = 34{:}55$:

$$[a, b] + (-1)^n(a, b) = 13,090 - 7 = 13,083$$
$$= (238 + 385) \cdot 21 = (a + b)F_8$$

2. Since $(a + b)F_8 = [a, b] + (-1)^9(a, b)$, it follows that

$$c = F_8 + \frac{F_9 F_7}{d - F_8}$$
$$= 21 + \frac{34 \cdot 13}{d - 21} = 21 + \frac{442}{d - 21}$$

Since $442 = 2 \cdot 13 \cdot 17$, 442 has eight positive factors: 1, 2, 13, 17, 26, 34, 221, and 442. So d has eight possible values: 22, 23, 34, 38, 47, 55, 242, and 463. Consequently, the various values of $c : d$ are $463 : 22, 242 : 23, 55 : 34, 47 : 38, 38 : 47, 34 : 55, 23 : 242$, and $463 : 22$. Since one-half of them are duplicates, the four distinct values of $c : d$ are $38 : 47, 34 : 55, 23 : 242$, and $22 : 463$, keeping the numerator to be smaller. Notice that one of the ratios is $34 : 55 = F_9 : F_{10}$, as expected. ■

Theorem 16.6 has a counterpart for Lucas numbers. Its proof requires the following lemma. We leave its proof as an exercise.

Lemma 16.3. $F_{2n-1} = F_{n+1}L_{n+2} - L_nL_{n+1}, n \geq 2$. ■

Theorem 16.8. (Freeman, 1967)

1. Let $a{:}b = L_n{:}L_{n+1}$. Then $(a + b)F_{n+1} = [a, b] + (a, b)F_{2n-1}, n \geq 2$.
2. Let $a{:}b = F_{n-2}{:}F_{n-1}$. Then $(a + b)F_{n+1} = [a, b] + (a, b)F_{2n-1}, n \geq 3$.
3. Let $(c, d) = 1$ such that $a{:}b = c{:}d$. If $(a + b)F_{n+1} = [a, b] + (a, b)F_{2n-1}$, where $n \geq 2$, then the ratios $c{:}d$ are determined by the positive factors of $F_{n+1}^2 - F_{2n-1}$, one of them being $L_n{:}L_{n+1}$. For $n \geq 3$, $F_{n-2}{:}F_{n-1}$ is also a solution.

Proof.

1. Let $a{:}b = L_n{:}L_{n+1}$. Since $(L_n, L_{n+1}) = 1, a = kL_n, b = kL_{n+1}, (a, b) = k$, and $[a, b] = L_n L_{n+1} k$ for some positive integer k. Then

$$\begin{aligned}(a+b)F_{n+1} &= (L_n + L_{n+1})kF_{n+1} = F_{n+1}L_{n+2}k \\ &= (F_{2n-1} + L_n L_{n+1})k \\ &= [a, b] + (a, b)F_{2n-1}\end{aligned}$$

as desired.

2. Suppose $a{:}b = F_{n-2}{:}F_{n-1}$. Then $a = kF_{n-2}, b = kF_{n-1}, (a, b) = k$, and $[a, b] = F_{n-1}{:}F_{n-2}k$ for some positive integer k. Then

$$\begin{aligned}(a+b)F_{n+1} &= (F_{n-2} + F_{n-1})kF_{n+1} = F_n F_{n+1} k \\ &= (F_{2n-1} + F_{n-1}F_{n-2})k\end{aligned}$$

since $$\begin{aligned}F_{2n-1} &= F_n F_{n+1} - F_{n-2}F_{n-1} \\ &= [a, b] + (a, b)F_{2n-1}\end{aligned}$$

again as desired.

3. Let $a{:}b = c{:}d$, where $(c, d) = 1$. As before, $a = ck, b = dk, (a, b) = k$, and $[a, b] = cdk$ for some positive integer k. Since $(a + b)F_{n+1} = [a, b] + (a, b)F_{2n-1}$,

$$\begin{aligned}(c+d)F_{n+1} &= cd + F_{2n-1} \\ c &= \frac{dF_{n+1} - F_{2n-1}}{d - F_{n+1}} \qquad (16.4) \\ &= F_{n+1} + \frac{F_{n+1}^2 - F_{2n-1}}{d - F_{n-1}}\end{aligned}$$

Since c and d are positive integers, it follows that the ratio $c{:}d$ is determined by the positive factors of $F_{n+1}^2 - F_{2n-1}$.

In particular, let $d = F_{n+1}$. Then, by Lemma 16.3,

$$\begin{aligned}c &= F_{n+1} + \frac{F_{n+1}^2 - F_{n+1}L_{n+2} + L_n L_{n+1}}{L_{n+1} - F_{n+1}} \\ &= \frac{F_{n+1}(L_{n+1} - L_{n+2}) + L_n L_{n+1}}{L_{n+1} - F_{n+1}} \\ &= \frac{L_n L_{n+1} - L_n F_{n+1}}{L_{n+1} - F_{n+1}} = L_n\end{aligned}$$

Thus L_n:L_{n+1} is a solution of the ratio c:d. (By symmetry, L_{n+1}:L_n is also a solution.)

Unlike Theorem 16.7, not all solutions are obtained by considering the case $d > F_{n+1}$. For instance, let $d = F_{n-1}$. Then, by Eq. (16.4),

$$\begin{aligned}
c &= \frac{F_{n-1}F_{n+1} - F_{2n-1}}{F_{n-1} - F_{n+1}}\\
&= \frac{F_{n-1}F_{n+1} - (F_nF_{n+1} - F_{n-2}F_{n-1})}{-F_n}\\
&= -\frac{F_{n+1}(F_{n-1} - F_n) + F_{n-2}F_{n-1}}{F_n}\\
&= -\frac{F_{n-2}F_{n+1} + F_{n-2}F_{n-1}}{F_n}\\
&= \frac{F_{n-2}(F_{n-1} - F_{n+1})}{F_n} = F_{n-2}
\end{aligned}$$

Thus F_{n-2}:F_{n-1} is also a solution of the ratio. ■

The next example illustrates this theorem.

Example 16.3. Let $n = 8$. We have $F_{n+1} = F_9 = 34$ and $F_{2n-1} = F_{15} = 610$.

1. Let a:$b = L_n$:$L_{n+1} = L_8$:$L_9 = 47$:76. Let $a = 235$ and $b = 380$. Then

$$\begin{aligned}
[a, b] + (a, b)F_{2n-1} &= [235, 380] + (235, 380)\cdot 610\\
&= 17,860 + 5\cdot 610 = 20,910\\
&= (235 + 380)\cdot 34\\
&= (a + b)F_{n+1}
\end{aligned}$$

2. Let a:$b = F_{n-2}$:$F_{n-1} = F_6$:$F_7 = 8$:13. Let $a = 96$ and $b = 156$. Then

$$\begin{aligned}
[a, b] + (a, b)F_{2n-1} &= [96, 156] + 12\cdot 610\\
&= 8568 = (96 + 156)\cdot 34\\
&= (a + b)F_{n+1}
\end{aligned}$$

3. Let a:b = 180:204 = 15:17, where c:d = 15:17 and $(15, 17) = 1$. Then, by part 3 of Theorem 16.8,

$$\begin{aligned}
c &= F_{n+1} + \frac{F_{n+1}^2 - F_{2n-1}}{d - F_{n+1}}\\
&= 34 + \frac{34^2 - 610}{d - 34} = 34 + \frac{546}{d - 34}
\end{aligned}$$

Since $546 = 2 \cdot 3 \cdot 7 \cdot 13$, 546 has 16 positive factors: 1, 2, 3, 6, 7, 13, 14, 21, 26, 39, 42, 78, 91, 182, 273, and 546. The corresponding ratios are 35:580, 36:307, 37:216, 40:125, 41:112, 47:76, 48:73, 55:60, 60:55, 73:48, 76:47, 112:41, 125:40, 216:37, 307:36, and 580:35. These yield the eight distinct ratios $c{:}d$ with $(c, d) = 1$, namely, 7:116, 8:25, 11:12, 36:307, 37:216, 41:112, 47:76, and 48:73. Notice that 8:13 is also a solution. Among these ratios we find $L_8{:}L_9 = 47{:}76$ and $F_6{:}F_7 = 8{:}13$, as expected. ■

AN ALTERED FIBONACCI SEQUENCE

In 1971, Underwood Dudley and Bessie Tucker of DePauw University in Indiana investigated a slightly altered Fibonacci sequence, defined by $G_n = F_n + (-1)^n$, where $n \geq 1$. They made an interesting observation, as Table 16.1 shows: The 1st, 3rd, 5th, ... entries (see the circled numbers) in the (G_n, G_{n+1})-row are the 2nd, 4th, 6th, ... Fibonacci numbers; and the 2nd, 4th, 6th, ... entries are the 3rd, 5th, 7th, ... Lucas numbers.

TABLE 16.1.

n	1	2	3	4	5	6	7	8	9	10	11	12	13	14	15
G_n	0	2	1	4	4	9	12	22	33	56	88	145	232	378	609
(G_n, G_{n+1})		1		4		3		11		8		29		21	

To establish these two results, we need the following theorem.

Theorem 16.9. (Dudley and Tucker, 1971)

(1) $F_{4n} + 1 = F_{2n-1}L_{2n+1}$ (2) $F_{4n} - 1 = F_{2n+1}L_{2n-1}$

(3) $F_{4n+1} + 1 = F_{2n+1}L_{2n}$ (4) $F_{4n+1} - 1 = F_{2n}L_{2n+1}$

(5) $F_{4n+2} + 1 = F_{2n+2}L_{2n}$ (6) $F_{4n+2} - 1 = F_{2n}L_{2n+2}$

(7) $F_{4n+3} + 1 = F_{2n+1}L_{2n+2}$ (8) $F_{4n+3} - 1 = F_{2n+2}L_{2n+1}$

Proof. The proof requires the following identities from Chapter 5:

$$F_{m+n} + F_{m-n} = \begin{cases} F_n L_m & \text{if } n \text{ is odd} \\ F_m L_n & \text{otherwise} \end{cases}$$

$$F_{m+n} - F_{m-n} = \begin{cases} F_m L_n & \text{if } n \text{ is odd} \\ F_n L_m & \text{otherwise} \end{cases}$$

Then

$$\begin{aligned} F_{4n} + 1 &= F_{4n} + F_2 = F_{(2n+1)+(2n-1)} + F_{(2n+1)-(2n-1)} \\ &= F_{2n-1} L_{2n+1} \end{aligned}$$

and

$$\begin{aligned} F_{4n+1} + 1 &= F_{4n+1} + F_1 = F_{(2n+1)+2n} + F_{(2n+1)-2n} \\ &= F_{2n+1} L_{2n} \end{aligned}$$

The other formulas can be established similarly (see Exercises 59–64). ■

The following corollary, observed in 1971 by Hoggatt, follows easily from this theorem.

Corollary 16.7. (Hoggatt, 1971)

(1) $(F_{4n+1} + 1, F_{4n+2} + 1) = L_{2n}$ (2) $(F_{4n+1} + 1, F_{4n+3} + 1) = F_{2n+1}$

(3) $(F_{4n+1} - 1, F_{4n+2} - 1) = F_{2n}$ (4) $(F_{4n+1} - 1, F_{4n+3} - 1) = L_{2n+1}$

(5) $(F_{4n-1} - 1, F_{4n+1} - 1) = F_{2n}$ (6) $(F_{4n-1} + 1, F_{4n+1} + 1) = L_{2n}$

(7) $(F_{4n+3} + 1, F_{4n} - 1) = F_{2n+1}$ (8) $(F_{4n+3} + 1, F_{4n+2} - 1) = F_{2n}$

(9) $(F_{4n+4} - 1, F_{4n+3} - 1) = L_{2n+1}$

■

Although it is not yet known whether or not the Fibonacci sequence contains infinitely many primes, this theorem establishes their finiteness in the sequences $\{F_n + 1\}$ and $\{F_n - 1\}$, as the next corollary shows.

Corollary 16.8. $F_n + 1$ is composite if $n \geq 4$, and $F_n - 1$ is composite if $n \geq 7$.

Proof. When $n = 1$, $F_{4n} + 1 = 4$ is composite. When $n \geq 2$, it follows from Theorem 16.9 that $F_{4n+1} + 1$, $F_{4n+2} + 1$, and $F_{4n+3} + 1$ have nontrivial factors. Thus $F_n + 1$ is composite if $n \geq 4$. Likewise, $F_n - 1$ is composite if $n \geq 7$. ■

Notice that $F_n + 1$ is a prime if $n < 4$ and $F_n - 1$ is a prime if $n < 7$. The next corollary confirms the observation we made earlier.

Corollary 16.9. $(G_{4n}, G_{4n+1}) = L_{2n+1}$, $(G_{4n+1}, G_{4n+3}) = L_{2n+1}$, and $(G_{4n+2}, G_{4n+3}) = F_{2n+2}$, where $n \geq 1$.

Proof. By Theorem 16.9,

$$\begin{aligned} (G_{4n}, G_{4n+1}) &= (F_{4n+1} + 1, F_{4n+1} - 1) \\ &= (F_{2n-1} L_{2n+1}, F_{2n} L_{2n+1}) \end{aligned}$$

$$= L_{2n+1}(F_{2n-1}, F_{2n})$$
$$= L_{2n+1}$$

We can establish the other two parts similarly (see Exercises 65 and 66). ■

The next result also follows from Theorem 16.9 (see Exercises 67–69).

Corollary 16.10. Let $H_n = F_n - (-1)^n$. Then $(H_{4n}, H_{4n+1}) = F_{2n+1}$, $(H_{4n+1}, H_{4n+3}) = F_{2n+1}$, and $(H_{4n+2}, H_{4n+3}) = L_{2n+2}$, where $n \geq 1$. ■

The following divisibility properties were discovered in 1974 by V. E. Hoggatt, Jr., and G. E. Bergum, except those noted otherwise, where p and q are odd primes, and $k, m, n, r, t \geq 1$. We omit their proofs in the interest of brevity.

- If $p|L_n$, then $p^k|L_{np}^{k-1}$ (Carlitz and Bergum, independently).
- Let $p|L_{2\cdot 3^k}$ and $n = 2 \cdot 3^k p^t$. Then $n|L_n$.
- Let $p \neq q$, $p|L_n$, and $q|L_m$, where m and n are odd. Then $(pq)^k|L_{mn(pq)^{k-1}}$.
- Let $p, q > 3$, $p \neq q$, $p|L_{2\cdot 3^k}$, $q|L_{2\cdot 3^k}$, and $n = 2 \cdot 3^k p^t q^r$, where $r, t \geq 0$. Then $n|L_n$.
- If $p|L_n$, then $p^k|F_{2np^{k-1}}$ (Carlitz and Bergum, independently).
- Let $p \neq q$, $p|L_n$, and $q|L_m$, where m and n are odd. Then $(pq)^k|F_{2mn(pq)^{k-1}}$ (Carlitz and Bergum, independently).
- If $p|F_n$, then $p^k|F_{np^{k-1}}$ (Carlitz and Bergum, independently).
- Let $p \neq q$, $p|F_n$, and $q|F_m$. Then $(pq)^k|F_{mn(pq)^{k-1}}$.
- If $n = 3^m 2^{r+1}$, then $n|F_n$.
- Let $n = 2^{r+1}3^m 5^k$. Then $n|F_n$.
- Let $p > 3$ such that $p|F_{2^{r+1}3^m}$. Let $n = 2^{r+1}3^m p^k$. Then $n|F_n$.
- Let $s = 2^{r+1}3^m$, $p \neq q$, $p|F_s$, and $q|F_s$, and $n = sp^k q^t$, where $k, t \geq 0$. Then $n|F_n$.
- $2^{k+2}|F_{3\cdot 2^k}$.
- If n is odd, then $L_n = 4^t M$, where M is odd and $t = 0$ or 1.
- Let n be odd. Then $L_n = 4^t M$, where $t = 0$ or 1, and the prime factors of M are of the form $10m \pm 1$ (Hoggatt).

EXERCISES 16

Verify each.

1. $F_7|F_{21}$
2. $F_6|F_{24}$

3. $(F_{12}, F_{18}) = F_{(12,18)}$
4. $(F_{10}, F_{21}) = F_{(10,21)}$
5. $(F_{144}, F_{1925}) = 1$
6. $L_5|F_{10}$
7. $L_6|F_{24}$
8. $L_4|L_{12}$

Find each.

9. (F_{144}, F_{440})
10. (F_{80}, F_{100})
11. Prove Theorem 16.1 using Binet's formula.

Prove that $(F_n, F_{n+1}) = 1$ using each method.

12. PMI
13. The well-ordering principle (WOP).
14. Prove that $(L_n, L_{n-1}) = 1$.

Disprove each.

15. $m|n$ implies $L_m|L_n$.
16. $(L_m, L_n) = L_{(m,n)}$
17. Let $m, n \geq 3$. Then $F_m F_n | F_{mn}$.
18. Let $m, n \geq 2$. Then $L_m L_n | L_{mn}$.
19. $[F_m, F_n] = F_{[m,n]}$
20. $[L_m, L_n] = L_{[m,n]}$
21. Compute (F_n, L_n) for $1 \leq n \leq 10$ and make a conjecture about (F_n, L_n).
22. Identify the integers n for which $(F_n, L_n) = 2$.

Compute each.

23. $F_{(F_5, F_{10})}$
24. $F_{(F_6, F_{18})}$
25. $L_{(F_5, F_{15})}$
26. $L_{(F_6, L_6)}$
27. $F_{(F_5, F_{10}, F_{15})}$
28. $F_{(F_6, F_{18}, F_{21})}$
29. $L_{(F_5, F_{10}, F_{15})}$
30. $L_{(F_6, L_6, L_9)}$
31. Disprove: If n is a prime, then F_n is a prime.

Prove each.

32. If F_n is a prime, then n is a prime, where $n \geq 5$.
33. $2|F_{3n}$
34. $3|n$ if and only if $2|F_n$.
35. $4|n$ if and only if $3|F_n$.

36. $6|n$ if and only if $4|F_n$.
37. $5|n$ if and only if $5|F_n$.
38. $(F_a, F_b, F_c) = F_{(a,b,c)}$
39. $2|L_{3n}$.
40. $2|F_{3n}$
41. $(F_n, L_n) = 2$ if and only if $3|n$.
42. $L_n|L_{3n}$
43. $L_n|L_{(2k-1)n}$, where $k \geq 1$.
44. $(F_n, L_n) = 1$ or 2, where $n \geq 1$.
45. Using Identity (32.3), prove that $F_m|F_{mn}$.
46. There are n consecutive composite Fibonacci numbers, $n \geq 1$ (Litvack, 1964).

Verify that $(a + b)F_{n-1} = [a, b] + (-1)^n(a, b)$ for each ratio $a{:}b$.

47. 21:34
48. 89:144

Prove Lemma 16.3 using:

49. PMI.
50. Binet's formula.

Verify that $(a + b)F_{n+1} = [a, b] + (a, b)F_{2n-1}$ for each ratio $a{:}b$.

51. 11:18
52. 21:34
53. 72:116
54. 65:105

Prove each.

55. $(F_m, F_n) = (F_m, F_{m+n}) = (F_n, F_{m+n})$ (Brown, 1967).
56. $\{a_n\}$ is an increasing sequence, where $a_n = \sum_{d|n} F_d$, $n \geq 1$ (Lind, 1967).
57. If $k > 4$, then $F_k \nmid L_n$ (Brousseau, 1968).
58. Let $F_m|L_n$, where $0 < m < n$. Then $m = 1, 2, 3$, or 4 (Lang, 1973).

59–64. Establish the identities 2, and 4–8 in Theorem 16.8 (Dudley and Tucker, 1971).

Prove each, where $G_n = F_n + (-1)^n$, $H_n = F_n - (-1)^n$, and $n \geq 1$ (Dudley and Tucker, 1971).

65. $(G_{4n+1}, G_{4n+3}) = L_{2n+1}$
66. $(G_{4n+2}, G_{4n+3}) = F_{2n+2}$
67. $(H_{4n}, H_{4n+1}) = F_{2n+1}$
68. $(H_{4n+1}, H_{4n+3}) = F_{2n+1}$
69. $(H_{4n+2}, H_{4n+3}) = L_{2n+2}$

Use the function $g_n = F_{4n-2} + F_{4n} + F_{4n+2}$ for Exercises 70–72 (Grassl, 1971a).

70. Define g_n recursively.
71. Prove that $12|g_n$ for every $n \geq 0$.

72. $168|(F_{8n-4} + F_{8n} + F_{8n+4})$
73. $(L_{2r} + 1)F_k|(F_{kn-2r} + F_{kn} + F_{kn+2r})$ (Hillman, 1971).
74. There are no even perfect Fibonacci numbers (Whitney, 1972).
75. Let $h = 5^k$, where $k \geq 1$. Prove that $h|F_h$ (Hoggatt, 1973).
76. Let $g = 2 \cdot 3^k$, where $k \geq 1$. Prove that $g|L_g$ (Hoggatt, 1973c).

GENERALIZED FIBONACCI NUMBERS REVISITED

In Chapter 5, we found that the sum of any 10 consecutive Fibonacci numbers is 11 times the seventh number in the sequence. Is this true for generalized Fibonacci numbers? To find out, notice that the first 10 terms of the generalized Fibonacci sequence are $a, b, a+b, a+2b, 2a+3b, 3a+5b, 5a+8b, 8a+13b, 13a+21b$, and $21a+34b$. Their sum is $55a+88b$, which is clearly 11 times the seventh term $5a+8b$. Interestingly enough, $11 = L_5$. Thus

$$\sum_1^{10} G_i = L_5 \cdot G_7$$

where $L_5 = (55, 88) = (55, 89-1) = (F_{10}, F_{11}-1)$

More generally, is $\sum_1^n G_i$ a multiple of some Lucas number L_m? To answer this question, recall that $G_i = aF_{i-2} + bF_{i-1}$. So

$$\begin{aligned}\sum_1^n G_i &= a\sum_1^n F_{i-2} + b\sum_1^n F_{i-1} \\ &= aF_n + b(F_{n+1}-1)\end{aligned}$$

When $n = 10$, this sum is divisible by L_5, as we just observed. Consequently, let us look for a way to factor this sum. Since a and b are arbitrary, we look for the common factors of F_n and $F_{n+1}-1$. [Although $(F_n, F_{n+1}) = 1$, F_n and $F_{n+1}-1$ need not be relatively prime.]

Table 17.1 shows a few specific values of F_n, $F_{n+1}-1$, and their factorizations; we have omitted the cases where $(F_n, F_{n+1}-1) = 1$.

TABLE 17.1.

n	F_n	$F_{n+1} - 1$	Factorization	
6	8	—	$4 \cdot$ 2	← Fibonacci numbers
		12	$4 \cdot$ 3	
8	21	—	$3 \cdot$ 7	← Lucas numbers
		33	$3 \cdot$ 11	
10	55	—	$11 \cdot$ 5	← Fibonacci numbers
		88	$11 \cdot$ 8	
12	144	—	$8 \cdot$ 18	← Lucas numbers
		232	$8 \cdot$ 29	
14	377	—	$29 \cdot$ 13	← Fibonacci numbers
		609	$29 \cdot$ 21	
16	987	—	$21 \cdot$ 47	← Lucas numbers
		1,596	$21 \cdot$ 76	
18	2,584	—	$76 \cdot$ 34	← Fibonacci numbers
		4,180	$76 \cdot$ 55	
20	6,765	—	$55 \cdot$ 123	← Lucas numbers
		10,945	$55 \cdot$ 199	

It is apparent from the table that when n is of the form $4k + 2$, $(F_n, F_{n+1} - 1)$ is a Lucas number and the various quotients are consecutive Fibonacci numbers; and when n is of the form $4k$, $(F_n, F_{n+1} - 1)$ is a Fibonacci number and the various quotients are consecutive Lucas numbers.

Next we proceed to confirm these two observations, for which we need the following facts from Theorem 16.9:

$$F_{4n+1} - 1 = L_{2n+1}F_{2n} \qquad \text{and} \qquad F_{4n+3} - 1 = L_{2n+1}F_{2n+2}$$

Case 1. Let n be of the form $4k+2$. Then

$$\begin{aligned}\sum_{1}^{4k+2} G_i &= aF_{4k+2} + b(F_{4k+3} - 1)\\ &= aL_{2k+1}F_{2k+1} + bL_{2k+1}F_{2k+2}\\ &= L_{2k+1}(aF_{2k+2} + bF_{2k+2})\\ &= L_{2k+1}G_{2k+3}\end{aligned}$$

Thus $\sum_{1}^{4k+2} G_i$ can be obtained by multiplying G_{2k+3} with L_{2k+1}.

In particular,

$$\sum_{1}^{10} G_i = L_5 \cdot G_7 = 11 \cdot G_7$$

as observed earlier. This is an interesting case, since multiplication by 11 is remarkably easy. Likewise, we can compute $\sum_{1}^{30} G_i$ by multiplying G_{17} with $L_{15} = 1364$.

Case 2. Let n be of the form $4k$. Then

$$\begin{aligned}\sum_{1}^{4k} G_i &= aF_{4k} + b(F_{4k+1} - 1)\\ &= aL_{2k}F_{2k} + bL_{2k+1}F_{2k}\\ &= F_{2k}(aL_{2k} + bL_{2k+1})\\ &= F_{2k}[a(F_{2k-1} + F_{2k+1}) + b(F_{2k} + F_{2k+2})]\\ &= F_{2k}[(aF_{2k-1} + bF_{2k}) + (aF_{2k+1} + bF_{2k+2})]\\ &= F_{2k}(G_{2k+1} + G_{2k+3})\end{aligned}$$

Thus we can realize $\sum_{1}^{4k} G_i$ by multiplying the sum $G_{2k+1} + G_{2k+3}$with F_{2k}.

For instance, we can obtain $\sum_{1}^{4k} G_i$ by multiplying the sum $G_{11} + G_{13}$ with 55.

EXERCISES 17

Prove each, where G_k denotes the kth generalized Fibonacci number.

1. $G_{m+n} + G_{m-n} = \begin{cases} G_m L_n & \text{if } n \text{ is even} \\ (G_{m+1} + G_{m-1})F_n & \text{otherwise (Koshy, 1998)} \end{cases}$
2. $G_{m+n} - G_{m-n} = \begin{cases} (G_{m+1} + G_{m-1})F_n & \text{if } n \text{ is even} \\ G_m L_n & \text{otherwise (Koshy, 1998)} \end{cases}$

3. $G_{m+n}^2 - G_{m-n}^2 = (G_{m+1} + G_{m-1})G_m F_{2n}$ (Koshy, 1998)

Consider the sequence $\{a_n\}$ defined by $a_{2n+1} = a_{2n} + a_{2n-1}$ and $a_{2n} = a_n$, where $a_1 = a, a_2 = b$, and $n \geq 1$. Verify each (Lind, 1968).

4. $\sum_1^n a_k = a_{2n+1} - a$
5. $\sum_1^n a_{2k-1} = a_{4n+1} - a_{2n+1}$

Prove each (Koshy, 1999).

6. $G_{4m} + b = (G_{2m} + G_{2m-1})F_{2m-1}$
7. $G_{4m+1} + a = G_{2m+1}L_{2m}$
8. $G_{4m+2} + b = G_{2m+2}L_{2m}$
9. $G_{4m+3} + a = (G_{2m+3} + G_{2m+1})F_{2m+1}$
10. $G_{4m} - b = G_{2m+1}L_{2m-1}$
11. $G_{4m+1} - a = (G_{2m+2} + G_{2m})F_{2m}$
12. $G_{4m+2} - b = (G_{2m+3} + G_{2m+1})F_{2m}$
13. $G_{4m+3} - a = G_{2m+2}L_{2m+1}$
14. $(G_{4m+1} + a, G_{4m+2} + b) = L_{2m}$
15. $(G_{4m+1} - a, G_{4m+2} - b) = F_{2m}$

GENERATING FUNCTIONS

Generating functions provide a powerful tool for solving linear homogeneous recurrence relations with constant coefficients (LHRRWCCs), as will be seen shortly. In 1718, the French mathematician Abraham De Moivre (1667–1754) invented generating functions in order to solve the Fibonacci recurrence relation.

First, notice that the polynomial $1 + x + x^2 + x^3 + x^4 + x^5$ can be written $(x^6 - 1)/(x - 1)$. We can verify this by either cross-multiplication or the familiar long-division method. Accordingly, $f(x) = (x^6 - 1)/(x - 1)$ is called the *generating function* of the sequence of coefficients 1, 1, 1, 1, 1, 1 in the polynomial.

More generally, we make the following definition.

GENERATING FUNCTION

Let $a_0, a_1, a_2, \ldots$ be a sequence of real numbers. Then the function

$$g(x) = a_0 + a_1 x + a_2 x^2 + \cdots + a_n x^n + \cdots \tag{18.1}$$

is called the *generating function* for the sequence $\{a_n\}$. We can also define generating functions for the finite sequence $a_0, a_1, \ldots, a_n$ by letting $a_i = 0$ for $i > n$; thus $g(x) = a_0 + a_1 x + a_2 x^2 + \cdots + a_n x^n$ is the *generating function* for the finite sequence $a_0, a_1, \ldots, a_n$.

For example, $g(x) = 1 + 2x + 3x^2 + \cdots + (n + 1)x^n + \cdots$ is the generating function for the sequence of positive integers and

$$f(x) = 1 + 3x + 6x^2 + \cdots + \frac{n(n+1)}{2}x^n + \cdots$$

is the generating function for the sequence of triangular numbers 1, 3, 6, 10, … . Since

$$\frac{x^n - 1}{x - 1} = 1 + x + x^2 + \cdots + x^{n-1} \qquad g(x) = \frac{x^n - 1}{x - 1}$$

is the generating function for the sequence of n ones.

A word of caution. The right-hand side (RHS) of Eq. (18.1) is a *formal power series* in x. The letter x does not represent anything. We use the various powers x^n of x simply to keep track of the corresponding terms a_n of the sequence. In other words, we think of the powers x^n as place-holders. Consequently, we are not interested in the convergence of the series.

Equality of Generating Functions

Two generating functions $f(x) = \sum_0^\infty a_n x^n$ and $g(x) = \sum_0^\infty b_n x^n$ are *equal* if $a_n = b_n$ for every $n \geq 0$.

For example, let

$$f(x) = 1 + 3x + 6x^2 + 10x^3 + \cdots$$

and

$$g(x) = 1 + \frac{2 \cdot 3}{2}x + \frac{3 \cdot 4}{2}x^2 + \frac{4 \cdot 5}{2}x^3 + \cdots$$

Then $f(x) = g(x)$.

A generating function we will use frequently is

$$\frac{1}{1 - ax} = 1 + ax + a^2x^2 + \cdots + a^n x^n + \cdots \tag{18.2}$$

Then

$$\frac{1}{1 - x} = 1 + x + x^2 + \cdots + x^n + \cdots \tag{18.3}$$

Can we add and multiply generating functions? Yes. Such operations are performed exactly the same way as polynomials are combined.

Addition and Multiplication of Generating Functions

Let $f(x) = \sum_0^\infty a_n x^n$ and $g(x) = \sum_0^\infty b_n x^n$ be two generating functions. Then

$$f(x) + g(x) = \sum_0^\infty (a_n + b_n)x^n \qquad \text{and} \qquad f(x)g(x) = \sum_{n=0}^\infty \left(\sum_{i=0}^n a_i b_{n-i} \right) x^n$$

For example,

$$\begin{aligned}
\frac{1}{(1-x)^2} &= \frac{1}{1-x}\cdot\frac{1}{1-x}\\
&= \left(\sum_0^{\infty} x^i\right)\left(\sum_0^{\infty} x^i\right) = \sum_{n=0}^{\infty}\left(\sum_0^{n} 1\cdot 1\right)x^n\\
&= \sum_{n=0}^{\infty}(n+1)x^n\\
&= 1+2x+3x^2+\cdots+(n+1)x^n+\cdots
\end{aligned} \tag{18.4}$$

and

$$\begin{aligned}
\frac{1}{(1-x)^3} &= \frac{1}{1-x}\cdot\frac{1}{(1-x)^2}\\
&= \left(\sum_0^{\infty} x^n\right)\left[\sum_0^{\infty}(n+1)x^n\right]\\
&= \sum_{n=0}^{\infty}\left[\sum_{i=0}^{n} 1\cdot(n+1-i)\right]x^n\\
&= \sum_{n=0}^{\infty}[(n+1)+n+\cdots+1]x^n\\
&= \sum_{n=0}^{\infty}\frac{(n+1)(n+2)}{2}x^n\\
&= 1+3x+6x^2+10x^3+\cdots
\end{aligned} \tag{18.5}$$

Before exploring how valuable generating functions are in solving LHRRWCCs, we examine how the technique of *partial fraction decomposition*, used in integral calculus, enables us to express the quotient $p(x)/q(x)$ of two polynomials $p(x)$ and $q(x)$ as a sum of proper fractions, where $\deg p(x) < \deg q(x)$.* For example,

$$\frac{6x+1}{(2x-1)(2x+3)} = \frac{1}{2x-1}+\frac{2}{2x+3}$$

(Verify this.)

*deg $f(x)$ denotes the degree of the polynomial $f(x)$.

PARTIAL FRACTION DECOMPOSITION RULE FOR $p(x)/q(x)$ WHERE deg $p(x) < \deg q(x)$

If $q(x)$ has a factor of the form $(ax+b)^m$, then the decomposition contains a sum of the form

$$\frac{A_1}{ax+b}+\frac{A_2}{(ax+b)^2}+\cdots+\frac{A_m}{(ax+b)^m}$$

where $A_i \in \mathbb{R}$.

Examples 18.1–18.3 illustrate the partial fraction decomposition technique. We will use their results to solve three recurrence relations in Examples 18.4–18.6.

Example 18.1. Express $x/((1-x)(1-2x))$ as a sum of partial fractions.

Solution. Since the denominator contains two linear factors, we let

$$\frac{x}{(1-x)(1-2x)}=\frac{A}{1-x}+\frac{B}{1-2x}$$

To find the constants A and B, multiply both sides by $(1-x)(1-2x)$:

$$x = A(1-2x)+B(1-x)$$

Now give convenient values to x. Setting $x=1$ yields $A=-1$ and setting $x=1/2$ yields $B=1$. (We can also find the values of A and B by equating coefficients of like terms from either side of the equation, and solving the resulting linear system.)

$$\frac{x}{(1-x)(1-2x)}=\frac{-1}{1-x}+\frac{1}{1-2x}$$

(We can verify this by combining the sum on the RHS into a single fraction.) We use this result in Example 18.4. ■

Example 18.2. Express $x/(1-x-x^2)$ as a sum of partial fractions.

Solution. First, factor $1-x-x^2$:

$$1-x-x^2=(1-\alpha x)(1-\beta x)$$

Let

$$\frac{x}{1-x-x^2}=\frac{A}{1-\alpha x}+\frac{B}{1-\beta x}$$

$$\therefore\quad x = A(1-\beta x)+B(1-\alpha x)$$

Equating coefficients of like terms, we get

$$A+B=0$$

$$-\beta A-\alpha B=1$$

Solving this linear system yields $A = 1/\sqrt{5} = -B$. (Verify this.) Thus

$$\frac{x}{1 - x - x^2} = \frac{1}{\sqrt{5}}\left[\frac{1}{1 - \alpha x} - \frac{1}{1 - \beta x}\right]$$

We use this result in Example 18.3. ∎

Example 18.3. Express $(2 - 9x)/(1 - 6x + 9x^2)$ as a sum of partial fractions.

Solution. Again, factor the denominator:

$$1 - 6x + 9x^2 = (1 - 3x)^2$$

By the decomposition rule, let

$$\frac{2 - 9x}{1 - 6x + 9x^2} = \frac{A}{1 - 3x} + \frac{B}{(1 - 3x)^2}$$

Then

$$2 - 9x = A(1 - 3x) + B$$

This yields $A = 3$ and $B = -1$. (Verify this.) Thus

$$\frac{2 - 9x}{1 - 6x + 9x^2} = \frac{3}{1 - 3x} - \frac{1}{(1 - 3x)^2}$$

We use this result in Example 18.6. ∎

Now we are ready to use partial fraction decomposition and generating functions to solve recurrence relations in the next three examples.

Example 18.4. Use generating functions to solve the recurrence relation $b_n = 2b_{n-1} + 1$, where $b_1 = 1$.

Solution. First, notice that the condition $b_1 = 1$ yields $b_0 = 0$. To find the sequence $\{b_n\}$ that satisfies the recurrence relation, we consider the corresponding generating function

$$g(x) = b_0 + b_1 x + b_2 x^2 + b_3 x^3 + \cdots + b_n x^n + \cdots$$

Then

$$2xg(x) = 2b_1 x^2 + 2b_2 x^3 + \cdots + 2b_{n-1} x^n + \cdots$$

Also,

$$\frac{1}{1-x} = 1 + x + x^2 + x^3 + \cdots + x^n + \cdots$$

$$\begin{aligned}\therefore\quad g(x) - 2xg(x) - \frac{1}{1-x} &= -1 + (b_1 - 1)x + (b_2 - 2b_1 - 1)x^2 + \cdots \\ &\qquad + (b_n - 2b_{n-1} - 1)x^n + \cdots \\ &= -1\end{aligned}$$

since $b_1 = 1$ and $b_n = 2b_{n-1} + 1$ for $n \geq 2$. That is,

$$\begin{aligned}(1-2x)g(x) &= \frac{1}{1-x} - 1 = \frac{x}{1-x} \\ \therefore\quad g(x) &= \frac{x}{(1-x)(1-2x)} \\ &= -\frac{1}{1-x} + \frac{1}{1-2x} \qquad \text{by Example 18.1} \\ &= -\left(\sum_0^\infty x^n\right) + \left(\sum_0^\infty 2^n x^n\right) \\ &= \sum_0^\infty (2^n - 1)x^n\end{aligned}$$

But $g(x) = \sum_0^\infty b_n x^n$, so $b_n = 2^n - 1$, $n \geq 1$. ■

Example 18.5. Use generating functions to solve the Fibonacci recurrence relation $F_n = F_{n-1} + F_{n-2}$, where $F_1 = 1 = F_2$.

Solution. Notice that the two initial conditions yield $F_0 = 0$. Let

$$g(x) = F_0 + F_1x + F_2x^2 + \cdots + F_nx^n + \cdots$$

be the generating function of the Fibonacci sequence. Since the orders of F_{n-1} and F_{n-2} are 1 and 2 less than the order of F_n, respectively, find $xg(x)$ and $x^2g(x)$:

$$\begin{aligned}xg(x) &= F_1x^2 + F_2x^3 + F_3x^4 + \cdots + F_{n-1}x^n + \cdots \\ x^2g(x) &= \qquad\qquad F_1x^3 + F_2x^4 + F_3x^5 + \cdots + F_{n-2}x^n + \cdots \\ \therefore\quad g(x) - xg(x) - x^2g(x) &= F_1x + (F_2 - F_1)x^2 + (F_3 - F_2 - F_1)x^3 + \cdots \\ &\qquad + (F_n - F_{n-1} - F_{n-2})x^n + \cdots \\ &= x\end{aligned}$$

since $F_2 = F_1$ and $F_n = F_{n-1} + F_{n-2}$. Thus

$$
\begin{aligned}
(1 - x - x^2)g(x) &= x \\
g(x) &= \frac{x}{1 - x - x^2} \\
&= \frac{1}{\sqrt{5}}\left[\frac{1}{1 - \alpha x} - \frac{1}{1 - \beta x}\right] \qquad \text{by Example 18.2} \\
\therefore \quad \sqrt{5}g(x) &= \frac{1}{1 - \alpha x} - \frac{1}{1 - \beta x} \\
&= \sum_0^\infty \alpha^n x^n - \sum_0^\infty \beta^n x^n = \sum_0^\infty (\alpha^n - \beta^n)x^n \\
g(x) &= \sum_0^\infty \frac{(\alpha^n - \beta^n)}{\sqrt{5}} x^n
\end{aligned}
$$

Thus, by the equality of generating functions, we get the *Binet formula* for F_n:

$$F_n = \frac{\alpha^n - \beta^n}{\sqrt{5}} = \frac{\alpha^n - \beta^n}{\alpha - \beta}$$

■

Example 18.6. Use generating functions to solve the recurrence relation $a_n = 6a_{n-1} - 9a_{n-2}$, where $a_0 = 2$ and $a_1 = 3$.

Solution. Let $g(x) = a_0 + a_1 x + a_2 x^2 + \cdots + a_n x^n + \cdots$. Then

$$
\begin{aligned}
6xg(x) &= 6a_0 x + 6a_1 x^2 + 6a_2 x^3 + \cdots + 6a_{n-1} x^n + \cdots \\
9x^2 g(x) &= 9a_0 x^2 + 9a_1 x^3 + 9a_2 x^4 + \cdots + 9a_{n-2} x^n + \cdots \\
\therefore \quad g(x) - 6xg(x) + 9x^2 g(x) &= a_0 + (a_1 - 6a_0)x + (a_2 - 6a_1 + 9a_0)x^2 + \cdots \\
&\quad + (a_n - 6a_{n-1} + 9a_{n-2})x^n + \cdots \\
&= 2 - 9x
\end{aligned}
$$

using the given conditions. Thus

$$
\begin{aligned}
\therefore g(x) &= \frac{2 - 9x}{1 - 6x + 9x^2} \\
&= \frac{3}{1 - 3x} - \frac{1}{(1 - 3x)^2} \qquad \text{by Example 18.3} \\
&= 3\left(\sum_0^\infty 3^n x^n\right) - \sum_0^\infty (n + 1)3^n x^n
\end{aligned}
$$

$$= \sum_{0}^{\infty}[3^{n+1} - (n+1)3^n]x^n$$

$$= \sum_{0}^{\infty} 3^n(2-n)x^n$$

Thus

$$a_n = (2-n)3^n \qquad n \geq 0$$

■

The following example presents an identity linking binomial coefficients and Fibonacci numbers. The identity, developed in 1968 by Hoggatt, is an application of Example 18.5. The proof given here is due to L. Carlitz.

Example 18.7. Prove that

$$\sum_{2j\leq n} j\binom{n-j}{j} = \sum_{0}^{n} F_{n-j}F_j.$$

Solution. Let

$$C_n = \sum_{2j\leq n} j\binom{n-j}{j}$$

$$\sum_{0}^{\infty} C_n x^n = \sum_{n=0}^{\infty} x^n \sum_{2j\leq n} j\binom{n-j}{j}$$

$$= \sum_{j=0}^{\infty} jx^{2j} \sum_{n=0}^{\infty}\binom{n+j}{j}x^n = \sum_{j=0}^{\infty} jx^{2j}(1-x)^{-j-1}$$

$$= \frac{x^2}{(1-x)^2}\sum_{j=0}^{\infty}(j+1)x^{2j}(1-x)^{-j}$$

$$= \frac{x^2}{(1-x)^2}\left(1 - \frac{x^2}{1-x}\right)^{-2}$$

$$= \frac{x^2}{(1-x-x^2)^2}$$

$$= \left(\sum_{0}^{\infty} F_n x^n\right)\left(\sum_{0}^{\infty} F_j x^j\right)$$

$$= \sum_{0}^{\infty} \left(\sum_{j=0}^{n} F_j F_{n-j} \right) x^n$$

$$\therefore C_n = \sum_{j=0}^{n} F_j F_{n-j}$$

Suppose we left-justify Pascal's triangle, multiply each column by j, and then add the rising diagonals, where $j \geq 0$. The resulting sum on the nth diagonal is C_n (see Fig. 18.1). ■

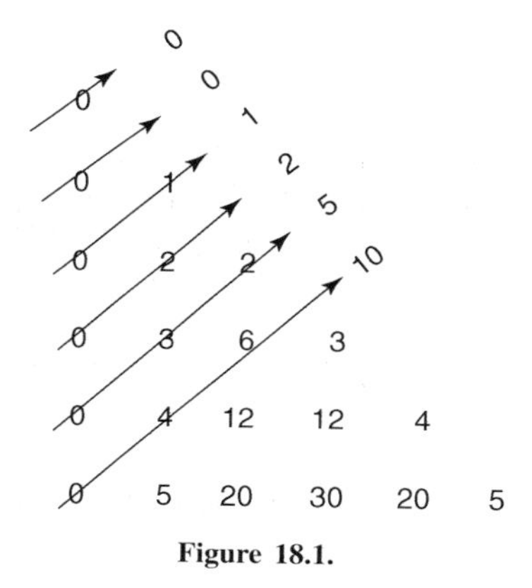

Figure 18.1.

A *lattice point* on the Cartesian plane is a point (x, y) such that both coordinates x and y are integers. The next example, proposed as a problem in 1970 by R. C. Drake of North Carolina A & T University at Greensboro, deals with paths connecting lattice points. The solution given here is based on one given by L. Carlitz of Duke University.

Example 18.8. Let $f(n)$ denote the number of paths on the Cartesian plane from $(0, 0)$ to $(n, 0)$. Each path is made up of directed line segments of one or more of the following types:

Type	1	2	3	4
Initial point Endpoint	$(k, 0)$ $(k, 1)$	$(k, 0)$ $(k + 1, 0)$	$(k, 1)$ $(k + 1, 1)$	$(k, 1)$ $(k + 1, 0)$

Thus, the next point on the path from the point $(k, 0)$ can be $(k, 1)$ or $(k + 1, 0)$; and that, from the point $(k, 1)$, can be $(k + 1, 1)$ or $(k + 1, 0)$. Find a formula for $f(n)$.

Solution. There are $f(1) = 2 = F_3$ paths from $(0, 0)$ to $(1, 0)$; $f(2) = 5 = F_5$ paths from $(0, 0)$ to $(2, 0)$; and $f(3) = 13 = F_7$ paths from $(0, 0)$ to $(3, 0)$. Figure 18.2 depicts all of them.

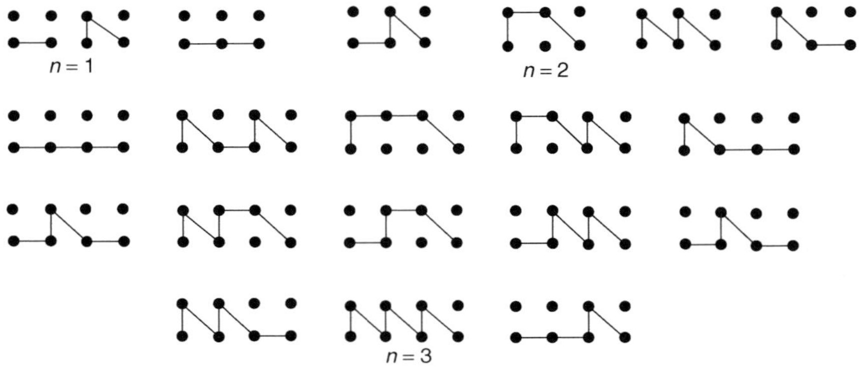

Figure 18.2.

Let $f_2(n)$ denote the number of paths ending with a line segment of type 2 and $f_4(n)$ the number of paths ending with a line segment of type 4. Then

$$f_2(n+1) = f_2(n) + f_4(n) = f(n)$$

$$f_2(n+1) = f(0) + f(1) + \cdots + f(n) = \sum_0^n f(k)$$

$$\therefore \quad f(n+1) = f_2(n+1) + f_4(n+1) = f(n) = \sum_0^n f(k)$$

Then $f(1) = f(0) + f(0) = 2f(0)$. But $f(1) = 2$, so $f(0) = 1$. Consider the power series

$$F(x) = \sum_0^\infty f(n)x^n = f(0) + \sum_1^\infty f(n)x^n = 1 + \sum_0^\infty f(n+1)x^{n+1}$$

$$= 1 + \sum_{n=0}^\infty \left[f(n) + \sum_{k=0}^n f(k) \right] x^{n+1}$$

$$= 1 + xF(x) + \frac{x}{1-x} F(x)$$

$$\therefore \quad F(x) = \frac{1-x}{1-3x+x^2}$$

But

$$\frac{1-x}{1-3x+x^2} = \sum_0^\infty F_{2n+1} x^n$$

(See the list of generating functions on p. 230.) Thus $f(n) = F_{2n+1}$, where $n \geq 0$. ■

EXERCISES 18

Express each quotient as a sum of partial fractions.

1. $\dfrac{x+7}{(x-1)(x+3)}$
2. $\dfrac{4x^2-3x-25}{(x+1)(x-2)(x+3)}$
3. $\dfrac{5}{1-x-6x^2}$
4. $\dfrac{2+4x}{1+8x+15x^2}$
5. $\dfrac{x(x+2)}{(2+3x)(x^2+1)}$
6. $\dfrac{-2x^2-2x+2}{(x-1)(x^2+2x)}$
7. $\dfrac{x^3+x^2+x+3}{x^4+5x^2+6}$
8. $\dfrac{-x^3+2x^2+x}{x^4+x^3+x+1}$
9. $\dfrac{3x^3-x^2+4x}{x^4-x^3+2x^2-x+1}$

*10. $\dfrac{x^3+x^2+5x-2}{x^4-x^2+x-1}$

Use generating functions to solve each LHRRWCC.

11. $a_n = 2a_{n-1}, a_0 = 1$
12. $a_n = a_{n-1} + 1, a_1 = 1$
13. $a_n = a_{n-1} + 2, a_1 = 1$
14. $a_n = a_{n-1} + 2a_{n-2}, a_0 = 3, a_1 = 0$
15. $a_n = 4a_{n-2}, a_0 = 2, a_1 = -8$
16. $a_n = a_{n-1} + 6a_{n-2}, a_0 = 5, a_1 = 0$
17. $a_n = 5a_{n-1} - 6a_{n-2}, a_0 = 4, a_1 = 7$
18. $a_n = a_{n-1} + a_{n-2}, a_0 = 1, a_1 = 2$
19. $a_n = a_{n-1} + a_{n-2}, a_0 = 2, a_1 = 3$
20. $L_n = L_{n-1} + L_{n-2}, L_1 = 1, L_2 = 3$
21. $a_n = 4a_{n-1} - 4a_{n-2}, a_0 = 3, a_1 = 10$
22. $a_n = 6a_{n-1} - 9a_{n-2}, a_0 = 2, a_1 = 3$
23. $a_n = 3a_{n-1} + 4a_{n-2} - 12a_{n-3}, a_0 = 3, a_1 = -7, a_2 = 7$
24. $a_n = 8a_{n-1} - 21a_{n-2} + 18a_{n-3}, a_0 = 0, a_1 = 2, a_2 = 13$
25. $a_n = 7a_{n-1} + 16a_{n-2} - 12a_{n-3}, a_0 = 0, a_1 = 5, a_2 = 19$

26. $a_n = 3a_{n-1} + 4a_{n-2} - 12a_{n-3}$, $a_0 = 3$, $a_1 = -7$, $a_2 = 7$
27. $a_n = 6a_{n-1} - 12a_{n-2} - 8a_{n-3}$, $a_0 = 0$, $a_1 = 2$, $a_2 = -2$
28. $a_n = 13a_{n-2} - 36a_{n-4}$, $a_0 = 7$, $a_1 = -6$, $a_2 = 38$, $a_3 = -84$
29. $a_n = -a_{n-1} + 3a_{n-2} + 5a_{n-3} + 2a_{n-4}$, $a_0 = 0$, $a_1 = -8$, $a_2 = 4$, $a_3 = -42$

GENERATING FUNCTIONS REVISITED

Generating functions, as we saw in the preceding chapter, can be employed to derive additional Fibonacci and Lucas identities, using the identities developed earlier.

A GENERATING FUNCTION FOR F_{3n}

First, we develop a generating function $g(x)$ for F_{3n}. To this end, we let

$$\begin{aligned} g(x) &= F_0 + F_3x + F_6x^2 + F_9x^3 + \cdots + F_{3n}x^n + \cdots \\ 4xg(x) &= \quad 4F_0x + 4F_3x^{2a} + 4F_6x^3 + \cdots + 4F_{3n-3}x^n + \cdots \\ x^2g(x) &= \quad F_0x^2 + F_3x^3 + \cdots + F_{3n-6}x^n + \cdots \\ \therefore \quad (1 - 4x - x^2)g(x) &= 2x \qquad \text{since } F_{3n} = 4F_{3n-3} + F_{3n-6} \end{aligned}$$

Thus

$$g(x) = \frac{2x}{1 - 4x - x^2}$$

A GENERATING FUNCTION FOR F_n^3

Next we derive a generating function for F_n^3. Let

$$\begin{aligned} g(x) &= F_0^3 + F_1^3x + F_2^3x^2 + F_3^3x^3 + \cdots \\ 3xg(x) &= \quad 3F_0^3x + 3F_1^3x^2 + 3F_2^3x^3 + 3F_3^3x^4 + \cdots \end{aligned}$$

$$\begin{aligned}
6x^2g(x) &= \qquad\qquad 6F_0^3x^2 + 6F_1^3x^3 + 6F_2^3\,x^4 + \cdots \\
3x^3g(x) &= \qquad\qquad\qquad 3F_0^3x^3 + 3F_1^3\,x^4 + \cdots \\
x^4g(x) &= \qquad\qquad\qquad\qquad F_0^3x^4 \;+ \cdots \\
(1 - 3x - 6x^2 + 3x^3 + x^4)g(x) &= x + x^2 - 3x^2 + 2x^3 - 3x^3 - 6x^3 \\
&\qquad\qquad \text{by identity 1 on p. 87} \\
&= x - 2x^2 - 7x^3
\end{aligned}$$

Thus

$$g(x) = \frac{x - 2x^2 - 7x^3}{1 - 3x - 6x^2 + 3x^3 + x^4}$$

In 1948, J. Ginzburg employed generating functions to prove Formula (5.1) that $\sum_1^n F_i = F_{n+2} - 1$. To see how this was done, first we derive a generating function for the sum $s_n = \sum_1^n F_i$, where $s_0 = 0$: Let $g(x) = \sum_0^\infty s_n x^n$. Then

$$2xg(x) = \sum_1^\infty 2s_{n-1}x^n \qquad \text{and} \qquad x^3g(x) = \sum_3^\infty s_{n-3}x^n$$

Since

$$s_n - 2s_{n-1} + s_{n-3} = 0$$

it follows that

$$(1 - 2x + x^3)g(x) = x$$

Thus

$$g(x) = \frac{x}{1 - 2x + x^3}$$

is the desired generating function; s_n is the coefficient of x^n in the power series expansion of this function.

Since $1 - 2x + x^3 = (1 - x - x^2)(1 - x)$, we can convert this into partial fractions:

$$g(x) = \frac{1 + x}{1 - x - x^2} - \frac{1}{1 - x}$$

Since

$$\frac{x}{1 - x - x^2} = \sum_0^\infty F_n x^n$$

this yields

$$\begin{aligned}
\sum_0^\infty s_n x^n &= (1 + x)\sum_1^\infty F_n x^{n-1} - \sum_0^\infty x^n \\
&= \sum_1^\infty (F_n x^{n-1} + F_n x^n) - \sum_0^\infty x^n
\end{aligned}$$

$$= \sum_{0}^{\infty} (F_{n+1} + F_n - 1)x^n$$

$$= \sum_{0}^{\infty} (F_{n+2} - 1)x^n$$

Thus,

$$s_n = \sum_{1}^{n} F_i = F_{n+2} - 1,$$

as desired.

A LIST OF GENERATING FUNCTIONS

In 1967, V. E. Hoggatt, Jr., and D. A. Lind compiled the following list of 18 generating functions that generate the various powers and products of Fibonacci and Lucas numbers. Some are their own creations.

1. $\dfrac{x}{1 - x - x^2} = \sum_{0}^{\infty} F_n x^n$

2. $\dfrac{1}{1 - x - x^2} = \sum_{0}^{\infty} F_{n+1} x^n$

3. $\dfrac{2 - x}{1 - x - x^2} = \sum_{0}^{\infty} L_n x^n$

4. $\dfrac{1 + 2x}{1 - x - x^2} = \sum_{0}^{\infty} L_{n+1} x^n$

5. $\dfrac{x - x^2}{1 - 2x - 2x^2 + x^3} = \sum_{0}^{\infty} F_n^2 x^n$

6. $\dfrac{1 - x}{1 - 2x - 2x^2 + x^3} = \sum_{0}^{\infty} F_{n+1}^2 x^n$

7. $\dfrac{1 + 2x - x^2}{1 - 2x - 2x^2 + x^3} = \sum_{0}^{\infty} F_{n+2}^2 x^n$

8. $\dfrac{x}{1 - 2x - 2x^2 + x^3} = \sum_{0}^{\infty} F_n F_{n+1} x^n$

9. $\dfrac{4 - 7x - x^2}{1 - 2x - 2x^2 + x^3} = \sum_0^\infty L_n^2 x^n$

10. $\dfrac{1 + 7x - 4x^2}{1 - 2x - 2x^2 + x^3} = \sum_0^\infty L_{n+1}^2 x^n$

11. $\dfrac{9 - 2x - x^2}{1 - 2x - 2x^2 + x^3} = \sum_0^\infty L_{n+2}^2 x^n$

12. $\dfrac{x - 2x^2 - x^3}{1 - 3x - 6x^2 + 3x^3 + x^4} = \sum_0^\infty F_n^3 x^n$

13. $\dfrac{1 - 2x - x^2}{1 - 3x - 6x^2 + 3x^3 + x^4} = \sum_0^\infty F_{n+1}^3 x^n$

14. $\dfrac{1 + 5x - 3x^2 - x^3}{1 - 3x - 6x^2 + 3x^3 + x^4} = \sum_0^\infty F_{n+2}^3 x^n$

15. $\dfrac{8 + 3x - 4x^2 - x^3}{1 - 3x - 6x^2 + 3x^3 + x^4} = \sum_0^\infty F_{n+3}^3 x^n$

16. $\dfrac{2x}{1 - 3x - 6x^2 + 3x^3 + x^4} = \sum_0^\infty F_n F_{n+1} F_{n+2} x^n$

17. $\dfrac{F_k x}{1 - L_k x + (-1)^k x^2} = \sum_0^\infty F_{kn} x^n$ (Hoggatt, 1971)

18. $\dfrac{F_r + (-1)^r F_{k-r} x}{1 - L_k x + (-1)^k x^2} = \sum_0^\infty F_{kn+r} x^n$

The following four generating functions were derived by V. E. Hoggatt, Jr., in 1971, and the fifth was discovered in the following year.

1. $\dfrac{x}{1 - 3x + x^2} = \sum_0^\infty F_{2n} x^n$

2. $\dfrac{1 - x}{1 - 3x + x^2} = \sum_0^\infty F_{2n+1} x^n$

3. $\dfrac{3 - 2x}{1 - 3x + x^2} = \sum_0^\infty L_{2n+2} x^n$

4. $$\frac{x+x^2}{1-3x+x^2} = \sum_0^\infty L_{2n+1}x^{n+1}$$

5. $$\frac{1}{1-2x-2x^2+x^3} = \sum_0^\infty F_{n+1}F_{n+2}x^n$$

GENERATING FUNCTIONS FOR F_{m+n} AND L_{m+n}

In 1972, R. T. Hansen of Montana State University also employed generating functions in his investigation of Fibonacci and Lucas numbers. For example, the generating function of F_{m+n} is given by:

$$\begin{aligned}
\sum_{n=0}^\infty F_{m+n}x^n &= \sum_{n=0}^\infty \frac{\alpha^{m+n}-\beta^{m+n}}{\alpha-\beta}x^n \\
&= \frac{1}{\alpha-\beta}\left[\alpha^m\sum_0^\infty \alpha^n x^n - \beta^m \sum_0^\infty \beta^n x^n\right] \\
&= \frac{1}{\alpha-\beta}\left[\frac{\alpha^m}{1-\alpha x} - \frac{\beta^m}{1-\beta x}\right] \\
&= \frac{(\alpha^m-\beta^m)+(\alpha^{m-1}-\beta^{m-1})x}{(\alpha-\beta)(1-\alpha x)(1-\beta x)} \\
&= \frac{F_m+F_{m-1}x}{1-x-x^2} \qquad (19.1)
\end{aligned}$$

Likewise, it can be shown that

$$\sum_{n=0}^\infty L_{m+n}x^n = \frac{L_m+L_{m-1}x}{1-x-x^2} \qquad (19.2)$$

See Exercise 15.

IDENTITIES USING GENERATING FUNCTIONS

These two generating functions can be applied to derive a host of identities. For example, notice that

$$\sum_0^\infty F_{n+1}x^n = \frac{1}{D} \qquad \sum_0^\infty F_{n-1}x^n = \frac{1-x}{D} \qquad \text{and} \qquad \sum_0^\infty L_n x^n = \frac{2-x}{D}$$

where $D = 1-x-x^2$. Since

$$\frac{2-x}{D} = \frac{1}{D} + \frac{1-x}{D}$$

it follows that

$$\sum_0^\infty L_n x^n = \sum_0^\infty F_{n+1} x^n + \sum_0^\infty F_{n-1} x^n$$
$$= \sum_0^\infty (F_{n+1} + F_{n-1}) x^n$$
$$\therefore \quad F_{n+1} + F_{n-1} = L_n$$

a fact already known from Chapter 5.

Next, we shall prove that $F_m L_n + F_{m-1} L_{n-1} = L_{m+n-1}$. We have

$$\sum_{m=0}^\infty (F_m L_n + F_{m-1} L_{n-1}) x^m = L_n \sum_{m=0}^\infty F_m x^m + L_{n-1} \sum_{m=0}^\infty F_{m-1} x^m$$
$$= L_{n-1} \frac{x}{D} + L_{n-1} \frac{1-x}{D}$$
$$= \frac{L_{n-1} + (L_n - L_{n-1})x}{D}$$
$$= \frac{L_{n-1} + L_{n-2}x}{D}$$
$$= \sum_{m=0}^\infty L_{m+n-1} x^m$$
$$\therefore \quad F_m L_n + F_{m-1} L_{n-1} = L_{m+n-1}$$

Similarly, it can be shown that $F_m F_n + F_{m-1} F_{n-1} = F_{m+n-1}$ and $L_m L_n + L_{m-1} L_{n-1} = 5L_{m+n-1}$ (see Exercises 6 and 7).

Next, we develop a generating function for $F_n/n!$ and then for $L_n/n!$.

EXPONENTIAL GENERATING FUNCTIONS

Since $e^t = \sum_0^\infty (t^n/n!)$, it follows that

$$e^{\alpha x} = \sum_0^\infty \frac{\alpha^n x^n}{n!} \quad \text{and} \quad e^{\beta x} = \sum_0^\infty \frac{\beta^n x^n}{n!}$$

$$\therefore \quad \frac{e^{\alpha x} - e^{\beta x}}{(\alpha - \beta)} = \sum_{n=0}^\infty \left(\frac{\alpha^n - \beta^n}{\alpha - \beta} \right) \frac{x^n}{n!} = \sum_{n=0}^\infty \frac{F_n}{n!} x^n \tag{19.3}$$

Thus, the exponential function $(e^{\alpha x} - e^{\beta x})/(\alpha - \beta)$ generates the numbers $F_n/n!$.

More generally, we can show that

$$\frac{e^{\alpha^k x} - e^{\beta^k x}}{\alpha - \beta} = \sum_{n=0}^{\infty} \frac{F_{nk}}{n!} x^n$$

Likewise,

$$e^{\alpha^k x} + e^{\beta^k x} = \sum_{n=0}^{\infty} \frac{L_{nk}}{n!} x^n$$

We can employ the generating functions for $F_n/n!$ and $L_n/n!$ to derive three combinatorial identities. It follows from Eq. (19.3) that

$$e^{x/2}(e^{\sqrt{5}x/2} - e^{-\sqrt{5}x/2}) = \sqrt{5} \sum_{n=0}^{\infty} \frac{F_n}{n!} x^n$$

$$2e^{x/2} \sinh(\sqrt{5}x/2) = \sqrt{5} \sum_{n=0}^{\infty} \frac{F_n}{n!} x^n$$

Thus

$$2e^x \sinh(\sqrt{5}x) = \sqrt{5} \sum_{n=0}^{\infty} \frac{2^n F_n}{n!} x^n \tag{19.4}$$

Likewise, the generating function $e^{\alpha x} + e^{\beta x} = \sum_{n=0}^{\infty} (L_n/n!) x^n$ can be employed to derive the formula

$$2e^x \cosh(\sqrt{5}x) = \sqrt{5} \sum_{n=0}^{\infty} \frac{2^n L_n}{n!} x^n \tag{19.5}$$

We can employ the exponential generating functions for F_n and L_n to develop a host of identities, as C. A. Church and M. Bicknell showed in 1973. To see this, let

$$A(t) = \sum_{0}^{\infty} a_n \frac{t^n}{n!} \quad \text{and} \quad B(t) = \sum_{0}^{\infty} b_n \frac{t^n}{n!}$$

Then

$$A(t)B(t) = \sum_{n=0}^{\infty} \left[\sum_{k=0}^{n} \binom{n}{k} a_k b_{n-k} \right] \frac{t^n}{n!} \tag{19.6}$$

and

$$A(t)B(-t) = \sum_{n=0}^{\infty} \left[\sum_{k=0}^{n} (-1)^{n-k} \binom{n}{k} a_k b_{n-k} \right] \frac{t^n}{n!} \tag{19.7}$$

In particular, let $A(t) = (e^{\alpha t} - e^{\beta t})/(\alpha - \beta)$ and $B(t) = e^t$. Then

$$\frac{e^t (e^{\alpha t} - e^{\beta t})}{\alpha - \beta} = \sum_{n=0}^{\infty} \left[\sum_{k=0}^{n} \binom{n}{k} F_k \right] \frac{t^n}{n!} \qquad \text{by Eq. (19.3)}$$

That is,

$$\frac{e^{(\alpha+1)t} - e^{(\beta+1)t}}{\alpha - \beta} = \sum_{n=0}^{\infty}\left[\sum_{k=0}^{n}\binom{n}{k}F_k\right]\frac{t^n}{n!}$$

$$\frac{e^{\alpha^2 t} - e^{\beta^2 t}}{\alpha - \beta} = \sum_{n=0}^{\infty}\left[\sum_{k=0}^{n}\binom{n}{k}F_k\right]\frac{t^n}{n!}$$

That is,

$$\sum_{n=0}^{\infty}F_{2n}\frac{t^n}{n!} = \sum_{n=0}^{\infty}\left[\sum_{k=0}^{n}\binom{n}{k}F_k\right]\frac{t^n}{n!} \tag{19.8}$$

Equating the coefficients of $t^n/n!$ yields the combinatorial identity (12.2) we derived in Chapter 12:

$$\sum_{k=0}^{n}\binom{n}{k}F_k = F_{2n} \tag{12.2}$$

Using $B(t) = e^{-t}$, Property (19.7), and the preceding steps, it follows that

$$\begin{aligned}\sum_{n=0}^{\infty}\left[\sum_{k=0}^{n}(-1)^{n-k}\binom{n}{k}F_k\right]\frac{t^n}{n!} &= \frac{e^{(\alpha-1)t} - e^{(\beta-1)t}}{\alpha - \beta}\\ &= \frac{e^{-\beta t} - e^{-\alpha t}}{\alpha - \beta}\\ &= \sum_{n=0}^{\infty}(-1)^{n-1}F_n\frac{t^n}{n!}\end{aligned}$$

This yields yet another combinatorial identity:

$$\sum_{k=0}^{n}(-1)^{n-k}\binom{n}{k}F_k = (-1)^{n-1}F_n \tag{12.4}$$

Obviously, by selecting $A(t)$ and $B(t)$ as suitable exponential functions, we can apply this method to derive an array of Fibonacci and Lucas identities. For example, choosing

$$A(t) = \frac{e^{\alpha^2 t} - e^{\beta^2 t}}{\alpha - \beta} \qquad \text{and} \qquad B(t) = e^{-t}$$

we can show that

$$\sum_{k=0}^{n}(-1)^{n-k}\binom{n}{k}F_{2k} = F_n$$

See Exercise 21.

HYBRID IDENTITIES

Again, we choose $A(t)$ and $B(t)$ strategically to develop a family of hybrid identities that contain both Fibonacci and Lucas numbers. Let

$$A(t) = \frac{e^{\alpha t} - e^{\beta t}}{\alpha - \beta} \qquad \text{and} \qquad B(t) = e^{\alpha t} + e^{\beta t}$$

Then, by Eq. (19.6),

$$\sum_{n=0}^{\infty} \left[\binom{n}{k} F_k L_{n-k} \right] \frac{t^n}{n!} = \frac{e^{2\alpha t} - e^{2\beta t}}{\alpha - \beta}$$

$$= \sum_{n=0}^{\infty} 2^n F_{2n} \frac{t^n}{n!}$$

This yields the combinatorial identity

$$\sum_{k=0}^{n} \binom{n}{k} F_k L_{n-k} = 2^n F_{2n} \tag{19.9}$$

Likewise, we can show that

$$\sum_{k=0}^{n} \binom{n}{k} F_k F_{n-k} = \frac{2^n L_n - 2}{5} \tag{19.10}$$

and

$$\sum_{k=0}^{n} \binom{n}{k} L_k L_{n-k} = 2^n L_n + 2 \tag{19.11}$$

See Exercises 9 and 10.

In fact, Identities (19.9) through (19.11) can be generalized as follows:

$$\sum_{k=0}^{n} \binom{n}{k} F_{mk} L_{mn-mk} = 2^n F_{mn} \tag{19.12}$$

$$\sum_{k=0}^{n} \binom{n}{k} F_{mk} F_{mn-mk} = \frac{2^n L_{mn} - 2L_m^n}{5} \tag{19.13}$$

$$\sum_{k=0}^{n} \binom{n}{k} L_{mk} L_{mn-mk} = 2^n L_{mn} + 2L_m^n \tag{19.14}$$

IDENTITIES USING THE DIFFERENTIAL OPERATOR d/dt

We can realize more generalized families of identities by using the differential operator d/dt. Since $A(t) = \sum_0^\infty a_n(t^n/n!)$, it follows that

$$\frac{d^r}{dt^r}A(t) = \sum_{n=0}^{\infty} a_{n+r}\frac{t^n}{n!}$$

Let $A(t) = (e^{\alpha t} - e^{\beta t})/(\alpha - \beta)$ and $B(t) = e^{\alpha t}$. Then, by Eq. (19.6),

$$\begin{aligned}\sum_{n=0}^{\infty}\left[\sum_{k=0}^{n}\binom{n}{k}F_{k+r}\right]\frac{t^n}{n!} &= e^t\frac{d^r}{dt^r}\left(\frac{e^{\alpha t} - e^{\beta t}}{\alpha - \beta}\right)\\ &= \frac{\alpha^r e^{(\alpha+1)t} - \beta^r e^{(\beta+1)t}}{\alpha - \beta}\\ &= \frac{\alpha^r e^{\alpha^2 t} - \beta^r e^{\beta^2 t}}{\alpha - \beta}\\ &= \sum_{n=0}^{\infty} F_{2n+r}\frac{t^n}{n!}\end{aligned}$$

This yields the identity

$$\sum_{k=0}^{n}\binom{n}{k}F_{k+r} = F_{2n+r} \tag{19.15}$$

Similarly, we can show that

$$\sum_{k=0}^{n}\binom{n}{k}F_{4mk+r} = L_{2m}^{n}F_{2mn+4mr} \tag{19.16}$$

$$\sum_{k=0}^{n}\binom{n}{k}F_{m-1}^{n-k}F_m^kF_k = F_{2n+r} \tag{19.17}$$

and

$$\sum_{k=0}^{n}\binom{n}{k}F_{m-1}^{n-k}F_m^kF_{k+rm} = F_{mn+rm} \tag{19.18}$$

COMPOSITIONS WITH 1s AND 2s REVISITED

In Chapter 4, we proved that the number of compositions C_n of a positive integer n, using 1s and 2s only, is given by $C_n = F_{n+1}$. We now reestablish this fact using generating functions. Let

$$\begin{aligned}
C(x) &= C_1x + C_2x^2 + C_3x^3 + \cdots + C_nx^n + \cdots \\
xC(x) &= \qquad\quad C_1x^2 + C_2x^3 + \cdots + C_{n-1}x^n + \cdots \\
x^2C(x) &= \qquad\qquad\quad C_1x^3 + \cdots + C_{n-2}x^n + \cdots \\
(1 - x - x^2)C(x) &= x + x^2 \\
\therefore \quad C(x) &= \frac{x + x^2}{1 - x - x^2} \\
&= \sum_0^\infty F_nx^n + \sum_0^\infty F_nx^{n+1} \\
&= \sum_1^\infty (F_n + F_{n-1})x^n = \sum_1^\infty F_{n+1}x^n
\end{aligned}$$

Thus $C_n = F_{n+1}$, as expected.

EXERCISES 19

1. Let $f(x) = \sum_0^\infty \frac{F_n}{n!}x^n$. Show that $f(x) = -e^x f(-x)$. (Lehmer, 1938)
2. Show that $e^{\alpha x} + e^{\beta x} = \sum_0^\infty \frac{L_n}{n!}x^n$.
3. Let $g(x) = \sum_0^\infty \frac{L_n}{n!}x^n$. Show that $g(x) = e^x g(-x)$.

Verify each.

4. $\dfrac{x}{1 + x - x^2} = \sum_0^\infty (-1)^{n+1}F_nx^n$ (Hoggatt, Jr., 1964)
5. $\dfrac{x + 2x^2}{(1 - x - x^2)^2} = \sum_0^\infty nF_{n+1}x^n$ (Grassl, 1974)

Prove each, using generating functions.

6. $F_mF_n + F_{m-1}F_{n-1} = F_{m+n-1}$ (Hansen, 1972)
7. $L_mL_n + L_{m-1}L_{n-1} = 5F_{m+n-1}$ (Hansen, 1972)
8. $\sum_{k=0}^n (-1)^{n-k}\binom{n}{k}F_{2k} = F_n$
9. $\sum_{k=0}^n \binom{n}{k}F_kF_{n-k} = \dfrac{2^nL_n - 2}{5}$
10. $\sum_{k=0}^n \binom{n}{k}L_kL_{n-k} = 2^nL_n + 2$

Use the function $A_n(x) = \sum_1^n F_ix^i$ to answer Exercises 11–14 (Lind, 1967).

11. Show that $A_n(x) = \frac{F_n x^{n+2} + F_{n+1} x^{n+1} - x}{1 - x - x^2}$.
12. Deduce the value of $\sum_{1}^{n} F_i$.
13. Derive a formula for $B(x) = \sum_{1}^{\infty} \frac{A_n(x)}{n!}$.
14. Deduce the value of $B(1)$.
15. Show that $\sum_{m=0}^{\infty} L_{m+n} x^m = \frac{L_n + L_{n-1} x}{1 - x - x^2}$ (Hansen, 1972)

Let $C_{n+2} = C_{n+1} + C_n + F_{n+2}$, where $C_1 = 1$ and $C_2 = 2$. Verify each. (Hoggatt, 1964)

16. $C_n = \sum_{i=1}^{n-1} F_i F_{n-i}$
17. $C_{n+1} = \sum_{i=0}^{\lfloor n/2 \rfloor} (n - i + 1) \binom{n-i}{i}$
18. $C_n = \frac{nL_{n+1} + 2F_n}{5}$, $n \geq 0$.

Prove each using exponential generating functions. (Church and Bicknell, 1973)

19. $\sum_{k=0}^{n} \binom{n}{k} L_k = L_{2n}$
20. $\sum_{k=0}^{n} (-1)^{n-k} \binom{n}{k} L_k = (-1)^{n-1} L_n$
21. $\sum_{k=0}^{n} (-1)^{n-k} \binom{n}{k} F_{2k} = F_n$
22. $\sum_{k=0}^{n} (-1)^{n-k} \binom{n}{k} L_{2k} = L_n$
23. $\sum_{k=0}^{n} \binom{n}{k} F_{mk} L_{mn-mk} = 2^n F_{mn}$
24. $\sum_{k=0}^{n} \binom{n}{k} F_{mk} F_{mn-mk} = \frac{2^n L_{mn} - 2L_m^n}{5}$
25. $\sum_{k=0}^{n} \binom{n}{k} L_{mk} L_{mn-mk} = 2^n L_{mn} + 2L_m^n$
26. $\sum_{k=0}^{n} \binom{n}{k} F_{4mk+r} = L_{2m}^n F_{2mn+4mr}$
27. $\sum_{k=0}^{n} \binom{n}{k} F_{m-1}^{n-k} F_m^k F_k = F_{mn}$
28. $\sum_{k=0}^{n} \binom{n}{k} F_{m-1}^{n-k} F_m^k F_{k+rm} = F_{mn+rm}$

THE GOLDEN RATIO

He that holds fast the golden mean,
And lives contentedly between
The little and the great,
Feels not the wants that pinch the poor
Nor plagues that haunt the rich man's door,
Embittering all his state.

—William Cowper, English poet (1731–1800)

What can we say about the sequence of ratios (F_{n+1}/F_n) of consecutive Fibonacci numbers? Does it converge? If it does, what is its limit? If the limit exists, does it have any geometric significance? These are a few interesting questions which, along with their counterparts, we shall pursue in this chapter.

First, let us compute the ratios F_{n+1}/F_n and L_{n+1}/L_n of the first 20 Fibonacci and Lucas numbers, and then examine them for a possible pattern:

TABLE 20.1.

F_{n+1}/F_n	L_{n+1}/L_n
$\frac{1}{1} = 1.0000000000$	$\frac{3}{1} = 3.0000000000$
$\frac{2}{1} = 2.0000000000$	$\frac{4}{3} \approx 1.3333333333$
$\frac{3}{2} = 1.5000000000$	$\frac{7}{4} = 1.7500000000$
$\frac{5}{3} \approx 1.6666666667$	$\frac{11}{7} \approx 1.5714285714$
$\frac{8}{5} = 1.6000000000$	$\frac{18}{11} \approx 1.6363636364$
$\frac{13}{8} = 1.6250000000$	$\frac{29}{18} \approx 1.6111111111$
$\frac{21}{13} \approx 1.6153846154$	$\frac{47}{29} \approx 1.6206896551$
$\frac{34}{21} \approx 1.6190476191$	$\frac{76}{47} \approx 1.6170212766$

TABLE 20.1. (*Continued*)

F_{n+1}/F_n	L_{n+1}/L_n
$\frac{55}{34} \approx 1.6176470588$	$\frac{123}{76} \approx 1.6184210526$
$\frac{89}{55} \approx 1.6181818182$	$\frac{199}{123} \approx 1.6178861788$
$\frac{144}{89} \approx 1.6179775281$	$\frac{322}{199} \approx 1.6180904523$
$\frac{233}{144} \approx 1.6180555556$	$\frac{521}{322} \approx 1.6180124224$
$\frac{377}{233} \approx 1.6180257511$	$\frac{843}{521} \approx 1.6180422265$
$\frac{610}{377} \approx 1.6180371353$	$\frac{1,364}{843} \approx 1.6180308422$
$\frac{987}{610} \approx 1.6180327869$	$\frac{2,207}{1,364} \approx 1.6180351906$
$\frac{1,597}{987} \approx 1.6180344478$	$\frac{3,571}{2,207} \approx 1.6180335297$
$\frac{2,584}{1,597} \approx 1.6180338134$	$\frac{5,778}{3,571} \approx 1.6180341641$
$\frac{4,181}{2,584} \approx 1.6180340557$	$\frac{9,349}{5,778} \approx 1.6180339218$
$\frac{6,765}{4,181} \approx 1.6180339632$	$\frac{15,127}{9,349} \approx 1.6180340143$
$\frac{10,946}{6,765} \approx 1.6180339985$	$\frac{24,476}{15,127} \approx 1.6180339789$

As n gets larger and larger, it appears that F_{n+1}/F_n approaches a limit, namely, $1.618033\ldots$.

This phenomenon was observed by the German astronomer and mathematician, Johannes Kepler (1571–1630). It appears that L_{n+1}/L_n also approaches the same magic number, as $n \to \infty$.

Interestingly enough, $\alpha = (1+\sqrt{5})/2 = 1.61803398875\ldots$. So it is reasonable to predict that both ratios converge to the same limit α, the positive root of the quadratic equation $x^2 - x - 1 = 0$.

To confirm this, let $x = F_{n+1}/F_n$. From the Fibonacci recurrence relation, we have

$$\begin{aligned}\frac{F_{n+1}}{F_n} &= 1 + \frac{F_{n-1}}{F_n}\\ &= 1 + \frac{1}{(F_n/F_{n-1})}\end{aligned}$$

As $n \to \infty$, this yields the equation $x = 1 + (1/x)$; that is, $x^2 - x - 1 = 0$. Thus

$$x = \frac{1 \pm \sqrt{5}}{2}$$

Since the limit is positive, it follows that

$$\lim_{n\to\infty} \frac{F_{n+1}}{F_n} = \frac{1+\sqrt{5}}{2} = \alpha$$

as was predicted.

Let α and β be the solutions of the quadratic equation $x^2 - x - 1 = 0$. The numbers α and β are the only numbers such that the reciprocal of each is obtained by subtracting 1 from it, that is, $x - 1 = 1/x$, where $x = \alpha$ or β. Thus α is the only positive number that has this property.

Since $\alpha - 1 = 1/\alpha$, it follows that α and $1/\alpha$ have the same infinite decimal portion:

$$\alpha = 1.61803398875\ldots$$
$$\frac{1}{\alpha} = 0.61803398875\ldots$$

An *interesting observation from Table 20.1*: When n is even, $F_{n+1}/F_n > \alpha$ and when it is odd, $F_{n+1}/F_n < \alpha$; the same behavior holds for L_{n+1}/L_n.

It follows from the preceding discussion that

$$\lim_{n\to\infty} \frac{L_{n+1}}{L_n} = \frac{1+\sqrt{5}}{2} = \alpha$$

This number α is so intriguing a number that it was known to the ancient Greeks at least sixteen centuries before Fibonacci. They called it the *Golden Section*, for reasons that will be clear shortly.

Before the Greeks, the ancient Egyptians used it in the construction of their great pyramids. The *Papyrus of Ahmes*, written hundreds of years before ancient Greek civilization existed and now kept in the British Museum, contains a detailed account of how the number was used in the building of the Great Pyramid of Giza around 3070 B.C. Ahmes refers to this number as a "sacred ratio."

The height of the Great Pyramid (Fig. 20.1*a*) is 484.4 feet, which is about 5813 inches; notice the three consecutive Fibonacci numbers in the height: 5, 8, and 13.

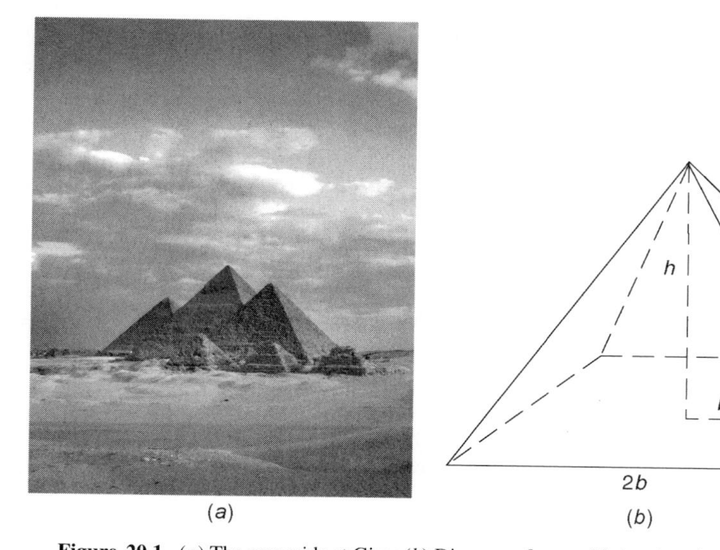

Figure 20.1. (*a*) The pyramids at Giza; (*b*) Diagram of pyramid showing the Golden Section.

Herodotus, a Greek historian of the fifth century B.C., wrote that he was told by the Egyptian priests that the proportions of the Great Pyramid were chosen in such a way that "the area of a square with a side of length equaling the height of the Pyramid is the same as the area of a slanted [triangular] face."

To confirm this, let $2b$ (Figure 20.1b) denote the base of the Pyramid, a the altitude of a slanted (triangular) face, and h the height of the Pyramid. According to Herodotus formula, $h^2 = (2b \cdot a)/2 = ab$. But, by the Pythagorean theorem, $h^2 = a^2 - b^2$, so $a^2 - b^2 = ab$, and hence $(a/b)^2 = 1 + (a/b)$. Thus a/b satisfies the quadratic equation $x^2 = x + 1$, so $x = a/b = \alpha$.

The actual measurements are $a = 188.4$ meters, $b = 116.4$ meters, and $h = 148.2$ meters. So $a/b = 188.4/116.4 \approx 1.618$.

In 1938, L. Hogben observed that the ratio of the base perimeter $8b$ of the Pyramid to its vertical height equals that of the circumference of a circle to its radius, that is, $8b = 2\pi h$. Thus

$$\pi = \frac{4b}{h} = \frac{4b}{\sqrt{ab}} = 4\sqrt{\frac{b}{a}} = \frac{4}{\sqrt{\alpha}}$$
$$\approx 3.1446$$

This estimate is accurate for two decimal places.

From Herodotus' statement, it follows that the ratio of the sum of the areas of the lateral faces of the Great Pyramid to the base area is also α (see Exercise 18).

The golden ratio is often denoted by φ, the Greek letter *phi*. It was given this name about a century ago by the American mathematician Mark Barr, who chose *phi* because it is the first Greek letter in the name of Phidias (490?–420? B.C.), the greatest of Greek sculptors, who employed the Golden Section constantly in his work. However, here we shall continue to denote the Golden Section by α for consistency.

The Golden Ratio is also often denoted by another popular name, τ, the Greek letter *tau*. According to H. S. M. Coxeter of the University of Toronto, this usage comes from the fact that τ is the first letter of the Greek word "$\tau o \mu \eta$," which means "the section."

The great German astronomer and mathematician Johannes Kepler referred to α as "*sectio divina*" (divine section) and Leonardo da Vinci (1452–1519), the great Italian artist, who employed the number in many of his great works, called it "*sectio aurea*" (the golden chapter), a term still in popular use.

Kepler singles out α in his *Mysterium Cosmographicum de Admirabile Proportine Orbium Celestium* as one of the two "great treasures" of geometry, the other being the Pythagorean theorem:

> Geometry has two great treasures: one is the Theorem of Pythagoras; the other, the division of a line into extreme and mean ratio. The first we may compare to a measure of gold, the second we may name a precious jewel.

The mysterious number α was the principal character of a book, *De Divina Proportione*, by Fra Luca Pacioli de Borgo, published in Venice in 1509. Pacioli describes the properties of α, stopping at thirteen "for the sake of our salvation." Another edition of the book appeared in Milan in 1956.

Today, the magical number α is variously called the *golden mean*, the *golden ratio*, the *golden proportion*, the *divine section*, or the *divine proportion*.

The concept of a golden mean has its origin in plane geometry. It stems from locating a point on a line segment such that it divides the line segment into two in a certain ratio. To explain this more clearly, we define the concept of *mean proportional*.

MEAN PROPORTIONAL

Let a, b, and c be any three positive integers such that $a^2 = bc$. Then a is called the *mean proportional* of b and c. Notice that $a^2 = bc$ if and only if $a/b = c/a$; that is, $a^2 = bc$ if and only if $a{:}b = c{:}a$. For example, since $6^2 = 4 \cdot 9$, 6 is the mean proportional of 4 and 9. Likewise, $\sqrt{6}$ is the mean proportional of 2 and 3.

A GEOMETRIC INTERPRETATION

Geometrically, we would like to find a point C on a line segment $\overline{AB}$ such that the length of the greater part $\overline{AC}$ is the mean proportional of the whole length AB and the length BC of the smaller part (see Fig. 20.2). Thus, we would like to find C such that $AC/BC = AB/AC$; then C divides $\overline{AB}$ in the Golden Ratio.

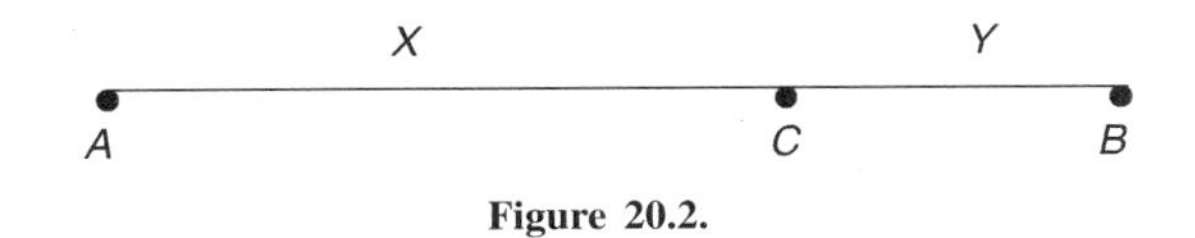

Figure 20.2.

To locate C, let $AC = x$ and $BC = y$. Then the equation $AC/BC = AB/AC$ yields

$$\frac{x}{y} = \frac{x+y}{x}$$

$$= 1 + \frac{y}{x} = 1 + \frac{1}{(x/y)}$$

That is, $(x/y)^2 - (x/y) - 1 = 0$. So x/y satisfies our well-known quadratic equation $t^2 - t - 1 = 0$. Since the ratio $x/y > 0$, this implies $x/y = (1+\sqrt{5})/2 = \alpha$, that is, $x{:}y = \alpha{:}1$. Thus we must choose the point C in such a way that $AC/BC = \alpha$, that is, C divides $\overline{AB}$ in the Golden Ratio.

RULER AND COMPASS CONSTRUCTION

Although this process defines the point C algebraically, how do we locate it geometrically? In other words, how do we locate it using a ruler and compass?

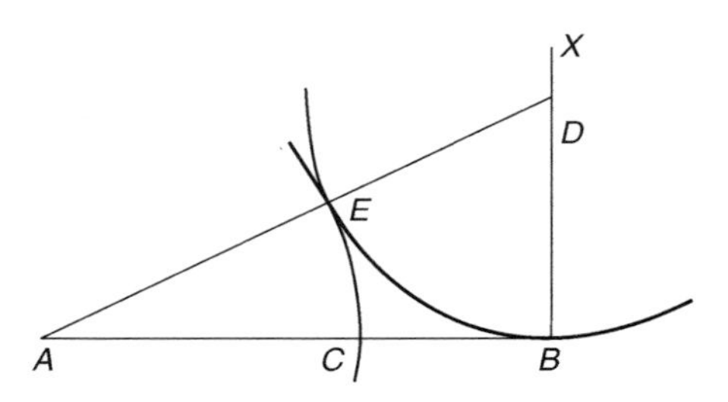

Figure 20.3.

To this end, draw $\overline{BX} \perp \overline{AB}$ (see Fig. 20.3). Select a point D on $\overline{BX}$ such that $AB = 2BD$. With D as center, draw an arc of radius DB to intersect $\overline{AD}$ at E. Now, with A as center, draw an arc of radius AE to meet $\overline{AB}$ at C. We now claim that C divides $\overline{AB}$ in the desired ratio.

To verify this, we have $AC = AE$, $BD = ED$, and $AB = AC + BC = 2BD = 2ED$. Since $\triangle ABD$ is a right triangle, by the Pythagorean theorem, $AD = \sqrt{5}BD = \sqrt{5}ED$. Therefore $AC = AE = AD - ED = (\sqrt{5} - 1)ED$. Thus

$$\frac{AB}{AC} = \frac{2ED}{(\sqrt{5} - 1)ED} = \frac{2}{(\sqrt{5} - 1)} = \frac{1 + \sqrt{5}}{2} = \alpha$$

Then $AB/(AB - AC) = \alpha$. This yields $1 - (BC/AB) = 1/\alpha$, that is,

$$\frac{BC}{AB} = \frac{\alpha}{\alpha - 1} = -\frac{\alpha}{\beta} = \alpha^2$$

$$\therefore \quad \frac{AC}{BC} = \frac{AC}{AB} \cdot \frac{AB}{BC} = \frac{1}{\alpha} \cdot \alpha^2 = \alpha$$

Thus $AC/BC = AB/AC = \alpha$, so C is the desired point.

EULER'S CONSTRUCTION

Next, we present Euler's method for locating the point C. This construction has in fact been attributed to the Pythagoreans, since Euler included it along with the theorems and constructions the Pythagoreans developed.

To locate the point C that divides $\overline{AB}$ in the Golden Ratio, first complete the square $ABDE$. Let F bisect $\overline{AE}$. With F as center, draw an arc of radius FB to cut ray $\overrightarrow{FA}$ at G. Now, with A as center and AG as radius, draw an arc to intersect $\overline{AB}$ at C. Then C is the desired point (see Fig. 20.4).

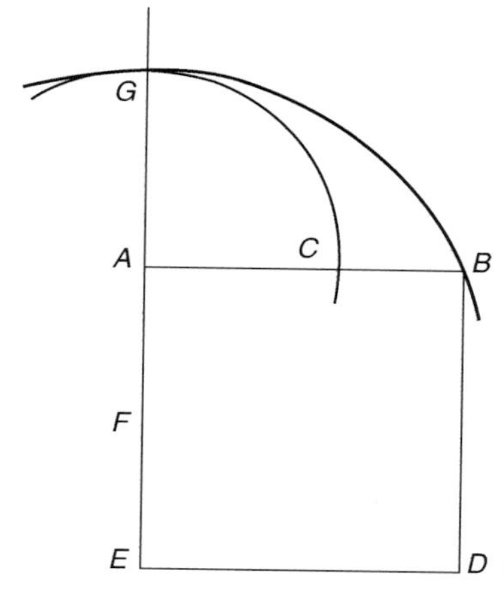

Figure 20.4.

To confirm this, we have $AB = AE = 2AF$. So, by the Pythagorean theorem, $FB = FG = \sqrt{5}AF$. Then $FA + AG = FG = \sqrt{5}AF$, so $AG = AC = (\sqrt{5} - 1)AF$.

$$\therefore \quad \frac{AB}{AC} = \frac{2AF}{(\sqrt{5} - 1)AF} = \frac{2}{\sqrt{5} - 1} = \alpha$$

Moreover, since $BC = AB - AC$,

$$\frac{BC}{AC} = \frac{AB}{AC} - 1 = \alpha - 1 = -\beta \qquad \text{so} \quad \frac{AC}{BC} = -\frac{1}{\beta} = \alpha$$

Thus

$$\frac{AB}{AC} = \frac{AC}{BC} = \alpha$$

so C is indeed the desired point.

GENERATING THE GOLDEN RATIO BY NEWTON'S METHOD

In 1999, J. W. Roche of LaSalle High School at Wyndmoor, Pennsylvania, tried to estimate the Golden Ratio using Newton's method of approximation and the function $f(x) = x^2 - x - 1$. In the process, he found a spectacular relationship between the various approximations and Fibonacci numbers.

Using $x_1 = 2$ as the seed and the recursive formula $x_{n+1} = x_n - f(x_n)/f'(x_n)$, he found the next three approximations to be $x_2 = 5/3$, $x_3 = 34/21$, and $x_4 = 1597/987$. Noticing that they are all ratios of consecutive Fibonacci numbers, he conjectured that $x_n = F_{2^n+1}/F_{2^n}$, where $n \geq 1$.

The validity of Roche's conjecture can be established using the principle of mathematical induction (PMI). Since $F_3/F_2 = 2 = x_1$, the formula works when $n = 1$.

Assume it is true for an arbitrary positive integer $n{:}x_n = F_{m+1}/F_m$, where $m = 2^n$. Since $f'(x) = 2x - 1$,

$$x_{n+1} = x_n - \frac{x_n^2 - x_n - 1}{2x_n - 1} = \frac{x_n^2 + 1}{2x_n - 1} = \frac{F_{m+1}^2/F_m^2 + 1}{2F_{m+1}/F_m - 1} = \frac{F_{m+1}^2 + F_m^2}{F_m(2F_{m+1} - F_m)}$$

$$= \frac{F_{2m+1}}{F_m(F_{m+1} + F_{m-1})} = \frac{F_{2m+1}}{F_m L_n} = \frac{F_{2m+1}}{F_{2m}}$$

Thus, by PMI, the formula holds for all $n \geq 1$.

As a by-product, since $\lim_{m\to\infty} F_{m+1}/F_m = \alpha$, it follows that the sequence of approximations $\{x_n\}$ approaches α as $n \to \infty$, as expected.

EXERCISES 20

1. Is $\alpha{:}1 = 1{:}\frac{1}{\alpha}$?
2. Is $\alpha{:}1 = 1{:} -\beta$?
3. Is $\alpha{:}1 = 1{:}\alpha - 1$?
4. Let C divide $\overline{AB}$ in the Golden Ratio, $\overline{AC}$ being the larger segment. Let $AC = 1$. Show that $BC = 1/\alpha$ and $AB = \alpha$.
5. Let C divide $\overline{AB}$ of unit length in the Golden Ratio, $\overline{AC}$ being the larger segment. Show that $BC = 1/\alpha^2$ and $AC = 1/\alpha$.
6. Suppose $BD = 1$ in Figure 20.3. Find BC.

Let C divide the line segment $\overline{AB}$ in the Golden Ratio, where $AB = 1$ and $AC = t$.

7. Find the quadratic equation satisfied by t.
8. Solve the equation.
9. Find the value of t.
10. Show that $t = -\beta$.
11. Evaluate the sum $\sqrt{1 - \sqrt{1 - \sqrt{1 - \sqrt{1 - \cdots}}}}$.
12. Let $x^2 = 1 - x$. Show that $x = \sqrt{1 - \sqrt{1 - \sqrt{1 - \sqrt{1 - \cdots}}}}$.

Let $a/b = c/d$. Prove each.

13. $\dfrac{b}{a} = \dfrac{d}{c}$
14. $\dfrac{a+b}{b} = \dfrac{c+d}{d}$
15. $\dfrac{a-b}{b} = \dfrac{c-d}{d}$
16. $\dfrac{a+b}{a-b} = \dfrac{c+d}{c-d}$
17. Suppose a side of the Great Pyramid is $2b$. Show that the altitude of a lateral face is $b\alpha$.

18. The base of the Great Pyramid is square. Show that the ratio of the sum of the areas of its lateral faces to the base area is α.

Prove each.

19. $\alpha = 1 + \frac{1}{\alpha}$
20. $\alpha = \frac{1}{\alpha - 1}$
21. $\alpha^n = \alpha^{n-1} + \alpha^{n-2}, n \geq 2$
22. $\frac{1}{\alpha^n} = \frac{1}{\alpha^{n+1}} + \frac{1}{\alpha^{n+2}}$
23. $\sum_{n=1}^{\infty} \frac{1}{\alpha^n} = \alpha$
24. $\sum_{n=0}^{\infty} \frac{1}{\alpha^n} = \alpha^2$
25. $\sum_{n=1}^{\infty} \frac{1}{\alpha^{2n}} = 3\alpha + 2$
26. $\sum_{n=0}^{\infty} \frac{1}{\alpha^{kn}} = \alpha F_{2k} + F_{2k-1}$

Verify each.

27. $\alpha\sqrt{3 - \alpha} = \sqrt{\alpha + 2}$
28. $\sqrt{3 - \alpha} = \frac{\sqrt{10 - 2\sqrt{5}}}{2}$
29. $\sqrt{\alpha + 2} = \frac{\sqrt{10 + 2\sqrt{5}}}{2}$

Using the fact that $\cos \pi/5 = \alpha/2$, express each in terms of α.

30. $\sin \pi/5$
31. $\cos \pi/10$
32. $\sin \pi/10$
33. Let ν (lowercase Greek letter *nu*) be a solution of the equation $x^2 = x + 1$. Show that $\nu + (1/\nu^2) = 2$.

Evaluate each limit, where G_n denotes the nth generalized Fibonacci number.

34. $\lim_{n \to \infty} \frac{F_n}{F_{n+1}}$
35. $\lim_{n \to \infty} \frac{L_n}{L_{n+1}}$
36. $\lim_{n \to \infty} \frac{L_n}{F_n}$
37. $\lim_{n \to \infty} \frac{G_{n+1}}{G_n}$
38. $\lim_{n \to \infty} \frac{2n}{n + 1 + \sqrt{5n^2 - 2n + 1}}$ (Hoggatt and Lind, 1967)

21 THE GOLDEN RATIO REVISITED

On January 21, 1911, W. Schooling wrote in the *Daily Telegraph* that there is a "very wonderful number which may be called by the Greek letter *phi*, of which nobody has heard much as yet, but of which, perhaps, a great deal is likely to be heard in the course of time." It is intriguing to note that his prediction has come true.

According to R. Fisher, Saint Thomas Aquinas (1225–1274), the greatest of the medieval philosophers and theologians, "described one of the basic rules of aesthetics—man's senses enjoy objects that are properly proportioned. He referred to the direct relationship between beauty and mathematics, which is often measurable and can be found in nature." St. Thomas Aquinas was of course referring to the Golden Ratio.

The Golden Ratio α can occur in extremely unlikely places, as we will see throughout this chapter.

URANIUM AND THE GOLDEN RATIO

Uranium, an important source of nuclear energy, enjoys a unique place among the chemical elements. The ratio of the number of neutrons to that of protons is maximum for uranium; curiously enough, this ratio is approximately α:

$$\frac{\text{Number of neutrons}}{\text{Number of protons}} = \frac{146}{92} \approx 1.5869565 \approx \alpha$$

π AND THE GOLDEN RATIO

M. J. Zerger observed two fascinating relationships between π and α:

- The first 10 digits of α can be permuted to obtain the first 10 digits of $1/\pi$: $\alpha = 1.618033988\ldots$ and $1/\pi = 0.3183098861\ldots$.
- The first nine digits of $1/\alpha$ can be permuted to form the first nine digits of $1/\pi$: $1/\alpha = 0.618033988\ldots$ and $1/\pi = 0.318309886\ldots$.

ILLINOIS AND THE GOLDEN RATIO

Zerger made yet another striking observation about the state of Illinois. Both the telephone area code 618 and the Zip Code prefix 618 are assigned to Illinois. (Recall that 618 are the first three digits in α after the decimal point.)

THE GOLDEN RATIO AND THE HUMAN BODY

Studies have shown that several proportions of the human body exemplify the Golden Ratio. For instance, consider the drawing of a typical athlete in Figure 21.1.

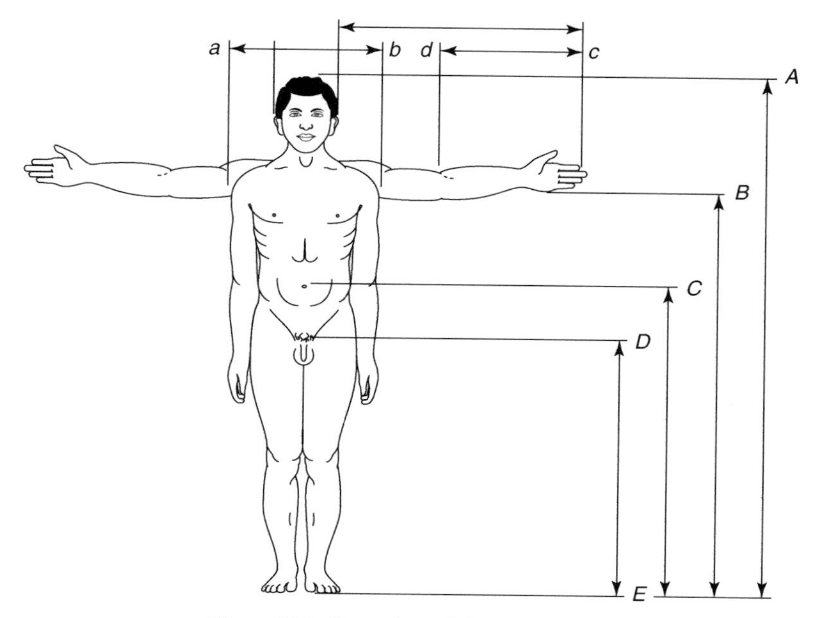

Figure 21.1. Proportions of the human body.

Then

$$\frac{AE}{CE} = \frac{\text{Height}}{\text{Navel height}} \approx \alpha$$

and

$$\frac{CE}{AC} = \frac{\text{Navel height}}{\text{Distance from the navel to the top of the head}} \approx \alpha$$

Thus height $= \alpha$ (navel height).

Moreover,

$$\frac{bc}{ab} = \frac{\text{Arm length}}{\text{Shoulder width}} \approx \alpha$$

In fact, using the figure, we can find several other remarkable ratios that approximate the Golden Ratio.

Certain bones in our body also show a relationship to the magical ratio. Figure 21.2, for instance, illustrates such a relationship between the hand and forearm. Because of the Golden Ratio's close association with the human body, the Golden Ratio is often referred to as "the number of our physical body."

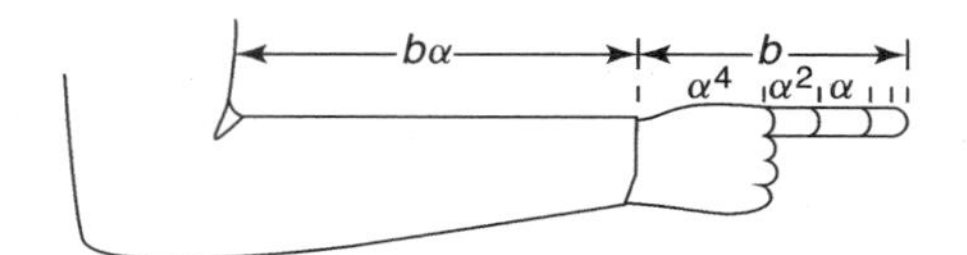

Figure 21.2. Reprinted with permission from M. H. Holt, 1964, *The Pentagon*.

According to S. Vajda of the University of Sussex in England, J. Gordon has detected (1938) the Golden Ratio in the English landscape *The Cornfield* by John Constable (1776–1837), in *Portrait of a Lady* by the Dutch painter and graphic artist Rembrandt Harmenszoon van Rijn (1606–1669), and in *Venus and Adonnis* by the Venetian painter Titian (Tiziano Vecellio, 1487?–1576). *The Cornfield* and *Portrait of a Lady* are displayed in the National Gallery in London.

MEXICAN PYRAMIDS

Just as the Egyptian pyramids exemplify the basic principles of aesthetics and perfect proportion, so do the Mexican pyramids. Both appear to have been built by people of common ancestry and both seem to have incorporated the magic ratio in their construction.

For instance, the cross section of a Mexican pyramid shown in Figure 21.3 clearly reveals the incorporation of the Golden Ratio into its architecture. The cross section depicts a staircase-like structure. There are 16 steps in the first set, 42 in the second,

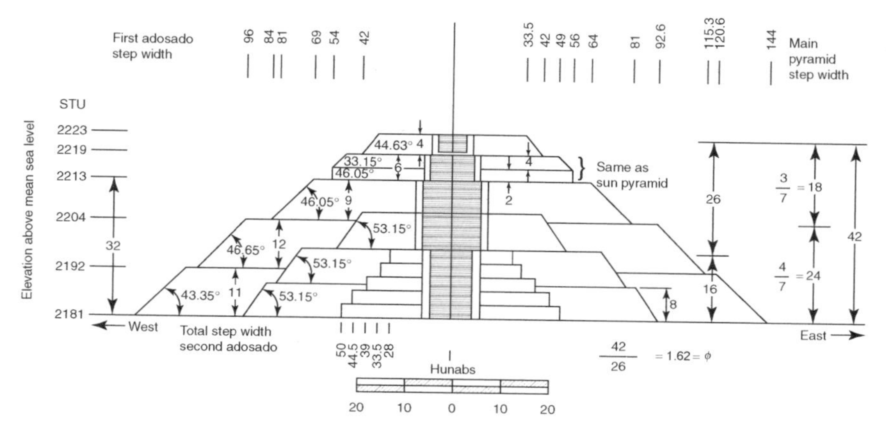

Figure 21.3. A cross-section of a Mexican pyramid.

and 68 in the third. These innocent-looking numbers, amazingly enough, are linked by the sacred ratio α:

$$16\alpha \approx 26$$
$$16 + 26 = 42$$
$$26\alpha \approx 42$$
$$42\alpha \approx 68$$

These numbers—16, 26, 42, and 68—have interesting relationships to the body of a well-proportioned man 68 inches tall. First, taking 10 inches as one unit of height, his height is about α^4; his navel height is 42 inches, which is about α^3; the height of the top of the head from the navel is 26 inches, that is, about α^2; the height of the vortex from his breast line is 16 inches, which is about α; and, finally, his breast line is 10 inches above the navel, that is, one unit of measurement, which is α^0.

In fact, $68 = \lfloor 10\alpha^4 \rfloor = \lceil 42\alpha \rceil$, $42 = \lfloor 10\alpha^3 \rfloor = \lfloor 26\alpha \rfloor$, and $26 = \lceil 16\alpha \rceil$.

VIOLIN AND THE GOLDEN TRIANGLE

The Golden Ratio plays an important role in the making of the violin, one of the most beautiful of orchestral instruments. The point B, where the two lines through the centers of the f-holes intersect, divides the body in the Golden Ratio: $AB/BC = \alpha$ (see Fig. 21.4). Besides, $AC/CD = \alpha$, so the body and the neck are in the golden proportion. It now follows that

$$\frac{AD}{AC} = \frac{AC}{AB} = \frac{CD}{BC} = \alpha$$

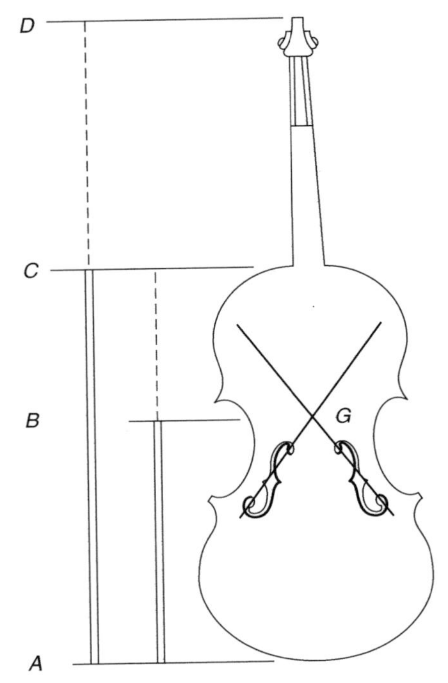

Figure 21.4. The point, B, on the violin where two lines drawn through the centers of the f holes intersect divide the instrument in the Golden Ratio; the body and neck are likewise in the Golden Ratio (*Source*: Trudi Hammel Garland, *Fascinating Fibonaccis: Mystery and Magic in Numbers*, Palo Alto, CA: Seymour, 1987. Copyright © 1987 by Seymour Publications. Used by permission of Pearson Learning.)

ANCIENT FLOOR MOSAICS AND THE GOLDEN RATIO

In 1970, R. E. M. Moore of Guy's Hospital Medical School, London, after studying numerous two-thousand-year-old floor mosaics from Syria, Greece, and Rome, observed an interesting phenomenon: all the mosaic patterns in these cultures showed the exact same dimensions. Therefore, the mosaicists in all these cultures must have

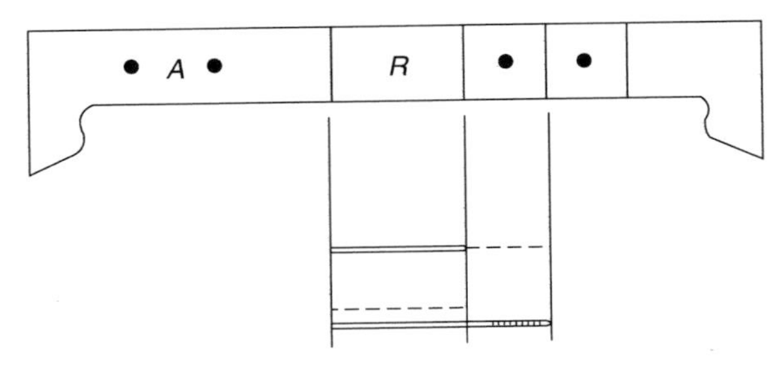

Figure 21.5. Calibrations on a ruler used in ancient mosaics. (*Source*: The Fibonacci Association)

used the same measuring technique and device. In fact, the calibrations on rulers employed by the mosaicists clearly and convincingly underscore the application of the golden proportions in the mosaic patterns (see Fig. 21.5).

THE GOLDEN RATIO IN AN ELECTRICAL NETWORK

Figure 21.6 represents an infinite network consisting of resistors, each with resistence r. We would like to compute the resistence between the points A and B. (This problem was posed at the 1967 International Physics Olympiad, Poland.)

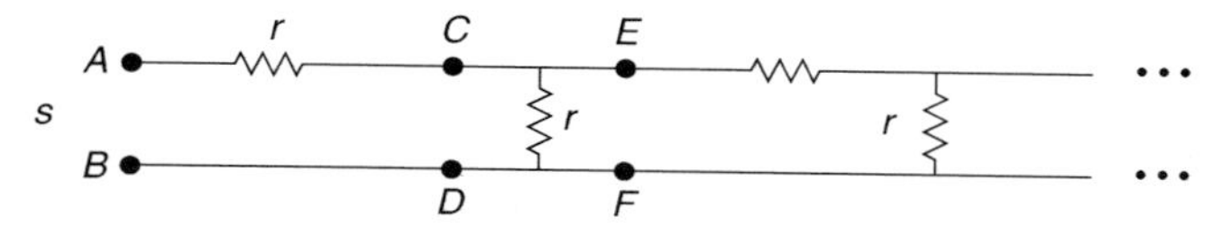

Figure 21.6. An infinite network consisting of resistors.

Let s denote the resistence between the points E and F of the infinite network to their right side. Then the resistence r_{CD} between the points C and D is given by

$$\frac{1}{r_{CD}} = \frac{1}{r} + \frac{1}{s}$$

Now add the resistence r to this. The resistence r_{AB} between A and B of the given network is given by

$$\begin{aligned} r_{AB} &= r + r_{CD} \\ &= r + \frac{rs}{r+s} \end{aligned}$$

Since the resulting network is again infinite, $r_{AB} = s$. Thus

$$s = r + \frac{rs}{r+s}$$

Solving, $s = r\alpha$.

Suppose the network in Figure 21.6 consists of n resistors (see Fig. 21.7). It follows from Chapter 3 that the resistence between A and B is given by

$$Z_i(n) = r + \frac{1}{(1/r) + (1/Z_i(n-1))}$$

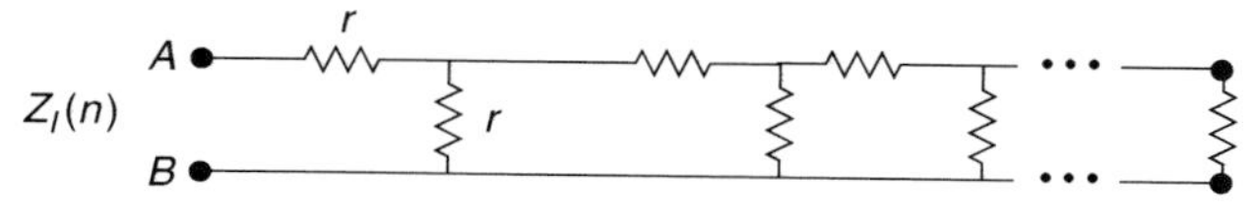

r
A
$Z_i(n)$
r
B

Figure 21.7.

where $Z_i(1) = 2r$. As $n \to \infty$, this recurrence relation yields

$$s = r + \frac{1}{(1/r) + (1/s)}$$

that is, $s = r + (rs/(r + s))$ as we just found.

Thus $\lim_{n\to\infty} Z_i(n) = r\alpha$; so when $r = 1$, $\lim_{n\to\infty} Z_i(n) = \alpha$. This we already knew, since, from Chapter 3, $Z_i(n) = (F_{2n+1}/F_{2n})$, so $\lim_{n\to\infty} Z_i(n) = \lim_{n\to\infty} (F_{2n+1}/F_{2n}) = \alpha$.

THE GOLDEN RATIO IN ELECTROSTATICS

The following problem in electrostatics, the branch of physics that deals with the properties and effects of static electricity, was studied in 1972 by B. Davis, then a student at the Indian Statistical Institute:

> A positive charge $+e$ and two negative charges $-e$ are to be placed on a line in such a way that the potential energy of the whole system is zero.

Suppose the charges are at points A, B, and C; and let $AB = x$ and $BC = y$ (see Fig. 21.8). The potential energy of a system of static charges is the work done in bringing the charges from infinity to these points. The potential energy between two charges is the product of the charges divided by the distance between them.

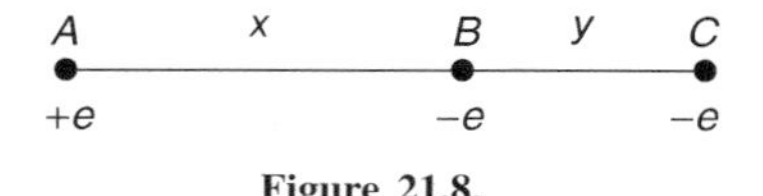

Figure 21.8.

The potential energy due to the charges at A and B is

$$\frac{(+e)(-e)}{x} = -\frac{e^2}{x}$$

The potential energy due to the charges at A and C is

$$\frac{(+e)(-e)}{x+y} = -\frac{e^2}{x+y}$$

and the potential energy due to the charges at A and B is

$$\frac{(-e)(-e)}{y} = \frac{e^2}{y}$$

For the potential energy of the system to be zero, we must have

$$-\frac{e^2}{x} - \frac{e^2}{x+y} + \frac{e^2}{y} = 0$$

$$-y(x+y) - xy + x(x+y) = 0$$

$$x^2 - xy - y^2 = 0$$

$$(x/y)^2 - (x/y) - 1 = 0$$

So $x/y = \alpha$. Thus x/y must be the Golden Ratio for the potential energy to be zero.

THE GOLDEN RATIO BY ORIGAMI

In 1999, P. Glaister of Reading University in England employed the Japanese art of *origami* (folding paper into make decorative shapes) to illustrate yet another mysterious occurrence of our ubiquitous friend α.

Take a 2×1 rectangular piece of paper and fold it in half both ways, as Figure 21.9 shows. Make a crease along $\overline{AD}$ (see Fig. 21.10). Place $\overline{AD}$ along $\overline{AB}$ and form a crease along the fold $\overline{AQ}$ so that $\overline{AQ}$ bisects $\angle DAB$.

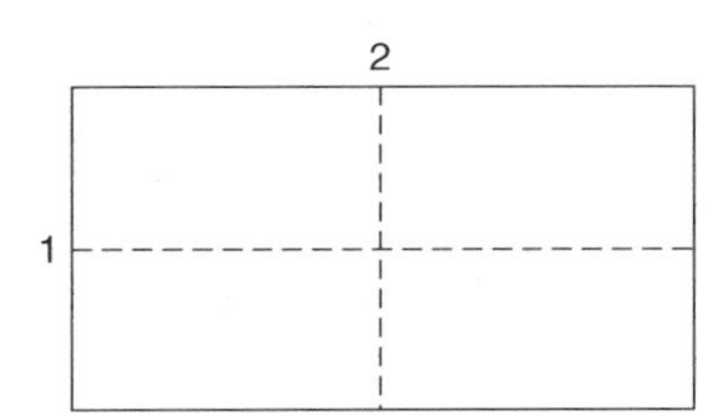

Figure 21.9.

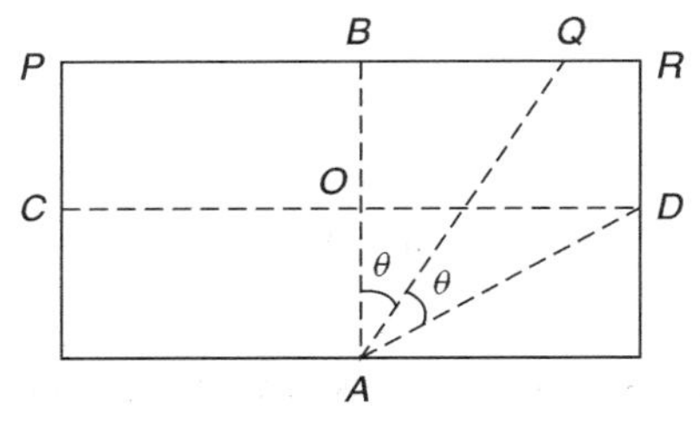

Figure 21.10.

Let $\angle DAB = 2\theta$. From ΔAOD, $\tan 2\theta = 2$. Using the double-angle formula,

$$\tan 2\theta = \frac{2\ \tan\theta}{1 - \tan^2\theta}$$

this yields

$$\frac{2\tan\theta}{1 - \tan^2\theta} = 2$$

That is,

$$\tan^2\theta + \tan\theta - 1 = 0$$

$$\tan\theta = -\beta$$

Therefore, $PQ = PB + BQ = 1 + AB\tan\theta = 1 - \beta = \alpha$, and hence $QR = 2 - \alpha = \beta$.

The following example is based on a calendar problem that appeared in the October 1999 issue of *Mathematics Teacher*.

Example 21.1. The points A and C on the axes are each one unit away from the origin. The point B lies one unit away from both axes in the first quadrant. Find the value of x such that the y-axis bisects the area $ABCD$, where D is the point $(-x,-x)$ and $x > 0$ (see Fig. 21.11).

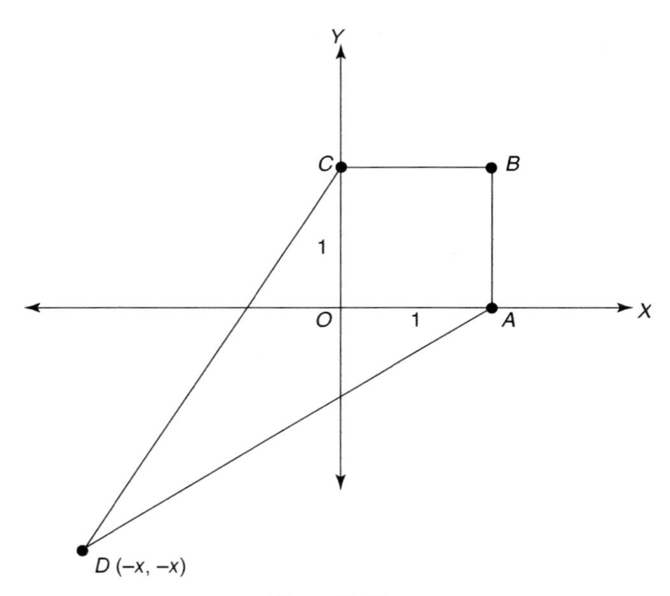

Figure 21.11.

Solution. Since $OABC$ is a square of unit area, the problem is to find x such that area $CDE = 1 +$ area OAE (see Fig. 21.12). The slope of the line $\overleftrightarrow{AD}$ is $x/(x+1)$,

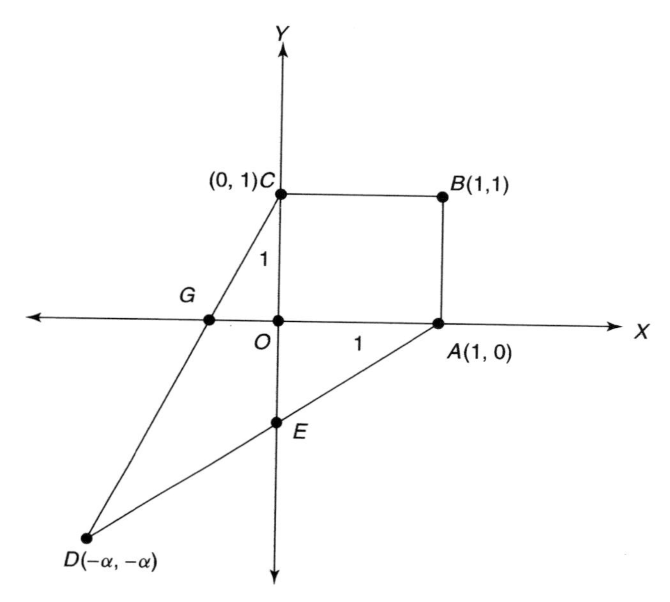

Figure 21.12.

so its equation is $y = [x/(x+1)](x-1)$. Therefore, the point E is $(0, -x/(x+1))$. Thus $OE = x/(x+1)$ and hence

$$CE = 1 + \frac{x}{x+1} = \frac{2x+1}{x+1}$$

$$\therefore \quad \text{Area } CDE = \frac{1}{2} \cdot CE \cdot x = \frac{x(2x+1)}{2(x+1)}$$

and area

$$OAE = \frac{1}{2} \cdot OA \cdot OE = \frac{1}{2} \cdot 1 \cdot \frac{x}{x+1} = \frac{x}{2(x+1)}$$

Thus

$$\frac{x(2x+1)}{2(x+1)} = 1 + \frac{x}{2(x+1)}$$

This yields $x^2 = x + 1$, so $x = \alpha$. Then

$$CE = \frac{2\alpha+1}{\alpha+1} = \frac{\alpha + (\alpha+1)}{\alpha+1} = \frac{\alpha+\alpha^2}{\alpha+1} = \alpha$$

That is, the point O divides $\overline{CE}$ in the Golden Ratio. Besides, $BD{:}OD = \sqrt{2}\alpha^2{:}\sqrt{2}\alpha = \alpha{:}1$, so O also divides $\overline{BD}$ in the Golden Ratio.

In addition, since E is the point $(0, \beta)$, $DA^2 = \alpha^2(1+\alpha^2)$ and $DE^2 = \alpha^2 + (\alpha+\beta)^2 = 1+\alpha^2$; therefore, $DA^2{:}DE^2 = \alpha^2{:}1$, so $DA{:}DE = \alpha{:}1$. Thus E divides

$\overline{DA}$ in the same magic ratio. By symmetry, it follows that the point G divides $\overline{DC}$ in the same ratio. ■

Example 21.2. Consider an equilateral triangle ABC inscribed in a circle. Let Q and R be the midpoints of the sides $\overline{AB}$ and $\overline{BC}$. Let $\overleftrightarrow{QR}$ meet the circle at P and S, as Figure 21.13 shows.

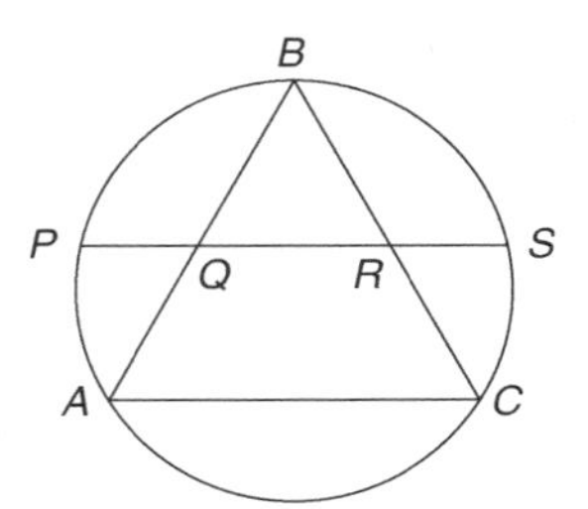

Figure 21.13.

Let $PQ = RS = 1$ and $QR = x$ (see Fig. 21.14). By the intersecting chord theorem, $PR \cdot RS = BR \cdot RC$; that is, $1 + x = x^2$. Therefore, $x = \alpha$. ■

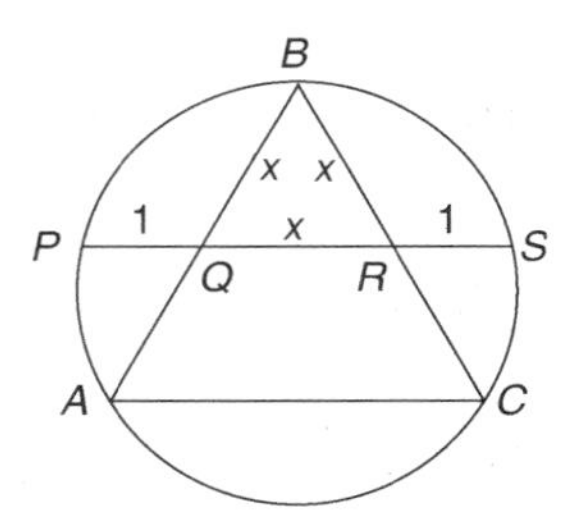

Figure 21.14.

This example was originally proposed as a problem in 1983 by G. Odom of Poughkeepsie, New York, in *The American Mathematical Monthly*. It resurfaced five years later in an article by J. F. Rigby of University College at Cardiff, with an intriguing by-product: The ratio of the length of a side of one of the four large triangles in Figure 21.15 to that of a side of one of the three small triangles is indeed the Golden Ratio.

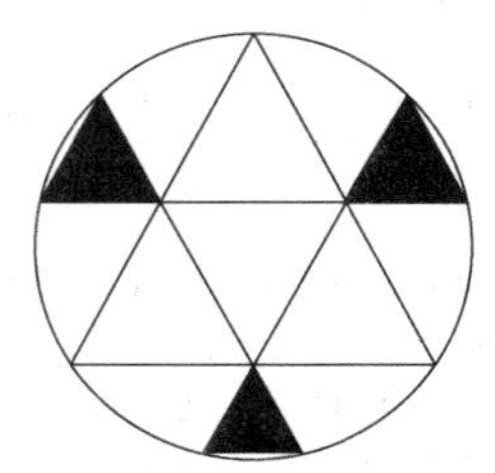

Figure 21.15.

Now we turn to an interesting problem on the congruence and similarity of triangles.

In 1965, M. H. Holt of Minnesota studied an interesting problem proposed by V. E. Hoggatt, Jr: *Do there exist triangles ΔABC and ΔPQR that have five of their six parts (three sides and three angles) congruent, but still not congruent*? This problem also appears in a high school geometry book by E. Moise and F. Downs, who gave two such triangles as a solution, both of which are shown in Figure 21.16.

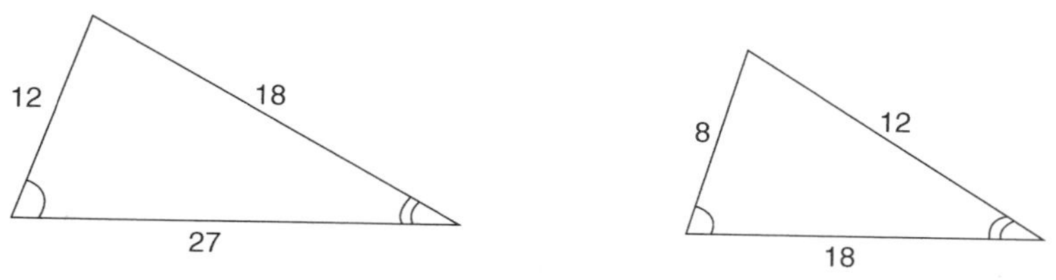

Figure 21.16.

Are there other solutions? If there are, how are they related?

To answer these questions, first notice that the five equal parts cannot include the three sides, since the triangles would then be congruent. Consequently, the five parts must consist of three angles and two sides, so the triangles are indeed similar. But the equal sides cannot be in the same order; otherwise, the triangles would be congruent by the side-angle-side (SAS) theorem or the angle-side-angle (ASA) theorem.

This yields two possibilities for ΔPQR, as Figure 21.17 shows. Since $\Delta ABC \cong \Delta P'Q'R'$, $a/b = b/d = 1/k$ (say), where $k > 0$ and $k \neq 1$. Then $b = ak$ and $d = bk = ak^2$. Since $\Delta ABC \cong \Delta P''Q''R''$, $a/b = b/d = 1/k$, so $b = ak$ and $d = bk = ak^2$. In both cases, the lengths of the sides are in the same ratio $a{:}b{:}d = a{:}ak{:}ak^2 = 1{:}k{:}k^2$. So, if there are triangles whose sides are in the ratio $1{:}k{:}k^2$, their parts would be congruent, but the triangles still would not be congruent.

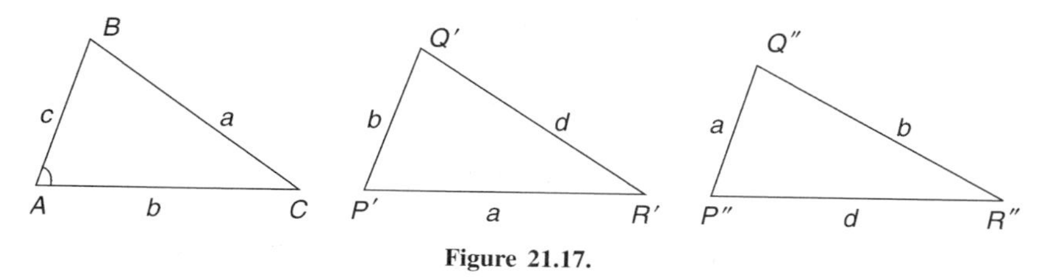

Figure 21.17.

To determine the values of k that yield such triangles, suppose such a triangle exists. Then, by the triangle inequality, $1 + k > k^2$, $1 + k^2 > k$, and $k + k^2 > 1$.

Case 1. Suppose $k > 1$. Then $k^2 > k$, so $1 + k^2 > 1 + k > k$. Also, $k + k^2 > k > 1$. Thus, if $k > 1$, then $1 + k^2 > k$ and $k + k^2 > 1$. So it suffices to identify the values of k for which $1 + k > k^2$, that is, $k^2 - k - 1 < 0$.

Since $k^2 - k - 1 = (k - \alpha)(k - \beta)$, $k^2 - k - 1 < 0$ if and only if $\beta < k < \alpha$. But $k > 1$, so $1 < k < \alpha$.

Graphically, k is the value of x for which the line $1 + x = y$ lies above the parabola $y = x^2$, where $x > 1$, as Figure 21.18 shows.

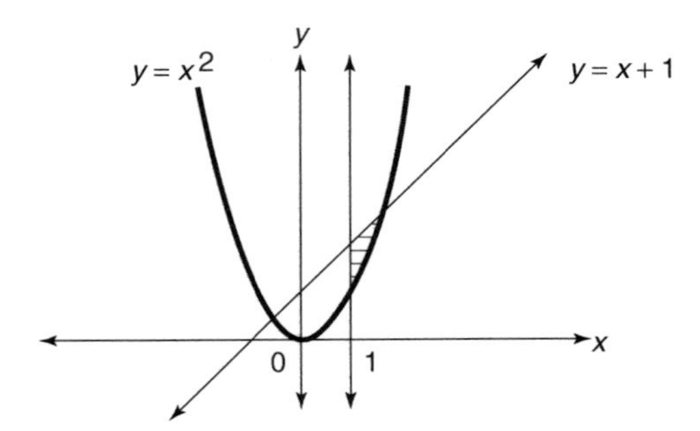

Figure 21.18.

Thus, if k is a number such that $1 < k < a$, then every ΔPQR with sides a, ak, and ak^2 will meet the desired conditions.

Case 2. Suppose $k < 1$. Then $k > k^2$; so $1 + k > k^2$. Also, since $1 > k$, $1 + k^2 > k$. Thus, $1 + k > k^2$ and $1 + k^2 > k$. So it suffices to look for values of k for which $k^2 + k > 1$, that is, $k^2 + k - 1 > 0$, where $k < 1$.

Since $k^2 + k - 1 = (k + \alpha)(k + \beta)$, $k^2 + k - 1 > 0$ if and only if either $k < -\alpha$ or $k > \beta$. Since $k < 1$, this yields $\beta < k < 1$.

Graphically, k is the value of x for which the parabola $y = x + x^2$ lies above the line $y = 1$, where $x < 1$, as Figure 21.19 shows.

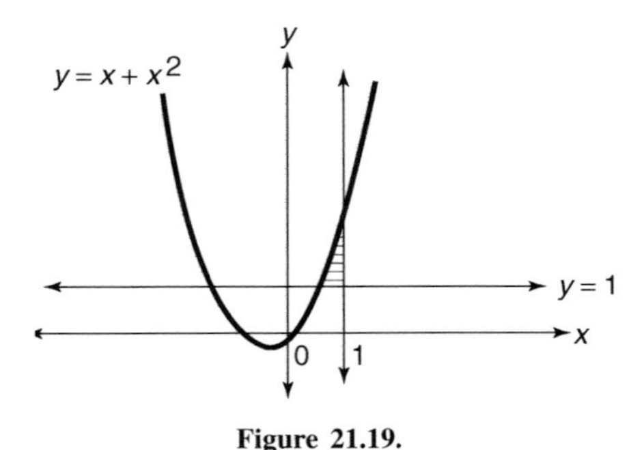

Figure 21.19.

Thus, if k is such that $\beta < k < 1$, then every ΔPQR with sides a, ak, and ak^2 will satisfy the given conditions.

To sum up, there are infinitely many triangles PQR whose five parts are congruent to those of ΔABC, but still not congruent to ΔABC.

DIFFERENTIAL EQUATIONS

Strange as it may seem, the Golden Ratio occurs in the solution of differential equations.

For instance, consider the second-degree differential equation $y'' - y' - y = 0$, where $y' = dy/dx$. Its characteristic equation is $t^2 - t - 1 = 0$, so the characteristic roots are α and β. Thus the basic solutions of the differential equation are $e^{\alpha x}$ and $e^{\beta x}$, so the general solution is $y = Ae^{\alpha x} + Be^{\beta x}$, where A and B are arbitrary constants.

GATTEI'S DISCOVERY OF THE GOLDEN RATIO

When P. Gattei was just a sixth former at Queen Elizabeth's Grammar School, Blackburn, England, he stumbled across a problem involving the inverse f^{-1} of a real-valued function f. Accidently, he dropped the minus sign and ended up taking the derivative f' of f. This led him to investigate if there were real functions f such that

$$f'(x) = f^{-1}(x) \qquad x \geq 0. \tag{21.1}$$

Gattei discovered an interesting solution: $f(x) = Ax^n$, where A is a constant. Then $f'(x) = Anx^{n-1}$ and $f^{-1}(x) = (x/A)^{1/n}$. Condition (21.1) yields $Anx^{n-1} = (x/A)^{1/n}$; that is, $A^{n+1}n^n x^{n(n-1)-1} = 1$; $n^2 - n - 1 = 0$ and $A^{n+1}n^n = 1$. Then $n = \alpha, \beta$ and $A = n^{-n/(n+1)}$.

Suppose $n = \beta$. Then $n < 0$, but $n + 1 > 0$. Thus $A = (\text{a negative number})^{(\text{a positive irrational number})}$, so A is not a real number. Consequently, f is not a real function. But when $n = \alpha$, f is a real function. Thus, the only solution to Eq. (21.1) is $f(x) = (\alpha x)^{\alpha}$.

More generally, consider the equation

$$f^{-1}(x) = f^{(m)}(x) \tag{21.2}$$

Let $f(x) = Ax^n$, so $f^{(m)}(x) = An(n-1)\cdots(n-m+1)x^{n-m}$. Then Eq. (21.2) yields $(x/A)^{1/n} = An(n-1)\cdots(n-m+1)x^{n-m}$. Thus $A^{n+1}[n(n-1)\cdots(n-m+1)]^n x^{n(n-m)-1} = 1$, so $A^{n+1}[n(n-1)\cdots(n-m+1)]^n = 1$ and $n^2 - mn - 1 = 0$. Thus

$$n = \frac{m + \sqrt{m^2+4}}{2}$$

and $A = [n(n-1)\cdots(n-m+1)]^{-n/(n+1)}$. (We shall revisit these two values of m in Chapter 38.)

Let m be odd. If

$$n = \frac{m - \sqrt{m^2+4}}{2}$$

then as before, it can be shown that A is not a real number; but

$$n = \frac{m + \sqrt{m^2+4}}{2}$$

leads to a valid solution:

$$f(x) = [n(n-1)\cdots(n-m+1)]^{-n/(n+1)}x^n$$

On the other hand, let m be even. Then $n(n-1)\cdots(n-m+1)$ is positive, whether n is positive or negative. Thus, we have two solutions:

$$f(x) = [n(n-1)\cdots(n-m+1)]^{-n/(n+1)}x^n$$

where

$$n = \frac{m \pm \sqrt{m^2+4}}{2}$$

THE GOLDEN RATIO AND SNOW PLOWING

The following snowplowing problem was discussed by T. Ratliff of Wheaton College in Massachusetts at the 1996 Fall Meeting of the Northeast Section the Mathematical Association of America, held at the University of Massachusetts in Boston. A quite similar problem appeared in 1984 in *Mathematical Spectrum* (Problem 16.6). The solutions of both versions involve the Golden Ratio:

> On one wintry morning, it started snowing at a constant and heavy rate. A snowplow started plowing at 8 A.M.; by 9 A.M., it plowed two miles; and by 10 A.M., it plowed another mile. Assuming that the snowplow removes a constant volume of snow per hour, what time did it start snowing?

Suppose, it started snowing at time t (in hours) and the plow began T hours before 8 o'clock. Let $x = x(t)$ denote the distance traveled by the snowplow in time t. Since the speed of the plow is inversely proportional to the depth of the snow, it follows that

$$\begin{aligned}\frac{dx}{dt} &= \frac{k}{\text{depth at time } t}\\ &= \frac{k}{ct}\\ &= \frac{K}{t}\end{aligned}$$

where k, c, and K are constants.

Solving this differential equation, we get $x = K\ln t + C$, where $x(T) = 0$, $x(T+1) = 2$, $x(T+2) = 3$, and C is a constant. The condition $x(T) = 0$ yields $C = -k\ln T$, so $x = K\ln(t/T)$. The other two conditions yield:

$$2 = K\ln\left(\frac{T+1}{T}\right) \qquad \text{and} \qquad 3 = K\ln\left(\frac{T+2}{T}\right)$$

Then

$$\begin{aligned}\frac{2}{3} &= \frac{\ln(T+1)/T}{\ln(T+2)/T}\\ 2\ln(T+2) - 2\ln T &= 3\ln(T+1) - 3\ln T\\ T(T+2)^2 &= (T+1)^3\end{aligned}$$

This yields $T^2 + T - 1 = 0$, so $T = -\beta \approx 0.61803398875$. Thus $T \approx 37$ minutes, 5 seconds. Consequently, it started snowing at 7:22:55 A.M.

THE GOLDEN RATIO IN ALGEBRA

In 1936, Eric T. Bell (1883–1960), a well-known Scottish-American mathematician, proved that the only polynomial, symmetric function $\varphi(s, t)$ that satisfies the associativity condition $\varphi(x, \varphi(x, y)) = \varphi(\varphi(x, y), z)$ is $\varphi(s, t) = s * t = a + b(s + t) + cst$, where a, b, and c are arbitrary constants such that $b^2 - b - ac = 0$, and s and t are complex numbers. In particular, let $ac = 1$. Then $b^2 - b - 1 = 0$, so $b = \alpha$ or β.

Thus the binary operation $*$, defined by $s * t = a + b(s + t) + cst$, where $ac = 1$, is associative only if $b = \alpha$ or β. We can confirm this (see Exercise 5).

BILINEAR TRANSFORMATION

In 1964, V. E. Hoggatt, Jr., discovered a close relationship between the bilinear transformation $w = (az + b)/(cz + d)$ and Fibonacci numbers. This is the essence of the following theorem.

Theorem 21.1. The bilinear transformation $w = (az + b)/(cz + d)$ has two distinct fixed points α and β if and only if $a - d = b = c \neq 0$, where a, b, c, and d are integers; $a, d > 0$; and $ad - bc = 1$.

Proof. Suppose the bilinear transformation has a fixed point. It is the solution of the equation $z = (az + b)/(cz + d)$; that is, $cz^2 - (a - d)z - b = 0$. Since there are two fixed points α and β, $c \neq 0$ and

$$
\begin{aligned}
z^2 - \frac{a-d}{c}z - \frac{b}{c} &= (z - \alpha)(z - \beta) \\
&= z^2 - z - 1
\end{aligned}
$$

Equating coefficients of like terms, we get $a - d = b = c$. Thus $a - d = b = c \neq 0$.

Conversely, let $a - d = b = c \neq 0$. Then

$$
w = \frac{az + b}{bz + (a - b)}
$$

Its fixed points are given by

$$
z = \frac{az + b}{bz + (a - b)}
$$

that is, $z^2 - z - 1 = 0$. So the fixed points are α and β. ■

We now turn to yet another occurrence of the Golden Ratio in geometry, discovered in 1966 by J. A. H. Hunter. A triangle in which the square of one side equals the product

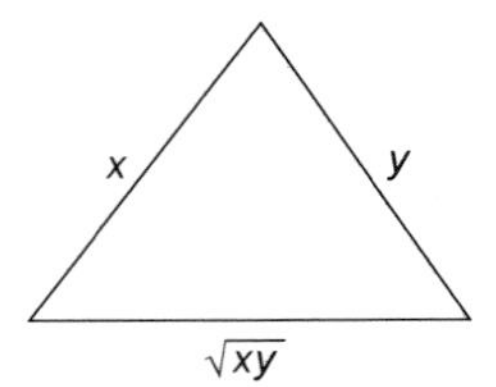

Figure 21.20.

of the other two sides. Suppose the sides are x, y, and $\sqrt{xy}$ units long, where $x > y$ (see Fig. 21.20). Let $x = a^2$ and $y = b^2$, so the three sides a^2, b^2, and ab. Then, by the triangle inequality, $ab + b^2 > a^2$; that is, $(a/b)^2 - (a/b) - 1 < 0$, so $\beta < a/b < \alpha$.

THE GOLDEN RATIO AND CENTROIDS OF CIRCLES

Consider two circles, A and B, one inside the other, but tangential to each other at a point O (see Fig. 21.21). Let their radii be a and b ($< a$), respectively, so their areas are πa^2 and πb^2. Let C_A and C_B be the centroids of the circles, so the points O, C_A, and C_B are collinear. Then the centroid C of the remnant $A - B$ is the endpoint of the diameter of C_B through O.

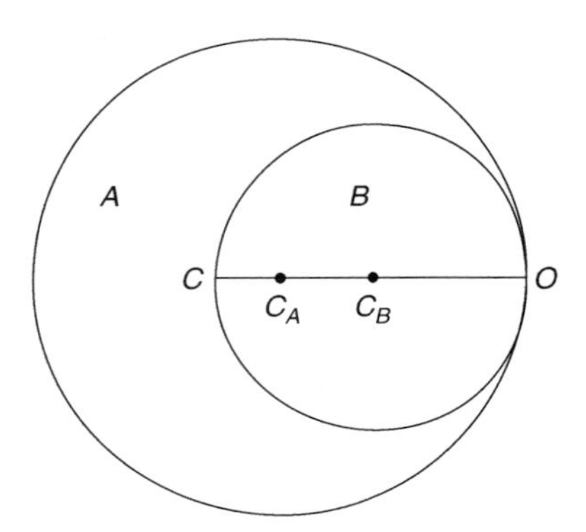

Figure 21.21.

Taking moments about O,

$$\pi b^2 \cdot OC_B + \pi(a^2 - b^2) = \pi a^2 \cdot OC_A$$

That is,

$$1 + \left(\frac{a^2}{b^2} - 1\right)\left(\frac{2b}{b}\right) = \left(\frac{a^2}{b^2}\right)\left(\frac{a}{b}\right)$$

$$2\left(\frac{a^2}{b^2} - 1\right) = \frac{a^3}{b^3} - 1$$

since $a \neq b$, this yields

$$2(a/b + 1) = a^2/b^2 + a/b + 1$$

That is,

$$(a/b)^2 - (a/b) - 1 = 0$$

Since $a > b$, it follows that $a/b = \alpha$, the Golden Ratio.

We can extend this discussion to any planar figure, as was done by H. E. Huntley in 1974.

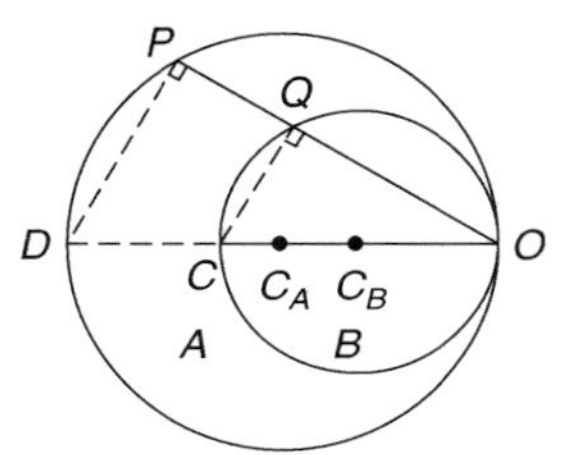

Figure 21.22.

As an additional exercise, let a chord $\overline{OP}$ of circle C_A intersect circle C_B at Q (see Fig. 21.22). Since the angle in a semicircle is a right angle, it follows that $\Delta\ OPD \approx \Delta\ OQC$. Therefore,

$$\frac{OP}{OQ} = \frac{OD}{OC} = \frac{2a}{2b} = \alpha$$

Thus Q divides the chord $\overline{OP}$ in the Golden Ratio.

EXERCISES 21

The semivertical angle of a right-circular cone is 54°, and its lateral side is one unit long. Compute each.

1. Base circumference.
2. Base area.
3. Volume of the cone.
4. Lateral surface area.
5. Show that the binary operation $*$, defined by $s * t = a + b(s + t) + cst$, where $ac = 1$ is associative only if $b = \alpha$ or β.
6. Show that the equation $x^n - xF_n - F_{n-1} = 0$ has no solution $> \alpha$, where $n \geq 2$ (Wall, 1964).
7. Let $x_{k+1} = \sqrt[n]{F_{n-1} + x_k F_n}$, where $x_0 \geq 0$ and $n \geq 2$. Find $\lim_{k\to\infty} x_k$, if it exists (Wall, 1964).
8. Find the value of x such that $n^n + (n + x)^n = (n + 2x)^n$, where $n \geq 1$ (Alfred, 1964).

9. Evaluate $\sum_0^{\infty} |\beta|^n$.
10. Let $I_n = \int_0^1 x^{n-1} dx$, where $n \geq 2$. Evaluate $\lim_{n\to\infty} I_n$.
11. Let t be a number such that $t = \int_0^1 x^t dx$. Find the value of t.
12. Derive a formula for $\sum_1^n \lfloor \alpha^i \rfloor$.
13. Let k be a positive integer. Evaluate $\lim_{n\to\infty} (F_{n+k}/L_n)$ (Dence, 1968).
14. Let k be a positive integer. Evaluate $\lim_{n\to\infty} (L_{n+k}/F_n)$ (Koshy, 1998).
15. Let $a_n = a_{n-1} + a_{n-2} + k$, where $a_0 = 0$, $a_1 = 1$, and k is a constant. Find $\lim_{n\to\infty} (a_n/F_n)$ (Shallit, 1976).

Let $b_n = b_{n-1} + b_{n-2} + k$, where $b_0 = 2$, $b_1 = 1$, and k is a constant. Find each (Koshy, 1999).

16. $\lim_{n\to\infty} (b_n/F_n)$
17. $\lim_{n\to\infty} (b_n/L_n)$
18. Let $c_n = c_{n-1} + c_{n-2} + k$, where $c_1 = a$, $c_2 = b$, and k is a constant. Find $\lim_{n\to\infty} (c_n/G_n)$ (Koshy, 1999).
19. Evaluate $\sum_0^n \binom{n}{i} \alpha^{3i-2n}$ (Freitag, 1975)
20. Evaluate the infinite product

$$\left(1 + \frac{1}{2}\right)\left(1 + \frac{1}{13}\right)\left(1 + \frac{1}{610}\right)\left(1 + \frac{1}{1346269}\right)\cdots$$

(Shallit, 1981).

21. Consider the real sequence $\{x_n\}_0^{\infty}$, defined by $x_{n+1} = 1/(x_n + 1)$. Find x_0 such that $\lim_{n\to\infty} x_n$ exists and find the value of the limit (Neumer, 1993).

Consider the vector space $V = \{v = (v_1, v_2, \ldots, v_n, \ldots) | v_n = v_{n-1} + v_{n-2}, n \geq 3, \text{ and } v_i \in \mathbb{R}\}$ with the usual operations. Do the vectors $u = (1, 0, 1, 1, 2, 3, 5, \ldots)$ and $v = (0, 1, 1, 2, 3, 5, 8, \ldots)$ belong to V? (Barbeau, 1993).

22. Show that V is 2-dimensional.
23. Show that $(r, r^2, r^3, \ldots) \in V$ if and only if $r^2 = r + 1$.
24. Find r if $(r, r^2, r^3, \ldots) \in V$.
25. Let $F = (1, 1, 2, 3, 5, 8, \ldots) \in V$. Let $u = (\alpha, \alpha^2, \alpha^3, \ldots)$ and $v = (\beta, \beta^2, \beta^3, \ldots)$. Find the constants a and b such that $F = au + bv$.
26. With F, u, and v as in Exercise 25, deduce Binet's formula.
27. Let $f(x) = Ax^n$ and $f^{-1}(x) = [f^{(m)}(x)]^p$. Show that $n = \frac{mp \pm \sqrt{p^2m^2+4p}}{2p}$ (Gattei, 1999).

GOLDEN TRIANGLES

According to *Scientific American* columnist Martin Gardner, "Pi (π) is the best known of all irrational numbers. The irrational number α is not so well-known, but it expresses a fundamental ratio that is almost as ubiquitous as *pi*, and it has the same amusing habit of popping up where least expected." Gardner made this trenchant observation in 1959. The magical number α makes some interesting appearances in plane and solid geometry.

Some triangles are linked to this ubiquitous number in a mysterious way. This is true of the golden triangle, so we begin with the following definition.

GOLDEN TRIANGLE

An isosceles triangle is a *golden triangle* if the ratio of one its lateral sides to the base is α.

Next we pursue a few properties of a golden triangle.

Theorem 22.1. Let $\triangle ABC$ be a golden triangle with base $\overline{AC}$. Let D divide $\overline{BC}$ in the Golden Ratio, $\overline{BD}$ being the larger segment. Then $\overline{AD}$ bisects $\angle A$.

Proof. Let D divide $\overline{BC}$ in the Golden Ratio such that $BD = \alpha CD$, $BD + CD = \alpha CD + CD = (\alpha+1)CD = \alpha^2 CD$; that is, $BC = \alpha^2 CD$. Thus $BC = \alpha^2 CD = \alpha AC$, so $\alpha CD = AC$ (see Fig. 22.1).

Thus $\angle BCA = \angle ACD$ and $(AB/AC) = (AC/CD)$ $\therefore$ $\triangle BAC \sim \triangle ACD$. Consequently, $\angle ACD = \angle ADC$ and $\angle ABC = \angle CAD$. So $\triangle ACD$ is an isosceles triangle with $AD = AC = \alpha CD$. Thus $BD = AD$, so $\triangle ABD$ is also an isosceles triangle. Hence $\angle BAD = \angle ABD$. Thus $\angle BAD = \angle CAD$; that is, $\overline{AD}$ bisects $\angle BAC$.

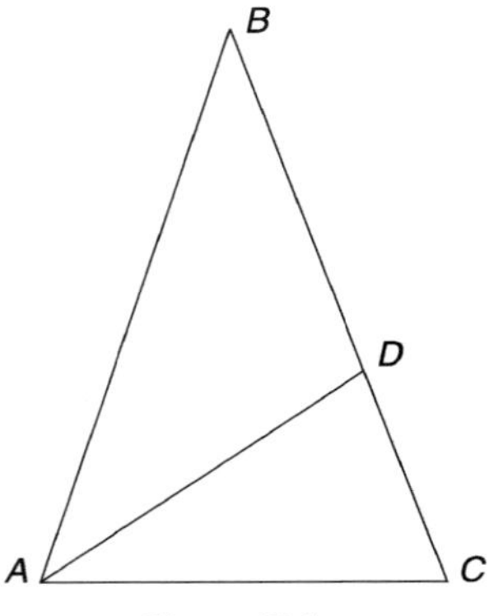

Figure 22.1.

■

Corollary 22.1. Let $\triangle ABC$ be a golden triangle with base $\overline{AC}$. Let D divide $\overline{BC}$ in the Golden Ratio, $\overline{BD}$ being the larger segment. Then $\triangle CAD$ is also a golden triangle.

Proof. Using Figure 22.1 and the preceding proof, $AD = AC = \alpha CD$, so $\triangle CAD$ is a golden triangle. ■

Thus $\overline{AD}$ cuts the golden triangle ABC into two isosceles triangles, $\triangle ABD$ and $\triangle CAD$, the latter being similar to $\triangle ABC$.

Theorem 22.2. The included angle between the equal sides of a golden triangle is 36°.

Proof. Let $\triangle ABC$ be a golden triangle with $AB = BC = \alpha AC$. Let D divide $\overline{BC}$ in the Golden Ratio, as in Figure 22.1. By Corollary 22.1, $\triangle CAD$ is a golden triangle similar to $\triangle ABC$.

Let $\angle ACD = 2x$. Then, from $\triangle ACD$, $2x + 2x + x = 180^\circ$, so $x = 36^\circ$. Thus $\angle A = \angle C = 72^\circ$ and $\angle B = 36^\circ$. ■

Is the converse true? Yes. If the nonrepeating angle in an isosceles triangle is 36°, then the triangle is a golden triangle, as the next theorem demonstrates.

Theorem 22.3. If the nonrepeating angle in an isosceles triangle is 36°, then the triangle is a golden triangle.

Proof. Let $\triangle ABC$ be an isosceles triangle with $AB = AC$ and $\angle A = 36^\circ$. Then $\angle B = \angle C = 72^\circ$ (see Fig. 22.2).

Let $\overline{AD}$ bisect $\angle A$. Then $\angle ADC = 72^\circ$, so $AC = AD = BD = y$ and $\triangle ABC \sim \triangle ACD$. Then $(AB/AC) = (BC/CD)$; that is, $x/y = y(x - y)$. Thus, $(x/y)^2 = (x/y) + 1$. Consequently, $x/y = \alpha$

Thus, $AB{:}\,AC = \alpha{:}1$, so $\triangle ABC$ is a golden triangle.

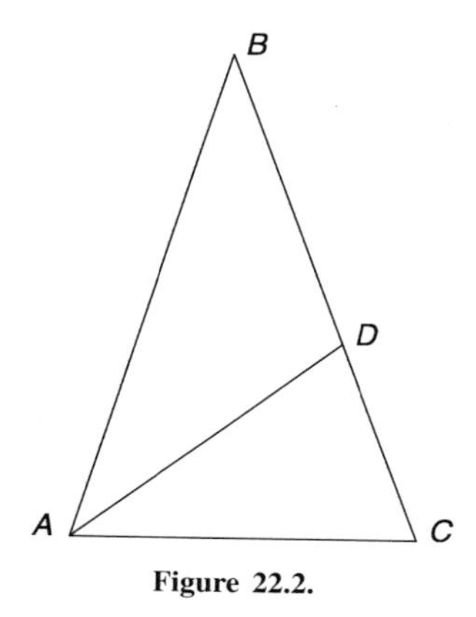

Figure 22.2.

■

Corollary 22.2. An isosceles triangle is a golden triangle if and only if its angles are 36°, 72°, and 72°. ■

A golden triangle can also be characterized by areas, as the next theorem shows. We shall leave its proof as an exercise.

Theorem 22.4. Let D be a point on side $\overline{BC}$ of an isosceles triangle ABC such that $\triangle ABC \sim \triangle CAD$, where $AB = BC$. Then $\triangle ABC$ is a golden triangle if and only if area of $\triangle ABC$:area of $\triangle BDA = \alpha:1$. ■

Theorem 22.5. Let the ratio a/b of two sides a and b of $\triangle ABC$ be greater than one. Remove a triangle with side b from $\triangle ABC$. The remaining triangle is similar to the original triangle if and only if $a/b = \alpha$.

Proof. Remove $\triangle ABD$ from $\triangle ABC$ (see Fig. 22.3). Since $\triangle ADC \sim \triangle BAC$,

$$\frac{AC}{BC} = \frac{DC}{AC},$$

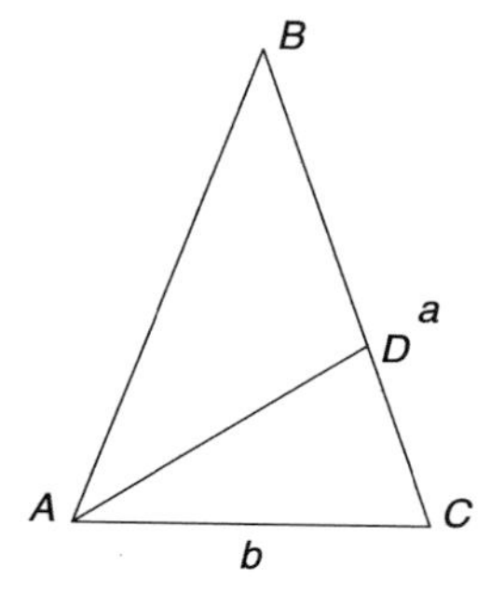

Figure 22.3.

that is,

$$\frac{b}{a} = \frac{a-b}{b}$$

$$\frac{b}{a} = \frac{a}{b} - 1$$

Since $a > b$, it follows that $a/b = \alpha$.

Conversely, let $a/b = \alpha$. Then

$$\frac{DC}{AC} = \frac{a-b}{b} = \frac{a}{b} - 1 = \alpha - 1 = -\beta = \frac{1}{\alpha} = \frac{b}{a} = \frac{AC}{BC}$$

Since $\angle C$ is common to triangles ADC and BAC, it follows that $\triangle ADC \sim \triangle BAC$. ■

The next theorem is closely related to this.

Theorem 22.6. Let the ratio of two sides of a triangle be $k > 1$. A triangle similar to the triangle can be removed from it in such a way that the ratio of the area of the original triangle and that of the remaining triangle is also k if and only if $k = \alpha$.

Proof. (see Figure 22.4) Let $\triangle ADC \sim \triangle BAC$ such that $(AC/BC) = (DC/AC) = k$. Let

$$\frac{\text{Area } \triangle BAC}{\text{Area } \triangle ABD} = k$$

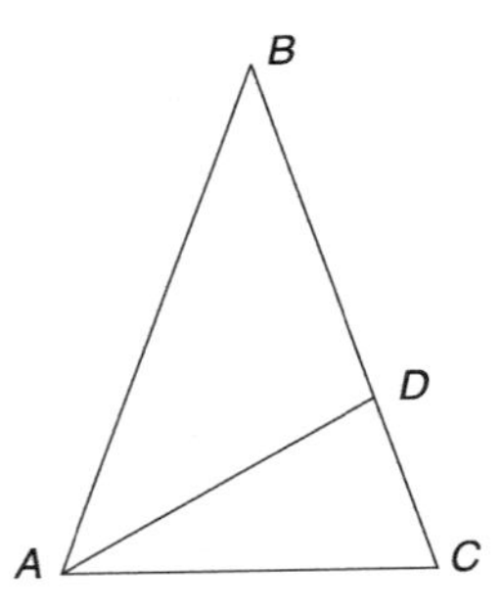

Figure 22.4.

Since $\triangle ABC$ and $\triangle ADC$ have the same altitude h from A,

$$\frac{\text{Area } \triangle BAC}{\text{Area } \triangle ABD} = \frac{1/2BC \cdot h}{1/2BD \cdot h} = \frac{BC}{BD} = \frac{BC}{BC - CD} = \frac{BC/AC}{BC/AC - CD/AC}$$

That is, $k = k/(k - 1/k)$, so $k = \alpha$.

Conversely, let

$$\frac{BC}{AC} = k = \alpha = \frac{\text{Area } \triangle BAC}{\text{Area } \triangle ADC}$$

Then

$$\alpha = \frac{BC/AC}{BC/AC - DC/AC} = \frac{\alpha}{\alpha - DC}$$

$$\therefore \quad \frac{DC}{AC} = \alpha - 1 = -\beta = \frac{1}{\alpha}$$

that is, $AC/DC = \alpha$. Thus $\triangle BAC \sim \triangle ADC$. ■

Since the central angle of a regular decagon (10-gon) is 360°, each side subtends an angle of 36° at the center (see Fig. 22.5). It now follows from Corollary 22.2 that each of the triangles AOB is a golden triangle.

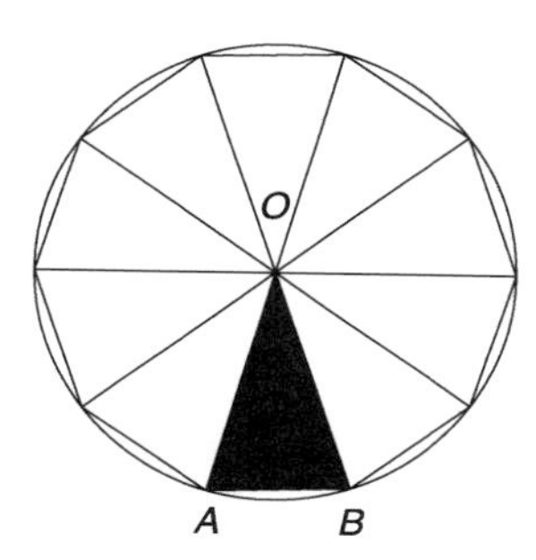

Figure 22.5.

Consider the regular pentagram in Figure 22.6. Since the angle at a vertex is 108°, it follows that $\angle BAC = 36°$, so $\triangle ABC$ is a golden triangle. The pentagon contains five golden triangles.

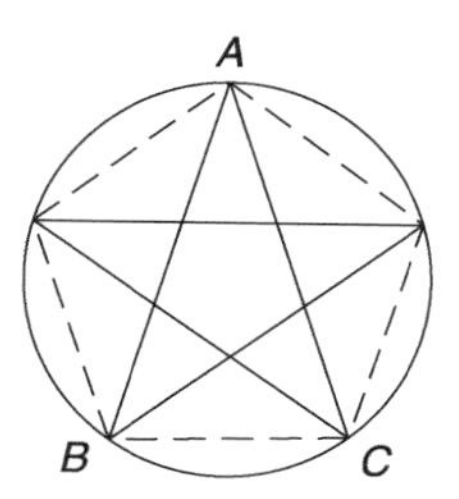

Figure 22.6.

EXERCISES 22

1. Let ABC be an isosceles triangle, where the nonrepeating angle $\angle B = 36°$. Let the bisector of $\angle A$ intersect $\overline{BC}$ at D. Prove that $\triangle CAD$ is a golden triangle.

Let D be a point on side $\overline{BC}$ of an isosceles triangle ABC such that $\triangle ABC \sim \triangle BDC$, where $AB = BC$. Prove each.

2. If $\triangle ABC$ is a golden triangle, then area $\triangle ABC$: area $\triangle BDA = \alpha : 1$.
3. If area $\triangle ABC$: area $\triangle BDA = \alpha : 1$, then $\triangle ABC$ is a golden triangle.
4. Let $\triangle ABC$ be a golden triangle, and D a point on $\overline{BC}$ such that $\triangle ABC : \triangle BDA = \alpha : 1$. Prove that $\triangle ABC : \triangle CAD = \alpha^2 : 1$.
5. The lengths of the sides of a right triangle form a geometric sequence with common ratio r. Prove that $r = \sqrt{\alpha}$.

GOLDEN RECTANGLES

In *Der goldene Schnitt* (1884), Adolf Zeising's 457-page classic work on the Golden Section, Zeising argued that "the golden ratio is the most artistically pleasing of all proportions and the key to the understanding of all morphology (including human anatomy), art, architecture, and even music."

Take a good look at the four picture frames of various proportions, represented in Figure 23.1. Which is aesthetically most appealing? Most pleasing to the eyes? Frame (*a*) is too square; frame (*b*) looks too narrow; and frame (*c*) appears too wide! So if we picked frame (*d*) as our top choice, we are right; it has aesthetically more pleasing proportions.

In fact, this choice puts us in good company. German psychologists Gustav Theodor Fechner (1801–1887) and Wilhelm Max Wundt (1832–1920) provide ample empirical support to Zeising's claims. They measured thousands of windows, picture frames, playing cards, books, mirrors, and other rectangular objects, and even checked the points where graveyard crosses were divided. They concluded that most people unconsciously select rectangular shapes in the Golden Ratio when selecting such

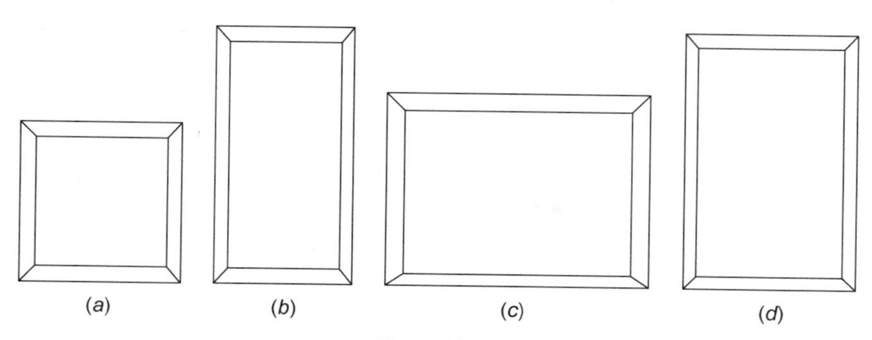

Figure 23.1.

objects. And, of course, such pleasing proportions were the basis of most ancient Greek art and architecture.

The American artist Jay Hambidge (1867–1924) of Yale University, in his extensive writings on *dynamic symmetry*, highlighted the prominent role the Golden Ratio has played in numerous Greek artworks, as well as modern art, architecture, and furniture design.

More recently, Frank A. Lone of New York confirmed one of Zeising's favorite theories. He measured the heights of 65 women and compared them to the heights of their navels. The ratio was found to be about 1.618, which he called the *lone relativity constant*. He also found a fascinating relationship between α and π:

$$\frac{6\alpha^2}{5} \approx \pi$$

GOLDEN RECTANGLE

Figure 23.1*d*, represented in Figure 23.2, has the fascinating property that the ratio of the length x of the longer side to the length y of the shorter side equals the ratio of their sum to the length of the longer side, that is,

$$\frac{x}{y} = \frac{x+y}{x}.$$

This yields the equation $x/y = 1 + y/x$, so x/y satisfies the familiar equation $t = 1 + 1/t$. Thus $x/y = \alpha$, as we could have conjectured. Such a rectangle is called a *golden rectangle*.

M. J. Zerger devised a clever method for constructing a large rectangle that approximates a golden rectangle. Place 20 ordinary $8\frac{1}{2} \times 11$ sheets of paper, in four rows of five each, as in Figure 23.3. The resulting shape is a 34×55 rectangle, which is a pretty good approximation to a golden rectangle.

As another example, consider the picture* in Figure 23.4. The lighthouse in the picture is drawn at a pivotal position. It divides the picture into two rectangular parts in such a way that if a denotes its distance from the left side and b that from the right side, then $a/b = (a+b)/a$. This is the Golden Ratio, so the rectangle in Figure 23.5 is indeed a golden rectangle.

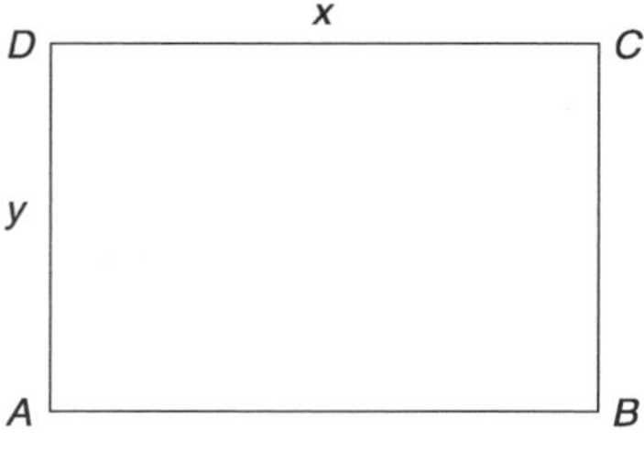

Figure 23.2.

*Based on F. Land, *The Language of Mathematics*, Murray, London, 1960.

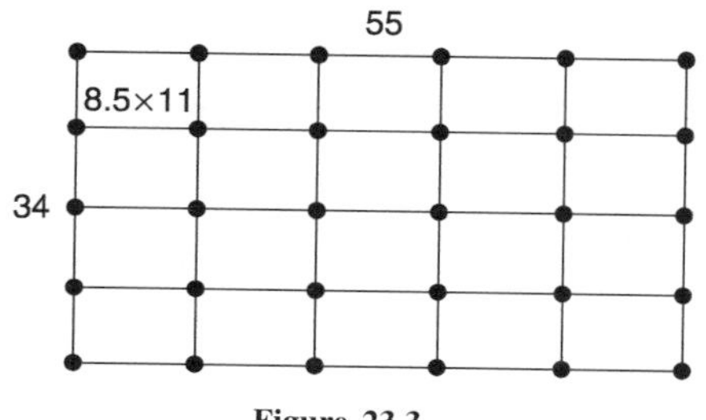

Figure 23.3.

Figure 23.4. The Lighthouse divides the picture in a way that creates a golden rectangle (*Source*: Trudi Hammel Garland, *Fascinating Fibonaccis: Mystery and Magic in Numbers*, Palo Alto, CA: Seymour, 1987. Copyright © 1987 by Dale Seymour Publications. Used by permission of Pearson Learning.).

Since the golden rectangle is the most pleasing rectangle, countless artists have used golden rectangles and their magnificent properties in their work.

The *Holy Family* by Michelangelo Buonarroti (1475–1564), and *Madonna of the Magnificat* by Sandro Botticelli (1444–1510), and more recently, *Corpus Hipercubus* and *The Sacrament of the Last Supper* by Spanish surrealist Salvador Dali (1904–1989) are fine illustrations of the visual power and beauty of the golden rectangle.

Dali originally entitled his masterpiece *Corpus hipercubus* (Hypercubic Body), according to *Time* magazine of January 24, 1955. His painting is based on "the harmonious division of a specific golden rectangle."

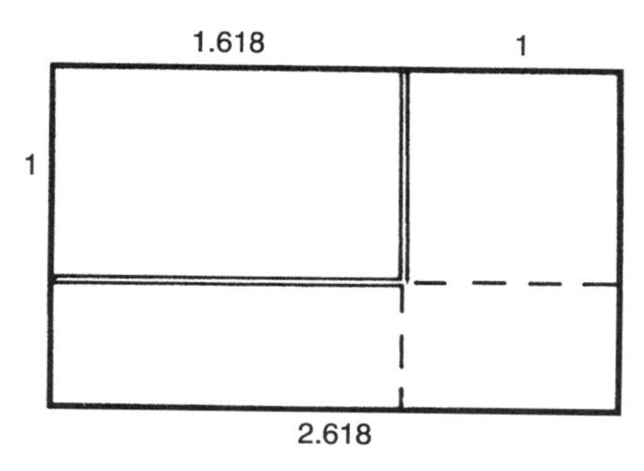

Figure 23.5.

Figure 23.6. (*a*) St. Jerome by da Vinci fits into a golden rectangle; (*b*) Michelangelo's David also illustrates a golden rectangle (*Source*: Both images from Scala/Art Resouce, New York.).

Leonardo da Vinci (1452–1519) painted *St. Jerome* to fit very nicely into a golden rectangle; art historians believe that da Vinci deliberately painted the figure according to the classical proportions he inherited from the Greeks. Michelangelo's *David* also illustrates a golden rectangle (see Fig. 23.6).

According to Sr. M. Stephen of Rosary College in Illinois, "da Vinci used [the Golden ratio] in laying out canvases in such a manner that the points of interest would be at the intersections of the diagonals and perpendicular from the vertices." (See Fig. 23.7.)

The Golden rectangles are also evident in the work of Albrecht Dürer (1471–1528), the foremost German painter, engraver, and designer of the Renaissance. Golden

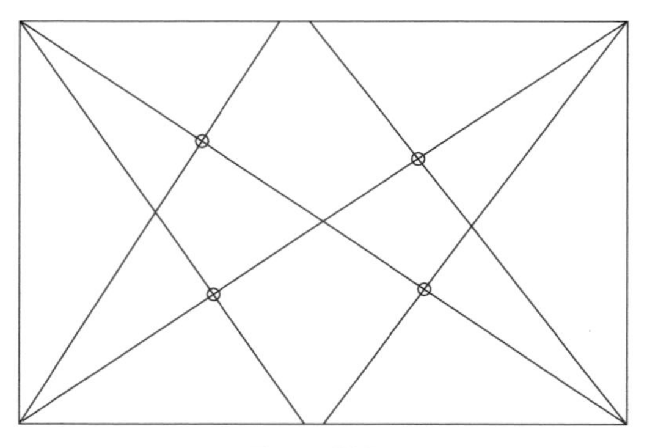

Figure 23.7.

(a)

(b)

Figure 23.8. (*a*) Georges Seurat's *La Parade* (*b*) Contains a golden rectangle [*Source*: The Metropolitan Museum of Art, Bequest of Stephen C. Clark, 1960 (61.101.17).].

rectangles appear in modern abstract art such as *La Parade* by the French impressionist Georges Seurat (1859–1891) (see Fig. 23.8). Seurat is said to have approached every canvas with the magical ratio in mind. The same can be said about much of the work by the Dutch abstractionist Pieter Cornelis Mondriaan (1872–1944). Juan Gris (1887–1927), the Spanish-born cubist who was greatly influenced by Pablo Picasso and Georges Braque, lavishly applied the golden ratio in his work and promoted its beauty.

THE PARTHENON

The *Parthenon*, the magnificent building erected by the ancient Athenians in honor of Athena Parthenos, the patron goddess of Athens, stands on the Acropolis. It is a monument to the ancients' worship of the golden rectangle (see Fig. 23.9). The

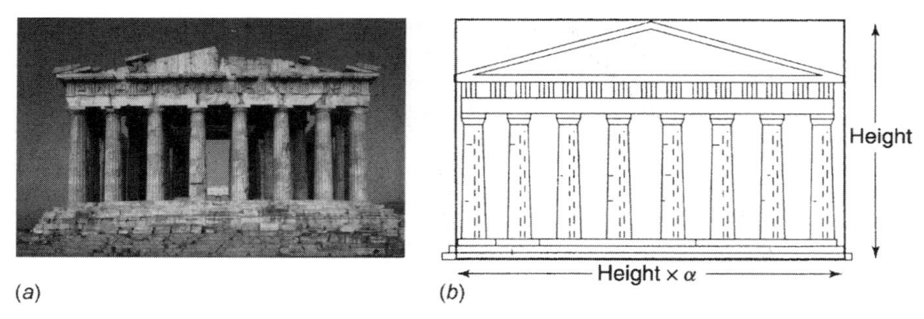

Figure 23.9. (*a*) View of the Parthenon at Athens; (*b*) This magnificent building fits into a golden rectangle (*Source*: Photo Researchers. © Marcello Bertinetti, Photo Researchers, Inc., New York.).

Figure 23.10. The Parthenon in Nashville (Photo: Gary Layda. © Metro Government of Nashville, 2000.).

whole shape fits nicely into a golden rectangle. Even the reconstruction of the original Parthenon in Nashville, Tennessee, vividly illustrates the aesthetic power of the golden rectangle (see Fig. 23.10).

According to R. F. Graesser of the University of Arizona, the Golden Ratio was used in the facade and floor plan of the Parthenon, as it was used in facades and floor plans of other Greek temples. The various occurrences of the golden rectangle in the architecture are depicted beautifully by Walt Disney's animation film *Donald Duck in Mathemagicland*.

Architect Le Corbusier (1887–1965) (Charles Edouard Jeanneret-Gris) one of the most influential designers of the twentieth century, developed a scale of proportions called the *modulator*. This unit was based on a human body, whose height is divided by the navel into the Golden Ratio, as Figure 23.11 shows.

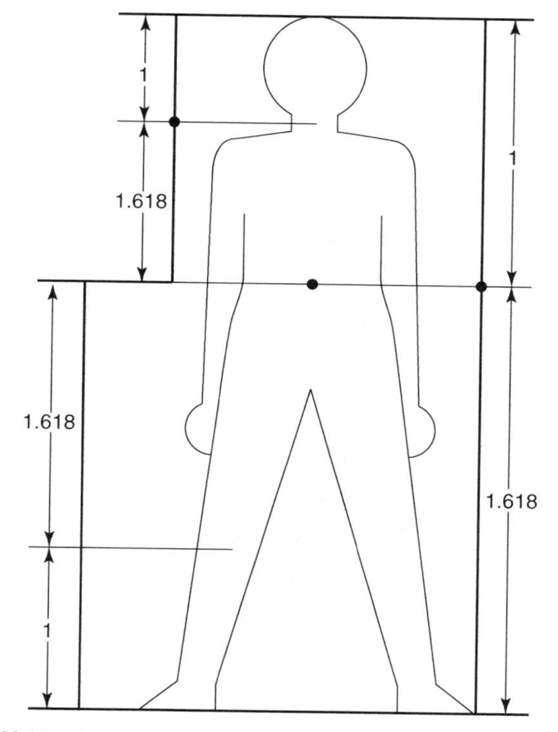

Figure 23.11. *The modulator*, a scale of proportions developed by Le Corbusier.

The golden rectangle is also used extensively in *Cathedral of Chartres* and the *Tower of Saint Jacques* in Paris. The royal doorway of the cathedral vividly illustrates a golden rectangle (see Fig. 23.12).

Figure 23.12. The Doorway of the Cathedral of Chartes (Reprinted with permission from MATHEMATICS TEACHER, copyright 1956, by the National Council of Teachers of Mathematics. All rights reserved.).

According to Sr. Marie Stephen of Rosary College, River Forest, Illinois, the tower of Saint Jacques in Paris illustrates

> the architectural leitmotif [α] in inverse progressions. At the corners, the buttresses rise in four superimposed layers, which diminish in size as they rise. The ratio thus established is exactly 1.618. The buttresses, like a human hand, whose proportions we shall see are the same, point toward the sky, while the three stories of windows which illumine the interior of the tower appear as a hand pointing down from the sky to the ground.

See Figure 23.13.

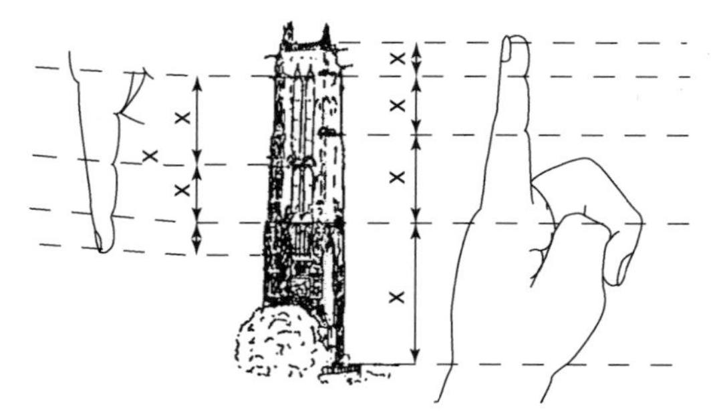

Figure 23.13. The Tower of Saint Jacques. (Reprinted with permission from MATHEMATICS TEACHER, copyright 1956, by the National Council of Teachers of Mathematics. All rights reserved.).

According to Stephen, "Dr. Christian Jacob of Buenos Aires has discovered the interesting proportion in the human brain."

THE HUMAN BODY AND THE GOLDEN RECTANGLE

As the ancient Greeks knew, the human body exemplifies the golden proportion. The head fits nicely into a golden rectangle, as Figure 23.14 demonstrates. In addition, the face provides visual examples of the Golden Ratio:

$$\frac{AC}{CD} = \frac{CD}{BC} = \frac{AD}{BD} = \alpha$$

So do the fingers, as Figures 23.15 and 23.16 illustrate:

$$\frac{b}{a} = \frac{c}{d} = \frac{d}{c} = \alpha.$$

According to T. H. Garland, most of the ancient graveyard crosses in Europe exemplify the golden proportion: the point where the two arms meet, divides the cross

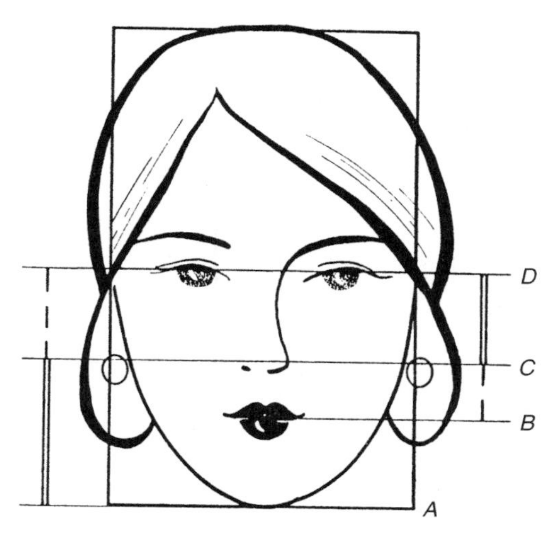

Figure 23.14. The golden proportions in a human head and face (*Source*: Trudi Hammel Garland, *Fascinating Fibonaccis: Mystery and Magic in Numbers*, Palo Alto, CA: Seymour, 1987. Copyright © 1987 by Dale Seymour Publications. Used by permission of Pearson Learning.).

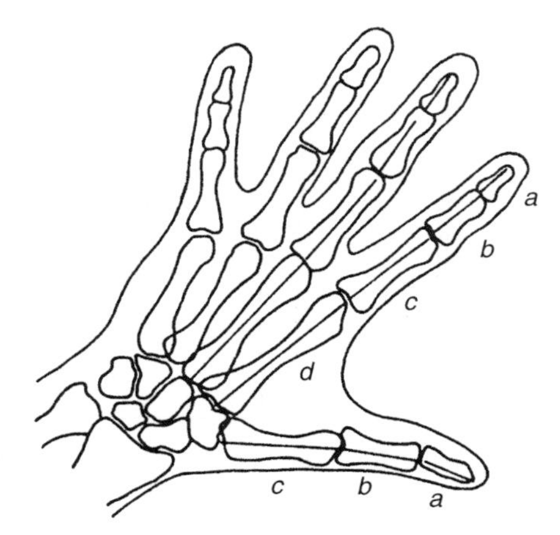

Figure 23.15. The golden proportions in a human hand.

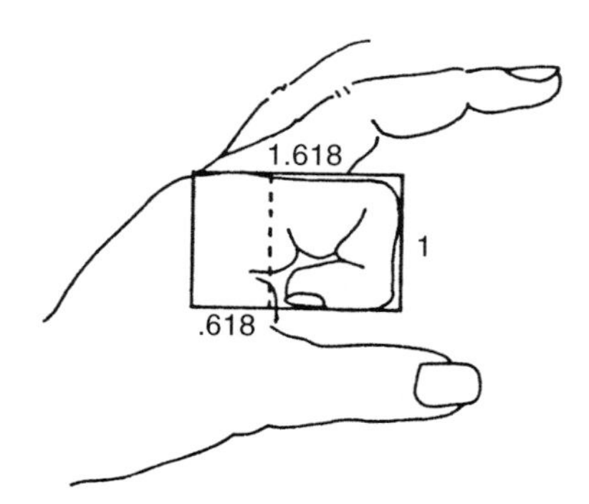

Figure 23.16. A personal golden rectangle formed by a pointer finger.

Figure 23.17. An Ancient Graveyard Cross (*Source*: Trudi Hammel Garland, *Fascinating Fibonaccis: Mystery and Magic in Numbers*, Palo Alto, CA: Seymour, 1987. Copyright © 1987 by Dale Seymour Publications. Used by permission of Pearson Learning.).

Figure 23.18. A Modern Cross (*Source*: Trudi Hammel Garland, *Fascinating Fibonaccis: Mystery and Magic in Numbers*, Palo Alto, CA: Seymour, 1987. Copyright © 1987 by Dale Seymour Publications. Used by permission of Pearson Learning.).

Figure 23.19. A Prostate Cancer Awareness stamp.

in the Golden Ratio (see Fig. 23.17). Although many modern crosses do not display this magnificent characteristic, some still fit into a golden rectangle (see Fig. 23.18).

Postage stamps, interestingly enough, often remind us of the golden rectangle. For example, the 1999 Prostate Cancer Awareness stamp in Figure 23.19. Its outer size is $4\,\text{cm} \times 2.5\,\text{cm}$ and $4/2.5 \approx \alpha$; the size of the inner rectangle is $3.5\,\text{cm} \times 2.1\,\text{cm}$, and $3.5/2.1 \approx \alpha$.

The statue of a seated Buddha (563?–483? B.C.) in Figure 23.20, also displays the golden proportions; it fits magnificently into a golden rectangle. So do the Chinese bowl in Figure 23.21 that belongs to the Ching dynasty, and the Greek urn in Figure 23.22.

Figure 23.20. Statue of Buddha (*Source*: Trudi Hammel Garland, *Fascinating Fibonaccis: Mystery and Magic in Numbers*, Palo Alto, CA: Seymour, 1987. Copyright © 1987 by Dale Seymour Publications. Used by permission of Pearson Learning.).

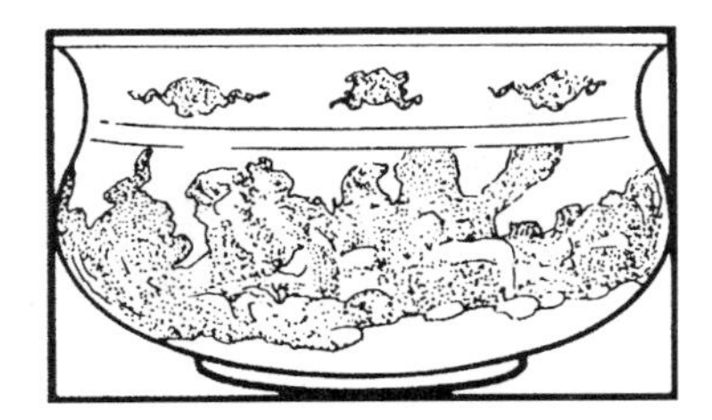

Figure 23.21. Chinese bowl (*Source*: Trudi Hammel Garland, *Facinating Fibonaccis: Mystery and Magic in Numbers*, Palo Alto, CA: Seymour, 1987. Copyright © 1987 by Dale Seymour Publications. Used by permission of Pearson Learning.).

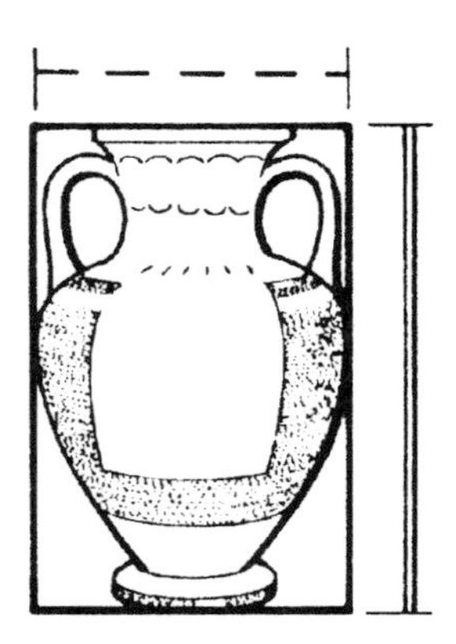

Figure 23.22. Greek urn (*Source*: Trudi Hammel Garland, *Fascinating Fibonaccis: Mystery and Magic in Numbers*, Palo Alto, CA: Seymour, 1987. Copyright © 1987 by Dale Seymour Publications. Used by permission of Pearson Learning.).

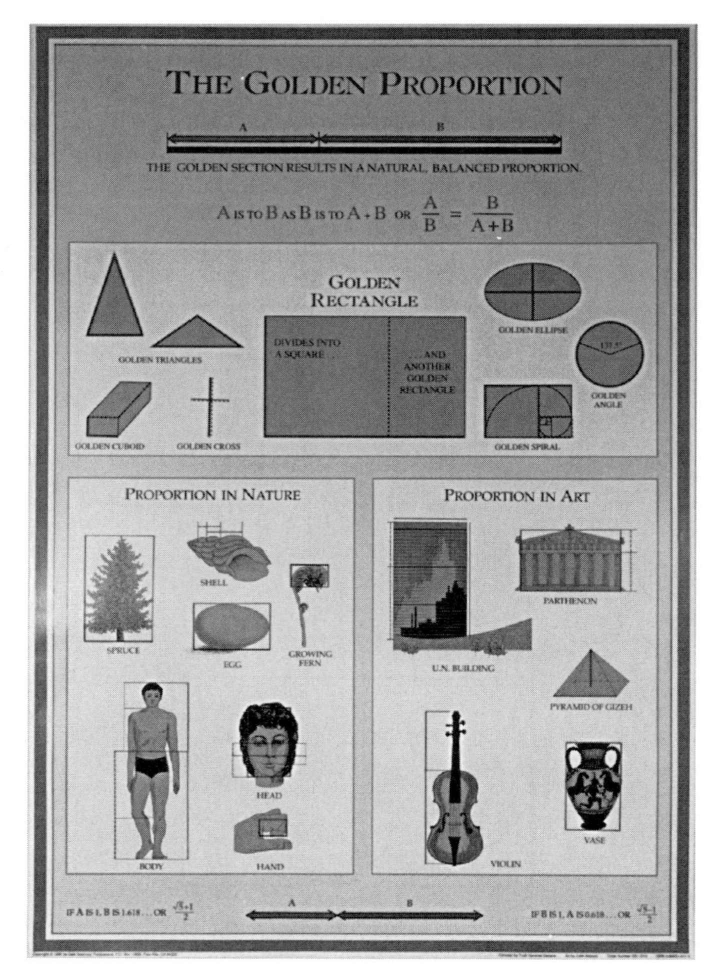

From *The Golden Proportion (poster)*. Copyright © 1990 by Dale Seymour Publications. Used with permission of Pearson Education.

THE GOLDEN RECTANGLE AND THE CLOCK

It is common knowledge that the positions of the hour and minute hands on an analog wristwatch or clock in store displays, or in newspaper and magazine advertisements tend to be approximately 10:09 or 8:18 (see Fig. 23.23).

One myth concerning the time 8:18 is that it was precisely the time Abraham Lincoln died by an assassin's bullet on April 15, 1865. Another misconception is that such a setting of the hands gives more space on the face of the clock to show the name of the manufacturer clearly.

In any case, in 1983, M. G. Monzingo of Southern Methodist University in Dallas argued convincingly that such a setting is related to the golden rectangle, and hence appealing aesthetically. He showed that the angle θ in Figure 23.24 is about 58.3°. Suppose $OE = 1$. Then $EB \approx \tan 58.3^\circ \approx \alpha$. So $AB : AD \approx 2\alpha{:}2 = \alpha{:}1$. In other words, such a setting pleases the eye, since it creates an imaginary golden rectangle $OEBF$ on the face of the clock.

Suppose the points A, B, C, and D in Figure 23.25 divide the respective sides of the square $PQRS$ in the Golden Ratio. Then $PA = PB$, $QB = QC$, and $PB/BQ = \alpha$.

Figure 23.23. Wristwatch with hands set at 10:09.

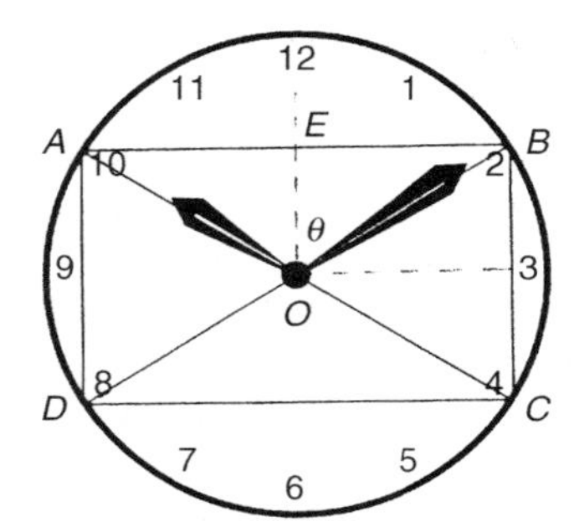

Figure 23.24. The 10:09 setting on a watch is related to the golden rectangle.

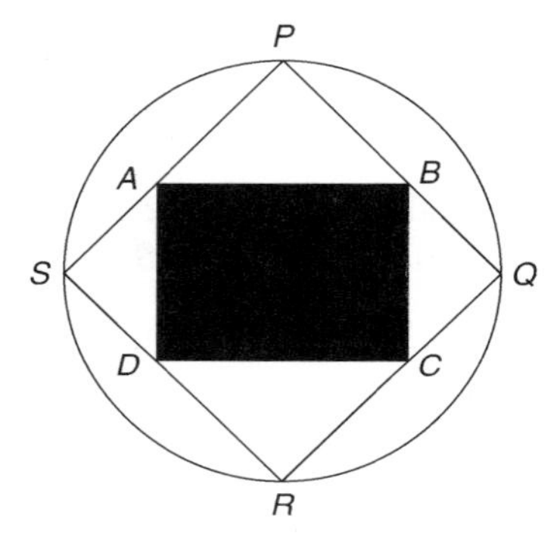

Figure 23.25.

Then

$$\frac{AB^2}{BC^2} = \frac{PA^2 + PB^2}{QB^2 + QC^2} = \frac{2PB^2}{2QB^2} = \alpha^2$$

Thus, $AB/BC = \alpha$, so $ABCD$ is indeed a golden rectangle.

STRAIGHTEDGE AND COMPASS CONSTRUCTION

How do we construct a golden rectangle with a straightedge and a compass? To this end, consider a line segment $\overline{AB}$ with C dividing it in the golden ratio: $AC/CB = AB/AC = \alpha$. Now with C as the center, draw an arc of radius CB; let the perpendicular $\overline{CH}$ to $\overline{CB}$ meet the arc at D. Complete the rectangle $ACDE$, as Figure 23.26 shows. It is a golden rectangle since $AC/CD = AC/CB = \alpha$.

Using the golden rectangle $ACDE$, we can draw another rectangle. With A as the center, draw an arc of radius AC so as to intersect the perpendicular $\overline{AB}$ at D. Complete the rectangle $ABGF$, as Figure 23.26 shows. It is also a golden rectangle, because

$$\frac{AB}{BG} = \frac{AB}{AC} = \alpha$$

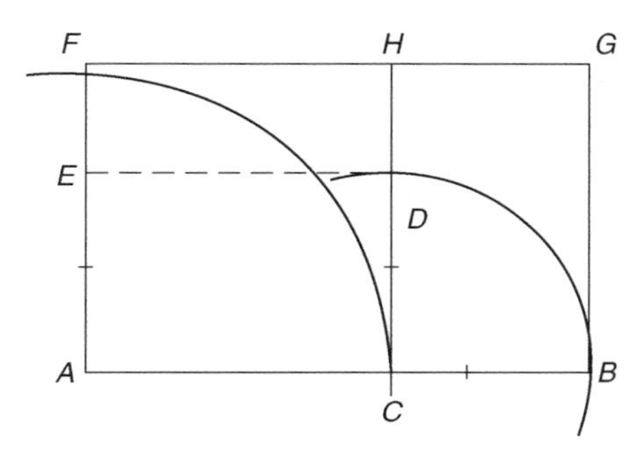

Figure 23.26.

In fact, we have gained a third golden rectangle, namely, $BCHG$. It is so since

$$\frac{BG}{BC} = \frac{AC}{BC} = \alpha$$

As a by-product, we can show that the ratio of the area of rectangle $ABGF$ to that of rectangle $BCHG$ is $\alpha + 1$ (see Exercise 2).

In Figure 23.26, Suppose we remove the square $ACHF$ from the golden rectangle $ABGF$; then the resulting rectangle $BCHG$ is also a golden rectangle. That is, if the ratio of the length to width of a rectangle $ABGF$ is the Golden Ratio, then that of the rectangle $BCHG$ obtained by removing a square with one side equal to the width of the original rectangle is also the golden rectangle.

Conversely, suppose the ratio of length to width of a rectangle $ABCD$ is k; that is, $AB/BC = l/w = k$ (see Fig. 23.27). Let $BEFC$ be the rectangle obtained by deleting the square $AEFD$ from the rectangle $ABCD$. The ratio of length to width of the rectangle $BEFC$ is $BC/BE = w/(l - w)$. Suppose $AB/BC = BC/EC$. Then $l/w = w/(l - w)$, that is, $k = 1/(k - 1)$; so $k = \alpha$. Thus, if removing the square yields a rectangle similar to the original square, then $k = \alpha$; that is, the original rectangle must be a golden rectangle.

On the other hand, suppose the ratio of length to width in a rectangle $ABCD$ is $k > 1$. Remove from rectangle $ABCD$ a rectangle $BEFC$ similar to it. Then the

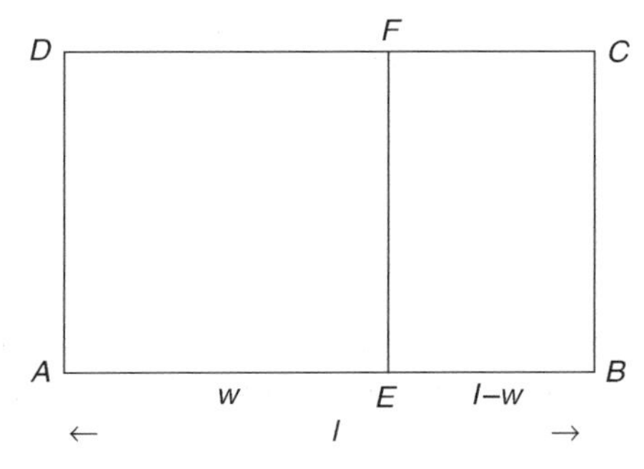

Figure 23.27.

ratio of the area of the original rectangle to that of the remaining rectangle $AEFC$ is k if and only if $k = \alpha$. In fact, $AEFC$ is a square (see Exercise 3).

RECIPROCAL OF A RECTANGLE

Let us now look at the golden rectangle $ABGF$ from a slightly different angle. Suppose we remove square $ACHF$ from the golden rectangle $ABGF$. The resulting rectangle $BCHG$ is also a golden rectangle; it is called the *reciprocal* of rectangle ABGF.*

Thus, the reciprocal of a rectangle is a smaller, similar rectangle such that one side of the original rectangle becomes a side of the new rectangle.

It now follows that the area of the reciprocal rectangle $BCHG$ in Figure 23.26 is (area $ABGF)/(\alpha + 1)$.

In Figure 23.26, square $ACHF$ is the smallest figure that when added to the golden rectangle $CBGH$ to yield a similar shape, another rectangle. Accordingly, square $ACHF$ is called the *gnomon* of the reciprocal rectangle $CBGH$, a term introduced by Sir D'Arcy W. Thompson.

Suppose the diagonals $\overline{BF}$ and $\overline{CG}$ of the reciprocal rectangles meet at P, as Figure 23.28 illustrates. We can show that they are perpendicular (see Exercise 4). Let $\overline{BF}$ intersect $\overline{CH}$ at Q. Let $\overline{QR} \perp \overline{BG}$. Then $BRQC$ is the reciprocal of $BGHC$ and is also a golden rectangle. This gives a systematic way of constructing the reciprocal of a (golden) rectangle.

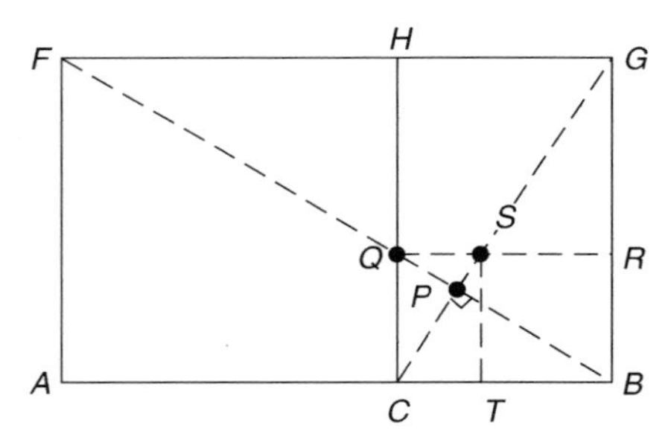

Figure 23.28.

Suppose we continue this procedure to draw the golden reciprocal of $BRQC$. Let $\overline{CG}$ meet $\overline{QR}$ at S and draw $\overline{ST} \perp \overline{BC}$. Then $CTSQ$ is the (golden) reciprocal of $BRQC$.

LOGARITHMIC SPIRAL

Obviously, we can continue this algorithm indefinitely, producing a sequence of smaller and smaller golden rectangles, as Figure 23.29 shows. The points that divide

*The term *reciprocal rectangle* was introduced by J. Hambidge.

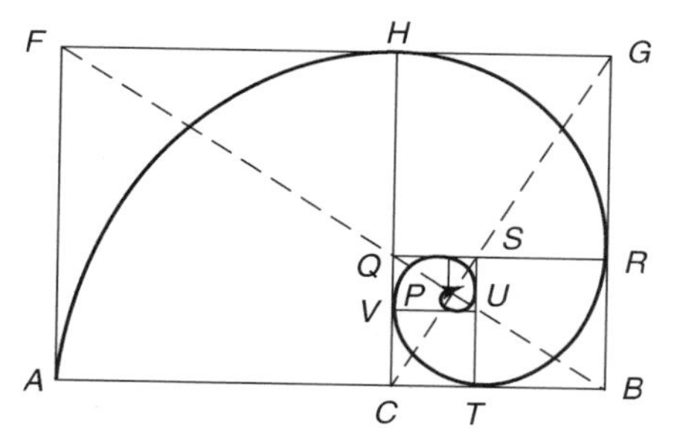

Figure 23.29.

the sides of the various golden rectangles spiral inward to the point P, the intersection of the two original diagonals. They lie on the *logarithmic spiral*, as Figure 23.29 demonstrates. The spiral, with its pole at P, touches the various golden rectangles at the golden sections.

The nautilus in Figure 23.30 is one of the most gorgeous examples of the logarithmic spiral in nature. Figure 23.31 also shows shells that display this beautiful logarithmic spiral.

Figure 23.30. A Nautilus photograph courtesy of Chip Clark

"The Chambered Nautilis," a poem written in 1858 by the American writer and physician Oliver Wendell Holmes (1809–1894) describes the creation of the spiral:

> Year after year beheld the silent toll
> That spread his lustrous coil;
> Still, as the spiral grew,
> He left the past year's dwelling for the new.

Consider the diagonals of the various squares that are snipped off, namely, $\overline{AH}$, $\overline{HR}$, $\overline{RT}$, $\overline{TV}$, ... Their lengths form a decreasing geometric sequence and their sum is $\sqrt{2}a\alpha^2$, where $AC = a$ (see Exercise 8).

Figure 23.31.

Interestingly enough, we can employ golden triangles to generate the logarithmic spiral. Bisect a 72°-angle in the golden triangle ABC in Figure 23.32. The point where the bisector meets the opposite side divides it in the Golden Ratio. The bisector produces a new similar golden rectangle, as we saw in Chapter 22. Now divide this triangle by a 72°-angle bisector to yield another golden triangle. Continuing this algorithm indefinitely generates a sequence of whirling golden triangles and hence

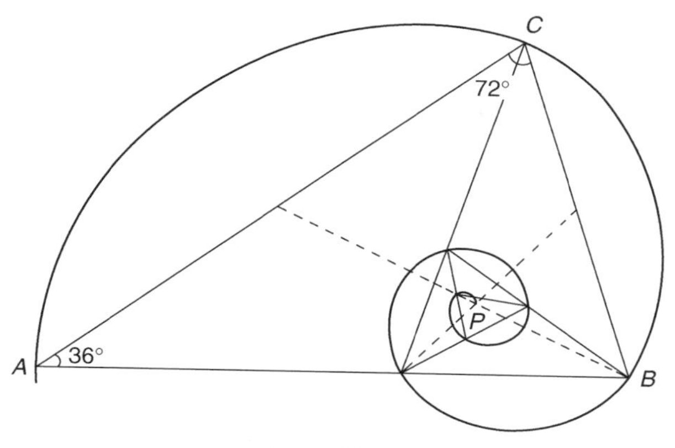

Figure 23.32.

the logarithmic spiral shown in the figure. Its pole P is the intersection of the two medians, indicated by broken segments.

GOLDEN RECTANGLE REVISITED

Suppose we remove a $t \times t$ square from one of the corners of a unit square lamina, as Figure 23.33 shows. We would like to find the value of t such that the center of gravity of the remaining gnomon is the corner G of the square removed. Taking moments about the side $\overline{AD}$, we have

$$\begin{pmatrix} \text{Moment of the} \\ \text{removed square} \end{pmatrix} + \begin{pmatrix} \text{Moment of the} \\ \text{gnomon} \end{pmatrix} = \begin{pmatrix} \text{Moment of the} \\ \text{original square} \end{pmatrix}$$

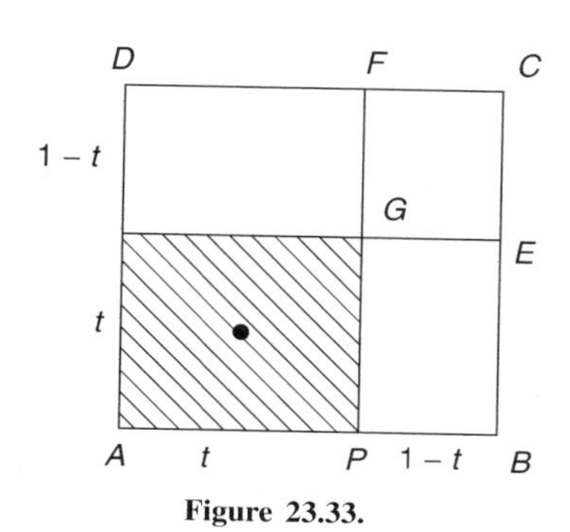

Figure 23.33.

That is,

$$\frac{1}{2}t \cdot t^2 + t(1 - t^2) = \frac{1}{2} \cdot 1$$

This yields the equation $t^3 - 2t + 1 = 0$; that is, $(t - 1)(t^2 + t - 1) = 0$. But $t \neq 1$, so $t^2 + t - 1 = 0$. Since $t > 0$, it follows that

$$t = \frac{-1 + \sqrt{5}}{2} = \frac{1}{\alpha}$$

Then

$$1 - t = 1 - \frac{1}{\alpha} = \frac{\alpha - 1}{\alpha}$$

$$\therefore \quad t : 1 - t = 1 : \alpha - 1 = \alpha : \alpha^2 - \alpha = \alpha : 1$$

Thus P divides $\overline{AB}$ in the Golden Ratio, and hence both legs of the gnomon are golden rectangles, as proved in 1995 by Nick Lord of Tonbridge School, Kent, England.

EXERCISES 23

1. Find the ratio of the length of a shorter side of golden rectangle to that of its longer side.
2. Using Figure 23.26, prove that the ratio of the area of rectangle $ABGF$ to that of rectangle $BCHG$ is $\alpha + 1$.
3. Let the ratio of length to width in a rectangle $ABCD$ be $k > 1$. From rectangle $ABCD$, remove a rectangle $BEFC$ similar to it. Prove that the ratio of rectangle $ABCD$ to the remaining area $AEFD$ is k if and only if $k = \alpha$. Besides, the remaining rectangle is a square.
4. Prove that the diagonals of two reciprocal rectangles are perpendicular.
5. Let $ABGF$ and $BGHC$ be two reciprocal golden rectangles. Let P be the point of intersection of the diagonals $\overline{BF}$ and $\overline{CG}$. Prove that $FP/GP = BP/CP = \alpha$.
6. Let $ABGF$ be a golden rectangle and $BGHC$ its reciprocal, as in Figure 23.26. Prove that $ACHF$ is a square.
7. Let $BGHC$ be a golden rectangle. Complete the square $ACHF$ on its left. Prove that $ABGF$ is a golden rectangle.
8. Show that the sum of the lengths of the diagonals of the various "whirling squares" in Figure 23.28 is $\sqrt{2}a\alpha^2$, where $AC = a$.
9. Let P, Q, R, and S be points on the sides of a square $ABCD$, dividing each in the Golden Ratio. Prove that $PQRS$ is a golden rectangle.

Consider the sequence of decreasing smaller reciprocal golden rectangles in Figure 23.34, beginning with the golden rectangle $ABCD$. Let $DE = a$ and $EC = b$, where $a = \alpha b$.*

10. Complete the following table.

*Based on G. E. Runion, *The Golden Section and Related Curiosa*, Scott, Foresman, Glenview, IL., 1972.

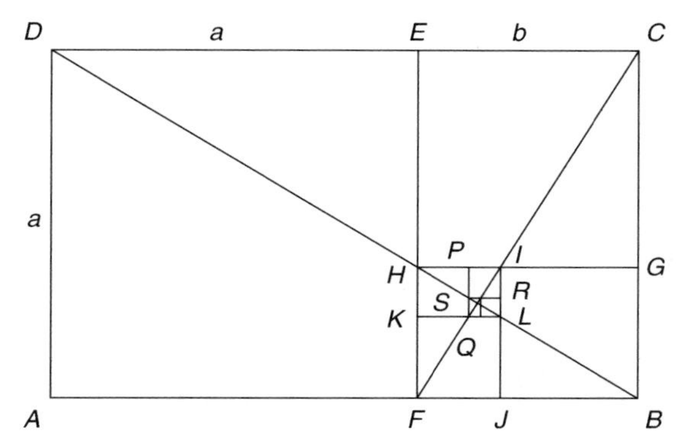

Figure 23.34.

Rectangle	*ABCD*	*BCEF*	*BGHF*	*FJIH*	*HKLI*	*PQLI*	*QLRS*
Shorter side	a						
Longer side	$a+b$						

11. Predict the size of the nth reciprocal rectangle in the sequence, $n \geq 0$.

FIBONACCI GEOMETRY

This chapter features some additional delightful properties of the Golden Ratio in Euclidean geometry. We begin with a problem proposed and solved by J. A. H. Hunter in 1963.

Example 24.1. Locate points P and Q on two adjacent sides of a rectangle $ABCD$, as in Figure 24.1, such that the areas of triangles APQ, BQC, and CDP are equal.

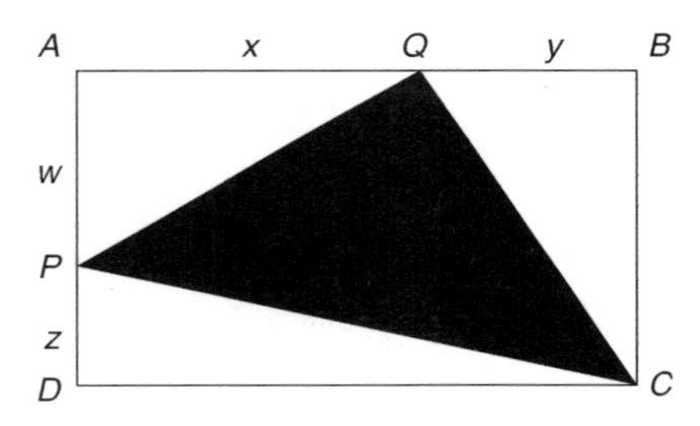

Figure 24.1.

Solution. Let $AQ = x$, $QB = y$, $AP = w$, and $PD = z$. Since the areas of $\triangle APQ$, $\triangle QBC$, and $\triangle CDP$ are equal, we have $xw/2 = y(w+z)/2 = z(x+y)/2$; that is, $xw = y(w+z) = z(x+y)$. Equation $y(w+z) = z(x+y)$ yields $yw = zx$; that is, $x/y = w/z$. From the equation $xw = z(x+y)$, we have $w/z = (x+y)/x$; that is,

$$\frac{x}{y} = 1 + \frac{y}{x}$$

Thus x/y satisfies the quadratic equation $t^2 = t + 1$. Since $x/y > 0$, we choose the positive root α for x/y:

$$\frac{x}{y} = \alpha = \frac{1+\sqrt{5}}{2}$$

Thus

$$\frac{w}{z} = \frac{x}{y} = \alpha \qquad (24.1)$$

Consequently, we must choose the points P and Q in such a way that P divides $\overline{AD}$ in the ratio $AP{:}PB = w{:}z = \alpha{:}1$. Likewise, we must locate Q on $\overline{AB}$ such that $AQ{:}QD = x{:}y = \alpha{:}1$. Thus $\overline{PQ}$ divides the two sides in the golden section. ■

In 1964, H. E. Huntley of Somerset, England, pursued this problem further. He proved that if $ABCD$ is a golden rectangle, then $\triangle PQC$ is an isosceles right triangle with right angle at Q, as the following example shows.

Example 24.2. Suppose the rectangle $ABCD$ in Figure 24.1 is a golden rectangle. Prove that $\triangle PQC$ is an isosceles right triangle and $\angle Q = 90°$.

Solution. Since $BCAD$ is a golden rectangle,

$$\frac{AB}{BC} = \frac{x+y}{w+z} = \alpha.$$

From Eq. (24.1), $x = y\alpha$ and $w = z\alpha$;

$$\therefore \quad \frac{y(1+\alpha)}{z(1+\alpha)} = \alpha$$

Thus, $y = z\alpha$. But $z\alpha = w$, so $y = w$. Thus $AP = BQ$.

Then $x = y\alpha = (z\alpha)\alpha = z\alpha^2 = z(\alpha + 1) = z\alpha + z = w + z$; that is, $AQ = BC$. Therefore, by the side-angle-side (SAS) theorem, $\triangle APQ \cong \triangle BQC$. Consequently, $PQ = QC$, so $\triangle PQC$ is an isosceles triangle.

Since $\triangle APQ \cong \triangle BQC$, $\angle AQP = \angle BCQ$. But $\angle BCQ + \angle BQC = 90°$; that is, $\angle AQP + \angle BQC = 90°$, so $\angle PQC = 90°$; $\therefore \angle PQC = 90°$. Thus $\triangle PQC$ is an isosceles right triangle (see Fig. 24.2).

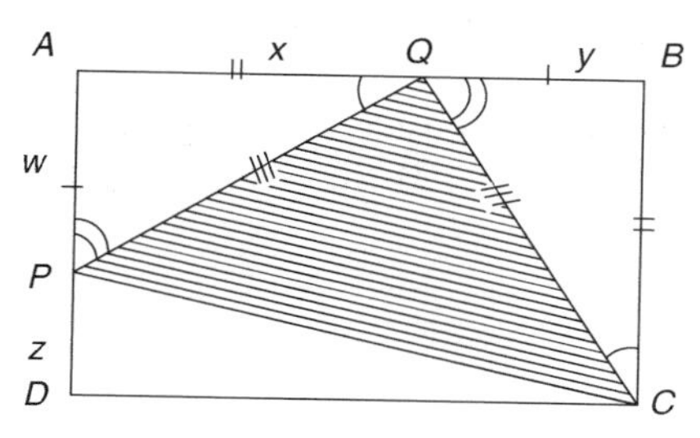

Figure 24.2.

■

As a byproduct, it follows that $\angle CPQ = 45°$, so $\angle APQ = 45°$; $\therefore \overleftrightarrow{PQ} \| \overleftrightarrow{BD}$. Besides, we can derive a formula for the area of $\triangle PQC$:

$$\begin{aligned}
\triangle PQC &= \frac{1}{2} PQ \cdot QC \\
&= \frac{1}{2}\sqrt{AP^2 + AQ^2} \cdot \sqrt{BQ^2 + BC^2} \\
&= \frac{1}{2}\sqrt{w^2 + x^2} \cdot \sqrt{y^2 + (w+z)^2} \\
&= \frac{1}{2}\sqrt{x^2 + y^2} \cdot \sqrt{y^2 + x^2} \\
&= \frac{1}{2}(x^2 + y^2) \\
&= \frac{1}{2}(y^2\alpha^2 + y^2) = \frac{y^2}{2}(1 + \alpha^2) \\
&= \frac{1}{2}(\alpha + 2)y^2
\end{aligned}$$

Notice that the area of the golden rectangle is given by

$$x(x+y) = y\alpha(y\alpha + 1) = (\alpha^2 + \alpha)y^2 = (2\alpha + 1)y^2$$

$$\begin{aligned}
\therefore \quad \triangle APQ + \triangle QBC + \triangle CDP &= (2\alpha + 1)y^2 - \frac{1}{2}(\alpha + 2)y^2 \\
&= \frac{(4\alpha + 2 - \alpha - 2)y^2}{2} = \frac{3}{2}\alpha y^2
\end{aligned}$$

In 1964, Hunter also proved that the ratios of the dimensions of a special rectangular prism are closely linked to the Golden Section, as the next example demonstrates.

Example 24.3. (Golden Cuboid) Consider a rectangular prism with unit volume and a diagonal of 2 units long (see Fig. 24.3). Suppose the edges are a, b, and c units long. Then $abc = 1$ and $a^2 + b^2 + c^2 = 4$.

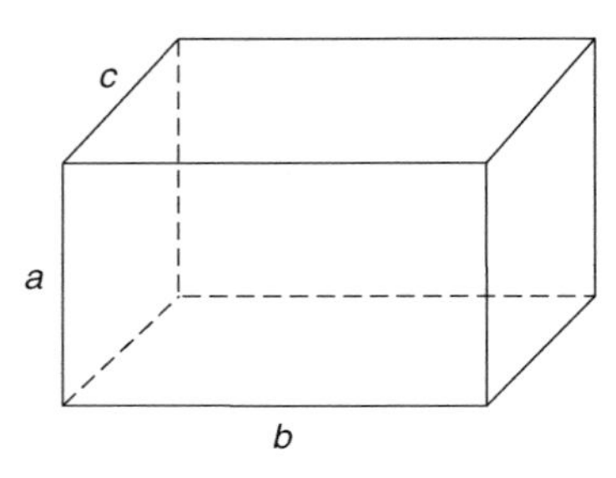

Figure 24.3.

Since we are interested in the ratios of the dimensions of this cuboid, assume, for convenience, that $a = 1$. Then $bc = 1$ and $b^2 + c^2 = 3$. Substituting for c, this equation yields $b^2 + 1/b^2 = 3$; that is, $b^4 - 3b^2 + 1 = 0$. Then

$$b^2 = \frac{3+\sqrt{5}}{2} = 1 + \alpha = \alpha^2$$

Therefore, $b = \alpha$ and hence $c = 1/\alpha$. Thus $a{:}b{:}c = 1{:}\alpha{:}1/\alpha$. ■

Notice that $a^2 + b^2 + c^2 = 1 + \alpha^2 + \alpha^{-2} = 1 + (\alpha^2 + \beta^2) = 1 + L_2 = 4$, as expected.

Example 24.3 leads to several interesting properties of the cuboid:

- Ratios of the areas of the three different faces $= ab{:}bc{:}ca = \alpha{:}1{:}1/\alpha$.
- Total surface area of the cuboid $= 2(ab + bc + ca) = 2(\alpha + 1 + 1/\alpha) = 2(2\alpha) = 4\alpha$
- Since $a{:}b{:}c = \alpha{:}1{:}1/\alpha$, it follows that the faces of the cuboid are indeed golden rectangles. For example, consider the face $ABCD$ in Figure 24.4. We have $AB{:}BC = b{:}a = \alpha{:}1$.

$$\frac{\text{Surface area of the golden cuboid}}{\text{Surface area of the circumscribing sphere}} = \frac{4\alpha}{4\pi} = \frac{\alpha}{\pi}.$$

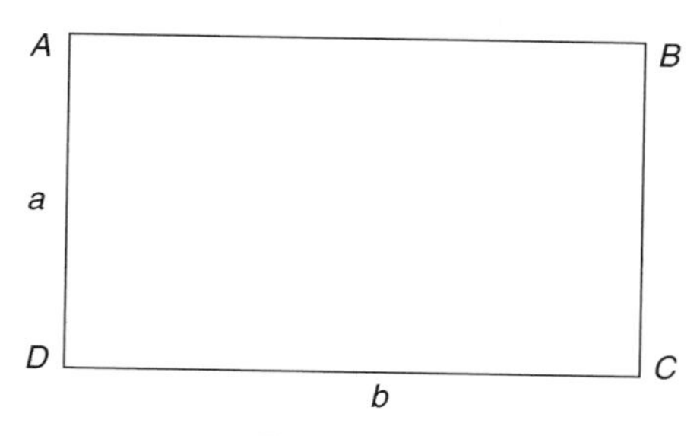

Figure 24.4.

The following example,* although elementary in nature, is certainly interesting in its own right. It also manifests the omnipresence of our marvelous number α.

Example 24.4. Let P be a point on a chord $\overline{AB}$ of a circle and $\overrightarrow{PT}$ be a tangent to it at T such that $PT = AB$ (see Fig. 24.5). Compute the ratio $PB{:}AB$.

*Based on H. E. Huntley, *The Divine Proportion*, Dover, New York, 1970.

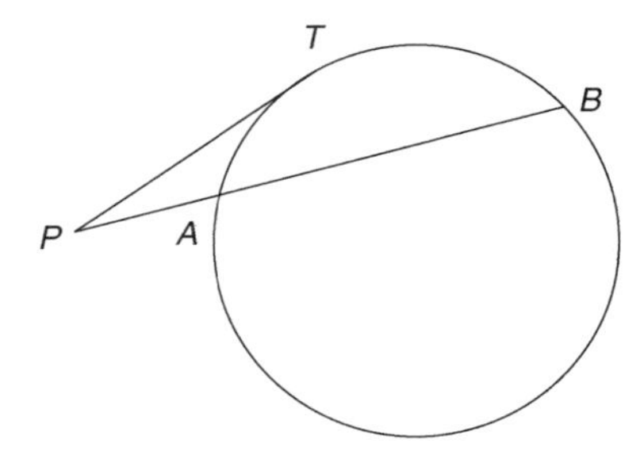

Figure 24.5.

Solution. It follows by elementary geometry that $PT^2 = PA \cdot PB$, that is,

$$AB^2 = PA \cdot PB$$
$$AB^2 = PB(PB - AB)$$
$$(PB/AB)^2 = 1 + PB/AB$$

So it follows that $PB{:}AB = \alpha{:}1$; that is, A divides $\overline{PB}$ in the golden ratio.

As a bonus, it follows that $PA{:}AB = 1/\alpha{:}1$. ■

To continue this example a bit further, let C be a point on $\overline{AB}$ such that $PT = AB = PC$ (see Fig. 24.6). Since $PC = AB$, $PA + AC = AC + CB$; thus $PA = CB$. Since $PB/AB = \alpha$, $(PC + CB)/AB = \alpha$. That is,

$$\frac{AB + CB}{AB} = \alpha$$

$$1 + \frac{CB}{AB} = \alpha$$

$$\therefore \quad \frac{AB}{CB} = \alpha$$

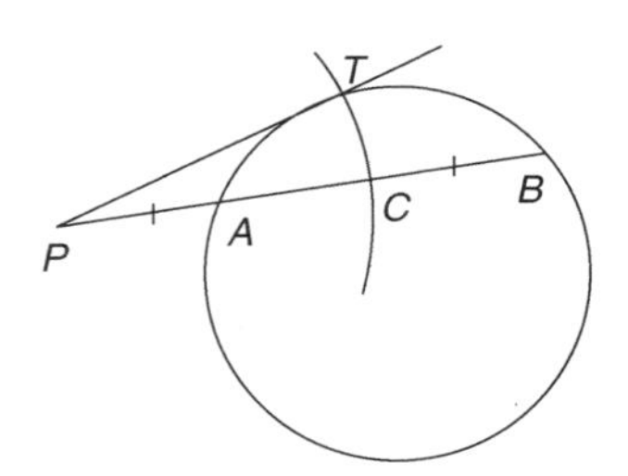

Figure 24.6.

Thus C divides $\overline{AB}$ in the Golden Ratio. Moreover,

$$\frac{PB}{PC} = \frac{PC + CB}{PC} = 1 + \frac{CB}{AB} = 1 + \frac{1}{\alpha} = 1 - \beta = \alpha$$

$$\frac{CB}{CA} = \frac{CB}{AB - CB} = \frac{1}{AB/CB - 1} = \frac{1}{\alpha - 1} = \alpha$$

$$\frac{PC}{PA} = \frac{PA + AC}{PA} = 1 + \frac{AC}{CB} = 1 + \frac{1}{\alpha} = \alpha$$

$$\frac{AB}{AC} = \frac{PC}{AC} = \frac{PC}{PA} \cdot \frac{PA}{AC} = \alpha \cdot \alpha = \alpha^2$$

$$\frac{PB}{AC} = \frac{PA + AB}{AB - CB} = \frac{PA/AB + 1}{1 - CB/AB} = \frac{1/\alpha + 1}{1 - 1/\alpha} = \frac{\alpha + 1}{\alpha - 1} = \frac{\alpha^2}{-\beta} = \alpha^3$$

CANDIDO'S IDENTITY

Candido's identity,* namely,

$$[x^2 + y^2 + (x + y)^2]^2 = 2[x^4 + y^4 + (x + y)^4]$$

provides an interesting application to Fibonacci numbers, where x and y are arbitrary real numbers. In particular, let $x = F_n$ and $y = F_{n+1}$. Then $(F_n^2 + F_{n+1}^2 + F_{n+2}^2)^2 = 2(F_n^4 + F_{n+1}^4 + F_{n+2}^4)$. This result has an interesting geometric interpretation.

To see this, consider a line segment $\overline{AD}$ such that $AB = F_n^2$, $BC = F_{n+1}^2$, and $CD = F_{n+2}^2$. Complete the square $ADEF$, as Figure 24.7 shows. Then

$$\begin{aligned} \text{Area ADEF} &= (F_n^2 + F_{n+1}^2 + F_{n+2}^2)^2 \\ &= 2(F_n^4 + F_{n+1}^4 + F_{n+2}^4) \\ &= 2(\text{sum of the areas of the three squares}) \end{aligned}$$

For example, let $m = 4$. Then $AB = 9$, $BC = 25$, and $CD = 64$ (see Fig. 24.8). Then

$$\begin{aligned} \text{Area ADEF} &= (9 + 25 + 64)^2 = 9604 \\ &= 2(81 + 625 + 4096) \\ &= 2(\text{sum of the areas of the three smaller squares}) \end{aligned}$$

*Named after Italian mathematician, Giacomo Candido (1871–1941).

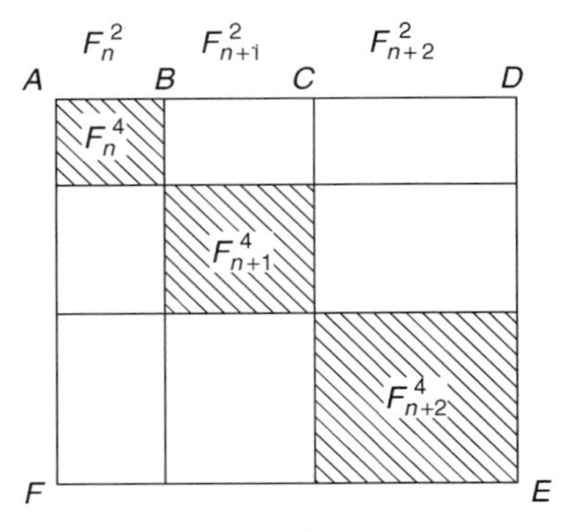

Figure 24.7.

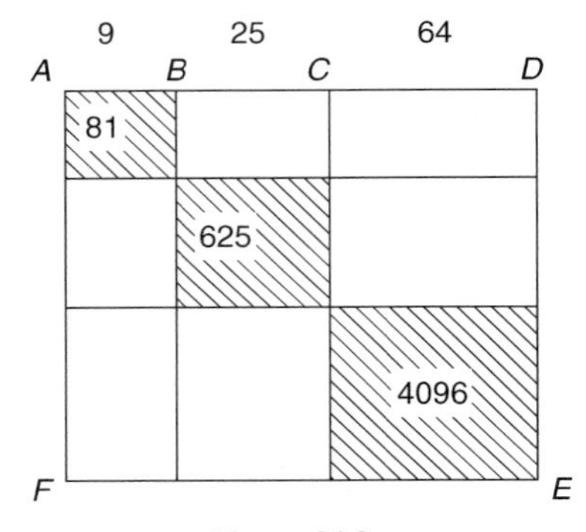

Figure 24.8.

Candido's identity for Fibonacci numbers can be extended to generalized Fibonacci numbers:

$$(G_n^2 + G_{n+1}^2 + G_{n+2}^2)^2 = 2(G_n^4 + G_{n+1}^4 + G_{n+2}^4)$$

See Figure 24.9.

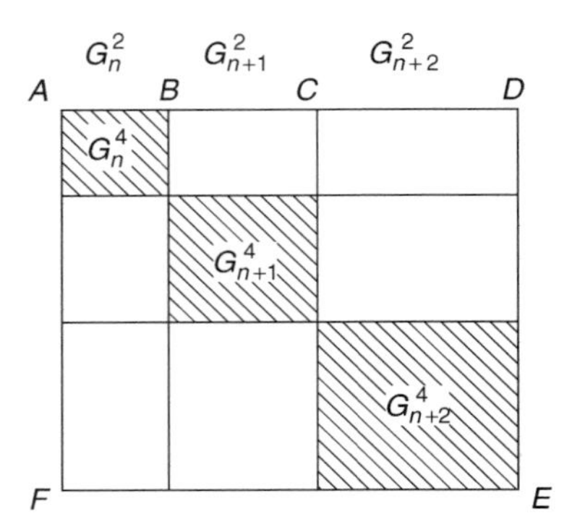

Figure 24.9.

FIBONACCI MEETS APPOLLONIUS

A mathematical giant of the third century B.C., Appollonius (262?–190? B.C.) proposed the following problem: *Given three fixed circles, find a circle that touches each of them.* In total, there are eight solutions. But, if the given circles are mutually tangential, then there are exactly two solutions.

Assume, for convenience, that the given circles are not only tangential to each other; but their centers form the vertices of a Pythagorean triangle (as studied in 1973 by W. H. Horner of Pittsburgh) (see Fig. 24.10). Let r_1, r_2, and r_3 denote the radii of the given circles, and R and r those of the solutions. Assume that $r_1 < r_2 < r_3$ and $r < R$.

Let $a = F_n, b = F_{n+1}, c = F_{n+2}$, and $d = F_{n+3}$. Since $(c^2 - b^2)^2 + (2bc)^2 = (c^2 + b^2)^2$, we can assume that the lengths of the sides of the Pythagorean triangle are $c^2 - b^2$, $2bc$, and $c^2 + b^2$. So, since the original circles are mutually tangential, it follows that:

$$\begin{aligned} r_1 + r_2 &= c^2 - b^2 \\ r_2 + r_3 &= c^2 + b^2 \\ r_3 + r_1 &= 2bc \end{aligned}$$

Solving this linear system, we get $r_1 = b(c - b) = ab, r_2 = c(c - b) = ac$, and $r_3 = b(c + b) = bd$. Then $r_1r_2r_3 = a^2b^2cd, r_1r_2 = a^2bc, r_2r_3 = abcd$, and $r_3r_1 = ab^2d$.

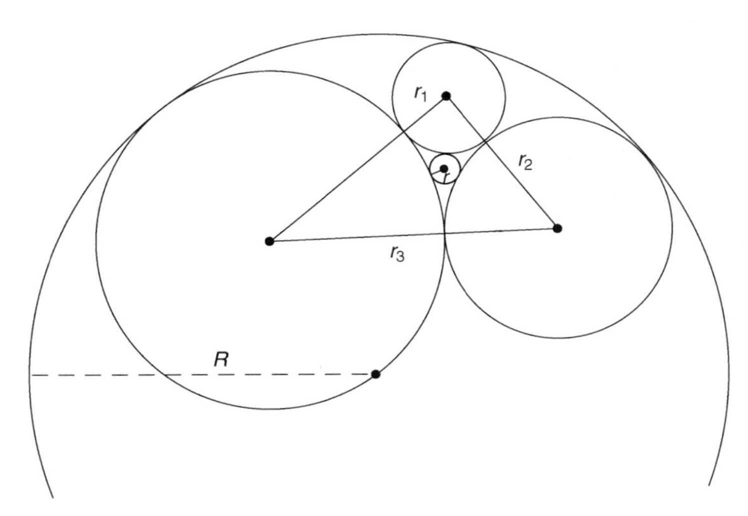

Figure 24.10.

In 1955, Col. R. S. Beard showed that

$$R \text{ or } r = \frac{r_1 r_2 r_3}{r_1 r_2 + r_2 r_3 + r_3 r_1 \mp 2\sqrt{r_1 r_2 r_3 (r_1 + r_2 + r_3)}}$$

where the negative root gives R and the positive root gives r.

Substituting for r_1, r_2, and r_3, we get

$$R \text{ or } r = \frac{a^2 b^2 cd}{a^2 bc + ab^2 d + abcd \mp 2\sqrt{a^2 b^2 c^2 d^2}}$$

So

$$\begin{aligned} R &= \frac{abcd}{ac + bd - cd} = \frac{abcd}{ac - d(c - b)} \\ &= \frac{abcd}{ac - ad} = \frac{abcd}{ab} = cd \end{aligned}$$

Similarly,

$$r = \frac{abcd}{4cd - ab}$$

Substituting for a, b, c, and d, we have

$$r_1 = F_n F_{n+1} \qquad r_2 = F_n F_{n+2} \qquad r_3 = F_{n+1} F_{n+3},$$

$$R = F_{n+2} F_{n+3},$$

$$r = \frac{F_n F_{n+1} F_{n+2} F_{n+3}}{4F_{n+2} F_{n+3} - F_n F_{n+1}}$$

Clearly, similar formulas hold for Lucas numbers also.

A FIBONACCI SPIRAL

We can arrange a series of $F_n \times F_n$ Fibonacci squares to form a *Fibonacci spiral*, as Figure 24.11 shows, where $n \geq 1$. Moreover, their centers appear to lie on two lines, and the two lines appear perpendicular. This is in fact the case.

To confirm this, suppose we choose the center of the first square as the origin, and the horizontal and vertical lines through it as the axes. Let n be odd. Then the change in the y-values in going from the $(n-4)$th square to the nth square is $\pm(F_n + F_{n-4})/2$, and that in the corresponding x-values is $\pm(F_{n-2} - 2F_{n-3} - F_{n-4})/2$. Therefore, the slope of the line passing through their centers is

$$\frac{F_n + F_{n-4}}{F_{n-2} - 2F_{n-3} - F_{n-4}} = \frac{3F_{n-2}}{F_{n-2}} = 3$$

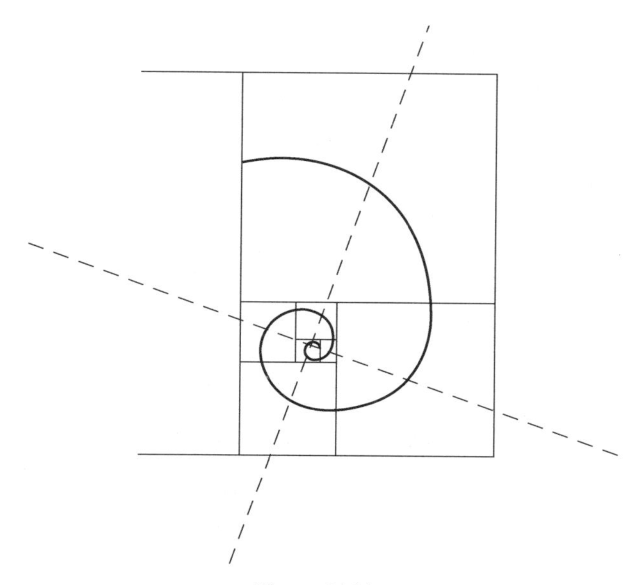

Figure 24.11.

But this line passes through the origin. Thus the center of all odd-numbered Fibonacci squares lies on the line $y = 3x$, as proved in 1983 by T. Gardiner of the University of Birmingham, England. [In particular, the line containing the centers (0,0) and $(1/2, 3/2)$ is $y = 3x$.] Similarly, the centers of all even-numbered Fibonacci squares lie on the line $x + 3y - 1 = 0$. Notice, in particular, that the centers (1,0) and $(-2, 1)$ lie on it.

Example 24.5. Suppose we inscribe a square *BDEF* in a semicircle such that one side of the square lies along its diameter (see Fig. 24.12). Since the right triangles *AFE* and *CFE* are similar,

$$\frac{AE}{FE} = \frac{FE}{CE} = \frac{AF}{CF}$$

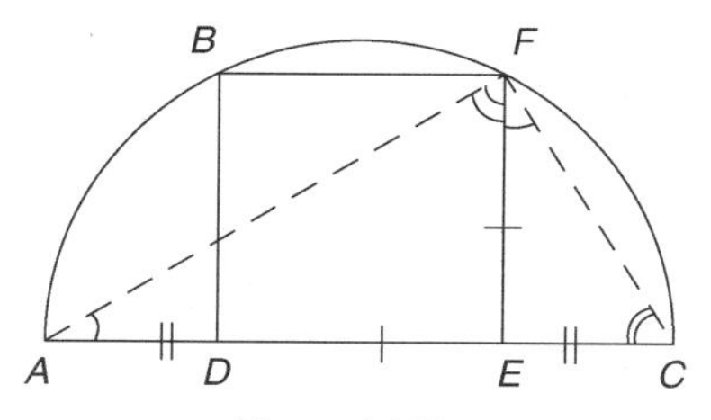

Figure 24.12.

But

$$\frac{AE}{FE} = \frac{AD + DE}{FE} = \frac{AD + FE}{FE} = \frac{AD}{FE} + 1$$

Since $CE = AD$, this implies $AD/FE + 1 = FE/AD$. Let $x = FE/AD$. Then this yields the equation $1/x + x = 1$, so $x = \alpha$. Thus $AE/DE = \alpha$, so D divides $\overline{AE}$ in the Golden Ratio. Moreover,

$$\frac{AE}{FE} = \frac{FE}{CE} = \frac{AF}{CF} = \alpha$$

that is, the ratio of the corresponding sides of the similar triangles is the Golden Ratio. ■

Example 24.6. In Figure 24.13, A is the midpoint of the side $\overline{PQ}$ of the square $PQRS$. Let $\overline{AR}$ be the tangent to the circle with center O. Since $AD^2 = AB \cdot AC$, it follows that $AD/AB = AC/AD$. In fact, we can be show that $AD/AB = AC/AD = \alpha$.

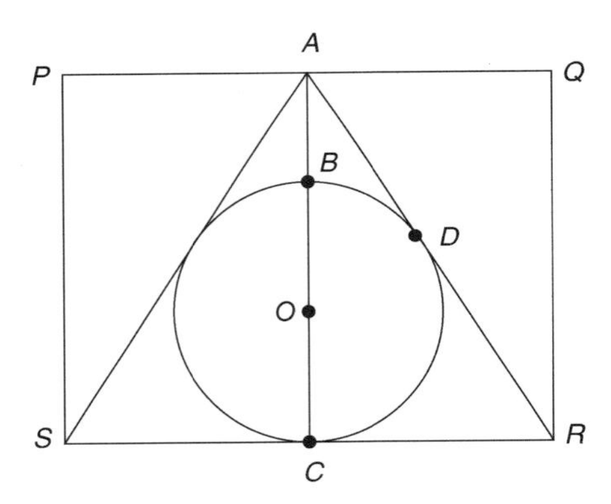

Figure 24.13.

■

RIGHT TRIANGLES AND THE GOLDEN RATIO

Suppose the lengths of the sides of a right triangle form a geometric sequence. What can we say about its common ratio r?

To answer this, suppose the three lengths are a, ar, and ar^2. Clearly, $r \neq 1$.

Case 1. Let $r < 1$. Then $ar^2 < ar < a$ (see Fig. 24.14). Then

$$a^2 = a^2r^2 + a^2r^4$$

$$r^4 + r^2 = 1$$

$$\therefore \quad r = 1/\sqrt{\alpha}$$

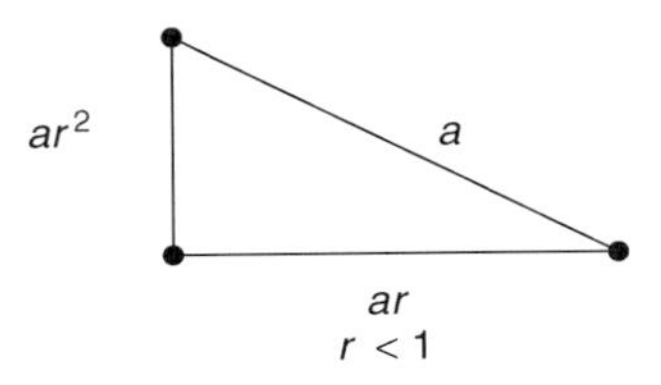

Figure 24.14.

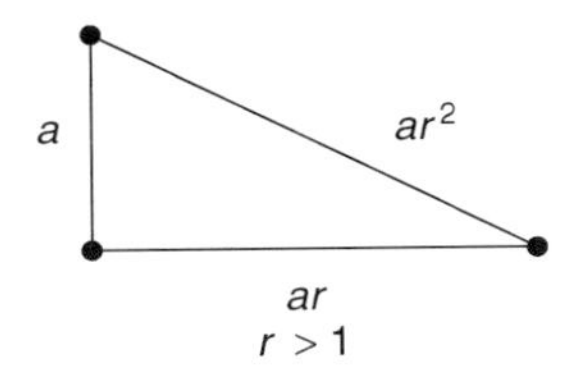

Figure 24.15.

Case 2. On the other hand, let $r > 1$. Then $a < ar < ar^2$ (see Fig. 24.15). Then

$$a^2r^4 = a^2r^2 + a^2$$
$$r^4 = r^2 + 1$$
$$\therefore \quad r = \sqrt{\alpha}$$

Thus, the common ratio is either $1/\sqrt{\alpha}$ or $\sqrt{\alpha}$.

THE CROSS OF LORRAINE

An interesting problem, related to the *Cross of Lorraine** or the *Patriarchal Cross*, brought to light by Martin Gardner. This ancient emblem reintroduced in modern times by General Charles de Gaulle (1890–1970) of France, consists of three beams—two horizontal and one vertical—and covers an area of $13 = F_7$ square units (see Fig. 24.16). We would like to cut the cross through C into two pieces of equal area, namely, 6.5 each.

Suppose the line segment $\overline{PQ}$ has the desired property. Then the area of $\triangle PQR$ is 2.5 units. Let $BP = x$ and $DQ = y$. Since $\triangle BPC \sim \triangle DQC$, $x/1 = 1/y$; that is, $xy = 1$.

$$\text{Area } \triangle PQR = \frac{1}{2}PR \cdot QR = \frac{1}{2}(x+1)(y+1)$$
$$\therefore \quad (x+1)(y+1) = 5$$
$$x^2 - 3x + 1 = 0$$
$$x = \frac{3 \pm \sqrt{5}}{2}$$

*Lothringen or Lorraine is a province on the border between France and Germany.

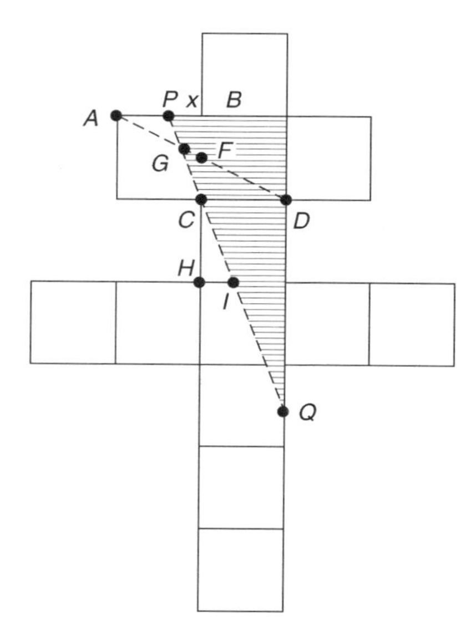

Figure 24.16.

Since $0 < x < 1$, it follows that $x = (3 - \sqrt{5})/2 = 1 + \beta$, so $y = 1/x = 1 + \alpha$. Thus $AP = 1 - x = -\beta$ and $EQ = y - 1 = \alpha$. So

$$\frac{AP}{PB} = \frac{-\beta}{1+\beta} = \alpha \qquad \text{and} \qquad \frac{DE}{EQ} = \frac{1}{\alpha}$$

Thus P and E divide $\overline{AB}$ and $\overline{DQ}$ in the Golden Ratio, respectively.

LOCATING *P* AND *Q* GEOMETRICALLY

To locate the points P and Q geometrically, let $\overline{AD}$ meet $\overline{BC}$ at F, so $AF = FD$. With F as center and FB as radius, draw an arc to intersect $\overline{AF}$ at G. With G as center and AG as radius, draw an arc to intersect $\overline{AB}$ at P. Let $\overline{PC}$ meet $\overline{DE}$ at Q. Then P divides $\overline{AB}$ in the Golden Ratio and E divides $\overline{DQ}$ in the Golden Ratio; moreover, $\overline{PQ}$ divides the cross into two equal areas.

To confirm this, $\triangle PBC \cong \triangle CHI$ and $\triangle PBC \sim \triangle DQC$, so $xy = 1$; $\therefore$ shaded area $= 2 +$ area $\triangle EQI$. Since $\triangle PBC \sim \triangle EQI$, $x/1 = EI/(y-1)$, so

$$EI = x(y-1) = xy - x = 1 - x$$

$$\therefore \quad \text{Area } \triangle EQI = \frac{1}{2} EI \cdot EQ = \frac{1}{2}(1-x)(y-1)$$

$$= \frac{1}{2}(1-x)\left(\frac{1}{x} - 1\right) = \frac{(1-x)^2}{2x}$$

But

$$\frac{1-x}{x} = \frac{-\beta}{1+\beta} = \alpha$$

$$\therefore \quad \text{Area } \triangle EQI = \frac{\alpha(-\beta)}{2} = \frac{1}{2}$$

Thus the shaded area = 2.5 units, so $\overline{PQ}$ partitions the cross equally as desired.

In 1959, M. Gardner, in his famous column in the *Scientific American*, invited his readers to compute the length BC. The same puzzle appeared five years later in an article by M. H. Holt in *The Pentagon*.

EXERCISES 24

Verify each.

1. $(F_n^2 + F_{n+1}^2 + F_{n+2}^2)^2 = 2(F_n^4 + F_{n+1}^4 + F_{n+2}^4)$
2. $(L_n^2 + L_{n+1}^2 + L_{n+2}^2)^2 = 2(L_n^4 + L_{n+1}^4 + L_{n+2}^4)$
3. $(G_n^2 + G_{n+1}^2 + G_{n+2}^2)^2 = 2(G_n^4 + G_{n+1}^4 + G_{n+2}^4)$
4. Compute the area of an equilateral trapezoid with bases F_{n-1} and F_{n+1}, and with lateral side F_n (Woodlum, 1968).
5. Show that the lengths $L_{n-1}L_{n+2}, 2L_nL_{n+1}$, and $L_{2n}L_{2n+2}$ form the sides of a Pythagorean triangle (Freitag, 1975).
6. In $\triangle ABC, AB = AC$. Let D be a point on side $\overline{AB}$ such that $AD = CD = BC$. Prove that $2\cos A = AB/BC = \alpha$ (Source unknown).

REGULAR PENTAGONS

Regular pentagons provide us with many examples of the Golden Ratio in everyday life. Some flowers have pentagonal shape; so do the starfish and the former Chrysler logo (see Fig. 25.1).

In 1948, H. V. Baravalle of Adelphi University observed, "Outstanding among the mathematical facts connected with the (regular) pentagon are the manifold implications of the irrational ratio of the Golden Section." This chapter investigates some of these implications.

Example 25.1. The diagonals $\overline{AC}$ and $\overline{BE}$ of the regular pentagon $ABCDE$ in Figure 25.2 meet at F. Prove that F divides both diagonals in the Golden Ratio.

*Proof:** Let $AB = a$, $BF = b$, and $FE = c$ (see Fig. 25.3). By the side-angle-side (SAS) theorem, $\triangle ABC \cong \triangle ABE$. Since $\angle ABC = 108^\circ$, $\angle BAC = \angle ABE = 36^\circ$. Therefore, $\angle CAE = 72^\circ = \angle AFE$. Then $AF = BF = b$ and $AE = AF$, so $a = c$.

By drawing the perpendicular $\overline{AR}$ to $\overline{BE}$, we can be show that

$$ER = a\cos 36^\circ = BR$$

$$\therefore\ BE = BR + RE = b + c = 2a\cos 36^\circ$$

Likewise,

$$b = \frac{a}{2\cos 36^\circ}$$

*Based on J. A. H. Hunter and J. S. Madachy, *Mathematical Diversions*, Dover, New York, 1975.

Figure 25.1. (*a*) A starfish.

Figure 25.1. (*b*) The former Chrysler logo are pentagonal shapes (*Source*: The photo image of the Dodge Intrepid® is used with permission from the DaimlerChrysler Corporation.).

$$\therefore\ c = 2a\cos 36^\circ - \frac{a}{2\cos 36^\circ}$$

$$= \frac{a(4\cos^2 36^\circ - 1)}{2\cos 36^\circ}$$

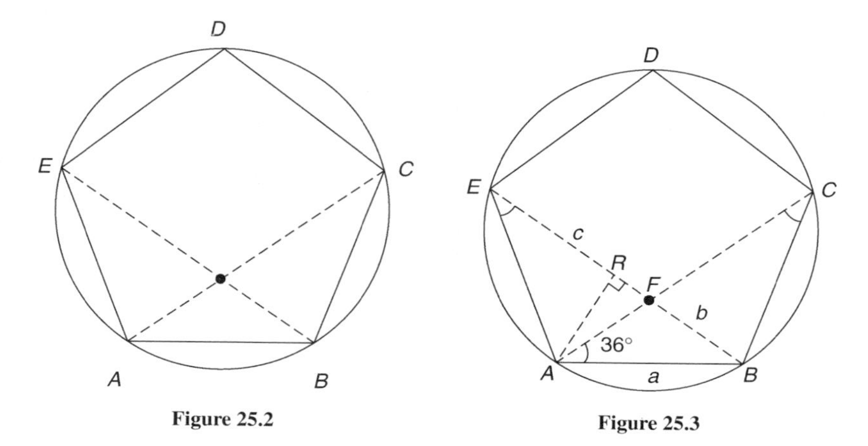

Figure 25.2

Figure 25.3

Since $c = a$, this yields the quadratic equation $4\cos^2 36^\circ - 2\cos 36^\circ - 1 = 0$. Solving this, we get

$$\cos 36^\circ = \frac{1+\sqrt{5}}{4} = \frac{\alpha}{2}$$

(Since $\cos 36^\circ > 0$, we take the positive root.)

$$\therefore\ BE = b + c = \alpha a = AC$$
$$BE{:}FE = \alpha a{:}a = \alpha{:}1 = AC{:}FC$$

Thus F divides both diagonals in the golden ratio. ■

This is the ninth property of the divine proportion delineated by Pacioli in his book.

Since $\cos \pi/5 = \alpha/2$, it follows that $\sin \pi/5 = (\sqrt{3-\alpha})/2$, $\cos \pi/10 = (\sqrt{\alpha+2})/2$, and $\sin \pi/10 = 1/2\alpha$.

AN ALTERNATE PROOF THAT cos $\pi/5 = \alpha/2$

We can derive the fact that $\cos \pi/5 = \alpha/2$ in a shorter, more elegant way. Let $\theta = \pi/10$. Then $2\theta + 3\theta = \pi/2$, so 2θ and $\pi/2 - 3\theta$ are complimentary angles. Since the values of cofunctions of complementary angles are equal, it follows that

$$\sin 2\theta = \cos 3\theta$$

That is,

$$\begin{aligned} 2\sin\theta\cos\theta &= 4\cos^3\theta - 3\cos\theta \\ 4\sin^3\theta + 2\sin\theta - 1 &= 0 \\ \sin\theta &= \frac{-1 \pm \sqrt{5}}{2} \end{aligned}$$

Since $\sin\theta > 0$, it follows that

$$\sin\theta = -\frac{\beta}{2} = \frac{1}{2\alpha}$$

$$\begin{aligned} \therefore\ \cos\pi/5 &= 1 - 2\sin^2\pi/10 = 1 - 2\cdot\frac{1}{4\alpha^2} = 1 - \frac{1}{2\alpha^2} \\ &= \frac{2-\beta^2}{2} = \frac{2-(1+\beta)}{2} = \frac{1-\beta}{2} = \frac{\alpha}{2} \end{aligned}$$

Knowing the values of $\sin\pi/10$ and $\cos\pi/10$, we can compute the exact values of sines and cosines of several acute angles that are multiples of $\pi/20 = 9^\circ$, as Table 25.1 shows.

TABLE 25.1.

Angle θ	Sin θ	Cos θ
$\frac{\pi}{20}$	$\frac{2-\alpha}{4}$	$\frac{2+\alpha}{4}$
$\frac{\pi}{10}$	$\frac{1}{2\alpha}$	$\frac{\sqrt{2+\alpha}}{2}$
$\frac{3\pi}{20}$	$\frac{11-4\alpha}{16}$	$\frac{7\alpha-11}{16}$
$\frac{\pi}{5}$	$\frac{\sqrt{3-\alpha}}{2}$	$\frac{\alpha}{2}$
$\frac{\pi}{4}$	$1-2\alpha$	$1-2\alpha$
$\frac{3\pi}{10}$	$\frac{\alpha}{2}$	$\frac{\sqrt{3-\alpha}}{2}$
$\frac{7\pi}{20}$	$\frac{7\alpha-11}{16}$	$\frac{11-7\alpha}{16}$
$\frac{2\pi}{5}$	$\frac{\sqrt{2+\alpha}}{2}$	$\frac{1}{2\alpha}$
$\frac{9\pi}{20}$	$\frac{2+\alpha}{4}$	$\frac{2-\alpha}{4}$

We pursue Example 25.1 a bit further in the following example.

Example 25.2. Suppose the perpendicular $\overline{AN}$ at A meets $\overline{ED}$ at N (see Fig. 25.4). Show that N divides $\overline{ED}$ in the Golden Ratio.

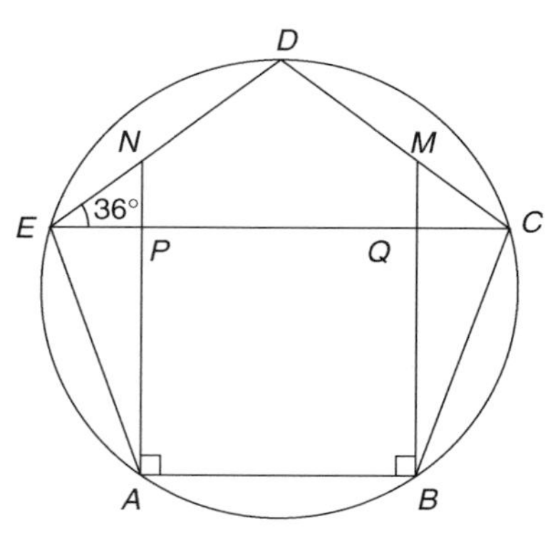

Figure 25.4.

Proof. Draw $\overline{BM} \perp \overline{AB}$. We have $CE = a\alpha$, $PQ = AB = a$, $EP = QC$, and $\angle DEC = \pi/5$. Then

$$EP = QC = \frac{a\alpha - a}{2} = \frac{a(\alpha - 1)}{2}$$

$$EN = \frac{EP}{\cos \pi/5} = \frac{2EP}{\alpha} = \frac{a(\alpha - 1)}{\alpha}$$

$$\therefore\ ND = DE - EN = a - \frac{a(\alpha - 1)}{\alpha} = \frac{a}{\alpha}$$

Thus $DE{:}DN = a{:}a/\alpha = 1{:}1/\alpha = \alpha{:}1$; that is, N divides $\overline{DE}$ in the Golden Ratio. ■

Example 25.3. Compute the area of the regular pentagon in Figure 25.5 with a side a units long.

Solution. Notice that $\triangle CEE$ is an isosceles triangle. (In fact, $\overleftrightarrow{CE} \parallel \overleftrightarrow{AB}$.) Let $\overline{DN} \perp \overline{CE}$. Then $CN = EN = a\cos\pi/5 = a\alpha/2$, so $CE = a\alpha$. Likewise, $PQ = a\alpha/2$, where P and Q are the midpoints of $\overline{CD}$ and $\overline{DE}$, respectively.

Let R be the circumradius and r the inradius of the regular pentagon. Then $R = (a/2)\csc \pi/5$ and $r = (a/2)\cot \pi/5$:

$$\begin{aligned}\text{Area of } \triangle AOB &= (1/2)a \cdot (a/2)\cot \pi/5 \\ &= (a^2/4)\cot \pi/5\end{aligned}$$

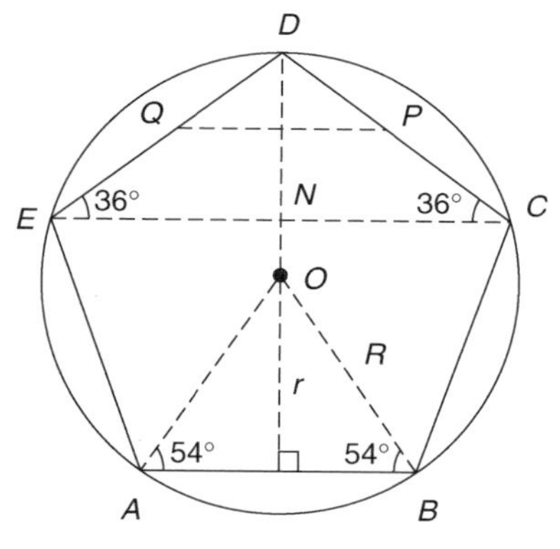

Figure 25.5.

$$= \frac{a^2}{4} \cdot \frac{\alpha}{\sqrt{3-\alpha}} \qquad \text{since } \cos \pi/5 = \alpha/2$$

$$= \frac{a^2\alpha}{4\sqrt{3-\alpha}}$$

$$\therefore \text{ Area of the pentagon } = \frac{5a^2\alpha}{4\sqrt{3-\alpha}}$$

■

THE BEE AND THE REGULAR PENTAGON

Using the fact that $\cos \pi/5 = \alpha/2$, we can derive a trigonometric formula for F_n:

$$\begin{aligned}\cos 3\pi/5 &= 4\cos^3 \pi/5 - 3\cos \pi/5 \\ &= 4(\alpha/5)^3 - 3(\alpha/2) = \alpha(\alpha^2 - 3)/2 \\ &= \alpha(-\beta^2)/2 = \beta/2\end{aligned}$$

$$\therefore\ F_n = \frac{\alpha^n - \beta^n}{\alpha - \beta} = \frac{2^n(\cos^n \pi/5 - \cos^n 3\pi/5)}{\sqrt{5}}$$

This formula was discovered in 1921 by W. Hope-Jones.

THE PENTAGRAM

Let us return to the regular pentagon $ABCDE$ in Figure 25.2. Drawing all its diagonals yields the *star polygon APBQCRDSET*, called a *pentagram*, as Figure 25.6 shows. It

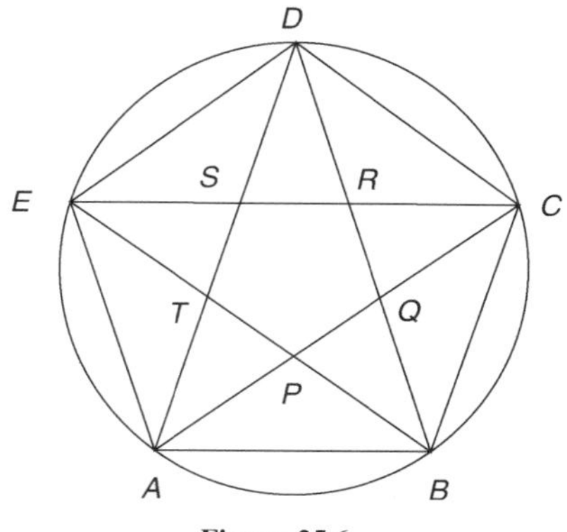

Figure 25.6.

follows from Example 25.1 that the points P, Q, R, S, and T divide the diagonals in the Golden Ratio.

The pentagram was the logo of the Pythagorean School of the sixth century B.C. The flags of many countries in the world contain one or more five-pointed stars. For example, the Australian flag contains 6 stars, the Chinese flag contains 5, and the United States flag contains 50. The diameter of every star on the U.S. flag is $0.616 \approx 1/\alpha$. Even the flags of several cities contain five-pointed stars; Chicago's flag contains four stars, and the flags of Dallas, Houston, and San Antonio each contain one.

Returning to Figure 25.6, the polygon $PQRST$ is also a regular polygon (see Exercise 1). Draw its diagonals to produce a new pentagram and a smaller regular pentagon; the points V, W, X, Y, and Z divide them in the Golden Ratio. Obviously, this process can be continued indefinitely. See Figure 25.7.

Figure 25.7 contain many angles of various sizes, namely, $\pi/5, 2\pi/5, 3\pi/5, 4\pi/5, \pi, 6\pi/5, 7\pi/5, 8\pi/5, 9\pi/5$, and 2π; they form a finite arithmetic sequence with a common difference of $\pi/5$.

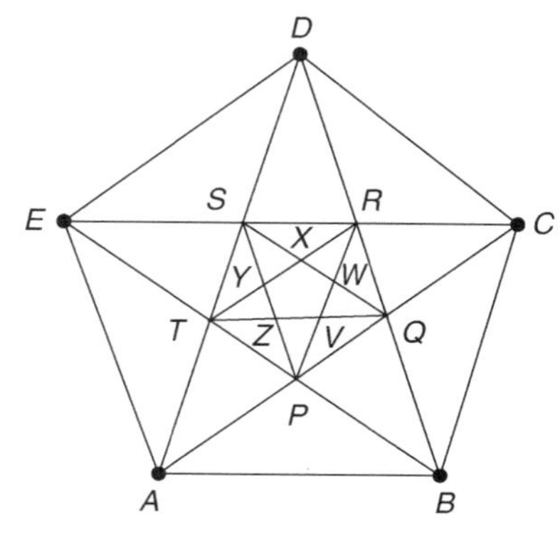

Figure 25.7.

We can partition the various line segments in Figure 25.7 into six different classes with representatives $\overline{BD}$, $\overline{AB}$, $\overline{BQ}$, $\overline{PQ}$, $\overline{QW}$, and $\overline{VW}$; the lengths of these line segments are different. Clearly, the longest of them is $\overline{BD}$.

There are 5 line segments of the same size as $\overline{BD}$, 15 line segments of the same size as $\overline{AB}$, 15 line segments of the same size as $\overline{BQ}$, 15 line segments of the same size as $\overline{PQ}$, 10 line segments of the same size as $\overline{QW}$, and 5 line segments of the same size as $\overline{VW}$, a total of 65 line segments.

Let $BD = a$. Since R divides $\overline{BD}$ in the Golden Ratio, it follows that $AB = BR = BD/\alpha = a/\alpha$.

Notice that $\triangle BER \sim \triangle BCQ$, so $BR/BQ = BE/BC$; that is, $BR/BQ = a/(a/\alpha) = \alpha$. Therefore, $BQ = BR/\alpha = a/\alpha^2$. Thus Q divides $\overline{BR}$ in the Golden Section. Similarly, R divides $\overline{DQ}$ in the Golden Ratio. Obviously, we can extend this property to other diagonals as well.

Consider the triangles SPQ and BPQ. Clearly, they are congruent, so $SQ = BQ$;

$$\therefore\ PQ = \frac{SQ}{\alpha} = \frac{BQ}{\alpha} = \frac{BR}{\alpha^2} = \frac{a}{\alpha^3}$$

$$QW = \frac{PQ}{\alpha} = \frac{a}{\alpha^4}$$

$$VW = \frac{QW}{\alpha} = \frac{a}{\alpha^5}$$

Thus, $BD = a$, $AB = a/\alpha$, $BQ = a/\alpha^2$, $PQ = a/\alpha^3$, $QW = a/\alpha^4$, and $VW = a/\alpha^5$. They form a decreasing geometric sequence with first term a and common ratio $1/\alpha$; so

$$BD{:}AB{:}BQ{:}PQ{:}QW{:}VW = 1{:}\frac{1}{\alpha}{:}\frac{1}{\alpha^2}{:}\frac{1}{\alpha^3}{:}\frac{1}{\alpha^4}{:}\frac{1}{\alpha^5}.$$

Thus every number in the sequence is α times the following element.

Suppose this procedure is continued indefinitely. Then the sum of the resulting geometric sequence of the different lengths of the various line segments is given by

$$\frac{a}{1-1/\alpha} = \frac{a\alpha}{\alpha-1} = a\alpha^2$$
$$\approx 2.61803398875a$$

Here is an interesting observation: $\triangle APT$ is a golden triangle. In fact, Figure 25.7 contains several golden rectangles, which become apparent if we search for them (see Exercise 14).

REGULAR DECAGON

Consider a regular decagon of side l and circumradius R (see Fig. 25.8). The central angle subtended by a side is $2\pi/10 = \pi/5$. Then

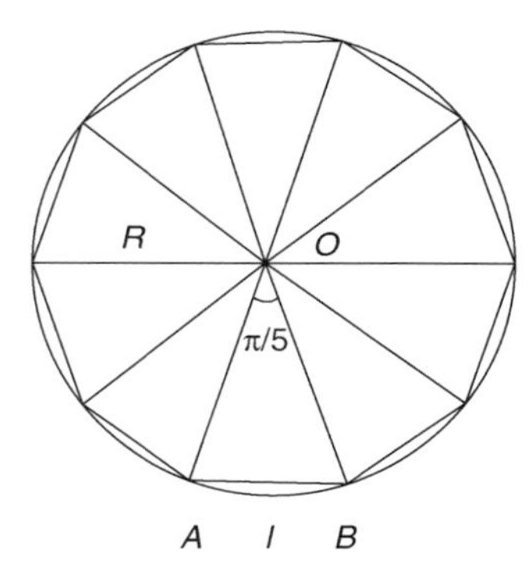

Figure 25.8.

$$
\begin{aligned}
R &= \left(\frac{l}{2}\right)\csc\frac{\pi}{5} = \frac{l}{2}\sqrt{\frac{2}{1-\cos\pi/5}} \\
&= \frac{l}{2}\sqrt{\frac{2}{1-\alpha/2}} = \frac{l}{\sqrt{2-\alpha}} \\
&= \frac{l}{\sqrt{\beta^2}} = -\frac{l}{\beta} \qquad \text{since } \beta < 0 \\
&= l\alpha
\end{aligned}
$$

In particular, let $l = 1$. Then $R = \alpha$; that is, the circumradius of a decagon of unit side is α. This is the seventh property of α described by Pacioli in his classic book.

Returning to Figure 25.8, we have:

$$
\begin{aligned}
\text{Area of } \triangle AOB &= 2lR\cos\frac{\pi}{10} = 2l\cdot l\cos\frac{\pi}{10} \\
&= 2l^2\alpha\frac{\sqrt{\alpha+2}}{2} = \alpha\sqrt{\alpha+2}\,l^2 \\
\therefore \text{ Area of the decagon} &= 10\alpha\sqrt{\alpha+2}\,l^2
\end{aligned}
$$

Consequently, the area enclosed by the decagon of unit side is $10\alpha\sqrt{\alpha+2}$.

Next we employ the fifth roots of unity to explore the various properties of the divine proportion related to the regular pentagon.

THE REGULAR PENTAGON AND THE FIFTH ROOTS OF UNITY

By DeMoivre's theorem, the complex nth roots of unity are given by $z = \text{cis } 2k\pi/n$, where $0 \le k < n$ and $\text{cis } \theta = \cos\theta + i \sin\theta$. They are equally spaced on the unit circle $|z| = 1$ on the complex plane.

In particular, the fifth roots of unity are given by $z = \text{cis } 2k\pi/5$, where $0 \le k < 5$. They are $z_0 = \text{cis } 0 = 1$, $z_1 = \text{cis } 2\pi/5$, $z_2 = \text{cis } 4\pi/5$, $z_3 = \text{cis } 6\pi/5 = \text{cis } (-4\pi/5)$, and $z_4 = \text{cis } 8\pi/5 = \text{cis } (-2\pi/5)$.

Since $\cos \pi/5 = \alpha/2$, it follows that

$$\cos\frac{2\pi}{5} = 2\cos^2\frac{\pi}{5} - 1 = 2 \cdot \frac{\alpha^2}{4} - 1$$

$$= \frac{\alpha^2 - 2}{2} = \frac{\alpha - 1}{2}$$

and

$$\sin\frac{2\pi}{5} = 2\sin\frac{\pi}{5}\cos\frac{\pi}{5} = 2 \cdot \frac{\sqrt{3-\alpha}}{2} \cdot \frac{\alpha}{2} = \frac{\alpha\sqrt{3-\alpha}}{2}$$

$$\therefore\ z_1 = \frac{\alpha-1}{2} + \frac{\alpha\sqrt{3-\alpha}}{2}i \quad \text{and} \quad z_4 = \frac{\alpha-1}{2} - \frac{\alpha\sqrt{3-\alpha}}{2}i$$

Now

$$\cos\frac{4\pi}{5} = 2\cos^2\frac{2\pi}{5} - 1 = 2 \cdot \left(\frac{\alpha-1}{2}\right)^2 - 1 = -\frac{\alpha}{2}$$

and

$$\sin\frac{4\pi}{5} = 2\sin\frac{2\pi}{5}\cos\frac{2\pi}{5} = 2 \cdot \frac{\alpha\sqrt{3-\alpha}}{2} \cdot \frac{\alpha-1}{2} = \frac{\alpha(\alpha-1)\sqrt{3-\alpha}}{2}$$

$$= \frac{\sqrt{3-\alpha}}{2}$$

$$\therefore\ z_2 = -\frac{\alpha}{2} + \frac{\sqrt{3-\alpha}}{2}i \quad \text{and} \quad z_3 = -\frac{\alpha-1}{2} - \frac{\sqrt{3-\alpha}}{2}i$$

Thus the five roots of unity are 1, $\frac{\alpha-1}{2} \pm \frac{\alpha\sqrt{\alpha+2}}{2}i$ and $-\frac{\alpha}{2} \pm \frac{\sqrt{3-\alpha}}{2}i$. We can verify this without resorting to DeMoivre's theorem (see Exercise 19). The roots are represented by the points C, D, B, E, and A, respectively, in Figure 25.9.

We can extract many properties of the regular pentagon using the coordinates of its vertices. For example, we can be show that $BD = \sqrt{\alpha+2}$ and $AB = \sqrt{\alpha+2}/\alpha$, so $BD = \alpha AB$, as expected! (See Exercises 20 and 21.)

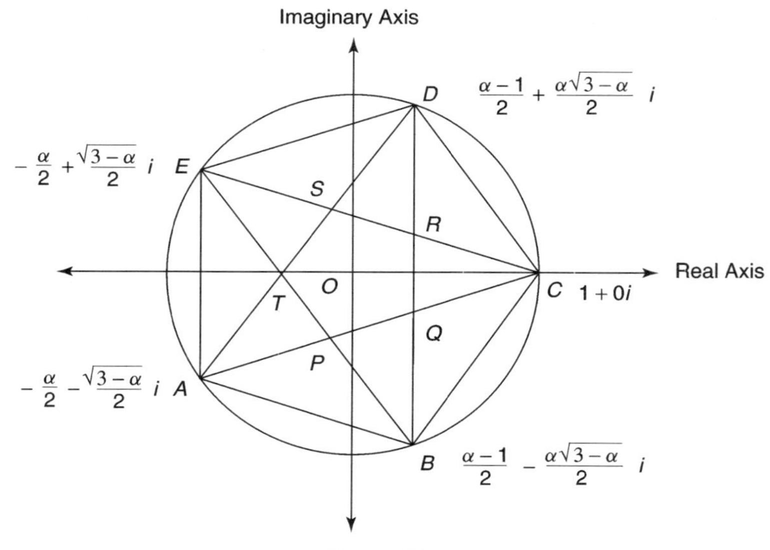

Figure 25.9.

Interestingly enough, our mysterious number α appears even in geometric figures not connected to pentagons or decagons. For example, consider the square ABFE inscribed in a semicircle, as Figure 25.10 shows. It can be shown that

$$\frac{AE}{AC} = \frac{AD}{AE} = \alpha$$

See Exercise 29.

The Golden Ratio α also appears in a circle inscribed in an isosceles triangle which is in turn inscribed in a square (see Fig. 25.11). Since the angle in a semicircle is a right angle, it follows that $\angle AEB = \angle CED$. But $\angle CED = \angle CDE$, since $\triangle CDE$ is isosceles. Therefore, $\angle AEB = \angle CDE$.

It now follows that $\triangle ABE \sim \triangle AED$, so $AE/AB = AD/AE$. Let $AE/AB = AD/AE = x$.

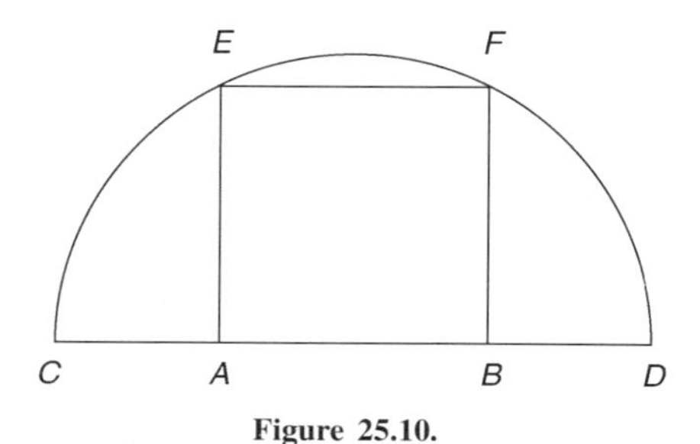

Figure 25.10.

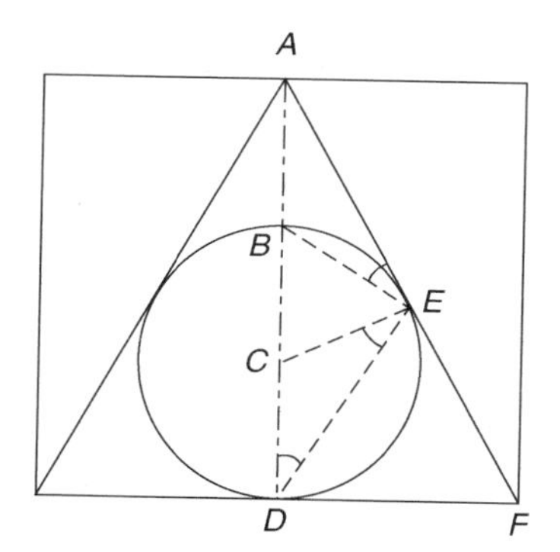

Figure 25.11.

Now consider $\triangle ADF$ and $\triangle ACE$. They are similar, so $AD/DF = AE/CE$. But $AD = 2DF$, so $AE = 2CE = BD$;

$$\therefore \frac{AD}{AE} = \frac{AB + BD}{AE} = \frac{AB + AE}{AE} = 1 + \frac{AB}{AE}$$

That is,

$$x = 1 + \frac{1}{x} \qquad \text{so } x = \alpha$$

Thus

$$\frac{AD}{AE} = \frac{AE}{AB} = \alpha$$

Suppose the lengths of the sides of a right triangle form a geometric sequence with common ratio r. Then we can show that $r = \sqrt{\alpha}$ (see Exercise 15).

REGULAR ICOSAHEDRON

Golden rectangles occur in solid geometry also. A *regular icosahedron* is one of the five Platonic solids. It has 12 vertices, 20 equilateral triangular faces, and 30 edges. Five faces meet at each vertex and they form a pyramid with a regular pentagonal base (see Fig. 25.12). We can place three mutually perpendicular and symmetrically placed golden rectangles (see Fig. 25.13) inside the icosahedron in such a way that their 12 corners will coincide with those of the icosahedron (see Fig. 25.14).

The length of a longer side of the golden rectangle equals the length of a diagonal of the pentagon. As we saw in Chapter 24, the length of a diagonal is α times that of a side of the pentagon. Thus, the length of the golden rectangle is α times the length of an edge between any two adjacent vertices of the icosahedron. In particular, if the adjacent vertices are one unit away, then the length of a longer side of the golden rectangle is α. This is the essence of the "twelfth incomprehensible" property described by Pacioli in his book.

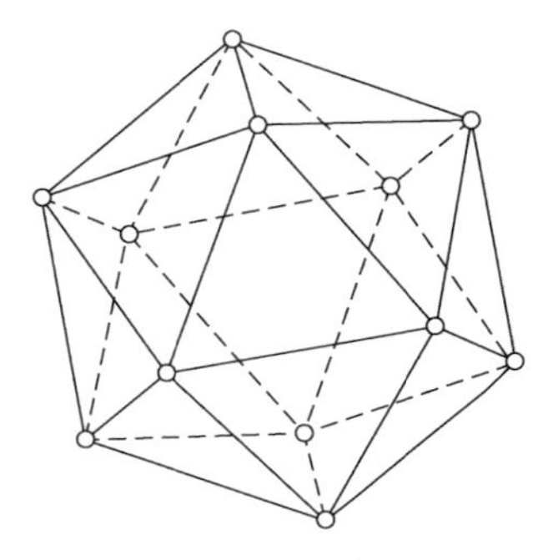

Figure 25.12. A regular icosahedron (*Source*: H. S. M. Coxeter, *Introduction to Geometry*, 2nd ed., Wiley, New York, 1969. Copyright © 1969, reproduced by permission of John Wiley & Sons, Inc.).

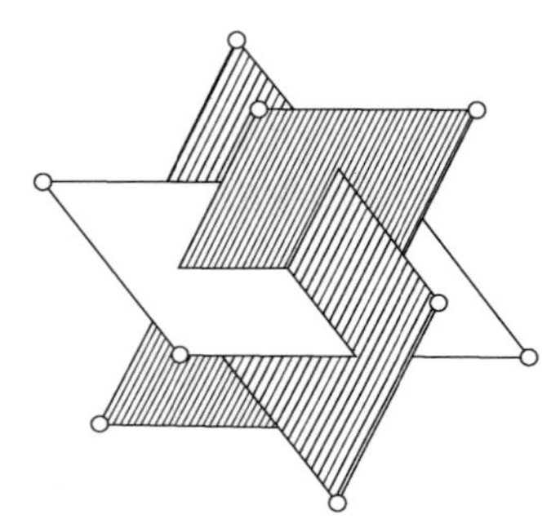

Figure 25.13. Three golden rectangles (*Source*: H. S. M. Coxeter, *Introduction to Geometry*, 2nd ed., Wiley, New York, 1969. Copyright © 1969, reproduced by permission of John Wiley & Sons, Inc.).

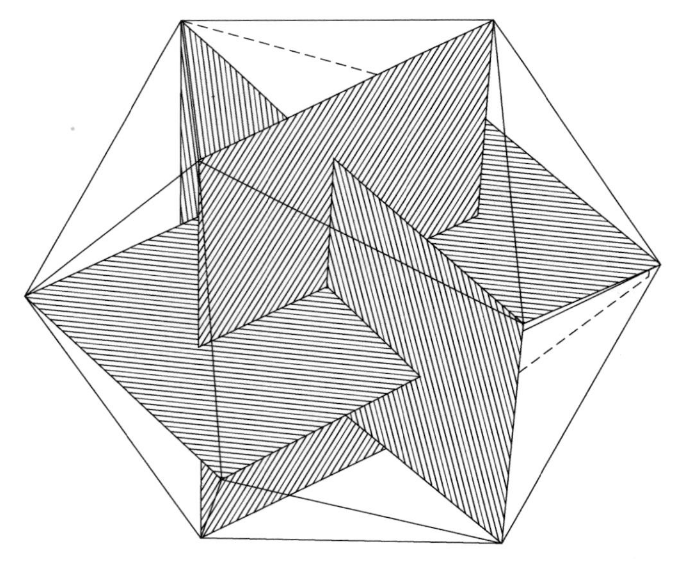

Figure 25.14. The corners of three golden rectangles meet at the corners of a regular icosahedron (*Source*: H. S. M. Coxeter, *Introduction to Geometry*, 2nd ed., Wiley, New York, 1969. Copyright © 1969, reproduced by permission of John Wiley & Sons, Inc.).

REGULAR DODECAHEDRON

We can also place three mutually perpendicular and symmetrically placed golden rectangles in another Platonic solid, the *regular dodecahedron*, which has 12 pentagonal faces, 20 vertices, and 30 edges. The various corners of the rectangles meet the faces at their centers, as Figure 25.15 shows.

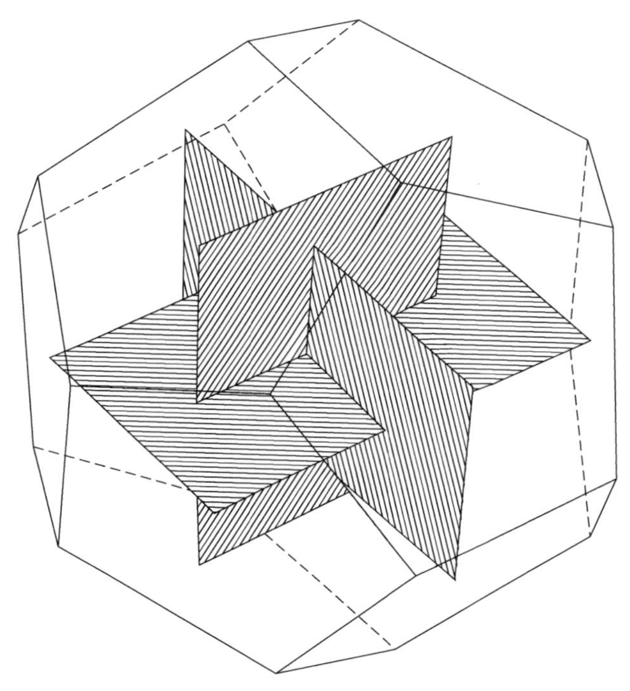

Figure 25.15. The corners of the same rectangles coincide with the centers of the sides of a regular dodecahedron (*Source*: H. S. M. Coxeter, *Introduction to Geometry*, 2nd ed., Wiley, New York, 1969. Copyright © 1969, reproduced by permission of John Wiley & Sons, Inc.).

A PENTAGONAL ARCH

In 1974, D. W. DeTemple of Washington State University studied the pentagonal arch formed by rolling a regular pentagon along a line (see Fig. 25.16). As the leftmost pentagon is rolled to the right, the vertex A moves toward B, then to C, D, and finally to E as the successive sides touch the base line. Connecting these five points, we generate the pentagonal arch $ABCDE$. Surprisingly enough, this arch is also related to the magic ratio.

To see this, let s denote the length of a side of the regular pentagon. Since a vertex angle of the pentagon is $108°$, it follows that $\angle APB = 72°$. Since $AP = BP = s$,

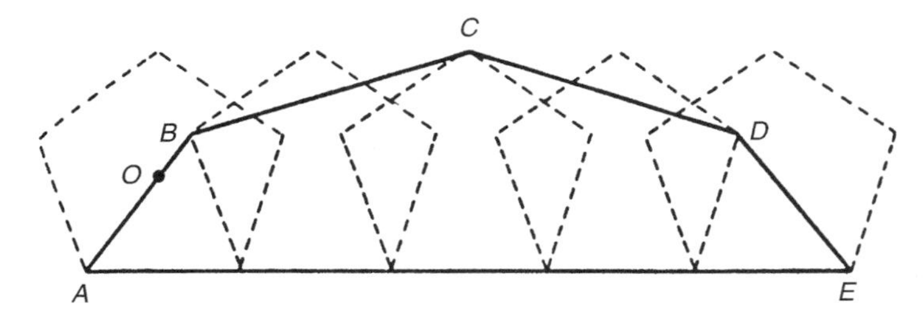

Figure 25.16. Pentagonal Arch. *Source*: D. W. Temple, "A Pentagonal Arch," *The Fibonacci Quaterly*, 1974.

by the law of cosines,

$$\begin{aligned} AB^2 &= AP^2 + BP^2 - 2AP \cdot BP \cos 72^\circ \\ &= 2s^2(1 - \cos 72^\circ) = 4s^2 \sin^2 36^\circ \\ &= 4s^2 \left(\frac{3-\alpha}{4}\right) = (3-\alpha)s^2 \\ \therefore\ AB &= s\sqrt{3-\alpha} \end{aligned}$$

Likewise, we can be show that $BC = s\sqrt{2+\alpha}$. Moreover,

$$\frac{BC}{AB} = \sqrt{\frac{2+\alpha}{3-\alpha}}$$

But

$$\alpha^2(3-\alpha) = 3\alpha^2 - \alpha^3 = 3(\alpha+1) - (2\alpha+1) = 2+\alpha$$

$$\therefore\ \alpha = \sqrt{\frac{2+\alpha}{3-\alpha}}$$

Thus $BC/AB = \alpha$, the Golden Ratio.

A TRIGONOMETRIC FORMULA FOR F_n

In Example 25.1 we found that $\cos \pi/5 = \alpha/2$. So $2\sin^2 \pi/10 = 1 - \cos \pi/5 = 1 - \alpha/2 = (2-\alpha)/2 = \beta^2/4$, and hence $\sin \pi/10 = |\beta|/2 = -\beta/2$; that is, $\beta = -2\sin \pi/10$. Thus $\alpha = 2\cos \pi/5$ and $\beta = -2\sin \pi/10$.

We can employ these trigonometric values of α and β to develop a trigonometric summation formula for F_n, derived in 1964 by J. L. Brown, Jr., of Pennsylvania State University.

By Binet's formula, we have

$$\begin{aligned}
F_n &= \frac{\alpha^n - \beta^n}{\alpha - \beta} = \sum_{k=0}^{n-1} \alpha^{n-k-1}\beta^k \\
&= \sum_{k=0}^{n-1} \left(2\cos\frac{\pi}{5}\right)^{n-k-1} \left(-2\sin\frac{\pi}{10}\right)^k \\
&= 2^{n-1} \sum_{k=0}^{n-1} (-1)^k \cos^{n-k-1}\frac{\pi}{5} \sin^k \frac{\pi}{10} \qquad (25.1)
\end{aligned}$$

For example,

$$\begin{aligned}
F_3 &= 4\sum_0^2 (-1)^k \cos^{2-k}\frac{\pi}{5}\sin^k\frac{\pi}{10} = 4\sum_0^2 (-1)^k \left(\frac{\alpha}{2}\right)^{2-k}\left(\frac{-\beta}{2}\right)^k \\
&= 4\left[(-1)^0\left(\frac{\alpha}{2}\right)^2\left(\frac{-\beta}{2}\right)^0 + (-1)\left(\frac{\alpha}{2}\right)\left(\frac{-\beta}{2}\right) + (-1)^2\left(\frac{\alpha}{2}\right)^0\left(\frac{-\beta}{2}\right)^2\right] \\
&= 4\left[\frac{\alpha^2}{4} + \frac{\alpha\beta}{4} + \frac{\beta^2}{4}\right] = \alpha^2 + \alpha\beta + \beta^2 \\
&= (\alpha+\beta)^2 - \alpha\beta = 1 - (-1) = 2
\end{aligned}$$

as expected.

Next we derive two additional trigonometric formulas for F_n.

TWO ADDITIONAL TRIGONOMETRIC FORMULA FOR F_n

Since $\cos\pi/5 = \alpha/2$, it follows that $\sin\pi/5 = (\sqrt{3-\alpha})/2$.

$$\begin{aligned}
\therefore\ \sin 3\pi/5 &= \sin(2\pi/5 + \pi/5) \\
&= \sin 2\pi/5 \cos\pi/5 + \cos 2\pi/5 \sin\pi/5 \\
&= 2\sin\pi/5\cos^2\pi/5 + (2\cos^2\pi/5 - 1)\sin\pi/5 \\
&= 4\sin\pi/5\cos^2\pi/5 - \sin\pi/5 \\
&= 4\cdot\frac{\sqrt{3-\alpha}}{2}\cdot\frac{\alpha^2}{4} - \frac{\sqrt{3-\alpha}}{2} = \frac{\alpha\sqrt{3-\alpha}}{2}
\end{aligned}$$

and hence

$$\sin\pi/5\,\sin 3\pi/5 = \frac{\sqrt{3-\alpha}}{2}\cdot\frac{\alpha\sqrt{3-\alpha}}{2} = \frac{\alpha(3-\alpha)}{4} = \frac{\sqrt{5}}{4}.$$

Since $\sin 9\pi/5 = -\sin \pi/5$, it also follows that $\sin 3\pi/5 \sin 9\pi/5 = -\sqrt{5}/4$;

$$\begin{aligned}
\therefore\ F_n &= \frac{\alpha^n - \beta^n}{\sqrt{5}} \\
&= \frac{(2\cos\pi/5)^n - (2\cos 3\pi/5)^n}{\sqrt{5}} \\
&= \frac{2^{n+2}}{5}\left(\frac{\sqrt{5}}{4}\cos^n\frac{\pi}{5} - \frac{\sqrt{5}}{4}\cos^n\frac{3\pi}{5}\right) \\
&= \frac{2^{n+2}}{5}(\cos^n \pi/5 \sin \pi/5 \sin 3\pi/5 + \cos^n 3\pi/5 \sin 3\pi/5 \sin 9\pi/5)
\end{aligned} \tag{25.2}$$

It follows from this formula that

$$F_n = \frac{(-2)^{n+2}}{5}(\cos^n 2\pi/5 \sin 2\pi/5 \sin 6\pi/5 + \cos^n 4\pi/5 \sin 4\pi/5 \sin 12\pi/5) \tag{25.3}$$

See Exercise 32.

These two formulas were discovered in 1979 by F. Stern of San Jose State University, California.

Since $\sqrt{5} = 4\sin\pi/5 \sin 3\pi/5$, we can also write Binet's formula as

$$\begin{aligned}
F_n &= \frac{(2\cos\pi/5)^n - (2\cos 3\pi/5)^n}{4\sin\pi/5\sin 3\pi/5} \\
&= \frac{2^{n-2}(\cos^n \pi/4 - \cos^n 3\pi/5)}{\sin\pi/5 \sin 3\pi/5}
\end{aligned} \tag{25.4}$$

EXERCISES 25

1. Show that the polygon $PQRST$ in Figure 25.7 is a regular pentagon.

Using Figure 25.7, compute the area of each polygon, where $BD = a$.

2. $\triangle APB$
3. $\triangle APT$
4. $\triangle CDR$
5. $\triangle CDS$
6. Rhombus $CDEP$
7. Rhombus $SPRD$

Use Figure 25.7 to compute each ratio.

8. $\triangle CDS : \triangle CDR$
9. Rhombus $CDEP$: Rhombus $SPRD$

Compute the shaded area in each figure.

10. Figure 25.17
11. Figure 25.18

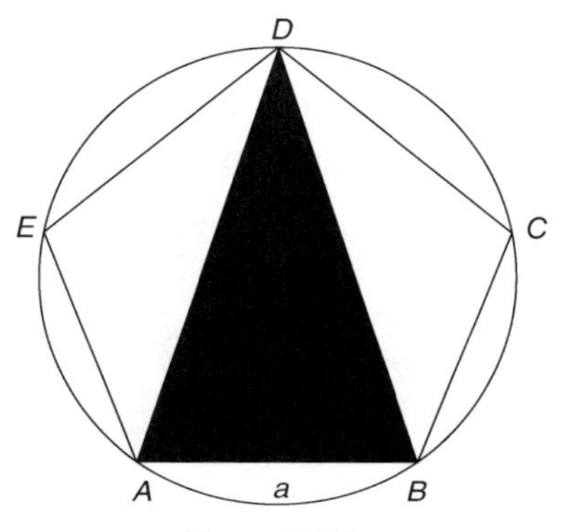

Figure 25.17.

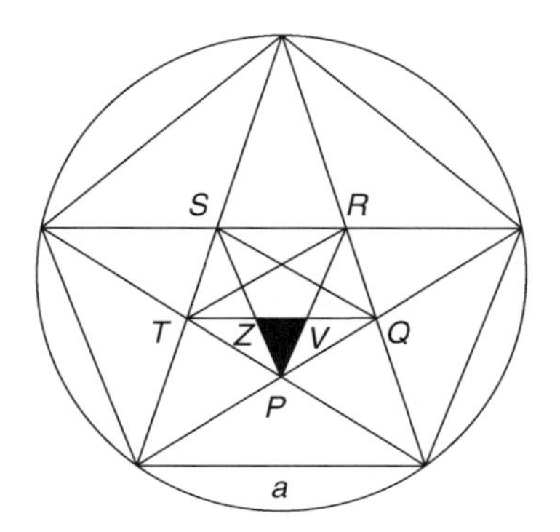

Figure 25.18.

Using Figure 25.7, compute the area of each polygon, where $BD = a$.

12. Pentagon $PQRST$
13. Pentagram $APBQCRDSET$
14. Find the number of golden rectangles in Figure 25.7.
15. The lengths of the sides of a Pythagorean triangle form a geometric sequence with common ratio r. Show that $r = \sqrt{\alpha}$.
16. Prove that the length of a side of a regular decagon with circumradius r is r/α.

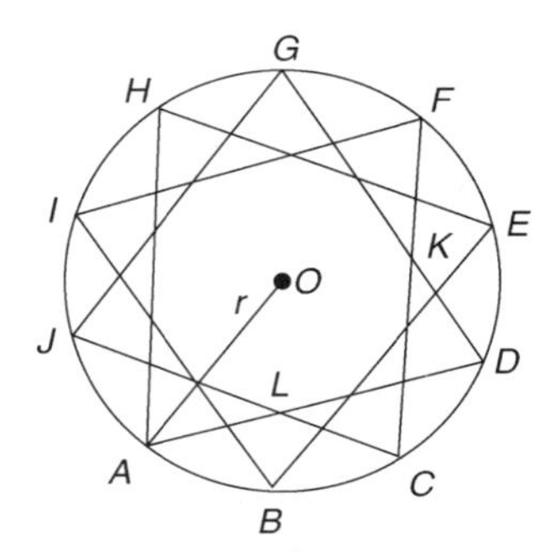

Figure 25.19.

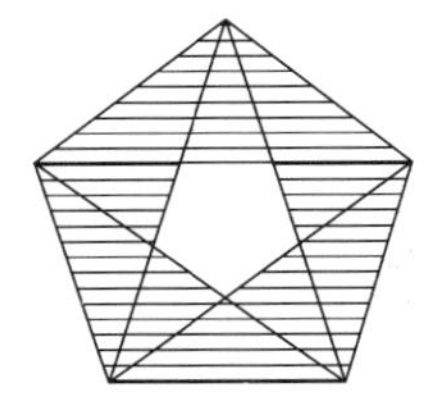

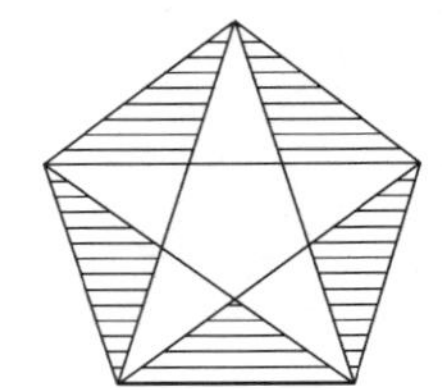

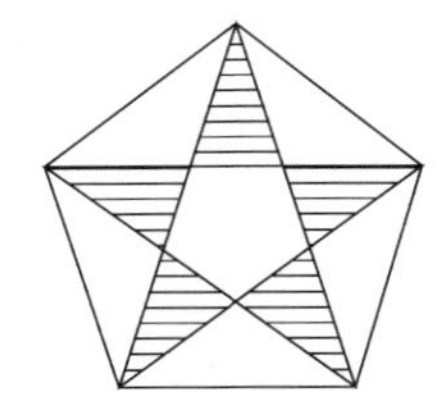

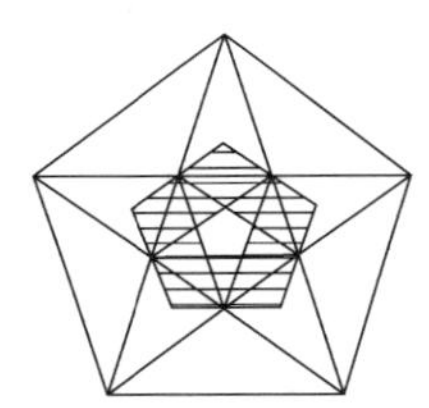

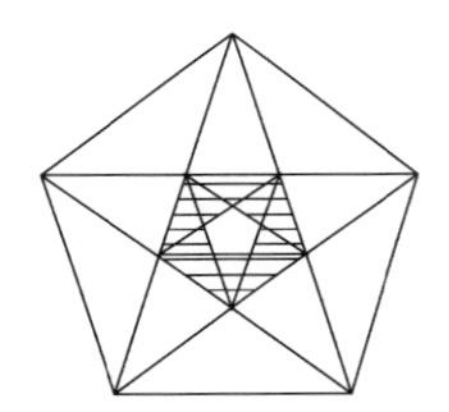

Figure 25.20.

17. The circumradius of the 10-pointed star in Figure 25.19 is R. Prove that $AD = r\alpha$.
18. Show that the shaded areas in Figure 25.20 form a geometric sequence with common ratio $1/\alpha$ (Baravalle, 1948).
19. Solve the equation $x^5 - 1 = 0$ algebraically.

Use Figure 25.9 to answer Exercises 20–30.

20. Find BD.
21. Find AB.
22. Find BD:AB.
23. Compute area $ABDE$.
24. Using the fact that P and Q divide $\overline{AC}$ in the Golden Ratio, determine their coordinates.

25. Using Exercise 24, compute PQ, AP, and QC.
26. Find the ratio $\triangle ABQ:\triangle ABP$.
27. Find the ratio $BD:AB:BQ:PQ$.
28. Find the inradius of the circle inscribed in the pentagon.
29. Using Figure 25.10, show that $AE/AC = AD/AE = \alpha$.
30. A regular pentagon of side p, a regular hexagon of side h, and a regular decagon of side d are inscribed in the same circle. Prove that these lengths can be used to form the sides of a Pythagorean triangle (Bicknell, 1974).
31. Using Eq. (25.1), compute F_2 and F_4.
32. Prove Eq. (25.3) (Stern, 1979).
33. Prove that

$$\frac{2^{n+1}}{5}\sum_{k=1}^{4}\cos^n k\pi/5 \sin k\pi/5 \sin 3k\pi/5 = \begin{cases} F_n & \text{if } n \text{ is even} \\ 0 & \text{otherwise} \end{cases}$$

(Hoggatt, 1979).

26 THE GOLDEN ELLIPSE AND HYPERBOLA

The concept of a golden ellipse was introduced in 1974 by H. E. Huntley of England. Huntley investigated the properties of the golden ellipse in detail. We shall pursue some of them in this chapter.

THE GOLDEN ELLIPSE

The ratio of the major axis to the minor axis of a *golden ellipse* is the magic ratio α.

Let $2a$ denote the length of the major axis and $2b$ that of its minor axis. Then it is well-known that $b^2 = a^2(1 - e^2)$. So for a golden ellipse,

$$\frac{b^2}{a^2} = 1 - e^2 = \frac{1}{\alpha^2} = \beta^2$$

Therefore,

$$e^2 = 1 - \beta^2 = -\beta$$

Thus we define the eccentricity of the golden ellipse by $e = \sqrt{-\beta}$ (see Fig. 26.1). Consequently, one-half of the minor axis is given by $b^2 = a^2\beta^2$, so $b = a|\beta|$.

If we inscribe the golden ellipse in a rectangle with its sides parallel to the axes, the rectangle would be a golden rectangle.

Let F and F' denote the foci of the golden ellipse. Then $OF = ae = a/\sqrt{\alpha} = a\sqrt{-\beta}$ and $BF = \sqrt{b^2 + a^2e^2} = a$.

Let $\angle OBF = \theta$. Then $\sec\theta = BF/OB = a/b = \alpha$. Let $\overleftrightarrow{ON}$ be perpendicular to the directrix $\overleftrightarrow{ND}$. Then $ON = a/e = a\sqrt{\alpha}$.

$$FN = ON - OF = a\sqrt{\alpha} - \frac{a}{\sqrt{\alpha}}$$

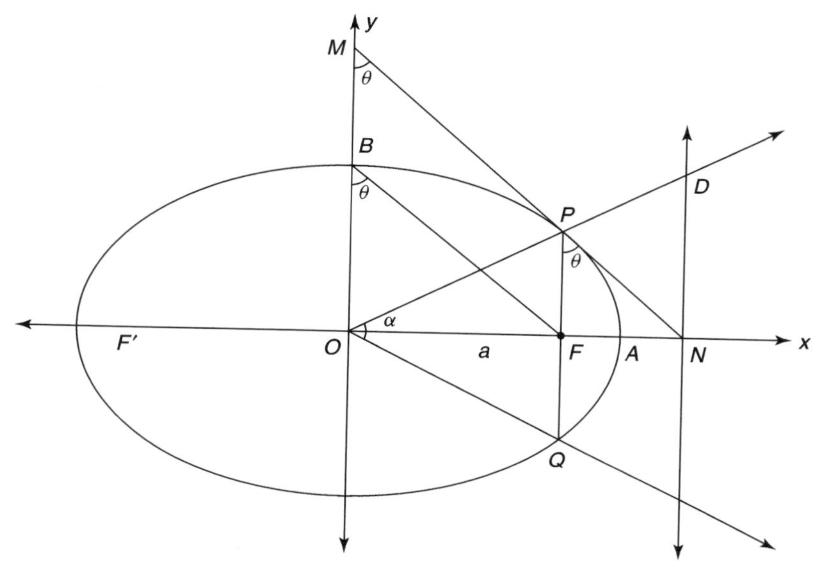

Figure 26.1.

$$= \frac{a(\alpha - 1)}{\sqrt{\alpha}} = -\frac{a\beta}{\sqrt{\alpha}} = a|\beta|^{3/2}$$

For any ellipse, the minor axis is the geometric mean of the major axis and the lactus rectum l; that is, $b^2 = al$. So, for the golden ellipse,

$$PQ = l = \frac{b^2}{a} = b\left(\frac{b}{a}\right) = \frac{b}{\alpha}$$

Thus $a : b : l = b\alpha : b : b/\alpha = \alpha : 1 : 1/\alpha = \alpha^2 : \alpha : 1$.

Notice that

$$\frac{ON}{FN} = \frac{a\sqrt{\alpha}}{a/\sqrt{\alpha}} = \alpha \qquad \text{and} \qquad \frac{OF}{FN} = \frac{a/\sqrt{\alpha}}{a(\alpha - 1)\sqrt{\alpha}} = \frac{1}{\alpha - 1} = \alpha$$

so the focus F divides $\overline{ON}$ in the Golden Ratio.

Let $\overline{PQ}$ denote the latus rectum. Then

$$\begin{aligned} OP^2 &= OF^2 + FP^2 = \frac{a^2}{\alpha} + \frac{b^2}{\alpha^2} \\ &= b^2\alpha + \frac{b^2}{\alpha^2} = \frac{b^2(\alpha^3 + 1)}{\alpha^2} \\ &= \frac{b^2(2\alpha + 2)}{\alpha^2} = 2b^2 \\ \therefore \quad OP &= \sqrt{2}b \end{aligned}$$

An ellipse has the property that the tangent at P passes through N and $\cot \angle FPN = e$. Since $\cot \theta = OB/OF = b/ae$, it follows in the case of the golden ellipse that

$$\cot \angle FPN = \frac{1}{\sqrt{\alpha}} = \sqrt{-\beta} \qquad \text{and} \qquad \cot \theta = \frac{1}{e\alpha} = \frac{1}{\sqrt{\alpha}} = \sqrt{-\beta}$$

so $\angle FPN = \theta$. Thus $MPFB$ is a parallelogram and hence $MP = BF = a$.

In addition, since $\triangle OMN \sim \triangle FPN$,

$$\frac{MN}{PN} = \frac{ON}{FN}$$

That is,

$$\frac{MP}{PN} + 1 = \frac{OF}{FN} + 1$$

$$\therefore \quad \frac{MP}{PN} = \frac{OF}{FN} = \alpha$$

Thus P divides $\overline{MN}$ in the Golden Ratio.

Finally, let $\overrightarrow{OP}$ intersect the directrix $\overleftrightarrow{ND}$ at D. Since $\triangle OND \sim \triangle OFP$, $OD/OP = ON/OF = \alpha$, so P divides $\overline{OD}$ in the Golden Ratio.

THE GOLDEN HYPERBOLA

The *golden hyperbola* (Fig. 26.2) was also studied by Huntley, who gives a fairly extensive account of its properties in his fascinating book, *The Divine Proportion*.

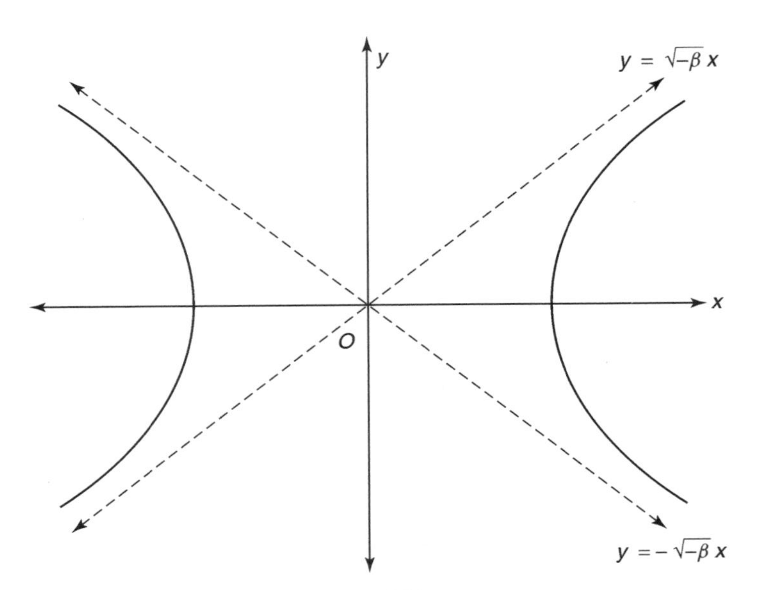

Figure 26.2.

The eccentricity of the golden hyperbola is defined by $e^2 = \alpha$. Then

$$b^2 = a^2(e^2 - 1) = a^2(\alpha - 1) = -a^2\beta$$

$$\therefore \quad \frac{a}{b} = \sqrt{a} \quad \text{and} \quad b = a\sqrt{-\beta}$$

The asymptotes of the golden hyperbola are given by $y = \pm(b/a)x$; that is, $y = \pm\sqrt{-\beta}x$.

Huntley also studied the parabola $y^2 = 4ax$ using $(a\alpha^2, 2a\alpha)$ as a point on it. If we draw the parabola and the golden hyperbola to the same scale with the same origin, then we can easily verify that the asymptotes of the hyperbola would intersect the parabola at the origin and at the points $(4a\alpha, \pm 4a\sqrt{\alpha})$ (see Exercise 1).

EXERCISES 26

1. Show that the asymptotes of the golden hyperbola intersect the parabola $y^2 = 4ax$ at the origin and at the points $(4a\alpha, \pm 4a\sqrt{\alpha})$.
2. One endpoint P of the focal chord of the parabola $y^2 = 4ax$ is $(a\alpha^2, 2a\alpha)$. Find the other endpoint Q.
3. Compute the length of the focal chord $\overline{PQ}$ in Exercise 2.
4. Find the equations of the tangent and the normal to the parabola $y^2 = 4ax$ at the point $P(a\alpha^2, 2a\alpha)$.
5. Find the equations of the tangent and the normal to the parabola $y^2 = 4ax$ at the endpoint Q of the focal chord $\overline{PQ}$ in Exercise 2.
6. Find the point of intersection of the tangents at the ends of the focal chord $\overline{PQ}$ in Exercise 2.
7. Find the angle between the tangents in Exercises 4 and 5.
8. Find the point of intersection of the normals to the parabola $y^2 = 4ax$ at the ends of the focal chord $\overline{PQ}$ in Exercise 2.
9. Find the angle between the normals at P and Q.
10. Suppose the focal chord $\overline{PQ}$ in Exercise 2 intersects the y-axis at R. Show that the focus S divides $\overline{PR}$ in the Golden Ratio.
11. With S, Q, and R as in Exercise 10, show that Q divides $\overline{SR}$ in the Golden Ratio.

CONTINUED FRACTIONS

In Chapter 20, we found that the sequence of ratios F_{n+1}/F_n of consecutive Fibonacci numbers approaches the Golden Ratio α as $n \to \infty$. Interestingly, we can employ these ratios to generate rational numbers of a very special nature, called *continued fractions*.* So we begin with a few characterizations of continued fractions.

FINITE CONTINUED FRACTIONS

A *finite continued fraction* is an expression of the form

$$x = a_1 + \cfrac{1}{a_2 + \cfrac{1}{a_3 + \cfrac{1}{\ddots \, a_{m-2} + \cfrac{1}{a_{m-1} + \cfrac{1}{a_m}}}}} \tag{27.1}$$

where $a_1 \geq 0$, and a_i is a positive integer and $i \geq 2$. Since this notation is a bit cumbersome, this fraction is often written as

$$a_1 + \frac{1}{a_2 +} \frac{1}{a_3 +} \cdots + \frac{1}{a_m}$$

*The Italian mathematician Pietro Antonio Cataldi (1548–1626) has been credited with laying the foundation for the theory of continued fractions.

Since the numerator of each fraction is 1, we refine this notation further as

$$[a_1; a_2, a_3, \ldots, a_m]$$

where $a_1 = \lfloor x \rfloor$ and the semicolon separates the fractional part from the integral part. For example,

$$\begin{aligned}
[1; 2, 3, 4, 5, 6] &= 1 + \frac{1}{2+}\frac{1}{3+}\frac{1}{4+}\frac{1}{5+}\frac{1}{6} \\
&= 1 + \cfrac{1}{2 + \cfrac{1}{3 + \cfrac{1}{4 + \cfrac{1}{5 + \cfrac{1}{6}}}}} \\
&= \frac{1393}{972}
\end{aligned}$$

On the other hand, finding the continued fraction of this rational number involves the repeated application of the Euclidean algorithm:

$$\begin{aligned}
1393 &= 1 \cdot 972 + 421 \\
972 &= 2 \cdot 421 + 130 \\
421 &= 3 \cdot 130 + 31 \\
130 &= 4 \cdot 31 \quad + 6 \\
31 &= 5 \cdot 6 \quad + 1
\end{aligned}$$

Now divide each dividend by the corresponding divisor, save the fractional remainder, and then apply substitution for the fractional remainder:

$$\begin{aligned}
\frac{1393}{972} &= 1 + \frac{421}{972} = 1 + \frac{1}{972/421} \\
&= 1 + \frac{1}{2 + 130/421} \qquad = 1 + \cfrac{1}{2 + \cfrac{1}{421/130}} \\
&= 1 + \cfrac{1}{2 + \cfrac{1}{3 + \cfrac{31}{130}}} \qquad = 1 + \cfrac{1}{2 + \cfrac{1}{3 + \cfrac{1}{130/31}}} \\
&= 1 + \cfrac{1}{2 + \cfrac{1}{3 + \cfrac{1}{4 + \cfrac{6}{31}}}} = 1 + \cfrac{1}{2 + \cfrac{1}{3 + \cfrac{1}{4 + \cfrac{1}{31/6}}}}
\end{aligned}$$

$$= 1 + \cfrac{1}{2 + \cfrac{1}{3 + \cfrac{1}{4 + \cfrac{1}{5 + \cfrac{1}{6}}}}}$$

$$= [1; 2, 3, 4, 5, 6]$$

CONVERGENTS OF A CONTINUED FRACTION

By chopping off the continued fraction for x in Eq. (27.1) at the various plus signs, we get a sequence $\{C_k\}$ of approximations of x, where $1 \leq k \leq m$:

$$a_1, a_1 + \frac{1}{a_1}, \quad a_1 + \cfrac{1}{a_2 + \cfrac{1}{a_3}}, \ldots$$

Each $C_k = [a_1; a_2, a_3, \ldots, a_k]$ is a *convergent* of x, where $k \geq 1$ and $C_1 = [a_1] = a_1$.

For example, consider the Fibonacci ratio $21/13$. As a finite continued fraction,

$$\frac{21}{13} = [1; 1, 1, 1, 1, 1, 1]$$

We can verify this. The various convergents are:

$$\begin{aligned}
C_1 &= [1] && = 1\\
C_2 &= [1; 1] && = 2\\
C_3 &= [1; 1, 1] && = 1.5\\
C_4 &= [1; 1, 1, 1] && \approx 1.6666666667\\
C_5 &= [1; 1, 1, 1, 1] && = 1.6\\
C_6 &= [1; 1, 1, 1, 1, 1] && = 1.625\\
C_7 &= [1; 1, 1, 1, 1, 1, 1] = \frac{21}{13} && \approx 1.6153846154
\end{aligned}$$

Obviously, these convergents, C_k approach the actual value $21/13$, as k increases, where $1 \leq k \leq 7$. In fact, the convergents with odd subscripts approach it from below, whereas those with even subscripts approach it from above. The convergents are alternately less than and greater than $21/13$, except the last one (see Fig. 27.1).

These convergents display a remarkable pattern:

$$C_1 = \frac{1}{1} \quad C_2 = \frac{2}{1} \quad C_3 = \frac{3}{2} \quad C_4 = \frac{5}{3} \quad C_5 = \frac{8}{5} \quad C_6 = \frac{13}{8} \quad C_7 = \frac{21}{13}$$

These ratios look familiar. They are, in fact, the ratios of consecutive Fibonacci numbers. (We shall return to them a bit later.) It is possible to conjecture and prove the value of C_n (see Exercise 9).

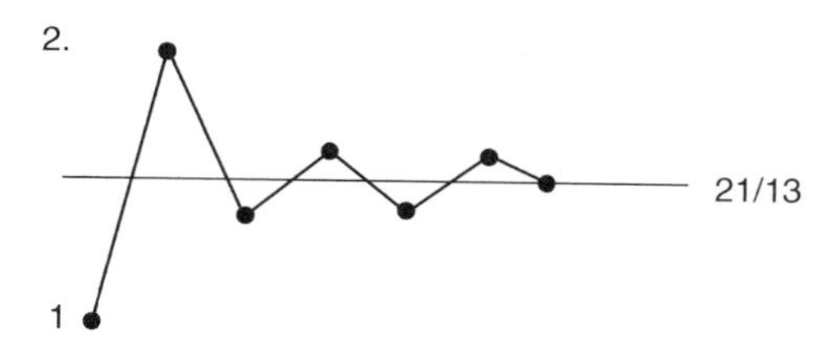

Figure 27.1.

Evaluating each convergent may seem to be a tedious job. This is where recursion becomes useful.

RECURSIVE DEFINITION OF C_n

Let $C_n = p_n/q_n$ denote nth convergent of the continued fraction (Eq. 27.1). Then we can show that

$$p_n = a_n p_{n-1} + p_{n-2}$$

and

$$q_n = a_n q_{n-1} + q_{n-2}$$

where

$$\frac{p_1}{q_1} = a_1, \qquad \frac{p_2}{q_2} = a_1 + \frac{1}{a_2}$$

and $n \geq 3$. Thus, using the convergents C_{n-2} and C_{n-1}, we can easily compute C_n.

For example, consider the continued fraction $21/13 = [1;\ 1, 1, 1, 1, 1, 1]$, where $a_i = 1$ for every i. We have

$$C_3 = \frac{p_3}{q_3} = \frac{5}{3} \qquad \text{and} \qquad C_4 = \frac{p_4}{q_4} = \frac{8}{5}$$

$$\therefore \quad C_5 = \frac{p_5}{q_5} = \frac{a_5 p_4 + p_3}{a_5 q_4 + q_3} = \frac{1 \cdot 8 + 5}{1 \cdot 5 + 3} = \frac{13}{8}$$

as expected.

In fact, we can use a table such as Table 27.1 to compute C_n. The table shows the numerators and denominators of all convergents p_n/q_n of the continued fraction $[2;\ 1, 3, 4, 2, 3, 5]$. By direct computation, we can verify that $1915/693 = [2;\ 1, 3, 4, 2, 3, 5]$.

TABLE 27.1.

n	1	2	3	4	5	6	7
a_n	2	1	3	4	2	3	5
p_n	1	3	11	47	105	362	1915
q_n	1	1	4	17	38	131	693

INFINITE CONTINUED FRACTION

Suppose we have infinitely many terms in the expression $[a_1; a_2, a_3, a_4, \ldots]$, where $a_1 \geq 0$ and $a_i \geq 1$ for $i \geq 2$. The resulting fraction is an *infinite continued fraction.* In particular, $[1; 1, 1, 1, 1, \ldots]$ is an infinite continued fraction, the simplest of them all.

It appears from our early analysis of corresponding finite continued fractions that the nth convergent C_n of the infinite continued fraction $[1; 1, 1, 1, 1, \ldots]$ is the Fibonacci ratio F_{n+1}/F_n. This is indeed the case and can be established using the principle of mathematical induction (PMI) (see Exercise 9). Thus

$$C_n = \frac{p_n}{q_n} = \frac{F_{n+1}}{F_n} \qquad n \geq 1$$

This relationship was first observed in 1753 by R. Simson.

Since

$$\lim_{n\to\infty} C_n = \lim_{n\to\infty} \frac{F_{n+1}}{F_n} = \alpha$$

it follows that the infinite continued fraction $[1; 1, 1, 1, \ldots]$ converges to the Golden Ratio. This yields a remarkably beautiful formula for α:

$$\begin{aligned}
\alpha &= [1; 1, 1, 1, 1, \ldots] \\
&= 1 + \frac{1}{1+}\,\frac{1}{1+}\,\frac{1}{1+\cdots} \\
&= 1 + \cfrac{1}{1 + \cfrac{1}{1 + \cfrac{1}{1 + \cfrac{1}{1 + \cdots}}}}
\end{aligned}$$

This is consistent with the fact that the value of every infinite continued fraction is an irrational number.

We can be establish the fact that $[1; 1, 1, 1, \ldots] = \alpha$ by using an alternate route, without employing convergents. To confirm this, let $x = [1; 1, 1, 1, \ldots]$. It is fairly obvious that the infinite continued fraction converges to a limit, so

$$[1; 1, 1, 1, \ldots] = [1; [1; 1, 1, 1, \ldots]].$$

That is,

$$\begin{aligned}
x &= [1; x] \\
x &= 1 + \frac{1}{x}
\end{aligned}$$

Therefore, $x = \alpha$, since $x > 0$. Thus

$$\lim_{n\to\infty} C_n = \lim_{n\to\infty} \frac{F_{n+1}}{F_n} = \alpha = [1; 1, 1, 1, \ldots]$$

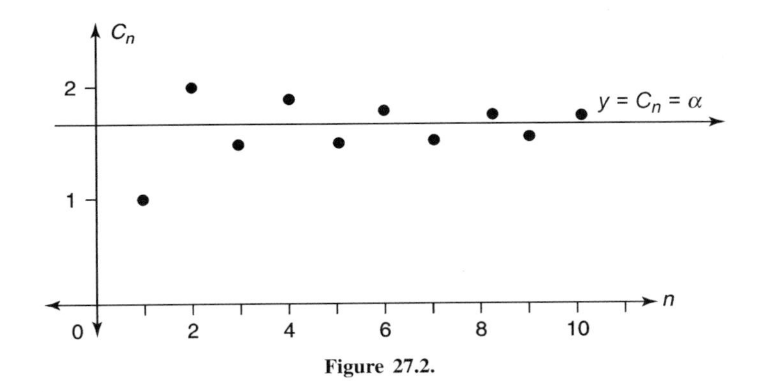

Figure 27.2.

It follows from the preceding discussion and Chapter 20 that when n is even, $C_n = F_{n+1}/F_n$ approaches α from above; and when n is odd, it approaches α from below. Figure 27.2 exhibits this marvelous behaviour for $1 \leq n \leq 10$.

AN INFINITE CONTINUED FRACTION FOR $-\beta$

In 1951, as a response to J. C. Pierce's article on the Fibonacci series in *The Scientific Monthly*, F. C. Ogg of Bowling Green State University, in a letter to the editor, provided a fancy way of converting $\sqrt{5}-1$ into an infinite continued fraction, which has an interesting by-product:

$$\begin{aligned}\sqrt{5}-1 &= 1+\sqrt{5}-2 = 1+\frac{1}{\sqrt{5}+2} = 1+\frac{1}{4+\sqrt{5}-2}\\ &= 1+\cfrac{1}{4+\cfrac{1}{\sqrt{5}+2}}\\ &= 1+\cfrac{1}{4+\cfrac{1}{4+\sqrt{5}-2}}\\ &= 1+\cfrac{1}{4+\cfrac{1}{4+\cfrac{1}{4+\cdots}}}\\ &= [1;4,4,4,\ldots]\end{aligned}$$

The first few convergents are $1, \frac{5}{4}, \frac{21}{17}, \frac{89}{72}, \ldots$. Now, divide each by 2. The resulting numbers are $\frac{1}{2}, \frac{5}{8}, \frac{21}{34}, \frac{89}{144}, \ldots$; so the nth convergent of the infinite continued fraction is F_n/F_{n+1}, where $n \neq 3k+1$. Since

$$\lim_{n\to\infty} \frac{F_n}{F_{n+1}} = \frac{1}{\alpha} = -\beta = \frac{\sqrt{5}-1}{2}$$

it follows that

$$-\beta = \frac{\sqrt{5}-1}{2} = [1; 4, 4, 4, \ldots]$$

PELL'S EQUATION

C. T. Long and J. H. Jordan of Washington State University, in their 1967 study of continued fractions, discovered a close relationship among the Golden Ratio, F_n, L_n, and a special class of Pell's equation. *Pell's equation*, named after the English mathematician John Pell (1611–1685), is an equation of the form $x^2 - dy^2 = N$, where x, y, d, and N are integers. (Although the equation bears his name, Pell added little to the study of such equations. It is so-called due to a mistake by Euler.)

The following theorems, discovered by Long and Jordan, "provide unusual characterizations of both Fibonacci and Lucas numbers." We omit their proofs in the interest of brevity.

Theorem 27.1. The Pell's equation $x^2 - 5y^2 = -4$ is solvable in positive integers if and only if $x = L_{2n-1}$ and $y = F_{2n-1}$, where $n \geq 1$. ■

Theorem 27.2. The Pell's equation $x^2 - 5y^2 = 4$ is solvable in positive integers if and only if $x = L_{2n}$ and $y = F_{2n}$, where $n \geq 1$. ■

EXERCISES 27

Represent each number as a continued fraction.

1. 51/35
2. 68/89

Represent each continued fraction as a rational number.

3. [2; 3, 1, 5]
4. [3; 1, 3, 2, 4, 7]

Find the convergents of each continued fraction.

5. [1; 2, 3, 4, 5]
6. [1; 1, 1, 1, 1, 1, 1, 1]
7. The second and third convergents of the continued fraction [1; 2, 3, 4, 5, 6] are 3/2 and 10/7. Find its fourth and fifth convergents.
8. The eighth and the ninth convergents of the continued fraction [1; 1, 1, 1, 1, 1, 1, 1, 1] are 34/21 and 55/34. Compute the tenth convergent.

9. Let C_n denote the nth convergent of the finite continued fraction $[1; 1, 1, 1, \ldots, 1]$. Prove that $C_n = F_{n+1}/F_n$, $n \geq 1$.
10. Let p_n/q_n denote the nth convergent of the continued fraction $[1; 1, 1, 1, \ldots, 1]$. Prove that $p_n q_{n-1} - q_n p_{n-1} = (-1)^n$, $n \geq 1$.
11. Using Cassini's formula, prove that $\lim_{n \to \infty} (C_n - C_{n-1}) = 0$, where C_n denotes the nth convergent of the infinite continued fraction $[1; 1, 1, 1, \ldots]$.

WEIGHTED FIBONACCI AND LUCAS SUMS

In Chapter 5, we found the following summation formulas:

$$\sum_{1}^{n} F_i = F_{n+2} - 1 \tag{5.1}$$

$$\sum_{1}^{n} L_i = L_{n+2} - 3 \tag{5.6}$$

More generally, we would like to find formulas for $\sum_{1}^{n} w_i F_i$ and $\sum_{1}^{n} w_i F_i$, where the *weights* w_i are positive integers.

To begin with, we would like to find a formula for each, when $w_i = i$, that is, a formula for $\sum_{1}^{n} iF_i$ and one for $\sum_{1}^{n} iL_i$. To derive the formula for the Fibonacci sum, let $A_n = \sum_{1}^{n} F_i$ and $B_n = \sum_{1}^{n} iF_i$. Then

$$\begin{aligned} B_n &= F_1 + 2F_2 + 3F_3 + \cdots + nF_n \\ &= \sum_{1}^{n} F_i + \sum_{2}^{n} F_i + \sum_{3}^{n} F_i + \cdots + \sum_{n}^{n} F_i \\ &= A_n + (A_n - A_1) + (A_n - A_2) + \cdots + (A_n - A_{n-1}) \end{aligned}$$

$$= nA_n - \sum_1^{n-1} A_i = n(F_{n+2} - 1) - \sum_1^{n-1}(F_{i+2} - 1) \qquad \text{by (5.1)}$$

$$= nF_{n+2} - n - (F_{n+3} - 3) + (n - 1) = nF_{n+2} - F_{n+3} + 2$$

Thus

$$\sum_1^n iF_i = nF_{n+2} - F_{n+3} + 2 \tag{28.1}$$

Similarly,

$$\sum_1^n iL_i = nL_{n+2} - L_{n+3} + 4 \tag{28.2}$$

For example, $\sum_1^7 iF_i = 7F_9 - F_{10} + 2 = 7 \cdot 34 - 55 + 2 = 185$; by direct computation, the sum equals $1 \cdot 1 + 2 \cdot 1 + 3 \cdot 2 + 4 \cdot 3 + 5 \cdot 5 + 6 \cdot 8 + 7 \cdot 13 = 185$. Also, $\sum_1^6 iL_i = 6L_8 - L_9 + 4 = 6 \cdot 47 - 76 + 4 = 210$, and by direct computation, $\sum_1^6 iL_i = 1 \cdot 1 + 2 \cdot 3 + 3 \cdot 4 + 4 \cdot 7 + 5 \cdot 11 + 6 \cdot 18 = 210$.

Now that we have formulas for $B_n = \sum_1^n iF_i$ and $C_n = \sum_1^n iL_i$, we can ask if there are formulas for:

$$B_n^* = \sum_1^n (n - i + 1)F_i = nF_1 + (n - 1)F_2 + \cdots + 2F_{n-1} + F_1$$

$$C_n^* = \sum_1^n (n - i + 1)L_i = nL_1 + (n - 1)L_2 + \cdots + 2L_{n-1} + L_1$$

Notice that B_n^* is the sum in Formula (28.1) with the coefficients in reverse order and similarly for C_n^*.

For example, $B_1^* = F_1 = 1$, and $B_2^* = 2F_1 + F_1 = 2 + 1 = 3$. Similarly, $B_3^* = 7$, $B_4^* = 14$, $B_5^* = 26$, and $B_6^* = 46$. Although these values do not seem to follow an obvious pattern, we can easily derive a formula for B_n^* as follows:

$$B_n + B_n^* = \sum_1^n iF_i + \sum_1^n (n - i + 1)F_i$$

$$= \sum_1^n [i + (n - i + 1)]F_i$$

$$\begin{aligned}
&= \sum_{1}^{n}(n+1)F_i = (n+1)\sum_{1}^{n} F_i \\
&= (n+1)(F_{n+2}-1) \qquad \text{by Formula (5.1)} \\
\therefore\quad B_n^* &= (n+1)(F_{n+2}-1) - B_n \\
&= (n+1)(F_{n+2}-1) - (nF_{n+2} - F_{n+3} + 2) \quad \text{by Formula (28.1)} \\
&= F_{n+4} - n - 3
\end{aligned}$$

Thus

$$\sum_{1}^{n}(n-i+1)F_i = F_{n+4} - n - 3 \tag{28.3}$$

Using this formula, $B_6^* = F_{10} - 6 - 3 = 55 - 9 = 46$, as expected. Using the same technique, we can be show that

$$C_n^* = \sum_{1}^{n}(n-i+1)L_i = L_{n+4} - 3n - 7 \tag{28.4}$$

For example, by direct computation, $\sum_{1}^{5}(n-i+1)L_i = 5L_1 + 4L_2 + 3L_3 + 2L_4 + L_5 = 5\cdot 1 + 4\cdot 3 + 3\cdot 4 + 2\cdot 7 + 1\cdot 11 = 54$, and using Formula (28.4), the sum equals $L_9 - 15 - 7 = 76 - 22 = 54$.

Formula (28.1) tempts us to investigate Fibonacci sums with odd integer coefficients and subscripts, and even integer coefficients and subscripts, that is, the sums $\sum_{1}^{n}(2i-1)F_{2i-1}$ and $\sum_{1}^{n}(2i)F_{2i}$, and the same sums with coefficients reversed.

To derive a formula for $C_n = \sum_{1}^{n}(2i-1)F_{2i-1}$, we employ Identity (5.2). Let $E_n = \sum_{1}^{n} F_{2i-1}$. Then

$$\begin{aligned}
C_n &= F_1 + 3F_3 + 5F_5 + \cdots + (2n-1)F_{2n-1} \\
&= \sum_{1}^{n} F_{2i-1} + 2\sum_{2}^{n} F_{2i-1} + 2\sum_{3}^{n} F_{2i-1} + \cdots + 2\sum_{n}^{n} F_{2i-1} \\
&= E_n + 2(E_n - E_1) + 2(E_n - E_2) + \cdots + 2(E_n - E_{n-1}) \\
&= E_n + 2(n-1)E_n - 2\sum_{1}^{n-1} E_i = (2n-1)E_n - 2\sum_{1}^{n-1} F_{2j} \\
&= (2n-1)E_n - 2(F_{2n-1} - 1) = (2n-1)F_{2n} - 2(F_{2n-1} - 1) \\
&= (2n-1)F_{2n} - 2F_{2n-1} + 2
\end{aligned} \tag{28.5}$$

For example, $C_4 = \sum_1^4 (2i-1)F_{2i-1} = 7F_8 - 2F_7 + 2 = 7\cdot 21 - 2\cdot 13 + 2 = 123$, which can be verified by direct computation.

Using Formulas 5.2 and 28.5,

$$\begin{aligned} C_n + C_n^* &= \sum_1^n (2i-1)F_{2i-1} + \sum_1^n (2n-2i+1)F_{2i-1} \\ &= \sum_1^n [(2i-1) + (2n-2i+1)]F_{2i-1} \\ &= \sum_1^n (2n)F_{2i-1} = 2n\sum_1^n F_{2i-1} = 2nF_{2n} \\ \therefore\quad C_n^* &= 2nF_{2n} - C_n = 2nF_{2n} - [(2n-1)F_{2n} - 2F_{2n-1} + 2] \\ &= F_{2n} + 2F_{2n-1} - 2 = (F_{2n} + F_{2n-1}) + F_{2n-1} - 2 \\ &= F_{2n+1} + F_{2n-1} - 2 \end{aligned}$$

That is,

$$\sum_1^n (2n-2i+1)F_{2i-1} = F_{2n+1} + F_{2n-1} - 2. \tag{28.6}$$

For example, $C_5^* = F_{11} + F_9 - 2 = 89 + 34 - 2 = 121$ and by direct computation, $C_5^* = 9F_1 + 7F_3 + 5F_5 + 3F_7 + F_9 = 9\cdot 1 + 7\cdot 2 + 5\cdot 5 + 3\cdot 13 + 1\cdot 34 = 121$.

Using Identity 5.3 and the same technique as in the proof of Formula (28.5), we can show that

$$\sum_1^n (2i)F_{2i} = 2(nF_{2n+1} - F_{2n}) \tag{28.7}$$

and as in the proof of Formula (28.6),

$$\sum_1^n (2n-2i+2)F_{2i} = 2F_{2n+2} - 2n - 2 \tag{28.8}$$

For example, $\sum_1^5 (2i)F_{2i} = 2(5F_{11} - F_{10}) = 2(5\cdot 89 - 55) = 780$ and $\sum_1^4 (10-2i)F_{2i} = 2F_{10} - 8 - 2 = 2\cdot 55 - 10 = 100$. We can verify both by direct calculation.

Interestingly enough, Formulas (28.5)–(28.8), have analogous results for Lucas numbers; we can derive and verify them as illustrated before:

$$\sum_{1}^{n}(2i-1)L_{2i-1} = (2n-1)L_{2n} - 2L_{2n-1} \tag{28.9}$$

$$\sum_{1}^{n}(2n-2i+1)L_{2i-1} = L_{2n+1} + L_{2n-1} - 4n \tag{28.10}$$

$$\sum_{1}^{n}(2i)L_{2i} = 2(nL_{2n+1} - L_{2n} + 2) \tag{28.11}$$

$$\sum_{1}^{n}(2n-2i+2)L_{2i} = 2L_{2n+2} - 2n - 6 \tag{28.12}$$

Their proofs employ Identities 5.7, 5.8, and 28.1, and we can establish them using induction.

Interestingly enough, we can extend Identity (28.1) to any Fibonacci sum where the coefficients form an arbitrary arithmetic sequence with first term a and common difference d. Let

$$\begin{aligned} S_n &= \sum_{1}^{n}[a+(i-1)d]F_i \\ &= a\sum_{1}^{n}F_i + d\left(\sum_{1}^{n}iF_i\right) - d\left(\sum_{1}^{n}F_i\right) \\ &= a(F_{n+2}-1) + d(nF_{n+2} - F_{n+3} + 2) - d(F_{n+2}-1) \\ &= (a+nd-d)F_{n+2} - d(F_{n+3}-3) - a \end{aligned}$$

Thus

$$\sum_{1}^{n}[a+(i-1)d]F_i = (a+nd-d)F_{n+2} - d(F_{n+3}-3) - a \tag{28.13}$$

Formula (28.13) has an analogous result for Lucas numbers also:

$$\sum_{1}^{n}[a+(i-1)d]L_i = (a+nd-d)L_{n+2} - d(L_{n+3}-7) - 3a \tag{28.14}$$

In particular, $\sum_{1}^{n} L_i = L_{n+2} - 3$ and $\sum_{1}^{n} iL_i = nL_{n+2} - L_{n+3} + 4$.

Let S_n^* denote the Fibonacci sum in Formula (28.13) with the coefficients reversed:

$$S_n^* = \sum_1^n [a + (n-i)d]F_i$$

Then

$$\begin{aligned}
S_n + S_n^* &= \sum_1^n \{[a + (i-1)d] + [a + (n-i)d]\}F_i \\
&= [(2a + (n-1)d] \sum_1^n F_i = [(2a + (n-1)d](F_{n+2} - 1) \\
\therefore \quad S_n^* &= [(2a + (n-1)d](F_{n+2} - 1) \\
&\qquad -[(a + nd - d)F_{n+2} - d(F_{n+3} - 3) - a] \\
&= [2a + (n-1)d - (a + nd - d)]F_{n+2} \\
&\qquad -[2a + (n-1)d] + d(F_{n+3} - 3) + a \\
&= aF_{n+2} + d(F_{n+3} - 3) - a - (n-1)d
\end{aligned}$$

Thus

$$\sum_1^n [a + (n-i)d]F_i = aF_{n+2} + d(F_{n+3} - 3) - a - (n-1)d \qquad (28.15)$$

When $a = 1 = d$, this reduces to the identity

$$\sum_1^n (n - i + 1)F_i = F_{n+4} - n - 3$$

Using the same technique, we can show that

$$\sum_1^n [a + (n-i)d]L_i = aL_{n+2} + d(L_{n+3} - 7) - 3[a + (n-1)d] \qquad (28.16)$$

Using the facts $\sum_1^n F_i^2 = F_n F_{n+1}$ and

$$\sum_1^n F_i F_{i+1} = \begin{cases} F_n^2 & \text{if } n \text{ is even} \\ F_n^2 - 1 & \text{otherwise} \end{cases}$$

we can show that

$$\sum_1^n [a + (i-1)d]F_i^2 = (a + nd - d)F_n F_{n+1} - d(F_n^2 - \gamma) \qquad (28.17)$$

where

$$\gamma = \begin{cases} 1 & \text{if } n \text{ is odd} \\ 0 & \text{otherwise} \end{cases}$$

In particular, this yields the identities

$$\sum_1^n F_i^2 = F_n F_{n+1}$$

$$\sum_1^n i F_i^2 = n F_n F_{n+1} - F_n^2 + \gamma \tag{28.18}$$

For example,

$$\sum_1^5 i F_i^2 = 5F_5F_6 - F_5^2 + 1 = 5 \cdot 5 \cdot 8 - 25 + 1 = 176$$

and

$$\sum_1^6 i F_i^2 = 6F_6F_7 - F_6^2 + 0 = 6 \cdot 8 \cdot 13 - 64 + 0 = 560$$

Let $D_n^* = \sum_1^n [a + (n-i)d]F_i^2$, the same sum (28.17) with the coefficients in the reverse order. Then

$$\begin{aligned} D_n + D_n^* &= [2a + (n-1)d] \sum_1^n F_i^2 = [2a + (n-1)d] F_n F_{n+1} \\ D_n^* &= [2a + (n-1)d] F_n F_{n+1} - (a + nd - d) F_n F_{n+1} + d(F_n^2 - \gamma) \\ &= a F_n F_{n+1} + d(F_n^2 - \gamma) \end{aligned}$$

Thus

$$\sum_1^n [a + (n-i)d] F_i^2 = a F_n F_{n+1} + d(F_n^2 - \gamma) \tag{28.19}$$

In particular,

$$\sum_1^n (n - i + 1)] F_i^2 = F_n F_{n+1} + F_n^2 - \gamma \tag{28.20}$$

For example, $\sum_1^5 (6-i)]F_i^2 = F_5F_6 + F_5^2 - 1 = 5 \cdot 8 + 25 - 1 = 64$

Fortunately, Formulas 28.17 and 28.19 have analogous counterparts to Lucas numbers:

$$\sum_{1}^{n}[a+(i-1)d]L_i^2 = (a+nd-d)(L_nL_{n+1}-2) - d(L_n^2-2n-\nu) \quad (28.21)$$

$$\sum_{1}^{n}[a+(n-i)d]L_i^2 = a(L_nL_{n+1}-2) + d(L_n^2-\nu) \quad (28.22)$$

where

$$\nu = \begin{cases} -1 & \text{if } n \text{ is odd} \\ 4 & \text{otherwise} \end{cases}$$

Identity Formula 28.21 yields $\sum_{1}^{n} L_i^2 = L_nL_{n+1} - 2$ and

$$\sum_{1}^{n} iL_i^2 = n(L_nL_{n+1}-2) - L_n^2 + \nu \quad (28.23)$$

Identity 28.22 yields

$$\sum_{1}^{n}(n-i+1)]L_i^2 = L_nL_{n+1} - \nu - 2 \quad (28.24)$$

For example, $\sum_{1}^{6} iL_i^2 = 6(L_6L_7-2) - L_6^2 + 6 = 6(18\cdot 29-2) - 18^2 + 6 = 2802$. Likewise, $\sum_{1}^{5}(7-i)L_i^2 = 2(L_5L_6-2) + L_5^2 - 1 = 512$. We can verify both by direct computation.

EXERCISES 28

1. Verify Identity 28.1 for $n = 7$.
2. Verify Identity 28.2 for $n = 7$.

Prove each.

3. $\sum_{1}^{n}(2i)F_{2i} = 2(nF_{2n+1} - F_{2n})$
4. $\sum_{1}^{n}(2n-2i+2)F_{2i} = 2F_{2n+2} - 2n - 2$
5. $\sum_{1}^{n}(2i-1)L_{2i-1} = (2n-1)L_{2n} - 2L_{2n-1}$
6. $\sum_{1}^{n}(2n-2i+1)L_{2i-1} = L_{2n+1} + L_{2n-1} - 4n$

7. $\sum_{1}^{n}(2i)L_{2i} = 2(nL_{2n+1} - L_{2n} + 2)$

8. $\sum_{1}^{n}(2n - 2i + 2)L_{2i} = 2L_{2n+2} - 2n - 6$

9. $\sum_{1}^{n}[a + (i - 1)d]L_i = (a + nd - d)L_{n+2} - d(L_{n+3} - 7) - 3a$

10. $\sum_{1}^{n}[a + (n - i)d]L_i = aL_{n+2} + d(L_{n+3} - 7) - 3[a + (n - 1)d]$

11. $\sum_{1}^{n}[a + (i - 1)d]F_i^2 = (a + nd - d)F_n F_{n+1} - d(F_n^2 - \gamma)$, where γ is defined as in Formula 28.17.

In Exercises 12–15, the number ν is defined as in Formula 28.22.

12. $\sum_{1}^{n} iL_i^2 = nL_n L_{n+1} - L_n^2 + \nu$

13. $\sum_{1}^{n}(n - i + 1)]L_i^2 = L_n L_{n+2} - 2(n + 1) - \nu$

14. $\sum_{1}^{n}[a + (i - 1)d]L_i^2 = (a + nd - d)(L_n L_{n+1} - 2) - d(L_n^2 - 2n - \nu)$

15. $\sum_{1}^{n}[a + (n - i)d]L_i^2 = a(L_n L_{n+1} - 2) + d(L_n^2 - 2n - \nu)$

Let G_i denote the ith term of the generalized Fibonacci sequence. Derive a formula for each sum.

16. $\sum_{1}^{n} G_i$

17. $\sum_{1}^{n} iG_i$

18. $\sum_{1}^{n}(n - i + 1)G_i$

19. $\sum_{1}^{n} G_{2i-1}$

20. $\sum_{1}^{n} G_{2i}$

21. $\sum_{1}^{n}(2i - 1)G_{2i-1}$

22. $\sum_{1}^{n}(2n - 2i + 1)G_{2i-1}$

29

FIBONACCI AND LUCAS SUMS REVISITED

This chapter continues to explore explicit formulas for $S(m) = \sum_1^n i^m F_i$ and $T(m) = \sum_1^n i^m L_i$, where $m \geq 0$.

We developed the formulas corresponding to $m = 0$ in Chapter 5 and those corresponding to $m = 1$ in Chapter 28. In fact, from Chapter 7, we have

$$\sum_1^n iG_i = nG_{n+2} - G_{n+3} + a + b$$

The formulas corresponding to $m = 2$ and $m = 3$ were developed algebraically by P. Glaister of the University of Reading, England, and N. Gauthier of The Royal Military College of Canada:

$$\sum_1^n i^2 F_i = (n+1)^2 F_{n+2} - (2n+3)F_{n+4} + 2F_{n+6} - 8 \tag{29.1}$$

$$\sum_1^n i^2 L_i = (n+1)^2 L_{n+2} - (2n+3)L_{n+4} + 2L_{n+6} - 18 \tag{29.2}$$

Formulas for $S(1)$ and $S(2)$ were rediscovered by Gauthier using a fascinating method involving the differential operator $x(d/dx)$, which, for the sake of brevity, we shall denote by ∇. In addition to giving a general method for computing $S(m)$, Gauthier gives an explicit formula for S_3:

$$\sum_1^n i^3 F_i = (n+1)^3 F_{n+2} - (3n^2+9n+7)F_{n+4} + (6n+12)F_{n+6} - 6F_{n+8} + 50 \tag{29.3}$$

Interestingly enough, we can employ Gauthier's differential approach to derive a formula for $T(m)$. To see this, we need to use Binet's formulas. Also, we will need the facts that $1/(1-\alpha)^i = (-\alpha)^i$ and $1/(1-\beta)^i = (-\beta)^i$. Suppose we have a formula for $T(m)$. Since we can obtain F_i from L_i by changing β^i to $-\beta^i$ and then dividing the difference by $\sqrt{5}$, we can find a formula for $S(m)$ from $T(m)$. To arrive at a formula for $T(m)$, notice that $L_i = \alpha^i + \beta^i = (x^i)_{x=\alpha} + (x^i)_{x=\beta}$, which we shall abbreviate as $L_i = (x^i)_\alpha + (x^i)_\beta$. Let

$$f(x) = \sum_1^n x^i = x + x^2 + \cdots + x^n = \frac{1 - x^{n+1}}{1 - x} - 1$$

where $x \neq 1$. Then $\frac{df}{dx} = \sum_1^n ix^{i-1}$, so

$$x\frac{df}{dx} = \sum_1^n ix^i$$

that is, $\nabla f = \sum_1^n ix^i$. Similarly, $\nabla^2 f = \nabla(\nabla f) = \sum_1^n i^2 x^i$. More generally, we have:

$$\nabla^m f = \sum_1^n i^m x^i \tag{29.4}$$

where $m \geq 0$ and $\nabla^0 f = f$.

By Formula (29.4),

$$\begin{aligned}
\sum_1^n L_i &= \left(\frac{1-\alpha^{n+1}}{1-\alpha} - 1\right) + \left(\frac{1-\beta^{n+1}}{1-\beta} - 1\right) \\
&= \left(\frac{1}{1-\alpha} + \frac{1}{1-\beta}\right) - \left(\frac{\alpha^{n+1}}{1-\alpha} + \frac{\beta^{n+1}}{1-\beta}\right) - 2 \\
&= -(\alpha+\beta) + (\alpha^{n+1} + \beta^{n+1}) - 2 \qquad (29.5) \\
&= -L_1 + L_{n+2} - 2 \\
&= L_{n+2} - 3 = L_{n+2} - L_2 \qquad (29.6)
\end{aligned}$$

which is Formula (5.6).

We can rewrite this formula as $\sum_1^n L_i = (\alpha^{n+2} + \beta^{n+2}) - (\alpha^2 + \beta^2)$. Now change β^i to $-\beta^i$ and then divide both sides by $\sqrt{5}$. This yields $\sum_1^n F_i = F_{n+2} - F_2 = F_{n+2} - 1$. In other words, it suffices to change L_i to F_i in Formula (29.6).

Let $m \geq 1$ and $g(x) = (1 - x^{n+1})/(1 - x)$. Then $\nabla g = \nabla f$, and hence $\nabla^m g = \sum_1^n i^m x^i$. Thus $(\nabla^m g)_\alpha = \sum_1^n i^m \alpha_i$, so by Binet's formula,

$$\sum_1^n i^m L_i = (\nabla^m g)_\alpha + (\nabla^m g)_\beta \tag{29.7}$$

This gives us the general formula for computing $T(m)$. Notice that

$$g(x) = \frac{1}{1-x} - \frac{x^{n+1}}{1-x} = g_0(x) - g_{n+1}(x)$$

where $g_t = x^t/(1 - x)$. Then $\nabla g = \nabla(g_0 - g_{n+1}) = \nabla g_0 - \nabla g_{n+1}$, and more generally $\nabla^m g = \nabla^m g_0 - \nabla^m g_{n+1}$, which we can find from $\nabla^m g_t$. Thus, we can modify Formula (29.7) as

$$\sum_1^n i^m L_i = [(\nabla^m g_0)_\alpha + (\nabla^m g_0)_\beta] - [(\nabla^m g_{n+1})_\alpha + (\nabla^m g_{n+1})_\beta] \tag{29.8}$$

Since

$$\nabla g_t = t\frac{x^t}{1-x} + \frac{x^{t+1}}{(1-x)^2}$$

$\nabla g_0 = x/(1-x)^2$ and

$$\nabla g_{n+1} = (n+1)\frac{x^n}{1-x} + \frac{x^{n+2}}{(1-x)^2}$$

So, when $m = 1$, by Formula (29.6),

$$\begin{aligned}
\sum_1^n iL_i &= \left[\frac{\alpha}{(1-\alpha)^3} + \frac{\beta}{(1-\beta)^2}\right] - (n+1)\left[\frac{\alpha^{n+1}}{1-\alpha} + \frac{\beta^{n+1}}{1-\beta}\right] \\
&\quad - \left[\frac{\alpha^{n+2}}{(1-\alpha)^2} + \frac{\beta^{n+2}}{(1-\beta)^2}\right] \\
&= (\alpha^3 + \beta^3) + (n+1)(\alpha^{n+2} + \beta^{n+2}) - (\alpha^{n+4} + \beta^{n+4}) &(29.9)\\
&= L_3 + (n+1)L_{n+2} - L_{n+4} &(29.10)\\
&= (n+1)L_{n+2} - L_{n+4} + 4
\end{aligned}$$

which is Identity (28.2). Changing L_i to F_i in Eq. (29.10), we get

$$\begin{aligned}
\sum_1^n iF_i &= (n+1)F_{n+2} - F_{n+4} + F_3 \\
&= (n+1)F_{n+2} - F_{n+4} + 2
\end{aligned}$$

which is Identity (28.1).

For the case $m = 2$, notice that

$$\nabla^2 g_t = \nabla(\nabla g_t) = t^2 \frac{x^t}{1-x} - (2t+1)\frac{x^{t+1}}{(1-x)^2} + 2\frac{x^{t+2}}{(1-x)^3}$$

$$\begin{aligned}\sum_1^n i^2 L_i &= \left[\frac{\alpha}{(1-\alpha)^2} + \frac{\beta}{(1-\beta)^2}\right] + 2\left[\frac{\alpha^2}{(1-\alpha)^3} + \frac{\beta^2}{(1-\beta)^3}\right] \\ &\quad - (n+1)^2\left[\frac{\alpha^{n+1}}{1-\alpha} + \frac{\beta^{n+1}}{1-\beta}\right] \\ &\quad - (2n+3)\left[\frac{\alpha^{n+2}}{(1-\alpha)^2} + \frac{\beta^{n+2}}{(1-\beta)^2}\right] - 2\left[\frac{\alpha^{n+3}}{(1-\alpha)^3} + \frac{\beta^{n+3}}{(1-\beta)^3}\right] \\ &= (\alpha^3+\beta^3) - 2(\alpha^5+\beta^5) + (n+1)^2(\alpha^{n+2}+\beta^{n+2}) \\ &\quad - (2n+3)(\alpha^{n+4}+\beta^{n+4}) + 2(\alpha^{n+6}+\beta^{n+6}) \\ &= (n+1)^2 L_{n+2} - (2n+3)L_{n+4} + 2L_{n+6} + L_3 - 2L_5\end{aligned}$$

which is Identity (29.2).

Replacing L_j with F_j, this yields

$$\sum_1^n i^2 F_i = (n+1)^2 F_{n+2} - (2n+3)F_{n+4} + 2_{Fn+6} + F_3 - 2F_5$$

which is Identity (29.1).

For the case $m = 3$, it may be verified that

$$\nabla^3 g_t = \nabla(\nabla^2 g_t) = t^3 \frac{x^t}{1-x} + (3t^2+3t+1)\frac{x^{t+1}}{(1-x)^2} + (6t+6)\frac{x^{t+2}}{(1-x)^3} + \frac{x^{t+3}}{(1-x)^4}$$

Then

$$\begin{aligned}(\nabla^3 g_0)_\alpha + (\nabla^3 g_0)_\beta &= \left[\frac{\alpha}{(1-\alpha)^2} + \frac{\beta}{(1-\beta)^2}\right] + 6\left[\frac{\alpha^2}{(1-\alpha)^3} + \frac{\beta^2}{(1-\beta)^3}\right] \\ &\quad + 6\left[\frac{\alpha^3}{(1-\alpha)^4} + \frac{\beta^3}{(1-\beta)^4}\right] \\ &= (\alpha^3+\beta^3) - 6(\alpha^5+\beta^5) + 6(\alpha^7+\beta^7) \\ &= L_3 - 6L_5 + 6L_7\end{aligned}$$

Likewise,

$$\begin{aligned}(\nabla^3 g_{n+1})_\alpha + (\nabla^3 g_{n+1})_\beta &= -(n+1)^3 L_{n+2} \\ &\quad + (3n^2+9n+7)L_{n+4} - (6n+12)L_{n+6} + 6L_{n+8}\end{aligned}$$

Thus,

$$\sum_{1}^{n} i^3 L_i = (n+1)^3 L_{n+2} - (3n^2 + 9n + 7)L_{n+4}$$
$$+ (6n + 12)L_{n+6} - 6L_{n+8} + L_3 - 6L_5 + 6L_7 \quad (29.11)$$
$$= (n+1)^3 L_{n+2} - (3n^2 + 9n + 7)L_{n+4}$$
$$+ (6n + 12)L_{n+6} - 6L_{n+8} + 112 \quad (29.12)$$

Changing L_j to F_j yields the identity

$$\sum_{1}^{n} i^3 F_i = (n+1)^3 F_{n+2} - (3n^2 + 9n + 7)F_{n+4}$$
$$+ (6n + 12)F_{n+6} - 6F_{n+8} + F_3 - 6F_5 + 6F_7 \quad (29.13)$$
$$= (n+1)^3 F_{n+2} - (3n^2 + 9n + 7)F_{n+4}$$
$$+ (6n + 12)F_{n+6} - 6F_{n+8} + 50 \quad (29.14)$$

Clearly, we can continue this procedure for an arbitrary positive integer m. For the curious-minded, we give the formulas for $T(4)$ and $S(4)$:

$$\sum_{1}^{n} i^4 L_i = (n+1)^4 L_{n+2} - (4n^3 + 18n^2 + 28n + 15)L_{n+4}$$
$$+ (12n^2 + 48n + 50)L_{n+6}$$
$$- (24n + 60)L_{n+8} + 24L_{n+10} + L_3 - 14L_5$$
$$+ 36L_7 - 24L_9 \quad (29.15)$$
$$= (n+1)^4 L_{n+2} - (4n^3 + 18n^2 + 28n + 15)L_{n+4}$$
$$+ (12n^2 + 48n + 50)L_{n+6}$$
$$- (24n + 60)L_{n+8} + 24L_{n+10} - 930 \quad (29.16)$$

Consequently,

$$\sum_{1}^{n} i^4 F_i = (n+1)^4 F_{n+2} - (4n^3 + 18n^2 + 28n + 15)F_{n+4}$$
$$+ (12n^2 + 48n + 50)F_{n+6}$$
$$- (24n + 60)F_{n+8} + 24F_{n+10}$$
$$+ F_3 - 14F_5 + 36F_7 - 24F_9 \quad (29.17)$$
$$= (n+1)^4 F_{n+2} - (4n^3 + 18n^2 + 28n + 15)F_{n+4}$$
$$+ (12n^2 + 48n + 50)F_{n+6}$$
$$- (24n + 60)F_{n+8} + 24F_{n+10} - 416 \quad (29.18)$$

For example, by Formula (29.16),

$$\sum_1^5 i^4 L_i = 1296L_7 - 1105L_9 + 590L_{11} - 180L_{13} + 24L_{15} - 930 = 9040$$

which we can verify by direct computation.

A few interesting observations about the formulas for $S(m)$ and $T(m)$:

- Both $S(m)$ and $T(m)$ contain $m + 2$ terms.
- The coefficients in $S(m)$ and $T(m)$ alternate in signs and the corresponding coefficients in them are identical.
- The leading term in $S(m)$ is $(n+1)^m F_{n+2}$, and that in $T(m)$ is $(n+1)^m L_{n+2}$. The subscripts in the Fibonacci and Lucas sums increase by 2, while the exponent of n in each coefficient decreases by one.
- We can obtain the formula for $S(m)$ from that of $T(m)$ and vice versa by switching F_j and L_j.
- Except for the trailing constant term, we can obtain the formula for $S(m-1)$ from that of $S(m)$. The same is true for $T(m)$ also.

For example, consider Formula (29.7) for $S(4)$. The nonconstant coefficients on the right-hand side are $(n+1)^4$, $-(4n^3+18n^2+28n+15)$, $12n^2+48n+50$, $-(24n+60)$, and 24. Their derivatives with respect to n are $4(n+1)^3$, $-4(3n^2+9n+7)$, $4(6n+12)$, $4(-6)$, and 0. The derivative of i^4 with respect to i is $4i^3$. Dividing them by 4, we get the nonconstant coefficients in $S(3)$:

$$\begin{aligned}\sum_1^n i^3 F_i &= (n+1)^3 F_{n+2} - (3n^2+9n+7)F_{n+4} \\ &\quad - (6n+12)F_{n+6} + 6F_{n+8} + k\end{aligned}$$

where k is a constant, which is consistent with Formula (29.13).

In fact, $k = (\nabla^m g_0)_\alpha - (\nabla^m g_0)_\beta$ in the case of $S(m)$ and $k = (\nabla^m g_0)_\alpha + (\nabla^m g_0)_\beta$ in the case of $T(m)$. For example, when $m = 3$, $k = (\nabla^3 g_0)_\alpha - (\nabla^3 g_0)_\beta = F_3 - 6F_5 + 6F_7 = 50$, as obtained earlier.

On the other hand, if we could use the coefficients in $S(m-1)$ to determine those in $S(m)$, it would be a tremendous advantage in the study of weighted Fibonacci and Lucas sums. The same would hold for $T(m)$ as well.

EXERCISES 29

Compute each sum.

1. $\sum_1^{10} i^2 F_i$
2. $\sum_1^{10} i^2 L_i$

3. $\sum_{1}^{5} i^3 F_i$

4. $\sum_{1}^{5} i^3 L_i$

5. $\sum_{1}^{5} i^4 F_i$

6. $\sum_{1}^{5} i^4 L_i$

Verify each identity for $n = 6$.

7. Identity (29.1)
8. Identity (29.2)
9. Identity (29.3)
10. Identity (29.12)
11. Identity (29.14)
12. Identity (29.16)
13. Establish Identity (29.1) algebraically.
14. Establish Identity (29.2) algebraically.

THE KNAPSACK PROBLEM

In this chapter, we investigate the well-known knapsack problem, with Fibonacci and Lucas numbers as weights.

Let G_i denote the ith generalized Fibonacci number. Then, recall that

$$\sum_{1}^{n} G_i = G_{n+2} - b$$

that is,

$$\sum_{1}^{n} G_i = \begin{cases} F_{n+2} - 1 & \text{if } G_i = F_i \\ F_{n+2} - 3 & \text{if } G_i = L_i \end{cases}$$

Consequently,

$$G_i + G_{i+1} + \cdots + G_{i+n-1} = \sum_{1}^{i+n-1} G_j - \sum_{1}^{i-1} G_j = G_{i+n+1} - G_{i+1} \tag{30.1}$$

THE KNAPSACK PROBLEM

Given a knapsack of volume S and n items of various volumes, $a_1, a_2, \ldots, a_n$, which of the items can fill the knapsack? In other words, given the positive integers $a_1, a_2, \cdots, a_n$, called *weights*, and a positive integer S, solve the linear diophantine equation (LDE) $a_1x_1 + a_2x_2 + \cdots + a_nx_n = S$, where $x_i = 0$ or 1. This is the celebrated *knapsack problem*.

In particular, consider the knapsack problem

$$G_i x_1 + G_{i+1} x_2 + \cdots + G_{i+n-1} x_n = G_{i+n} \tag{30.2}$$

where $i \geq 2$. By virtue of the Fibonacci recurrence relation, this LDE is solvable with $(0, 0, \ldots, 0, 1, 1)$ as a solution. In fact, since $G_i + G_{i+1} + \cdots + G_{i+n-2} = G_{i+n} - G_{i-1} < G_{i+n}$, no sum of $G_i, G_{i+1}, \ldots, G_{i+n-2}$ can add up to G_{i+n}. Thus $(0, 0, \ldots, 0, 1, 1)$ is the unique solution, with $x_{n-1} = 1$.

For example, the only solution of $x_2 + 2x_3 + 3x_4 + 5x_5 + 8x_6 + 13x_7 = 21$, with $x_6 = 1$, is (0,0,0,0,1,1).

So, is Eq. (30.2) solvable with $x_{n-1} = 0$? If yes, how many such solutions does the problem have? First, notice that:

$$\begin{aligned}
G_{i+n} &= G_{i+n-1} + G_{i+n-2} \\
&= G_{i+n-1} + G_{i+n-3} + G_{i+n-4} \\
&= G_{i+n-1} + G_{i+n-3} + G_{i+n-5} + G_{i+n-6} \\
&= G_{i+n-1} + G_{i+n-3} + G_{i+n-5} + G_{i+n-7} + G_{i+n-8} \\
&\ \ \vdots \\
&= G_{i+n-1} + G_{i+n-3} + G_{i+n-5} + G_{i+n-7} + \cdots + G_i
\end{aligned} \tag{30.3}$$

Since $i = (i + n) - n$, the right-hand side (RHS) of Eq. (30.3) contains $\lfloor n/2 \rfloor$ additions. Thus, there are $\lfloor n/2 \rfloor$ number of ways of expressing G_{i+n} as a sum of its predecessors through G_i. In other words, the knapsack problem (Eq. 30.2) has $\lfloor n/2 \rfloor$ solutions, one of which corresponds to $x_{n-1} = 1$.

Theorem 30.1. The knapsack problem $G_i x_1 + G_{i+1} x_2 + \cdots + G_{i+n-1} x_n = G_{i+n}$ has $\lfloor n/2 \rfloor$ solutions, where $i \geq 1$. ■

For example, $F_5 x_1 + F_6 x_2 + \cdots + F_{10} x_6 = F_{11}$ has $\lfloor 6/2 \rfloor = 3$ solutions $(x_1, \ldots, x_6)$. They correspond to the internal nodes in the binary tree in Figure 30.1 and to the three

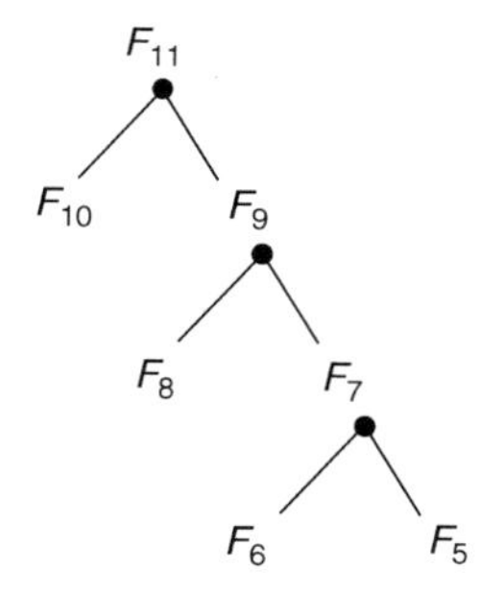

Figure 30.1.

different ways of expressing F_{11} in terms of its predecessors through F_5:

$$\begin{aligned} F_{11} &= F_{10} + F_9 \\ &= F_{10} + F_8 + F_7 \\ &= F_{10} + F_8 + F_6 + F_5 \end{aligned}$$

Correspondingly, the three solutions are (0,0,0,0,1,1), (0,0,1,1,0,1), and (1,1,0,1,0,1).

Theorem 30.1 yields the next result.

Corollary 30.1. The knapsack problem $G_1x_1 + G_2x_2 + \cdots + G_nx_n = G_{n+1}$ has $\lfloor n/2 \rfloor$ solutions.

For example, $F_1x_1 + F_2x_2 + \cdots + F_{10}x_{10} = F_{11}$ has $\lfloor n/2 \rfloor = 5$ solutions $(x_1, \ldots, x_{10})$. They correspond to the five different ways of expressing F_{11} in terms of its predecessors:

$$\begin{aligned} F_{11} &= F_{10} + F_9 \\ &= F_{10} + F_8 + F_7 \\ &= F_{10} + F_8 + F_6 + F_5 \\ &= F_{10} + F_8 + F_6 + F_4 + F_3 \\ &= F_{10} + F_8 + F_6 + F_4 + F_2 + F_1 \end{aligned}$$

They are represented by the internal nodes in the binary tree in Figure 30.2. Correspondingly, the five solutions are (0, 0, 0, 0, 0, 0, 0, 0, 1, 1), (0, 0, 0, 0, 0, 0, 1, 1, 0, 1), (0, 0, 0, 0, 1, 1, 0, 1, 0, 1), (0, 0, 1, 1, 0, 1, 0, 1, 0, 1), and (1,1,0,1,0,1,0,1,0,1).

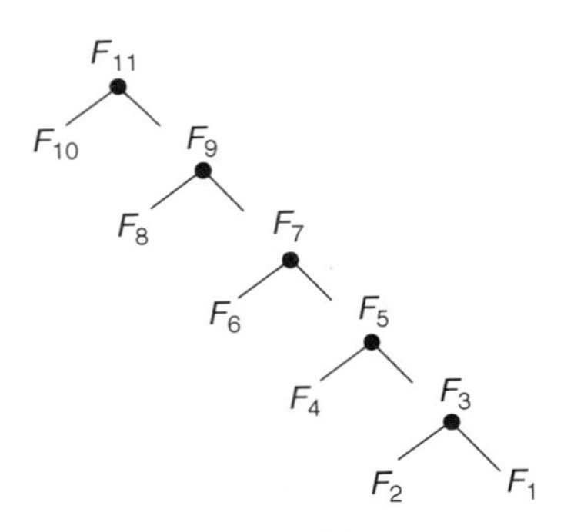

Figure 30.2.

Suppose the RHS of Eq. 30.2 is G_j, where $j \neq i + n$:

$$G_ix_1 + G_{i+1}x_2 + \cdots + G_{i+n-1}x_n = G_j \qquad j \neq i + n \tag{30.4}$$

If $i \leq j \leq i + n - 1$, the knapsack problem (Eq. 30.4) is solvable with a solution $(0, \ldots, 0, 1, 0, \ldots, 0)$, where the 1 occurs in position j; it need not be unique. If $j < i$, then no solution is possible. Suppose $j > i + n$. Since $G_i + G_{i+1} + \cdots + G_{i+n-1} = G_{i+n+1} - G_{i+1} < G_{i+n+1} \leq G_j$, the knapsack problem has no solutions.

Next we establish that the knapsack problem

$$G_1x_1 + G_2x_2 + \cdots + G_nx_n = S \tag{30.5}$$

is solvable, where S is any positive integer $\leq G_{n+1}$. This, in fact, follows by Theorems 11.1 and 11.2.

Theorem 30.2. The knapsack problem $G_1x_1 + G_2x_2 + \cdots + G_nx_n = S$ is solvable, where S is a positive integer $\leq G_{n+1}$. ■

For example, $x_1 + x_2 + 2x_3 + 3x_4 + 5x_5 + 8x_6 + 13x_7 = 7$ with Fibonacci weights is solvable, and since $7 = 2 + 5$, (0,0,1,0,1,0,0) is a solution. Likewise, the knapsack problem $x_1 + x_2 + 2x_3 + 3x_4 + 5x_5 + 8x_6 + 13x_7 + 21x_8 + 34x_9 = 48$ is solvable; since $48 = 1 + 5 + 8 + 34$, $(1, 0, 0, 0, 1, 1, 0, , 0, 1)$ is a solution. Both problems have more than one solution.

The knapsack problem $2x_1 + x_2 + 3x_3 + 4x_4 + 7x_5 + 11x_6 = 15$ with Lucas weights is solvable; since $15 = 4 + 11$, $(0, 0, 0, 1, 0, 1)$ is a solution; so is (0,1,1,0,0,1). The problem $2x_1 + x_2 + 3x_3 + 4x_4 + 7x_5 + 11x_6 = 10$ is also solvable, (0,0,1,0,1,0) being a solution.

FIBONACCI MAGIC SQUARES

For centuries, magic squares were a source of entertainment in royal courts around the world. Today, they are still popular with both amateurs and professionals.

A *magic square* is a square array of distinct positive integers such that the sum of the numbers along each row, column, and diagonal is a constant k; k is the *magic constant* of the magic square. The oldest known magic square is the Chinese magic square, *lo-shu*, shown in Figure 31.1. According to legend, the array was discovered around 2200 B.C. on the back of a divine tortoise along the banks of the Yellow River. The array was displayed using knots and strings. Lo-shu's magic constant is 15.

4	9	2
3	5	7
8	1	6

Figure 31.1.

In 1964, Br. U. Alfred of St. Mary's College, California, initiated an investigation of magic squares using Fibonacci numbers, to discover if such magic squares exist. Unfortunately, in the following year, J. L. Brown, Jr., of Pennsylvania State University proved that there are no magic squares with only Fibonacci entries.

We shall now confirm this by contradiction. Suppose there are 2×2 Fibonacci magic squares (FMSs), as Figure 31.2 shows, where the entries are all distinct. Then $a + b = a + c$, so $b = c$, which is a contradiction. Thus there are no 2×2 FMSs.

a	b
c	d

Figure 31.2.

Let us now assume that there are $n \times n$ FMSs, where $n \geq 3$. Let $F_{i_1}, F_{i_2}, \ldots,$ and F_{i_n}; $F_{j_1}, F_{j_2}, \ldots,$ and F_{j_n}; and $F_{k_1}, F_{k_2}, \ldots,$ and F_{k_n} denote the elements of the first three columns. Then

$$\begin{aligned} F_{i_1} + F_{i_2} + \cdots + F_{i_n} &= F_{j_1} + F_{j_2} + \cdots + F_{j_n} \\ &= F_{k_1} + F_{k_2} + \cdots + F_{k_n} = S \text{ (say)} \end{aligned}$$

Since they are all distinct, without loss of generality, we can assume that

$$F_{i_1} > F_{i_2} > \cdots > F_{i_n}, \qquad F_{j_1} > F_{j_2} > \cdots > F_{j_n}, \qquad \text{and}$$
$$F_{k_1} > F_{k_2} > \cdots > F_{k_n}.$$

Again, without loss of generality, we can assume that $F_{i_1} > F_{j_1} > F_{k_1}$, so $F_{i_1} \geq F_{k_1+2}$. Then $F_{i_1} + F_{i_2} + \cdots + F_{i_n} > F_{i_1}$. Thus $S \geq F_{k_1+2}$. Since $F_{k_n} < \cdots < F_{k_2} < F_{k_1}$, $F_{k_1} + F_{k_2} + \cdots + F_{k_n} < \sum_{1}^{k_1} F_i$. That is, $S = F_{k_1+2} - 1$, by Identity (5.1). Thus $F_{k_1+2} \leq S < F_{k_1+2} - 1$, which is a contradiction. Consequently, there are no $n \times n$ FMSs, where $n \geq 2$.

FIBONACCI MATRICES

The application of matrices to the theory of Fibonacci and Lucas numbers yields excellent dividends.

THE *Q*-MATRIX

Using the properties of matrices presented in the Appendix, we can demonstrate a close link between matrices and Fibonacci numbers. To this end, consider the matrix

$$\mathbf{Q} = \begin{bmatrix} 1 & 1 \\ 1 & 0 \end{bmatrix}$$

This matrix, called the **Q**-*matrix*, was studied by Charles H. King in 1960 for his Master's thesis at what was then San Jose State College, California. Notice that $|\mathbf{Q}| = -1$. In addition, we have:

$$\mathbf{Q}^2 = \begin{bmatrix} 1 & 1 \\ 1 & 0 \end{bmatrix}\begin{bmatrix} 1 & 1 \\ 1 & 0 \end{bmatrix} = \begin{bmatrix} 2 & 1 \\ 1 & 1 \end{bmatrix}$$

$$\mathbf{Q}^3 = \begin{bmatrix} 1 & 1 \\ 1 & 0 \end{bmatrix}\begin{bmatrix} 2 & 1 \\ 1 & 1 \end{bmatrix} = \begin{bmatrix} 3 & 2 \\ 2 & 1 \end{bmatrix}$$

Likewise,

$$\mathbf{Q}^4 = \begin{bmatrix} 5 & 3 \\ 3 & 2 \end{bmatrix}$$

We can see a pattern emerging. More generally, we have the following intriguing result.

Theorem 32.1. Let $n \geq 1$. Then

$$\mathbf{Q}^n = \begin{bmatrix} F_{n+1} & F_n \\ F_n & F_{n-1} \end{bmatrix}$$

Proof. [by the principle of mathematical induction (PMI)]. When $n = 1$,

$$\mathbf{Q}^1 = \begin{bmatrix} F_2 & F_1 \\ F_1 & F_0 \end{bmatrix} = \begin{bmatrix} 1 & 1 \\ 1 & 0 \end{bmatrix} = \mathbf{Q}$$

so the result is true. Now, assume it is true for an arbitrary positive integer k:

$$\mathbf{Q}^k = \begin{bmatrix} F_{k+1} & F_k \\ F_k & F_{k-1} \end{bmatrix}$$

Then

$$\begin{aligned} \mathbf{Q}^{k+1} = \mathbf{Q}^k\mathbf{Q}^1 &= \begin{bmatrix} F_{k+1} & F_k \\ F_k & F_{k-1} \end{bmatrix}\begin{bmatrix} 1 & 1 \\ 1 & 0 \end{bmatrix} \\ &= \begin{bmatrix} F_{k+1} + F_k & F_k \\ F_k + F_{k-1} & F_k \end{bmatrix} \\ &= \begin{bmatrix} F_{k+2} & F_{k+1} \\ F_{k+1} & F_k \end{bmatrix} \end{aligned}$$

Thus the result follows by PMI. ∎

CASSINI'S FORMULA REVISITED

Theorem 32.1 provides an alternate proof of Cassini's formula (Identity 5.4), as the next corollary shows.

Corollary 32.1. Let $n \geq 1$. Then $F_{n-1}F_{n+1} - F_n^2 = (-1)^n$.

Proof. Since $|\mathbf{Q}| = -1$, it follows, by Theorem A.25, that $|\mathbf{Q}^n| = (-1)^n$. But, by Theorem 32.1, $|\mathbf{Q}^n| = F_{n+1}F_{n-1} - F_n^2$. Thus $F_{n-1}F_{n+1} - F_n^2 = (-1)^n$. ∎

We can apply Theorem 32.1 to derive four new Fibonacci identities, as the next corollary shows. They are basically the same.

Corollary 32.2.

$$F_{m+n+1} = F_{m+1}F_{n+1} + F_mF_n \tag{32.1}$$

$$F_{m+n} = F_{m+1}F_n + F_mF_{n-1} \tag{32.2}$$

$$F_{m+n} = F_mF_{n+1} + F_{m-1}F_n \tag{32.3}$$

$$F_{m+n-1} = F_mF_n + F_{m-1}F_{n-1} \tag{32.4}$$

Proof. Since $\mathbf{Q}^m\mathbf{Q}^n = \mathbf{Q}^{m+n}$, we have

$$\begin{bmatrix} F_{m+1} & F_m \\ F_m & F_{m-1} \end{bmatrix} \begin{bmatrix} F_{n+1} & F_n \\ F_n & F_{n-1} \end{bmatrix} = \begin{bmatrix} F_{m+n+1} & F_{m+n} \\ F_{m+n} & F_{m+n-1} \end{bmatrix}$$

That is,

$$\begin{bmatrix} F_{m+1}F_{n+1} + F_mF_n & F_{m+1}F_n + F_mF_{n-1} \\ F_mF_{n+1} + F_{m-1}F_n & F_mF_n + F_{m-1}F_{n-1} \end{bmatrix} = \begin{bmatrix} F_{m+n+1} & F_{m+n} \\ F_{m+n} & F_{m+n-1} \end{bmatrix}$$

Equating the corresponding entries, the identities follow. ■

In particular, let $m = n$. Then Identity 32.1 yields the well-known formula $F_n^2 + F_{n+1}^2 = F_{2n+1}$ (Identity 5.11), and Identity 32.2 yields $F_{2n} = F_{n+1}F_n + F_nF_{n-1} = F_n(F_{n+1} + F_{n-1}) = F_nL_n$ (Identity 5.13).

Corollary 32.3.

$$F_{m+1}L_n + F_mL_{n-1} = L_{m+n} \tag{32.5}$$

Proof. Replace n with $n + 1$ in Identity (32.1); and add the resulting formula and Identity (32.2):

$$\begin{aligned} F_{m+1}F_{n+2} + F_mF_{n+1} &= F_{m+n+2} \\ F_{m+1}F_n + F_mF_{n-1} &= F_{m+n} \end{aligned}$$

We then get $F_{m+1}(F_{n+2} + F_n) + F_m(F_{n+1} + F_{n-1}) = F_{m+n+2} + F_{m+n}$. Using Identity (5.14), we get $F_{m+1}L_{n+1} + F_mL_n = L_{m+n+1}$. Changing n to $n - 1$ yields the desired result. ■

We can use Identities (32.2) and (32.3) to derive an identity that links both Fibonacci and Lucas numbers. To derive it, add the two identities:

$$F_m(F_{n-1} + F_{n+1}) + F_n(F_{m-1} + F_{m+1}) = 2F_{m+n}$$

Using Identity 5.14, this yields $F_mL_n + F_nL_m = 2F_{m+n}$. This has an analogous formula for L_{m+n} also: $2L_{m+n} = L_mL_n + 5F_mF_n$. We invite you to confirm this (see Exercise 8). Accordingly, we have the following results.

Corollary 32.4.

$$2F_{m+n} = F_mL_n + F_nL_m \tag{32.6}$$

$$2L_{m+n} = L_mL_n + 5F_mF_n \tag{32.7}$$

■

THE M-MATRIX

In lieu of the **Q**-matrix, consider the closely related *M-matrix* M, studied in 1983 by Sam Moore of the Community College of Allegheny County, Pennsylvania:

$$M = \begin{bmatrix} 1 & 1 \\ 1 & 2 \end{bmatrix}$$

We can show by PMI that

$$M^n = \begin{bmatrix} F_{2n-1} & F_{2n} \\ F_{2n} & F_{2n+1} \end{bmatrix}$$

where $n \geq 1$ (see Exercise 19). Then

$$\frac{M^n}{F_{2n-1}} = \begin{bmatrix} 1 & F_{2n}/F_{2n-1} \\ F_{2n}/F_{2n-1} & F_{2n+1}/F_{2n-1} \end{bmatrix}$$

Since $\lim_{k\to\infty} (F_k/F_{k-1}) = \alpha$, it follows that

$$\lim_{n\to\infty} \frac{M^n}{F_{2n-1}} = \begin{bmatrix} 1 & \alpha \\ \alpha & \alpha^2 \end{bmatrix} = \begin{bmatrix} 1 & \alpha \\ \alpha & 1+\alpha \end{bmatrix}$$

That is, the sequence $\{M^n/F_{2n-1}\}$ of Fibonacci matrices with leading entries 1 converges to the matrix

$$\begin{bmatrix} 1 & \alpha \\ \alpha & 1+\alpha \end{bmatrix}$$

Likewise, the sequence $\{\mathbf{Q}^n/F_{2n-1}\}$ converges to the matrix

$$\begin{bmatrix} 1+\alpha & \alpha \\ \alpha & 1 \end{bmatrix}$$

CHARACTERISTIC EQUATION

Let $A = (a_{ij})_{n\times n}$ and I the $n \times n$ identity matrix. Then the equation $|A - xI| = 0$ is the *characteristic equation* of matrix A. Its roots are the *characteristic roots* of A.

To determine the characteristic roots of $\mathbf{Q}^n$, let us first find its characteristic equation:

$$\begin{aligned} |\mathbf{Q}^n - xI| &= \begin{vmatrix} F_{n+1} - x & F_n \\ F_n & F_{n-1} - x \end{vmatrix} \\ &= (F_{n+1} - x)(F_{n-1} - x) - F_n^2 \\ &= x^2 - (F_{n+1} + F_{n-1})x + F_{n-1}F_{n+1} - F_n^2 \\ &= x^2 - L_n x + (-1)^n \end{aligned}$$

by Identities 5.4 and 5.14. Thus the characteristic equation is

$$x^2 - L_n x + (-1)^n = 0. \tag{32.8}$$

Using the quadratic formula, we arrive at the characteristic roots

$$x = \frac{L_n \pm \sqrt{L_n^2 - 4(-1)^n}}{2}$$

But $L_n^2 - 4(-1)^n = 5F_n^2$, by Exercise 39 in Chapter 5. So $x = (L_n \pm \sqrt{5}F_n)/2$. Since $\alpha^n - \beta^n = \sqrt{5}F_n$ and $\alpha^n + \beta^n = L_n$. Consequently,

$$\frac{L_n + \sqrt{5}F_n}{2} = \alpha^n \qquad \text{and} \qquad \frac{L_n - \sqrt{5}F_n}{2} = \beta^n$$

Thus we have the following result.

Theorem 32.2. The characteristic roots of $\mathbf{Q}^n$ are α^n and β^n. ■

Corollary 32.5. The characteristic roots of $\mathbf{Q}$ are α and β. ■

When $n = 1$, Eq. (32.8) becomes $x^2 - x - 1 = 0$, which is the characteristic equation of $\mathbf{Q}$. But notice that $\mathbf{Q}^2 - \mathbf{Q} - I = O$ (see Exercise 2). Thus $\mathbf{Q}$ satisfies its characteristic equation, illustrating the well-known *Cayley–Hamilton Theorem*, which states that *every square matrix satisfies its characteristic equation*.

In 1963, I. D. Ruggles and V. E. Hoggatt, Jr., established Identity (5.1) using the $\mathbf{Q}$-matrix. To see this, we can use PMI to establish that

$$(I + \mathbf{Q} + \mathbf{Q}^2 + \cdots + \mathbf{Q}^N)(\mathbf{Q} - I) = \mathbf{Q}^{N+1} - I \tag{32.9}$$

(see Exercise 3). Since $|\mathbf{Q} - I| = -1 \neq 0$, $\mathbf{Q} - I$ is invertible. Since $\mathbf{Q}^2 = \mathbf{Q} + I$, $\mathbf{Q}^2 - \mathbf{Q} = I$; that is, $\mathbf{Q}(\mathbf{Q} - I) = I$. Thus $(\mathbf{Q} - I)^{-1} = \mathbf{Q}$. Now multiply both sides of Eq. (32.9) by $(\mathbf{Q} - I)^{-1}$:

$$\begin{aligned} I + \mathbf{Q} + \mathbf{Q}^2 + \cdots + \mathbf{Q}^N &= (\mathbf{Q}^{N+1} - I)\mathbf{Q} \\ &= \mathbf{Q}^{n+2} - \mathbf{Q} \end{aligned}$$

Equating the upper right-hand elements in this matrix equation yields the desired formula, $F_1 + F_2 + F_3 + \cdots + F_n = F_{n+2} - 1$.

R-MATRIX

Consider the **R**-*matrix*, which corresponds to the **Q**-matrix:

$$\mathbf{R} = \begin{bmatrix} 1 & 2 \\ 2 & -1 \end{bmatrix}$$

The **R**-matrix was introduced by Hoggatt and Ruggles in 1963. Recall that $L_{n+1} = F_{n+1} + 2F_n$, $L_n = 2F_{n+1} - F_n$, $5F_{n+1} = L_{n+1} + 2L_n$, and $5F_n = 2L_{n+1} - L_n$. Using these formulas, we have

$$\mathbf{RQ}^n = \begin{bmatrix} 1 & 2 \\ 2 & -1 \end{bmatrix} \begin{bmatrix} F_{n+1} & F_n \\ F_n & F_{n-1} \end{bmatrix} = \begin{bmatrix} L_{n+1} & L_n \\ L_n & L_{n-1} \end{bmatrix}$$

Using Theorem A.25, this implies

$$\begin{vmatrix} 1 & 2 \\ 2 & -1 \end{vmatrix} \begin{vmatrix} F_{n+1} & F_n \\ F_n & F_{n-1} \end{vmatrix} = \begin{vmatrix} L_{n+1} & L_n \\ L_n & L_{n-1} \end{vmatrix}$$

That is, $L_{n+1}L_{n-1} - L_n^2 = (-5)(F_{n+1}F_{n-1} - F_n^2) = 5(-1)^{n+1}$, by Identity (5.4). Thus

$$L_{n+1}L_{n-1} - L_n^2 = 5(-1)^{n+1} \tag{32.10}$$

See Exercise 38 in Chapter 5.

CASSINI'S FORMULA AND CRAMER'S RULE

Next we show how we can employ Cramer's* rule for 2×2 linear systems to derive Cassini's formula. We first review the rule.

The 2×2 linear system

$$\begin{aligned} ax + by &= e \\ cx + dy &= f \end{aligned}$$

has a unique solution if and only if $ad - bc \neq 0$. It is given by

$$x = \frac{\begin{vmatrix} e & b \\ f & e \end{vmatrix}}{\begin{vmatrix} a & b \\ c & d \end{vmatrix}} \qquad y = \frac{\begin{vmatrix} a & e \\ c & f \end{vmatrix}}{\begin{vmatrix} a & b \\ c & d \end{vmatrix}}$$

In particular, consider the system:

$$\begin{aligned} F_n x + F_{n-1} y &= F_{n+1} \\ F_{n+1} x + F_n y &= F_{n+2} \end{aligned}$$

Since $(F_k, F_{k+1}) = 1$, by virtue of the Fibonacci recurrence relation (FRR), $x = 1 = y$ is the unique solution of this system. Therefore, by Cramer's rule,

$$y = \frac{\begin{vmatrix} F_n & F_{n+1} \\ F_{n+1} & F_{n+2} \end{vmatrix}}{\begin{vmatrix} F_n & F_{n-1} \\ F_{n+1} & F_n \end{vmatrix}} = 1$$

Thus, $F_nF_{n+2} - F_{n+1}^2 = F_n^2 - F_{n-1}F_{n+1}$. That is, $F_nF_{n+2} - F_{n+1}^2 = -(F_{n-1}F_{n+1} - F_n^2)$.

*Named after the Swiss mathematician Gabriel Cramer (1704–1752).

Let $p_n = F_{n-1}F_{n+1} - F_n^2$. Then this equation yields the recurrence relation $p_n = -p_{n-1}$, where $p_1 = F_0F_2 - F_1^2 = -1$. Solving this recurrence relation, we get $p_n = (-1)^n$ (see Exercise 22). Thus, $F_{n-1}F_{n+1} - F_n^2 = (-1)^n$, where $n \geq 1$.

Now we turn to vectors formed by adjacent Fibonacci and Lucas numbers.

FIBONACCI AND LUCAS VECTORS*

Consider the vectors $\mathbf{U}_n = (F_{n+1}, F_n)$ and $\mathbf{V}_n = (L_{n+1}, L_n)$. Their magnitudes are given by

$$|\mathbf{U}_n|^2 = F_{n+1}^2 + F_n^2 = F_{2n+1}$$

and

$$\begin{aligned} |\mathbf{V}_n|^2 &= L_{n+1}^2 + L_n^2 = [5F_{n+1}^2 + 4(-1)^{n+1}] + [5F_n^2 + 4(-1)^n] \\ &= 5(F_n^2 + F_{n+1}^2) = 5F_{2n+1} \end{aligned}$$

Their directions are given by

$$\tan\theta = \frac{F_n}{F_{n+1}} \qquad \text{and} \qquad \tan\theta' = \frac{L_n}{L_{n+1}}$$

We shall show later that

$$\frac{F_n}{F_{n+1}} \approx \frac{\sqrt{5}-1}{2} = \frac{1}{\alpha} \qquad \text{and} \qquad \frac{L_n}{L_{n+1}} \approx \frac{\sqrt{5}-1}{2} = \frac{1}{\alpha}$$

You may notice that

$$\mathbf{U}_0\mathbf{Q}^{n+1} = (1, 0)\begin{bmatrix} F_{n+2} & F_{n+1} \\ F_{n+2} & F_{n+1} \end{bmatrix} = (F_{n+2}, F_{n+1}) = \mathbf{U}_{n+1} = \mathbf{U}_n\mathbf{Q}$$

Likewise, $\mathbf{V}_0\mathbf{Q}^{n+1} = \mathbf{V}_{n+1} = \mathbf{V}_n\mathbf{Q}$. Besides,

$$\begin{aligned} \mathbf{U}_m\mathbf{Q}^n &= (F_{m+1}, F_m)\begin{bmatrix} F_{n+2} & F_{n+1} \\ F_{n+1} & F_n \end{bmatrix} \\ &= \begin{bmatrix} F_{m+1}F_{n+2} & + & F_mF_{n+1} \\ F_{m+1}F_{n+1} & + & F_mF_n \end{bmatrix} = (F_{m+n+1}, F_{m+n}) \\ &= \mathbf{U}_{m+n+1} \end{aligned}$$

by Identity (32.1). Likewise, $\mathbf{V}_m\mathbf{Q}^n = \mathbf{V}_{m+n+1}$ (see Exercise 29).

*Throughout this chapter, the ordered pair (x, y) denotes a vector and *not* the greatest common divisor (gcd) of x and y.

Let us now return to the **R**-matrix:

$$\mathbf{R} = \begin{bmatrix} 1 & 2 \\ 2 & -1 \end{bmatrix}$$

Notice that $|\mathbf{R}| = -5 \neq 0$, so **R** is invertible and

$$\mathbf{R}^{-1} = \frac{1}{5}\begin{bmatrix} 1 & 2 \\ 2 & -1 \end{bmatrix}$$

We have

$$\begin{aligned}
\mathbf{V}_n\mathbf{R} &= (L_{n+1}, L_n)\begin{bmatrix} 1 & 2 \\ 2 & -1 \end{bmatrix} \\
&= (L_{n+1} + 2L_n, 2L_{n+1} - L_n) = (5F_{n+1}, 5F_n) \\
&= 5\mathbf{U}_n \\
\therefore \quad \mathbf{V}_n &= (5\mathbf{U}_n)\mathbf{R}^{-1} = 5(\mathbf{U}_n\mathbf{R}^{-1}) \\
&= 5 \cdot \frac{1}{5}(F_{n+1}, F_n)\begin{bmatrix} 1 & 2 \\ 2 & -1 \end{bmatrix} = (F_{n+1} + 2F_n, 2F_{n+1} - F_n) \\
&= (L_{n+1}, L_n)
\end{aligned}$$

as expected. Likewise, $\mathbf{U}_n\mathbf{R} = \mathbf{V}_n$ and $\mathbf{U}_n = \mathbf{R}^{-1}\mathbf{V}_n$.

What is the effect of **R** on any nonzero vector $\mathbf{U} = (x, y)$? To see this, observe that:

$$\begin{aligned}
\mathbf{UR} &= (x, y)\begin{bmatrix} 1 & 2 \\ 2 & -1 \end{bmatrix} = (x + 2y, 2x - y) \\
\therefore \quad |\mathbf{UR}|^2 &= (x + 2y)^2 + (2x - y)^2 \\
&= 5(x^2 + y^2) = 5|\mathbf{U}|^2
\end{aligned}$$

Thus, the Fibonacci matrix **R** magnifies every nonzero vector by a factor of $\sqrt{5}$.

To find the effect of **R** on the slope of **U**, suppose that the acute angles made with the x-axis by the directions of the vectors **U** and **UR** are θ and θ', respectively. Then $\tan\theta = y/x$ and $\tan\theta' = (2x - y)/(x + 2y)$.

$$\begin{aligned}
\therefore \quad \tan(\theta + \theta') &= \frac{\tan\theta + \tan\theta'}{1 - \tan\theta\tan\theta'} \\
&= \frac{(y)/(x) + (2x - y)/(x + 2y)}{1 - (y/x)\cdot(2x - y)/(x + 2y)} = \frac{y(x + 2y) + x(2x - y)}{x(x + 2y) - y(2x - y)} \\
&= \frac{2(x^2 + y^2)}{x^2 + y^2} = 2
\end{aligned}$$

(Note that $\mathbf{U} \neq 0$.)

Let 2γ be the angle between the vectors **U** and **UR**. Then $\theta + \gamma = \theta' - \gamma$, so $2\gamma = \theta + \theta'$. Since

$$\tan 2\gamma = \frac{2\tan\gamma}{1-\tan^2\gamma}$$

it follows that

$$\frac{2\tan\gamma}{1-\tan^2\gamma} = 2$$

that is, $\tan^2\gamma + \tan\gamma - 1 = 0$.

$$\therefore \quad \tan\gamma = \frac{-1\pm\sqrt{5}}{2}$$

Since $2\gamma = \tan^{-1} 2 \approx 63.43^\circ$, $\gamma \approx 31.7175^\circ$; so, we choose $\tan\gamma = (\sqrt{5}-1)/2 = -\beta$, which is the negative of an eigenvalue for **Q**. Thus the vector that bisects the angle between the vectors **U** and **UR** has slope $-\beta$; it is a vector of the form $\mathbf{W} = (\alpha x, x)$.

Consequently, we have the following result.

Theorem 32.3. (Hoggatt and Ruggles) The **R**-matrix transforms the nonzero vector $\mathbf{U} = (x, y)$ into a vector **UR** such that $|\mathbf{UR}| = \sqrt{5}|\mathbf{U}|$ and the bisector of the angle between them is the vector of the form $(\alpha x, x)$ with slope $-\beta$. ∎

Corollary 32.6. The **R**-matrix maps the vector $\mathbf{U}_n$ into $\mathbf{V}_n$ and $\mathbf{V}_n$ into $\sqrt{5}\mathbf{U}_n$. ∎

AN INTRIGUING FIBONACCI MATRIX

In 1996, David M. Bloom of Brooklyn College, New York, proposed the following problem in *Math Horizons*:
Determine the sum

$$\sum_{\substack{i,j,k>0 \\ i+j+k=n}} F_i F_j F_k$$

The solution provided by C. Libis of the University of West Alabama in the February 1997 issue involved an intriguing, infinite-dimensional Fibonacci matrix:

$$H = \begin{bmatrix} H_{0,n} \\ H_{1,n} \\ \vdots \\ H_{m,n} \\ \vdots \end{bmatrix}$$

where each element $h_{i,j}$ is defined recursively as follows:

$$
\begin{aligned}
h_{0,j} &= 0 && \text{if } j \geq 0 \\
h_{j,j} &= 1 && \text{if } j \geq 1 \\
h_{i,j} &= 0 && \text{if } i > j \\
h_{i,j} &= h_{i,j-2} + h_{i,j-1} + h_{i-1,j-1} && \text{if } i \geq 1 \text{ and } j \geq 2
\end{aligned}
\tag{32.11}
$$

As an example:

$$
\begin{aligned}
h_{2,5} &= h_{2,3} + h_{2,4} + h_{1,4} \\
&= (h_{2,1} + h_{2,2} + h_{1,2}) + (h_{2,2} + h_{2,3} + h_{1,3}) + F_4 \\
&= (0 + 1 + F_2) + [1 + (h_{2,1} + h_{2,2} + h_{1,2}) + F_3] + F_4
\end{aligned}
$$

By the recurrence relation (Eq. 32.11), it follows that

$$h_{1,j} = h_{1,j-2} + h_{1,j-1} + h_{0,j-1} = h_{1,j-2} + h_{1,j-1}$$

where $h_{1,0} = 0$ and $h_{1,1} = 1$. Consequently, $h_{1,n} = F_n$. Thus

$$
\begin{aligned}
h_{2,5} &= 2 + [1 + (0 + 1 + F_2) + 2] + 3 \\
&= 10
\end{aligned}
$$

The condition, $h_{0,j} = 0$ for every $j \geq 0$ implies that the top row of matrix H consists of zeros; $h_{i,i} = 1$ means, every element on the main diagonal is 1; and $h_{i,j} = 0$ for $i > j$ means the matrix H is upper triangular; that is, every element below the main diagonal is zero. The recurrence relation (32.11) implies that we can obtain every element $h_{i,j}$ by adding the two previous elements $h_{i,j-2}$ and $h_{i,j-1}$ in the same row, and the element $h_{i-1,j-1}$, which lies just above $h_{i,j-1}$, where $i \geq 1$ and $j \geq 2$.

Using these straightforward observations, we can determine the various elements of H. Thus:

$H =$

i \ j	0	1	2	3	4	5	6	7	8	...		
0	0	0	0	0	0	0	0	0	0	0	0	
1	0	1	1	2	3	5	8	13	21	34	55	$\leftarrow F_n$
2	0	0	1	2	5	10 ↑	↘ 20	38	71	130	235	
3	0	0	0	1	3 →	9	22	51	111	233	474	
4	0	0	0	0	1	4	14	40	105	255	593	
5	0	0	0	0	0	1	15	56	176	487	918	
							⋮					

Notice that $h_{3,6} = 22 = 3 + 9 + 10 = h_{3,4} + h_{3,5} + h_{2,4}$. See the arrows in the array. Notice also that:

$$\begin{aligned} h_{2,7} &= 38 \\ &= 1 \cdot 8 + 1 \cdot 5 + 2 \cdot 3 + 3 \cdot 2 + 5 \cdot 1 + 8 \cdot 1 \\ &= F_1 h_{1,6} + F_2 h_{1,5} + F_3 h_{1,4} + F_4 h_{1,3} + F_5 h_{1,2} + F_6 h_{1,1} \\ &= \sum_{j=1}^{7} F_j h_{1,7-j} = \sum_{j=1}^{7} F_j F_{7-j} \\ &= \sum_{\substack{j,k \geq 1 \\ j+k=7}} F_j F_k \end{aligned}$$

More generally, we have the following result.

Theorem 32.4.

$$h_{2,n} = \sum_{\substack{j,k \geq 1 \\ j+k=n}} F_j F_k$$

Proof. (by PMI) When $n = 1$, the left-hand side (LHS)$= h_{2,1} = 0 = \sum_{\substack{j,k \geq 1 \\ j+k=1}} F_j F_k =$ RHS. Thus the result is true when $n = 1$.

Now assume it is true for all positive integers $\leq m$, where $m \geq 2$:

$$h_{2,m} = \sum_{\substack{j,k \geq 1 \\ j+k=m}} F_j F_k$$

Then

$$\begin{aligned} \sum_{\substack{j,k \geq 1 \\ j+k=m+1}} F_j F_k &= \sum_{j=1}^{m} F_j F_{m+1-j} \\ &= \sum_{j=1}^{m} F_j (F_{m-j} + F_{m-j-1}) \\ &= \sum_{j=1}^{m} F_j F_{m-j} + \sum_{j=1}^{m} F_j F_{m-1-j} \\ &= \sum_{j=1}^{m} F_j F_{m-j} + \sum_{j=1}^{m-1} F_j F_{m-1-j} + F_m F_{-1} \end{aligned}$$

$$\begin{aligned} &= h_{2,m} + h_{2,m-1} + F_m \qquad \text{by the IH} \\ &= h_{2,m} + h_{2,m-1} + h_{1,m} \\ &= h_{2,m+1} \end{aligned}$$

by the recurrence relation (32.11). Thus, by the strong version of PMI, the formula is true for $n \geq 1$. ■

Since $h_{1,k} = F_k$, this theorem yields the following result.

Corollary 32.7.

$$h_{2,n} = \sum_{i=1}^{n} F_i h_{1,n-i}$$ ■

That is, we can obtain every element $h_{2,n}$ by multiplying the elements $h_{1,n-1}$, $h_{1,n-2}, \ldots, h_{1,1}$ with weights $F_1, F_2, \ldots, F_{n-1}$, respectively, and then by adding up the products, as we observed earlier. (Recall that $h_{1,0} = 0$.)

Corresponding to this corollary, we have a similar result for row 3 of matrix H also. It can also be established by PMI, so we omit its proof.

Theorem 32.5.

$$h_{3,n} = \sum_{i=1}^{n} F_i h_{2,n-i}$$ ■

By this theorem, we can obtain every element $h_{3,n}$ by multiplying the elements $h_{2,n-1}, h_{2,n-2}, \ldots, h_{2,1}$ with weights $F_1, F_2, \ldots, F_{n-1}$ in that order, and then by summing them up. For example,

$$\begin{aligned} h_{3,n} &= \sum_{1}^{7} F_i h_{2,7-i} \\ &= F_1 h_{2,6} + F_2 h_{2,5} + F_3 h_{2,4} + F_4 h_{2,3} + F_5 h_{2,2} + F_6 h_{2,1} \\ &= 1 \cdot 20 + 1 \cdot 10 + 2 \cdot 5 + 3 \cdot 2 + 5 \cdot 1 + 8 \cdot 0 \\ &= 51 \end{aligned}$$

The next corollary provides the answer to the problem proposed earlier.

Corollary 32.8.

$$h_{3,n} = \sum_{\substack{i,j,k \geq 1 \\ i+j+k=n}} F_i F_j F_k$$ ■

Proof. By Theorem 32.5,

$$\begin{aligned} h_{3,n} &= \sum_{i=1}^{n} F_i h_{2,n-i} \\ &= \sum_{i=1}^{n} F_i \Big(\sum_{\substack{j,k \geq 1 \\ j+k=n-i}} F_j F_k \Big) \\ &= \sum_{\substack{i,j,k \geq 1 \\ i+j+k=n}} F_i F_j F_k \end{aligned}$$

■

For example:

$$\begin{aligned} h_{3,5} &= \sum_{\substack{i,j,k \geq 1 \\ i+j+k=5}} F_i F_j F_k \\ &= F_1 \sum_{\substack{j,k \geq 1 \\ j+k=4}} F_j F_k + F_2 \sum_{\substack{j,k \geq 1 \\ j+k=3}} F_j F_k + F_3 \sum_{\substack{j,k \geq 1 \\ j+k=2}} F_j F_k \\ &= F_1(F_1 F_3 + F_2 F_2 + F_3 F_1) + F_2(F_1 F_2 + F_2 F_1) + F_3(F_1 F_1) \\ &= 1(1 \cdot 2 + 1 \cdot 1 + 2 \cdot 1) + 1(1 \cdot 1 + 1 \cdot 1) + 2(1 \cdot 1) \\ &= 9 \end{aligned}$$

as expected. In fact, we can generalize Corollaries 32.7 and 32.8 as follows.

Theorem 32.6. (Libis, 1997)

$$h_{m,n} = \sum_{i=1}^{n} F_i h_{m-1,n-i} \qquad m \geq 2$$

■

EXPLICIT FORMULAS FOR $h_{2,n}$ AND $h_{3,n}$

In the same issue of *Math Horizons* (1997), the editor, M. Klamkin of the University of Alberta, Canada, presented explicit formulas for $h_{2,n}$ and $h_{3,n}$ by introducing an operator E:

$$E h_{m,n} = h_{m,n+1}$$

Then

$$\begin{aligned} E^2 h_{m,n} &= E h_{m,n+1} = h_{m,n+2} \\ \therefore \quad (E^2 - E - 1) h_{m,n} &= h_{m,n+2} - h_{m,n+1} - h_{m,n} \\ &= h_{m-1,n+1} \qquad \text{by Eq. (32.11)} \end{aligned}$$

Consequently, $(E^2 - E - 1)^m h_{m,n} = 0$.

It now follows that $h_{2,n}$ must be of the form

$$h_{2,n} = (an + b)F_n + (cn + d)F_{n-1} \tag{32.12}$$

and $h_{3,n}$ must be of the form

$$h_{3,n} = (an^2 + bn + c)F_n + (dn^2 + en + f)F_{n-1}$$

where a, b, c, d, e, and f are constants to be determined.

Since $h_{2,1} = 0, h_{2,2} = 1, h_{2,3} = 2$, and $h_{2,4} = 5$, Eq. (32.12) yields the linear system:

$$\begin{aligned} a + b &= 0 & 2a + b + 2c + d &= 1 \\ 6a + 2b + 3c + d &= 2 & 12a + 3b + 8c + 2d &= 5 \end{aligned}$$

Solving this system, we get $a = 1/5 = -b, c = 2/5$, and $d = 0$. Thus:

$$h_{2,n} = \frac{(n-1)F_n + 2nF_{n-1}}{5} \tag{32.13}$$

For example,

$$h_{2,7} = \frac{6F_7 + 14F_6}{5} = \frac{6 \cdot 13 + 14 \cdot 8}{5} = 38$$

Likewise, it would be a good exercise to verify that

$$h_{3,n} = \frac{(5n^2 - 3n - 2)F_n - 6nF_{n-1}}{50} \tag{32.14}$$

For example,

$$\begin{aligned} h_{3,5} &= \frac{(5 \cdot 5^2 - 3 \cdot 5 - 2)F_5 - 6 \cdot 5F_4}{50} = \frac{108 \cdot 5 - 30 \cdot 3}{50} \\ &= 9 \end{aligned}$$

Since $h_{2,n}$ and $h_{3,n}$ are integers, it follows that $(n-1)F_n + 2nF_{n-1} \equiv 0 \pmod 5$ and $(5n^2 - 3n - 2)F_n \equiv 6nF_{n-1} \pmod{50}$.

AN INFINITE-DIMENSIONAL LUCAS MATRIX

A similar study of Lucas numbers L_n yields some interesting and rewarding dividends. To see this, consider the infinite-dimensional matrix $K = (k_{i,j})$, where we define recursively each element $k_{i,j}$ as follows and $i, j \geq 0$:

(1) $k_{0,j} = 0$

(2) $k_{1,1} = 1$

(3) $k_{j,j-1} = 2, j \geq 1$

(4) $k_{i,j} = 0$ if $j < i - 1$.

(5) $k_{i,j} = k_{i,j-2} + k_{i,j-1} + k_{i-1,j-1}, i \geq 1$ and $j \geq 2$.

Condition (1) implies that row 0 consists of zeros; conditions (2) and (3) imply the first two elements in row 1 are 2 and 1; by condition (3), the diagonal below the main diagonal consists of 2s; and by condition (4), every element below this diagonal is zero. We can now employ condition (5) to compute the remaining elements of K: add the two previous elements $k_{i,j-2}$ and $k_{i,j-1}$, and then add the element $k_{i-1,j-1}$ just above $k_{i,j-1}$:

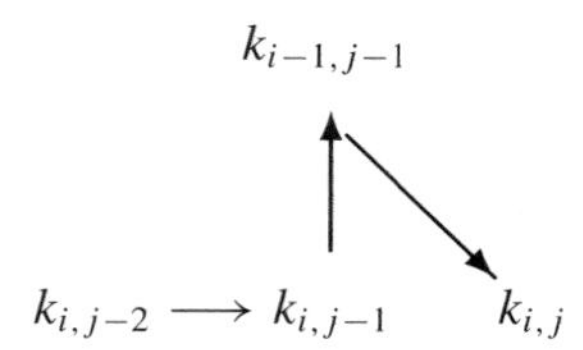

Thus

$K =$

i \ j	0	1	2	3	4	5	6	7	8	...	
0	0	0	0	0	0	0	0	0	0		
1	2	1	3	4	7	11	18	29	47		← Lucas numbers
2	0	2	3	8 →	15	30	56	104	189		
3	0	0	2	5	15	35	80	171	355		
4	0	0	0	2	7	24	66	170	407		
5	0	0	0	0	2	9	35	110	315		
6	0	0	0	0	0	2	11	48	169		
⋮											

All twos

Using the recursive formula,

$$\begin{aligned} k_{1,n} &= k_{1,n-2} + k_{1,n-1} + k_{0,n-2} \\ &= k_{1,n-2} + k_{1,n-1} + 0 \\ &= k_{1,n-2} + k_{1,n-1} \end{aligned}$$

where $k_{1,0} = 2$ and $k_{1,1} = 1$. Thus $k_{1,n} = L_n$, so row 1 consists entirely of Lucas numbers.

Here is an interesting observation:

$$\begin{aligned}
k_{2,7} &= 104 \\
&= 1 \cdot 18 + 1 \cdot 11 + 2 \cdot 7 + 3 \cdot 4 + 5 \cdot 3 + 8 \cdot 1 + 13 \cdot 2 \\
&= \sum_{1}^{7} F_j k_{1,7-j} = \sum_{\substack{j,k \geq 0 \\ j+k=7}} F_j L_k
\end{aligned}$$

More generally, we have the following result, which we prove by strong induction.

Theorem 32.7. (Koshy, 1999)

$$k_{2,n} = \sum_{\substack{j,k \geq 0 \\ j+k=n}} F_j L_k \tag{32.15}$$

Proof. When $n = 0$, each side equals 0, so the result is true.
Now assume it is true for every nonnegative integer $\leq m$:

$$k_{2,m} = \sum_{\substack{j,k \geq 0 \\ j+k=m}} F_j L_k$$

Then

$$\begin{aligned}
\sum_{\substack{j,k \geq 0 \\ j+k=m+1}} F_j L_k &= \sum_{j=0}^{m+1} F_j L_{m+1-j} \\
&= \sum_{0}^{m} F_j (L_{m-j} + L_{m-j-1}) + F_{m+1} L_0 \\
&= \sum_{0}^{m} F_j L_{m-j} + \sum_{0}^{m} F_j L_{m-j-1} + F_{m+1} L_0 \\
&= \sum_{0}^{m} F_j L_{m-j} + \sum_{0}^{m-1} F_j L_{m-j-1} + F_m L_{-1} + F_{m+1} L_0 \\
&= k_{2,m} + k_{2,m-1} + 2F_{m+1} - F_m \\
&= k_{2,m} + k_{2,m-1} + k_{1,m} \qquad \text{since } 2F_{m+1} - F_m = L_m. \\
&= k_{2,m+1}
\end{aligned}$$

by the recurrence relation. Thus, by strong induction, the formula holds for all $n \geq 0$.

■

Since $k_{1,n} = L_n$, we can rewrite Formula (32.15) as

$$k_{2,n} = \sum_{\substack{j,t \geq 0 \\ j+t=n}} F_j k_{1,t} \tag{32.16}$$

In words, every element $k_{2,n}$ can be obtained by multiplying the elements $k_{1,n-1}$, $k_{1,n-2}, \ldots, k_{1,1}$ in the previous row with weights $F_1, F_2, \ldots, F_{n-1}$, respectively, and then by adding up the products, as we observed earlier.

As in Theorem 32.7, we can prove that

$$k_{3,n} = \sum_{0}^{n} F_i k_{2,n-i} = \sum_{\substack{i,j,k \geq 0 \\ i+j+k=n}} F_i F_j L_k \tag{32.17}$$

For example,

$$\begin{aligned} k_{3,5} &= F_0 k_{2,5} + F_1 k_{2,4} + F_2 k_{2,3} + F_3 k_{2,2} + F_4 k_{2,1} + F_5 k_{2,0} \\ &= 0 \cdot 30 + 1 \cdot 15 + 1 \cdot 8 + 2 \cdot 3 + 3 \cdot 2 + 5 \cdot 0 = 35 \end{aligned}$$

Formulas (32.15) and (32.17) are in fact special cases of the following result, which we can establish also using strong induction.

Theorem 32.8. (Koshy, 1999)

$$k_{m,n} = \sum_{i=0}^{n} F_i k_{m-1,n-i} \qquad m \geq 2 \tag{32.18}$$

Proof. Assuming that the result is true for all m, we shall first prove that it is true for all $n \geq 0$. Since $k_{m,0} = 0 = \sum_{0}^{0} F_i k_{m-1,-i}$, the result is true when $n = 0$.

When $n = 1$, LHS $= k_{m,1}$ and RHS $= \sum_{0}^{1} F_i k_{m-1,1-i} = F_0 k_{m-1,1} + F_1 k_{m-1,0} = 0 + k_{m-1,0} = k_{m-1,0}$. Since $k_{2,1} = 2 = k_{1,0}$ and $k_{i,1} = 0 = k_{i-1,0}$ for $i > 2$, it follows that $k_{m,1} = k_{m-1,0}$ for $m \geq 2$.

Now assume the result is true for all integers t, where $t \geq 2$:

$$k_{m,t} = \sum_{i=0}^{t} F_i k_{m-1,t-i}$$

Then:

$$\begin{aligned} \sum_{0}^{t+1} F_i k_{m-1,t+1-i} &= \sum_{0}^{t+1} F_i (k_{m-1,t-i-1} + k_{m-1,t-i} + k_{m-2,t-i}) \\ &= \sum_{0}^{t+1} F_i k_{m-1,t-i-1} + \sum_{0}^{t+1} F_i k_{m-1,t-i} + \sum_{0}^{t+1} F_i k_{m-2,t-i} \end{aligned}$$

$$
\begin{aligned}
&= \sum_{0}^{t-1} F_i k_{m-1,t-i-1} + \sum_{0}^{t} F_i k_{m-1,t-i} + \sum_{0}^{t} F_i k_{m-2,t-i} \\
&= k_{m,t-1} + k_{m,t} + k_{m-1,t} \\
&= k_{m,t+1}
\end{aligned}
$$

Thus the result is true for all $n \geq 0$.

On the other hand, assume that Eq. (32.18) is true for all $n \geq 0$. We shall prove that it is true for all $m \geq 2$. It is true for $m = 2$ by Eq. (32.15), and for $m = 3$ by Eq. (32.17). Assume it is true for all integers $\leq t$, where $t \geq 2$:

$$k_{t,n} = \sum_{0}^{n} F_i k_{t-1,n-i}$$

Then

$$
\begin{aligned}
\sum_{0}^{n} F_i k_{t,n-i} &= \sum_{0}^{n} F_i (k_{t,n-i-2} + k_{t,n-1-i} + k_{t-1,n-1-i}) \\
&= \sum_{0}^{n} F_i k_{t,n-i-2} + \sum_{0}^{n} F_i k_{t,n-1-i} + \sum_{0}^{n} F_i k_{t-1,n-1-i} \\
&= \sum_{0}^{n-2} F_i k_{t,n-i-2} + \sum_{0}^{n-1} F_i k_{t,n-1-i} + \sum_{0}^{n-1} F_i k_{t-1,n-1-i} \\
&= k_{t+1,n-2} + k_{t+1,n-1} + k_{t,n-1} \\
&= k_{t+1,n}
\end{aligned}
$$

Thus the result is true for all $m \geq 2$ also. ■

For example,

$$
\begin{aligned}
k_{4,5} &= \sum_{0}^{5} F_i k_{3,5-i} = \sum_{1}^{4} F_i k_{3,5-i} \\
&= F_1 k_{3,4} + F_2 k_{3,3} + F_3 k_{3,2} + F_4 k_{3,1} = 15 + 5 + 4 + 0 = 24
\end{aligned}
$$

EXPLICIT FORMULAS FOR $k_{2,n}$ AND $k_{3,n}$

Row 2 of matrix K contains an intriguing pattern:

$$
\begin{aligned}
k_{2,0} &= 0 = 1 \cdot 0 \\
k_{2,1} &= 2 = 2 \cdot 1 \\
k_{2,2} &= 3 = 3 \cdot 1
\end{aligned}
$$

$$\begin{aligned} k_{2,3} &= 8 = 4 \cdot 2 \\ k_{2,4} &= 15 = 5 \cdot 3 \\ &\vdots \quad \overset{\uparrow}{\text{Fibonacci numbers}} \end{aligned}$$

So we conjecture that $k_{2,n} = (n+1)F_n$. The following theorem in fact confirms it using strong induction.

Theorem 32.9. (Koshy, 1999)

$$k_{2,n} = (n+1)F_n \tag{32.19}$$

Proof. Since $k_{2,0} = 0 = (0+1)F_0$, the result is true for $n = 0$. Now assume it is true for all integers $\leq t$, where $t \geq 0$. Then

$$\begin{aligned} (t+2)F_{t+1} &= (t+2)(F_t + F_{t-1}) \\ &= tF_{t-1} + (t+1)F_t + F_t + 2F_{t-1} \end{aligned}$$

But $F_t + 2F_{t-1} = F_{t+1} + F_{t-1} = L_t$;

$$\begin{aligned} \therefore \quad (t+2)F_{t+1} &= tF_{t-1} + (t+1)F_t + L_t \\ &= k_{2,t-1} + k_{2,t} + k_{1,t} \\ &= k_{2,t+1} \end{aligned}$$

Thus, by strong induction, Formula (32.19) is true for all $n \geq 0$. ■

We can also establish Formula (32.19) by assuming that $k_{2,n}$ is of the form $(an+b)F_n + (cn+d)F_{n-1}$. Thus

$$k_{2,n} = (n+1)F_n = \sum_{0}^{n} F_j L_{n-j} \tag{32.20}$$

Again, $k_{3,n}$ must be of the form $(an^2+bn+c)F_n + (dn^2+en+f)F_{n-1}$. Using the initial values of $k_{3,0}$ through $k_{3,5}$, we see that $a = 1/10 = b$, $c = -1/5 = -d$, $e = 2/5$, and $f = 0$. This yields

$$k_{3,n} = \frac{(n^2+n-2)F_n + 2n(n+2)F_{n-1}}{10}$$

Since $F_n + 2F_{n-1} = L_n$, we can rewrite this as

$$k_{3,n} = \frac{(n+2)(nL_n - F_n)}{10} \tag{32.21}$$

For example,

$$k_{3,7} = \frac{9(7L_7 - F_7)}{10} = \frac{9(7 \cdot 29 - 13)}{10} = 171$$

as expected. It follows from Eq. (32.21) that $(n+2)(nL_n - F_n) \equiv 0 \pmod{10}$.

More generally, suppose we construct a new matrix G that also satisfies conditions (1), (4), and (5), and two new conditions:

$$(2')\ G_{1,1} = a$$

$$(3')\ G_{1,2} = b$$

where a and b are arbitrary integers, and $G_{1,0} = b - a$. Then

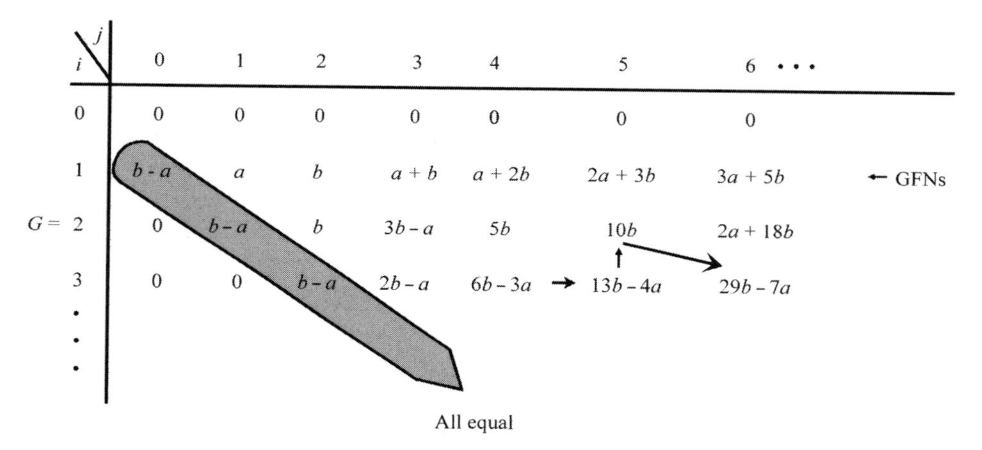

Row 1 of G consists of the generalized Fibonacci numbers (GFNs) G_n; when $a = 1 = b$, $G_n = F_n$; and when $a = 1$ and $b = 3$, $G_n = L_n$.

We can extend Formula (32.18) to G, as the next theorem shows. Its proof follows the same lines as in Theorem 32.8, so we skip it.

Theorem 32.10. (Koshy, 1999)

$$G_{m,n} = \sum_{i=0}^{n} F_i G_{m-1,n-i} \qquad m \geq 2 \tag{32.22}$$

■

For example,

$$\begin{aligned} G_{3,5} &= \sum_{0}^{5} F_i G_{2,5-i} = \sum_{1}^{4} F_i G_{2,5-i} \\ &= F_1 G_{2,4} + F_2 G_{2,3} + F_3 G_{2,2} + F_4 G_{2,1} \\ &= 5b + (3b - a) + 2b + 3(b - a) = 13b - 4a \end{aligned}$$

In particular, when $m = 2$ and $m = 3$, Formula (32.22) yields

$$G_{2,n} = \sum_{0}^{n} F_i G_{1,n-i} = \sum_{0}^{n} F_i G_{n-i}$$

and

$$G_{3,n} = \sum_{0}^{n} F_i G_{2,n-i} = \sum_{\substack{i,j,k \ge 0 \\ i+j+k=n}} F_j F_j G_k$$

THE LAMBDA FUNCTION

The *lambda function* λ of a matrix was studied extensively by Fenton S. Stancliff, a professional musician. We can use the lambda function coupled with Fibonacci matrices to derive a host of new Fibonacci identities.

Let $A = (a_{ij})_{n\times n}$. Let $A^* = (a_{ij}+1)_{n\times n}$. Thus A^* denotes the matrix obtained by adding 1 to every element of A. Then $\lambda(A) = |A^*| - |A|$, the change in the value of the determinant. For example, let

$$A = \begin{bmatrix} a & b \\ c & d \end{bmatrix}$$

Then

$$A^* = \begin{bmatrix} a+1 & b+1 \\ c+1 & d+1 \end{bmatrix}$$

$$\begin{aligned} |A^*| &= \begin{vmatrix} a+1 & b+1 \\ c+1 & d+1 \end{vmatrix} \\ &= (a+1)(d+1) - (b+1))(c+1) \\ &= (ad-bc) + (a+d-b-c) \\ \therefore \quad \lambda(A) &= a+d-b-c \end{aligned}$$

Suppose we add a constant k to each element in A. Then

$$\begin{aligned} |A^*| &= \begin{vmatrix} a+k & b+k \\ c+k & d+k \end{vmatrix} \\ &= (ad-bc) + k(a+d-b-c) \\ \therefore \quad |A^*| &= |A| + k\lambda(A) \end{aligned}$$

In particular, let $A = \mathbf{Q}^n$. Then $|(\mathbf{Q}^n)^*| = |\mathbf{Q}^n| + k\lambda(\mathbf{Q}^n)$. But $\lambda(\mathbf{Q}^n) = F_{n+1} + F_{n-1} - 2F_n = F_{n-1} - F_{n-2} = F_{n-3}$. Therefore, by Cassini's formula, $|(\mathbf{Q}^n)^*| = (-1)^n + kF_{n-3}$.

Now let $k = F_n$. Then

$$\begin{vmatrix} F_{n+1}+F_n & F_n+F_n \\ F_n+F_n & F_{n-1}+F_n \end{vmatrix} = (-1)^n + F_nF_{n-3}$$

That is,

$$\begin{vmatrix} F_{n+2} & 2F_n \\ 2F_n & F_{n+1} \end{vmatrix} = (-1)^n + F_nF_{n-3}$$

This yields the Identity

$$4F_n^2 = F_{n+2}F_{n+1} - F_nF_{n-3} + (-1)^{n+1} \tag{32.23}$$

THE P-MATRIX

The Fibonacci matrix

$$P = \begin{bmatrix} 0 & 0 & 1 \\ 0 & 1 & 2 \\ 1 & 1 & 1 \end{bmatrix}$$

was studied by Marjorie Bicknell and V. E. Hoggatt, Jr., of San Jose State College, and Terry Brennan of Lockheed Missiles and Space Company. As in the case of the **Q**-matrix, let us study its powers and look for any pattern:

$$P^2 = \begin{bmatrix} 1 & 1 & 1 \\ 2 & 3 & 4 \\ 1 & 2 & 4 \end{bmatrix} \qquad P^3 = \begin{bmatrix} 1 & 2 & 4 \\ 4 & 7 & 12 \\ 4 & 6 & 9 \end{bmatrix} \qquad P^4 = \begin{bmatrix} 4 & 6 & 9 \\ 12 & 19 & 30 \\ 9 & 15 & 25 \end{bmatrix}$$

Is a pattern observable? Can we conjecture P^n? The pattern is not that obvious, so keep trying before reading any further.

Notice that:

$$P = \begin{bmatrix} F_0^2 & F_0F_1 & F_1^2 \\ 2F_0F_1 & F_2^2 - F_0F_1 & 2F_1F_2 \\ F_1^2 & F_1F_2 & F_2^2 \end{bmatrix} \qquad P^2 = \begin{bmatrix} F_1^2 & F_1F_2 & F_2^2 \\ 2F_1F_2 & F_3^2 - F_1F_2 & 2F_2F_3 \\ F_2^2 & F_2F_3 & F_3^2 \end{bmatrix}$$

$$P^3 = \begin{bmatrix} F_2^2 & F_2F_3 & F_3^2 \\ 2F_2F_3 & F_4^2 - F_2F_3 & 2F_3F_4 \\ F_3^2 & F_3F_4 & F_4^2 \end{bmatrix}$$

and so on.

Clearly, a pattern emerges. Can we now predict P^n?

We can show by PMI that

$$P^n = \begin{bmatrix} F_{n-1}^2 & F_{n-1}F_n & F_n^2 \\ 2F_{n-1}F_n & F_{n+1}^2 - F_{n-1}F_n & 2F_nF_{n+1} \\ F_n^2 & F_nF_{n+1} & F_{n+1}^2 \end{bmatrix}$$

Let

$$A = \begin{bmatrix} a & b & c \\ d & e & f \\ g & h & i \end{bmatrix}$$

Then

$$\lambda(A) = \begin{vmatrix} a+e-b-d & b+f-c-e \\ d+h-g-e & e+i-h-f \end{vmatrix}$$

(see Exercise 31). In particular, let $A = P^n$. Then

$$\lambda(P^n) = \begin{vmatrix} F_{n-1}^2 + F_{n+1}^2 - 4F_{n-1}F_n & 2F_{n-1}F_n + 2F_nF_{n+1} - F_n^2 - F_{n+1}^2 \\ 3F_{n-1}F_n + F_nF_{n+1} - F_n^2 - F_{n+1}^2 & 2F_{n+1}^2 - 3F_{n-1}F_n - F_nF_{n-1} \end{vmatrix}$$

Notice, for example, that

$$\begin{aligned} 2F_{n-1}F_n + 2F_nF_{n+1} - F_n^2 - F_{n+1}^2 &= 2F_n(F_{n-1} + F_{n+1}) - (F_n^2 + F_{n+1}^2) \\ &= 2F_nL_n - F_{2n+1} \\ &= 2F_{2n} - F_{2n+1} \\ &= F_{2n} - (F_{2n+1} - F_{2n}) \\ &= F_{2n} - F_{2n-1} \\ &= F_{2n-2} \end{aligned}$$

We can simplify the other expression entries in $\lambda(P^n)$ likewise (see Exercises 40–42). The resulting determinant is

$$\begin{aligned} \lambda(P^n) &= \begin{vmatrix} F_{2n-3} & F_{2n-2} \\ -F_{n-2}^2 & (-1)^n - F_{n-2}F_{n-1} \end{vmatrix} \\ &= F_{2n-3}[(-1)^n - F_{n-2}F_{n-1}] + F_{n-2}^2F_{2n-2} \\ &= (-1)^n(F_{n-1}^2 - F_{n-3}F_{n-2}) \qquad \text{after simplification} \\ &= (-1)^n(\text{center element in } P^{n-2}) \end{aligned}$$

EXERCISES 32

1. Let $\mathbf{Q}$ denote the $\mathbf{Q}$-matrix. Prove that $\mathbf{Q}^n = F_n\mathbf{Q} + F_{n-1}I$, where I denotes the 2×2 identity matrix. (Notice the similarity between this result and the formula $\alpha^n = F_n\alpha + F_{n-1}$.)
2. Show that $\mathbf{Q}^2 - \mathbf{Q} - I = 0$.
3. Prove that $(I + \mathbf{Q} + \mathbf{Q}^2 + \cdots + \mathbf{Q}^n)(\mathbf{Q} - I) = \mathbf{Q}^{n+1} - I$, where $n \geq 1$.
4. Using Identity (32.6), prove Identity (32.7).

Prove each.

5. $F_{m+n} = F_{m+1}F_n + F_mF_{n-1}$
6. $L_{m+n} = F_{m+1}L_n + F_mL_{n-1}$
7. $2F_{m+n} = F_mL_n + F_nL_m$
8. $2L_{m+n} = L_mL_n + 5F_mF_n$
9. $2F_{m-n} = (-1)^n(F_mL_n - F_nL_m)$
10. $2L_{m-n} = (-1)^n(L_mL_n - 5F_mF_n)$
11. $5(L_mL_n + F_mF_n) = 6L_{m+n} + 4(-1)^nL_{m-n}$

12. $5(L_m L_n - F_m F_n) = 4L_{m+n} + 6(-1)^n L_{m-n}$
13. $F_{m-n} = (-1)^n (F_m F_{n-1} - F_{m-1} F_n)$
14. $L_{m-n} = (-1)^n (F_{m+1} L_n - F_m L_{n+1})$
15. $F_{m+n} + F_{m-n} = \begin{cases} L_m F_n & \text{if } n \text{ is odd} \\ F_m L_n & \text{otherwise} \end{cases}$
16. $F_{m+n} - F_{m-n} = \begin{cases} F_m L_n & \text{if } n \text{ is odd} \\ L_m F_n & \text{otherwise} \end{cases}$
17. $L_{m+n} + L_{m-n} = \begin{cases} 5F_m F_n & \text{if } n \text{ is odd} \\ F_m L_n & \text{otherwise} \end{cases}$
18. $L_{m+n} - L_{m-n} = \begin{cases} L_m F_n & \text{if } n \text{ is odd} \\ F_m L_n & \text{otherwise} \end{cases}$
19. Let $M = \begin{bmatrix} 1 & 1 \\ 1 & 2 \end{bmatrix}$. Prove that $M^n = \begin{bmatrix} F_{2n-1} & F_{2n} \\ F_{2n} & F_{2n+1} \end{bmatrix}$.

Let $A_n = \begin{bmatrix} F_n & L_n \\ L_n & F_n \end{bmatrix}$ (Rabinowitz, 1998).

20. Express A_{2n} in terms of A_n and A_{n+1}.
*21. Express A_{2n} in terms of A_n and A_{n+1} only.
22. Let $p_{n+1} = -p_n$, where $p_1 = -1$. Prove that $p_n = (-1)^n$.
23. Consider the linear system

$$G_n x + G_{n-1} y = G_{n+1}$$
$$G_{n+1} x + G_n y = G_{n+2}$$

where $G_1 = a$ and $G_2 = b$. Use Cramer's rule to prove that $G_{n-1}G_{n+1} - G_n^2 = \mu(-1)^n$.

24. Use Exercise 23 to deduce a formula for $L_{n-1}L_{n+1} - L_n^2$.

Let A be a 2×2 matrix and $\mathbf{V}_n$ a 2×1 matrix such that $\mathbf{V}_{n+1} = A\mathbf{V}_n$ (Thoro, 1963). Find $\mathbf{V}_n$ in each case.

25. $A = \begin{bmatrix} 1 & 1 \\ 1 & 0 \end{bmatrix} \quad \mathbf{V}_1 = \begin{bmatrix} 1 \\ 1 \end{bmatrix}$
26. $A = \begin{bmatrix} 2 & 1 \\ 1 & 1 \end{bmatrix} \quad \mathbf{V}_1 = \begin{bmatrix} 2 \\ 1 \end{bmatrix}$
27. $A = \begin{bmatrix} 0 & 1 \\ 1 & -1 \end{bmatrix} \quad \mathbf{V}_1 = \begin{bmatrix} 0 \\ 1 \end{bmatrix}$

Let $\mathbf{U}_m = (F_{m+1}, F_m)$ and $\mathbf{V}_m = (L_{m+1}, L_m)$. Verify each. (Ruggles and Hoggatt, 1963)

28. $\mathbf{U}_0\mathbf{Q}^{n+1} = \mathbf{U}_{n+1} = \mathbf{U}_n\mathbf{Q}$
29. $\mathbf{V}_m\mathbf{Q}^n = \mathbf{V}_{m+n+1}$
30. Find $\lambda(\mathbf{Q}^n)$.

31. Let $M = \begin{bmatrix} a & b & c \\ d & e & f \\ g & h & i \end{bmatrix}$. Find $\lambda(M)$ (Bicknell and Hoggatt, 1963).
32. Compute $|P|$.
33. Find $\lambda(P)$.

Use the matrix $\mathbf{R} = \begin{bmatrix} L_{n+1} & L_n \\ L_n & L_{n-1} \end{bmatrix}$ to answer Exercises 34 and 35.

34. Compute $|\mathbf{R}|$.
35. Find $\lambda(\mathbf{R})$.

Let $A = \begin{bmatrix} G_{n+k} & G_n \\ G_n & G_{n-k} \end{bmatrix}$ and $B = \begin{bmatrix} G_{n+k}+i & G_n+i \\ G_n+i & G_{n-k}+i \end{bmatrix}$, where G_n denotes the nth generalized Fibonacci number. Compute each.

36. $|A|$
37. $\lambda(A)$
38. $|B|$
39. Using PMI, establish the formula for P^n, where P denotes the P-matrix.

Prove each.

40. $F_{n-1}^2 + F_{n+1}^2 - 4F_{n-1}F_n = F_{2n-3}$
41. $F_n^2 + F_{n+1}^2 - 3F_{n-1}F_n - F_nF_{n+1} = F_{n-2}^2$
42. $3F_{n-1}F_n + F_nF_{n-1} - 2F_{n+1}^2 = F_{n-2}F_{n-1} - (-1)^n$

FIBONACCI DETERMINANTS

In Chapter 3, we found that Fibonacci numbers and Lucas numbers occur in graph theory; specifically, they occur in the study of paraffins and cycloparaffins. We now turn our attention to an additional occurrence of Lucas numbers in graph theory, and then to some Fibonacci and Lucas determinants.

AN APPLICATION TO GRAPH THEORY

In 1975, K. R. Rebman of California State University at Hayward showed the occurrence of Lucas numbers in the study of spanning trees of wheel graphs. Before we can present the main result, we need to lay groundwork with two lemmas and some basic vocabulary.

Lemma 33.1. Let A_n denote the $n \times n$ matrix

$$\begin{vmatrix} 3 & -1 & 0 & 0 & \cdots & & 0 \\ -1 & 3 & -1 & 0 & \cdots & & 0 \\ 0 & -1 & 3 & -1 & \cdots & & 0 \\ & & & & \vdots & & -1 \\ 0 & 0 & 0 & 0 & \cdots & -1 & 3 \end{vmatrix}$$

Then $|A_n| = F_{2n+2}$.

Proof. [by the principle of mathematical induction (PMI)] Since $|A_1| = 3 = F_4$ and $|A_2| = \begin{vmatrix} 3 & -1 \\ -1 & 3 \end{vmatrix} = 8 = F_6$, the result is true when $n = 1$ and $n = 2$.

Assume it is true for every positive integer $k < n$. Expanding $|A_n|$ by the first row,

$$\begin{aligned}
|A_n| &= 3|A_{n+1}| + \begin{vmatrix} -1 & -1 & 0 & \cdots & & 0 \\ 0 & 3 & -1 & \cdots & & 0 \\ & & & \cdots & & -1 \\ 0 & 0 & 0 & \cdots & -1 & 3 \end{vmatrix} \\
&= 3|A_{n-1}| - |A_{n-2}| \\
&= 3F_{2(n-1)+2} - F_{2(n-2)+2} \qquad \text{by the inductive hypothesis} \\
&= 3F_{2n} - F_{2n-2} = F_{2n+2}
\end{aligned}$$

Therefore, the result is true for every $n \geq 1$. ■

Lemma 33.2. Let B_n denote the $n \times n$ matrix:

$$\begin{vmatrix} 3 & -1 & 0 & 0 & \cdots & 0 & -1 \\ -1 & 3 & -1 & 0 & \cdots & 0 & 0 \\ 0 & -1 & 3 & -1 & \cdots & 0 & 0 \\ & & & & \vdots & 3 & -1 \\ -1 & 0 & 0 & 0 & \cdots & -1 & 3 \end{vmatrix}$$

where $B_1 = [1]$ and $B_2 = \begin{vmatrix} 3 & -2 \\ -2 & 3 \end{vmatrix}$. Then $|B_n| = L_{2n} - 2$.

Proof. Since $|B_1| = 1 = L_2 - 2$ and $|B_2| = 5 = L_4 - 2$, the result is true when $n = 1$ and $n = 2$. So assume that $n \geq 3$.

Expanding $|B_n|$ by the first row, $|B_n| = 3|A_{n-1}| + |R_{n-1}| + (-1)^{n+2}|S_{n-1}|$, where

$$R_m = \begin{vmatrix} -1 & -1 & 0 & 0 & \cdots & 0 & 0 \\ 0 & 3 & -1 & 0 & \cdots & 0 & 0 \\ 0 & -1 & 3 & -1 & \cdots & 0 & 0 \\ & & & & \vdots & 3 & -1 \\ 0 & 0 & 0 & 0 & \cdots & -1 & 3 \end{vmatrix}$$

and

$$S_m = \begin{vmatrix} -1 & 3 & -1 & 0 & \cdots & 0 & 0 \\ 0 & -1 & 3 & -1 & \cdots & 0 & 0 \\ 0 & 0 & -1 & 3 & \cdots & 0 & 0 \\ & & & & \vdots & 3 & -1 \\ 0 & 0 & 0 & 0 & \cdots & -1 & 3 \\ -1 & 0 & 0 & 0 & \cdots & 0 & -1 \end{vmatrix}$$

Expanding $|R_m|$ by the first column,

$$|R_m| = -|A_{m-1}| + (-1)^m \begin{vmatrix} -1 & 0 & 0 & \cdots & 0 & 0 \\ 3 & -1 & 0 & \cdots & 0 & 0 \\ -1 & 3 & -1 & \cdots & 0 & 0 \\ & & & \vdots & & \\ 0 & 0 & 0 & \cdots & 0 & -1 \end{vmatrix}$$

$$= -|A_{m-1}| + (-1)^m(-1)^{m-1} = -|A_{m-1}| - 1$$

$$= -F_{2m} - 1$$

Expanding $|S_m|$ by the first column,

$$|S_m| = (-1)^m + (-1)(-1)^{m-1}|A_{m-1}|$$

$$= (-1)^m(F_{2m} + 1)$$

$$\therefore \quad |B_n| = 3F_{2n} + (-F_{2m} - 1) + (-1)^n(-1)^{n-1}(F_{2n-2} + 1)$$

$$= 3F_{2n} - 2F_{2n-2} - 2$$

$$= L_{2n} - 2$$

■

A FEW BASIC FACTS FROM GRAPH THEORY

At this point, we need to introduce a few basic terms and some fundamental results from graph theory for clarity and consistency.

We can represent algebraically a graph G with n vertices by the incidence matrix $A(G) = (a_{ij})_{n\times n}$, where

$$a_{ij} = \begin{cases} 1 & \text{if there is an edge from vertex } i \text{ to vertex } j \\ 0 & \text{otherwise} \end{cases}$$

For example, the incidence matrix of the graph in Figure 33.1 is

$$\begin{array}{c} \\ 1 \\ 2 \\ 3 \\ 4 \end{array} \begin{array}{c} \begin{array}{cccc} 1 & 2 & 3 & 4 \end{array} \\ \begin{bmatrix} 0 & 1 & 0 & 0 \\ 1 & 0 & 1 & 1 \\ 0 & 1 & 0 & 1 \\ 0 & 1 & 1 & 0 \end{bmatrix} \end{array}$$

The *degree* of a vertex v, denoted by $\deg(v)$, is the number of edges meeting at v. For instance, the degree of vertex 2 in Figure 33.1 is three.

Let $D(G) = (d_{ij})_{n\times n}$ denote the matrix defined by

$$d_{ij} = \begin{cases} \deg(i) & \text{if } i = j \\ 0 & \text{otherwise} \end{cases}$$

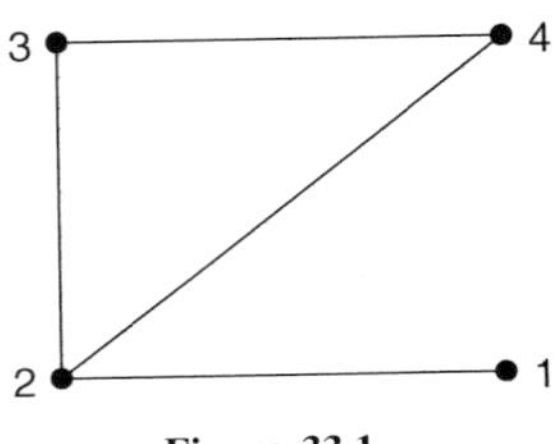

Figure 33.1.

For example, for the graph in Figure 33.1,

$$D(G) = \begin{array}{c} \\ 1 \\ 2 \\ 3 \\ 4 \end{array} \begin{array}{c} \begin{array}{cccc} 1 & 2 & 3 & 4 \end{array} \\ \begin{bmatrix} 1 & 0 & 0 & 0 \\ 0 & 3 & 0 & 0 \\ 0 & 0 & 2 & 0 \\ 0 & 0 & 0 & 2 \end{bmatrix} \end{array}$$

Recall from Chapter 5 that a spanning tree of a graph G is a subgraph of G that is a tree containing every vertex of G and its complexity $k(G)$ is the number of distinct spanning trees of the graph. For any graph G, $k(G)$ equals the determinant of any one of the n principal $(n-1)$-rowed minors of the matrix $D(G) - A(G)$. This remarkable result was established by the outstanding German physicist Gustav Robert Kirchhoff (1824–1887).

For example, using the graph in Figure 33.1,

$$D(G) - A(G) = \begin{bmatrix} 1 & -1 & 0 & 0 \\ -1 & 3 & -1 & -1 \\ 0 & -1 & 2 & -1 \\ 0 & -1 & -1 & 2 \end{bmatrix}$$

Since

$$\begin{vmatrix} 3 & -1 & -1 \\ -1 & 2 & -1 \\ -1 & -1 & 2 \end{vmatrix} = 3$$

it follows that the complexity of the graph is three, as we found in Chapter 5.

THE WHEEL GRAPH

Let $n \geq 3$. The *wheel graph* W_n is a graph with $n + 1$ vertices; n of them lie on a cycle(the rim) and the remaining vertex (the hub) is connected to every rim vertex. Figure 33.2 shows the wheel graphs W_3, W_4, and W_5.

We are now ready for the surprise.

Theorem 33.1. $k(G) = L_{2n} - 2$.

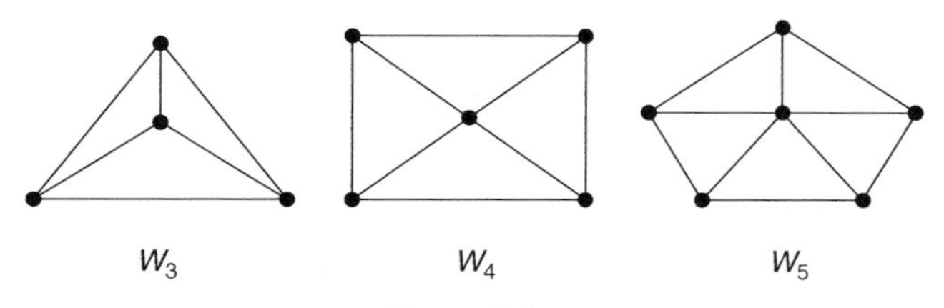

Figure 33.2.

Proof. Let us denote the rim vertices by v_1 through v_n, and the hub vertex by v_{n+1}. Then

$$\deg(v_i) = \begin{cases} 3 & \text{if } i \neq n+1 \\ n & \text{otherwise} \end{cases}$$

Consequently,

$$D(W_n) - A(W_n) = \begin{vmatrix} 3 & & & & 0 \\ . & 3 & & & \\ . & . & 3 & & \\ & & & & \\ & 0 & & & \\ & & & & n \end{vmatrix} - \begin{vmatrix} 0 & 1 & 0 & 0 & \cdots & 0 & 1 \\ 1 & 0 & 1 & 0 & \cdots & 0 & 0 \\ 0 & 1 & 0 & 1 & \cdots & 0 & 0 \\ & & & \cdots & & & \\ 0 & & & \cdots & & 0 & 1 \\ 0 & 0 & 0 & 0 & \cdots & 1 & 0 \end{vmatrix}$$

$$= \left| \begin{array}{cccc|c} & & & & -1 \\ & & & & -1 \\ & & A_n & & . \\ & & & & . \\ & & & & . \\ & & & & -1 \\ \hline -1 & -1 & \cdots & -1 & n \end{array} \right|$$

To compute $k(W_n)$, any principal $(n-1)$-rowed minor will suffice. So deleting row $(n+1)$ and column $(n+1)$, we get $k(W_n) = |A_n| = L_{2n} - 2$. ■

For example, W_3 has $L_6 - 2 = 18 - 2 = 16$ spanning trees. Figure 33.3 shows all of them.

Theorem 33.1 was originally discovered in 1969 by J. Sedlacek of The University of Calgary, Canada, and then rediscovered two years later by B. R. Myers of the University of Notre Dame. At the 1969 Calgary International Conference of Combinatorial Structures and their Applications, Sedlacek stated the formula as

$$k(W_n) = \left(\frac{3+\sqrt{5}}{2}\right)^n - \left(\frac{3-\sqrt{5}}{2}\right)^n - 2$$

In contrast, Myers gave it as $k(W_n) = F_{2n+2} - F_{2n-2} - 2$ in a problem he proposed in 1972 in *The American Mathematical Monthly*. Obviously, either formula can be rewritten in terms of the Lucas number L_{2n}.

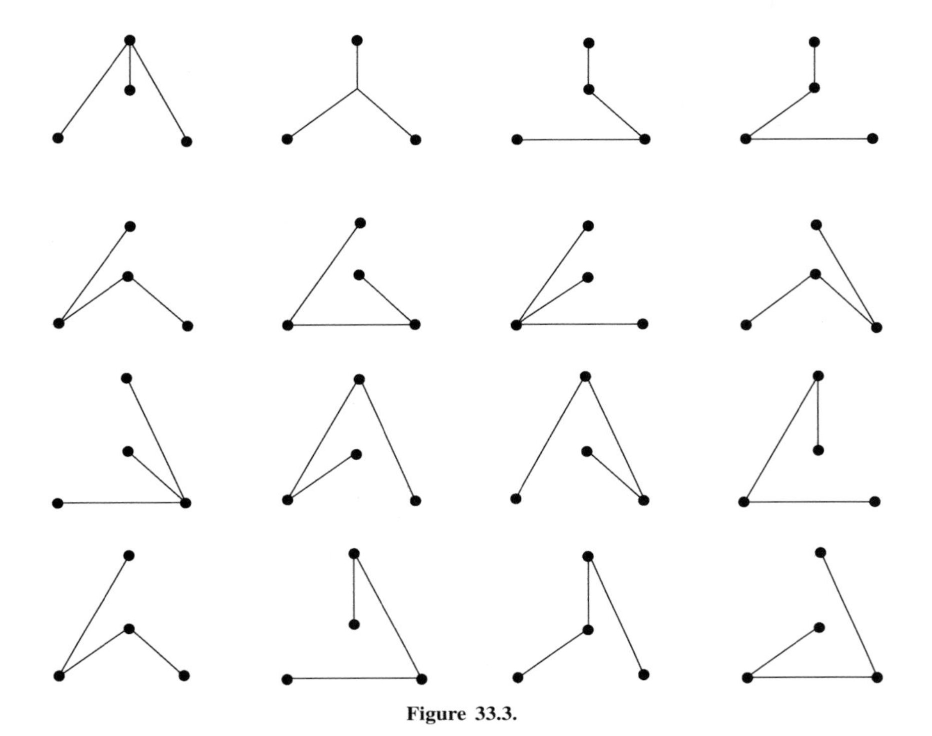

Figure 33.3.

THE SINGULARITY OF FIBONACCI MATRICES

Now we turn our attention to the singularity of a class of Fibonacci matrices, but first a definition. A square matrix A is *singular* if $|A| = 0$. For example, the matrix $\begin{bmatrix} 3 & 4 \\ 6 & 8 \end{bmatrix}$ is singular since $|A| = 3 \cdot 8 - 6 \cdot 4 = 0$.

A matrix $M_n = (a_{ij})_{n \times n}$ is called the *Fibonacci matrix* if it contains the first n^2 Fibonacci numbers such that $a_{11} = F_1, a_{12} = F_2, \ldots, a_{1n} = F_n$, $a_{21} = F_{n+1}, \ldots,$ and $a_{nn} = F_{n^2}$. Thus $a_{ij} = F_{(i-1)n+j}$, where $1 \leq i,\ j \leq n$.

For example, $M_1 = [1]$,

$$M_2 = \begin{bmatrix} 1 & 1 \\ 2 & 3 \end{bmatrix} \quad M_3 = \begin{bmatrix} 1 & 1 & 2 \\ 3 & 5 & 8 \\ 13 & 21 & 34 \end{bmatrix} \quad \text{and} \quad M_4 = \begin{bmatrix} 1 & 1 & 2 & 3 \\ 5 & 8 & 13 & 21 \\ 34 & 55 & 89 & 144 \\ 233 & 377 & 610 & 987 \end{bmatrix}$$

Clearly, $|M_1| \neq 0$ and $|M_2| \neq 0$. But

$$|M_3| = \begin{vmatrix} 1 & 0 & 0 \\ 3 & 2 & 2 \\ 13 & 8 & 8 \end{vmatrix} = 0$$

Using elementary column operations, it is easy to verify that $|M_4| = 0$. Thus both M_3 and M_4 are singular matrices.

More generally, we have the following theorem, which was observed in 1995 by Graham Fisher, then a student at Bournemouth School, England. Before we can present its proof, we need the next lemma, which was proposed as a problem in 1969 by D. V. Jaiswal of Kolkar Science College, Indore, India.

Lemma 33.3. Let m, n, p, q, and r be positive integers, and G_n the nth generalized Fibonacci number. Then

$$\begin{vmatrix} G_p & G_{p+m} & G_{p+m+n} \\ G_q & G_{q+m} & G_{q+m+n} \\ G_r & G_{r+m} & G_{r+m+n} \end{vmatrix} = 0$$

Proof. Since $G_{m+n} = G_m F_{n+1} + G_{m-1} F_n$ (see Chapter 7, Exercise 27), it follows that

$$G_{k+m+n} = G_{k+m} F_{n+1} + G_{k+m-1} F_n \tag{33.1}$$

Using Eq. (33.1), we can write the given determinant D as

$$\begin{aligned} D &= F_{n+1} \begin{vmatrix} G_p & G_{p+m} & G_{p+m} \\ G_q & G_{q+m} & G_{q+m} \\ G_r & G_{r+m} & G_{r+m} \end{vmatrix} + F_n \begin{vmatrix} G_p & G_{p+m} & G_{p+m-1} \\ G_q & G_{q+m} & G_{q+m-1} \\ G_r & G_{r+m} & G_{r+m-1} \end{vmatrix} \\ &= F_{n+1} \cdot 0 + F_n \begin{vmatrix} G_p & G_{p+m} & G_{p+m-1} \\ G_q & G_{q+m} & G_{q+m-1} \\ G_r & G_{r+m} & G_{r+m-1} \end{vmatrix} \\ &= F_n \begin{vmatrix} G_p & G_{p+m} & G_{p+m-1} \\ G_q & G_{q+m} & G_{q+m-1} \\ G_r & G_{r+m} & G_{r+m-1} \end{vmatrix} \end{aligned}$$

Subtract the third column from the second:

$$= F_n \begin{vmatrix} G_p & G_{p+m-2} & G_{p+m-1} \\ G_q & G_{q+m-2} & G_{q+m-1} \\ G_r & G_{r+m-2} & G_{r+m-1} \end{vmatrix}$$

Subtract the second column from the third:

$$= F_n \begin{vmatrix} G_p & G_{p+m-2} & G_{p+m-3} \\ G_q & G_{q+m-2} & G_{q+m-3} \\ G_r & G_{r+m-2} & G_{r+m-3} \end{vmatrix}$$

Continuing like this, we can reduce the subscripts of the elements in columns 2 and 3 further. At a certain stage, when m is even, columns 1 and 2 would be identical; when m is odd, columns 1 and 3 would be identical. In both cases, $D = 0$. ■

This lemma has several by-products (see Exercises 2–5).

Theorem 33.2. $|M_n| = 0$, where $n \geq 3$.

Proof. (by PMI) Since the result for $n = 3$, assume it is true for $n = k$: $|M_k| = 0$, where $k \geq 3$.

Consider M_{k+1}. Expanding $|M_{k+1}|$ by the first row and the resulting cofactors, we can express $|M_{k+1}|$ as a linear combination of a host of 3×3 matrices. Let

$$M = \begin{vmatrix} F_m & U_1 & U_2 \\ U_3 & U_4 & U_5 \\ U_6 & U_7 & U_8 \end{vmatrix}$$

where each U_i is a Fibonacci element belonging to M. Since we expanded $|M_{k+1}|$ and its successive cofactors by their first rows, M preserves the same row order as in M_{k+1}, so $U_3 = F_{m+r}$ and $U_6 = F_{m+2r}$, where $r \geq 1$. Thus

$$M = \begin{vmatrix} F_m & F_{m+s} & F_{m+t} \\ F_{m+r} & F_{m+r+s} & F_{m+r+t} \\ F_{m+2r} & F_{m+2r+s} & F_{m+2r+t} \end{vmatrix}$$

where $s, t \geq 1$. It follows by Lemma 33.3 that $|M| = 0$. Since every 3×3 matrix M is singular, it follows that M_{k+1} is singular. Thus, by PMI, M_n is singular for every $n \geq 3$. ■

For example,

$$\begin{aligned} |M_4| &= \begin{vmatrix} F_6 & F_7 & F_8 \\ F_{10} & F_{11} & F_{12} \\ F_{14} & F_{15} & F_{16} \end{vmatrix} - \begin{vmatrix} F_5 & F_7 & F_8 \\ F_9 & F_{11} & F_{12} \\ F_{13} & F_{15} & F_{16} \end{vmatrix} + 2\begin{vmatrix} F_5 & F_6 & F_8 \\ F_9 & F_{10} & F_{12} \\ F_{13} & F_{14} & F_{16} \end{vmatrix} \\ &\quad -3\begin{vmatrix} F_5 & F_6 & F_7 \\ F_9 & F_{10} & F_{11} \\ F_{13} & F_{14} & F_{15} \end{vmatrix} \\ &= 0 - 0 + 2 \cdot 0 - 3 \cdot 0 \qquad \text{by Lemma 33.3} \\ &= 0 \end{aligned}$$

Therefore, M_4 is singular.

FIBONACCI AND ANALYTIC GEOMETRY

Next we pursue a few results in analytic geometry that involve Fibonacci determinants. They were developed in 1974 by Jaiswal.

Theorem 33.3. The area of the triangle with vertices (G_n, G_{n+r}), (G_{n+p}, G_{n+p+r}), and (G_{n+q}, G_{n+q+r}) is independent of n.

Proof. Twice the area of the area of the triangle equals the absolute value of the determinant

$$\Delta = \begin{vmatrix} G_n & G_{n+r} & 1 \\ G_{n+p} & G_{n+p+r} & 1 \\ G_{n+q} & G_{n+q+r} & 1 \end{vmatrix}$$

Using the identity $G_{m+n} = G_m F_{n+1} + G_{m-1} F_n$ for the second column, we can write this determinant as

$$\Delta = F_{r+1} \begin{vmatrix} G_n & G_n & 1 \\ G_{n+p} & G_{n+p} & 1 \\ G_{n+q} & G_{n+q} & 1 \end{vmatrix} + F_r \begin{vmatrix} G_n & G_{n-1} & 1 \\ G_{n+p} & G_{n+p-1} & 1 \\ G_{n+q} & G_{n+q-1} & 1 \end{vmatrix}$$

$$\Delta = 0 + F_r \begin{vmatrix} G_n & G_{n-1} & 1 \\ G_{n+p} & G_{n+p-1} & 1 \\ G_{n+q} & G_{n+q-1} & 1 \end{vmatrix}$$

Add the negative of column 2 to column 1:

$$\Delta = F_r \begin{vmatrix} G_{n-2} & G_{n-1} & 1 \\ G_{n+p-2} & G_{n+p-1} & 1 \\ G_{n+q-2} & G_{n+q-1} & 1 \end{vmatrix}$$

Now add the negative of column 1 to column 2:

$$\Delta = F_r \begin{vmatrix} G_{n-2} & G_{n-3} & 1 \\ G_{n+p-2} & G_{n+p-3} & 1 \\ G_{n+q-2} & G_{n+q-3} & 1 \end{vmatrix}$$

Continuing in this way, we get

$$\Delta = \pm F_r \begin{vmatrix} G_1 & G_2 & 1 \\ G_{p+1} & G_{p+2} & 1 \\ G_{q+1} & G_{q+2} & 1 \end{vmatrix}$$

according as n is odd or even. Expanding by column 3,

$$\Delta = \pm F_r[(G_{p+1}G_{q+2} - G_{p+2}G_{q+1}) - (G_1 G_{q+2} - G_2 G_{q+1}) + (G_1 G_{p+2} - G_2 G_{p+1})]$$

But $G_n G_{m+k} - G_{n+k} G_m = (-1)^{n+1} F_k F_{m-n} \mu$;

$$\begin{aligned} \therefore \quad \Delta &= \pm F_r[(-1)^p F_{q-p} - F_q + F_p]\mu \\ &= \pm F_r[(-1)^p F_{q-p} + F_p - F_q]\mu \end{aligned}$$

Since this is independent of n, it follows that the desired area is independent of n. ■

Corollary 33.1. The area of the triangle with vertices (F_n, F_{n+h}), (F_{n+2h}, F_{n+3h}), and (F_{n+4h}, F_{n+5h}) is $F_h(F_{4h} - 2F_{2h})/2$.

Proof. The proof follows from the theorem, since $r = h$, $p = 2h$, $q = 4h$, and $a = b = \mu = 1$. ■

Theorem 33.4. The lines through the origin with direction ratios G_n, G_{n+p}, G_{n+q}, where p and q are constants, are coplanar for every n.

Proof. The direction ratios of three such lines can be taken as G_i, G_{i+p}, G_{i+q}; G_j, G_{j+p}, G_{j+q}; and G_k, G_{k+p}, G_{k+q}. The lines are coplanar if and only if

$$D = \begin{vmatrix} G_i & G_{i+p} & G_{i+q} \\ G_j & G_{j+p} & G_{j+q} \\ G_k & G_{k+p} & G_{k+q} \end{vmatrix} = 0$$

Using the identity $G_{m+n} = G_m F_{n+1} + G_{m-1} F_n$, we can express D as the sum of four determinants, each of which is zero. Therefore, $D = 0$, and hence the three lines are coplanar, as desired. ■

Theorem 33.5. The plane containing the family of points (G_n, G_{n+p}, G_{n+q}), where p and q are arbitrary constants, contains the origin.

Proof. By Theorem 33.4, the given points are coplanar. An equation of the plane containing any three points of the family of points is

$$\begin{vmatrix} x & y & z & 1 \\ G_i & G_{i+p} & G_{i+q} & 1 \\ G_j & G_{j+p} & G_{j+q} & 1 \\ G_k & G_{k+p} & G_{k+q} & 1 \end{vmatrix} = 0$$

Expanding this determinant with respect to row 1, we get the equation $Ax + By + Cx + D = 0$, where A, B, C, and D are constants. In fact,

$$\begin{aligned} D &= -\begin{vmatrix} G_i & G_{i+p} & G_{i+q} \\ G_j & G_{j+p} & G_{j+q} \\ G_k & G_{k+p} & G_{k+q} \end{vmatrix} \\ &= 0 \end{aligned}$$

by Theorem 33.4. Thus the equation of the plane is $Ax + By + Cz = 0$, which clearly contains the origin. ■

In fact, we can show that the equation of the plane is $(-1)^p F_{q-p} x - F_q y + F_p z = 0$. In particular, the plane containing the points (G_k, G_{k+3}, G_{k+8}) is $(-1)^3 F_5 x - F_8 y + F_3 z = 0$; that is, $5x + 2y - 2z = 0$.

Theorem 33.6. The family of planes $G_n x + G_{n+p} y + G_{n+q} z + G_{n+r} = 0$, where p, q, and r are arbitrary constants, intersect along a line whose equation is independent of n.

The proof is a bit complex, so we omit it. But we can show that the planes intersect along the line

$$\frac{(-)^p F_p x - F_{r-p}}{F_{q-p}} = \frac{F_p y + F_r}{-F_q} = \frac{z}{F_p}$$

For example, the planes $G_n x + G_{n+2}y + G_{n+3}z + G_{n+4} = 0$ intersect along the line

$$\frac{(-1)^2 F_2 x - F_3}{F_1} = \frac{F_2 y + F_5}{-F_3} = \frac{z}{F_2}$$

that is,

$$\frac{x-2}{1} = \frac{y+2}{-2} = \frac{z}{1}$$

In 1966, D. A. Klarner of the University of Alberta, Canada, developed a fascinating formula for computing the determinants whose elements satisfy the recurrence relation

$$a_{n+2} = pa_{n+1} - qa_n \tag{33.2}$$

where p and q are fixed complex numbers. It is given by the following theorem. Since the proof is complicated, we omit it.

Theorem 33.7. (Klarner, 1966) Let a_n be defined by the recurrence relation $a_{n+2} = pa_{n+1} - qa_n$, where p and q are fixed complex numbers and $n \geq 0$. Let

$$A_k(a_n) = \begin{vmatrix} a_n^k & a_{n+1}^k & \cdots & a_{n+k}^k \\ a_{n+1}^k & a_{n+2}^k & \cdots & a_{n+k+1}^k \\ & & \vdots & \\ a_{n+k}^k & a_{n+k+1}^k & \cdots & a_{n+2k}^k \end{vmatrix}$$

Then $A_k(a_{mn+r}) = q^{nk(k+1)/2} A_k(a_r)$. ■

When $p = 1$ and $q = -1$, the recurrence relation (Eq. 33.2) yields the Fibonacci recurrence relation (FRR), so we can employ this theorem to evaluate determinants of the form $A_k(F_n)$, as the next example illustrates.

Example 33.1. Evaluate each determinant D:

(1)
$$\begin{vmatrix} F_n & F_{n+1} \\ F_{n+1} & F_{n+2} \end{vmatrix}$$

Solution. Here $k = 1 = m$, $r = 0$, and $a_n = F_n$. By Theorem 33.7,

$$\begin{aligned} D &= A_1(F_n) = (-1)^{n\cdot 1\cdot 2/2} A_1(F_0) \\ &= (-1)^n \begin{vmatrix} F_0 & F_1 \\ F_1 & F_2 \end{vmatrix} = (-1)^n \begin{vmatrix} 0 & 1 \\ 1 & 1 \end{vmatrix} = (-1)^{n+1} \end{aligned}$$

That is, $F_n F_{n+2} - F_{n+1}^2 = (-1)^{n+1}$, which, we recall, is Cassini's rule.

(2)
$$\begin{vmatrix} F_n^2 & F_{n+1}^2 & F_{n+2}^2 \\ F_{n+1}^2 & F_{n+2}^2 & F_{n+3}^2 \\ F_{n+2}^2 & F_{n+3}^2 & F_{n+4}^2 \end{vmatrix}$$

Solution. Here $k = 2$, $m = 1$, $r = 0$, and $a_n = F_n$.

$$D = (-1)^{n \cdot 2 \cdot 3/2} A_2(F_0) = (-1)^{3n} \begin{vmatrix} F_0^2 & F_1^2 & F_2^2 \\ F_1^2 & F_2^2 & F_3^2 \\ F_2^2 & F_3^2 & F_4^2 \end{vmatrix}$$

$$= (-1)^{3n} \begin{vmatrix} 0 & 1 & 1 \\ 1 & 1 & 4 \\ 1 & 4 & 9 \end{vmatrix} = (-1)^{3n} \begin{vmatrix} 0 & 1 & 1 \\ 1 & 1 & 4 \\ 0 & 3 & 5 \end{vmatrix}$$

$$= (-1)^{3n}(-1)(2) = 2(-1)^{n+1} \tag{33.3}$$

(This example was proposed as a problem in 1963 by Br. Alfred.)

(3) $$\begin{vmatrix} F_n^3 & F_{n+1}^3 & F_{n+2}^3 & F_{n+3}^3 \\ F_{n+1}^3 & F_{n+2}^3 & F_{n+3}^3 & F_{n+4}^3 \\ F_{n+2}^3 & F_{n+3}^3 & F_{n+4}^3 & F_{n+5}^3 \\ F_{n+3}^3 & F_{n+4}^3 & F_{n+5}^3 & F_{n+6}^3 \end{vmatrix}$$

Solution. Here $k = 3$, $m = 1$, $r = 0$, and $a_n = F_n$, so by Theorem 33.7,

$$D = (-1)^{n \cdot 3 \cdot 4/2} A_3(F_0) = (-1)^{6n} \begin{vmatrix} F_0^3 & F_1^3 & F_2^3 & F_3^3 \\ F_1^3 & F_2^3 & F_3^3 & F_4^3 \\ F_2^3 & F_3^3 & F_4^3 & F_5^3 \\ F_3^3 & F_4^3 & F_5^3 & F_6^3 \end{vmatrix}$$

$$= \begin{vmatrix} 0 & 1 & 1 & 8 \\ 1 & 1 & 8 & 27 \\ 1 & 8 & 27 & 125 \\ 8 & 27 & 125 & 512 \end{vmatrix} = \begin{vmatrix} 0 & 1 & 1 & 8 \\ 1 & 1 & 8 & 27 \\ 0 & 7 & 19 & 98 \\ 0 & 19 & 61 & 296 \end{vmatrix}$$

$$= -\begin{vmatrix} 1 & 1 & 8 \\ 7 & 19 & 98 \\ 19 & 61 & 296 \end{vmatrix} = -\begin{vmatrix} 1 & 1 & 8 \\ 0 & 12 & 42 \\ 0 & 42 & 144 \end{vmatrix}$$

$$= -(12 \cdot 144 - 42 \cdot 42) = 36$$

(This was proposed as an advanced problem in 1963 by J. Erbacker et al.) ■

Since Theorem 33.7 holds for any sequence that satisfies the recurrence relation (Eq. 33.1), we can apply it when $a_n = G_n$, the nth generalized Fibonacci number, as the next example demonstrates.

Example 33.1. Evaluate the determinant

$$D = \begin{vmatrix} G_n & G_{n+1} & G_{n+2} \\ G_{n+1} & G_{n+2} & G_{n+3} \\ G_{n+2} & G_{n+3} & G_{n+4} \end{vmatrix}$$

Solution. Here $k = 1 = m$, $r = 0$, and $a_n = G_n$. So, by Theorem 33.7,

$$D = (-1)^{n\cdot 2\cdot 3/2} A_2(a_0) = (-1)^{3n} A_2(G_0)$$

$$= (-1)^n \begin{vmatrix} G_0 & G_1 & G_2 \\ G_1 & G_2 & G_3 \\ G_2 & G_3 & G_4 \end{vmatrix} = (-1)^n \begin{vmatrix} b-a & a & b \\ a & b & a+b \\ b & a+b & a+2b \end{vmatrix} = 0$$

(This follows obviously, since column 3 of D is the sum of columns 1 and 2.) ■

EXERCISES 33

Evaluate each determinant.

1. $\begin{vmatrix} L_n & L_{n+1} \\ L_{n+1} & L_{n+2} \end{vmatrix}$

2. $\begin{vmatrix} F_p & F_{p+m} & F_{p+m+n} \\ F_q & F_{q+m} & F_{q+m+n} \\ F_r & F_{r+m} & F_{r+m+n} \end{vmatrix}$ (Ivanoff, 1968)

3. $\begin{vmatrix} F_a & F_{a+d} & F_{a+2d} \\ F_{a+3d} & F_{a+4d} & F_{a+5d} \\ F_{a+6d} & F_{a+7d} & F_{a+8d} \end{vmatrix}$ (Finkelstein, 1969)

4. $\begin{vmatrix} L_a & L_{a+d} & L_{a+2d} \\ L_{a+3d} & L_{a+4d} & L_{a+5d} \\ L_{a+6d} & L_{a+7d} & L_{a+8d} \end{vmatrix}$ (Finkelstein, 1969)

5. $\begin{vmatrix} G_a & G_{a+d} & G_{a+2d} \\ G_{a+3d} & G_{a+4d} & G_{a+5d} \\ G_{a+6d} & G_{a+7d} & G_{a+8d} \end{vmatrix}$

6. $\begin{vmatrix} F_p + k & F_{n+1} + k & F_{n+2} + k \\ F_{n+1} + k & F_{n+2} + k & F_{n+3} + k \\ F_{n+2} + k & F_{n+3} + k & F_{n+4} + k \end{vmatrix}$ (Alfred, 1963)

7. $\begin{vmatrix} G_p + k & G_{p+m} + k & G_{p+m+n} + k \\ G_q + k & G_{q+m} + k & G_{q+m+n} + k \\ G_r + k & G_{r+m} + k & G_{r+m+n} + k \end{vmatrix}$ (Jaiswal, 1969)

8. $\begin{vmatrix} F_{n+3} & F_{n+2} & F_{n+1} & F_n \\ F_{n+2} & F_{n+3} & F_n & F_{n+1} \\ F_{n+1} & F_n & F_{n+3} & F_{n+2} \\ F_n & F_{n+1} & F_{n+2} & F_{n+3} \end{vmatrix}$ (Ledin, 1967)

9. $\begin{vmatrix} G_{n+3} & G_{n+2} & G_{n+1} & G_n \\ G_{n+2} & G_{n+3} & G_n & G_{n+1} \\ G_{n+1} & G_n & G_{n+3} & G_{n+2} \\ G_n & G_{n+1} & G_{n+2} & G_{n+3} \end{vmatrix}$ (Jaiswal, 1969)

10. Show that $\begin{vmatrix} L_{n+3} & L_{n+2} & L_{n+1} & L_n \\ L_{n+2} & L_{n+3} & L_n & L_{n+1} \\ L_{n+1} & L_n & L_{n+3} & L_{n+2} \\ L_n & L_{n+1} & L_{n+2} & L_{n+3} \end{vmatrix} = 25 \begin{vmatrix} F_{n+3} & F_{n+2} & F_{n+1} & F_n \\ F_{n+2} & F_{n+3} & F_n & F_{n+1} \\ F_{n+1} & F_n & F_{n+3} & F_{n+2} \\ F_n & F_{n+1} & F_{n+2} & F_{n+3} \end{vmatrix}$
    (Jaiswal, 1969)

Evaluate each.

11. $\begin{vmatrix} L_n^2 & L_{n+1}^2 & L_{n+2}^2 \\ L_{n+1}^2 & L_{n+2}^2 & L_{n+3}^2 \\ L_{n+2}^2 & L_{n+3}^2 & L_{n+4}^2 \end{vmatrix}$

12. $\begin{vmatrix} G_n^2 & G_{n+1}^2 & G_{n+2}^2 \\ G_{n+1}^2 & G_{n+2}^2 & G_{n+3}^2 \\ G_{n+2}^2 & G_{n+3}^2 & G_{n+4}^2 \end{vmatrix}$

13. $\begin{vmatrix} L_n^3 & L_{n+1}^3 & L_{n+2}^3 & L_{n+3}^3 \\ L_{n+1}^3 & L_{n+2}^3 & L_{n+3}^3 & L_{n+4}^3 \\ L_{n+2}^3 & L_{n+3}^3 & L_{n+4}^3 & L_{n+5}^3 \\ L_{n+3}^3 & L_{n+4}^3 & L_{n+5}^3 & L_{n+6}^3 \end{vmatrix}$

14. Let A be the $n \times n$ matrix

$$\begin{bmatrix} 3 & i & 0 & 0 & \cdots & 0 & 1 \\ i & 1 & i & 0 & \cdots & 0 & 0 \\ 0 & i & 1 & 0 & \cdots & 0 & 0 \\ 0 & 0 & i & 1 & \cdots & 0 & 0 \\ & & & & \vdots & & \\ 0 & 0 & 0 & 0 & \cdots & 1 & i \\ 0 & i & 1 & 0 & \cdots & i & 1 \end{bmatrix}$$

    where $i = \sqrt{-1}$. Prove that $|A| = L_{n+1}$. (Byrd, 1963)

Consider the $n \times n$ determinant

$$g_{n+1}(x) = \begin{bmatrix} 2x & i & 0 & 0 & \cdots & 0 & 1 \\ i & 2x & i & 0 & \cdots & 0 & 0 \\ 0 & i & 2x & i & \cdots & 0 & 0 \\ 0 & 0 & i & 2x & \cdots & 0 & 0 \\ & & & & \vdots & & \\ 0 & 0 & 0 & 0 & \cdots & 2x & i \\ 0 & i & 1 & 0 & \cdots & i & 2x \end{bmatrix}$$

    where $g_0(x) = 0$, $g_1(x) = 1$, $i = \sqrt{-1}$, and x is any real number (Byrd, 1963).

15. Find the recurrence relation satisfied by g.
16. Deduce the value of $g_{n+1}(1/2)$.
17. Show that $|e^{Q^n}| = e^{L_n}$, where e denotes the base of the natural logarithm and Q the Q-matrix (Hoggatt and King, 1963).

18. Compute the following determinant (Lucas).

$$\begin{vmatrix} 1 & 1 & 0 & 0 & 0 & \cdots & 0 & 0 & 0 & 0 \\ -1 & 1 & 1 & 0 & 0 & \cdots & 0 & 0 & 0 & 0 \\ 0 & -1 & 1 & 1 & 0 & \cdots & 0 & 0 & 0 & 0 \\ & & & & & \vdots & & & & \\ 0 & 0 & 0 & 0 & 0 & \cdots & 0 & -1 & 1 & 1 \\ 0 & 0 & 0 & 0 & 0 & \cdots & 0 & 0 & -1 & 1 \end{vmatrix}$$

19. Let g_n denote the number of nonzero terms in the expansion of the determinant

$$\begin{vmatrix} a_1 & b_1 & 0 & 0 & 0 & \cdots & 0 & 0 & 0 & 0 \\ -1 & a_2 & b_2 & 0 & 0 & \cdots & 0 & 0 & 0 & 0 \\ 0 & -1 & a_3 & b_3 & 0 & \cdots & 0 & 0 & 0 & 0 \\ & & & & & \vdots & & & & \\ 0 & 0 & 0 & 0 & 0 & \cdots & 0 & -1 & a_{n-1} & b_{n-1} \\ 0 & 0 & 0 & 0 & 0 & \cdots & 0 & 0 & -1 & a_n \end{vmatrix}$$

where $a_i b_i \neq 0$. Show that $g_n = F_{n+1}$ (Bridger, 1967).

20. Show that

$$\begin{vmatrix} 2 & 2 & -1 & 0 & 0 & \cdots & 0 & 0 & 0 & 0 \\ -1 & 2 & 2 & -1 & 0 & \cdots & 0 & 0 & 0 & 0 \\ 0 & -1 & 2 & 2 & -1 & \cdots & 0 & 0 & 0 & 0 \\ 0 & 0 & -1 & 2 & 2 & \cdots & 0 & 0 & 0 & 0 \\ & & & & & \vdots & & & & \\ 0 & 0 & 0 & 0 & 0 & \cdots & 0 & -1 & 1 & 1 \\ 0 & 0 & 0 & 0 & 0 & \cdots & 0 & 0 & -1 & 1 \end{vmatrix} = F_{n+1}F_{n+2}$$

(Lind, 1971).

21. Evaluate the determinant

$$D_n = \begin{vmatrix} a+b & ab & 0 & 0 & \cdots & 0 & 0 \\ 1 & a+b & ab & 0 & \cdots & 0 & 0 \\ 0 & 1 & a+b & ab & \cdots & 0 & 0 \\ & & & & \vdots & & \\ 0 & & & & \cdots & 1 & a+b \end{vmatrix} \qquad \text{(Church, 1964).}$$

22. Prove that

$$\begin{vmatrix} F_n^2 & F_{n+1}^2 & F_{n+2}^2 \\ F_{n+1}^2 & F_{n+2}^2 & F_{n+3}^2 \\ F_{n+2}^2 & F_{n+3}^2 & F_{n+4}^2 \end{vmatrix} = 2(-1)^{n+1} \qquad \text{(Alfred, 1963).}$$

(*Hint*: $F_{n+3}^2 = 2F_{n+2}^2 + 2F_{n+1}^2 - F_n^2$.)

FIBONACCI AND LUCAS CONGRUENCES

We can employ congruence relation to extract many interesting properties of both Fibonacci and Lucas numbers. For instance, in Chapter 16, we found that $F_m|F_n$ if and only if $m|n$; that is, $F_n \equiv 0 \pmod{F_m}$ if and only if $n \equiv 0 \pmod{m}$. In particular, $F_n \equiv 0 \pmod{F_5}$ if and only if $n \equiv 0 \pmod{5}$; that is, $F_n \equiv 0 \pmod{5}$ if and only if $n \equiv 0 \pmod{5}$. Thus, *beginning with F_0, every fifth Fibonacci number is divisible by* 5, *a fact we already knew.*

Furthermore, $F_3|F_n$ if and only if $3|n$; that is, $F_n \equiv 0 \pmod{2}$ if and only if $n \equiv 0 \pmod{3}$. Thus, *beginning with F_0, every third Fibonacci number is divisible by* 2, another fact we already knew.

Consequently, $F_n \equiv 0 \pmod{5}$ if and only if $n \equiv 0 \pmod{5}$, and $F_n \equiv 0 \pmod{2}$ if and only if $n \equiv 0 \pmod{3}$. Also, $F_n \equiv 0 \pmod{10}$ if and only if $n \equiv 0 \pmod{15}$. In other words, *beginning with F_0, every fifteenth Fibonacci number ends in a zero, and conversely, if a Fibonacci number ends in a zero, then* n *is divisible by* 15. For example, $F_{15} = 610$, $F_{30} = 832{,}040$, and $F_{45} = 1{,}134{,}903{,}170$ end in a zero.

Are there Lucas numbers ending in a zero? To answer this, let us see if there are Lucas numbers ending in a 5. Suppose $L_n \equiv 0 \pmod{5}$ for some integer n. Then, by the binomial theorem, we have

$$\begin{aligned}
2^n L_n &= (1+\sqrt{5})^n + (1-\sqrt{5})^n \\
&= \sum_0^n \binom{n}{i} (\sqrt{5})^n + \sum_0^n \binom{n}{i} (-\sqrt{5})^n \\
&= 2 \sum_{j=0}^{\lfloor n/2 \rfloor} \binom{n}{2j} 5^j = 2(1+5m)
\end{aligned}$$

for some integer m. This yields $2^{n-1}L_n = 1 + 5m$. Since $L_n \equiv 0 \pmod 5$, this equation implies $1 \equiv 0 \pmod 5$, which is a contradiction. Thus, *no Lucas number is divisible by* 5. *It now follows that no Lucas numbers end in a zero.*

Next we turn to a few additional Fibonacci and Lucas congruences. We shall prove a few of them and keep the others as routine exercises.

Theorem 34.1.

(1) $L_n \equiv 0 \pmod 2$ if and only if $n \equiv 0 \pmod 3$ (34.1)

(2) $L_n \equiv 0 \pmod 3$ if and only if $n \equiv 2 \pmod 4$ (34.2)

(3) If $n \equiv 0 \pmod 2$ and $n \not\equiv 0 \pmod 3$, then $L_n \equiv 3 \pmod 4$ (34.3)

(4) $L_{n+2k} \equiv -L_n \pmod{L_k}$, where $k \equiv 0 \pmod 2$ and $k \not\equiv 0 \pmod 3$ (34.4)

(5) $F_{n+2k} \equiv -F_n \pmod{L_k}$, where $k \equiv 0 \pmod 2$ and $k \not\equiv 0 \pmod 3$ (34.5)

(6) $L_{n+12} \equiv L_n \pmod 8$ (34.6)

Proof.

(1) By Exercise 39 in Chapter 5, $5F_n^2 = L_n^2 - 4(-1)^n$. Suppose $L_n \equiv 0 \pmod 2$. Then $5F_n^2 \equiv 0 \pmod 2$, so $F_n^2 \equiv 0 \pmod 2$. Then $F_n \equiv 0 \pmod 2$; that is, $F_n \equiv 0 \pmod{F_3}$. Therefore, $n \equiv 0 \pmod 3$.

Conversely, let $n \equiv 0 \pmod 3$. Then $F_n \equiv 0 \pmod{F_3}$; that is, $F_n \equiv 0 \pmod 2$. This implies $L_n^2 - 4(-1)^n \equiv 0 \pmod 2$, so $L_n \equiv 0 \pmod 2$. Thus, $L_n \equiv 0 \pmod 2$ if and only if $n \equiv 0 \pmod 3$.

(5) By Identity 83 on p. 91, $2F_{m+n} = F_m L_n + F_n L_m$;

$$\begin{aligned}
\therefore \quad 2F_{n+2k} &= F_n L_{2k} + F_{2k} L_n \\
&= F_n[L_k^2 + 2(-1)^{k-1}] + F_{2k} L_n \\
&= 2(-1)^{k-1} F_n + F_{2k} L_n \pmod{L_k} \\
&= -2F_n + F_k L_k L_n \pmod{L_k} \qquad \text{since } F_{2k} = F_k L_k \\
&= -2F_n \pmod{L_k}
\end{aligned}$$

Since $k \not\equiv 0 \pmod 3$, L_k is odd, so $(2, L_k) = 1$. Thus $F_{n+2k} \equiv -F_n \pmod{L_k}$.

(6) Since $2L_{m+n} = 5F_mF_n + L_nL_m$, (see Identity 84 on p. 91) we have

$$\begin{aligned} 2L_{n+12} &= 5F_nF_{12} + L_nL_{12} \\ &= 5 \cdot 144F_n + 322L_n \\ L_{n+12} &= 360F_n + 161L_n \\ &\equiv 0 + L_n \equiv L_n \pmod 8 \end{aligned}$$

■

For example, let $n = 15$ and $k = 8$. Clearly, k is even, and k is not divisible by 3. Then $L_k = L_8 = 47$. We have

$$\begin{aligned} (1) \qquad L_{n+2k} &= F_{31} = 3{,}010{,}349 \equiv 46 \pmod{47} \quad \text{and} \\ -L_n &= -L_{15} = -1364 \equiv 46 \pmod{47} \\ \therefore \quad L_{31} &\equiv -L_{15} \pmod{47} \\ (2) \qquad F_{n+2k} &= F_{31} = 1{,}346{,}269 \equiv 1 \pmod{47} \quad \text{and} \\ -F_{15} &= -610 \equiv 1 \pmod{47} \\ \therefore \quad F_{31} &\equiv -F_{15} \pmod{47} \end{aligned}$$

Property (3) can be stated in words: If n is an even integer, not divisible by 3, that is, if n is of the form $6k \pm 2$, then $L_n \equiv 3 \pmod 4$. For example, let $n = 20 = 6 \cdot 3 + 2$. Then $L_{20} = 15{,}127 \equiv 3 \pmod 4$. Likewise, $L_{16} = 2207 \equiv 3 \pmod 4$.

LUCAS SQUARES

Are there Lucas numbers that are perfect squares? Clearly, $L_1 = 1$ and $L_3 = 4$ are squares. Are there any others? In fact, in April 1964, J. H. E. Cohn of the University of London, UK, established that there are no other such Lucas numbers. His proof hinges on the following formulas:

$$L_{2n} = 2L_n^2 + 2(-1)^{n-1} \tag{34.7}$$

$$\text{If } k \equiv 0 \pmod 2 \text{ and } k \not\equiv 0 \pmod 3, \text{ then } L_k \equiv 0 \pmod 4 \tag{34.8}$$

$$\text{If } k \equiv 0 \pmod 2 \text{ and } k \not\equiv 0 \pmod 3, \text{ then } L_{n+2k} \equiv -L_n \pmod{L_k} \tag{34.9}$$

Theorem 34.2. The only Lucas numbers that are perfect squares are 1 and 4.

Proof. Let L_n be a perfect square x^2 for some positive integer $n : L_n = x^2$.

Suppose n is even, say, $n = 2r$. Then $L_n = L_{2r} = L_r^2 \pm 2$, by Identity 34.7. Since L_r^2 is a square, $L_r^2 \pm 2$ cannot be a perfect square. This is a contradiction.

Suppose n is odd. Let $n \equiv 1 \pmod 4$. If $n = 1$, then $L_n = 1$ is a perfect square. So assume $n > 1$. Then we can write n as $n = 1 + 2 \cdot 3^i k$, where $i \geq 0$ and k is an even integer not divisible by 3. Therefore, by Identity (34.9), $L_n \equiv -L_1 \equiv -1 \pmod{L_k}$. Since -1 is a quadratic residue of L_k by identity (34.8), it follows that L_n cannot be a square.

On the other hand, let $n \equiv 3 \pmod 4$. If $n = 3$, then $L_n = L_3 = 4$ is a perfect square. Suppose $n \neq 3$. Then $n = 3 + 2 \cdot 3^i k$, where $i \geq 0$ and k is an even integer not divisible by 3. Then, by Identity (34.9), $L_n \equiv -L_3 \equiv -4 \pmod{L_k}$. Consequently, L_n cannot be a perfect square.

Thus, the only Lucas numbers that are perfect squares are 1 and 4. ■

The next theorem, also discovered by Cohn in 1964, identifies Lucas numbers of the form $2x^2$. Its proof uses the following formulas:

$$L_n \equiv 0 \pmod 2 \text{ if and only if } n \equiv 0 \pmod 3 \tag{34.1}$$

$$L_n \equiv 0 \pmod 3 \text{ if and only if } n \equiv 2 \pmod 4 \tag{34.2}$$

$$F_{-n} = (-1)^{n-1} F_n \tag{5.17}$$

$$L_{-n} = (-1)^n L_n \tag{5.18}$$

$$L_{n+12} \equiv L_n \pmod 8 \tag{34.6}$$

Theorem 34.3. Let x be an integer such that $L_n = 2x^2$. Then $n = 0$ or ± 6.

Proof. Since $x^2 \equiv 0$, 1, or 4 (mod 8), $L_n = 2x^2 \equiv 0$ or 2 (mod 8).

Since L_n is even, by Identity (34.1), $n \equiv 0 \pmod 3$. So $3|n$.

Suppose n is odd. Then n is of the form $12q + r$, where $0 \leq r < 12$. Since n is odd and is a multiple of 3, $r = 3$ or 9. Thus, n is of the form $12q + 3$ or $12q + 9$. If $n = 12q + 3$, by Identity (34.6), $L_n = L_{12q+3} \equiv L_3 \equiv 4 \pmod 8$. This is a contradiction, since $L_n \equiv 0$ or 2 (mod 8).

If $n = 12q + 9$, $L_n = L_{12q+9} \equiv L_9 \equiv 4 \pmod 8$. This, again, is a contradiction. Thus n cannot be odd.

Suppose n is even. Then n is of the form $8t$, $8t \pm 2$, or $8t + 4$. If $n = 8t$ or $8t = 4$, $n \equiv 0 \pmod 4$. If $n = 0$, then $L_n = L_0 = 2 = 2 \cdot 1^2$ has the desired form. If $n \neq 0$, n has the form $n = 2 \cdot 3^i \cdot k$. Then, $2L_n \equiv -2L_0 \equiv -4 \pmod{L_k}$. Therefore, $2L_n$ cannot be a perfect square y^2. Consequently, L_n cannot be of the form $2x^2$.

Suppose $n \equiv -2 \equiv 6 \pmod 8$. If $n = 6$, then $L_6 = 2 \cdot 3^2$ has the desired property. On the other hand, if $n \neq 6$, then n is of the form $n = 6 + 2 \cdot 3^i \cdot k$, where $4|k$ and $3 \nmid k$. Therefore, $2L_n \equiv -2L_6 \equiv -36 \pmod{L_k}$. It follows, by Identities (34.2) and (34.3), that -36 is a quadratic nonresidue of L_k. Thus, as earlier, L_n cannot be of the form $2x^2$.

Finally, if $n \equiv 2 \pmod 8$, then, by Identity 5.18, $L_{-n} = L_n$. Therefore, $-n \equiv 6 \pmod 8$. This yields $-n = 6$, so $n = -6$.

Thus, if L_n is of the form $2x^2$, then $n = 0$ or ± 6. ■

SQUARE FIBONACCI NUMBERS

Historically, one of the oldest conjectures in the theory of Fibonacci numbers is that 0, 1, and 144 are the only perfect squares. In 1963, M. Wunderlich conducted an extensive computer search among the first one million Fibonacci numbers, but did not unearth any new ones. Surprisingly, the conjecture was confirmed by Cohn the following year.

Theorem 34.4. If F_n is a perfect square x^2, then $n = 0, \pm 1, 2$, or 12.

Proof.

Case 1. Let n be odd. Then $n \equiv \pm 1 \pmod 4$.

Suppose $n \equiv 1 \pmod 4$. If $n = 1$, then $F_1 = 1 = 1^2$ is a perfect square. If $n \neq 1$, then n must be of the form $n = 1 + 2 \cdot 3^i \cdot k$. So, by Identity (34.5), $F_n \equiv -F_1 \equiv -1 \pmod{L_k}$. Consequently, F_n cannot be a perfect square.

On the other hand, suppose $n \equiv -1 \pmod 4$; that is, $-n \equiv 1 \pmod 4$. Then $F_{-n} = F_n$, so, by the preceding subcase, $-n = 1$, that is, $n = -1$.

Case 2. Let n be even, say, $n = 2s$. Then $F_n = F_{2s} = F_s L_s = x^2$.

Suppose $3|n$. Then $F_3|F_n$, that is, $2|F_n$. So $F_s = 2y^2$ and $L_s = 2z^2$ for some integers y and z. Then, by Theorem 34.3, $n/2 = s = 0$ or ± 6; that is, $n = 0$ or ± 12.

When $n = 0$, $F_s = F_0 = 0 = 2 \cdot 0^2$; when $n = 12$, $F_s = F_6 = 8 = 2 \cdot 2^2$, but, when $n = -12$, $F_s = F_{-6} = (-1)^5 F_6 = -8$, which is not of the form $2y^2$. Therefore, $n = 0$ or 12.

Suppose $3 \nmid n$, so F_n is not even. Then $F_s = y^2$ and $L_s = z^2$ for some integers y and z. By Theorem 34.2, $n/2 = s = 1$ or 3, so $n = 2$ or 6.

When $n = 2$, $F_s = F_1 = 1^2$; but when $n = 6$, $F_s = F_3 = 2$ is not a perfect square. Thus $n = 0, \pm 1, 2$, or 12, as desired. ∎

It follows from this theorem that the only distinct positive Fibonacci numbers that are perfect squares are 1 and 144.

Cohn also proved that 0, 2, and 8 are the only Fibonacci numbers that are of the form $2x^2$. This is the essence of the next theorem.

Theorem 34.5. If F_n is of the form $2x^2$, then $n = 0, \pm 3$, or 6.

Proof.

Case 1. Let n be odd. Then $n \equiv \pm 1 \pmod 4$.

Suppose $n \equiv -1 \equiv 3 \pmod 4$. If $n = 3$, then $F_n = F_3 = 2 = 2 \cdot 1^2$ has the desired form. If $n \neq 3$, n is of the form $n = 3 + 2 \cdot 3^i \cdot k$. Then, by formula (34.5), $2F_n \equiv -2F_3 \equiv -4 \pmod{L_k}$, that is, $x^2 \equiv -1 \pmod{L_k}$, so F_n cannot be of the given form.

Suppose $n \equiv 1 \equiv -3 \pmod 4$. Then $-n \equiv 3 \pmod 4$ and $F_{-n} = F_n$, so $-n = 3$ by the preceding paragraph. This yields $n = -3$.

Case 2. Let n be even, say, $n = 2s$. Then $F_n = F_{2s} = F_s L_s = 2x^2$. So either $F_s = y^2$ and $L_s = 2z^2$, or $F_s = 2y^2$ and $L_s = z^2$ for some integers y and z.

By Theorems 34.3 and 34.4, the only value of s that satisfies the equations $F_s = y^2$ and $L_s = 2z^2$ is $s = 0$. Then $n = 0$.

Suppose $F_s = y^2$ and $L_s = 2z^2$. By Theorem 34.2, $s = 1$ or 3. But F_1 is not of the form $2y^2$, so $s \neq 1$. But $F_3 = 2 \cdot 1^2$, so $F_s = 2y^2$ is solvable when $s = 3$. Then $n = 6$.

Thus, collecting all the possible values of n, we have $n = 0, \pm 3$, or 6. ■

A GENERALIZED FIBONACCI CONGRUENCE

In 1963, J. A. Maxwell proposed the following problem:

Generalize the congruences

$$F_{n+1}2^n + F_n 2^{n+1} \equiv 1 \pmod 5 \qquad F_{n+1}3^n + F_n 3^{n+1} \equiv 1 \pmod{11}$$

$$F_{n+1}5^n + F_n 5^{n+1} \equiv 1 \pmod{29}$$

where $n \geq 0$. Their generalization, although not quite obvious, is

$$F_{n+1}p^n + F_n p^{n+1} \equiv 1 \pmod{p^2 + p - 1} \tag{34.10}$$

where p is an arbitrary prime (see Exercise 25).

Interestingly enough, we can generalize Congruence 34.10 even further to include the generalized Fibonacci numbers G_n, where $n \geq 2$.

The following theorem gives the desired generalization.

Theorem 34.6. (Koshy, 1999). Let $m \geq 2$ and $n \geq 0$. Then

$$G_{n+1}m^n + G_n m^{n+1} \equiv a(1-m) + bm \pmod{m^2 + m - 1} \tag{34.11}$$

Proof. We shall prove this using the strong version of induction on n.

Since $G_1 m^0 + G_0 m = a + (b-a)m = a(1-m) + bm \equiv a(1-m) + bm \pmod{m^2 + m - 1}$ and $G_2 m + G_1 m^2 = bm + am^2 \equiv a(1-m) + bm \pmod{m^2 + m - 1}$, the result is true when $n = 0$ and 1.

Now assume it is true for all integers i, where $0 \leq i \leq k$ and $k \geq 1$. Then

$$G_k m^{k-1} + G_{k-1}m^k \equiv a(1-m) + bm \pmod{m^2 + m - 1}$$

$$G_{k+1}m^k + G_k m^{k+1} \equiv a(1-m) + bm \pmod{m^2 + m - 1}$$

$$\begin{aligned}
\therefore \quad G_{k+2}m^{k+1} + G_{k+1}m^{k+2} &= (G_k + G_{k+1})m^{k+1} + (G_{k-1} + G_k)m^{k+2} \\
&= m(G_{k+1}m^k + G_k m^{k+1}) + m^2(G_k m^{k-1} + G_{k-1}m^k) \\
&\equiv m[a(1-m) + bm] + m^2[a(1-m) + bm]
\end{aligned}$$

$$
\begin{aligned}
&(\bmod\ m^2 + m - 1)\\
&= (m + m^2)[a(1-m) + bm]\ (\bmod\ m^2 + m - 1)\\
&= 1[a(1-m) + bm]\ (\bmod\ m^2 + m - 1)\\
&= [a(1-m) + bm]\ (\bmod\ m^2 + m - 1)
\end{aligned}
$$

Thus Congruence 34.11 is true for all integers $n \geq 0$. ■

Corollary 34.1.

$$F_{n+1}m^n + F_n m^{n+1} \equiv 1 (\bmod\ m^2 + m - 1) \tag{34.12}$$

$$L_{n+1}m^n + L_n m^{n+1} \equiv 1 + 2m (\bmod\ m^2 + m - 1) \tag{34.13}$$

■

Formula (34.12) follows from Eq. (34.11) when $a = 1 = b$, and Eq. (34.13) when $a = 1$ and $b = 3$.

For example, let $n = 12$ and $m = 15$, so $m^2 + m - 1 = 239$. Then

$$
\begin{aligned}
F_{13}15^{12} + F_{12}15^{13} &= 233 \cdot 15^{12} + 144 \cdot 15^{13}\\
&\equiv 233 \cdot 80 + 144 \cdot 5\ (\bmod\ 239)\\
&\equiv 1\ (\bmod\ 55)
\end{aligned}
$$

Likewise,

$$
\begin{aligned}
L_{17}7^{16} + L_{16}7^{17} &= 3571 \cdot 7^{16} + 2207 \cdot 7^{17}\\
&\equiv 3571 \cdot 26 + 2207 \cdot 17\ (\bmod\ 55)\\
&\equiv 15 \equiv 1 + 2 \cdot 7\ (\bmod\ 55)
\end{aligned}
$$

Interestingly enough, Theorem 34.6 can be extended to negative subscripts. To establish the generalization, we need the following lemma, which we shall establish using strong induction.

Lemma 34.1. (Koshy, 1999). Let $m \geq 2$ and $n \geq 0$. Then

$$F_{n-1} - mF_n \equiv (-1)^n m^n\ (\bmod\ m^2 + m - 1) \tag{34.14}$$

Proof. Since $F_{-1} - mF_0 = 1 - 0 = 1 \equiv (-1)^0 m^0\ (\bmod\ m^2 + m - 1)$ and $F_0 - mF_1 = 0 - m \equiv (-1)^1 m^1\ (\bmod\ m^2 + m - 1)$, the result is true for $n = 0$ and $n = 1$.

Now, assume it is true for every nonnegative integer $\le k$, where $k \ge 2$. Then

$$\begin{aligned}
F_k - mF_{k+1} &= (F_{k-2} + F_{k-1}) - m(F_{k-1} + F_k) \\
&= (F_{k-2} - mF_{k-1}) + (F_{k-1} - mF_k) \\
&\equiv (-1)^{k-1}m^{k-1} + (-1)^k m^k \pmod{m^2 + m - 1} \\
&\equiv (-1)^{k-1}m^{k-1}(1 - m) \pmod{m^2 + m - 1} \\
&\equiv (-1)^{k-1}m^{k-1}m^2 \pmod{m^2 + m - 1} \\
&\equiv (-1)^{k+1}m^{k+1} \pmod{m^2 + m - 1}
\end{aligned}$$

Thus, by strong induction, Formula 34.14 is true for all $n \ge 0$. ■

For instance, $F_{12} - 5F_{13} = 144 - 5 \cdot 233 \equiv 23 \equiv (-1)^{13}5^{13} \pmod{29}$.

Curiously enough, Formula 34.14 does not hold for Lucas numbers, that is, $L_{n-1} - mL_n \not\equiv (-1)^n m^n \pmod{m^2 + m - 1}$. For instance, $L_9 - 6L_{10} = 76 - 6 \cdot 123 \equiv 35 \pmod{41}$, whereas $(-1)^{10}6^{10} \equiv 32 \pmod{41}$.

The lemma yields a fascinating by-product, as the next corollary reveals.

Corollary 34.2. $L_n \equiv (-1)^n 2^{n+1} \equiv 2 \cdot 3^n \pmod 5$.

Proof. By Lemma 34.1, $F_n - 2F_{n+1} \equiv (-1)^{n+1}2^{n+1} \pmod 5$. But $2F_{n+1} - F_n = F_{n+1} + (F_{n+1} - F_n) = F_{n+1} + F_{n-1} = L_n$. Therefore, $-L_n \equiv (-1)^{n+1}2^{n+1} \pmod 5$; that is, $L_n \equiv (-1)^n 2^{n+1} \pmod 5$. ■

For example, $L_{10} = 123 \equiv 3 \equiv (-1)^{11}2^{11} \pmod 5$.

The next corollary follows from Corollary 34.2 since $3^4 \equiv 1 \pmod 5$.

Corollary 34.3. $L_{4n} \equiv 2 \pmod 5$, $L_{4n+1} \equiv 1 \pmod 5$, $L_{4n+2} \equiv 3 \pmod 5$, and $L_{4n+3} \equiv 4 \pmod 5$. ■

For example, $L_{12} = 322 \equiv 2 \pmod 5$, $L_{13} = 521 \equiv 1 \pmod 5$, $L_{14} = 843 \equiv 3 \pmod 5$, and $L_{14} = 1364 \equiv 4 \pmod 5$.

With Lemma 34.1 and the fact that

$$G_{-n} = (-1)^{n+1}(aF_{n+2} - bF_{n+1}) \tag{34.15}$$

we are now ready to generalize Theorem 34.6.

Theorem 34.7. (Koshy, 1999). Let $m \ge 2$ and n be any integer. Then

$$G_{n+1}m^n + G_n m^{n+1} \equiv a(1 - m) + bm \pmod{m^2 + m - 1}$$

Proof. By virtue of Theorem 34.6, it suffices to show that Formula (34.11) holds when n is negative. We have:

$$\begin{aligned}
G_{-n+1}m^{-n} + G_{-n}m^{-n+1} &= m^{-n}G_{-(n-1)} + m^{-n+1}G_{-n} \\
&= m^{-n}[(-1)^n(aF_{n+1} - bF_n)] \\
&\quad + m^{-n+1}[(-1)^{n+1}(aF_{n+2} - bF_{n+1})] \\
&= a(-1)^n m^{-n}(F_{n+1} - mF_{n+2}) \\
&\quad - b(-1)^n m^{-n}(F_n - mF_{n+1}) \\
&\equiv a(-1)^n m^{-n}[(-1)^{n+2}m^{n+2}] \\
&\quad - b(-1)^n m^{-n}[(-1)^{n+1}m^{n+1}] \pmod{m^2 + m - 1} \\
&\equiv am^2 + bm \pmod{m^2 + m - 1} \\
&\equiv a(1-m) + bm \pmod{m^2 + m - 1}
\end{aligned}$$

as desired. ■

For example, let $n = -8$ and $m = 6$. Then

$$\begin{aligned}
L_{-7}6^{-8} + L_{-8}6^{-7} &= (-29) \cdot 6^{-8} + 47 \cdot 6^{-7} \\
&\equiv 12 \cdot 6^{-8} + 6 \cdot 6^{-7} \equiv 12 \cdot 6^{-8} + 6^{-6} \pmod{41} \\
&\equiv (2 \cdot 6^{-1} + 1) \cdot 6^{-6} \equiv (2 \cdot 7 + 1) \cdot 7^6 \pmod{41} \\
&\equiv 15 \cdot 20 \equiv 13 \equiv 1 + 2 \cdot 6 \pmod{41}
\end{aligned}$$

Likewise, $F_{-7}5^{-8} + F_{-8}5^{-7} = 13 \cdot 5^{-8} + (-21) \cdot 5^{-7} \equiv 1 \pmod{29}$.

The next theorem shows that every prime p divides some Fibonacci number. Its proof employs the number-theoretic Legendre symbol, so we omit the proof. (See Hardy and Wright, p. 150, for a proof.)

Theorem 34.8. Let p be a prime. Then $F_{p-1} \equiv 0 \pmod{p}$ if $p \equiv \pm 1 \pmod{5}$ and $F_{p+1} \equiv 0 \pmod{p}$ if $p \equiv \pm 2 \pmod{5}$. ■

For example, let $p = 19 \equiv -1 \pmod{5}$. Then $F_{p-1} = F_{18} = 2584 \equiv 0 \pmod{19}$. Likewise, $p = 23 \equiv -2 \pmod{5}$ and $F_{p+1} = F_{24} = 46{,}368 \equiv 0 \pmod{23}$.

The next result was discovered in 1967 by Martin Pettet of Toronto. We will need it in the proof of the following theorem, so we label it a lemma. The theorem was discovered in 1970 by J. E. Desmond of Florida State University.

Lemma 34.2. (Pettet, 1967) Let p be a prime. Then $L_p \equiv 1$ (mod p).

Proof. By the binomial theorem,

$$L_p = \frac{1}{2^{p-1}} \sum_{i=0}^{\lfloor n/2 \rfloor} \binom{p}{2i} 5^i$$

Since p is a prime, $\binom{p}{j} \equiv 0 \pmod{p}$ for $0 < j < p$. Therefore,

$$L_p \equiv \frac{1}{2^{p-1}} \pmod{p}$$

But, by Fermat's little theorem, $2^{p-1} \equiv 1 \pmod{p}$. Thus $L_p \equiv 1$ (mod p). ■

For instance, let $L_{43} = 969{,}323{,}029 \equiv 1 \pmod{43}$, as expected.

Theorem 34.9. (Desmond, 1970) Let p be a prime. Then $F_{np} \equiv F_n F_p \pmod{p}$ and $L_{np} \equiv L_n L_p \equiv L_n \pmod{p}$.

Proof. [by the principle of mathematical induction (PMI)] We shall prove the first result and leave the second as an exercise. The statement is clearly true when $n = 0$ and $n = 1$. So assume it is true for every integer $n \leq k$, where $k \geq 1$.

Since $F_{r+s} = F_r L_s + (-1)^s F_{r-s}$ (this identity was established in 1963 by I. D. Ruggles), $F_{np+p} = F_{np} L_p + (-1)^{p+1} F_{np-p}$. So $F_{(n+1)p} \equiv F_{np} + F_{(n-1)p} \pmod{p}$, by Lemma 34.1. Then

$$\begin{aligned} F_{(k+1)p} &\equiv F_{kp} + F_{(k-1)p} \equiv F_k F_p + F_{k-1} F_p \pmod{p} \\ &\equiv (F_k + F_{k-1}) F_p \equiv F_{k+1} F_p \pmod{p} \end{aligned}$$

Thus $F_{np} \equiv F_n F_p \pmod{p}$ for every prime p and for every integer $n \geq 0$. ■

For example, let $n = 4$ and $p = 7$. Then $F_n = F_4 = 3$, $F_p = F_7 = 13$, and $F_{np} = F_{28} = 317{,}811 \equiv 4 \equiv 3 \cdot 13 = F_4 F_7 \pmod{7}$. Likewise, $L_{28} = 710{,}647 \equiv 7 \cdot 29 = L_4 L_7 \pmod{7}$.

The next result was developed in 1977 as an advanced problem by L. Carlitz of Duke University.

Theorem 34.10. Then $L_{p^2} \equiv 1 \pmod{p^2}$ if and only if $L_p \equiv 1 \pmod{p^2}$, where p is a prime.

Proof. We have

$$1 = (\alpha + \beta)^n = L_n + \sum_{1}^{n-1} \binom{n}{k} \alpha^k \beta^{n-k}$$

$$\therefore \quad L_p = 1 - \sum_{1}^{p-1} \binom{p}{k} \alpha^k \beta^{p-k} \qquad \text{and} \qquad L_{p^2} = 1 - \sum_{1}^{p^2-1} \binom{p^2}{k} \alpha^k \beta^{p^2-k}$$

Since $\binom{p}{k} = \frac{p}{k} \binom{p-1}{k-1}$ and $\binom{p-1}{k-1} \equiv (-1)^{k-1} \pmod{p}$, it follows that $L_p \equiv 1 \pmod{p^2}$ if and only if

$$\sum_{1}^{p-1} \frac{(-1)^{k-1}}{k} \alpha^k \beta^{p-k} \equiv 0 \pmod{p}$$

Since $p \nmid k$, $\binom{p^2}{k} \equiv 0 \pmod{p^2}$ and

$$\binom{p^2}{pk} = \frac{p}{k}\binom{p^2-1}{pk-1} \equiv (-1)^{k-1}\frac{p}{k} \pmod{p^2}$$

Thus $L_{p^2} \equiv 1 \pmod{p^2}$ if and only if

$$\sum_1^{p-1} \frac{(-1)^{k-1}}{k}\alpha^{pk}\beta^{p^2-pk} \equiv 0 \pmod{p}$$

Since

$$\left(\sum_1^{p-1} \frac{(-1)^{k-1}}{k}\alpha^k\beta^{p-k}\right)^p \equiv \sum_1^{p-1} \frac{(-1)^{k-1}}{k}\alpha^{pk}\beta^{p^2-pk} \pmod{p}$$

it follows that $L_{p^2} \equiv 1 \pmod{p^2}$ if and only if $\sum_1^{p-1}((-1)^{k-1}/k)\alpha^k\beta^{p-k} \equiv 0 \pmod{p}$. Thus $L_{p^2} \equiv 1 \pmod{p^2}$ if and only if $L_p \equiv 1 \pmod{p^2}$. ■

More generally, we have the following result.

Theorem 34.11. (Carlitz, 1977) $L_{p^m} \equiv 1 \pmod{p^2}$ if and only if $L_p \equiv 1 \pmod{p^2}$, where p is a prime and $m \geq 2$. ■

The next example was proposed as a problem in 1977 by G. Berzsenyi of Lamar University in Texas.

Example 34.1. Prove that $L_{2mn}^2 \equiv 3 \pmod{10}$, where $m, n \geq 1$.

Solution. Since $L_{(2k+1)m} = (\alpha^m)^{2k+1} + (\beta^m)^{2k+1}$, it follows that $L_m | L_{(2k+1)m}$.

Case 1. Let $n = 2k + 1$ be odd. Then, by Exercise 41 in Chapter 5,

$$L_{2nm} = L_{nm}^2 - 2(-1)^{nm} = L_{(2k+1)m}^2 - 2(-1)^m$$

$$\therefore \quad L_{(2k+1)m}^2 = \begin{cases} L_{2nm} + 2 & \text{if } m \text{ is even} \\ L_{2nm} - 2 & \text{otherwise} \end{cases}$$

Consequently, $(L_{2nm} + 2)(L_{2nm} - 2) = L_{2nm}^2 - 4$ is divisible by $L_{(2k+1)m}^2$, and hence by L_m^2. Thus $L_{2nm}^2 \equiv 4 \pmod{L_m^2}$, where m is odd.

Case 2. Let $n = 2k$ be even. We shall prove this half by PMI on k. When $k = 1$,

$$\begin{aligned} L_{2nm} &= L_{4m} = L_{2m}^2 - 2 \\ &= [L_m^2 - 2(-1)^m]^2 - 2 = L_m^4 - 4(-1)^m L_m^2 + 2 \end{aligned}$$

Thus $L_{2nm} \equiv 2 \pmod{L_m^2}$, so $L_{2nm}^2 \equiv 4 \pmod{L_m^2}$.

Assume the result holds for all positive even integers less than $n = 2k$. Then

$$L_{2nm} = L_{4km} = L_{2km}^2 - 2$$
$$L_{2nm} - 2 = L_{2km}^2 - 4$$

By Case 1 and the inductive hypothesis, $L_{2km}^2 \equiv 4 \pmod{L_m^2}$. Thus $L_{2nm} \equiv 2 \pmod{L_m^2}$, so $L_{2nm}^2 \equiv 4 \pmod{L_m^2}$.

Thus the result holds in both cases. ■

An intriguing theorem that combines Fibonacci numbers with *Euler's phi function* φ. $\varphi(n)$ denotes the number of positive integers $\le n$ and relatively prime to it. For example, $\varphi(1) = 1$, $\varphi(5) = 4$, and $\varphi(6) = 2$.

The theorem was proposed originally in 1965 as an advanced problem in *The Fibonacci Quarterly* by D. Lind of Falls Church, Virginia. An incomplete proof by J. L. Brown, Jr., of Pennsylvania State University appeared in the *Quarterly* in the following year. It resurfaced as a problem in 1976 by C. Kimberling of the University of Evansville, Indiana, in *The American Mathematical Monthly*. In the following year, P. L. Montgomery of Huntsville, Alabama, provided an elegant solution using group theory. In 1980, V. E. Hoggatt, Jr., and H. Edgar of San Jose State University provided an alternate proof of the theorem.

Theorem 34.12. $\varphi(F_n) \equiv 0 \pmod 4$, where $n \ge 5$. ■

For example, $\varphi(F_{10}) = \varphi(55) = \varphi(5\cdot 11) = \varphi(5)\cdot\varphi(11) = 4\cdot 10 \equiv 0 \pmod 4$ and $\varphi(F_{17}) = \varphi(1597) = 1596 \equiv 0 \pmod 4$.

EXERCISES 34

Verify Corollary 34.1 for the given values of m and n.

1. $m = 13, n = 5$
2. $m = 20, n = 11$

Verify Corollary 34.2 for the given values of m and n.

3. $m = 11, n = 5$
4. $m = 18, n = 7$

Prove each, where $m, n, k \ge 1$ and p is a prime.

5. $F_n \equiv 0 \pmod 3$ if and only if $n \equiv 0 \pmod 4$.
6. $F_n \equiv 0 \pmod 4$ if and only if $n \equiv 0 \pmod 6$.
7. $F_n \equiv 0 \pmod 5$ if and only if $n \equiv 0 \pmod 5$.
8. $L_n^2 \equiv L_{2n} \pmod 2$
9. $L_n^2 \equiv F_n^2 \pmod 4$
10. $L_n^2 \equiv L_{n-1}L_{n+1} \pmod 5$, $n \ge 2$
11. $2L_{m+n} \equiv L_m L_n \pmod 5$
12. $F_{15n} \equiv 0 \pmod{10}$

13. $L_{(2k-1)n} \equiv 0 \pmod{L_n}$
14. $F_{2n} \equiv (-1)^n n \pmod 5$
15. $F_{n+24} \equiv F_n \pmod 9$ (Householder, 1963)
16. $F_{n+3} \equiv F_n \pmod 2$
17. $F_{3n} \equiv 0 \pmod 2$
18. $F_{n+5} \equiv 3F_n \pmod 5$
19. $F_{5n} \equiv 0 \pmod 5$
20. Identity (34.2)
21. Identity (34.3)
22. Identity (34.4)
23. $L_{3n} \equiv 0 \pmod 2$
24. $L_{3n} \equiv 0 \pmod{L_n}$
25. $(F_n, L_n) = 2$ if and only if $n \equiv 0 \pmod 3$
26. $L_n \equiv 3L_{n-1} \pmod 5$
27. $3L_{2n-2} \equiv (-1)^{n-1} \pmod 5$
28. $nL_n \equiv F_n \pmod 5$ (Wall, 1964)
29. $L_p \equiv 1 \pmod p$ (Pettet, 1967)
30. $2^n L_n \equiv 2 \pmod 5$ (Wall, 1968)
31. $2^n F_n \equiv 2n \pmod 5$ (Wall, 1968)
32. $L_{np} \equiv L_n \pmod p$ (Desmond, 1970)
33. $F_{5^n} \equiv 5^n \pmod{5^{n+3}}$ (Bruckman, 1980)
34. $L_{5^n} \equiv L_{5^{n+1}} \pmod{5^{n+3}}$ (Bruckman, 1980)
35. $(5F_n^2)^2 + 4^2 \equiv (L_n^2)^2 \pmod{5F_n^2}$ (Freitag, 1982)
36. Let $L(n) = L_n$ and $t_n = n(n+1)/2$. Then $L(n) \equiv (-1)^{t_n - 1} \pmod 5$ (Freitag, 1982).
37. $\sum_{i=1}^{20} F_{n+i} \equiv 0 \pmod{F_{10}}$ (Ruggles, 1963)
38. The Lucas numbers L_{2^n} end in the digit 7; that is, $L_4, L_8, L_{16}, L_{32}, \ldots$ end in 7.
39. Compute $F_{F_5}, F_{L_5}, L_{F_5}, L_{L_5}$.

Prove each, where $n \geq 1$.

40. $F_{L_n} \not\equiv -0 \pmod 5$
41. $F_{F_n} \equiv 0 \pmod 5$ if and only if $n \equiv 0 \pmod 5$
42. $F_n \equiv L_n \pmod 2$
43. $F_m F_n \equiv L_m L_n \pmod 2$

FIBONACCI AND LUCAS PERIODICITY

A cursory examination of the units digits of the Fibonacci numbers F_0 through F_{59} reveals no obvious or interesting pattern. But then take a look at the ones digits in F_{60} and F_{61}; they are the same as those in F_1 and F_2. That is, $F_{60} \equiv F_0 \pmod{10}$ and $F_{61} \equiv F_1 \pmod{10}$. So, by virtue of the Fibonacci recurrence relation (FRR), the pattern continues: $F_{60+i} \equiv F_i \pmod{10}$.

More generally, we have the following result.

Theorem 35.1. Let $i, n \geq 0$. Then $F_{60n+i} \equiv F_i \pmod{10}$.

Proof. [by the principle of mathematical induction (PMI)] The statement is clearly true when $n = 0$. So, assume it is true for an arbitrary integer $k \geq 0$: $F_{60k+i} \equiv F_i \pmod{10}$. Notice from the Fibonacci table that $F_{60} \equiv 0 \pmod{10}$ and $F_{59} \equiv 1 \pmod{10}$. Then:

$$\begin{aligned} F_{60(k+1)+i} &= F_{(60k+i)+60} \\ &= F_{60k+i+1}F_{60} + F_{60k+i}F_{59}, \qquad \text{by Identity 5.22} \\ &\equiv F_{60k+i} \cdot 0 + F_i \cdot 1 \pmod{10} \\ &\equiv F_i \pmod{10} \end{aligned}$$

Thus, by PMI, the statement is true for every $n \geq 0$. ■

For example, $F_{14} = 377 \equiv 7 \pmod{10}$, so $F_{74} \equiv F_{14} \equiv 7 \pmod{10}$. To confirm this, notice that $F_{74} = 1,304,969,544,928,657$ ends in 7.

Let p be the smallest positive integer such that $F_{p+i} \equiv F_i \pmod{10}$ for every i. Then p is called the *period* of the Fibonacci sequence modulo 10. By virtue of

Theorem 35.1, $p \le 60$. But when we examine the ones digits in F_0 through F_{59}, we see no repetitive pattern, so $p \ge 60$. Thus $p = 60$; that is, the *period of the Fibonacci sequence modulo 10 is 60.*

In 1963, using an extensive computer search, S. P. Geller of the University of Alaska established that the last two digits of F_n repeat every 300 times, the last three every 1500, the last four every 15,000, the last five every 150,000, and the last six digits every 1,500,000 times. That is,

$$\begin{aligned} F_{n+300} &\equiv F_n(\text{mod } 300), \qquad F_{n+1500} \equiv F_n(\text{mod } 1500), \\ F_{n+15,000} &\equiv F_n(\text{mod } 15,000), \\ F_{n+150,000} &\equiv F_n(\text{mod } 100,000), \qquad \text{and} \\ F_{n+1,500,000} &\equiv F_n(\text{mod } 1,000,000) \end{aligned}$$

Thus $F_{n+300} \equiv F_n$ (mod 300) and $F_{n+1.5\times 10^k} \equiv F_n$ (mod 10^k), where $3 \le k \le 6$.

It is reasonable to ask if the Lucas sequence is also periodic modulo 10. If it is, what is the period? To answer these questions, let us extract the ones digits in L_0 through L_{11}. They are 2, 1, 3, 4, 7, 1, 8, 9, 7, 6, 3, and 9. There is no pattern so far, so the period is at least 12. But, by virtue of the Lucas recurrence relation (LRR), we obtain every residue modulo 10 by the sum of the two previous residues modulo 10. So the next two units digits are 2 and 1. Clearly, a pattern begins to emerge. Thus, *the Lucas sequence modulo 10 is also periodic*; *its period is 12*; *that is*, $L_{12+i} \equiv L_i$ (*mod* 10).

More generally, we have the following theorem. We leave its proof as an exercise (see Exercise 10).

Theorem 35.2. Let $i, n \ge 0$. Then $L_{12n+i} \equiv L_i$ (mod 10). ■

For example, $L_{11} = 199 \equiv 9$ (mod 10), so, by Theorem 35.2, $L_{47} \equiv 9$ (mod 10). This is true since $L_{47} \equiv 6,643,838,879$ ends in 9.

SQUARE LUCAS NUMBERS REVISITED

In February 1964, Br. Alfred published a neat and simple proof of the fact that 1 and 4 are the only Lucas numbers that are perfect squares. His proof hinges on the periodicity of the Lucas sequence modulo 8.

Since $L_0 \equiv 2$ (mod 8) and $L_1 \equiv 1$ (mod 8), it follows that $L_i \equiv L_{i-1} + L_{i-2}$ (mod 8), where $i \ge 2$. For example, $L_8 \equiv 7$ (mod 8) and $L_9 \equiv 4$ (mod 8), so $L_{10} \equiv 7 + 4 \equiv 3$ (mod 8). Table 35.1 shows the residues of the Lucas numbers modulo 8, where $0 \le i \le 11$.

It follows from the table that $L_{12} \equiv 2$ (mod 8) and $L_{13} \equiv 1$ (mod 8). Consequently, $L_0 \equiv L_{12}$ (mod 8) and $L_1 \equiv L_{13}$ (mod 8). By virtue of the LRR, the Lucas residues continue repeating. Thus there are exactly 12 distinct Lucas residues

TABLE 35.1.

i	0	1	2	3	4	5	6	7	8	9	10	11
L_i (mod 8)	2	1	3	4	7	3	2	5	7	4	3	7

modulo 8, as the table shows. *The period of the Lucas sequence modulo 8 is twelve*, that is, $L_{i+12} \equiv L_i \pmod 8$. More generally, $L_{12n+i} \equiv L_i \pmod 8$, where $n \geq 0$.

Since the least residue of a perfect square modulo 8 is 0, 1, or 4, it follows from Table 35.1 that the only Lucas numbers that can be squares are those of the form L_{12k+i}, where $i = 1, 3$, or 9. This observation narrows considerably our search for Lucas squares.

In order to identify the Lucas squares, we need the results in the following lemma.

Lemma 35.1. Let $m, n \geq 0$, $r \geq 1$, and $t = 2^r$. Then

1. $L_{2t} = L_t^2 - 2$
2. $(L_t, 2) = 1$
3. $L_{2t+i} \equiv -L_i \pmod{L_t}$
4. $L_{6n\pm2} \equiv 3 \pmod 4$

Proof.

1. Since $L_{2n} = L_n^2 - 2(-1)^n$, it follows that $L_{2t} = L_t^2 - 2$.
2. By Identity (34.1), $L_n \equiv 0 \pmod 2$ if and only if $n \equiv 0 \pmod 3$. Since $3 \nmid t$, $2 \nmid L_t$. Therefore, $(L_t, 2) = 1$.
3. Since $2L_{m+n} = 5F_mF_n + L_mL_n$, we have

$$\begin{aligned} 2L_{2t+i} &= 5F_{2t}F_i + L_{2t}L_i \\ &= 5F_tL_tF_i + L_i(L_t^2 - 2) \\ &\equiv -2L_i \pmod{L_t} \end{aligned}$$

But $(L_t, 2) = 1$, so $L_{2t+i} \equiv -L_i \pmod{L_t}$.

We shall leave the proof of part 4 as an exercise (see Exercises 12 and 15). ■

We are now in a position to establish the theorem. The following proof is essentially the one given by Br. Alfred.

Theorem 35.3. The only Lucas numbers that are perfect squares are 1 and 4.

Proof. Clearly, $L_1 = 1$ and $L_3 = 4$ are perfect squares. Let $i, k \geq 1$. Then $12k = 2mt$ for some odd integer m and $t = 2^r$, where $r \geq 1$. By the repeated application of Lemma 35.1, we have

$$\begin{aligned} L_{12k+i} &= L_{2mt+i} \\ &\equiv -L_{2(m-1)t+i} \pmod{L_t} \\ &\equiv (-1)^2 L_{2(m-2)t+i} \pmod{L_t} \\ &\vdots \\ &\equiv (-1)^m L_i \pmod{L_t} \\ &\equiv -L_t, \qquad \text{since } m \text{ is odd} \end{aligned}$$

Case 1. Let $i = 1$. Then $L_{12k+1} \equiv -L_1 \equiv -1 \pmod{L_t}$. Since $2|t$ and $3 \nmid t$, it follows by Theorem 35.1 that $L_t \equiv 3 \pmod 4$. Consequently, -1 is a quadratic nonresidue of L_t; that is, $x^2 \equiv -1 \pmod{L_t}$ has no solutions. Thus L_{12k+1} is not a perfect square.

Case 2. Let $i = 3$. Then $L_{12k+3} \equiv -L_3 \equiv -4 \pmod{L_t}$. Again, -4 is a quadratic nonresidue of L_t, so L_{12k+3} cannot be a perfect square.

Case 3. Let $i = 9$. Then L_{12k+9} can be factored as $L_{12k+9} = L_{4k+3}(L_{4k+3}^2 + 3)$ (see Exercise 17).

Suppose $d|L_{4k+3}$ and $d|(L_{4k+3}^2 + 3)$. Then $d|3$, so $d = 1$ or 3. But the only Lucas numbers divisible by 3 are of the form L_{4k+2} (see Exercises 19–21). So $d \nmid (4k + 3)$. Thus $d = 1$, so $(L_{4k+3}, L_{4k+3}^2 + 3) = 1$. Therefore, if L_{12k+9} is to be a perfect square, both factors must be squares.

Clearly, L_{4k+3} is not a perfect square when $k = 1$ or 2. By the division algorithm, we have $k = 3s$, $3s + 1$, or $3s + 2$ with $s \geq 1$. By Table 35.1, the corresponding Lucas numbers L_{12s+3}, L_{12s+7}, and L_{12s+11} are not squares.

Thus, the only Lucas numbers that are perfect squares are 1 and 4. ■

FIBONACCI AND LUCAS PERIODICITY

In 1960, D. D. Wall of the IBM Corporation investigated the periodicity of the Fibonacci sequence modulo a positive integer $m \geq 2$. He established that if $m = \prod_1^k p_i^{e_i}$ and h_i denotes the period of the sequence modulo $p_i^{e_i}$, then the period of the sequence modulo m is $[h_1, h_2, \ldots, h_k]$.

Twelve years later, J. Kramer and V. E. Hoggatt, Jr., both of San Jose State University, continued the investigation and established the periodicity of both Fibonacci and Lucas numbers modulo 10^n. To demonstrate this, we need the following theorems.

Lemma 35.2. $L_{3n} \equiv 0 \pmod 2$. ■

This follows by Exercise 39 in Chapter 16.

Theorem 35.4. (Kramer and Hoggatt, 1972). The period of the Fibonacci sequence modulo 2^n is $3 \cdot 2^{n-1}$.

Proof. (by PMI) By virtue of the Fibonacci recurrence relation, it suffices to prove that $F_{3 \cdot 2^{n-1}} \equiv F_0 \pmod{2^n}$ and $F_{3 \cdot 2^{n-1}+1} \equiv F_1 \pmod{2^n}$.

1. *To prove that* $F_{3 \cdot 2^{n-1}} \equiv F_0 \pmod{2^n}$ *for* $n \geq 1$:

   When $n = 1$, $F_{3 \cdot 2^{n-1}} = F_3 = 2 \equiv 0 \pmod 2$, so the result is true when $n = 1$.

   Now assume it is true for an arbitrary integer $k \geq 1$:

   $$F_{3 \cdot 2^{k-1}} \equiv F_0 \pmod{2^k}$$

   Then $F_{3 \cdot 2^k} = F_{3 \cdot 2^{k-1}} L_{3 \cdot 2^{k-1}} \equiv 0 \pmod{2^{k+1}}$, by Lemma 35.1 and the inductive hypothesis. Thus $F_{3 \cdot 2^{n-1}} \equiv 0 \pmod{2^n}$ for every $n \geq 1$.

2. *To prove that* $F_{3 \cdot 2^{n-1}+1} \equiv F_1 \pmod{2^n}$:

   Using the identity $F_{2m+1} = F_{m+1}^2 + F_m^2$,

   $$F_{3 \cdot 2^{n-1}+1} = \left(F_{3 \cdot 2^{n-2}+1}\right)^2 + \left(F_{3 \cdot 2^{n-2}}\right)^2$$

   Since $F_{3 \cdot 2^{n-2}} \equiv 0 \pmod{2^{n-1}}$ by Part 1, it follows that $(F_{3 \cdot 2^{n-2}})^2 \equiv 0 \pmod{2^n}$. Therefore, by Cassini's formula,

   $$\begin{aligned}\left(F_{3 \cdot 2^{n-2}+1}\right)^2 &= F_{3 \cdot 2^{n-2}+2} F_{3 \cdot 2^{n-2}} - (-1)^{3 \cdot 2^{n-2}+1} \\ &\equiv 0 + 1 \equiv \pmod{2^n}\end{aligned}$$

Thus $F_{3 \cdot 2^{n-1}+1} \equiv 0 + 1 \equiv F_1 \pmod{2^n}$. ■

The next theorem generalizes this result to the generalized Fibonacci sequence.

Theorem 35.5. (Kramer and Hoggatt, 1972) The period of the generalized Fibonacci sequence modulo 2^n is $3 \cdot 2^{n-1}$. ■

Its proof hinges on establishing that $G_{3 \cdot 2^{n-1}+1} \equiv G_1 \pmod{2^n}$ and $G_{3 \cdot 2^{n-1}+2} \equiv G_2 \pmod{2^n}$, Cassini's rule, and the following identities:

$$\begin{aligned}G_{m+n+1} &= G_{m+1} F_{n+1} + G_m F_n \\ G_{n+1} &= a F_{n-1} + b F_n\end{aligned}$$

See Exercise 26.

Next we need two simple lemmas. We can establish Lemma 35.3 using Binet's formula and Lemma 35.4 from Lemma 35.2 using PMI (see Exercises 27 and 28).

Lemma 35.3. $F_{5^{n+1}} = (L_{4\cdot 5^n} - L_{2\cdot 5^n} + 1)F_{5^n},\ n \geq 1$ ■

Lemma 35.4. $F_{5^n} \equiv 0 \pmod{5^n}$, $n \geq 1$ ■

Theorem 35.6. The period of the Fibonacci sequence modulo 5^n is $4 \cdot 5^n$.

Proof. Again, by virtue of the Fibonacci recurrence relation, it suffices to show that $F_{4\cdot 5^n} \equiv F_0 \pmod{5^n}$ and $F_{4\cdot 5^n+1} \equiv F_1 \pmod{5^n}$.

1. *To prove that* $F_{4\cdot 5^n} \equiv F_0$ *(mod* 5^n*)*:

   Since $F_{5^n} | F_{4\cdot 5^n}$, $F_{5^n} \equiv F_{4\cdot 5^n} \equiv 0$ (*mod* 5^n), by Lemma 35.3.

2. *To prove that* $F_{4\cdot 5^n+1} \equiv F_1 \pmod{5^n}$:

   Using the identity $F_{2m+1} = F_{m+1}^2 + F_m^2$, we have

$$\begin{aligned} F_{4\cdot 5^n+1} &= (F_{2\cdot 5^n+1})^2 + (F_{2\cdot 5^n})^2 \\ &\equiv (F_{2\cdot 5^n+1})^2 \pmod{5^n} \end{aligned}$$

   Using Cassini's formula,

$$\begin{aligned} (F_{2\cdot 5^n+1})^2 &= F_{2\cdot 5^n+2}F_{2\cdot 5^n} - (-1)^{2\cdot 5^n+1} \\ &\equiv 0 + 1 \equiv 1 \pmod{5^n} \end{aligned}$$

Thus $F_{4\cdot 5^n+1} \equiv F_1 \pmod{5^n}$. ■

The following theorem provides the corresponding result for Lucas numbers. We omit its proof in the interest of brevity.

Theorem 35.7. The period of the Lucas sequence modulo 5^n is $4 \cdot 5^{n-1}$. ■

Theorems 35.4 – 35.7 yield the periodicity of each sequence modulo 10^n.

Theorem 35.8.

1. The period of the Fibonacci sequence modulo 10^n is

$$\begin{cases} 60 & \text{if } n = 1 \\ 300 & \text{if } n = 2 \\ 15 \cdot 10^{n-1} & \text{otherwise} \end{cases}$$

2. The period of the Lucas sequence modulo 10^n is

$$\begin{cases} 12 & \text{if } n = 1 \\ 60 & \text{if } n = 2 \\ 3 \cdot 10^{n-1} & \text{otherwise} \end{cases}$$

Proof.

1. The period of the Fibonacci sequence modulo 10^n is given by

$$[3 \cdot 2^{n-1}, 4 \cdot 5^n] = \begin{cases} 60 & \text{if } n = 1 \\ 300 & \text{if } n = 2 \\ 15 \cdot 10^{n-1} & \text{otherwise} \end{cases}$$

2. The period of the Lucas sequence modulo 10^n is given by

$$[3 \cdot 2^{n-1}, 4 \cdot 5^{n-1}] = \begin{cases} 60 & \text{if } n = 1 \\ 300 & \text{if } n = 2 \\ 15 \cdot 10^{n-1} & \text{otherwise} \end{cases}$$

■

The next corollary follows immediately from this theorem.

Corollary 35.1.

1. The ones digit of a Fibonacci number repeats in a cycle of period of 60; the last two digits in a cycle of period of 300; and the last $n(\geq 3)$ digits in a period of $15 \cdot 10^{n-1}$.
2. The ones digit of a Lucas number repeats in a cycle of period of 12; the last two digits in a cycle of period of 60; and the last $n(\geq 3)$ digits in a period of $3 \cdot 10^{n-1}$. ■

Kramer and Hoggatt (1972) also established the following results.

Theorem 35.9.

$$\text{(1)} \quad L_{2\cdot 3^n} \equiv 0 \pmod{2 \cdot 3^n}$$

$$\text{(2)} \quad F_{4\cdot 3^n} \equiv 0 \pmod{4 \cdot 3^n}$$

■

For example, $L_{54} = 192{,}900{,}153{,}618 \equiv 0 \pmod{54}$ and $F_{36} = 14{,}930{,}352 \equiv 0 \pmod{36}$.

We can employ Theorem 35.8 to prove the following result, which was proposed as a problem in 1976 by H. T. Freitag of Virginia. The proof presented here is essentially the same as the one given by P. S. Bruckman of the University of Illinois.

Example 35.1. Prove that $L_{2p^k} \equiv 3 \pmod{10}$ for primes $p \geq 5$.

Proof. For primes $p \geq 5$, $p \equiv \pm 1 \pmod 6$, so $p^k \equiv \pm 1 \pmod 6$, and hence $2p^k \equiv \pm 2 \pmod{12}$. By Theorem 35.8, the period of the Lucas sequence modulo 10 is 12; that is, $L_{n+12} \equiv L_n \pmod{10}$. But $L_n \equiv 3 \pmod{10}$ if and only if $n \equiv \pm 2 \pmod{12}$. Therefore, $L_{2p^k} \equiv 3 \pmod{10}$ for primes $p \geq 5$. ■

The next theorem was discovered in 1974 by M. R. Turner of Regis University in Denver. It characterizes those Fibonacci numbers that terminate in the same last two digits as their subscripts. Its proof is fairly long, so we omit it.

Theorem 35.10. $F_n \equiv n \pmod{100}$ if and only if $n = 1, 5, 25, 29, 41$, or $49 \pmod{60}$ or $n = 0 \pmod{300}$. ■

For example, $F_{41} = 165{,}580{,}141 \equiv 41 \pmod{100}$ and $n = 41 \equiv 41 \pmod{60}$. On the other hand, $n = 85 \equiv 25 \pmod{60}$ and $F_{85} = 259{,}695{,}496{,}911{,}122{,}585 \equiv 85 \pmod{100}$.

This theorem has an interesting by-product.

Corollary 35.2. Let p be a prime ≥ 5. Then $F_{p^2} \equiv p^2 \pmod{100}$.

Proof. By the theorem, $F_{25} \equiv 25 \pmod{100}$. If $p > 5$, then $p \equiv 1, 3, 7, 9, 11, 13, 17$, or $19 \pmod{20}$. Then $p^2 \equiv 1$ or $9 \pmod{20}$, so $p^2 \equiv 1$ or $49 \pmod{20}$. Since $p^2 \equiv 1 \pmod 3$, it follows that $p^2 \equiv 1$ or $49 \pmod{60}$. Therefore, by Theorem 35.10, $F_{p^2} \equiv p^2 \pmod{100}$. ■

For example, let $p = 7$. Clearly, $F_{49} = 7{,}778{,}742{,}049 \equiv 49 \pmod{100}$.

EXERCISES 35

1. The ones digits in F_{31} is 9 and that in F_{32} is also 9. Compute the ones digit in F_{33}.
2. F_{38} ends in 9 and F_{39} in 6. Find the ones digit in F_{42}.
3. F_{43} ends in 7. Determine the ones digit in F_{703}.
4. L_{20} ends in 7 and F_{21} in 6. Find the ones digit in L_{22} and L_{25}.
5. L_{45} ends in 6. Find the ones digit in L_{93}.
6. Complete the following table.

Modulus m	Period of the Fibonacci Sequence Modulo m	Period of the Lucas Sequence Modulo m
2		
3		
4		
5		
6		
7		
8		
9		
10		

7. Given that $L_{23} \equiv 7 \pmod 8$ and $L_{24} \equiv 2 \pmod 8$, compute L_{25} modulo 8.

8. Let $L_i \equiv 5 \pmod 8$ and $L_{i+1} \equiv 7 \pmod 8$. Compute L_{i+2} modulo 8.
9. Let $L_i \equiv 3 \pmod 8$ and $L_{i-1} \equiv 4 \pmod 8$. Compute L_{i-2} modulo 8.

Prove each, where $n, i \geq 0$.

10. $L_{12n+i} \equiv L_i \pmod{10}$
11. $F_{6n+1} \equiv 1 \pmod 4$
12. $F_{6n-2} \equiv 3 \pmod 4$
13. $F_{6n-1} \equiv 1 \pmod 4$
14. $L_{6n-1} \equiv 3 \pmod 4$
15. $L_{6n+2} \equiv 3 \pmod 4$
16. $L_{6n+4} \equiv 3 \pmod 4$
17. $L_{12n+9} = L_{4n+3}(L_{4n+3}^2 + 3)$.
18. $F_{4n} \equiv 0 \pmod 3$
19. $L_{4n+2} \equiv 0 \pmod 3$
20. $L_{4n} \equiv \pm 1 \pmod 3$
21. $L_{4n+1} \equiv \pm 1 \pmod 3$
22. $L_{12n+i} \equiv L_i \pmod 3$
23. Let $2|t$ and $3 \not| t$. Then $L_t \equiv 3 \pmod 4$.
24. Use the fact that $2F_{m+n} = F_m L_n + F_n L_m$ to prove that $F_{60n+i} \equiv F_i \pmod{10}$.
25. Use the fact that $L_{m+2k} \equiv -L_m \pmod{L_k}$ to prove that $L_{4n+2} \equiv 0 \pmod 3$.
26. Theorem 35.5
27. Lemma 35.3
28. Lemma 35.4
29. $F_{60k} \equiv 20k \pmod{100}$ (Turner, 1974)
30. If $n \equiv 1 \pmod{60}$, then $F_n \equiv n \pmod{100}$ (Turner, 1974).
31. $F_{60k+n} \equiv 20kF_{n-1} + (60k+1)F_n \pmod{100}$ (Turner, 1974)
32. The sum of n consecutive Lucas numbers is divisible by 5 if and only if $4|n$ (Freitag, 1974).
33. $F_{(n+2)k} \equiv F_{nk} \pmod{L_k}$, where k is odd (Freitag, 1974).
34. $F_{(n+2)k} + F_{nk} \equiv 2F_{(n+1)k} \pmod{L_k - 2}$, where k is even (Freitag, 1974).
35. $L_{2m(2n+1)} \equiv L_{2m} \pmod{F_{2m}^2}$ (Bruckman, 1975)
36. $L_{(2m+1)(4n+1)} \equiv L_{2m+1} \pmod{F_{2m+1}}$ (Bruckman, 1975b)
37. $L_{(2m+1)(4n+1)} \equiv L_{2m+1} \pmod{F_{2n}F_{2n+1}}$ (Koshy, 1999)
38. $L_{2p^k} \equiv 3 \pmod{10}$, where p is a prime ≥ 5 (Freitag, 1976).
39. $F_{3n+1} + F_{n+3} \equiv 0 \pmod 3$ (Berzsenyi, 1979)
40. $F_{2n} \equiv n(-1)^{n+1} \pmod 5$ (Freitag, 1979)
41. $L_{2^n} \equiv 7 \pmod{10}$, where $n \geq 2$ (Shannon, 1979).
42. $F_{3 \cdot 2^n} \equiv 2^{n+2} \pmod{2^{n+3}}$, where $n \geq 1$ (Bruckman, 1979).
43. $L_{3 \cdot 2^n} \equiv 2 + 2^{n+2} \pmod{2^{n+4}}$, where $n \geq 1$ (Bruckman, 1979).

FIBONACCI AND LUCAS SERIES

If, beginning with F_0, we place successively every Fibonacci number F_n after a decimal point, so that its ones digit falls in the $(n+1)$st decimal place, then the resulting real number is the decimal expansion of the rational number $1/89 = 1/F_{11}$. Be sure to account for the carries:

$$\begin{array}{rlcccccccccccc}
 & 0. & 0 & 1 & 1 & 2 & 3 & 5 & 8 & & & & & \\
 & & & & & & & & 1 & 3 & & & & \\
 & & & & & & & & & 2 & 1 & & & \\
 & & & & & & & & & & 3 & 4 & & \\
 & & & & & & & & & & & 5 & 5 & \\
 & & & & & & & & & & & 1 & 4 & 4 \\
 & & & & & & & & & & & & \cdots & \\
= & 0. & 0 & 1 & 1 & 2 & 3 & 5 & 9 & 5 & 5 & 0 & 5 & \cdots \\
= & \frac{1}{89} & & & & & & & & & & & &
\end{array}$$

that is, $\sum_{i=0}^{\infty}(F_i/10^{i+1}) = 1/F_{11}$. This result was discovered in 1953 by F. Stancliff. But, how do we establish this fact? For that, we need to study the convergence of the Fibonacci series

$$S = \sum_{i=0}^{\infty} \frac{F_i}{k^{i+1}} \tag{36.1}$$

where k is a positive integer.

Suppose the series converges. Then, by Binet's formula,

$$\sum_{i=0}^{\infty} \frac{F_i}{k^{i+1}} = \frac{1}{\sqrt{5}k}\left[\sum_{0}^{\infty}\left(\frac{1+\sqrt{5}}{2k}\right)^i - \left(\frac{1-\sqrt{5}}{2k}\right)^i\right]$$

$$= \frac{1}{\sqrt{5}k}\left[\frac{1}{1-(1+\sqrt{5})/2k} - \frac{1}{1-(1+\sqrt{5})/2k}\right] \tag{36.2}$$

$$= \frac{2}{\sqrt{5}}\left(\frac{1}{2k-1-\sqrt{5}} - \frac{1}{2k-1+\sqrt{5}}\right)$$

Thus,

$$S = \frac{1}{k^2-k-1} \tag{36.3}$$

Notice that the denominator of the right-hand side (RHS) is the characteristic polynomial of the Fibonacci recurrence relation (FRR). Also, S is an integer if and only if $k = 2$.

Since the power series $1/(1-x) = \sum_{i=0}^{\infty} x^i$ converges if and only if $|x| < 1$, it follows from Eq. (36.2) that the Fibonacci power series (Eq. 36.1) converges if and only if $|\alpha| < k$ and $|\beta| < k$, that is, if and only if $k > \max(|\alpha|, |\beta|)$. But $\alpha = |\alpha| > |\beta|$. Thus S converges if and only $k > \alpha$, that is, if and only if $k \geq 2$, which was somewhat obvious.

Equation (36.3) yields the following results:

When $k = 2$ $\qquad \sum_0^{\infty} \frac{F_i}{2^{i+1}} = 1 = \frac{1}{F_1}$

When $k = 3$ $\qquad \sum_0^{\infty} \frac{F_i}{3^{i+1}} = \frac{1}{5} = \frac{1}{F_5}$

When $k = 8$ $\qquad \sum_0^{\infty} \frac{F_i}{8^{i+1}} = \frac{1}{55} = \frac{1}{F_{10}}$

When $k = 10$ $\qquad \sum_0^{\infty} \frac{F_i}{10^{i+1}} = \frac{1}{89} = \frac{1}{F_{11}}$

These values of k yield the value of the infinite sum to be of the form $1/F_t$ for some Fibonacci number F_t; in other words, they are such that $k^2 - k - 1 = F_t$; that is, $k(k-1) = 1 + F_t$.

Conversely, suppose $1 + F_t$ is the product $b(b-1)$ of two consecutive positive integers b and $b-1$. Solving the equation $k^2 - k - (1 + F_t) = 0$ for k,

$$k = \frac{1 \pm \sqrt{1 + 4(1 + F_t)}}{2}$$

$$= \frac{1 \pm \sqrt{4F_t + 5}}{2}$$

Since $k > 0$, this implies

$$k = \frac{1 + \sqrt{4F_t + 5}}{2}$$

But $4F_t + 5 = 4(1 + F_t) + 1 = 4b(b - 1) + 1 = (2b - 1)^2$;

$$\therefore \quad k = \frac{1 + 2b - 1}{2} = b$$

so $S = 1/F_t$. Thus $1/S$ is a Fibonacci number F_t if and only if $1 + F_t$ is the product of two consecutive positive integers.

It now follows that there are at least four such values of k:

$$1 \cdot 2 = 1 + F_1 \qquad 2 \cdot 3 = 1 + F_5 \qquad 7 \cdot 8 = 1 + F_{10} \qquad \text{and } 9 \cdot 10 = 1 + F_{11}$$

What, then, occurs when we turn to the convergence of the corresponding Lucas series,

$$S^* = \sum_{i=0}^{\infty} \frac{L_i}{k^{i+1}} \tag{36.4}$$

As before, we can show that this series converges to a finite sum if and only if $k > \alpha$, and the sum is

$$S^* = \frac{2k - 1}{k^2 - k - 1} \tag{36.5}$$

when $k = 2$, $S^* = 3$; and when $k = 3$, $S^* = 1$. In both cases, S^* is an integer.

So we wish to investigate the integral values of k for which S^* is an integer t. Notice that $t = 0$ implies $k = 1/2$, which is a contradiction. So $t \geq 1$. Let $(2k - 1)/(k^2 - k - 1) = t$. Then $tk^2 - (t + 2)k - (t - 1) = 0$, so

$$k = \frac{(t + 2) \pm \sqrt{(t + 2)^2 + 4t(t - 1)}}{2t}$$

$$= \frac{(t + 2) \pm \sqrt{5t^2 + 4}}{2t} \tag{36.6}$$

Since k is an integer, $\sqrt{5t^2 + 4}$ must be a perfect square. When $t = 1$, $k = 3$ or 0. Since $k \geq 1$, 0 is not acceptable, so $t = 1$ yields the case $k = 3$.

When $t > 1$, $\sqrt{5t^2 + 4} > t + 2$. Consequently, since $k > 0$, the negative root in Eq. (36.3) is not acceptable. Thus

$$k = \frac{(t + 2) + \sqrt{5t^2 + 4}}{2t} \tag{36.7}$$

Suppose $k \geq 4$. Then

$$\frac{(t + 2) + \sqrt{5t^2 + 4}}{2t} \geq 4$$

$$\sqrt{5t^2 + 4} \geq 7t - 2$$

$$5t^2 + 4 \geq 49t^2 - 28t + 4$$

$$11t^2 \leq 7t$$

$$t \leq 7/11$$

which is a contradiction.

Thus, the only positive integral values of k that yield an integral value for S^* are $k = 2$ and $k = 3$. The corresponding values of S^* are the Lucas numbers $L_2 = 3$ and $L_1 = 1$, respectively.

A similar argument shows that the only positive integral value of k that produces an integral value for $S = 1/(k^2 - k - 1)$ is $k = 2$, in which case $S = 1 = F_1$ (or F_2).

In 1981, Calvin T. Long of Washington State University at Pullman showed that the following summation results can be derived from a bizzare identity established in the following theorem:

$$\frac{1}{89} = \sum_{1}^{\infty} \frac{F_{n-1}}{10^n} \tag{36.8}$$

$$\frac{19}{89} = \sum_{1}^{\infty} \frac{L_{n-1}}{10^n} \tag{36.9}$$

$$\frac{1}{109} = \sum_{1}^{\infty} \frac{F_{n-1}}{(-10)^n} \tag{36.10}$$

$$-\frac{21}{109} = \sum_{1}^{\infty} \frac{L_{n-1}}{(-10)^n} \tag{36.11}$$

Theorem 36.1. (Long, 1981). Let a, b, c, d, and B be integers. Let $U_{n+2} = aU_{n+1} + bU_n$, where $U_0 = c$, $U_1 = d$, and $n \geq 2$. Let the integers m and N be defined by $B^2 = m + aB + b$ and $N = cm + dB + bc$. Then

$$B^n N = m \sum_{i=1}^{n+1} B^{n-i+1} U_{i-1} + BU_{n+1} + bU_n \tag{36.12}$$

for all $n \geq 0$.

Proof. (PMI) When $n = 0$, Eq. (36.12) yields $N = cm + dB + bc$. So the result is true when $n = 0$.

Assume it is true when $n = k$:

$$B^k N = m \sum_{i=1}^{k+1} B^{k-i+1} U_{i-1} + BU_{k+1} + bU_k$$

Then

$$\begin{aligned} B^{k+1} N &= m \sum_{i=1}^{k+1} B^{k-i+2} U_{i-1} + B^2 U_{k+1} + bBU_k \\ &= m \sum_{i=1}^{k+1} B^{k-i+2} U_{i-1} + (m + aB + b)U_{k+1} + bBU_k \end{aligned}$$

$$= m\sum_{i=1}^{k+2} B^{k-i+2}U_{i-1} + B(aU_{k+1} + bU_k) + bU_{k+1}$$

$$= m\sum_{i=1}^{k+2} B^{k-i+2}U_{i-1} + BU_{k+2} + bU_{k+1}$$

Thus the formula is true for all $n \geq 0$. ■

Formula (36.12) yields the following theorem.

Theorem 36.2. (Long, 1981) Let a, b, c, d, m, B, and N be integers as in Theorem 36.1. Let

$$r = \frac{a + \sqrt{a^2 + 4b}}{2} \qquad \text{and} \qquad s = \frac{a - \sqrt{a^2 + 4b}}{2}$$

where $|r| < |B|$ and $|s| < |B|$. Then

$$\frac{N}{mB} = \sum_{1}^{\infty} \frac{U_{i-1}}{B^i} \tag{36.13}$$

Proof. Using the recurrence relation in Theorem 36.1, we can show (see Exercise 11) that

$$U_n = Pr^n + Qs^n \tag{36.14}$$

where

$$P = \frac{c}{2} + \frac{2d - ca}{2\sqrt{a^2 + 4b}} \qquad \text{and} \qquad Q = \frac{c}{2} - \frac{2d - ca}{2\sqrt{a^2 + 4b}}$$

Then, by Eq. (36.12),

$$\frac{N}{mB} = \sum_{i=1}^{n+1} \frac{U_{i-1}}{B^i} + \frac{BU_{n+1} + bU_n}{mB^{n+1}}$$

$$= \sum_{1}^{\infty} \frac{U_{i-1}}{B^i} + 0 \qquad \text{as } n \to \infty, \text{ since } |r|, |s| < 1$$

$$= \sum_{1}^{\infty} \frac{U_{i-1}}{B^i}$$

■

In particular, let $a = 1 = b$, $c = 0$, $d = 1$, and $B = 10$. Then $m = B^2 - aB - b = 100 - 10 - 1 = 89$ and $N = cm + dB + bc = 0 + 10 + 0 = 10$. So Eq. (36.13) yields

$$\frac{10}{10 \cdot 89} = \sum_{1}^{\infty} \frac{F_{i-1}}{10^i}$$

That is,

$$\frac{1}{89} = \sum_{1}^{\infty} \frac{F_{i-1}}{10^i}$$

We can derive summation formulas (36.9) through (36.11) similarly (see Exercises 12–14).

Corollary 36.1. Let $a = b = d = 1$ and $c = 0$:

1. If $B = 10^h$, then $1/(10^{2h} - 10^h - 1) = \sum_{1}^{\infty} \frac{F_{i-1}}{10^{ih}}$
2. If $B = (-10)^h$, then $1/(10^{2h} - (-10)^h - 1) = \sum_{1}^{\infty} \frac{F_{i-1}}{(-10)^{ih}}$

Proof. If $B = 10^h$, then $m = 10^{2h} - 10^h - 1$ and $N = 10^h$. If $B = (-10)^h$, then $m = 10^{2h} - (-10)^h - 1$ and $N = (-10)^h$. Both formulas now follow by substitution. ■

The following summation formulas follow from this corollary:

$$\frac{1}{89} = \sum_{1}^{\infty} \frac{F_{i-1}}{10^i} = 0.01123593505557\ldots$$

= 0.0112358
       13
        21
         34
          55
           89
            144
             233
              ...

$$\frac{1}{9899} = \sum_{1}^{\infty} \frac{F_{i-1}}{10^{2i}} = 0.000101020305081321\ldots$$

$$\frac{1}{998{,}999} = \sum_{1}^{\infty} \frac{F_{i-1}}{10^{3i}} = 0.000001001002003005008013\ldots$$

$$\frac{1}{109} = \sum_{1}^{\infty} \frac{F_{i-1}}{(-10)^i}$$

$$\frac{1}{10{,}099} = \sum_{1}^{\infty} \frac{F_{i-1}}{(-100)^i}$$

$$\frac{1}{1{,}000{,}999} = \sum_{1}^{\infty} \frac{F_{i-1}}{(-1000)^i}$$

We now turn our attention to the convergence of the Fibonacci power series

$$T = \sum_{i=0}^{\infty} F_i x^i \tag{36.15}$$

We have

$$\begin{aligned} T &= F_1 x + F_2 x^2 + \sum_{3}^{\infty} (F_{i-1} + F_{i-2}) x^2 \\ &= x + x^2 + x \sum_{2}^{\infty} F_r x^r + x^2 \sum_{1}^{\infty} F_s x^s \\ &= x + x^2 + x(T - x) + x^2 T \\ \therefore \quad T &= \frac{x}{1 - x - x^2} \qquad (36.16) \\ &= \frac{x}{(1 - \alpha x)(1 - \beta x)} \end{aligned}$$

Converting into partial fractions, we find

$$T = \frac{A}{1 - \alpha x} - \frac{A}{1 - \beta x}$$

where $A = 1/\sqrt{5}$. The series from the first term converges if and only if $|\alpha x| < 1$ and that from the second term converges if and only if $|\beta x| < 1$.

So the Fibonacci power series (36.15) converges if and only if $|x| < \min(1/|\alpha|, 1/|\beta|)$. Since $\alpha\beta = -1$, $1/|\alpha| = -\beta$, so $\min(1/|\alpha|, 1/|\beta|) = \min(-\beta, \alpha) = -\beta$. Thus, the series converges if and only if $|x| < -\beta$, that is, if and only if $\beta < x < -\beta$.

Next, we proceed to identify the rational values of x for which T is an integer $k \geq 1$, as P. Glaister did in 1995:

$$\begin{aligned} \frac{x}{1 - x - x^2} &= k \\ kx^2 + (k+1)x - k &= 0 \\ x &= \frac{-(k+1) \pm \sqrt{(k+1)^2 + (2k)^2}}{2k} \end{aligned}$$

This implies that there are two possible values of x.

Since we require x to be rational, $(k+1)^2 + (2k)^2$ must be a perfect square. This can be realized using Pythagorean triples, so we let

$$k + 1 = m^2 - n^2 \qquad \text{and} \qquad 2k = 2mn \tag{36.17}$$

for some integers m and n, where $m > n \geq 1$. Then:

$$\begin{aligned}(k+1)^2 + (2k)^2 &= (m^2 - n^2) + (2mn)^2 \\ &= (m^2 + n^2)^2 \\ \therefore \quad x &= \frac{-(m^2 - n^2) \pm (m^2 - n^2)}{2mn} \\ &= \frac{m}{n}, \frac{n}{m}\end{aligned}$$

Suppose both of these values lie within the interval of convergence $(\beta, -\beta)$, that is, $\beta < -m/n < -\beta$ and $\beta < n/m < -\beta$. From the second double inequality, $m/n > -1/\beta$, so $-m/n < -\alpha$. This is a contradiction, since $-m/n > \beta > -\alpha$. Thus, both values of x, $-m/n$ and n/m, cannot lie within the interval at the same time.

From Eq. (36.17), we have $m^2 - n^2 = mn + 1$;

$$\therefore \quad (m - n/2)^2 = m^2 - mn + n^2/4 = n^2 + 1 + n^2/4 = 1 + 5n^2/4$$

that is,

$$(2m - n)^2 = 4 + 5n^2$$

Thus, $4 + 5n^2$ must be a perfect square r^2 for some positive integer r. So $2m - n = r$.

Let us now look at the first three possible values of n and compute the corresponding values of r, m, and $T = k$:

Case 1. Let $n = 1$. Then $4 + 5n^2 = 3^2$, so $2m - 1 = 3$ and $m = 2$. Therefore, $T = k = mn = 2, n/m = 1/2$, and $-m/n = -2$. But $-2 \notin (\beta, -\beta)$. Thus $\sum_1^\infty F_i(1/2)^i = 2$.

Case 2. Let $n = 3$. Then $4 + 5n^2 = 7^2$, so $2m - 3 = 7$ and $m = 5$. Therefore, $T = mn = 15, n/m = 3/5$, and $-m/n = -5/3$. But $-5/3 \notin (\beta, -\beta)$. Thus $\sum_1^\infty F_i(3/5)^i = 15$.

Case 3. Let $n = 8$. Then $4 + 5n^2 = 18^2$, so $2m - 8 = 18$ and $m = 13$. Therefore, $T = mn = 104, n/m = 8/13$, and $-m/n = -13/8$. Again, $-m/n \notin (\beta, -\beta)$. Thus $\sum_1^\infty F_i(8/13)^i = 104$.

The next choice of n is 21. Surprisingly enough, a clear pattern begins to emerge: These four values of n are Fibonacci numbers with even subscripts: $F_2 = 1$, $F_4 = 3$, $F_6 = 8$, and $F_8 = 21$; the corresponding m-values are their immediate successors: $F_3 = 2$, $F_5 = 5$, $F_7 = 13$, and $F_9 = 34$; and

$$\sum_0^\infty F_i(1/2)^i = 1 \cdot 2, \qquad \sum_0^\infty F_i(3/5)^i = 3 \cdot 5$$

$$\sum_0^\infty F_i(8/13)^i = 8 \cdot 13, \qquad \sum_0^\infty F_i(21/34)^i = 21 \cdot 34$$

In 1996, Glaister established that this fascinating pattern does indeed hold:

To see this, let $n = F_{2k}$, where $k \geq 1$. Then

$$\begin{aligned} 4 + 5n^2 &= 4 + 5F_{2k}^2 = 4 + 5\left(\frac{\alpha^{2k} - \beta^{2k}}{\sqrt{5}}\right)^2 \\ &= 4 + (\alpha^{4k} + \beta^{4k} - 2) = \alpha^{4k} + \beta^{4k} + 2 \\ &= \alpha^{4k} + \beta^{4k} + 2(\alpha\beta)^{2k} = (\alpha^{2k} + \beta^{2k})^2 \end{aligned}$$

Therefore, $4 + 5n^2$ is a perfect square, as desired. Then

$$\begin{aligned} 2m - n &= \alpha^{2k} + \beta^{2k} \\ \therefore \quad m &= \frac{\alpha^{2k} + \beta^{2k} + n}{2} = \frac{1}{2}\left[\alpha^{2k} - \beta^{2k} + \left(\frac{\alpha^{2k} - \beta^{2k}}{\sqrt{5}}\right)\right] \\ &= \frac{\alpha^{2k+1} - \beta^{2k+1}}{\sqrt{5}} = F_{2k+1} \end{aligned}$$

Since $\beta < 0$, it follows that

$$\begin{aligned} \frac{n}{m} &= \frac{F_{2k}}{F_{2k+1}} = \frac{\alpha^{2k} - \beta^{2k}}{\alpha^{2k+1} - \beta^{2k+1}} \\ &< \frac{\alpha^{2k}}{\alpha^{2k+1}} \\ &= \frac{1}{\alpha} = -\beta \end{aligned}$$

Thus $0 < n/m < -\beta$ and hence n/m lies within the interval of convergence. Consequently, $T = k = mn = F_{2k}F_{2k+1}$.

Accordingly, we have the following theorem.

Theorem 36.3. The Fibonacci power series (Eq. 36.15) converges if $\beta < x < -\beta$ and

$$\sum_{i=1}^\infty F_i(F_{2k}/F_{2k+1})^i = F_{2k}F_{2k+1}$$

where $k \geq 1$. ■

For example, $F_{12} = 144$ and $F_{13} = 233$. Then

$$\sum_{i=1}^{\infty} F_i(144/233)^i = 144 \cdot 233 = 33{,}552$$

A similar study of the related Lucas power series

$$T = \sum_{i=0}^{\infty} L_i x^i \tag{36.18}$$

yields fascinating dividends. Suppose this series (36.18) converges. Then

$$\begin{aligned}
T &= L_0 + L_1 x + \sum_{2}^{\infty}(L_{i-1} + L_{i-2})x^i \\
&= 2 + x + x\sum_{1}^{\infty} L_i x^i + x^2 \sum_{0}^{\infty} L_i x^i \\
&= 2 + x + x(T-2) + x^2 T \\
(1 - x - x^2)T &= 2 - x \\
T &= \frac{2-x}{1-x-x^2}
\end{aligned} \tag{36.19}$$

Since $1 - x - x^2 = (1-\alpha x)(1-\beta x)$, we can convert this into partial fractions:

$$T = \frac{A}{1-\alpha x} + \frac{B}{1-\beta x}$$

where A and B are constants. Expanding the right-hand side yields

$$T = A\sum_{0}^{\infty}(\alpha x)^i + B\sum_{0}^{\infty}(\beta x)^i$$

These two series converge if and only if $|\alpha x| < 1$ and $|\beta x| < 1$, that is, if and only if $|x| < 1/|\alpha|$ and $|x| < 1/|\beta|$. But $|\alpha| = \alpha$ and $|\beta| = -\beta$. Thus the series (36.18) converges if and only if $|x| < \min(1/\alpha, 1/-\beta)$. Since $\alpha\beta = -1$, this implies $|x| < \min(-\beta, \alpha)$. But $-\beta < \alpha$, so $\min(-\beta, \alpha) = -\beta$. Thus, the Lucas power series (36.18) converges if and only if $|x| < -\beta$; that is, if and only $\beta < x < -\beta$. When the series converges, Formula (36.19) gives the value of the infinite sum.

Formula (36.19) generates a delightful question for the curious-minded: *Are there rational numbers* x *in the interval of convergence for which* T *is an integer*?

Before answering this question, let us study a few examples and look for any possible pattern:

$$\text{When } x = 1/3, \qquad T = \frac{2 - 1/3}{1 - 1/3 - (1/3)^2} = 3 = 3 \cdot 1$$

$$\text{When } x = 4/7, \qquad T = \frac{2 - 4/7}{1 - 4/7 - (4/7)^2} = 14 = 7 \cdot 2$$

$$\text{When } x = 11/18, \qquad T = \frac{2 - 11/18}{1 - 11/18 - (11/18)^2} = 90 = 18 \cdot 5$$

Clearly, a pattern emerges: In each case, x lies within the interval of convergence; x is of the form L_{2k-1}/L_{2k}, $k \geq 1$; and T is an integer of the form $L_{2k}F_{2k-1}$.

Fortunately, this is always the case. Before we can prove it, we need to lay some groundwork in the form of two lemmas; we can establish both using Binet's formula.

Lemma 36.1. $L_i + L_{i-2} = 5F_{i-1}, i \geq 2$. ■

Corollary 36.2. $L_i + L_{i-2} \equiv 0 \pmod 5$, $i \geq 2$; that is, the sum of two Lucas numbers with consecutive even (or odd) subscripts is divisible by 5. ■

For example, $L_{19} + L_{17} = 9349 + 3571 = 12{,}920 \equiv 0 \pmod 5$. Likewise, $L_{26} + L_{24} \equiv 0 \pmod 5$.

Lemma 36.2. Let $k \geq 1$. Then $L_{2k}^2 - L_{2k}L_{2k-1} - L_{2k-1}^2 = 5$. ■

We are now ready to establish the conjecture.

Theorem 36.4. (Koshy, 1999) Let k be any positive integer. Then

$$\sum_{i=0}^{\infty} L_i(L_{2k-1}/L_{2k})^i = L_{2k}F_{2k-1}$$

Proof. First, we show that $\beta < L_{2k-1}/L_{2k} < -\beta$. By Binet's formula,

$$\begin{aligned} \frac{L_{2k-1}}{L_{2k}} &= \frac{\alpha^{2k-1} + \beta^{2k-1}}{\alpha^{2k} + \beta^{2k}} \\ &< \frac{\alpha^{2k-1} + \beta^{2k-1}}{\alpha^{2k}} \\ &< \frac{\alpha^{2k-1}}{\alpha^{2k}} \\ &= \frac{1}{\alpha} = -\beta \end{aligned}$$

Thus $0 < L_{2k-1}/L_{2k} < -\beta$, so $\beta < L_{2k-1}/L_{2k} < -\beta$.

Since $x = L_{2k-1}/L_{2k} \in (\beta, -\beta)$, by Formula (36.19),

$$\begin{aligned}
\sum_{i=0}^{\infty} L_i (L_{2k-1}/L_{2k})^i &= \frac{2 - L_{2k-1}/L_{2k}}{1 - L_{2k-1}/L_{2k} - (L_{2k-1}/L_{2k})^2} \\
&= \frac{L_{2k}(2L_{2k} - L_{2k-1})}{L_{2k}^2 - L_{2k}L_{2k-1} - L_{2k-1}^2} \\
&= \frac{L_{2k}(2L_{2k} - L_{2k-1})}{5} \qquad \text{by Lemma 36.2} \\
&= \frac{L_{2k} + (L_{2k} - L_{2k-1})}{5} = \frac{L_{2k} + L_{2k-2}}{5} \\
&= \frac{L_{2k}(5F_{2k-1})}{5} \qquad \text{by Lemma 36.1} \\
&= L_{2k}F_{2k-1}
\end{aligned}$$

■

For example, let $k = 5$. Then $x = L_9/L_{10} = 76/123$, and $F_9 = 34$;

$$\therefore \quad \sum_{0}^{\infty} L_i (76/123)^i = 123 \times 34 = 4182$$

Interestingly enough, although when $x = L_{2k-1}/L_{2k}$, the Fibonacci power series (36.15) converges to a finite sum, it is not an integer. We can show that

$$\sum_{i=0}^{\infty} F_i (L_{2k-1}/L_{2k})^i = \frac{L_{2k}L_{2k-1}}{5}$$

where $L_{2k}L_{2k-1} \not\equiv 0 \pmod 5$. For example,

$$\sum_{0}^{\infty} F_i (L_9/L_{10})^i = \sum_{0}^{\infty} F_i (76/123)^i = \frac{123 \cdot 76}{5} = 1869.6$$

As in the proof of Theorem 36.1, we can show that $\beta < F_{2k}/F_{2k+1} < -\beta$, so the Lucas series (36.18) converges to a finite sum when $x = F_{2k}/F_{2k+1}$, where $k \geq 0$. Then:

$$\begin{aligned}
\sum_{i=0}^{\infty} L_i (F_{2k}/F_{2k+1})^i &= \frac{2 - F_{2k}/L_{2k+1}}{1 - F_{2k}/F_{2k+1} - (F_{2k}/F_{2k+1})^2} \\
&= \frac{F_{2k+1}(2F_{2k+1} - F_{2k})}{F_{2k+1}^2 - F_{2k}F_{2k+1} - F_{2k}^2} \\
&= \frac{F_{2k+1}(F_{2k+1} + F_{2k-1})}{1} = F_{2k+1}L_{2k}
\end{aligned}$$

again an integer. Again, for the sake of brevity, we have omitted the cumbersome details.

Accordingly, we have the following result.

Theorem 36.5. (Koshy, 1999) Let $k \geq 0$. Then

$$\sum_{i=0}^{\infty} L_i (F_{2k}/F_{2k+1})^i = F_{2k+1} L_{2k} \qquad \blacksquare$$

For example, let $k = 3$. Then $x = F_6/F_7 = 8/13$. Therefore,

$$\sum_{i=0}^{\infty} L_i (8/13)^i = F_7 L_6 = 13 \cdot 18 = 234$$

Likewise, $\sum_{i=0}^{\infty} L_i (F_{10}/F_{11})^i = F_{11} L_{10} = 89 \cdot 123 = 10{,}947$.

The next example, proposed in 1963 by L. Carlitz of Duke University, North Carolina, is an interesting telescoping sum involving a Fibonacci sum. The beautiful proof given here is by J. H. Avila of the University of Maryland.

A *telescoping sum* is a sum of the form $\sum_{1}^{n} (a_i - a_{i-1})$, where a_i is any number. When we expand such a sum, all terms get canceled, except the two ends.

Thus

$$\sum_{1}^{n} (a_i - a_{i-1}) = a_n - a_0$$

Example 36.1. Show that

$$\sum_{1}^{\infty} \frac{1}{F_n F_{n+1}^2 F_{n+2}} + \sum_{1}^{\infty} \frac{1}{F_n F_{n+2}^2 F_{n+3}} = \frac{1}{2}$$

Solution. For convenience, let $a = F_n$, $b = F_{n+1}$, $c = F_{n+2}$, and $d = F_{n+3}$. Then $a + b = c$ and $b + c = d$.

$$\begin{aligned}
\text{LHS} &= \sum_{1}^{\infty} \frac{1}{ac^2d} + \sum_{1}^{\infty} \frac{1}{ab^2d} \\
&= \sum_{1}^{\infty} \left(\frac{1}{ac^2d} + \frac{1}{ab^2d} \right) \quad = \sum_{1}^{\infty} \left(\frac{b}{ac^2d} + \frac{c}{ab^2d} \right) \\
&= \sum_{1}^{\infty} \left(\frac{c-a}{abc^2d} + \frac{d-b}{ab^2cd} \right) = \sum_{1}^{\infty} \left(\frac{1}{abcd} - \frac{1}{bc^2d} + \frac{1}{ab^2c} - \frac{1}{abcd} \right)
\end{aligned}$$

$$= \sum_{1}^{\infty} \left(\frac{1}{ab^2c} - \frac{1}{bc^2d} \right)$$

$$= \sum_{1}^{\infty} \left(\frac{1}{F_n F_{n+1}^2 F_{n+2}} - \frac{1}{F_{n+1} F_{n+2}^2 F_{n+3}} \right) \qquad \leftarrow \text{Telescoping sum}$$

$$= \frac{1}{F_1 F_2^2 F_3} = \frac{1}{2}$$

■

We can apply the same telescoping technique to derive the following formulas, developed in 1963 by R. L. Graham of Bell Telephone Laboratories, Murray Hill, New Jersey, now called Lucent Technologies:

$$\sum_{2}^{\infty} \frac{1}{F_{n-1} F_{n+1}} = 1 \qquad \text{and} \qquad \sum_{2}^{\infty} \frac{F_n}{F_{n-1} F_{n+1}} = 2$$

See Exercises 24 and 44.

Letting $x = 1/2$ in the power series

$$\frac{1}{1 - x - x^2} = \sum_{0}^{\infty} F_i x^{i-1} \tag{36.20}$$

we get $\sum_{1}^{\infty} F_i/2^{i-1} = 4$; that is, $\sum_{1}^{\infty} F_i/2^i = 2$.

Differentiating Eq. (36.20) with respect to x, we get

$$\frac{1 + 2x}{(1 - x - x^2)^2} = \sum_{1}^{\infty} (i - 1) F_i x^{i-2}$$

Let $x = 1/2$. Then

$$8 = \sum_{1}^{\infty} \frac{iF_i}{2^i} - 2$$

That is,

$$\sum_{1}^{\infty} \frac{iF_i}{2^i} = 10$$

The following result was established in 1974 by I. J. Good of Virginia Polytechnic Institute and State University at Blacksburg, Virginia.

Example 36.2. Prove that

$$\sum_{0}^{\infty} \frac{1}{F_{2^i}} = 4 - \alpha = 3 + \beta$$

Proof. First, we shall prove by PMI that

$$\sum_{0}^{n} \frac{1}{F_{2^i}} = 3 - \frac{F_{2^n-1}}{F_{2^n}} \tag{36.21}$$

When $n = 1$, LHS $= 2$ and RHS $= 3 - (F_1/F_2) = 3 - 1 = 2$. So the result is true when $n = 1$.

Assume Eq. (36.21) is true for all positive integers $\leq n$. Then

$$\begin{aligned}\sum_{0}^{n+1} \frac{1}{F_{2^i}} &= 3 - \frac{F_{2^n-1}}{F_{2^n}} + \frac{1}{F_{2^{n+1}}}\\ &= 3 - \frac{L_{2^n}F_{2^n-1}}{L_{2^n}F_{2^n}} + \frac{1}{F_{2^{n+1}}}\\ &= 3 - \frac{L_{2^n}F_{2^n-1} - 1}{F_{2^{n+1}}}\end{aligned} \tag{36.22}$$

With $m = 2^n$ and $k = 2^n - 1$ in Exercise 56 in Chapter 5, we have:

$$F_{2^{n+1}-1} + F_{-1} = L_{2^n}F_{2^n-1}$$

That is,

$$L_{2^n}F_{2^n-1} - 1 = F_{2^{n+1}-1}$$

Therefore, Eq. (36.22) becomes

$$\sum_{0}^{n+1} \frac{1}{F_{2^i}} = 3 - \frac{F_{2^{n+1}-1}}{F_{2^{n+1}}}$$

Thus, by the strong version of PMI, Eq. (36.21) is true for every $n \geq 1$;

$$\therefore \quad \lim_{n\to\infty} \sum_{0}^{n} \frac{1}{F_{2^i}} = 3 - \lim_{n\to\infty} \frac{F_{2^{n+1}-1}}{F_{2^{n+1}}}$$

That is,

$$\sum_{0}^{\infty} \frac{1}{F_{2^i}} = 3 - (-\beta) = 3 + \beta = 4 - \alpha \qquad \blacksquare$$

Next we pursue an interesting problem proposed in 1963 by H. W. Gould of West Virginia University.

Example 36.3. Show that $\dfrac{x(1-x)}{1-2x-2x^2+x^3} = \sum_{0}^{\infty} F_n^2 x^n$.

Solution. Let

$$\frac{x(1-x)}{1-2x-2x^2+x^3} = \sum_0^\infty a_n x^n. \tag{36.23}$$

Then $x - x^2 = \sum_0^\infty a_n(1 - 2x - 2x^2 + x^3)x^n$. Equating coefficients of like terms, $a_0 = 0$, $1 = -2a_0 + a_1$, $-1 = -2a_0 - 2a_1 + a_2$, and $0 = a_{n-3} - 2a_{n-2} - 2a_{n-1} + a_n$, where $n \geq 3$. Thus

$$a_0 = 0, \quad a_1 = 1 = a_2 \quad \text{and} \quad a_n = 2a_{n-1} + 2a_{n-2} - a_{n-3} \quad n \geq 3 \tag{36.24}$$

Since $a_n = F_n^2$ for $0 \leq n \leq 2$, it remains to show that $a_n = F_n^2$ for $n \geq 3$. To this end, notice that:

$$\begin{aligned} F_n^2 &= (F_{n-1} + F_{n-2})^2 \\ &= 2F_{n-1}^2 + 2F_{n-2}^2 - (F_{n-1} - F_{n-2})^2 \\ &= 2F_{n-1}^2 + 2F_{n-2}^2 - F_{n-3}^2 \end{aligned}$$

So F_n^2 satisfies the Recurrence Relation 36.24 and the three initial conditions. Thus $a_n = F_n^2$ for all $n \geq 0$, so

$$\frac{x(1-x)}{1-2x-2x^2+x^3} = \sum_0^\infty F_n^2 x^n \qquad \blacksquare$$

Let us take this example a bit further. The roots of the cubic equation $1 - 2x - 2x^2 + x^3 = 0$ are -1, α^2, and β^2, of which β^2 has the least absolute value. Therefore, the power series (36.23) converges if and only if $|x| < \beta^2$, where $\beta^2 \approx 0.38196601125$.

In particular, the series converges when $x = 1/4$ and $\sum_0^\infty F_n^2/4^n = \frac{12}{25}$.

A LIST OF SUMMATION FORMULAS

The following summation formulas were developed in 1969 by Br. Alfred Brousseau of St. Mary's College in California.

1. $\sum_1^\infty \frac{(-1)^{n-1}}{F_n F_{n+3}} = \frac{6\alpha - 9}{4}$

2. $\sum_1^\infty \frac{(-1)^{n-1} L_{2n+2}}{F_n F_{n+1} L_{n+1} L_{n+2}} = \frac{1}{3}$

3. $\sum_1^\infty \frac{(-1)^{n-1} F_{2n+2}}{L_n^2 L_{n+2}^2} = \frac{8}{45}$

4. $\sum_1^\infty \frac{1}{F_n F_{n+2} F_{n+3}} = \frac{1}{4}$

5. $\sum_{1}^{\infty} \frac{F_{n+3}}{F_n F_{n+2} F_{n+4} F_{n+6}} = \frac{17}{480}$

6. $\sum_{1}^{\infty} \frac{F_{4n+3}}{F_{2n} F_{2n+1} F_{2n+2} F_{2n+3}} = \frac{1}{2}$

7. $\sum_{1}^{\infty} \frac{L_{n+2}}{F_n F_{n+4}} = \frac{17}{6}$

8. $\sum_{1}^{\infty} \frac{(-1)^{n-1} F_{2n+1}}{F_n^2 F_{n+1}^2} = 1$

9. $\sum_{1}^{\infty} \frac{(-1)^{n-1} L_{n+1}}{F_n F_{n+1} F_{n+2}} = 1$

10. $\sum_{1}^{\infty} \frac{(-1)^{n-1}}{L_{3n} L_{3n+3}} = \frac{3-\alpha}{40(1+\alpha)}$

11. $\sum_{1}^{\infty} \frac{(-1)^{n-1} F_{6n+3}}{F_{3n}^3 F_{3n+3}^3} = \frac{1}{8}$

12. $\sum_{1}^{\infty} \frac{(-1)^{n-1}}{F_n F_{n+5}} = \frac{150 - 83\alpha}{105\alpha}$

13. $\sum_{1}^{\infty} \frac{(-1)^{n-1} F_{6n+3}}{F_{6n} F_{6n+6}} = \frac{1}{16}$

14. $\sum_{1}^{\infty} \frac{F_{2n+5}}{F_n F_{n+1} F_{n+2} F_{n+3} F_{n+4} F_{n+5}} = \frac{1}{15}$

15. $\sum_{1}^{\infty} \frac{(-1)^{n-1}}{F_{2n-1} F_{2n+3}} = \frac{1}{6}$

16. $\sum_{1}^{\infty} \frac{F_{2n}}{F_{n+2}^2 F_{n-2}^2} = \frac{85}{108}$

EXERCISES 36

Evaluate each sum.

1. $\sum_{1}^{\infty} \frac{F^i}{2^i}$
2. $\sum_{0}^{\infty} \frac{L^i}{3^i}$

Consider the recurrence relation $u_n - (\alpha + \beta)u_{n-1} + \alpha\beta u_{n-2} = 0$, where $u_0 = a$ and $u_1 = b$.

3. Solve the recurrence relation.
4. Evaluate the sum $\sum_{i=0}^{\infty} \frac{u_i}{k^{i+1}}$, where $k > \alpha$ (Cross, 1996).
5. Evaluate the sum $\sum_{i=0}^{\infty} \frac{F_i}{k^{i+1}}$, where $k > a$.
6. Suppose $k > \alpha$. Show that the Lucas series $\sum_{0}^{\infty} \frac{L_i}{k^{i+1}}$ converges to the limit

$$S^* = \frac{2k-1}{k^2 - k - 1}.$$

7. Show that $-m/n = F_{2k-1}/F_{2k} \notin (\beta, -\beta)$ (Glaister, 1996).
8. Show that the value of m corresponding to $n = -F_{2k}$ is F_{2k-1}, where $k \geq 1$ (Glaister, 1996).
9. Show that $F_{2k}/F_{2k-1} \notin (\beta, -\beta)$ (Glaister, 1996).
10. Show that $\beta < F_k/L_k < -\beta$, where $k \geq 0$.

11. Does the Fibonacci power series (36.15) converge when $x = F_k/L_k$, where $k \geq 0$?
12. Does the Lucas power series (36.18) converge when $x = F_k/L_k$, where $k \geq 0$?
13. Show that $U_n = Pr^n + Qs^n$, with P, Q, r, s, and n are defined as in Theorem 36.2.

Derive each.

14. Formula 36.9
15. Formula 36.10
16. Formula 36.11
17. Show that $\sum_1^\infty \frac{1}{F_n}$ converges (Lind, 1967a).
18. Show that $\sum_1^\infty \frac{1}{\ln F_n}$ converges (Lind, 1967b).
19. Show that $\sum_1^\infty \frac{1}{F_n} > \frac{803}{240}$ (Guillottee, 1972).
20. Let $M(x) = \sum_1^\infty \frac{L_n x^n}{n}$. Show that the Maclaurin expansion of $e^{M(x)}$ is $\sum_1^\infty F_n x^{n-1}$ (Hoggatt, 1976).

Evaluate each sum.

21. $\sum_1^\infty \frac{1}{\alpha F_{n+1} + F_n}$ (Guillotte, 1971a)
22. $\sum_1^\infty \frac{1}{F_n + \sqrt{5}F_{n+1} + F_{n+2}}$ (Guillotte, 1971b)

Prove each.

23. $\sum_1^\infty \frac{1}{F_n} = 3 + \sum_1^\infty \frac{(-1)^{n+1}}{F_n F_{n+1} F_{n+2}}$ (Graham, 1963a)
24. $\sum_2^\infty \frac{1}{F_{n-1} F_{n+2}}$ (Graham, 1963b)
25. $\sum_2^\infty \frac{F_n}{F_{n-1} F_{n+1}} = 2$ (Graham, 1963b)
26. $\sum_2^\infty \frac{1}{G_{n-1} G_{n+1}} = \frac{1}{ab}$ (Koshy, 1998)
27. $\sum_2^\infty \frac{G_n}{G_{n-1} G_{n+1}} = \frac{1}{a} + \frac{1}{b}$ (Koshy, 1998)
28. $\sum_1^\infty \frac{1}{G_n} = \frac{2}{a} + \frac{1}{b} + \sum_1^\infty \frac{(-1)^{n+1}\mu}{G_n G_{n+1} G_{n+2}}$ (Koshy, 1998)
29. $\frac{F_{n+1}}{F_n} = 1 + \sum_{i=2}^n \frac{(-1)^i}{F_i F_{i-1}}$ (Basin, 1964a)
30. $\sum_2^\infty \frac{(-1)^i}{F_i F_{i-1}} = |\beta|$ (Basin, 1964a)

31. $\sum_{1}^{\infty} \frac{F_{2n+1}}{L_n L_{n+1} L_{n+2}} = \frac{1}{3}$ (Ferns, 1967)

32. $\sum_{0}^{\infty} \frac{1}{F_{2n+1}} = \sqrt{5}\sum_{0}^{\infty} \frac{(-1)^n}{L_{2n+1}}$ (Carlitz, 1967a)

33. $\sum_{0}^{\infty} \frac{(-1)^n}{F_{4n+2}} = \sqrt{5}\sum_{0}^{\infty} \frac{1}{L_{4n+2}}$ (Carlitz, 1967a)

34. $\sum_{0}^{\infty} \frac{F_{n+1}}{2^n} = 4$ (Butchart, 1968)

35. $\sum_{1}^{\infty} \frac{1}{F_n F_{n+2}} = 1$ (Brousseau, 1969b)

36. $\sum_{1}^{\infty} \frac{1}{F_n F_{n+4}} = \frac{7}{18}$ (Brousseau, 1969b)

37. $\sum_{1}^{\infty} \frac{1}{F_n F_{n+2} F_{n+3}} = \frac{1}{4}$ (Brousseau, 1969b)

38. $\sum_{1}^{\infty} \frac{F_n}{F_{n+1} F_{n+2}} = 1$ (Brousseau, 1969b)

39. $\sum_{1}^{\infty} \frac{F_{n+1}}{F_n F_{n+3}} = \frac{5}{4}$ (Brousseau, 1969b)

40. $\sum_{1}^{\infty} \frac{(-1)^{n-1}}{L_{n-1} L_n} = \frac{\sqrt{5}}{10}$ (Brousseau, 1969b)

41. $\sum_{1}^{\infty} F_{2n-1} x^n = \frac{1-x}{1-3x+x^2}$, where $|x| < \beta^2$ (Hoggatt, 1971c)

42. Show that $\sum_{0}^{\infty} F_i (L_{2k-1}/L_{2k})^i = L_{2k} L_{2k-1}/5$, where $k \geq 1$ (Koshy, 1998).

43. Show that $\sum_{0}^{\infty} F_i (F_k/L_k)^i = F_{2k}/(L_k^2 - L_k F_k - F_k^2)$, where $k \geq 0$ (Koshy, 1998).

44. Show that $\sum_{0}^{\infty} L_i (F_k/L_k)^i = \frac{2L_k - F_{2k}}{L_k^2 - L_k F_k - F_k^2}$, where $k \geq 0$ (Koshy, 1998).

FIBONACCI POLYNOMIALS

Large classes of polynomials can be defined by Fibonacci-like recurrence relations, and yield Fibonacci numbers. Such polynomials, called the *Fibonacci polynomials*, were studied in 1883 by the Belgian mathematician Eugene Charles Catalan (1814–1894) and the German mathematician E. Jacobsthal. The polynomials $f_n(x)$ studied by Catalan are defined by the recurrence relation

$$f_n(x) = xf_{n-1}(x) + f_{n-2}(x) \tag{37.1}$$

where $f_1(x) = 1$, $f_2(x) = x$, and $n \geq 3$. They were further investigated in 1966 by M. N. S. Swamy of the University of Saskatchwan in Canada.

The Fibonacci polynomials studied by Jacobsthal were defined by $J_n(x) = J_{n-1}(x) + xJ_{n-2}(x)$, where $J_1(x) = 1 = J_2(x)$. We shall pursue them in Chapter 39.

We now turn to the class of Fibonacci polynomials introduced by Catalan.

CATALAN'S FIBONACCI POLYNOMIALS

The first ten members of this Fibonacci family are:

$$
\begin{aligned}
f_1(x) &= 1 \\
f_2(x) &= x \\
f_3(x) &= x^2 + 1 \\
f_4(x) &= x^3 + 2x \\
f_5(x) &= x^4 + 3x^2 + 1
\end{aligned}
$$

$$\begin{aligned}
f_6(x) &= x^5 + 4x^3 + 3x \\
f_7(x) &= x^6 + 5x^4 + 6x^2 + 1 \\
f_8(x) &= x^7 + 6x^5 + 10x^3 + 4x \\
f_9(x) &= x^8 + 7x^6 + 15x^4 + 10x^2 + 1 \\
f_{10}(x) &= x^9 + 8x^7 + 21x^5 + 20x^3 + 5x
\end{aligned}$$

Here we make an interesting observation: $f_i(1) = F_i$ for $1 \le i \le 10$. In fact, $f_n(1) = F_n$ for all n; this follows directly from the recurrence relation (37.1). Besides, the degree of $f_n(x)$ is $n - 1$, where $n \ge 1$.

We make yet another interesting observation. Notice that $f_1(2) = 1$, $f_2(2) = 2$, and $f_n(2) = 2f_{n-1}(2) + f_{n-2}(2)$, where $n \ge 3$. So $P_n = f_n(2)$ defines the well-known *Pell numbers* 1, 2, 5, 12, 29, ...

We can extend the definition of the Fibonacci polynomials to negative subscripts also:

$$f_{-n}(x) = (-)^{n+1} f_n(x)$$

Notice that $f_0(x) = 0$.

TABLE 37.1.

n	x^0	x^1	x^2	x^3	x^4	x^5	x^6	x^7	x^8		
1	1										
2	0	1									
3	1	0	1								A Pascal row
4	0	2	0	1							Diagonal sum = 2^4
5	1	0	3	0	1						
6	0	3	0	4	0	1					
7	1	0	6	0	5	0	1				← Row sum = F_7
8	0	4	0	10	0	6	0	1			
9	1	0	10	0	15	0	7	0	1		
10	0	5	0	20	0	21	0	8	0	1	

Table 37.1 shows the various coefficients of the first ten Fibonacci polynomials, when arranged in increasing exponents. Three noteworthy observations:

- The elements on every rising diagonal beginning on row $2n$ are zero.
- The alternate rising diagonals form the various Pascal rows.
- The sum of the elements on the nth rising diagonal is $2^{(n-1)/2} = 2 \cdot 2^{(n-3)/2}$, where n is odd. For example, the sum of the numbers on row 7 is $1+3+3+1 = 8$.

TABLE 37.2.

n	Expansion of $(x+1)^n$
0	1
1	$x+1$
2	x^2+2x+1 ← $f_5(x)$
3	x^3+3x^2+3x+1 ← $f_7(x)$
4	$x^4+4x^3+6x^2+4x+1$
5	$x^5+5x^4+10x^3+10x^2+5x+1$
6	$x^6+6x^5+15x^4+20x^3+15x^2+6x+1$

In 1970, Marjorie Bicknell of Wilcox High School in California showed that the Fibonacci polynomials can be constructed using the binomial expansions of $(x+1)^n$, where $n \geq 0$. The sums of the elements along the rising diagonals in Table 37.2 yield the various Fibonacci polynomials. For instance, the sum of the elements along the diagonal beginning at row 4 is $x^4 + 3x^2 + 1$, which is $f_5(x)$; similarly, the diagonal beginning at row 6 yields $f_7(x)$.

AN EXPLICIT FORMULA FOR $f_n(x)$

More generally, the sum of the elements along the diagonal beginning at row n is $f_{n+1}(x)$; that is,

$$f_{n+1}(x) = \sum_{j=0}^{\lfloor n/2 \rfloor} \binom{n-j}{j} x^{n-2j} \qquad n \geq 0$$

$$\therefore\ f_n(x) = \sum_{j=0}^{\lfloor (n-1)/2 \rfloor} \binom{n-j-1}{j} x^{n-2j-1} \qquad n \geq 0 \qquad (37.2)$$

For example,

$$\begin{aligned} f_5(x) &= \sum_{0}^{2} \binom{4-j}{j} x^{4-2j} \\ &= \binom{4}{0} x^4 + \binom{3}{1} x^2 + \binom{2}{0} x^0 \\ &= x^4 + 3x^2 + 1 \end{aligned}$$

as we obtained earlier.

Using Formula (37.2), it is easy to verify that $f_1(x) = 1$ and $f_2(x) = x$, so $f_n(x)$ satisfies both initial conditions. We can now confirm that $f_n(x)$ is indeed the Fibonacci polynomial (see Exercise 11).

ANOTHER EXPLICIT FORMULA FOR $f_n(x)$

There is yet another explicit formula for $f_n(x)$:

$$f_n(x) = \frac{\alpha^n(x) - \beta^n(x)}{\alpha(x) - \beta(x)} \tag{37.3}$$

where

$$\alpha(x) = \frac{x + \sqrt{x^2+4}}{2} \quad \text{and} \quad \beta(x) = \frac{x - \sqrt{x^2+4}}{2}$$

See Chapter 38.

For example, let us compute $f_5(x)$. It is easy to verify that

$$(x + \sqrt{x^2+4})^5 - (x - \sqrt{x^2+4})^5 = 32(x^4 + 3x^2 + 1)\sqrt{x^2+4},$$

so $f_5(x) = x^4 + 3x^2 + 1$, as expected.

We can confirm Formula (37.3) using the recurrence relation (see Exercise 19).

Next we establish a few properties of Fibonacci polynomials, which are generalizations of some formulas we derived in Chapter 5.

Theorem 37.1.

$$x \sum_1^n f_i(x) = f_{n+1}(x) + f_n(x) - 1 \tag{37.4}$$

Proof. Using the recurrence relation (37.1),

$$\sum_1^n f_{i+1}(x) = x \sum_1^n f_i(x) + \sum_1^n f_{i-1}(x)$$

That is,

$$f_n(x) + f_{n+1}(x) = x \sum_1^n f_i(x) + f_0(x) + f_1(x)$$

Since $f_0(x) = 0$, it follows that

$$x \sum_1^n f_i(x) = f_{n+1}(x) + f_n(x) - 1$$

■

For example,

$$x\sum_{1}^{4} f_i(x) = x[1 + x + (x^2 + 1) + (x^3 + 2x)]$$

$$= x^4 + x^3 + 3x^2 + x$$

$$f_5(x) + f_4(x) - 1 = (x^4 + 3x^2 + 1) + (x^3 + 2x) - 1$$

$$= x^4 + x^3 + 3x^2 + x$$

$$= x\sum_{1}^{4} f_i(x)$$

Corollary 37.1.

$$\sum_{1}^{n} F_i = F_{n+2} - 1$$ ■

This corollary follows from the theorem, since $f_i(1) = F_i$.

A GENERATING FUNCTION FOR $f_n(x)$

Next, we will find a generating function for $f_n(x)$. To this end, we let

$$g(t) = \sum_{0}^{\infty} f_n(x)t^n$$

$$xtg(t) = \sum_{0}^{\infty} xf_n(x)t^{n+1}$$

$$t^2g(t) = \sum_{0}^{\infty} f_n(x)t^{n+1}$$

Then

$$(1 - xt - t^2)g(t) = f_0(x) + tf_1(x) - xtf_0(x) = t$$

Thus

$$g(t) = \frac{t}{1 - xt - t^2}$$

generates $f_n(x)$.

Theorem 37.2.

$$f_{m+n+1}(x) = f_{m+1}(x)f_{n+1}(x) + f_m(x)f_n(x)$$

Proof.

$$\frac{y}{1-xy-y^2}=\sum_{n=0}^{\infty} f_n(x)y^n$$

Therefore,

$$\frac{yf_m(x)}{1-xy-y^2}=\sum_{n=0}^{\infty} f_m(x)f_n(x)y^n \tag{37.5}$$

Then

$$\frac{f_{m+1}(x)}{1-xy-y^2}=\sum_{n=0}^{\infty} f_{m+1}(x)f_{n+1}(x)y^n \tag{37.6}$$

In 1964, D. Zeitlin showed that

$$\frac{f_{m+1}(x)+f_m(x)y}{1-xy-y^2}=\sum_{n=0}^{\infty} f_{m+n+1}(x)y^n \tag{37.7}$$

The desired result now follows from Eqs. (37.5), (37.6), and (37.7) by equating the coefficients of y^n from both sides. ■

This theorem illustrates an alternate method for constructing new members of the family of Fibonacci polynomials.

For example, let $m = 3$ and $n = 4$. Then

$$\begin{aligned} f_{m+1}(x)f_{n+1}(x)+f_m(x)f_n(x) &= f_4(x)f_5(x)+f_3(x)f_4(x)\\ &= (x^3+2x)(x^4+3x^2+1)+(x^2+1)(x^3+2x)\\ &= x^7+6x^5+10x^3+4x\\ &= f_8(x)=f_{m+n+1}(x)\end{aligned}$$

Theorem 37.2 yields the following Fibonacci identity without much effort.

Corollary 37.2. $F_{m+n}=F_{m+1}F_n+F_mF_{n-1}$ ■

Next we derive Formula (37.2) for $f_n(x)$ by an alternate method.

Theorem 37.3.

$$f_n(x)=\sum_{j=0}^{\lfloor (n-1)/2\rfloor}\binom{n-j-1}{j}x^{n-2j-1} \qquad n\ge 0 \tag{37.2}$$

Proof. We have

$$\frac{y}{1-xy-y^2}=\sum_{n=0}^{\infty} f_n(x)y^n \tag{37.8}$$

But

$$\frac{1}{1-2tz+z^2} = \sum_{n=0}^{\infty}\left[\sum_{j=0}^{\lfloor n/2\rfloor}(-1)^j\binom{n-j}{j}(2t)^{n-2j}\right]z^n \tag{37.9}$$

$$U_n(t) = \sum_{j=0}^{\lfloor n/2\rfloor}(-1)^j\binom{n-j}{j}(2t)^{n-2j}$$

is the *Chebyshev polynomial of the second kind*.* Let $z = iy$ and $t = x/2i$, where $i^2 = -1$. Then Eq. (37.9) yields

$$\frac{1}{1-xy-y^2} = \sum_{n=0}^{\infty} i^n U_n(x/2i)y^n$$

$$\therefore \quad \frac{y}{1-xy-y^2} = \sum_{n=0}^{\infty} i^n U_n(x/2i)y^{n+1}$$

From Eqs. (37.8) and (37.9), it follows that

$$\begin{aligned} f_{n+1}(x) &= i^n U_n(x/2i) \\ &= i^n \sum_{j=0}^{\lfloor n/2\rfloor}(-1)^j\binom{n-j}{j}(x/i)^{n-2j} \\ &= \sum_{j=0}^{\lfloor n/2\rfloor}\binom{n-j}{j}x^{n-2j} \\ \therefore \quad f_n(x) &= \sum_{j=0}^{\lfloor n/2\rfloor}\binom{n-j-1}{j}x^{n-2j-1} \end{aligned}$$ ■

The next result, which we derived in Chapter 12, follows from this formula.

Corollary 37.3. (Lucas, 1876)

$$F_n = \sum_{j=0}^{\lfloor (n-1)/2\rfloor}\binom{n-j-1}{j}$$ ■

*Named after the Russian mathematician Pafnuty Lvovich Chebyshev (1821–1894).

Theorem 37.4.

$$f_n'(x) = \sum_{i=1}^{n-1} f_i(x) f_{n-i}(x)$$

where $f_n'(x)$ denotes the derivative of $f_n(x)$ with respect to x and $n \geq 1$.

Proof. Differentiating Eq. (37.8) with respect to x,

$$\begin{aligned}
\sum_{n=0}^{\infty} f_n'(x) y^n &= \left(\frac{y}{1 - xy - y^2}\right)^2 \\
&= \left[\sum_{0}^{\infty} f_n(x) y^n\right]^2 \\
&= \sum_{n=0}^{\infty} \left[\sum_{i=0}^{n} f_i(x) f_{n-i}(x)\right] y^n
\end{aligned}$$

Equating the coefficients of y^n, we get

$$f_n'(x) = \sum_{i=0}^{n} f_i(x) f_{n-i}(x)$$

$$f_n'(x) = \sum_{i=1}^{n-1} f_i(x) f_{n-i}(x)$$

since $f_0(x) = 0$. ■

For example, we have:

$$\begin{aligned}
f_6(x) &= x^5 + 4x^3 + 3x \\
f_6'(x) &= 5x^4 + 12x^2 + 3 \\
\sum_{1}^{6} f_i(x) f_{6-i}(x) &= f_1(x) f_5(x) + f_2(x) f_4(x) \\
&\quad + f_3(x) f_3(x) + f_4(x) f_2(x) + f_5(x) f_1(x) \\
&= 2 f_1(x) f_5(x) + 2 f_2(x) f_4(x) + f_3^2(x) \\
&= 2(1)(x^4 + 3x^2 + 1) + 2x(x^3 + 2x) + (x^2 + 1)^2 \\
&= 5x^4 + 12x^2 + 3 \\
&= f_6'(x)
\end{aligned}$$

Using the Pascal-like array in Table 13.3, H. W. Gould in 1965 studied the polynomial

$$G_n(x) = \sum_{j=0}^{n} A(n, j)x^j$$

where

$$A(n, j) = \binom{n - \lfloor (j+1)/2 \rfloor}{\lfloor j/2 \rfloor}$$

denotes the jth entry in row n. Since $A(n, 2k) = \binom{n-k}{k}$ and $A(n, 2k+1) = \binom{n-k-1}{k}$, we can write $G_n(x)$ as

$$G_n(x) = \sum_{0}^{\lfloor n/2 \rfloor} \binom{n-k}{k} x^{2k} + \sum_{0}^{\lfloor (n-1)/2 \rfloor} \binom{n-k-1}{k} x^{2k+1} \tag{37.10}$$

Then

$$G_{n+1}(x) = \sum_{0}^{\lfloor (n+1)/2 \rfloor} \binom{n-k+1}{k} x^{2k} + \sum_{0}^{\lfloor n/2 \rfloor} \binom{n-k}{k} x^{2k+1}$$

and

$$\begin{aligned} x^2 G_n(x) &= \sum_{0}^{\lfloor n/2 \rfloor} \binom{n-k}{k} x^{2k+2} + \sum_{0}^{\lfloor (n-1)/2 \rfloor} \binom{n-k-1}{k} x^{2k+3} \\ &= \sum_{1}^{\lfloor (n+2)/2 \rfloor} \binom{n-k+1}{k} x^{2k} + \sum_{1}^{\lfloor (n+1)/2 \rfloor} \binom{n-k}{k-1} x^{2k+1} \end{aligned}$$

By virtue of Pascal's identity,

$$\begin{aligned} G_{n+1}(x) + x^2 G_n(x) &= \sum_{0}^{\lfloor (n+2)/2 \rfloor} \binom{n-k+2}{k} x^{2k} + \sum_{0}^{\lfloor (n+1)/2 \rfloor} \binom{n-k+1}{k} x^{2k+1} \\ &= G_{n+2}(x) \end{aligned}$$

Thus $G_n(x)$ satisfies the recurrence relation

$$G_{n+2}(x) = G_{n+1}(x) + x^2 G_n(x) \tag{37.11}$$

which yields the Fibonacci recurrence relation (FRR) when $x = 1$. Since $G_n(x) = \sum_{j=0}^{n} A(n, j)$, it follows, by Theorem 13.1, that $G_n(1) = F_{n+2}, n \geq 0$.

Using the polynomial $G_n(x)$, Gould also studied extensively a closely related polynomial $H_n(x)$:

$$H_n(x) = x^n G_n(1/x) = \sum_{j=0}^{n} A(n, j)x^{n-j}$$

Interestingly enough, this polynomial yields intriguing results. Notice that:

$$\begin{aligned} H_{n+2}(x) &= x^{n+2}G_{n+2}(1/x) \\ &= x^{n+2}[G_{n+1}(1/x) + x^{-2}G_n(1/x)] \\ &= x^{n+2}G_{n+1}(1/x) + x^n G_n(1/x) \\ &= xH_{n+1}(x) + H_n(x) \end{aligned}$$

where $H_0(x) = G_0(1/x) = 1$ and $H_1(x) = xG_1(1/x) = x + 1$. This is exactly the same Fibonacci polynomial, $f_n(x)$, studied by Catalan more than 80 years earlier.

Notice that $H_{n+2}(1) = H_{n+1}(1) + H_n(1)$, where $H_0(1) = 1$ and $H_1(1) = 2$. So $H_n(1) = F_{n+1}$, where $n \geq 0$.

A MATRIX GENERATOR FOR FIBONACCI POLYNOMIALS

Gould employed the Q-matrix technique that we studied in Chapter 32, to study the H-polynomials. To this end, he considered the generalized Q-matrix

$$Q(x) = \begin{bmatrix} x & 1 \\ 1 & 0 \end{bmatrix}$$

Then

$$Q^n(x) = \begin{bmatrix} f_{n+1}(x) & f_n(x) \\ f_n(x) & f_{n-1}(x) \end{bmatrix} \tag{37.12}$$

where $n \geq 1$ (see Exercise 22).

Since $|Q| = -1, |Q^n| = (-1)^n$. Accordingly, Eq. (37.12) yields a Cassini-like formula for $f_n(x)$:

$$f_{n+1}(x)f_{n-1}(x) - f_n^2(x) = (-1)^n \tag{37.13}$$

When $x = 1$, this reduces to the Cassini's rule.

It follows from Eq. (37.12) that

$$Q^{m+n}(x) = \begin{bmatrix} f_{m+n+1}(x) & f_{m+n}(x) \\ f_{m+n}(x) & f_{m+n-1}(x) \end{bmatrix}$$

But

$$\begin{aligned} Q^{m+n}(x) &= Q^m(x)Q^n(x) \\ &= \begin{bmatrix} f_{m+1}(x)f_{n+1}(x) + f_m(x)f_n(x) & f_{m+1}(x)f_n(x) + f_m(x)f_{n-1}(x) \\ f_m(x)f_{n+1}(x) + f_{m-1}(x)f_n(x) & f_m(x)f_n(x) + f_{m-1}(x)f_{n-1}(x) \end{bmatrix} \end{aligned}$$

Consequently,

$$f_{m+n}(x) = f_{m+1}(x)f_n(x) + f_m(x)f_{n-1}(x) \tag{37.14}$$

In particular, let $x = 1$. This yields an identity from Chapter 32:

$$F_{m+n} = F_{m+1}F_n + F_m F_{n-1} \tag{32.2}$$

Since $f_1(x) + f_0(x) = 1 + 0 = 1 = H_0(x)$ and $f_2(x) + f_1(x) = x + 1$, it follows that $H_n(x) = f_{n+1}(x) + f_n(x)$;

$$\begin{aligned}
\therefore\ (-1)^{i+1}H_i(x) &= (-1)^{i+1}f_{i+1}(x) - (-1)^i f_i(x) \\
\sum_{i=0}^{n}(-1)^{i+1}H_i(x) &= \sum_{i=0}^{n}[(-1)^{i+1}f_{i+1}(x) - (-1)^i f_i(x)] \\
&= (-1)^{n+1}f_{i+1}(x)
\end{aligned}$$

Thus, we can express the Fibonacci polynomials $f_n(x)$ in terms of the H-polynomials (or the G-polynomials) as

$$f_{n+1}(x) = \sum_{i=0}^{n}(-1)^{n+i}H_i(x)$$

Next, it follows from Eqs. (37.12) and (37.13) that

$$Q^{n+1}(x) + Q^n(x) = \begin{bmatrix} H_{n+1}(x) & H_n(x) \\ H_n(x) & H_{n-1}(x) \end{bmatrix}$$

$$\begin{aligned}
\begin{vmatrix} H_{n+1}(x) & H_n(x) \\ H_n(x) & H_{n-1}(x) \end{vmatrix} &= |Q^n(x)[Q(x) + I]| \\
&= |Q^n(x)| \cdot |Q + I|
\end{aligned}$$

That is,

$$H_{n+1}(x)H_{n-1}(x) - H_n^2(x) = x(-1)^{-n} \tag{37.15}$$

which is again a generalization of Cassini's rule.

In fact, we even have a further generalization of Eq. (37.15):

$$\begin{vmatrix} H_{n+a}(x) & H_{n+a+b}(x) \\ H_n(x) & H_{n+b}(x) \end{vmatrix} = (-1)^n \begin{vmatrix} H_a(x) & H_{a+b}(x) \\ H_0(x) & H_b(x) \end{vmatrix}$$

BYRD'S FIBONACCI POLYNOMIALS

Next we examine the Fibonacci polynomials $\varphi_n(x)$ studied extensively in 1963 by P. F. Byrd of San Jose State College. They are defined by the recurrence relation

$$\varphi_{n+2}(x) = 2x\ \varphi_{n+1}(x) + \varphi_n(x) \tag{37.16}$$

where $n \geq 0$, x is an arbitrary real number, $\varphi_0(x) = 0$, and $\varphi_1(x) = 1$.

Using the recurrence relation (37.16), we can extract various Fibonacci polynomials of this family:

$$\begin{aligned}
\varphi_2(x) &= 2x \cdot 1 - 0 = 2x \\
\varphi_3(x) &= 2x(2x) + 1 = 4x^2 + 1 \\
\varphi_4(x) &= 2x(4x^2 + 1) + 2x = 8x^3 + 4x \\
\varphi_5(x) &= 2x(8x^3 + 4x) + (4x^2 + 1) = 16x^4 + 12x^2 + 1
\end{aligned}$$

In particular, $\varphi_0(1/2) = 0$, $\varphi_1(1/2) = 1$, $\varphi_2(1/2) = 1$, $\varphi_3(1/2) = 2$, $\varphi_4(1/2) = 3$, and $\varphi_5(1/2) = 5$; they are all Fibonacci numbers. But, is this true always?

Notice that when $x = 1/2$, Eq. (37.16) yields the recurrence relation $\varphi_{n+2}(1/2) = \varphi_{n+1}(1/2)+\varphi_n(1/2)$, where $\varphi_0(1/2) = 0$ and $\varphi_1(1/2) = 1$. Consequently, $\varphi_n(1/2) = F_n$, as expected.

To derive a generating function for $\varphi_n(x)$, we let

$$g(t) = \sum_{n=0}^{\infty} \varphi_n(x)t^n$$

Then

$$2xtg(t) = \sum_{n=0}^{\infty} 2x\ \varphi_n\ (x)t^{n+1}$$

and

$$\begin{aligned}
t^2 g(t) &= \sum_{n=0}^{\infty} \varphi_n(x)t^{n+2} \\
\therefore\ t^2 g(t) - 2xtg(t) - g(t) &= -\ \varphi_1\ t + \sum_{n=0}^{\infty} [\varphi_n(x) - 2x\ \varphi_{n+1}\ (x) - \varphi_{n+2}]t^n \\
(t^2 - 2xt - 1)g(t) &= -t
\end{aligned}$$

by the recurrence relation (37.16). That is,

$$g(t) = \frac{t}{1 - 2xt - t^2}$$

So $g(t) = t/(1 - 2xt - t^2)$ is a generating function of the Fibonacci polynomials. Thus

$$\frac{t}{1 - 2xt - t^2} = \sum_{n=0}^{\infty} \varphi_n(x)t^n \tag{37.17}$$

Notice that when $x = 1/2$, this yields the generating function for F_n we obtained in Chapter 18.

The polynomials $\varphi_n(x)$ satisfy a fascinating property. To see this, change t to $-t$ and x to $-x$ in Eq. (37.17):

Thus

$$\frac{-t}{1-2xt-t^2} = \sum_{n=0}^{\infty} \varphi_n(-x)(-t)^n$$

That is,

$$-\sum_{n=0}^{\infty} \varphi_n(x)t^n = \sum_{n=0}^{\infty} \varphi_n(-x)(-t)^n$$

Equating coefficients of t^n from either side yields the property

$$\varphi_n(x) = (-1)^{n+1}\ \varphi_n\ (-x)$$

Consequently, $\varphi_n(x)$ is an even function if n is odd, and an odd function otherwise.

For example, notice that $\varphi_3(x)$ and $\varphi_5(x)$ are even functions, whereas $\varphi_2(x)$ and $\varphi_4(x)$ are odd.

Since $1/(1-s) = \sum_0^{\infty} s^i$, we can expand the left-hand side (LHS) of Eq. (37.17):

$$\begin{aligned}
\sum_{n=0}^{\infty} \varphi_n(x)t^n &= t\sum_{n=0}^{\infty}(2xt+t^2)^n \\
&= t\sum_{n=0}^{\infty}\sum_{i=0}^{n}\binom{n}{i}(2xt)^{n-i}(t^2)^i \\
&= \sum_{n=0}^{\infty}\sum_{i=0}^{n}\binom{n}{i}(2x)^{n-i}t^{n+i+1} \\
&= t + \sum_0^1\binom{1}{i}(2x)^{1-i}t^{i+2} + \sum_0^2\binom{2}{i}(2x)^{2-i}t^{3+i} \\
&\quad + \sum_0^3\binom{3}{i}(2x)^{3-i}t^{4+i} + \cdots \\
&= t + (2x)t^2 + [(2x)^2+1]t^3 + [(2x)^3 + 292x)]t^4 \\
&\quad + [(2x)^4 + 3(2x)^2 + 1]t^5 + \cdots
\end{aligned}$$

That is,

$$\sum_{n=0}^{\infty} \varphi_n(x)t^n = \sum_{n=0}^{\infty}\left[\sum_{i=0}^{\lfloor (n-1)/2\rfloor}\binom{n-i-1}{i}(2x)^{n-2i}\right]t^n$$

This yields the explicit formula for $\varphi_n(x)$:

$$\varphi_n(x) = \sum_{i=0}^{\lfloor (n-1)/2 \rfloor} \binom{n-i-1}{i} (2x)^{n-2i} \tag{37.18}$$

where $n \geq 1$.

When $x = 1/2$, this yields the Lucas formula for F_n, developed in Chapter 12:

$$F_n = \sum_{i=0}^{\lfloor (n-1)/2 \rfloor} \binom{n-i-1}{i} \tag{12.1}$$

EXERCISES 37

1. Verify Theorem 37.1 for $n = 5$ and $n = 6$.

Find each polynomial using Theorem 37.2.

2. $f_6(x)$
3. $f_{10}(x)$

Find each polynomial using Theorem 37.3.

4. $f_6(x)$
5. $f_{10}(x)$

Using Theorem 37.4, find $f_n'(x)$ for the given value of n.

6. $n = 4$
7. $n = 7$
8. Let $z_i = f_i(x) + f_i(y)$. Show that $z_{n+4} - (x+y)z_{n+3} + (xy-2)z_{n+2} + (x+y)z_{n+1} + z_n = 0$ (Swamy, 1966).
9. Let $z_i = f_i(x) \cdot f_i(y)$. Show that $z_{n+4} - (xy)z_{n+3} + (x^2 + y^2 + 2)z_{n+2} - xyz_{n+1} + z_n = 0$ (Swamy, 1968).
10. Using the principle of mathematical induction, prove that $f_{n+1}(x)f_{n-1}(x) - f_n^2(x) = (-1)^n, n \geq 1$.
11. Let $g_n(x) = \sum_{j=0}^{\lfloor (n-1)/2 \rfloor} \binom{n-j-1}{j} x^{n-2j-1}$. Show that $g_n(x) = f_n(x)$.
12. Let $A(n, j)$ denote the element in row n and column j of the array in Table 13.1. Define $A(n, j)$ recursively.
13. Show that $f_n(x)$ and $f_{n+1}(x)$ are relatively prime (Swamy, 1971).
14. Prove that $f_{n+1}(x)f_{n-2}(x) - f_n(x)f_{n-1}(x) + x(-1)^n = 0$ (Swamy, 1971).

Establish the following generating functions.

15. $$\frac{xt^2}{1-(x^2+2)t^2+t^4} = \sum_0^\infty f_{2n}(x)t^{2n}$$
16. $$\frac{t-t^3}{1-(x^2+2)t^2+t^4} = \sum_0^\infty f_{2n+1}(x)t^{2n+1}$$

17. $\dfrac{(x^2+1)-t^2}{1-(x^2+2)t^2+t^4} = \sum_0^\infty f_{2n+3}(x)t^{2n}$

18. Show that $\left[1+\sum_1^n \frac{1}{f_{2n-1}(x)f_{2n+1}(x)}\right]\left[1-x^2\sum_1^n \frac{1}{f_{2n}(x)f_{2n+2}(x)}\right]=1$ (Swamy, 1971a).

19. Verify that Formula (37.3) defines the Fibonacci polynomial $f_n(x)$.

20. *Chebyshev polynomials of the second kind* U_{n-1} are defined by $U_{n-1} = \dfrac{r^n - s^n}{2\sqrt{x^2-1}}$, where $r = x+\sqrt{x^2-1}$ and $s = x-\sqrt{x^2-1}$. Show that $U_{n-1}(3/2) = F_{2n}$ (Basin, 1963).

21. Let $g(x,n)$ denote the hypergeometric function

$$g(x,n) = \sum_{k=0}^{n-1} \frac{2^k(n+k)!}{(n-k-1)!(2k+1)!}(x-1)^k$$

Show that $g(3/2,n) = F_{2n}$. [*Note*: $g(x,n) = U_{n-1}(x)$.] (Basin, 1963).

22. Prove that $Q^n(x) = \begin{bmatrix} f_{n+1}(x) & f_n(x) \\ f_n(x) & f_{n-1}(x) \end{bmatrix}, n \geq 1.$

23. Prove that $H_n(x) = f_{n+1}(x) + f_n(x), n \geq 0$.

24. Find the Fibonacci polynomials $\varphi_6(x)$ and $\varphi_7(x)$.

Consider the polynomials $\psi_n(x)$ defined recursively by $\psi_{n+2}(x) = 2x\psi_{n+1}(x) + \psi_n(x)$, where $\psi_0(x) = 2$ and $\psi_1(x) = 1$.

25. Find the polynomials $\psi_n(x)$ for $2 \leq n \leq 5$.

26. Compute $\psi_n(1/2)$ for $0 \leq n \leq 5$.

27. Conjecture the value of $\psi_n(1/2)$.

28. Prove that $\psi_n(1/2) = L_n$.

29. Find a generating function for the polynomial $\psi_n(x)$.

Consider the polynomials $y_n(x)$ defined recursively by $y_{n+2}(x) = xy_{n+1}(x) + y_n(x)$, where $y_0(x) = 0$ and $y_1(x) = 1$.

30. Find the polynomials $y_n(x)$ for $2 \leq n \leq 5$.

31. Conjecture the value of $y_n(1)$.

32. Prove that $y_n(1) = F_n$.

33. Find a generating function for the polynomial $y_n(x)$.

34–35. Redo Exercises 30 and 31 if $y_0(x) = 2$.

36. Prove that $y_n(1) = L_n$.

37. Find a generating function for $y_n(x)$.

Consider the polynomials $z_n(x)$ defined recursively by $z_{n+2}(x) = z_{n+1}(x) + xz_n(x)$, where $z_0(x) = 0$ and $z_1(x) = 1$.

38. Find the polynomials $z_n(x)$ for $2 \leq n \leq 5$.

39. Conjecture the value of $z_n(1)$.
40. Establish the formula in Exercise 39.
41. Find a generating function for the polynomial $z_n(x)$.

42–45. Redo Exercises 38–41 if $z_0(x) = 2$.

46. Using Formula (37.16), compute the polynomials $\varphi_6(x)$ and $\varphi_7(x)$.
47. Let $D_n(x)$ denote the nth-order determinant:

$$\begin{vmatrix} 2x & i & 0 & 0 & \dots & 0 & 0 \\ i & 2x & i & 0 & \dots & 0 & 0 \\ 0 & i & 2x & i & \dots & 0 & 0 \\ 0 & 0 & i & 2x & \dots & 0 & 0 \\ \cdot & \cdot & \cdot & \cdot & \dots & & \\ \cdot & \cdot & \cdot & \cdot & & & \\ \cdot & \cdot & \cdot & \cdot & \dots & 2x & i \\ 0 & 0 & 0 & 0 & & i & 2x \end{vmatrix}$$

where $n \geq 2, i = \sqrt{-1}, x$ is an arbitrary real number, $D_0(x) = 0$, and $D_1(x) = 1$. Show that $D_n(x) = \varphi_n(x)$ (Byrd, 1963).

48. Evaluate the $n \times n$ determinant

$$\begin{vmatrix} 1 & i & 0 & 0 & \dots & 0 & 0 \\ i & 1 & i & 0 & \dots & 0 & 0 \\ 0 & i & 1 & i & \dots & 0 & 0 \\ 0 & 0 & i & 1 & \dots & 0 & 0 \\ \cdot & \cdot & \cdot & \cdot & \dots & & \\ \cdot & \cdot & \cdot & \cdot & & & \\ \cdot & \cdot & \cdot & \cdot & \dots & 1 & i \\ 0 & 0 & 0 & 0 & \dots & i & 1 \end{vmatrix}$$

where $n \geq 2$ and $i = \sqrt{-1}$ (Byrd, 1963).

38

LUCAS POLYNOMIALS

Lucas polynomials $l_n(x)$, originally studied in 1970 by Bicknell, are defined by

$$l_n(x) = xl_{n-1}(x) + l_{n-2}(x)$$

where $l_0(x) = 2$, $l_1(x) = x$, and $n \geq 2$.

The first ten Lucas polynomials are:

$$\begin{aligned}
l_1(x) &= x \\
l_2(x) &= x^2 + 2 \\
l_3(x) &= x^3 + 3x \\
l_4(x) &= x^4 + 4x^2 + 2 \\
l_5(x) &= x^5 + 5x^3 + 5x \\
l_6(x) &= x^6 + 6x^4 + 9x^2 + 2 \\
l_7(x) &= x^7 + 7x^5 + 9x^3 + 7x \\
l_8(x) &= x^8 + 8x^6 + 20x^4 + 16x^2 + 2 \\
l_9(x) &= x^9 + 9x^7 + 27x^5 + 30x^3 + 9x \\
l_{10}(x) &= x^{10} + 10x^8 + 35x^6 + 50x^4 + 25x^2 + 2
\end{aligned}$$

It follows from the recursive definition that $l_n(1) = L_n$ for $n \geq 0$; that is, the sum of the coefficients of $l_n(x)$ is L_n. We can verify this by computing $l_n(1)$ for these Lucas polynomials.

TABLE 38.1.

n	x^0	x^1	x^2	x^3	x^4	x^5	x^6	x^7	x^8	x^9	x^{10}
1	0	1									
2	2	0	1								
3	0	3	0	1							
4	2	0	4	0	1						
							Sum $= 48 = 3 \cdot 2^4$				
5	0	5	0	5	0	1					
6	2	0	9	0	6	0	1				
7	0	7	0	14	0	7	0	1	← Sum $= L_7 = 29$		
8	2	0	16	0	20	0	8	0	1		
9	0	9	0	30	0	27	0	9	0	1	
10	2	0	25	0	50	0	35	0	10	0	1

Lucas Polynomials satisfy three additional properties:

- $l_n(x) = f_{n+1}(x) + f_{n-1}(x) = xf_n(x) + 2f_{n-1}(x)$
- $xl_n(x) = f_{n+2}(x) - f_{n-2}(x)$
- $l_{-n}(x) = (-1)^n l_n(x)$

See Exercises 3–6.

In addition, $l_n(2) = f_{n+1}(2) + f_{n-1}(2) = P_{n+1} + P_{n-1}$, where P_n denotes the nth Pell number. For example, $l_5(2) = 82 = 70 + 12 = P_6 + P_4$.

Arranging the coefficients of the various polynomials in ascending order of exponents, we get the array in Table 38.1. The sum of the elements along the nth rising diagonal is $3 \cdot 2^{(n-1)/2}$, where n is even and is ≥ 2.

BINET'S FORMULAS FOR $f_n(x)$ AND $l_n(x)$

Next we find Binet's formulas for both Fibonacci and Lucas polynomials.

Let $\alpha(x)$ and $\beta(x)$ be the solutions of the quadratic equation $t^2 - xt - 1 = 0$:

$$\alpha(x) = \frac{x + \sqrt{x^2 + 4}}{2} \qquad \text{and} \qquad \beta(x) = \frac{x - \sqrt{x^2 + 4}}{2}$$

Notice that $\alpha(1) = \alpha$ and $\beta(1) = \beta$; $\alpha(2) = 1 + \sqrt{2}$ and $\beta(2) = 1 - \sqrt{2}$ are the characteristic roots of the *Pell recurrence relation* $x^2 - 2x - 1 = 0$. We can verify that

$$f_n(x) = \frac{\alpha^n(x) - \beta^n(x)}{\alpha(x) - \beta(x)} \qquad \text{and} \qquad l_n(x) = \alpha^n(x) + \beta^n(x)$$

See Exercises 8 and 9.

We can employ Formula (37.14) to derive identities linking Fibonacci and Lucas polynomials. Changing n to $-n$ in the formula, we get:

$$\begin{aligned} f_{m-n}(x) &= (-1)^n[-f_{m+1}(x)f_n(x) + f_m(x)f_{n+1}(x)] \\ \therefore \quad f_{m+n}(x) + (-1)^n f_{m-n}(x) &= f_m(x)f_{n-1}(x) + f_m(x)f_{n+1}(x) \\ &= f_m(x)l_n(x) \end{aligned}$$

This yields the formula

$$F_{m+n} + (-1)^n F_{m-n} = F_m L_n \tag{38.1}$$

Replacing n by k and m by $m - k$ in Eq. (37.14), we get yet another identity:

$$f_m(x) = l_k(x)f_{m-k}(x) + (-1)^{k+1} f_{m-2k}(x)$$

In particular,

$$F_m = L_k F_{m-k} + (-1)^{k+1} F_{m-2k} \tag{38.2}$$

Two new polynomials $g_n(x)$ and $h_n(x)$ correspond to the Fibonacci and Lucas polynomials:

$$\begin{aligned} g_0(x) &= 0 \qquad g_1(x) = 1 \\ g_n(x) &= xg_{n-1}(x) - g_{n-2}(x) \qquad n \geq 2 \end{aligned}$$

$$\begin{aligned} h_0(x) &= 2 \qquad h_1(x) = x \\ h_n(x) &= xh_{n-1}(x) - h_{n-2}(x) \qquad n \geq 2 \end{aligned}$$

Notice that we can obtain $g_n(x)$ and $h_n(x)$ from $f_n(x)$ and $l_n(x)$ by changing the plus sign to minus in the recurrence relations. These polynomials, studied extensively in 1971 by Hoggatt et al., and in 1972 by Hoggatt, are related to the Chebyshev polynomials of the first and second kind.

Let $\gamma(x)$ and $\delta(x)$ be the solutions of the quadratic equation $t^2 - xt + 1 = 0$:

$$\gamma(x) = \frac{x + \sqrt{x^2 - 4}}{2} \qquad \text{and} \qquad \delta(x) = \frac{x + \sqrt{x^2 - 4}}{2}$$

Notice that $\gamma(\sqrt{5}) = \alpha$ and $\delta(\sqrt{5}) = -\beta$. It is easy to verify that

$$g_n(x) = \frac{\gamma^n(x) - \delta^n(x)}{\gamma(x) - \delta(x)} \qquad \text{and} \qquad h_n(x) = \gamma^n(x) + \delta^n(x)$$

where $x \neq 2$ (see Exercises 19 and 20).

Besides, $h_n^2(x) - (x^2 - 4)g_n^2(x) = 4$ and $h_n(x) = g_{n+1}(x) - g_{n-1}(x)$ (see Exercises 21 and 22). We can see that the coefficients of $g_n(x)$ lie along the rising diagonals of Pascal's triangle.

DIVISIBILITY PROPERTIES

The polynomials $f_n(x), l_n(x), g_n(x)$, and $h_n(x)$ yield interesting divisibility dividends. To extract some of them, recall that

$$\alpha(x) = \frac{x + \sqrt{x^2+4}}{2} \qquad \text{and} \qquad \beta(x) = \frac{x - \sqrt{x^2+4}}{2}$$

Then

$$\begin{aligned} \alpha(L_{2m+1}(x)) &= \frac{L_{2m+1}(x) + \sqrt{x^2+4}F_{2m+1}}{2} \\ &= \alpha^{2m+1}(x) \end{aligned}$$

by Exercise 15. Similarly,

$$\begin{aligned} \beta(L_{2m+1}(x)) &= \beta^{2m+1}(x) \\ \therefore \quad f_n(L_{2m+1}(x)) &= \frac{\alpha^n(L_{2m+1}(x)) - \beta^n(L_{2m+1}(x))}{\alpha(L_{2m+1}(x)) - \beta(L_{2m+1}(x))} \\ &= \frac{\alpha^{(2m+1)n}(x) - \beta^{(2m+1)n}(x)}{\alpha^{2m+1}(x) - \beta^{2m+1}(x)} = \frac{f_{(2m+1)n}(x)}{f_{2m+1}(x)} \end{aligned}$$

Similarly, using the identity $l_{2m}^2(x) - 4 = (x^2+4)f_{2m}^2(x)$, we can show that:

$$\begin{aligned} \gamma(l_{2m}(x)) &= \frac{l_{2m}(x) + \sqrt{x^2+4}f_{2m}(x)}{2} = \alpha^{2m}(x) \\ \delta(l_{2m}(x)) &= \frac{l_{2m}(x) - \sqrt{x^2+4}f_{2m}(x)}{2} = \beta^{2m}(x) \\ g_n(L_{2m}(x)) &= \frac{\gamma^n(L_{2m}(x)) - \delta^n(L_{2m}(x))}{\gamma(L_{2m}(x)) - \delta(L_{2m}(x))} \\ &= \frac{\gamma^{2mn}(x) - \delta^{2mn}(x)}{\gamma^{2m}(x) - \delta^{2m}(x)} = \frac{f_{2mn}(x)}{f_{2m}(x)} \end{aligned}$$

Thus, we have established the following theorem.

Theorem 38.1. (Hoggatt, Jr., Bicknell, and King, 1972)

$$f_{nk}(x) = \begin{cases} f_k(x) \cdot f_n(l_k(x)) & \text{if } k \text{ is odd} \\ f_k(x) \cdot g_n(l_k(x)) & \text{otherwise} \end{cases}$$ ■

Corollary 38.1. $f_k(x) | f_{nk}(x)$ ■

$$\begin{aligned}\text{For example, } f_6(x) &= x^5 + 4x^3 + 3x \\ &= (x^2+1)(x^3+3x) \\ &= f_3(x) \cdot l_3(x) = f_3(x) \cdot f_2(l_3(x))\end{aligned}$$

Corollary 38.2. $F_k | F_{nk}$ ■

This follows from Corollary 38.1, since $f_n(1) = F_k$; note that we already knew this from Chapter 16.

Corollary 38.3.

$$f_{nk}(x) = f_k(x) \sum_{j=0}^{\lfloor (n-1)/2 \rfloor} \binom{n-j-1}{j} (-1)^{(k+1)j} l_k^{n-2j-1}(x)$$

Proof. We have

$$f_m(x) = \sum_{j=0}^{\lfloor (m-1)/2 \rfloor} \binom{m-j-1}{j} x^{m-2j-1}$$

$$g_m(x) = \sum_{j=0}^{\lfloor (m-1)/2 \rfloor} \binom{m-j-1}{j} (-1)^j x^{m-2j-1}$$

$$\therefore \quad f_{nk}(x) = \begin{cases} f_k(x) \displaystyle\sum_{j=0}^{\lfloor (n-1)/2 \rfloor} \binom{n-j-1}{j} l_k^{n-2j-1}(x) & \text{if } k \text{ is odd} \\ f_k(x) \displaystyle\sum_{j=0}^{\lfloor (n-1)/2 \rfloor} \binom{n-j-1}{j} (-1)^j L_k^{n-2j-1}(x) & \text{otherwise} \end{cases}$$

$$= f_k(x) \sum_{j=0}^{\lfloor (n-1)/2 \rfloor} \binom{n-j-1}{j} (-1)^{(k+1)j} l_k^{n-2j-1}(x)$$ ■

For example,

$$\begin{aligned} f_6(x) &= f_2(x) \sum_0^1 \binom{2-j}{j} (-1)^j l_2^{2-2j}(x) \\ &= x\left[\binom{2}{0} l_2^2(x) - \binom{1}{1} l_2^0(x)\right] = x[(x^2+2)^2 - 1] \\ &= x^5 + 4x^3 + 3x \end{aligned}$$

as we found in the preceding chapter.

We can also employ the polynomials $l_n(x)$ and $h_n(x)$ to derive similar divisibility properties. To this end, recall that

$$\alpha(l_{2m+1}(x)) = \alpha^{2m+1}(x) \quad \text{and} \quad \beta(l_{2m+1}(x)) = \beta^{2m+1}(x)$$
$$\therefore \quad l_n(l_{2m+1}(x)) = \alpha^{(2m+1)n}(x) + \beta^{(2m+1)n}(x) = l_{(2m+1)n}(x)$$

Since $\gamma(l_{2m}(x)) = \alpha^{2m}(x)$ and $\delta(l_{2m}(x)) = \beta^{2m}(x)$, it follows that

$$h_n(l_{2m}(x)) = \alpha^{2mn}(x) + \beta^{2mn}(x) = l_{2mn}(x)$$

Thus, we have the following theorem.

Theorem 38.2. $l_n(l_{2m-1}(x)) = l_{(2m-1)n}(x)$ and $h_n(l_{2m}(x)) = l_{2mn}(x)$. ■

Corollary 38.4. $l_n(x)|l_{(2m-1)n}(x)$.

Proof. Since $l_1(x) = x$ and $l_2(x) = x^2 + 2$, it follows from the recurrence relation that $x|l_{2k-1}(x)$ for every $k \geq 1$. Likewise, $x|h_{2k-1}(x)$ for every $k \geq 1$.

Since $l_{2m-1}(l_{2k-1}(x)) = l_{(2m-1)(2k-1)}(x)$ and $l_{2k-1}(x)|l_{2m-1}(l_{2k-1}(x))$, it follows that $l_{2k-1}(x)|l_{(2m-1)(2k-1)}(x)$. Likewise, $h_{2m-1}(l_{2k}(x)) = l_{(2m-1)(2k)}(x)$ and $l_{2k}(x)|h_{2m-1}(l_{2k}(x))$, so $l_{2k}(x)|l_{(2m-1)(2k)}(x)$. Thus, whether n is odd or even, $l_n(x)|l_{(2m-1)n}(x)$. ■

For example,

$$\begin{aligned} l_{3\cdot4}(x) &= l_{12}(x) = x^{12} + 12x^{10} + 54x^8 + 112x^6 + 105x^4 + 36x^2 + 2 \\ &= (x^4 + 4x^2 + 2)(x^8 + 8x^6 + 20x^4 + 16x^2 + 1) \\ &= l_4(x) \cdot [l_8(x) - 1] \end{aligned}$$

So $l_4(x)|l_{3\cdot4}(x)$.

Since $l_n(1) = L_n$, this corollary yields the following result.

Corollary 38.5. $L_n|L_{(2m-1)n}$ ■

CONVERGENCE OF A *Q*-LIKE MATRIX

In 1983, after studying various powers of the matrix

$$M = \begin{bmatrix} 1 & 1 \\ 1 & 1+x \end{bmatrix}$$

for several values of x, S. Moore conjectured that the powers with their leading entries scaled to 1 converge to the matrix

$$\begin{bmatrix} 1 & \alpha(x) \\ \alpha(x) & \alpha^2(x) \end{bmatrix}$$

In the same year, Hazel Perfect of Sheffield, England, furnished a neat proof of this using the well-known diagonalization technique in linear algebra.

The characteristic equation of M is

$$|M - \lambda I| = \begin{vmatrix} 1-\lambda & 1 \\ 1 & 1+x-\lambda \end{vmatrix} = \lambda^2 - (2+x)\lambda + x = 0$$

Its (distinct) roots are $r = 1 + \beta(x)$ and $s = 1 + \alpha(x)$.

The eigenvector $\begin{pmatrix} u \\ v \end{pmatrix}$ associated with r is given by $M\begin{pmatrix} u \\ v \end{pmatrix} = r\begin{pmatrix} u \\ v \end{pmatrix}$; that is,

$$\begin{bmatrix} 1 & 1 \\ 1 & 1+x \end{bmatrix}\begin{pmatrix} u \\ v \end{pmatrix} = r\begin{pmatrix} u \\ v \end{pmatrix}$$

Then $(1-r)u + v = 0$ and $u + (1+x-r)v = 0$. So we choose $u = 1$ and $v = r - 1 = \beta(x)$. The corresponding eigenvector is

$$\begin{pmatrix} u \\ v \end{pmatrix} = \begin{pmatrix} 1 \\ \beta(x) \end{pmatrix}$$

Likewise, the eigenvector corresponding to s is

$$\begin{pmatrix} u \\ v \end{pmatrix} = \begin{pmatrix} 1 \\ \alpha(x) \end{pmatrix}$$

Then, by matrix diagonalization,

$$M = \begin{bmatrix} 1 & 1 \\ \beta(x) & \alpha(x) \end{bmatrix}\begin{bmatrix} r & 0 \\ 0 & s \end{bmatrix}\begin{bmatrix} 1 & 1 \\ \beta(x) & \alpha(x) \end{bmatrix}^{-1}$$

$$\therefore\ M^n = \begin{bmatrix} 1 & 1 \\ \beta(x) & \alpha(x) \end{bmatrix}\begin{bmatrix} r & 0 \\ 0 & s \end{bmatrix}^n\begin{bmatrix} 1 & 1 \\ \beta(x) & \alpha(x) \end{bmatrix}^{-1}$$

$$= \begin{bmatrix} 1 & 1 \\ \beta(x) & \alpha(x) \end{bmatrix}\begin{bmatrix} r^n & 0 \\ 0 & s^n \end{bmatrix}\begin{bmatrix} 1 & 1 \\ \beta(x) & \alpha(x) \end{bmatrix}^{-1}$$

$$= \frac{1}{\alpha(x) - \beta(x)}\begin{bmatrix} 1 & 1 \\ \beta(x) & \alpha(x) \end{bmatrix}\begin{bmatrix} r^n & 0 \\ 0 & s^n \end{bmatrix}\begin{bmatrix} 1 & -1 \\ -\beta(x) & \alpha(x) \end{bmatrix}$$

$$= \frac{1}{\alpha(x) - \beta(x)}\begin{bmatrix} \alpha(x)r^n - \beta(x)s^n & s^n - r^n \\ s^n - r^n & \alpha(x)s^n - \beta(x)r^n \end{bmatrix} \tag{38.3}$$

$$\frac{\alpha(x)-\beta(x)}{\alpha(x)r^n-\beta(x)s^n}M^n = \begin{bmatrix} 1 & \dfrac{s^n-r^n}{\alpha(x)r^n-\beta(x)s^n} \\ \dfrac{s^n-r^n}{\alpha(x)r^n-\beta(x)s^n} & \dfrac{\alpha(x)s^n-\beta(x)r^n}{\alpha(x)r^n-\beta(x)s^n} \end{bmatrix} \tag{38.4}$$

$$= \begin{bmatrix} 1 & \dfrac{1-(r/s)^n}{\alpha(x)(r/s)^n-\beta} \\ \dfrac{1-(r/s)^n}{\alpha(x)(r/s)^n-\beta} & \dfrac{\alpha(x)-\beta(x)(r/s)^n}{\alpha(x)(r/s)^n-\beta} \end{bmatrix} \tag{38.5}$$

Suppose $x > -2$. Then $(r/s)^n \to 0$ as $n \to \infty$, so the matrix approaches the limit

$$\begin{bmatrix} 1 & -\dfrac{1}{\beta(x)} \\ -\dfrac{1}{\beta(x)} & -\dfrac{\alpha(x)}{\beta(x)} \end{bmatrix} = \begin{bmatrix} 1 & \alpha(x) \\ \alpha(x) & \alpha^2(x) \end{bmatrix}$$

as desired.

Suppose $x < -2$. Then $(s/r)^n \to 0$ as $n \to \infty$. Thus, as $n \to \infty$, Matrix (38.5) approaches the limit

$$\begin{bmatrix} 1 & -\dfrac{1}{\alpha(x)} \\ -\dfrac{1}{\alpha(x)} & -\dfrac{\beta(x)}{\alpha(x)} \end{bmatrix} = \begin{bmatrix} 1 & \beta(x) \\ \beta(x) & \beta^2(x) \end{bmatrix}$$

When $x = 1, \alpha(x) = \alpha, \beta(x) = \beta, r = 1+\beta = \beta^2$, and $s = 1+\alpha = \alpha^2$. Then the Matrix Equation (38.3) yields

$$\begin{aligned} M^n &= \frac{1}{\alpha-\beta}\begin{bmatrix} \alpha\beta^{2n}-\beta\alpha^{2n} & \alpha^{2n}-\beta^{2n} \\ \alpha^{2n}-\beta^{2n} & \alpha^{2n+1}-\beta^{2n+1} \end{bmatrix} \\ &= \frac{1}{\alpha-\beta}\begin{bmatrix} \alpha^{2n-1}-\beta^{2n-1} & \alpha^{2n}-\beta^{2n} \\ \alpha^{2n}-\beta^{2n} & \alpha^{2n+1}-\beta^{2n+1} \end{bmatrix} \\ &= \begin{bmatrix} F_{2n-1} & F_{2n} \\ F_{2n} & F_{2n+1} \end{bmatrix} \end{aligned}$$

as we saw in Chapter 32.

As $n \to \infty$, $(1/F_{2n-1})M^n$ approaches the limit $\begin{bmatrix} 1 & \alpha \\ \alpha & \alpha^2 \end{bmatrix}$.

When $x = -2$,

$$M = \begin{bmatrix} 1 & 1 \\ 1 & -1 \end{bmatrix}$$

so the sequence of scaled matrices is

$$\begin{bmatrix} 1 & 1 \\ 1 & -1 \end{bmatrix}, \begin{bmatrix} 1 & 0 \\ 0 & 1 \end{bmatrix}, \begin{bmatrix} 1 & 1 \\ 1 & -1 \end{bmatrix}, \begin{bmatrix} 1 & 0 \\ 0 & 1 \end{bmatrix}, \ldots.$$

EXERCISES 38

Find each Lucas polynomial.

1. $l_{11}(x)$
2. $l_{12}(x)$

Verify each.

3. $l_n(x) = f_{n+1}(x) + f_{n-1}(x)$
4. $l_n(x) = xf_n(x) + 2f_{n-1}(x)$
5. $xl_n(x) = f_{n+2}(x) - f_{n-2}(x)$
6. $l_{-n}(x) = (-1)^n l_n(x)$
7. Let $B(n, j)$ denote the element in row n and column j of the array in Table 38.1. Define $B(n, j)$ recursively.

Prove each

8. $f_n(x) = \dfrac{\alpha^n(x) - \beta^n(x)}{\alpha(x) - \beta(x)}$
9. $l_n(x) = \alpha^n(x) + \beta^n(x)$
10. $l_n^2(x) - (x^2+4)f_n^2(x) = 4(-1)^n$
11. $f_n^2(x) + f_{n+1}^2(x) = f_{2n+1}(x)$ (Koshy, 1999)
12. $l_n^2(x) + l_n^2(x) = (x^2+4)f_{2n+1}(x)$ (Koshy, 1999)
13. $l_{n+1}(x)l_{n-1}(x) - l_n^2(x) = (-1)^{n-1}(x^2+4)$ (Koshy, 1999)
14. $f_m(x)l_n(x) + f_n(x)l_m(x) = 2f_{m+n}(x)$ (Koshy, 1999)
15. $\alpha^n(x) = \dfrac{l_n(x) + \sqrt{x^2+4}\,f_n(x)}{2}$
16. $\beta^n(x) = \dfrac{l_n(x) - \sqrt{x^2+4}\,f_n(x)}{2}$
17. Find the polynomials $g_2(x)$, $g_3(x)$, $g_4(x)$, and $g_5(x)$.
18. Find the polynomials $h_2(x)$, $h_3(x)$, $h_4(x)$, and $h_5(x)$.

Verify each

19. $g_n(x) = \dfrac{\gamma^n(x) - \delta^n(x)}{\gamma(x) - \delta(x)}$ (Hoggatt, Bicknell, and King, 1972)
20. $h_n(x) = \gamma^n(x) + \delta^n(x)$ (Hoggatt, Bicknell, and King, 1972)
21. $h_n^2(x) - (x^2-4)g_n^2(x) = 4$
22. $h_n(x) = g_{n+1}(x) - g_{n-1}(x)$
23. $g_n(x) = \displaystyle\sum_{0}^{\lfloor (n-1)/2 \rfloor} \binom{n-i-1}{i} (-1)^i x^{n-2i-1}, n \ge 0.$

Establish the following generating functions.

24. $\dfrac{2 - xt}{1 - xt - t^2} = \displaystyle\sum_0^\infty l_n(x)t^n$ (Koshy, 1999)
25. $\dfrac{2 - (x^2+2)t^4}{1 - (x^2+2)t^2 + t^4} = \displaystyle\sum_0^\infty l_{2n}(x)t^{2n}$ (Koshy, 1999)

26. $\dfrac{(1+x)t - t^3}{1-(x^2+2)t^2+t^4} = \sum_0^{\infty} l_{2n+1}(x)t^{2n+1}$ (Koshy, 1999)
27. Show that the zeros of $f_n(x)$ are $2i \cos k\pi/n$, where $1 \le k \le n-1$ and $i = \sqrt{-1}$ (Webb and Parberry, 1969).

The *generalized Fibonacci polynomials* $t_n(x)$, studied in 1970 by Bicknell, are defined by $t_1(x) = a$, $t_2(x) = bx$ and $t_n(x) = xt_{n-1}(x) + t_{n-2}(x)$, $n \ge 3$.

28. Find $t_6(x)$ and $t_7(x)$.
29. Express $t_n(x)$ in terms of $f_{n-1}(x)$ and $f_{n-2}(x)$.
30. Use Exercise 29 to find $t_6(x)$.

JACOBSTHAL POLYNOMIALS

Jacobsthal polynomials, $J_n(x)$, named after the German mathematician E. Jacobsthal, are related to Fibonacci polynomials. They are defined by

$$J_n(x) = J_{n-1}(x) + xJ_{n-2}(x) \tag{39.1}$$

where $J_1(x) = 1 = J_2(x)$. Clearly, $J_n(1) = F_n$. The first 10 Jacobsthal polynomials are:

$$\begin{aligned}
J_1(x) &= 1\\
J_2(x) &= 1\\
J_3(x) &= x + 1\\
J_4(x) &= 2x + 1\\
J_5(x) &= x^2 + 3x + 1\\
J_6(x) &= 3x^2 + 4x + 1\\
J_7(x) &= x^3 + 6x^2 + 5x + 1\\
J_8(x) &= 4x^3 + 10x^2 + 6x + 1\\
J_9(x) &= x^4 + 10x^3 + 15x^2 + 7x + 1\\
J_{10}(x) &= 5x^4 + 20x^3 + 21x^2 + 8x + 1
\end{aligned}$$

We can make a few interesting observations:

- The Jacobsthal polynomials $J_{2n-1}(x)$ and $J_{2n}(x)$ have the same degree.
- The degree of $J_n(x)$ is $\lfloor (n-1)/2 \rfloor$.
- The leading coefficient of $J_{2n-1}(x)$ is one, whereas that of $J_{2n}(x)$ is n.
- The coefficients of $J_n(x)$ are the same of those of $f_n(x)$, but in the reverse order.
- The coefficients of the Jacobsthal polynomials lie on the rising diagonals of the left-justified Pascal's triangle, in the reverse order, as shown below:

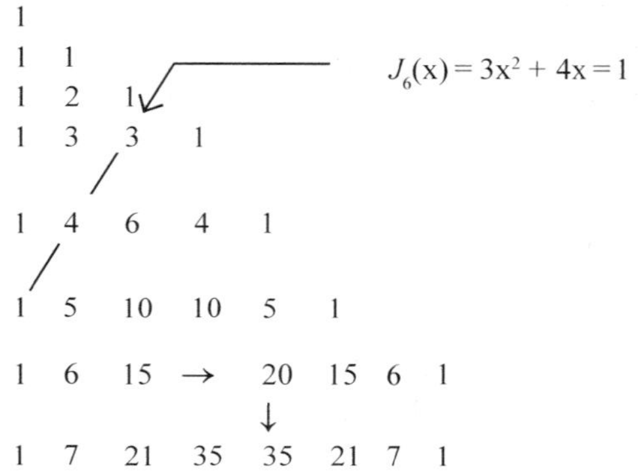

1
1 1
1 2 1 $J_6(x) = 3x^2 + 4x = 1$
1 3 3 1
1 4 6 4 1
1 5 10 10 5 1
1 6 15 → 20 15 6 1
1 7 21 35 35 21 7 1

AN EXPLICIT FORMULA FOR $J_n(x)$

Next, we can derive an explicit formula for $J_n(x)$. Since the coefficients of $J_n(x)$ are the same as those of $f_n(x)$ in reverse order, it follows that

$$J_n(x) = \sum_{j=0}^{\lfloor (n-1)/2 \rfloor} \binom{\lfloor n/2 \rfloor + j}{\lfloor (n-1)/2 \rfloor - j} x^{\lfloor (n-1)/2 \rfloor - j} \qquad (39.2)$$

See Exercise 3.

For example,

$$\begin{aligned} J_7(x) &= \sum_{0}^{3} \binom{3+j}{3-j} x^{3-j} \\ &= \binom{3}{3} x^3 + \binom{4}{2} x^2 + \binom{5}{1} x \binom{6}{0} x^0 \\ &= x^3 + 6x^2 + 5x + 1 \end{aligned}$$

Similarly,

$$J_8(x) = 4x^3 + 10x^2 + 6x + 1$$

BINET'S FORMULA FOR $J_n(x)$

Now, we find Binet's formula for $J_n(x)$. To this end, let r and s be the solutions of the characteristic equation $t^2 - t - x = 0$ of the recurrence relation Eq. (39.1). Then

$$r = \frac{1 + \sqrt{1+4x}}{2} \qquad s = \frac{1 - \sqrt{1+4x}}{2}$$

$$r + s = 1 \qquad rs = -x \qquad \text{and} \qquad r - s = \sqrt{1+4x}$$

It can be shown that $J_n(x)$ can also be defined by Binet's formula

$$J_n(x) = \frac{r^n - s^n}{\sqrt{1+4x}} \tag{39.3}$$

where $n \geq 1$

ANOTHER FAMILY OF POLYNOMIALS $K_n(x)$

Next, we introduce yet another family of polynomials, $K_n(x)$, which are closely related to Jacobsthal polynomials.

We define the polynomials $K_n(x)$ by

$$K_n(x) = K_{n-1}(x) + xK_{n-2}(x) \tag{39.4}$$

where $K_1(x) = 1$ and $K_2(x) = x$. The first 10 members of this family are:

$$\begin{aligned}
K_1(x) &= 1 \\
K_2(x) &= x \\
K_3(x) &= 2x \\
K_4(x) &= x^2 + 2x \\
K_5(x) &= 3x^2 + 2x \\
K_6(x) &= x^3 + 5x^2 + 2x \\
K_7(x) &= 4x^3 + 7x^2 + 2x \\
K_8(x) &= x^4 + 9x^3 + 9x^2 + 2x \\
K_9(x) &= 5x^4 + 16x^3 + 11x^2 + 2x \\
K_{10}(x) &= x^5 + 14x^4 + 25x^3 + 13^2 + 2x
\end{aligned}$$

The polynomials $K_n(x)$ have several interesting properties:

- The degree of $K_n(x)$ is $\lfloor n/2 \rfloor$, so $K_{2n}(x)$ and $K_{2n+1}(x)$ have the same degree.
- The leading coefficient of $K_n(x)$ is

$$\begin{cases} 1 & \text{if } n \text{ is even} \\ \lfloor (n+1)/2 \rfloor & \text{otherwise} \end{cases}$$

- $x|K_n(x)$ for every $n \geq 2$. (Assume that $x \neq 0$.)
- The coefficient of x is always 2, where $n \geq 3$.

Since $K_n(1) = F_n$, it follows that the sum of the coefficients in every polynomial $K_n(x)$ is a Fibonacci number. In other words, every row sum in the array of coefficients in Table 39.1 is a Fibonacci number.

TABLE 39.1.

n \ j	0	1	2	3	4	5	...
1	1						
2	1						
3	2						
4	1	2					
5	3	2					
6	1	5	2				
7	4	7	2			← Row sum = 13 = F_7	
8	1	9	9	2			
10	5	14	25	13	2		

Let $K(n, j)$ denote the element in row n and column j, where $n \geq 1$ and $j \geq 0$. It can be defined recursively as follows:

$$K(n, 0) = \begin{cases} 1 & \text{if } n \text{ is even} \\ \lfloor (n+1)/2 \rfloor & \text{otherwise} \end{cases}$$

$$K(n, j) = \begin{cases} K(n-1, j-1) + K(n-2, j) & \text{if } n \text{ is even} \\ K(n-1, j) + K(n-2, j) & \text{otherwise} \end{cases}$$

where $n \geq 3$ and $j \leq \lfloor (n-2)/2 \rfloor$. If $j > \lfloor (n-2)/2 \rfloor$, $K(n, j)$ can be considered 0.

Surprisingly enough, there is a close relationship between the polynomials $K_n(x)$ and $J_n(x)$, as the following theorem shows.

Theorem 39.1. (Koshy, 2000) $K_n(x) = x[J_{n-1}(x) + J_{n-2}(x)]$, where $n \geq 2$.

Proof. Since $K_n(x)$ satisfies the same recurrence relation as $J_n(x)$, it follows that $K_n(x) = Ar^n + Bs^n$, where the expressions A and B are to be determined subject to the initial conditions $K_1(x) = 1$ and $K_2(x) = x$. These two conditions yield the equations

$$Ar + Bs = 1$$
$$Ar^2 + Bs^2 = x$$

Solving this system, we get

$$A = \frac{x - s}{r\sqrt{1+4x}} \quad \text{and} \quad B = \frac{r - x}{s\sqrt{1+4x}}$$

$$\begin{aligned}
\therefore\ K_n(x) &= \frac{x - s}{r\sqrt{1+4x}} \cdot r^n + \frac{r - x}{s\sqrt{1+4x}} \cdot s^n \\
&= \frac{(x - s)r^{n-1} + (r - x)s^{n-1}}{\sqrt{1+4x}} \\
&= \frac{x(r^{n-1} - s^{n-1}) - (rs)(r^{n-2} - s^{n-2})}{\sqrt{1+4x}} \\
&= x[J_{n-1}(x) + J_{n-2}(x)] \qquad \text{by Formula (39.3)}
\end{aligned}$$

■

Thus, to find any polynomial $K_n(x)$, we need only multiply the sum of the consecutive numbers $J_{n-1}(x)$ and $J_{n-2}(x)$ of the Jacobsthal family by x, where $n \geq 3$.

For example,

$$\begin{aligned}
K_9(x) &= x[J_8(x) + J_7(x)] \\
&= x[(4x^3 + 10x^2 + 6x + 1) + (x^3 + 6x^2 + 5x + 1) \\
&= 5x^4 + 16x^3 + 11x^2 + 2x
\end{aligned}$$

Similarly, $K_8(x) = x^4 + 9x^3 + 9x^2 + 2x$.

Since $K_m(1) = F_m = J_m(1)$, Theorem 39.1 yields the familiar Fabonacci recurrence formula.

Corollary 39.1. $F_n = F_{n-1} + F_{n-2}, n \geq 3$. ■

A POLYNOMIAL EXPANSION FOR $K_n(x)$

We can employ Formula (39.2) and Theorem 39.1 to derive a polynomial formula for $K_n(x)$.

Case 1. Let $n = 2k + 1$ be odd. By Formula (39.2),

$$\begin{aligned}
J_{n-1}(x) + J_{n-2}(x) &= \sum_{0}^{k-1} \binom{k+j}{k-j-1} x^{k-j-1} + \sum_{0}^{k-1} \binom{k+j-1}{k-j-1} x^{k-j-1} \\
&= \sum_{0}^{k-1} \left(\binom{k+j}{k-j-1} + \binom{k+j-1}{k-j-1}\right) x^{k-j-1} \\
&= \sum_{0}^{k-1} \frac{k+3j+1}{2j+1} \binom{k+j-1}{k-j-1} x^{k-j-1}
\end{aligned}$$

Case 2. Let $n = 2k$ be even. Then:

$$\begin{aligned}
J_{n-1}(x) + J_{n-2}(x) &= \sum_{0}^{k-1} \binom{k+j-1}{k-j-1} x^{k-j-1} + \sum_{0}^{k-2} \binom{k+j-2}{k-j-2} x^{k-j-2} \\
&= \sum_{0}^{k-1} \binom{k+j-1}{k-j-1} x^{k-j-1} + \sum_{1}^{k-1} \binom{k+j-2}{k-j-1} x^{k-j-1} \\
&= \sum_{1}^{k-1} \left[\binom{k+j-1}{k-j-1} + \binom{k+j-2}{k-j-1}\right] x^{k-j-1} + x^{k-1} \\
&= x^{k-1} + \sum_{1}^{k-1} \frac{k+3j-1}{2j} \binom{k+j-2}{k-j-1} x^{k-j-1}
\end{aligned}$$

Thus, we have the following result.

Theorem 39.2. (Koshy, 2000)

$$K_n(x)/x = \begin{cases} x^{k-1} + \sum_{1}^{k-1} \dfrac{k+3j-1}{2j} \dbinom{k+j-2}{k-j-1} x^{k-j-1} & \text{if } n = 2k \text{ is even} \\ \sum_{0}^{k-1} \dfrac{k+3j+1}{2j+1} \dbinom{k+j-1}{k-j-1} x^{k-j-1} & \text{if } n = 2k+1 \text{ is odd} \end{cases}$$

■

For example,

$$\begin{aligned}
K_6(x)/x &= x^2 + \sum_{1}^{2} \frac{2+3j}{2j} \binom{1+j}{2-j} x^{2-j} \\
&= x^2 + \frac{5}{2}\binom{2}{1} x + \frac{8}{4}\binom{3}{0} \\
&= x^2 + 5x + 2 \\
\therefore\ K_6(x) &= x^3 + 5x^2 + 2x
\end{aligned}$$

Likewise,

$$K_7(x)/x = \sum_0^2 \frac{4+3j}{2j+1}\binom{2+j}{2-j}x^{2-j} = 4x^2 + 7x + 2,$$

so

$$K_7(x) = 4x^3 + 7x^2 + 2x$$

Since $K_n(1) = F_n$, the next result follows from Theorem 39.2.

Corollary 39.2.

$$(1) \quad F_{2n} = 1 + \sum_1^{n-1} \frac{n+3j-1}{2j}\binom{n+j-2}{n-j-1} \tag{39.5}$$

$$(2) \quad F_{2n+1} = \sum_0^{n-1} \frac{n+3j+1}{2j}\binom{n+j-1}{n-j-1} \tag{39.6}$$

■

For example,

$$\begin{aligned} F_{11} &= \sum_0^4 \frac{6+3j}{2j+1}\binom{4+j}{4-j} \\ &= \frac{6}{1}\binom{4}{4} + \frac{9}{3}\binom{5}{3} + \frac{12}{5}\binom{6}{2} + \frac{15}{7}\binom{7}{1} + \frac{18}{9}\binom{8}{0} \\ &= 6 + 30 + 36 + 15 + 2 = 89 \end{aligned}$$

and

$$F_{12} = 1 + \sum_1^5 \frac{5+3j}{2j}\binom{4+j}{5-j} = 144$$

Since $F_{2n} = F_n L_n$, it follows that the sum in Eq. (39.5) has nontrivial factors when $n \geq 3$. Besides, since $F_{2n+1} = F_n^2 + F_{n+1}^2$, it follows that the sum in Eq. (39.6) is the sum of two (Fibonacci) squares.

We now turn our attention to constructing a generating function for $K_n(x)$.

A GENERATING FUNCTION FOR $K_n(x)$

Since $1 - t - xt^2 = (1 - rt)(1 - st)$, we can show that

$$\frac{t}{1 - t - xt^2} = \sum_0^\infty J_n(x)t^n$$

Therefore,

$$\frac{t^2 + t^3}{1 - t - xt^2} = \sum_2^\infty [J_{n-1}(x) + J_{n-2}(x)]t^n$$
$$= \sum_2^\infty [K_n(x)/x]t^n, \quad \text{by Theorem 39.1}$$

That is,

$$\frac{x(1+t)t^2}{1 - t - xt^2} = \sum_2^\infty K_n(x)t^n$$

In other words,

$$\frac{t + (x-1)t^2}{1 - t - xt^2} = \sum_1^\infty K_n(x)t^n$$

Thus, the function on the left generates the polynomials $K_n(x)$ as coefficients of t^n, where $n \geq 1$.

EXERCISES 39

1. Using the recursive definition of $J_n(x)$, find $J_{11}(x)$ and $J_{12}(x)$.
2. Using the explicit formula for $J_n(x)$, find $J_9(x)$ and $J_{10}(x)$.
3. Prove the explicit Formula (39.2) for $J_n(x)$.
4. Show that $J_n(x) = \dfrac{r^n - s^n}{\sqrt{1+4x}}$, where $r = \dfrac{1+\sqrt{1+4x}}{2}$ and $s = \dfrac{1-\sqrt{1+4x}}{2}$.
5. Find a generating function for $J_n(x)$.

Consider the polynomial $k_n(x)$ defined by $k_n(x) = k_{n-1}(x) + xk_{n-2}(x)$, where $k_0(x) = 2$ and $k_1(x) = 1$.

6. Find $k_3(x)$ and $k_6(x)$.
7. Find $k_n(1)$.

8–9. Redo problems 6 and 7 if $k_1(x) = x$.

There is yet another family of polynomial functions, $Q_n(x)$, that are related to Jacobsthal polynomials. They are defined by $Q_n(x) = x[Q_{n-1}(x) + Q_{n-2}(x)]$, where $Q_1(x) = 1$ and $Q_2(x) = x$. Find each.

10. $Q_3(x)$ and $Q_7(x)$
11. $Q_n(1)$
12. The degree of $Q_n(x)$
13. The lowest power of x in $Q_n(x)$
14. The number of terms of $Q_n(x)$
15. Binet's formula for $Q_n(x)$
16. An explicit formula for $Q_n(x)$
17. A generating function for $Q_n(x)$

ZEROS OF FIBONACCI AND LUCAS POLYNOMIALS

In 1973, V. E. Hoggatt, Jr., and M. Bicknell investigated the zeros of the Fibonacci and Lucas polynomials using the hyperbolic functions

$$\sinh z = \frac{e^z - e^{-z}}{2} \qquad \text{and} \qquad \cosh z = \frac{e^z + e^{-z}}{2}$$

where $z = x + iy$ is a complex variable. These functions satisfy the identities $\cosh^2 z - \sinh^2 z = 1$, $\cosh iy = \cos y$, and $\sinh iy = i \sin y$.

Let $x = 2i \cosh z$. Then $\sqrt{x^2 + 4} = 2i \sinh z$, so $\alpha(x) = i(\cosh z + \sinh z) = ie^z$ and $\beta(x) = i(\cosh z - \sinh z) = ie^{-z}$;

$$\therefore \quad f_n(x) = \frac{(\alpha(x))^n - (\beta(x))^n}{\alpha - \beta} = i^{n-1}\left(\frac{e^{nz} - e^{-nz}}{e^z - e^{-z}}\right) = i^{n-1}\frac{\sinh nz}{\sinh z}$$

and $l_n(x) = (\alpha(x))^n + (\beta(x))^n = e^{nz} + e^{-nz} = 2i^n \cosh nz$.

Let $z = u + iv$. Then $|\sinh z|^2 = \sinh^2 u + \sin^2 v$ and $|\cosh z|^2 = \sinh^2 u + \cos^2 v$. Since u is real, $\sinh u = 0$ if and only if $u = 0$, so the zeros of $\sinh z$ are those of $\sinh iv = i \sin v$; and the zeros of $\cosh z$ are those of $\cosh iv = \cos v$.

Clearly, $f_n(x) = 0$ if and only if $\sinh nz = 0$ and $\sinh z \neq 0$. But $\sinh nz = 0$ if and only if $\sin ny = 0$, or $z = iy$. Therefore, $ny = \pm k\pi$, so $z = \pm k\pi/n$. Since $i \cosh iy = i \cos y$, $x = 2i \cosh z = 2i \cos k\pi/n$, where $1 \leq k \leq n - 1$.

Notice that $l_n(x) = 0$ if and only if $\cosh nz = 0$, that is, if and only if $\cos ny = 0$. Then $ny = (2k+1)\pi/2$ and $z = iy$. So $x = 2i \cosh z = 2i \cos(2k+1)\pi/2n$, where $0 \leq k \leq n - 1$.

Thus, the zeros of $f_n(x)$ are

$$x = 2i \cos k\pi/n \qquad 1 \leq k \leq n - 1$$

and those of $l_n(x)$ are

$$x = 2i\cos(2k+1)\pi/2n \qquad 0 \le k \le n-1$$

For example, the zeros of $f_6(x) = x^5+4x^3+3x$ are given by $x = 2i\cos k\pi/6$, $1 \le k \le 5$. When $k = 1$, $x = 2i\cos\pi/6 = \sqrt{3}i$; when $k = 2$, $x = 2i\cos\pi/3 = i$; similarly, $k = 3$, 4, and 5 yield the values 0, $-i$, and $-\sqrt{3}i$, respectively. Thus, the zeros of $f_6(x)$ are 0, $\pm i$, and $\pm\sqrt{3}i$. This should be obvious, since $f_6(x) = x(x^2+1)(x^2+3)$.

The zeros of $l_5(x) = x^5+5x^3+5x$ are given by $x = 2i\cos(2k+1)\pi/10$, where $0 \le k \le 4$; so they are 0, $\pm\frac{\sqrt{10\pm2\sqrt{5}}}{2}i$ since $\cos\pi/10 = (\sqrt{10+2\sqrt{5}})/4$. We can confirm this easily, since $l_5(x) = x(x^4+5x^2+5)$, and x^4+5x^2+5 is a quadratic in x^2.

FACTORING FIBONACCI AND LUCAS POLYNOMIALS

It now follows that both $f_n(x)$ and $l_n(x)$ can be factored:

$$f_n(x) = \prod_1^{n-1}(x - 2i\cos k\pi/n)$$

and

$$l_n(x) = \prod_0^{n-1}[x - 2i\cos(2k+1)\pi/2n]$$

It also follows from these two factorizations that

$$F_n = \prod_1^{n-1}(1 - 2i\cos k\pi/n) \tag{40.1}$$

and

$$L_n = \prod_0^{n-1}[1 - 2i\cos(2k+1)/2n] \tag{40.2}$$

Formula 40.1 was initially proposed as an advanced problem in 1965 by D. Lind of the University of Virginia at Charlottesville. Two years later, D. Zeitlin of Minnesota derived Formula 40.2 along with Formula 40.1 using trigonometric factorizations of Chebyshev polynomials of the first and the second kinds. We will elaborate Formula 40.1 further in Chapter 42.

For example,

$$\begin{aligned}
F_4 &= \prod_1^3(1 - 2i\cos k\pi/4) \\
&= (1 - 2i\cos\pi/4)(1 - 2i\cos 2\pi/4)(1 - 2i\cos 3\pi/4) \\
&= (1 - \sqrt{2}i)(1 - 0)(1 + \sqrt{2}i) = 3
\end{aligned}$$

and

$$\begin{aligned} L_3 &= \prod_0^2 [1 - 2i\cos(2k+1)\pi/6] \\ &= (1 - 2i\cos\pi/6)(1 - 2i\cos 3\pi/6)(1 - 2i\cos 5\pi/6) \\ &= (1 - 2i\cos\pi/6)(1 - 0)(1 + 2i\cos\pi/6) \\ &= 1 + 4\cos^2\pi/6 = 1 + 4\cdot 3/4 \\ &= 4 \end{aligned}$$

41

MORGAN-VOYCE POLYNOMIALS

In 1959, A. M. Morgan-Voyce of Convair, a division of General Dynamics Corporation, in his study of electrical ladder networks of resistors, discovered two large families of polynomials, $b_n(x)$ and $B_n(x)$. Closely related to Fibonacci polynomials, they are defined recursively as follows:

$$b_n(x) = xB_{n-1}(x) + b_{n-1}(x) \tag{41.1}$$

$$B_n(x) = (x+1)B_{n-1}(x) + b_{n-1}(x) \tag{41.2}$$

where $b_0(x) = 1 = B_0(x)$ and $n \geq 1$.

These polynomials were studied extensively in 1967 and 1968 by several investigators, including M. N. S. Swamy, S. L. Basin of Sylvania Electronic Systems, V. E. Hoggatt, Jr., and M. Bicknell.

To appreciate the origin of these polynomials, we must return to the ladder networks of n resistors that we examined in Chapter 4. Consider the case with $R_1 = x$ and $R_2 = 1$ (see Fig. 41.1). As before, Z_n denotes the resistance between the terminals C and D.

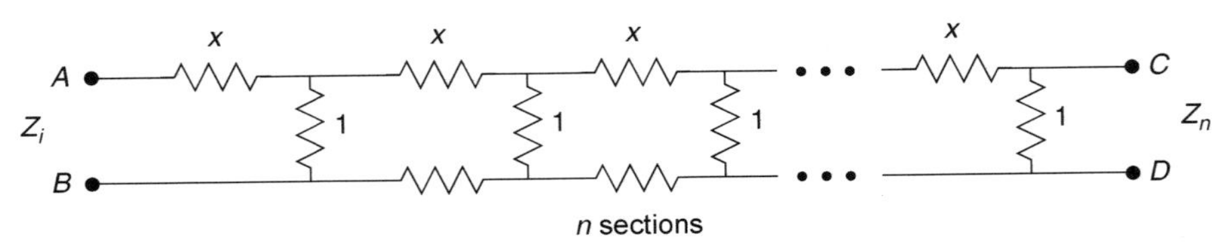

Figure 41.1.

Using Figure 41.2, since x and Z_n are connected in series, they yield a combined resistance of $R = x + Z_n$. Since R and 1 are connected in parallel,

$$\frac{1}{Z_{n+1}} = \frac{1}{x + Z_n} + \frac{1}{1}$$

$$= \frac{x + Z_n + 1}{x + Z_n}$$

$$\therefore \quad Z_{n+1} = \frac{x + Z_n}{x + Z_n + 1} \tag{41.3}$$

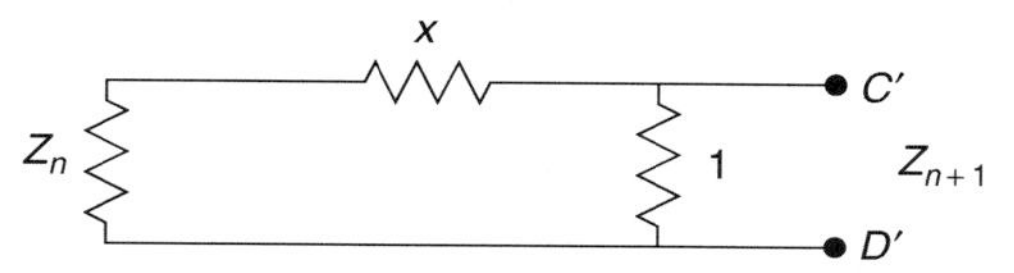

Figure 41.2.

Since Z_n is a polynomial in x, so are $x + Z_n$ and $x + Z_n + 1$. Thus Z_{n+1}, and hence Z_n, are the ratios of two related polynomials. Let

$$Z_n = \frac{b_n(x)}{B_n(x)}$$

By Eq. (41.3),

$$\frac{b_{n+1}(x)}{B_{n+1}(x)} = \frac{x + b_n(x)/B_n(x)}{x + 1 + b_n(x)/Bb_n(x)}$$

$$= \frac{xB_n(x) + b_n(x)}{(x + 1)B_n(x) + b_n(x)}$$

This equation yields the recurrence relations satisfied by $b_n(x)$ and $B_n(x)$:

$$b_n(x) = xB_{n-1}(x) + b_{n-1}(x) \tag{41.1}$$

$$B_n(x) = (x + 1)B_{n-1}(x) + b_{n-1}(x) \tag{41.2}$$

Since $Z_0 = 1$, we define $b_0(x) = 1 = B_0(x)$. Thus $b_n(x)$ and $B_n(x)$ are the *Morgan-Voyce polynomials* defined earlier.

Let us now study the resistance Z_n from the input end. It follows from Figure 41.3 that the resistence R resulting from the parallel resistors Z_n and 1 is given by

$$\frac{1}{R} = \frac{1}{Z_n} + \frac{1}{1}$$

$$R = \frac{Z_n}{Z_n + 1}$$

$$Z_{n+1} = x + \frac{Z_n}{Z_n + 1} = \frac{(x + 1)Z_n + x}{Z_n + 1}$$

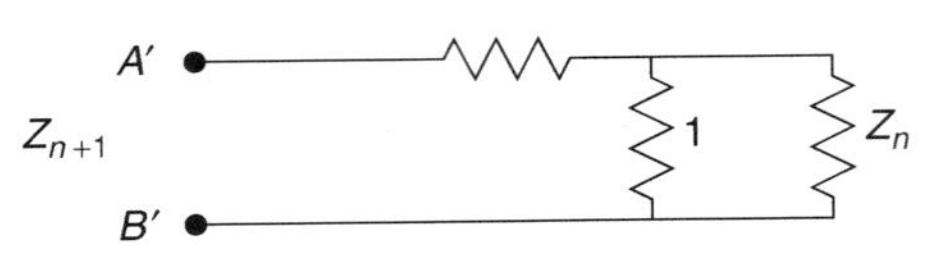

Figure 41.3.

As before, let $Z_n = P_n(x)/Q_n(x)$. Then

$$\begin{aligned}\frac{P_{n+1}(x)}{Q_{n+1}(x)} &= \frac{(x+1)P_n(x)/Q_n(x) + x}{P_n(x)/Q_n(x) + 1} \\ &= \frac{(x+1)P_n(x) + xQ_n(x)}{P_n(x) + Q_n(x)}\end{aligned}$$

Therefore, $P_n(x) = (x+1)P_{n-1}(x) + Q_{n-1}(x)$ and $Q_{n+1}(x) = P_n(x) + Q_n(x)$. Then

$$\begin{aligned}Q_n(x) - Q_{n-1}(x) &= P_{n-1}(x) \\ &= (x+1)P_{n-2}(x) + xQ_{n-2}(x) \\ &= (x+1)[Q_{n-1}(x) - Q_{n-2}(x)] + xQ_{n-2}(x) \\ \therefore \quad Q_n(x) &= (x+2)Q_{n-1}(x) - Q_{n-2}(x)\end{aligned}$$

Since $Q_1(x) = 1$ and $Q_2(x) = x + 2$, it follows that $Q_n(x) = B_n(x)$. Moreover, $P_n(x) = Q_{n+1}(x) - Q_n(x) = B_{n+1}(x) - B_n(x) = b_{n+1}(x)$. Thus $Z_n = b_{n+1}(x)/B_n(x)$ yields the resistance from the input end.

B_n AND b_n FAMILIES

The first five members of the B_n-family of polynomials are

$$\begin{aligned}B_0(x) &= 1 \\ B_1(x) &= x + 2 \\ B_2(x) &= x^2 + 4x + 3 \\ B_3(x) &= x^3 + 6x^2 + 10x + 4 \\ B_4(x) &= x^4 + 8x^3 + 21x^2 + 20x + 5\end{aligned}$$

More generally,

$$B_n(x) = \sum_{i=0}^{n} \binom{n+i+1}{n-i} x^i \tag{41.4}$$

For instance,

$$B_4(x) = \sum_{i=0}^{4} \binom{5+i}{4-i} x^i$$

$$= \binom{5}{4} + \binom{6}{3}x + \binom{7}{2}x^2 + \binom{8}{1}x^3 + \binom{9}{0}x^4$$
$$= x^4 + 8x^3 + 21x^2 + 20x + 5$$

Their cousins in the b_n-family of polynomials are:

$$\begin{aligned} b_0(x) &= 1 \\ b_1(x) &= x + 1 \\ b_2(x) &= x^2 + 3x + 1 \\ b_3(x) &= x^3 + 5x^2 + 6x + 1 \\ b_4(x) &= x^4 + 7x^3 + 15x^2 + 10x + 1 \end{aligned}$$

More generally,

$$b_n(x) = \sum_{i=0}^{n} \binom{n+i}{n-i} x^i \qquad (41.5)$$

For example,

$$\begin{aligned} b_4(x) &= \sum_{i=0}^{4} \binom{4+i}{4-i} x^i \\ &= \binom{4}{4} + \binom{5}{3}x + \binom{6}{2}x^2 + \binom{7}{1}x^3 + \binom{8}{0}x^4 \\ &= x^4 + 7x^3 + 15x^2 + 10x + 1 \end{aligned}$$

PROPERTIES OF MORGAN-VOYCE POLYNOMIALS

Both families enjoy a plethora of Fibonacci-like properties. We can examine a few of them here. First, notice an interesting pattern emerging:

$$B_0(1) = 1 = F_2 \qquad B_1(1) = 3 = F_4 \qquad B_2(1) = 8 = F_6$$
$$\text{and} \qquad B_3(1) = 21 = F_8$$

More generally, $B_n(1) = F_{2n+2}$, where $n \geq 0$ (see Exercise 8). Likewise, $b_n(1) = F_{2n+1}$, where $n \geq 0$ (see Exercise 9).

In fact, there is a close relationship between the Fibonacci polynomials $f_n(x)$ and the Morgan-Voyce polynomials $b_n(x)$ and $B_n(x)$:

$$f_n(x) = \begin{cases} b_{(n-1)/2}(x^2) & \text{if } n \text{ is odd} \\ xB_{(n-2)/2}(x^2) & \text{otherwise} \end{cases}$$

See Exercise 33.

Equations (41.1) and (41.2) yield two obvious properties:

$$b_n(x) = B_n(x) - B_{n-1}(x) \tag{41.6}$$

$$xB_n(x) = b_{n+1}(x) - b_n(x) \tag{41.7}$$

These equations together yield the Fibonacci recurrence relation when $x = 1$.

RECURSIVE DEFINITIONS

Substituting for $b_n(x)$ from Eq. (41.6) in Eq. (41.1) yields the recurrence relation satisfied by $B_n(x)$:

$$B_n(x) = (x+2)B_{n-1}(x) - B_{n-2}(x) \qquad n \geq 2 \tag{41.8}$$

where $B_0(x) = 1$ and $B_1(x) = x + 2$.

Using Eqs. (41.1), (41.2), and (41.7), we can show that $b_n(x)$ satisfies the very same recurrence relation (see Exercise 1):

$$b_n(x) = (x+2)b_{n-1}(x) - b_{n-2}(x) \qquad n \geq 2 \tag{41.9}$$

where $b_0(x) = 1$ and $b_1(x) = x + 1$.

PROPERTIES OF $B_n(x)$

The recurrence relation (41.6) can be employed to express $B_n(x)$ as the determinant of an $n \times n$ circulant matrix:

$$B_n(x) = \begin{vmatrix} x+2 & 1 & 0 & 0 & \dots & 0 \\ 1 & x+2 & 1 & 0 & \dots & 0 \\ 0 & 1 & x+2 & 1 & \dots & 0 \\ & & \dots & & & \\ & & \dots & & & 1 \\ 0 & & \dots & & & x+2 \end{vmatrix}$$

Consequently, we can extract many properties of $B_n(x)$ by studying this determinant.

For example,

$$B_{m+n}(x) = B_m(x)B_n(x) - B_{m-1}(x)B_{n-1}(x) \tag{41.10}$$

In particular,

$$B_{2n}(x) = B_n^2(x) - B_{n-1}^2(x) \tag{41.11}$$

and

$$B_{2n-1}(x) = [B_n(x) - B_{n-2}(x)]B_{n-1}(x) \tag{41.12}$$

For example,

$$\begin{aligned} B_2^2 - B_1^2 &= (x^2+4x+3)^2 - (x+2)^2 \\ &= x^4 + 8x^3 + 21x^2 + 20x + 5 \\ &= B_4(x) \end{aligned}$$

It now follows from Eq. (41.8) that

$$(x+2)B_{2n-1}(x) = B_n^2(x) - B_{n-2}^2(x) \tag{41.13}$$

See Exercise 12.

Since $B_n(1) = F_{2n+2}$, Identities (41.10) through (41.13) yield the following Fibonacci identities:

$$F_{m+n} = F_{m+2}F_n - F_m F_{n-2} \tag{41.14}$$

$$F_{2n+2} = F_{n+2}^2 - F_n^2 \tag{41.15}$$

$$F_{2n} = (F_{n+2} - F_{n-2})F_n \tag{41.16}$$

$$3F_{2n} = F_{n+2}^2 - F_{n-2}^2 \tag{41.17}$$

See Exercises 13–16.

The polynomial $B_n(x)$ also satisfies a Cassini-like formula:

$$B_{n+1}(x)B_{n-1}(x) - B_n^2(x) = -1 \tag{41.18}$$

In particular, let $x = 1$. Then Eq. (41.18) yields Cassini's formula.

PROPERTIES OF $b_n(x)$

We now turn our attention to the properties of $b_n(x)$. By Property (41.7), we have

$$x\sum_0^n B_i(x) = b_{n+1}(x) - b_0(x)$$

That is,

$$x\sum_0^n B_i(x) = b_{n+1}(x) - 1 \tag{41.19}$$

For example,

$$\begin{aligned} x\sum_0^3 B_i(x) &= x(x^3+7x^2+15x+10) \\ &= x^4 + 7x^3 + 15x^2 + 10x \\ &= b_4(x) - 1 \end{aligned}$$

In particular, let $x = 1$. Then Eq. (41.19) yields Identity (5.3):

$$\sum_{0}^{n} F_{2i} = F_{2n+1} - 1 \tag{5.3}$$

It follows from Eq. (41.6) that

$$\sum_{0}^{n} b_i(x) = B_n(x) - B_{-1}(x)$$

But $B_{-1}(x) = 0$ (see Exercise 20).

$$\therefore \quad \sum_{0}^{n} b_i(x) = B_n(x) \tag{41.20}$$

For example,

$$\begin{aligned} \sum_{0}^{3} b_i(x) &= 1 + (x+1) + (x^2 + 3x + 1) + (x^3 + 5x^2 + 6x + 1) \\ &= x^3 + 6x^2 + 10x + 4 \\ &= B_3(x) \end{aligned}$$

Letting $x = 1$ in Eq. (41.20) yields Identity (5.2):

$$\sum_{0}^{n} F_{2i-1} = F_{2n} \tag{5.2}$$

Using Identity (41.6), we have

$$\begin{aligned} b_{m+n}(x) &= B_{m+n}(x) - B_{m+n-1}(x) \\ &= [B_m(x)B_n(x) - B_{m-1}(x)B_{n-1}(x)] \\ &\quad -[B_m(x)B_{n-1}(x) - B_{m-1}(x)B_{n-2}(x)] \\ &= B_m(x)[B_n(x) - B_{n-1}(x)] - B_{m-1}(x)[B_{n-1}(x) - B_{n-2}(x)] \\ &= B_m(x)b_n(x) - B_{m-1}(x)b_{n-1}(x) \end{aligned} \tag{41.21}$$

Switching m and n, this yields

$$b_{m+n}(x) = b_m(x)B_n(x) - b_{m-1}(x)B_{n-1}(x) \tag{41.22}$$

In particular, $b_{2n}(x) = B_n(x)b_n(x) - B_{n-1}(x)b_{n-1}(x)$.

BINET'S FORMULA FOR $B_n(x)$

Next we derive Binet's formula for $B_n(x)$ by solving the recurrence relation (41.6). Its characteristic equation is $t^2 - (x+2)t + 1 = 0$ with roots

$$r(x) = \frac{x+2+\sqrt{x^2+4x}}{2} \quad \text{and} \quad s(x) = \frac{x+2-\sqrt{x^2+4x}}{2}$$

where $r + s = x + 2$, $r - s = \sqrt{x^2+4x}$, and $rs = 1$. So the general solution of the recurrence relation is $B_n(x) = Cr^n + Ds^n$, where the coefficients C and D are to be determined.

The initial conditions $B_0(x) = 1$ and $B_1(x) = x + 2 = r + s$ yield the following linear system:

$$C + D = 1$$

$$Cr + Ds = r + s$$

Solving,

$$C = \frac{r}{\sqrt{x^2+4x}} \quad \text{and} \quad D = -\frac{s}{\sqrt{x^2+4x}}$$

Thus, the desired Binet's formula is

$$B_n(x) = \frac{r^{n+1} - s^{n+1}}{r - s} \tag{41.23}$$

where $r = r(x)$, $s = s(x)$, and $n \geq 0$.

RELATIONSHIPS BETWEEN FIBONACCI AND MORGAN-VOYCE POLYNOMIALS

Binet's formulas for $B_n(x)$ and $f_n(x)$ can be employed to derive a close link between them. To this end, recall that

$$f_n(x) = \frac{\alpha^n(x) - \beta^n(x)}{\alpha(x) - \beta(x)},$$

where

$$\alpha(x) = \frac{x+\sqrt{x^2+4}}{2} \quad \text{and} \quad \beta(x) = \frac{x-\sqrt{x^2+4}}{2}$$

Then

$$\alpha^2(x) = \frac{x^2+2+\sqrt{x^2+4}}{2} \quad \text{and} \quad \beta^2(x) = \frac{x^2+2-\sqrt{x^2+4}}{2}.$$

We have

$$\begin{aligned}
f_{2n}(x) &= \frac{\alpha^{2n}(x) - \beta^{2n}(x)}{\alpha(x) - \beta(x)} = \frac{(\alpha^2(x))^n - (\beta^2(x))^n}{\alpha(x) - \beta(x)} \\
&= \frac{(r(x^2))^n - s(x^2))^n}{[r(x^2) - s(x^2)]/x} = xB_{n-1}(x^2)
\end{aligned}$$

Since $b_n(x) = B_n(x) - B_{n-1}(x)$, $xb_n(x^2) = xB_n(x^2) - xB_{n-1}(x^2) = f_{2n+2}(x) - f_{2n}(x) = xf_{2n+1}(x)$. Thus

$$b_n(x^2) = f_{2n+1}(x) \qquad \text{and} \qquad xB_{n-1}(x^2) = f_{2n}(x) \tag{41.24}$$

For example,

$$\begin{aligned}
b_3(x^2) &= x^6 + 5x^4 + 6x^2 + 1 = f_7(x) \\
xB_2(x^2) &= x(x^4 + 4x^2 + 3) = x^6 + 4x^3 + 3x = f_6(x)
\end{aligned}$$

GENERATING FUNCTIONS AND MORGAN-VOYCE POLYNOMIALS

Next, we show how generating functions can be employed to derive the properties in Eq. (41.24). To this end, first we derive the generating functions for both $B_n(x)$ and $b_n(x)$.

Since the roots of the characteristic equation $t^2 - (x+2)t + 1 = 0$ are r and s, it follows that

$$\begin{aligned}
\frac{1}{1 - (x+2)t + t^2} &= \frac{1}{(1 - rt)(1 - st)} \\
&= \frac{A}{1 - rt} + \frac{B}{1 - st}
\end{aligned}$$

where $A = r/(r-s)$, $B = -s/(r-s)$, $r = r(x)$, and $s = s(x)$.

$$\frac{1}{1 - (x+2)t + t^2} = \sum_{n=0}^{\infty} \frac{(r^{n+1} - s^{n+1})t^n}{r - s} = \sum_{n=0}^{\infty} B_n(x)t^n \tag{41.25}$$

$$\therefore \quad \frac{t}{1 - (x+2)t + t^2} = \sum_{n=0}^{\infty} B_n(x)t^{n+1} = \sum_{n=1}^{\infty} B_{n-1}(x)t^n \tag{41.26}$$

Then

$$\begin{aligned}
\frac{1 - t}{1 - (x+2)t + t^2} &= \sum_0^{\infty} B_n(x)t^n - \sum_1^{\infty} B_{n-1}(x)t^n \\
&= B_0(x) + \sum_1^{\infty} [B_n(x) - B_{n-1}(x)]t^n
\end{aligned}$$

$$= 1 + \sum_{1}^{\infty} b_n(x)t^n$$

$$= \sum_{0}^{\infty} b_n(x)t^n \tag{41.27}$$

Equations (41.25) and (41.27) provide us with the generating functions for $B_n(x)$ and $b_n(x)$.

It follows from Eq. (41.27) that

$$\frac{t(1-t^2)}{1-(x^2+2)t^2+t^4} = \sum_{0}^{\infty} b_n(x^2)t^{2n+1}$$

and

$$\frac{xt^2}{1-(x^2+2)t^2+t^4} = \sum_{0}^{\infty} xB_{n-1}(x^2)t^{2n}$$

Adding these two equations, we get

$$\frac{t(1+xt-t^2)}{(1+xt-t^2)(1-xt-t^2)} = \sum_{0}^{\infty} xB_{n-1}(x^2)t^{2n} + \sum_{0}^{\infty} b_n(x^2)t^{2n+1}$$

$$\frac{t}{1-xt-t^2} = \sum_{0}^{\infty} xB_{n-1}(x^2)t^{2n} + \sum_{0}^{\infty} b_n(x^2)t^{2n+1}$$

$$\sum_{0}^{\infty} f_n(x)t^n = \sum_{0}^{\infty} xB_{n-1}(x^2)t^{2n} + \sum_{0}^{\infty} b_n(x^2)t^{2n+1}$$

Thus $f_{2n}(x) = xB_{n-1}(x^2)$ and $f_{2n+1}(x) = b_n(x^2)$, as desired.

It follows from Eq. (41.24) also that $F_{2n} = B_{n-1}(1)$ and $F_{2n+1} = b_n(1)$. Consequently, the coefficients of $B_n(x)$ and $b_n(x)$ lie on the rising diagonals of Pascal's triangle, as shown below:

```
1
1  1                         ←—— Coefficients of B2 (x) = f6(x)
1  2  1                      ←—— Coefficients of B3 (x) = f7(x)
1  3  3  1
1  4  6  4  1
1  5  10 10 5  1
1  6  15 20 15 6  1
1  7  21 35 35 21 7  1
```

Coefficients of $B_2(x) = f_6(x)$

Coefficients of $B_3(x) = f_7(x)$

Moreover, $b_n(x)$ is irreducible if and only if $2n+1$ is a prime. For example, $b_9(x) = x^9 + 17x^8 + 120x^7 + 455x^6 + 1001x^5 + 128x^4 + 924x^3 + 330x^2 + 45x + 1$ has nontrivial factors.

VOLTAGE AND CURRENT

Next, we will see how the polynomials $B_n(x)$ and $b_n(x)$ are related to voltage and current in the ladder network.

Since the system is linear, assume that the output voltage is 1 volt. Let V_n denote the voltage across the nth unit resistance and I_n the current. Initially, that is, when $n = 0$, we have a no-resistance network, so there is no current and the voltage between the terminals is 1 volt. That is, $I_0 = 0$ and $V_0 = 1$ (see Fig. 41.4).

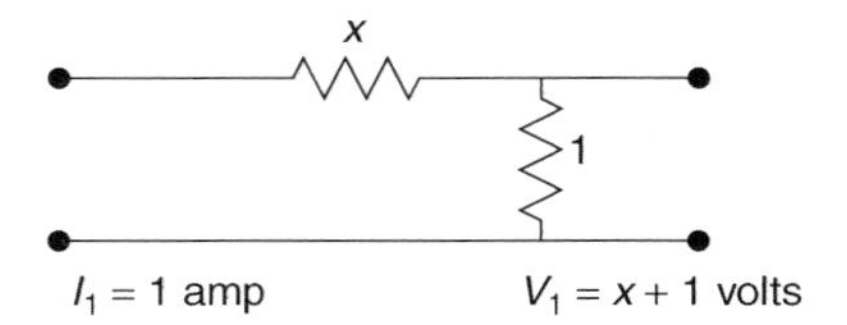

Figure 41.4.

Thus

$$\begin{aligned} I_0 &= 0 = B_{-1}(x) \qquad V_0 = 1 = b_0(x) \\ I_1 &= 1 = B_0(x) \qquad V_1 = x + 1 = b_1(x) \end{aligned}$$

Since $b_{n+1}(x) = xB_n(x) + b_n(x)$,

$$\begin{aligned} B_{n+1}(x) &= (x+1)B_n(x) + b_n(x) = B_n(x) + [xB_n(x) + b_n(x)] \\ &= B_n(x) + b_{n+1}(x) \end{aligned}$$

Using this result and the principle of mathematical induction, we now show that $I_n = B_{n-1}(x)$ and $V_n = b_{n-1}(x)$. Consider the ladder network in Figure 41.5. We have $V_{n+1} = xI_{n+1} + V_n$ and $I_{n+1} = V_n + I_n$. Assume that $I_n = B_{n-1}(x)$ and $V_n = b_n(x)$. Then

$$\begin{aligned} V_{n+1} &= xB_n(x) + b_n(x) = b_{n+1}(x) \\ I_{n+1} &= b_n(x) + B_{n-1}(x) = B_n(x) \end{aligned}$$

Since the results are true when $n = 0$ and $n = 1$, it follows by PMI that $I_n = B_{n-1}(x)$ and $V_n = b_{n-1}(x)$ for every $n \geq 0$.

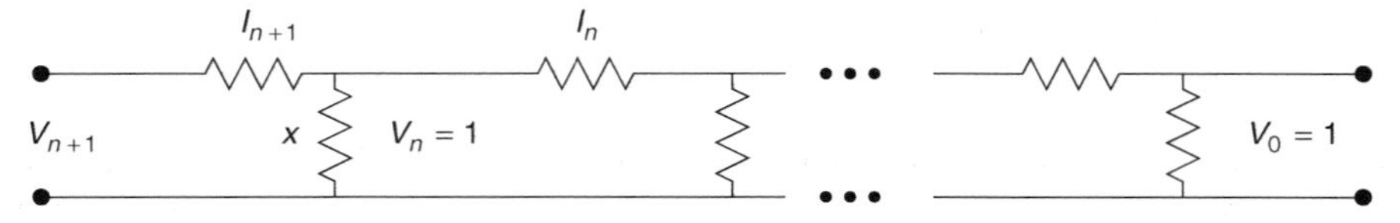

Figure 41.5.

As a by-product, notice that:

$$\begin{aligned} b_n(x) &= V_n = xB_{n-1}(x) + b_{n-1}(x) \\ &= xB_{n-1}(x) + xB_{n-2}(x) + b_{n-2}(x) \\ &\vdots \\ &= x[B_{n-1}(x) + B_{n-2}(x) + \cdots + B_0(x)] + 1 \end{aligned}$$

Likewise,

$$\begin{aligned} B_n(x) &= I_{n+1} = V_n + V_{n-1} + \cdots + V_0 \\ &= b_n(x) + b_{n-1}(x) + \cdots + b_0(x) \end{aligned}$$

Thus we can find the polynomial $b_n(x)$ by multiplying the sum of the polynomials $B_{n-1}(x)$, $B_{n-2}(x), \ldots, B_0(x)$ by x and then adding a 1 to the product; and $B_n(x)$ can be obtained by just adding the polynomials $b_n(x), b_{n-1}(x), \ldots, b_0(x)$.

For example,

$$\begin{aligned} b_4(x) &= x[B_3(x) + B_2(x) + B_1(x) + B_0(x)] + 1 \\ &= x[(x^3 + 6x^2 + 10x + 4) + (x^2 + 4x + 3) + (x + 2) + 1] + 1 \\ &= x^4 + 7x^3 + 15x^2 + 10x + 1 \end{aligned}$$

and

$$\begin{aligned} B_4(x) &= b_3(x) + b_2(x) + b_1(x) + b_0(x) \\ &= (x^3 + 5x^2 + 6x + 1) + (x^2 + 3x + 1) + (x + 1) + 1 \\ &= x^3 + 6x^2 + 10x + 4 \end{aligned}$$

MORGAN-VOYCE POLYNOMIALS AND THE S-MATRIX

Like the Q-matrix in Chapter 32, the matrix

$$S = \begin{bmatrix} x+2 & -1 \\ 1 & 0 \end{bmatrix}$$

can be employed to investigate the properties of $B_n(x)$ and $b_n(x)$. Notice that

$$S = \begin{bmatrix} B_1(x) & -B_0(x) \\ B_0(x) & B_{-1}(x) \end{bmatrix}$$

and

$$\begin{aligned} S^2 &= \begin{bmatrix} x+2 & -1 \\ 1 & 0 \end{bmatrix} \begin{bmatrix} x+2 & -1 \\ 1 & 0 \end{bmatrix} = \begin{bmatrix} x^2+4x+3 & -(x+2) \\ x+2 & -1 \end{bmatrix} \\ &= \begin{bmatrix} B_2(x) & -B_1(x) \\ B_1(x) & -B_0(x) \end{bmatrix} \end{aligned}$$

More generally, we can show that

$$S^n = \begin{bmatrix} B_n(x) & -B_{n-1}(x) \\ B_{n-1}(x) & B_{n-2}(x) \end{bmatrix} \tag{41.28}$$

See Exercise 34. Since $|S| = -1$, this yields the Cassini-like formula we found earlier: $B_{n+1}(x)B_{n-1}(x) - B_n^2(x) = -1$.

Using Formula (41.3), Eq. (41.28) implies

$$\begin{aligned} \begin{bmatrix} b_n(x) & -b_{n-1}(x) \\ b_{n-1}(x) & -b_{n-2}(x) \end{bmatrix} &= \begin{bmatrix} B_n(x) - B_{n-1}(x) & -[B_{n-1}(x) - B_{n-2}(x)] \\ B_{n-1}(x) - B_{n-2}(x) & -[B_{n-2}(x) - B_{n-3}(x)] \end{bmatrix} \\ &= S^n - S^{n-1} \\ &= S^{n-1}(S - I) \\ \therefore \quad \begin{vmatrix} b_n(x) & -b_{n-1}(x) \\ b_{n-1}(x) & -b_{n-2}(x) \end{vmatrix} &= |S - I| \\ &= \begin{vmatrix} x+1 & -1 \\ 1 & -1 \end{vmatrix} = x \end{aligned}$$

Thus

$$b_{n+1}(x)b_{n-1}(x) - b_n^2(x) = x \tag{41.29}$$

For example,

$$\begin{aligned} b_4(x)b_2(x) - b_3^2(x) &= (x^4 + 7x^3 + 15x^2 + 10x + 1)(x^2 + 3x + 1) \\ &\quad -(x^3 + 5x^2 + 6x + 1)^2 \\ &= x \end{aligned}$$

TRIGONOMETRIC FORMULAS FOR $B_n(x)$ AND $b_n(x)$

Next, we show that $B_n(x)$ and $b_n(x)$ can be expressed in terms of the sine and cosine functions. Since

$$\sin A + \sin B = 2 \sin \frac{A+B}{2} \cos \frac{A-B}{2}$$

it follows that

$$\sin(n+1)\theta + \sin(n-1)\theta = 2 \sin n\theta \cos \theta$$

Let $\cos \theta = (x+2)/2$, so $-4 \le x \le 0$. Then

$$\frac{\sin(n+1)\theta}{\sin \theta} + \frac{\sin(n-1)\theta}{\sin \theta} = (x+2)\frac{\sin n\theta}{\sin \theta} \tag{41.30}$$

Notice that

$$\frac{\sin(n+1)\theta}{\sin \theta} = \begin{cases} 1 & \text{if } n = 0 \\ x+2 & \text{if } n = 1 \end{cases}$$

Let $S_n(x) = [\sin(n+1)\theta]/\sin\theta$. Then $S_0(x) = 1$ and $S_1(x) = x + 2$. Besides, Eq. (41.29) shows that $S_{n+1}(x) = (x+2)S_n(x) - S_{n-1}(x)$. Thus, $S_n(x)$ satisfies the same initial conditions and the same recurrence relation as $B_n(x)$, so $S_n(x) = B_n(x)$. That is,

$$B_n(x) = \frac{\sin(n+1)\theta}{\sin\theta} \qquad -4 \le x \le 0 \tag{41.31}$$

Since $b_n = B_n - B_{n-1}$, this yields a trigonometric formula for $b_n(x)$:

$$b_n(x) = \frac{\cos(2n+1)\theta/2}{\cos\theta/2} \qquad -4 \le x \le 0 \tag{41.32}$$

See Exercise 35.

HYPERBOLIC FUNCTIONS FOR $B_n(x)$ AND $b_n(x)$

To derive a hyperbolic function for $B_n(x)$, we let $\cosh\varphi = (x+2)/2$. Using similar steps as before, we can show that

$$B_n(x) = \frac{\sinh(n+1)\varphi}{\sinh\varphi} \qquad x \ge 0 \tag{41.33}$$

See Exercise 36.

Since $b_n = B_n - B_{n-1}$, this implies

$$\begin{aligned} b_n(x) &= \frac{\sinh(n+1)\varphi}{\sinh\varphi} - \frac{\sinh(n-1)\varphi}{\sinh\varphi} \\ &= \frac{\sinh(n+1)\,\varphi - \sinh(n-1)\varphi}{\sinh(n+1)\varphi} \\ &= \frac{\cosh(2n+1)\,\varphi\,/2}{\cosh\varphi/2} \qquad x \ge 0 \end{aligned} \tag{41.34}$$

ZEROS OF $B_n(x)$ AND $b_n(x)$

Formulas (41.31) and (41.32) provide us with interesting bonuses, namely, the zeros of both $B_n(x)$ and $b_n(x)$.

Since $\sin m\theta = 0$ if and only if $\theta = k\pi/m$, it follows from Eq. (41.31) that $B_n(x) = 0$ if and only if $\theta = k\pi/(n+1)$, where $0 \le k < n$. Then

$$\begin{aligned} x + 2 &= 2\cos\theta = 2\cos k\pi/(n+1) \\ x &= 2[\cos k\pi/(n+1) - 1] = -4\sin^2 k\pi/(2n+2) \qquad 0 \le k < n \end{aligned}$$

Similarly, the zeros of $b_n(x)$ are given by $x = -4\sin^2(2k-1)\pi/(4n+2)$, where $0 \le k < n$ (see Exercise 37).

It now follows that the zeros of both $B_n(x)$ and $b_n(x)$ are real, negative, and distinct.

EXERCISES 41

1. Show that $b_n(x)$ satisfies the recurrence relation (41.7).
2. Find $B_5(x)$ using Formula (41.3).
3. Find $b_5(x)$ using Formula (41.5).
4. Verify Formula (41.3).
5. Verify Formula (41.5).
6. Verify that the explicit Formula (41.3) for $B_n(x)$ satisfies the Property (41.2).
7. Verify that the explicit Formula (41.4) for $b_n(x)$ satisfies the Property (41.1).
8. Prove that $B_n(1) = F_{2n+2}, n \geq 0$.
9. Prove that $b_n(1) = F_{2n+1}, n \geq 0$.
10. Using Binet's formula for $B_n(x)$, show that $B_n(1) = F_{2n+2}$, where $n \geq 0$.
11. Using Exercise 4, show that $b_n(1) = F_{2n+1}$, where $n \geq 0$.
12. Show that $(x + 2)B_{2n-1} = B_n^2 - B_{n-2}^2$.

13–16. Establish the Identities (41.14) through (41.17).

Prove each.

17. $xB_n(x) = (x + 1)b_n(x) - b_{n-1}(x)$ (Swamy, 1966)
18. $B_{n+1}(x) - B_{n-1}(x) = b_{n+1}(x) + b_n(x)$ (Swamy, 1966)
19. $x[B_n(x) + B_{n-1}(x)] = b_{n+1}(x) - b_{n-1}(x)$
20. $B_{-1}(x) = 0$ and $b_{-1}(x) = 1$.
21. $b_{2n}(x) = B_n(x)b_n(x) - B_{n-1}(x)b_{n-1}(x)$ (Swamy, 1966)
22. $b_{2n+1}(x) = B_n(x)b_{n+1}(x) - B_{n-1}(x)b_n(x)$ (Swamy, 1966)
23. $(x + 2)b_{2n+1}(x) = B_{n+1}(x)b_{n+1}(x) - B_{n-1}(x)b_{n-1}(x)$ (Swamy, 1966)
24. $(x + 2)B_{2n}(x) = B_{n+1}(x)B_n(x) - B_{n-1}(x)B_{n-2}(x)$ (Swamy, 1966)
25. $b_{2n}(x) - b_{2n-1}(x) = b_n^2(x) - b_{n-1}^2(x)$ (Swamy, 1966)
26. $(x + 2)b_{2n}(x) = B_{n+1}(x)b_n(x) - B_{n-1}(x)b_{n-2}(x)$ (Swamy, 1966)
27. $\sum_0^n B_{2i}(x) = B_n^2(x)$ (Swamy, 1966)
28. $\sum_1^n B_{2i-1}(x) = B_n(x)B_{n-1}(x)$ (Swamy, 1966)
29. $\sum_0^n b_{2i}(x) = B_n(x)b_n(x)$ (Swamy, 1966)
30. $\sum_1^n b_{2i-1}(x) = B_{n-1}(x)b_n(x)$ (Swamy, 1966)
31. $\sum_0^{2n}(-1)^i b_i(x) = b_n^2(x)$ (Swamy, 1966)

Let $g_n(x) = \begin{cases} b_{(n-1)/2}(x^2) & \text{if } n \text{ is odd} \\ xB_{(n-2)/2}(x^2) & \text{otherwise} \end{cases}$

32. Find the polynomials $g_5(x)$ and $g_6(x)$.

33. Show that $g_n(x) = f_n(x)$, the Fibonacci polynomial.

34. Let $S = \begin{bmatrix} x+2 & -1 \\ 1 & 0 \end{bmatrix}$. Prove that $S^n = \begin{bmatrix} B_n(x) & -B_{n-1}(x) \\ B_{n-1}(x) & -B_{n-2}(x) \end{bmatrix}$, $n \ge 1$.

35. Show that $b_n(x) = \dfrac{\cos(2n+1)\theta/2}{\cos\theta/2}$, where $-4 \le x \le 0$.

36. Let $\cosh\varphi = \dfrac{x+2}{2}$. Show that $B_n(x) = \dfrac{\sinh(2n+1)\,\varphi/2}{\sinh\varphi}$, where $x \ge 0$.

37. Show that the zeros of $b_n(x)$ are given by $-4\sin^2(2k-1)\pi/(4n+2), 0 \le k < n$.

The polynomial $C_n(x)$, defined by $C_n(x) = (x+2)C_{n-1}(x) - C_{n-2}(x)$, where $C_0(x) = 1$, $C_1(x) = (x+2)/2$, and $n \ge 2$, occurs in network theory. (Swamy, 1971)

38. Find $C_2(x)$, $C_3(x)$, and $C_4(x)$.

39. Show that $2C_n(x) = b_n(x) + b_{n-1}(x) = B_n(x) - B_{n-2}(x)$.

40. Let $S = \begin{bmatrix} x+2 & -1 \\ 1 & 0 \end{bmatrix}$. Show that $S^n - S^{n-2} = 2\begin{bmatrix} C_n(x) & -C_{n-1}(x) \\ C_{n-1}(x) & -C_{n-2}(x) \end{bmatrix}$, where $n \ge 2$.

41. Show that $C_{n+1}(x)C_{n-1}(x) - C_n^2(x) = x(x+4)/4$.

42. Show that $C_n(x) = \sum_{r=0}^{n} \dfrac{n}{n-r}\dbinom{n+r-1}{n-r-1}x^r$.

43. Using Exercise 42, find the polynomials $C_2(x)$, $C_3(x)$, and $C_4(x)$.

42 FIBONOMETRY

Several well-known trigonometric formulas relate Fibonacci and Lucas numbers with the trigonometric functions.

For example, a number of interesting relationships exist connecting Fibonacci and Lucas numbers with the inverse tangent function $\tan^{-1}$, and the ubiquitous irrational number π:

$$\begin{aligned}
\frac{\pi}{4} &= \tan^{-1}\frac{1}{1}\\
&= \tan^{-1}\frac{1}{2} + \tan^{-1}\frac{1}{3}\\
&= 2\tan^{-1}\frac{1}{3} + \tan^{-1}\frac{1}{7}\\
&= \tan^{-1}\frac{1}{2} + \tan^{-1}\frac{1}{5} + \tan^{-1}\frac{1}{8} \qquad \text{(Dase, 1844)}\\
&= 2\tan^{-1}\frac{1}{5} + \tan^{-1}\frac{1}{7} + 2\tan^{-1}\frac{1}{8}\\
&= \tan^{-1}\frac{1}{3} + \tan^{-1}\frac{1}{5} + \tan^{-1}\frac{1}{7} + \tan^{-1}\frac{1}{8}
\end{aligned}$$

THE GOLDEN RATIO AND THE INVERSE TRIGONOMETRIC FUNCTIONS

Several interesting relationships link the golden ratio and the inverse trigonometric functions.

Since $\cos^{-1} x = \sin^{-1}\sqrt{1-x^2}$ for $0 \le x \le 1$, it follows that $\cos^{-1}(1/\alpha) = \sin^{-1}\sqrt{1-\beta^2} = \sin^{-1}\beta$. Likewise, $\sin^{-1}(1/\alpha) = \cos^{-1}\sqrt{1-\beta^2} = \cos^{-1}(1/\sqrt{\beta})$.

Suppose that $\tan x = \cos x$. Then $\sin x = \cos^2 x$, so $\sin^2 x + \sin x - 1 = 0$. Since $\sin x \geq 0$, it follows that $\sin x = |\beta| = 1/\alpha$ and $x = \sin^{-1}(1/\alpha)$. Then $\tan(\sin^{-1}(1/\alpha)) = \cos(\sin 1/\alpha) = \cos(\cos^{-1}(1/\sqrt{\alpha}) = 1/\sqrt{\alpha}$. In addition, $\cot(\cos^{-1}(1/\alpha) = 1/\sqrt{\alpha} = \sin(\cos^{-1}(1/\alpha)$. These results, studied in 1970 by Br. L. Raphael of St. Mary's College, California, are summarized in Figure 42.1.

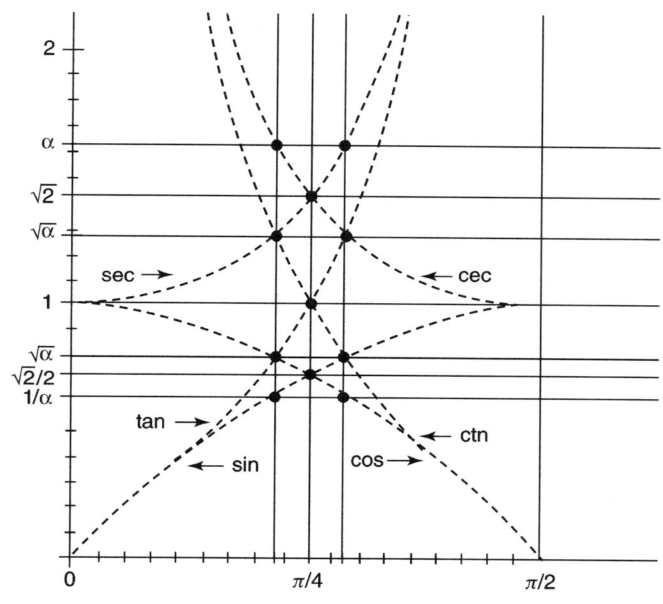

Figure 42.1.

THE GOLDEN TRIANGLE REVISITED

We can employ the golden triangle with vertex angle 36° to produce a surprising trigonometric result. The exact trigonometric values of some acute angles are known. The smallest such integral angle is $\theta = 3°$. Oddly enough, we can use the golden triangle to compute the exact value of sine of 3°. Once we know sin 3°, we can express the values of the remaining trigonometric functions of 3°, and hence those of the trigonometric values of multiples of 3° using the sum formulas.

Recall from Chapter 22 that the ratio of a lateral side to the base of the golden triangle with vertex angle 36° is the Golden Ratio α: $AB/AC = x/y = \alpha$ (see Fig. 42.2). In particular, let $x = 1$. Then $1/y = \alpha$, so $y = (\sqrt{5} - 1)/2 = -\beta = 1/\alpha$. Let $\overline{BN}$ be the perpendicular bisector of $\overline{AC}$ (see Fig. 42.3). Then

$$\sin 18^\circ = \frac{AN}{AB} = \frac{y}{2} = \frac{1}{2\alpha} = \frac{\sqrt{5} - 1}{4}$$

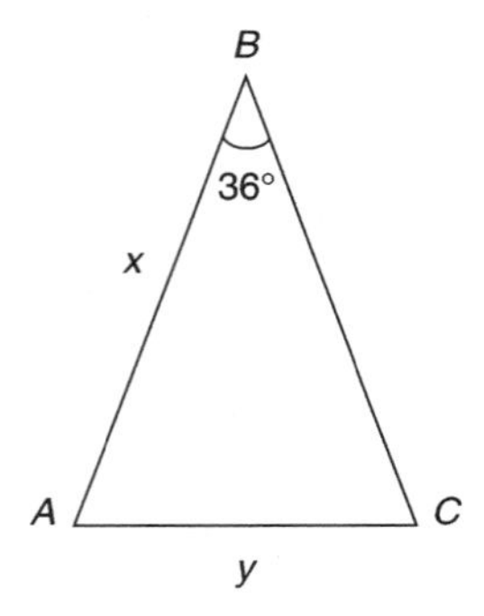

Figure 42.2.

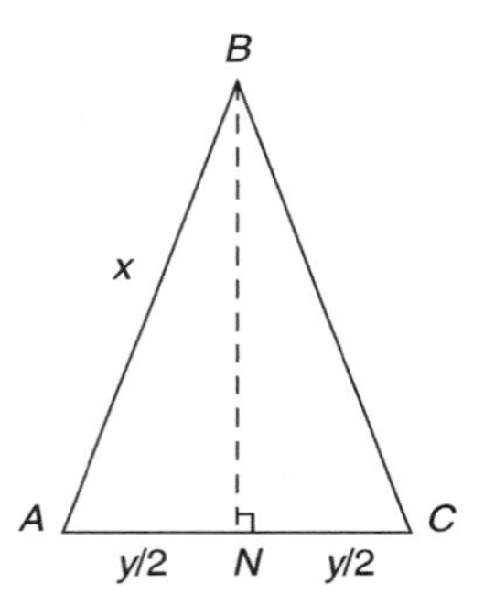

Figure 42.3.

Since $\sin^2 u + \cos^2 u = 1$, this implies

$$\cos 18^\circ = \frac{\sqrt{10 + 2\sqrt{5}}}{4} = \frac{\sqrt{\sqrt{5}\alpha}}{2}$$

Then

$$\begin{aligned}\sin 15^\circ &= \sin(45^\circ - 30^\circ)\\ &= \frac{\sqrt{2}}{2} \cdot \frac{\sqrt{3}}{2} - \frac{\sqrt{2}}{2} \cdot \frac{1}{2} = \frac{\sqrt{6} - \sqrt{2}}{4}\end{aligned}$$

Likewise,

$$\begin{aligned}\cos 15^\circ &= \frac{\sqrt{6} + \sqrt{2}}{4}\\ \therefore \quad \sin 3^\circ &= \sin(18^\circ - 15^\circ)\\ &= \sin 18^\circ \cos 15^\circ - \cos 18^\circ \sin 15^\circ\\ &= \frac{\sqrt{5} - 1}{4} \cdot \frac{\sqrt{6} + \sqrt{2}}{4} - \frac{\sqrt{6} - \sqrt{2}}{4} \cdot \frac{\sqrt{10 + 2\sqrt{5}}}{4}\end{aligned}$$

$$= \frac{(\sqrt{5}-1)(\sqrt{6}+\sqrt{2}) - (\sqrt{6}-\sqrt{2})\sqrt{10+2\sqrt{5}}}{16}$$

This formula was developed in 1959 by W. R. Ransom.

GOLDEN WEAVES

In 1978, W. E. Sharpe of the University of South Carolina at Columbia, while he was a member of the Norwegian Geological Survey, observed a set of remarkable weave patterns. Suppose the weave begins at the lower left-hand corner of a square loom of unit side. Suppose the first thread makes an angle θ with the base, where

$$\tan\theta = \frac{n+\alpha}{n+\alpha+1} \qquad n \geq 0$$

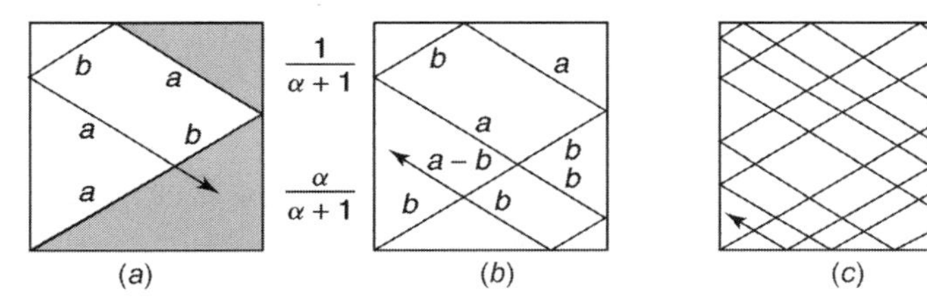

Figure 42.4. The Development of a Golden Weave on a Square Loom of Unit Side, for $n = 0$: (*a*) After 3 Reflections; (*b*) After 5 Reflections; (*c*) After 15 Reflections [*Source:* W. E. Sharpe, "Golden Weaves," *The Mathematical Gazette*, Vol. 62, 1978. Copyright(©) 1978 by The Mathematical Association (www.m-a.org.uk).].

Initially, $n = 0$ (see Fig. 42.4*a*). Every time the thread (or line) meets a side of the square, it is reflected in the same way as a ray of light and a new thread of the weave begins. After the third reflection, the thread crosses the original thread. Suppose the point of intersection divides the original thread into lengths a and b, as the figure shows. Using the properties of isosceles triangles and parallelograms, all the lengths marked a are equal and so are those marked b.

Since $\tan\theta = \alpha/(\alpha+1)$, it follows that the first thread meets the side of the square at the point that divides it into segments of lengths

$$\frac{\alpha}{\alpha+1} \qquad \text{and} \qquad 1 - \frac{\alpha}{\alpha+1} = \frac{1}{\alpha+1}$$

The shaded triangles in Figure 42.4*a* are congruent, so

$$\frac{a+b}{a} = \frac{\alpha/(\alpha+1)}{1/(\alpha+1)}$$

that is, $1 + (b/a) = \alpha$;

$$\therefore \quad \frac{a}{b} = \frac{1}{\alpha - 1} = \frac{\alpha}{\alpha^2 - \alpha} = \alpha$$

Thus the first point of intersection divides the first thread in the Golden Ratio.

From Figure 42.4*b*, the second point of intersection divides the line segment of length a into two parts of length b and $a - b$. Since

$$\frac{b}{a-b} = \frac{1}{a/b - 1} = \frac{1}{\alpha - 1} = \alpha$$

this line segment is also divided in the Golden Ratio at the point of intersection.

As additional crossovers occur, successive line segments are divided in the Golden Ratio at each intersection. For this reason, Sharpe called these weaves the *golden weaves*.

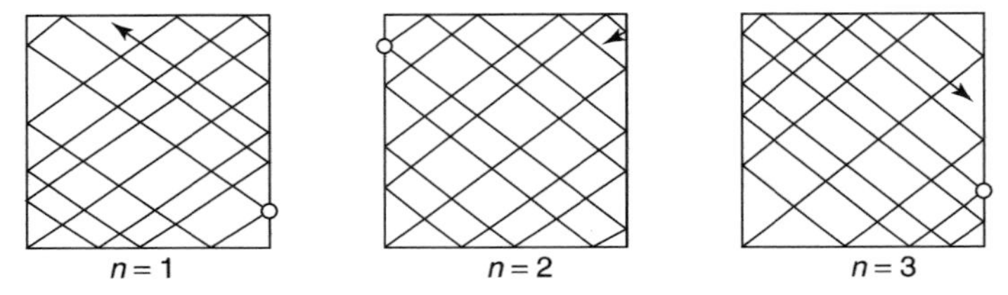

Figure 42.5. The Weaving Patterns for the First 15 Reflections [*Source:* W. E. Sharpe, "Golden Weaves," *The Mathematical Gazette*, Vol. 62, 1978. Copyright (©) 1978 by The Mathematical Association (www.m-a.org.uk).].

Figure 42.5 shows the weaving patterns for the first 15 reflections, for $n = 1$, 2, and 3. In fact, in all four cases, the first golden division occurs after $2n+3$ reflections, marked with a circle in these diagrams.

In Chapter 25, we found a trigonometric expansion of F_n:

$$F_n = 2^{n-1} \sum_{i=0}^{n-1} (-1)^i \cos^{n-i-1} \pi/5 \sin^k \pi/10 \tag{25.1}$$

Next we explore a host of additional relationships.

Theorem 42.1. Let G_n denote the nth generalized Fibonacci number. Then

$$\tan\left(\tan^{-1} \frac{G_n}{G_{n+1}} - \tan^{-1} \frac{G_{n+1}}{G_{n+2}}\right) = \frac{(-1)^{n+1}\mu}{G_{n+1}(G_n + G_{n+2})} \tag{42.1}$$

Proof.

$$\begin{aligned} \text{LHS} &= \frac{(G_n/G_{n+1}) - (G_{n+1}/G_{n+2})}{1 + (G_n/G_{n+1}) \cdot (G_{n+1}/G_{n+2})} = \frac{G_n G_{n+2} - G_{n+1}^2}{G_n G_{n+1} + G_{n+1} G_{n+2}} \\ &= \frac{(-1)^{n+1}\mu}{G_{n+1}(G_n + G_{n+2})} \end{aligned}$$

■

The following corollary follows directly from this theorem.

Corollary 42.1.

(1)
$$\tan\left(\tan^{-1}\frac{F_n}{F_{n+1}} - \tan^{-1}\frac{F_{n+1}}{F_{n+2}}\right) = \frac{(-1)^{n+1}}{F_{2n+2}} \tag{42.2}$$

(2)
$$\tan\left(\tan^{-1}\frac{L_n}{L_{n+1}} - \tan^{-1}\frac{L_{n+1}}{L_{n+2}}\right) = \frac{5(-1)^{n}}{L_{2n+1}+L_{2n+3}} \tag{42.3}$$

■

Theorem 42.2.

$$\tan^{-1}\frac{F_n}{F_{n+1}} = \sum_{i=1}^{n}(-1)^{i+1}\tan^{-1}\frac{1}{F_{2i}}$$

Proof. [by the principle of mathematical induction (PMI)] When $n = 1$, LHS $=$ $\tan^{-1} 1 = (-1)^2 \tan^{-1} 1/F_2 =$ RHS. Therefore, the formula works for $n = 1$.

Now, assume it works for $n = k$. Then:

$$\begin{aligned}
\sum_{i=1}^{k+1}(-1)^{i+1}\tan^{-1}\frac{1}{F_{2i}} &= \sum_{1}^{k}(-1)^{i+1}\tan^{-1}\frac{1}{F_{2i}} + (-1)^{k+2}\tan^{-1}\frac{1}{F_{2(k+1)}}\\
&= \tan^{-1}\frac{F_k}{F_{k+1}} + (-1)^k\tan^{-1}\frac{1}{F_{2k+2}}\\
&= \left(\tan^{-1}\frac{(-1)^{k+1}}{F_{2k+2}} + \tan^{-1}\frac{F_{k+1}}{F_{k+2}}\right)\\
&\quad + (-1)^k\tan^{-1}\frac{1}{F_{2k+2}} \qquad \text{By Corollary 42.1}\\
&= \tan^{-1}\frac{F_{k+1}}{F_{k+2}}
\end{aligned}$$

since $\tan^{-1}(-x) = -\tan^{-1} x$. So the formula works for $n = k+1$. Thus, by PMI, the formula holds for every $n \geq 1$. ■

Corollary 42.2.

$$\sum_{1}^{\infty}(-1)^{n+1}\tan^{-1}\frac{1}{F_{2n}} = \tan^{-1}(1/\alpha)$$

Proof. Since $\tan^{-1}$ is a continuous increasing function, $\tan^{-1}(1/F_{2n}) > \tan^{-1}(1/F_{2n+2})$. Also, $\lim_{n\to\infty}\tan^{-1}(1/F_{2n}) = \tan^{-1} 0 = 0$. Therefore, the series converges and

$$\sum_{1}^{\infty}(-1)^{n+1}\tan^{-1}\frac{1}{F_{2n}} = \lim_{m\to\infty}\sum_{n=1}^{m}(-1)^{n+1}\tan^{-1}\frac{1}{F_{2n}}$$

$$
\begin{aligned}
&= \lim_{m\to\infty} \tan^{-1} \frac{F_m}{F_{m+1}} \\
&= \tan^{-1}\left(\lim_{m\to\infty} \frac{F_m}{F_{m+1}}\right) = \tan^{-1}(1/\alpha)
\end{aligned}
\qquad \blacksquare
$$

This corollary has a companion result for Fibonacci numbers with odd subscripts. However, before we state and prove it, we need to lay the groundwork in the form of a few lemmas.

Lemma 42.1.

$$L_{2n}L_{2n+2} - 1 = 5F_{2n+1}^2.$$

This lemma follows by Identity 11 on p. 88.

Lemma 42.2.

$$\tan^{-1}\frac{1}{F_{2n+1}} = \tan^{-1}\frac{1}{F_{2n}} - \tan^{-1}\frac{1}{F_{2n+2}}$$

Proof. Let

$$
\begin{aligned}
\theta_n &= \tan^{-1}\frac{1}{F_{2n}} - \tan^{-1}\frac{1}{F_{2n+2}} \\
\tan\theta_n &= \frac{(1/F_{2n}) - (1/F_{2n+2})}{1 + (1/F_{2n})\cdot(1/F_{2n+2})} = \frac{F_{2n+2} - F_{2n}}{F_{2n}F_{2n+2} + 1} \\
&= \frac{F_{2n+1}}{F_{2n+1}^2} \quad \text{by Cassini's formula.} \\
&= \frac{1}{F_{2n+1}}
\end{aligned}
$$

Therefore,

$$\tan^{-1}\frac{1}{F_{2n+1}} = \tan^{-1}\frac{1}{F_{2n}} - \tan^{-1}\frac{1}{F_{2n+2}} \qquad \blacksquare$$

The proof of the next lemma is quite similar to this proof, so we leave it as an exercise (see Exercise 3).

Lemma 42.3.

$$\tan^{-1}\frac{1}{F_{2n+1}} = \tan^{-1}\frac{1}{L_{2n}} + \tan^{-1}\frac{1}{L_{2n+2}}$$

We are now ready to state and prove the celebrated theorem promised just before Lemma 42.1. It was discovered in 1936 by the American mathematician Derrick H. Lehmer (1905–1991) when he was at Lehigh University.

Theorem 42.3. (Lehmer, 1936)

$$\sum_{1}^{\infty} \tan^{-1} \frac{1}{F_{2n+1}} = \frac{\pi}{4}$$

Proof. By Lemma 42.2,

$$\begin{aligned}
\sum_{n=1}^{m} \tan^{-1} \frac{1}{F_{2n+1}} &= \sum_{n=1}^{m} \left(\tan^{-1} \frac{1}{F_{2n}} - \tan^{-1} \frac{1}{F_{2n+2}} \right) \\
&= \tan^{-1} \frac{1}{F_2} - \tan^{-1} \frac{1}{F_{2m+2}} \\
&= \frac{\pi}{4} - \tan^{-1} \frac{1}{F_{2m+2}} \\
\sum_{1}^{\infty} \tan^{-1} \frac{1}{F_{2n+1}} &= \lim_{m\to\infty} \sum_{1}^{m} \tan^{-1} \frac{1}{F_{2n+1}} \\
&= \lim_{m\to\infty} \left(\frac{\pi}{4} - \tan^{-1} \frac{1}{F_{2m+2}} \right) \\
&= \frac{\pi}{4} - \tan^{-1} 0 = \frac{\pi}{4} - 0 \\
&= \frac{\pi}{4}
\end{aligned}$$

■

The following theorem provides a similar result for Lucas numbers.

Theorem 42.4.

$$\sum_{1}^{\infty} \tan^{-1} \frac{1}{L_{2n}} = \tan^{-1}(-\beta)$$

Proof. By Lemma 42.3,

$$\begin{aligned}
\sum_{n=1}^{m} \tan^{-1} \frac{1}{F_{2n+1}} &= \sum_{n=1}^{m} \left(\tan^{-1} \frac{1}{L_{2n}} + \tan^{-1} \frac{1}{L_{2n+2}} \right) \\
&= \tan^{-1} \frac{1}{3} + \left(2 \sum_{2}^{m} \tan^{-1} \frac{1}{L_{2n}} + \tan^{-1} \frac{1}{L_{2m+2}} \right) \\
\sum_{n=1}^{\infty} \tan^{-1} \frac{1}{F_{2n+1}} &= \tan^{-1} \frac{1}{3} + 2 \sum_{2}^{\infty} \tan^{-1} \frac{1}{L_{2n}} + 0 \\
&= 2 \sum_{1}^{\infty} \tan^{-1} \frac{1}{L_{2n}} - \tan^{-1} \frac{1}{3}
\end{aligned}$$

Since $\pi/4 = \tan^{-1} 1$, by Theorem 42.3, this yields

$$\begin{aligned}
2\sum_{1}^{\infty} \tan^{-1} \frac{1}{L_{2n}} &= \tan^{-1} \frac{1}{3} + \tan^{-1} 1 \\
&= \tan^{-1} 2 \qquad \text{(See Exercise 4)} \\
\sum_{1}^{\infty} \tan^{-1} \frac{1}{L_{2n}} &= \frac{1}{2} \tan^{-1} 2 \\
&= \tan^{-1} \frac{\sqrt{5}-1}{2} \qquad \text{(See Exercise 5)} \\
&= \tan^{-1}(-\beta)
\end{aligned}$$

■

The next theorem was also proposed as a problem in 1936 by Lehmer. The proof given below is essentially the one derived two years later by M. A. Heaslet of San Jose State College, California. Alternate proofs were given by Hoggatt, Jr., in 1964 and 1968, and by C. W. Trigg of California in 1973.

Theorem 42.5. (Lehmer, 1936)

$$\cot^{-1} 1 = \sum_{1}^{\infty} \cot^{-1} F_{2n+1}$$

Proof.

$$\begin{aligned}
\cot^{-1} F_{2k} - \cot^{-1} F_{2k+1} &= \cot^{-1} \frac{F_{2k}F_{2k+1} + 1}{F_{2k+1} - F_{2k}} \\
&= \cot^{-1} \frac{F_{2k}F_{2k+1} + 1}{F_{2k-1}} \\
&= \cot^{-1} \frac{F_{2k-1}F_{2k+2}}{F_{2k-1}} \qquad \text{By Identity 2 on p. 87.} \\
&= \cot^{-1} F_{2k+2} \\
\therefore \quad \cot^{-1} F_{2k} - \cot^{-1} F_{2k+2} &= \cot^{-1} F_{2k+1} \\
\sum_{1}^{n} (\cot^{-1} F_{2k} - \cot^{-1} F_{2k+2}) &= \sum_{1}^{n} \cot^{-1} F_{2k+1} \\
\cot^{-1} F_2 - \cot^{-1} F_{2n+2} &= \sum_{1}^{n} \cot^{-1} F_{2k+1}
\end{aligned}$$

As $m \to \infty$, $\cot^{-1} F_m \to 0$. Therefore, as $n \to \infty$, this equation yields

$$\cot^{-1} 1 - 0 = \sum_{1}^{\infty} \cot^{-1} F_{2k+1}$$

That is,

$$\cot^{-1} 1 = \sum_{1}^{\infty} \cot^{-1} F_{2n+1}$$

■

The next theorem is a generalization of Lehmer's formula in Theorem 42.3.

Theorem 42.6. Let $f_n(x)$ denote the Fibonacci polynomial. Then

$$\tan^{-1} \frac{1}{x} = \sum_{1}^{\infty} \tan^{-1} \frac{x}{f_{2n+1}(x)}$$

Proof. Let $\tan \theta_n = 1/f_n(x)$. Then

$$\begin{aligned} \tan(\theta_n - \theta_{n+2}) &= \frac{f_{2n+2}(x) - f_{2n}(x)}{1 + f_{2n}(x) f_{2n+2}(x)} \\ &= \frac{x f_{2n+1}(x)}{1 + f_{2n}(x) f_{2n+2}(x)} \end{aligned} \tag{42.4}$$

Since $f_{k-1}(x) f_{k+1}(x) - f_k^2(x) = (-1)^k$, this yields

$$\tan(\theta_n - \theta_{n+2}) = \frac{x f_{2n+1}(x)}{f_{2n+1}^2(x)} = \frac{x}{f_{2n+1}(x)}$$

Then

$$\begin{aligned} \theta_n - \theta_{n+2} &= \tan^{-1} \frac{x}{f_{2n+1}(x)} \\ \tan^{-1} \frac{1}{f_{2n}(x)} - \tan^{-1} \frac{1}{f_{2n+2}(x)} &= \tan^{-1} \frac{x}{f_{2n+1}(x)} \\ \sum_{1}^{m} \tan^{-1} \frac{x}{f_{2n+1}(x)} &= \sum_{1}^{m} \left[\tan^{-1} \frac{1}{f_{2n}(x)} - \tan^{-1} \frac{x}{f_{2n+2}(x)} \right] \\ &= \tan^{-1} \frac{1}{f_2(x)} - \tan^{-1} \frac{1}{f_{2m+2}(x)} \end{aligned}$$

Since $f_2(x) = x$ and $\tan^{-1}(1/f_{2m+2}(x)) \to 0$ as $m \to \infty$, this yields the desired result:

$$\tan^{-1} \frac{1}{x} = \sum_{1}^{\infty} \tan^{-1} \frac{x}{f_{2n+1}(x)}$$

■

Lehmer's formula in Theorem 42.3 follows from this result since $f_k(1) = F_k$.

Corollary 42.3.

$$\sum_{1}^{\infty} \tan^{-1} \frac{1}{F_{2n+1}} = \frac{\pi}{4} \qquad \blacksquare$$

The next formula was developed in 1973 by J. R. Goggins of Glasgow, Scotland. It was rediscovered 22 years later by an alternate method by M. Harvey and P. Woodruff of St. Paul's School, London,.

Theorem 42.7. (Goggins, 1973)

$$\sum_{1}^{n} \tan^{-1} \frac{1}{2k+1} + \tan^{-1} \frac{1}{F_{2n+2}} = \frac{\pi}{4}$$

Proof. (by PMI) When $n = 1$, LHS $= \tan^{-1} \frac{1}{2} + \tan^{-1} \frac{1}{3} = \frac{\pi}{4} =$ RHS. So the result is true when $n = 1$.

Now, assume it is true for $n = m$, where $m \geq 1$:

$$\begin{aligned}
\sum_{1}^{m+1} \tan^{-1} \frac{1}{2k+1} + \tan^{-1} \frac{1}{F_{2m+4}} &= \sum_{1}^{m} \tan^{-1} \frac{1}{2k+1} \\
&\quad + \tan^{-1} \frac{1}{2m+3} + \tan^{-1} \frac{1}{F_{2m+4}} \\
&= \left(\frac{\pi}{4} - \tan^{-1} \frac{1}{F_{2m+2}}\right) \\
&\quad + \tan^{-1} \frac{1}{2m+3} + \tan^{-1} \frac{1}{F_{2m+4}} \qquad (42.5)
\end{aligned}$$

Let $\tan x = 1/F_{2m+2}$ and $\tan y = 1/F_{2m+4}$. Then

$$\begin{aligned}
\tan(x-y) &= \frac{1/F_{2m+2} - 1/F_{2m+4}}{1 + (1/F_{2m+2})(1/F_{2m+4})} = \frac{F_{2m+4} - F_{2m+2}}{1 + F_{2m+2}F_{2m+4}} \\
&= \frac{F_{2m+3}}{F_{2m+3}^2} = \frac{1}{F_{2m+3}} \\
\therefore \quad x - y &= \tan^{-1} \frac{1}{F_{2m+3}}
\end{aligned}$$

That is,

$$\tan^{-1} \frac{1}{F_{2m+2}} - \tan^{-1} \frac{1}{F_{2m+4}} = \tan^{-1} \frac{1}{F_{2m+3}}$$

So Eq. (42.5) becomes

$$\sum_{1}^{m+1} \tan^{-1} \frac{1}{2k+1} + \tan^{-1} \frac{1}{F_{2m+4}} = \frac{\pi}{4}$$

Thus the result is true for $n = m + 1$. So, by PMI, it is true for every $n \geq 1$. ■

This result also yields Theorem 42.3 and, hence, Lehmer's formula.

Corollary 42.1.

$$\sum_{1}^{\infty} \tan^{-1} \frac{1}{2k+1} = \frac{\pi}{4}$$ ■

FIBONACCI AND LUCAS FACTORIZATIONS REVISITED

In Chapter 40, we found that both F_n and L_n can be factored using complex numbers:

$$F_n = \prod_{1}^{n-1} (1 - 2i \cos k\pi/n) \tag{42.6}$$

$$L_n = \prod_{0}^{n-1} (1 - 2i \cos(2k+1)\pi/2n) \tag{42.7}$$

In 1967, D. A. Lind expressed the factorizations without the complex number i. They are given in the next theorem, which was posed as an advanced problem. The proof below is the one given by Swamy as a solution to the problem.

Theorem 42.8. (Lind, 1967)

$$F_n = \prod_{1}^{\lfloor (n-1)/2 \rfloor} (3 + 2\cos 2k\pi/n) \tag{42.8}$$

$$L_n = \prod_{0}^{\lfloor (n-2)/2 \rfloor} [3 + 2\cos(2k+1)\pi/n] \tag{42.9}$$

Proof.

Case 1. Let $n = 2m + 1$ be odd. Formula (42.6) becomes

$$\begin{aligned}
F_{2m+1} &= \prod_{1}^{2m}\left(1 - 2i\cos\frac{k\pi}{2m+1}\right) \\
&= \prod_{1}^{m}\left(1 - 2i\cos\frac{k\pi}{2m+1}\right)\prod_{m+1}^{2m}\left(1 - 2i\cos\frac{k\pi}{2m+1}\right) \\
&= \prod_{1}^{m}\left(1 - 2i\cos\frac{k\pi}{2m+1}\right)\prod_{m+1}^{2m}\left[1 + 2i\cos\left(\pi - \frac{k\pi}{2m+1}\right)\right]
\end{aligned}$$

Let $j = 2m + 1 - k$ in the second product. Then

$$\begin{aligned}
F_{2m+1} &= \prod_{1}^{m}\left(1 - 2i\cos\frac{k\pi}{2m+1}\right)\prod_{1}^{m}\left(1 + 2i\cos\frac{k\pi}{2m+1}\right) \\
&= \prod_{1}^{m}\left(1 + 4\cos^2\frac{j\pi}{2m+1}\right) \\
&= \prod_{1}^{m}\left(3 + 2\cos\frac{2j\pi}{2m+1}\right) \qquad (42.10)
\end{aligned}$$

Case 2. Let $n = 2m$ be even. Formula (42.6) becomes

$$F_{2m} = \prod_{1}^{2m-1}\left(1 - 2i\cos\frac{k\pi}{2m}\right)$$

As before, this yields

$$\begin{aligned}
F_{2m} &= \prod_{1}^{m-1}\left(1 - 2i\cos\frac{k\pi}{2m}\right)\prod_{1}^{m-1}\left(1 + 2i\cos\frac{j\pi}{2m}\right)\cdot(1 + 2i\cos\pi/2) \\
&= \prod_{1}^{m-1}\left(1 + 4\cos^2\frac{j\pi}{2m}\right) \\
&= \prod_{1}^{m-1}\left(3 + 2\cos\frac{2j\pi}{2m}\right) \qquad (42.11)
\end{aligned}$$

It follows from Formulas (42.10) and (42.11) that

$$F_n = \prod_{1}^{\lfloor (n-1)/2 \rfloor}(3 + 2\cos 2k\pi/n) \qquad (42.12)$$

To derive the formula for L_n, we have:

$$\begin{aligned} F_{2n} &= \prod_{1}^{\lfloor (2n-1)/2 \rfloor} (3 + 2\cos k\pi/n) \\ &= \prod_{\substack{\text{even integers} \\ i \le n-1}} (3 + 2\cos i\pi/n) \prod_{\substack{\text{odd integers} \\ j \le n-1}} (3 + 2\cos j\pi/n) \end{aligned}$$

Let $i = 2k$ and $j = 2k + 1$. Then

$$\begin{aligned} F_{2n} &= \prod_{1}^{\lfloor (n-1)/2 \rfloor} (3 + 2\cos k\pi/n) \prod_{0}^{\lfloor (n-2)/2 \rfloor} [3 + 2\cos(2k+1)\pi/n] \\ &= F_n \prod_{0}^{\lfloor (n-2)/2 \rfloor} [3 + 2\cos(2k+1)\pi/n] \end{aligned}$$

Since $F_{2n} = F_n L_n$, it follows that

$$L_n = \prod_{0}^{\lfloor (n-2)/2 \rfloor} [3 + 2\cos(2k+1)\pi/n] \tag{42.13}$$

■

For example,

$$\begin{aligned} F_6 &= \prod_{1}^{2} (3 + 2\cos k\pi/3) \\ &= (3 + 2\cos \pi/3)(3 + 2\cos 2\pi/3) \\ &= (3 + 2\cos \pi/3)(3 - 2\cos \pi/3) \\ &= 9 - 4\cos^2 \pi/3 = 9 - 1 \\ &= 8 \end{aligned}$$

and

$$\begin{aligned} L_6 &= \prod_{0}^{2} [3 + 2\cos(2k+1)\pi/6] \\ &= (3 + 2\cos \pi/6)(3 + 2\cos \pi/2)(3 + 2\cos 5\pi/6) \\ &= (3 + 2\cos \pi/6)(3 - 2\cos \pi/6)(3 + 0) \\ &= 3(9 - 4\cos^2 \pi/6) = 3(9 - 4 \cdot 3/4) \\ &= 18 \end{aligned}$$

EXERCISES 42

1. Deduce Formula (42.3) from Formula (42.1).
2. Prove Formula (42.1) using PMI.
3. Prove Lemma 42.3.
4. Show that $\tan^{-1} 1 + \tan^{-1} \frac{1}{3} = \tan^{-1} 2$.
5. Show that $\tan^{-1} 2 = 2\tan^{-1} \beta$.
6. Let θ be the angle between the vectors $\mathbf{u} = (B_0, B_1, B_2, \ldots, B_n)$ and $\mathbf{v} = (F_m, F_{m+1}, F_{m+2}, \ldots, F_{m+n})$ in the Euclidean $(n+1)$-space, where $B_i = \binom{n}{i}$. Find $\lim_{n\to\infty} \theta$ (Gootherts, 1966).
7. Show that $\pi = \tan^{-1}(1/F_{2n}) + \tan^{-1} F_{2n+1} + \tan^{-1} F_{2n+2}$ (Horner, 1969).
8. Prove that $\sum_{1}^{\infty} \dfrac{1}{\alpha F_{n+1} + F_n} = \sum_{1}^{\infty} \tan^{-1} \dfrac{1}{F_{2n+1}}$ (Guillottee, 1972).

43 FIBONACCI AND LUCAS SUBSCRIPTS

What are the characteristics of Fibonacci and Lucas numbers with Fibonacci and Lucas subscripts? That is, numbers of the form F_{F_n}, F_{L_n}, L_{F_n}, and L_{L_n}? For example, $F_{F_6} = F_8 = 21$, $F_{L_6} = F_{18} = 2584$, $L_{F_6} = L_8 = 47$, and $L_{L_6} = L_{18} = 5778$. In this discussion, we shall, for convenience, use the following notations:

$$U_n = F_{F_n} \qquad V_n = F_{L_n} \qquad X_n = L_{F_n} \qquad \text{and} \qquad W_n = L_{L_n}$$

Although Binet's formulas for F_n and L_n are not extremely useful, we can employ them to find explicit formulas for U_n, V_n, X_n, and W_n. For example,

$$U_n = \frac{\alpha^{F_n} - \beta^{F_n}}{\alpha - \beta} \qquad \text{and} \qquad V_n = \alpha^{F_n} + \beta^{F_n}$$

In 1966, R. E. Whitney of Lockhaven State College partially succeeded in defining U_n, V_n, X_n, and W_n recursively in terms of hybrid relations.

Theorem 43.1.

$$5U_nU_{n+1} = X_{n+2} - (-1)^{F_n}X_{n-1}$$

Proof. We have

$$\sqrt{5}F_n = \alpha^n - \beta^n \tag{43.1}$$

$$L_n = \alpha^n + \beta^n \tag{43.2}$$

Replacing n by F_n in Eq. (43.1):

$$\sqrt{5}U_n = \alpha^{F_n} - \beta^{F_n}$$

$$\sqrt{5}U_{n+1} = \alpha^{F_{n+1}} - \beta^{F_{n+1}}$$

Multiplying these, we get

$$\begin{aligned}
5U_nU_{n+1} &= \alpha^{F_{n+2}} - \beta^{F_{n+2}} - (\alpha^{F_n}\beta^{F_{n+1}} + \beta^{F_n}\alpha^{F_{n+1}}) \\
&= X_{n+2} - [\beta^{F_{n+1}}(-\alpha^{-1})^{F_n} + \alpha^{F_{n+1}}(-\beta^{-1})^{F_n}] \\
&= X_{n+2} - (-1)^{F_n}(\alpha^{F_{n+1}-F_n} + \beta^{F_{n+1}-F_n}) \\
&= X_{n+2} - (-1)^{F_n}(\alpha^{F_{n-1}} + \beta^{F_{n-1}}) \\
&= X_{n+2} - (-1)^{F_n}X_{n-1}
\end{aligned}$$ ■

For example, let $n = 7$. Then:

$$\begin{aligned}
5U_7U_8 = 5F_{13}F_{21} &= 5 \cdot 233 \cdot 10,946 = 12,752,090 \\
X_{7+2} - (-1)^{F_7}X_{7-1} &= X_9 - (-1)^{13}X_6 = L_{34} + L_8 = 12,752,043 + 47 \\
&= 12,752,090 \\
&= 5U_7U_8
\end{aligned}$$

Theorem 43.2.

$$X_nX_{n+1} = X_{n+2} + (-1)^{F_n}X_{n-1}$$ ■

Its proof follows along the same lines from Equation (43.2), so we leave it as an exercise (see Exercise 2).

For example, $X_6X_7 = L_8L_{13} = 47 \cdot 521 = 24,487$

$$\begin{aligned}
X_8 + (-1)^{F_6}X_5 &= L_{21} + (-1)^8L_5 = 24{,}476 + 11 = 24{,}487 \\
&= X_6X_7
\end{aligned}$$

Combining Theorems 43.1 and 43.2, we get the following result.

Corollary 43.1.

$$5U_nU_{n+1} + X_nX_{n+1} = 2X_{n+2}$$ ■

The following theorem follows from Eqs. 43.1 and 43.2 by replacing n by L_n (see Exercises 6 and 7).

Theorem 43.3.

$$(1) \qquad 5V_nV_{n+1} = W_{n+2} - (-1)^{L_n}W_{n-1}$$

$$(2) \qquad W_nW_{n+1} = W_{n+2} + (-1)^{L_n}W_{n-1}$$ ■

The following result follows from this theorem.

Corollary 43.2.

$$5V_nV_{n+1} + W_nW_{n+1} = 2W_{n+2} \qquad \blacksquare$$

The next theorem also follows from Eqs. (43.1) and (43.2).

Theorem 43.4.

$$\begin{aligned} &(1) \qquad X_{n-1}X_{n+1} = W_n + (-1)^{F_{n-1}}X_n \\ &(2) \qquad 5U_{n-1}U_{n+1} = W_n - (-1)^{F_{n-1}}X_n \end{aligned} \qquad \blacksquare$$

Corollary 43.3.

$$\begin{aligned} &(1) \qquad X_{n-1}X_{n+1} + 5U_{n-1}U_{n+1} = 2W_n \\ &(2) \qquad X_{n-1}X_{n+1} - 5U_{n-1}U_{n+1} = 2(-1)^{F_{n-1}}X_n \end{aligned} \qquad \blacksquare$$

In 1967, D. A. Lind of the University of Virginia succeeded in defining both $Y_n = F_{G_n}$ and $Z_n = L_{G_n}$ recursively, where G_n denotes the nth generalized Fibonacci number. We need the following identities from Chapter 5 to pursue those Fibonacci numbers:

$$2F_{n+1} = F_n + L_n \tag{43.3}$$

$$F_{n-1} = (L_n - F_n)/2 \tag{43.4}$$

$$L_n^2 - 5F_n^2 = 4(-1)^n \tag{43.5}$$

$$2L_{n+1} = 5F_n + L_n \tag{43.6}$$

It follows from Identities (43.3) and (43.5) that

$$F_{n+1} = \frac{1}{2}\left(\sqrt{5F_n^2 + 4(-1)^n} + F_n\right) \tag{43.7}$$

and from Identities (43.5) and (43.6) that

$$L_{n+1} = \frac{1}{2}\left(\sqrt{5L_n^2 - 20(-1)^n} + L_n\right) \tag{43.8}$$

For example,

$$F_{10} = \frac{1}{2}\left(\sqrt{5F_9^2 - 4} + F_9\right) = \frac{1}{2}\left(\sqrt{5 \cdot 1156 - 4} + 34\right) = 55$$

and

$$L_{10} = \frac{1}{2}\left(\sqrt{5L_9^2 + 20} + L_9\right) = \frac{1}{2}\left(\sqrt{5 \cdot 5776 + 20} + 76\right) = 123$$

Identity (43.7) implies that

$$F_{n-1} = \frac{1}{2}\left(\sqrt{5F_n^2 + 4(-1)^n} - F_n\right) \tag{43.9}$$

We require two more identities from Chapter 5:

$$F_{m+n+1} = F_m F_n + F_{m+1} F_{n+1} \tag{43.10}$$

$$L_{m+n+1} = F_m L_n + F_{m+1} L_{n+1} \tag{43.11}$$

Finally, let $s(n) = n^2 - 3\lfloor n^2/3 \rfloor$. Clearly,

$$s(n) = \begin{cases} 0 & \text{if } 3|n \\ 1 & \text{otherwise} \end{cases}$$

See Exercise 16. So $(-1)^{s(n)} = (-1)^{F_n} = (-1)^{L_n} = (-1)^{G_n}$ (see Exercise 18).

A RECURSIVE DEFINITION OF Y_n

Next we develop a recursive definition of Y_n. We have:

$$\begin{aligned} Y_{n+2} &= F_{G_{n+2}} = F_{(G_{n+1}-1)} F_{G_n} + F_{G_{n+1}} F_{(G_n+1)} \qquad \text{By Formula (43.10)} \\ &= \frac{1}{2}\left[Y_n \sqrt{5Y_{n+1}^2 + 4(-1)^{G_{n+1}}} + Y_{n+1}\sqrt{5Y_n^2 + 4(-1)^{G_n}}\right] \end{aligned}$$

Since $Y_1 = F_{G_1} = F_a$ and $Y_2 = F_{G_2} = F_b$, Y_n can be defined recursively as follows:

$$\begin{aligned} Y_1 &= F_a, Y_2 = F_b \\ Y_{n+2} &= \frac{1}{2}\left[Y_n \sqrt{5Y_{n+1}^2 + 4(-1)^{G_{n+1}}} + Y_{n+1}\sqrt{5Y_n^2 + 4(-1)^{G_n}}\right] \end{aligned} \tag{43.12}$$

where $n \geq 1$.

A RECURSIVE DEFINITION OF U_n

In particular, let $a = 1 = b$. Then $Y_n = U_n$. Accordingly, we have the following recursive definition of U_n:

$$\begin{aligned} U_1 &= 1 = U_2 \\ U_{n+2} &= \frac{1}{2}\left[U_n \sqrt{5U_{n+1}^2 + 4(-1)^{s(n+1)}} + U_{n+1}\sqrt{5U_n^2 + 4(-1)^{s(n)}}\right] \qquad n \geq 1 \end{aligned}$$

For example, let $n = 6$. Then $U_6 = F_8 = 21$ and $U_7 = F_{13} = 233$;

$$\begin{aligned}
\therefore\ U_8 &= \frac{1}{2}\left[U_6\sqrt{5U_7^2 - 4} + U_7\sqrt{5U_6^2 + 4}\right] \\
&= \frac{1}{2}\left[21\sqrt{5\cdot 54289 - 4} + 233\sqrt{5\cdot 441 + 4}\right] \\
&= \frac{1}{2}(10941 + 10951) = 10{,}946
\end{aligned}$$

A RECURSIVE DEFINITION OF V_n

Substituting $a = 1$ and $b = 3$ yields the following recursive definition of V_n:

$$V_1 = 1 \qquad V_2 = 4$$

$$V_{n+2} = \frac{1}{2}\left[V_n\sqrt{5V_{n+1}^2 + 4(-1)^{s(n+1)}} + V_{n+1}\sqrt{5V_n^2 + 4(-1)^{s(n)}}\right] \qquad n \geq 1$$

For example, let $n = 5$. Then $U_5 = F_{11} = 89$ and $V_6 = F_{18} = 2584$;

$$\begin{aligned}
\therefore\ V_7 &= \frac{1}{2}\left[V_5\sqrt{5V_6^2 + 4} + V_6\sqrt{5V_5^2 - 4}\right] \\
&= \frac{1}{2}\left[89\sqrt{5\cdot 6{,}677{,}056 + 4} + 2584\sqrt{5\cdot 7921 - 4}\right] \\
&= \frac{1}{2}(514{,}242 + 514{,}216) = 514{,}229
\end{aligned}$$

A RECURSIVE DEFINITION OF Z_n

Using Formulas (43.4), (43.5), (43.8), and (43.11), we can develop a recursive definition Z_n:

$$\begin{aligned}
Z_{n+2} &= L_{G_{n+2}} = L_{G_{n+1}+G_n} \\
&= F_{(G_{n+1}-1)}L_{G_n} + F_{G_{n+1}}L_{(G_n+1)} && \text{By Eq. (43.11)} \\
&= \frac{1}{2}\left(L_{G_{n+1}} - F_{G_{n+1}}\right)L_{G_n} + F_{G_{n+1}}L_{G_n+1} && \text{By Eq. (43.2)} \\
&= \frac{1}{2}L_{G_n}L_{G_{n+1}} - \frac{1}{2}L_{G_n}F_{G_{n+1}} + F_{G_{n+1}}L_{G_n+1} \\
&= \frac{1}{2}L_{G_n}L_{G_{n+1}} + \frac{1}{2}F_{G_n+1}(2L_{G_{n+1}} - L_{G_n}) \\
&= \frac{1}{2}\left\{Z_nZ_{n+1} + \sqrt{[Z_{n+1}^2 - 4(-1)^{G_{n+1}}][Z_n^2 - 4(-1)^{G_n}]}\right\}
\end{aligned}$$

Accordingly we can define Z_n as follows:

$$Z_1 = L_a \qquad Z_2 = L_b$$

$$Z_{n+2} = \frac{1}{2}\left\{Z_n Z_{n+1} + \sqrt{[Z_{n+1}^2 - 4(-1)^{G_{n+1}}][Z_n^2 - 4(-1)^{G_n}]}\right\} \tag{43.13}$$

A RECURSIVE DEFINITION OF X_n

When $a = 1 = b$, Formula (43.13) yields the recursive definition of X_n:

$$X_1 = 1 \qquad X_2 = 3$$

$$X_{n+2} = \frac{1}{2}\left\{X_n X_{n+1} + \sqrt{[X_{n+1}^2 - 4(-1)^{s(n+1)}][X_n^2 - 4(-1)^{s(n)}]}\right\} \qquad n \geq 1$$

For example, let $n = 5$. Then $X_5 = L_5 = 11$, and $X_6 = L_8 = 47$. Therefore,

$$\begin{aligned} X_7 &= \frac{1}{2}[X_5 X_6 + \sqrt{(X_6^2 - 4)(X_5^2 + 4)}] \\ &= \frac{1}{2}[11 \cdot 47 + \sqrt{(2209 - 4)(12 + 4)}] = 521 \end{aligned}$$

A RECURSIVE DEFINITION OF W_n

When $T_n = L_n$, Formula (43.11) yields the recursive definition of W_n:

$$W_1 = 1 \qquad W_2 = 4$$

$$W_{n+2} = \frac{1}{2}\left\{W_n W_{n+1} + \sqrt{[W_{n+1}^2 - 4(-1)^{s(n+1)}](W_n^2 - 4(-1)^{s(n)}]}\right\} \qquad n \geq 1$$

Using this definition, we can verify that $W_6 = 5778$.

EXERCISES 43

1. Verify Theorem 43.1 for $n = 5$.
2. Prove Theorem 43.2.
3. Verify Theorem 43.2 for $n = 5$.
4. Verify Corollary 43.1 for $n = 6$.
5. Verify Theorem 43.3 for $n = 5$.
6. Prove Part 1 of Theorem 43.3.
7. Prove Part 2 of Theorem 43.3.
8. Verify Part 1 of Theorem 43.4 for $n = 5$.

9. Verify Part 2 of Theorem 43.4 for $n = 5$.
10. Prove Part 1 of Theorem 43.4.
11. Prove Part 2 of Theorem 43.4.
12. Compute F_{13} using Identity (43.7).
13. Compute L_{13} using Identity (43.8).
14. Prove Identity (43.7).
15. Prove Identity (43.8).
16. Let $s(n) = n^2 - 3\lfloor n^2/3 \rfloor$, where n is an integer. Show that $s(n) = \begin{cases} 0 & \text{if } 3|n \\ 1 & \text{otherwise.} \end{cases}$
17. Prove that $2|L_n$ if and only if $3|n$.
18. Prove that $(-1)^{s(n)} = (-1)^{F_n} = (-1)^{L_n}$.

Compute each.

19. U_7
20. V_6
21. X_8
22. W_5

GAUSSIAN FIBONACCI AND LUCAS NUMBERS

In 1963, A. F. Horadam examined Fibonacci numbers on the complex plane and established some interesting properties about them. Two years later, J. H. Jordan of Washington State University followed up with a study of his own. We now briefly introduce these numbers.

GAUSSIAN NUMBERS

Gaussian numbers were investigated in 1832 by Gauss. A *Gaussian number* is a complex number $z = a + ib$, where a and b are integers. Its *norm*, $\|z\|$, is defined by $\|z\| = a^2 + b^2$; it is the square of the distance of z from the origin on the complex plane: $\|z\| = |z|^2$.

The norm function satisfies the following fundamental properties:

- $\|z\| \geq 0$
- $\|z\| = 0$ if and only if $z = 0$
- $\|wz\| = \|w\| \cdot \|z\|$

We leave it as an exercise to verify these (see Exercises 1–3).

GAUSSIAN FIBONACCI AND LUCAS NUMBERS

Gaussian Fibonacci numbers (GFNs) f_n are defined by $f_n = f_{n-1} + f_{n-2}$, where $f_0 = i$, $f_1 = 1$, and $n \geq 2$.

The first six GFNs are 1, $1+i, 2+i, 3+2i, 5+3i$, and $8+5i$. Clearly, $f_n = F_n + iF_{n-1}$. Consequently, $\|f_n\| = F_n^2 + F_{n-1}^2 = F_{2n-1}$.

Gaussian Lucas numbers (GLNs) l_n are defined by $l_n = l_{n-1} + l_{n-2}$, where $l_0 = 2-i, l_1 = 1+2i$, and $n \geq 2$. The first six GLNs are $1+2i, 3+i, 4+3i, 7+4i, 11+7i$, and $18+11i$. It is easy to see that $l_n = L_n + iL_{n-1}$. Also $\|l_n\| = L_n^2 + L_{n-1}^2$.

The identities we established in Chapter 5 can be extended to Gaussian Fibonacci and Lucas numbers, also. A few are given below, and others will be found as exercises. They can be validated using the principle of mathematics induction (PMI).

Theorem 44.1.

$$\sum_{0}^{n} f_i = f_{n+2} - 1$$

Proof. (by PMI) The formula is clearly true when $n = 0$, so assume it is true for an arbitrary integer $k \geq 0$. Then

$$\begin{aligned}\sum_{0}^{k+1} f_i &= \sum_{0}^{k} f_i + f_{k+1} \\ &= (f_{k+2} - 1) + f_{k+1} \\ &= f_{k+3} - 1\end{aligned}$$

Thus the result is true for every $n \geq 0$. ■

The next two theorems also can be established fairly easily using PMI, so we omit their proofs.

Theorem 44.2.

$$\sum_{0}^{n} l_i = l_{n+2} - (l + 2i)$$ ■

Theorem 44.3.

$$f_{n-1}f_{n+1} - f_n^2 = (2-i)(-1)^n \qquad n \geq 1$$ ■

We can extend the definitions of GFNs and GLNs to negative subscripts also:

$$f_{-n} = F_{-n} + iF_{-n-1} = (-1)^{n-1} + i(-1)^n F_{n+1} = (-1)^{n-1}(F_n - iF_{n+1})$$

Likewise,

$$l_{-n} = L_{-n} + iL_{-n-1} = (-1)^n + i(-1)^{n+1} L_{n+1} = (-1)^n (L_n - iL_{n+1})$$

For example, $f_{-3} = (-1)^2(F_3 - iF_4) = 2 - 3i$ and $l_{-4} = (-1)^4(L_4 - iL_5) = 7 - 11i$.

In Chapter 5, we found that $L_n^2 - 5F_n^2 = 4(-1)^n$. The next theorem shows the corresponding result for Gaussian Fibonacci and Lucas numbers.

Theorem 44.4.

$$l_n^2 - 5f_n^2 = 4(2-i)(-1)^n \qquad \blacksquare$$

This result has an interesting by-product. Since $5 = (2-i)(2+i)$, it follows that $2 - i|l_n^2$, so $2 - i|l_n$, where $n \geq 2$. Accordingly, we have the following result.

Corollary 44.1. (Jordan, 1965) l_n is composite for $n \geq 2$. $\blacksquare$

For example, $l_5 = 11 + 7_i = (2-i)(3+5i)$, so $2 - i|11 + 7i$.

Since $l_{-n} = (-1)^n(L_n - iL_{n+1})$, it follows that $2 - i|l_{-n}$, where $n \geq 2$. This gives the following result.

Corollary 44.2. l_n is composite for all integers $n \neq \pm 1$. $\blacksquare$

For instance, $l_{-5} = -11 + 18i = (2-i)(-8+5i)$, so $2 - i|l_{-5}$.

In Chapter 16, we found that $F_m|F_n$ if and only if $m|n$, where $m \geq 2$. There is a corresponding result for GFNs. To establish it, we need the next lemma.

Lemma 44.1. (Jordan, 1965) If $2m - 1|2n - 1$, then $2m - 1|m + n - 1$.

Proof. Suppose $2m - 1|2n - 1$. Then $2m - 1|[(2n-1) - (2m-1)]$; that is, $2m - 1|2n - 2m$. But $(2m - 1, 2) = 1$, so $2m - 1|(n - m)$. Therefore, $2m - 1|[(2n-1) - (n-m)]$; that is, $2m - 1|m + n - 1$. $\blacksquare$

Theorem 44.5. (Jordan, 1965) Let $m > 2$. Then $f_m|f_n$ if and only if $2m - 1|2n - 1$.

Proof. Suppose $f_m|f_n$. Then $\|f_m\|\,|\,\|f_n\|$; that is, $F_{2m-1}|F_{2n-1}$. So, by Corollary 16.2, $2m - 1|2n - 1$.

Conversely, let $2m - 1|2n - 1$. Then $F_{2m-1}|F_{2n-1}$, by Corollary 16.2. Therefore,

$$\left\|\frac{f_n}{f_m}\right\| = \frac{\|f_n\|}{\|f_m\|} = \frac{F_{2n-1}}{F_{2m-1}}$$

is a positive integer (see Exercise 4). But

$$\begin{aligned}
\frac{f_n}{f_m} &= \frac{F_n + iF_{n-1}}{F_m + iF_{m-1}} = \frac{F_mF_n + F_{m-1}F_{n-1} + i(F_{m-1}F_n - F_mF_{n-1})}{F_m^2 + F_{m-1}^2} \\
&= \frac{F_mF_n + F_{m-1}F_{n-1}}{F_{2m-1}} + i\frac{F_{m-1}F_n - F_mF_{n-1}}{F_{2m-1}} \qquad \text{By Identity (5.11).} \\
&= \frac{F_{m+n-1}}{F_{2m-1}} + i\frac{F_{m-1}F_n - F_mF_{n-1}}{f_{2m-1}} \qquad \text{By Identity (32.4).}
\end{aligned}$$

By Lemma 44.1 and Corollary 16.2, F_{m+n-1}/F_{2m-1} is a positive integer. Since $\|f_n\|/\|f_m\|$ is also an integer, it follows that $(F_{m-1}F_{n-1} - F_mF_{n-1})/F_{2m-1}$ also must be an integer. So f_n/f_m is a Gaussian integer. Thus $f_m|f_n$. ■

For example, let $m = 3$ and $n = 8$. Then $2m - 1|2n - 1$. Notice that $f_8 = 21 + 13i = (2+i)(11+i) = (11+i)f_3$. So $f_3|f_8$, as expected.

Theorem 44.5 has an interesting by-product.

Corollary 44.3. Let $m \geq 2$. Then $F_{2m-1}|(F_{m-1}F_n - F_mF_{n-1})$ if and only if $2m - 1|2n - 1$. ■

For example, with $m = 3$ and $n = 8$, $F_{2m-1} = F_5 = 5$ and $F_{m-1}F_n - F_mF_{n-1} = F_2F_8 - F_3F_7 = 1 \cdot 21 - 2 \cdot 13 = -5$, so clearly, $F_{2m-1}|(F_{m-1}F_n - F_mF_{n-1})$.

Before presenting the next result, we need to extend the definition of gcd to Gaussian integers.

GCD OF GAUSSIAN INTEGERS

The Gaussian integer y is the *gcd of the Gaussian integers* w and z if:

- $y|w$ and $y|z$
- If $x|w$ and $x|z$, then $\|x\| \leq \|y\|$

The gcd is denoted by $y = (w, z)$. For example, $(1 + 7i, 2 + 9i) = 1 + 2i$.

Theorem 44.6. (Jordan, 1965). $(f_m, f_n) = f_k$, where $2k - 1 = (2m - 1, 2n - 1)$.

Proof. Since $2k - 1|2m - 1$ and $2k - 1|2n - 1$, it follows by Theorem 44.5 that $f_k|f_m$ and $f_k|f_n$; therefore, $f_k|(f_m, f_n)$.

Suppose $x|f_m$ and $x|f_n$. Then $\|x\|\,|\,\|f_m\|$ and $\|x\|\,|\,\|f_n\|$; that is, $\|x\|\,|F_{2m-1}$ and $\|x\|\,|F_{2n-1}$. Therefore, $\|x\|\,|(F_{2m-1}, F_{2n-1})$; that is, $\|x\|\,|F_{(2m-1,2n-1)}$. In other words, $\|x\|\,|F_{2k-1}$; that is, $\|x\|\,|f_k$. Consequently, $\|x\| \leq \|f_k\|$. Thus $(f_m, f_n) = f_k$. ■

For example, let $m = 5$ and $n = 8$. Then $(2m - 1, 2n - 1) = 3 = 2k - 1$, where $k = 2$. We have $f_5 = 5 + 3i$, $f_8 = 21 + 13i$, and $f_2 = 1 + i$. Notice that $f_5 = (1+i)(4-i)$ and $f_8 = (1+i)(17-4i)$, so $(f_5, f_8) = 1+i = f_2$, as expected.

EXERCISES 44

Prove each, where w and z are Gaussian numbers.

1. $\|z\| \geq 0$
2. $\|z\| = 0$ if and only if $z = 0$.
3. $\|wz\| = \|w\| \cdot \|z\|$

4. $\|w/z\| = \|w\|/\|z\|$
5. $f_n = F_n + iF_{n-1}, n \geq 1$.
6. $l_n = L_n + iL_{n-1}, n \geq 1$.
7. $\|\bar{z}\| = \|z\|$, where $\bar{z}$ denotes the complex conjugate of z.
8. Compute f_{10} and l_{10}.
9. Compute f_{-10} and l_{-10}.
10. Verify Theorem 44.5 for $m = 4$ and $n = 7$.
11. Verify Corollary 44.3 for $m = 4$ and $n = 11$.

Prove each (Jordan, 1965).

12. $\sum_{0}^{n} l_i = l_{n+2} - (1 + 2i)$
13. $l_{n-1}l_{n+1} - l_n^2 = 5(2 - i)(-1)^{n+1}$
14. $f_{n+1} + f_{n-1} = l_n$
15. $f_n^2 + f_{n+1}^2 = (1 + 2i)f_{2n}$
16. $f_{n+1}^2 - f_{n-1}^2 = (1 + 2i)f_{2n-1}$
17. $f_n l_n = (1 + 2i)f_{2n-1}$
18. $f_{m+1}f_{n+1} + f_m f_n = (1 + 2i)f_{m+n}$
19. $l_n^2 - 5f_n^2 = 4(2 - i)(-1)^n$
20. $\sum_{1}^{n} f_i^2 = (1 + 2i)f_n^2 + i(-1)^n - i$
21. $\sum_{1}^{n} f_{2i-1} = f_{2n} - i$
22. $\sum_{1}^{n} f_{2i} = f_{2n+1} - 1$
23. $\sum_{1}^{2n} (-1)^i f_i = f_{2n-1} + i - 1$
24. $\sum_{1}^{n} (-1)^i f_i = (-1)^{n+1} f_n + i - 1$

Let $C_n = F_n + iF_{n+1}$. Prove each, where $\overline{C}_n$ denotes the complex conjugate of C_n.

25. $C_n\overline{C}_n = F_{2n+1}$
26. $C_n\overline{C}_{n+1} = F_{2n+2} + i(-1)^n$

ANALYTIC EXTENSIONS

In 1966, Whitney investigated analytic generalizations of both Fibonacci and Lucas numbers, by extending Binet's formulas to the complex plane:

$$f(z) = \frac{\alpha^z - \beta^z}{\alpha - \beta} \qquad \text{and} \qquad l(z) = \alpha^z + \beta^z$$

where z is an arbitary complex variable; $f(z)$ is the *complex Fibonacci function*; and $l(z)$ the *complex Lucas function*. Notice that $f(n) = F_n$ and $l(n) = L_n$, where n is an integer. Both functions possess several interesting properties. We will examine a few of them here.

PERIODICITY OF α^z AND β^z

Let p be the period of α^z. Then $\alpha^{z+p} = \alpha^z$, so $\alpha^p = 1 = e^{2\pi i}$. Thus $p = 2\pi i/\ln\alpha$.

On the other hand, let q be the period of β^z. Then $\beta^{z+q} = \beta^z$ implies $\beta^q = 1 = e^{2\pi i}$. Since $\beta < 0$, we rewrite $\beta = e^{\pi i}(-\beta)$. Then $e^{q\pi i}(-\beta)^q = e^{2\pi i}$, so $(-\beta)^q = e^{\pi(2-q)i}$. That is, $\alpha^q = e^{\pi(q-2)i}$, so $q\ln\alpha = \pi(q-2)i$;

$$\therefore \quad q = \frac{2\pi i}{\pi i - \ln\alpha} = \frac{2\pi(\pi - i\ln\alpha)}{\ln^2\alpha + \pi^2}$$

Thus α^z is periodic with period $(2\pi i/\ln\alpha)$ and β^z with periodic $[2\pi(\pi - i\ln\alpha)]/(\ln^2\alpha + \pi^2)$.

PERIODICITY OF $f(z)$ AND $l(z)$

Are $f(z)$ and $l(z)$ periodic? If yes, what are their periods? To answer these questions, suppose $f(z)$ is periodic with period w. Then $f(0) = 0 = f(w)$, which implies $\alpha^w = \beta^w$. So $f(z+w) = f(z)$ yields $\alpha^{z+w} - \beta^{z+w} = \alpha^z - \beta^z$; that is, $\alpha^w(\alpha^z - \beta^z) = \alpha^z - \beta^z$.

Therefore, $\alpha^w = 1$. This implies that the real part x of $w = x + iy$ must be zero. Then $\alpha^{yi} = 1$. But this is possible only if $y = 0$. Then $w = 0 + 0i = 0$, which is a contradiction. Thus $f(z)$ is not periodic. Likewise, we can show that $l(z)$ also is not periodic (see Exercise 1).

ZEROS OF $f(z)$ AND $l(z)$

Next we pursue the zeros of $f(z)$ and $l(z)$. First, notice that $f(z)$ has a real zero, namely, 0, and $l(z)$ has no real zeros.

To find the complex zeros of $f(z)$, let $f(z) = 0$. This yields $(\alpha/\beta)^z = 1 = e^{2k\pi i}$, where k is an arbitrary integer. Then $z \ln(\alpha/\beta) = 2k\pi i$. But $\beta = e^{\pi i}(-\beta)$, so $\alpha/\beta = \alpha^2 e^{-\pi i}$ and $\ln \alpha/\beta = 2\ln\alpha - i\pi$. Thus:

$$\begin{aligned} z(2\ln\alpha - i\pi) &= 2k\pi i \\ z &= \frac{2k\pi i}{2\ln\alpha - i\pi} \\ &= \frac{2k\pi i(2\ln\alpha + i\pi)}{4\ln^2\alpha + \pi^2} \\ &= \frac{2k\pi(-\pi + 2i\ln\alpha)}{4\ln^2\alpha + \pi^2} \end{aligned}$$

where k is an arbitrary integer. This equation gives the infinitely many complex zeros of $f(z)$.

Similarly, we can show that the complex zeros of $l(z)$ are given by

$$z = \frac{(2k+1)\pi(-\pi + 2i\ln\alpha)}{4\ln^2\alpha + \pi^2}$$

See Exercise 2.

BEHAVIOR OF $f(z)$ AND $l(z)$ ON THE REAL AXIS

To see how the two functions behave on the real axis, let $z = x$, an arbitrary real number. Then $\alpha^z = \alpha^x$ and $\beta^z = \beta^x = e^{x\ln\beta} = e^{x(\pi i - \ln\alpha)} = e^{-x\ln\alpha}(\cos\pi x + i\sin\pi x)$, since $\ln\beta = -\ln\alpha$. Since $\operatorname{Im} f(z) = 0 = \operatorname{Im} l(z)$, this yields $e^{-x\ln\alpha}\sin\pi x = 0$, so $\sin\pi x = 0$. This implies that x must be an integer n. Thus $f(z)$ is an integer if and only if z is an integer n. The same is true for $l(z)$.

IDENTITIES SATISFIED BY $f(z)$ AND $l(z)$

Many of the properties of F_n and L_n that we investigated in Chapter 5 have their counterparts on the complex plane. Some of these identities are listed below.

(1) $f(z+2) = f(z+1) + f(z)$

(2) $l(z+2) = l(z+1) + l(z)$

(3) $f(z-1)f(z+1) - f^2(z) = e^{\pi zi}$
(4) $l^2(z) - 5f^2(z) = 4e^{\pi zi}$
(5) $f(-z) = -f(z)e^{\pi zi}$
(6) $l(-z) = l(z)e^{\pi zi}$
(7) $f(2z) = f(z)l(z)$
(8) $f(z+w) = f(z)f(w+1) + f(z-1)f(w)$
(9) $f(3z) = f^3(z+1) + f^3(z) - f^3(z-1)$

We can establish these properties using Binet's formulas. For example,

$$\begin{aligned} 5[f(z-1)f(z+1) - f^2(z)] &= (\alpha^{z-1} - \beta^{z-1})(\alpha^{z+1} - \beta^{z+1}) - (\alpha^z - \beta^z)^2 \\ &= \alpha^{2z} + \beta^{2z} - (\alpha\beta)^z(\alpha^2 + \beta^2) \\ &\quad -[\alpha^{2z} + \beta^{2z} - 2(\alpha\beta)^z] \\ &= 3(-1)^z + 2(-1)^z = 5e^{\pi zi} \end{aligned}$$

Thus $f(z-1)f(z+1) - f^2(z) = e^{\pi zi}$, as expected. We leave the proofs of the others as routine exercises (see Exercises 3–10).

TAYLOR EXPANSIONS OF *f*(*z*) AND *l*(*z*)

Since both $f(z)$ and $l(z)$ are entire functions, both have Taylor expansions. Since $(d^k/dw^k)(\alpha^w) = \alpha^w \cdot \ln^k \alpha$, it follows that

$$\begin{aligned} f(z) &= \sum_{k=0}^{\infty} \frac{f^{(k)}(w)}{k!}(z-w)^k \\ &= \frac{1}{\sqrt{5}} \sum_{k=0}^{\infty} \frac{\alpha^w(\ln^k \alpha) - \beta^w(\ln^k \beta)}{k!}(z-w)^k \qquad (45.1) \end{aligned}$$

Likewise,

$$l(z) = \frac{1}{\sqrt{5}} \sum_{k=0}^{\infty} \frac{\alpha^w(\ln^k \alpha) + \beta^w(\ln^k \beta)}{k!}(z-w)^k \qquad (45.2)$$

In particular, let $z = n$ and $w = n-1$. Then Eq. (45.1) yields an interesting infinite series expansion:

$$F_n = \frac{1}{\sqrt{5}} \sum_{k=0}^{\infty} \frac{\alpha^{n-1}(\ln^k \alpha) - \beta^{n-1}(\ln^k \beta)}{k!}$$

Similarly,

$$L_n = \frac{1}{\sqrt{5}} \sum_{k=0}^{\infty} \frac{\alpha^{n-1}(\ln^k \alpha) + \beta^{n-1}(\ln^k \beta)}{k!}$$

EXERCISES 45

1. Prove that $l(z)$ is not periodic.
2. Find the complex zeros of $l(z)$.

3–10. Prove the Identities 1, 2, and 4–9.

Prove each.

11. $F_n = \frac{1}{\sqrt{5}} \sum_{k=0}^{\infty} \frac{(\ln^k \alpha - \ln^k \beta)n^k}{k!}$
12. $L_n = \frac{1}{\sqrt{5}} \sum_{k=0}^{\infty} \frac{(\ln^k \alpha + \ln^k \beta)n^k}{k!}$

TRIBONACCI NUMBERS

In the case of Fibonacci and Lucas numbers, every element, except for the first two, can be obtained by adding its two immediate predecessors. Now, suppose we are given three initial conditions and add the three immediate predecessors to compute their successor in a number sequence. Such a sequence is the *tribonacci sequence*, originally studied in 1963 by M. Feinberg when he was a 14-year-old ninth grader at Susquehanna Township Junior High School in Pennsylvania (1963a).

TRIBONACCI NUMBERS

The *tribonacci numbers* T_n are defined by the recurrence relation

$$T_n = T_{n-1} + T_{n-2} + T_{n-3} \tag{46.1}$$

where $T_1 = 1 = T_2$, $T_3 = 2$, and $n \geq 4$.

The first twelve tribonacci numbers are:

1, 1, 2, 4, 7, 13, 24, 44, 81, 149, 274, and 504

Just as the ratios of consecutive Fibonacci numbers and those of Lucas numbers converge to the Golden Ratio, α, the tribonacci ratios T_{n+1}/T_n converge to the irrational number $1.83928675521416\cdots$.

TRIBONACCI ARRAYS

In the same way that we can extract Fibonacci numbers from the rising diagonals of Pascal's triangle, so can we obtain the various tribonacci numbers by computing the sums of the elements on the rising diagonals of a similar triangular array, a *tribonacci array*. Every element $t(m,n)$ of the array is defined as follows, where $m, n \geq 0$:

$$t(m,n) = 0 \quad \text{if} \quad m > n \text{ or } \quad n < 0$$

$$t(m,m) = 1$$

$$t(m,n) = t(m-1,n-1) + t(m-1,n) + t(m-2,n-1) \quad \text{if} \quad m \geq 2$$

Using the recurrence relation, we can obtain every element of the array by adding the neighboring elements from the two preceding arrays. Figure 46.1 shows the resulting array. Notice that the rising diagonal sums of this array are indeed tribonacci numbers.

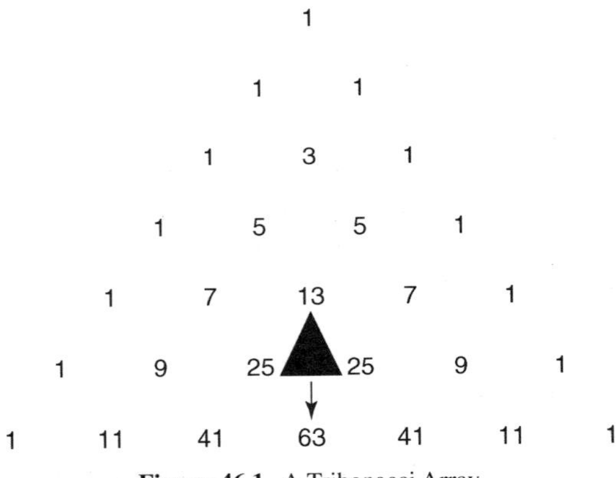

Figure 46.1. A Tribonacci Array.

There is yet another triangular array that also yields the various tribonacci numbers. To construct this array, first we find the trinomial expansions of $(1 + x + x^2)^n$ for several values of $n \geq 0$:

$$(1 + x + x^2)^0 = 1$$

$$(1 + x + x^2)^1 = 1 + x + x^2$$

$$(1 + x + x^2)^2 = 1 + 2x + 3x^2 + 2x^3 + x^4$$

$$(1 + x + x^2)^3 = 1 + 3x + 6x^2 + 7x^3 + 6x^4 + 3x^5 + x^6$$

$$(1 + x + x^2)^4 = 1 + 4x + 10x^2 + 16x^3 + 19x^4 + 16x^5 + 10x^6 + 4x^7 + x^8$$

Now arrange the coefficients in the various expansions to form a left-justified triangular array, as shown below:

$$\begin{array}{ccccccccccccc}
1 \\
1 & 1 & 1 \\
1 & 2 & 3 & 2 & 1 \\
1 & 3 & 6\text{—} & 7\text{—} & 6 & 3 & 1 \\
 & & & & \downarrow \\
1 & 4 & 10 & 16 & 19 & 16 & 10 & 4 & 1 \\
1 & 5 & 15 & 30 & 45 & 51 & 45 & 30 & 15 & 5 & 1 \\
1 & 6 & 21 & 50 & 90 & 126 & 141 & 126 & 90 & 50 & 21 & 6 & 1
\end{array}$$

Obviously, every row is symmetric. With the exception of the first two rows, every row can be obtained from the preceding row; see the arrow in the array. The rising diagonal sums of this trinomial coefficient array also yield the tribonacci numbers.

COMPOSITIONS WITH SUMMANDS 1, 2, AND 3

In Chapters 4 and 20, we found that the number of compositions of a positive integer n using the summands 1 and 2 is the Fibonacci number F_{n+1}. Suppose that we permit the numbers 1, 2, and 3 as summands. What can we say about the number of compositions C_n?

To answer this, let us first find the compositions of the integers 1 through 5, summarize the data in a table, and then look for any obvious patterns. It appears from Table 46.1 that $C_n = T_{n+1}$, where $n \geq 1$. Fortunately, this is indeed the case.

TABLE 46.1.

n	Compositions of n Using 1, 2, and 3	C_n
1	1	1
2	1 + 1, 2	2
3	1 + 1 + 1, 1 + 2, 2 + 1, 3	4
4	1 + 1 + 1 + 1, 1 + 1 + 2, 1 + 2 + 1, 2 + 1 + 1, 2 + 2, 1 + 3, 3 + 1	7
5	1 + 1 + 1 + 1 + 1, 1 + 1 + 1 + 2, 1 + 1 + 3, 1 + 1 + 2 + 1, 1 + 2 + 1 + 1, 2 + 1 + 1 + 1, 1 + 3 + 1, 3 + 1 + 1, 2 + 3, 3 + 2, 1 + 2 + 2, 2 + 1 + 2, 2 + 2 + 1	13

↑
Tribonacci numbers

RECURSIVE ALGORITHM

Using the recursive definition of T_n, we can fairly easily develop a recursive algorithm for computing T_n, as Algorithm 46.1 shows.

```
Algorithm tribonacci (n)
(* This algorithm computes the first n tribonacci
   numbers using recursion, where n ≥ 4. *)
Begin (* algorithm *)
   tribonacci (1) = 1
   tribonacci (2) = 1
   tribonacci (3) = 2
   while i ≤ n do
     tribonacci (i) = tribonacci (i − 1) + tribonacci
                      (i − 2) + tribonacci (i−3)
   endwhile
End (*algorithm *)
```

Algorithm 46.1.

Next, we explore an explicit formula for the number of additions a_n needed to compute T_n recursively. For example, it takes two additions to compute T_4; that is, $a_4 = 2$.

Using the recurrence relation Eq. (46.1), we can define a_n recursively:

$$a_n = a_{n-1} + a_{n-2} + a_{n-3} + 2$$

where $n \geq 4$, and $a_1 = a_2 = a_3 = 0$. Let $b_n = a_n + 1$. Then this yields

$$b_n - 1 = b_{n-1} + b_{n-2} + b_{n-3} - 3 + 2$$
$$b_n = b_{n-1} + b_{n-2} + b_{n-3}$$

where $b_1 = b_2 = b_3 = 1$. The first 12 elements of the sequence $\{b_n\}$ are 1, 1, 1, 3, 5, 9, 17, 31, 57, 105, 193, and 355.

Note an interesting characteristic: Write these values in a row, except the first three; then write the first 10 tribonacci numbers, except the first, in a row right below. Now add the two rows:

$$\begin{array}{rrrrrrrrrrl} & 3 & 5 & 9 & 17 & 31 & 57 & 105 & 193 & 355 & \\ + & 1 & 2 & 4 & 7 & 13 & 24 & 44 & 81 & 149 & \\ \hline & 4 & 7 & 13 & 24 & 44 & 81 & 149 & 274 & 504 & \leftarrow T_n \end{array}$$

See an intriguing pattern? The resulting sums are tribonacci numbers T_n, where $n \geq 4$.

So we conjecture that $b_n + T_{n-2} = T_n$; that is, $b_n = T_n - T_{n-2} = T_{n-1} + T_{n-3}$, where $n \geq 4$. Thus we predict that $a_n = T_{n-1} + T_{n-3} - 1$, where $n \geq 4$.

The next theorem confirms this using the strong version of induction.

Theorem 46.1. Let a_n denote the number of additions needed to compute T_n recursively. Then $a_n = T_{n-1} + T_{n-3} - 1$, where $n \geq 4$.

Proof. Since $T_3 + T_1 - 1 = 2 + 1 - 1 = 2 = a_4$, the formula works when $n = 4$. Now, assume it is true for all positive integers $k \leq n$, where $k \geq 4$. Then:

$$T_{n+1} = T_n + T_{n-1} + T_{n-2}$$

So

$$\begin{aligned} a_{n+1} &= a_n + a_{n-1} + a_{n-2} + 2 \\ &= (T_{n-1} + T_{n-3} - 1) + (T_{n-2} + T_{n-4} - 1) + (T_{n-3} + T_{n-5} - 1) + 2 \\ &= (T_{n-1} + T_{n-2} + T_{n-3}) + (T_{n-3} + T_{n-4} + T_{n-5}) - 1 \\ &= T_n + T_{n-2} - 1 \end{aligned}$$

Thus, by the strong version of the principle of mathematical induction, the formula holds for every $n \geq 4$. ■

It follows by the theorem that

$$a_n = \begin{cases} 0 & \text{if } 1 \leq n \leq 3 \\ T_{n-1} + T_{n-3} - 1 & \text{otherwise} \end{cases}$$

For example, $a_6 = T_5 + T_3 - 1 = 7 + 2 - 1 = 8$ additions are needed to compute T_6 recursively.

We can represent the recursive computation of T_n pictorially by a rooted tree, as Figure 46.2 illustrates. Each internal node (see the heavy dots in the figure) of the tree represents two additions, so $a_n = 2$ (number of internal nodes of the tree rooted at T_n). For example, the tree in Figure 46.2 has four internal nodes, so it takes eight additions to compute T_6 recursively, as we just discovered.

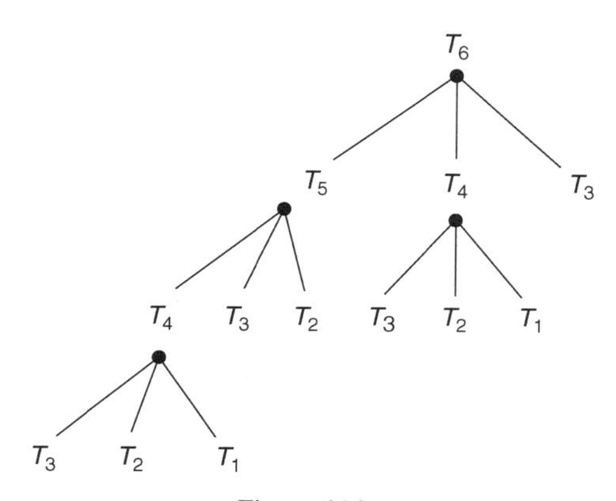

Figure 46.2.

A GENERATING FUNCTION FOR T_n

Using the recursive definition, we can develop a generating function for the tribonacci numbers. To this end, let $g(x) = \sum_0^\infty T_n x^n$. Since $T_0 = 0$, it follows that $(1 - x - x^2 - x^3)g(x) = T_1 x$; that is,

$$g(x) = \frac{x}{1 - x - x^2 - x^3}$$

Thus

$$\frac{x}{1 - x - x^2 - x^3} = \sum_0^\infty T_n x^n$$

Since

$$\frac{1}{1-u} = \sum_0^\infty u^n$$

it follows that

$$\frac{x}{1 - x - x^2 - x^3} = x + x^2 + 2x^3 + 4x^4 + \cdots$$

so the various tribonacci numbers appear as coefficients on the right-hand side.

EXERCISES 46

The Pascal-like array in Figure 46.3 can be employed to generate all tribonacci numbers.

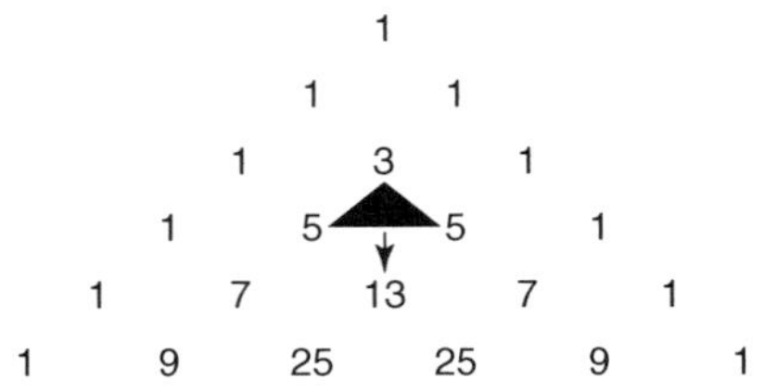

1
1 1
1 3 1
1 5 5 1
1 7 13 7 1
1 9 25 25 9 1

Figure 46.3.

1. Let $B(n, j)$ denote the element in row n and column j of the array. Define $B(n, j)$ recursively.
2. Prove that the sum of the elements on the nth rising diagonal is a tribonacci number, T_n; that is, $\sum_0^{\lfloor (n-1)/2 \rfloor} B(n, j) = T_n$.

TRIBONACCI POLYNOMIALS

In 1973, V. E. Hoggatt, Jr., and M. Bicknell generalized Fibonacci polynomials to *tribonacci polynomials* $t_n(x)$. These are defined by

$$t_n(x) = x^2 t_{n-1}(x) + x t_{n-2}(x) + t_{n-3}(x)$$

where $t_0(x) = 0$, $t_1(x) = 1$, and $t_2(x) = x^2$. Notice that $t_n(1) = T_n$, the nth tribonacci number.

The first ten tribonacci polynomials are:

$$\begin{aligned}
t_1(x) &= 1 \\
t_2(x) &= x^2 \\
t_3(x) &= x^4 + x \\
t_4(x) &= x^6 + 2x^3 + 1 \\
t_5(x) &= x^8 + 3x^5 + 3x^2 \\
t_6(x) &= x^{10} + 4x^7 + 6x^4 + 2x \\
t_7(x) &= x^{12} + 5x^9 + 10x^6 + 7x^3 + 1 \\
t_8(x) &= x^{14} + 6x^{11} + 15x^8 + 16x^5 + 6x^2 \\
t_9(x) &= x^{16} + 7x^{13} + 21x^{10} + 30x^7 + 19x^4 + 3x \\
t_{10}(x) &= x^{18} + 8x^{15} + 28x^{12} + 50x^9 + 45x^6 + 16x^3 + 1
\end{aligned}$$

TRIBONACCI ARRAY

Table 47.1 shows the corresponding left-justified array of tribonacci coefficients. As expected, the row sums yield the various tribonacci numbers.

Let $T(n, j)$ denote the element in row n and column j of this array, where $n > j \geq 0$. It satisfies the recurrence relation

$$T(n, j) = T(n-1, j) + T(n-2, j-1) + T(n-3, j-2)$$

where $n \geq 4$. See the arrows in the table.

TABLE 47.1. Tribonacci Array

n \ j								Row Sum
1	1							1
2	1							1
3	1	1						2
4	1	2	1					4
5	1	3	3					7
6	1	4	6	2				13
7	1	5	10	7	1			24
8	1	6	15	16	6			44
9	1	7	21	30	19	3		81
10	1	8	28	50	45	16	1	149

↑ tribonacci numbers

Interestingly enough, each row of the array in Table 47.1 is a rising diagonal of the triangular array of coefficients in the trinomial expansion of $(x + y + z)^n$, where $n \geq 0$. See the left-justified trinomial coefficient array in the following display.

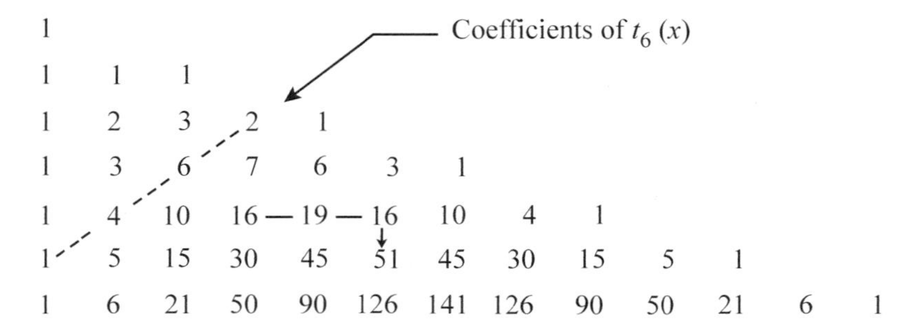

Obviously, every row is symmetric. With the exception of the first two rows, every row can be obtained from the preceding row. We can obtain the tribonacci array from this array by lowering each column one level more than the preceding column.

Consequently, the rising diagonal sums of the trinomial coefficient array also yield the tribonacci numbers, and therefore the sum of every rising diagonal is a tribonacci number.

A TRIBONACCI FORMULA

An explicit formula for the tribonacci polynomial $t_n(x)$ is given by

$$t_n(x) = \sum_{j=0}^{\lfloor (2n-2)/3 \rfloor} T(n, j)x^{2n-3j-2}$$

For example,

$$\begin{aligned} t_5(x) &= \sum_{0}^{2} T(5, j)x^{8-3j} \\ &= T(5, 0)x^8 + T(5, 1)x^5 + T(5, 2)x^2 \\ &= x^8 + 3x^5 + 3x^2 \end{aligned}$$

TRIBONACCI POLYNOMIALS AND THE Q-MATRIX

We can generate tribonacci polynomials by the Q-matrix

$$Q = \begin{bmatrix} x^2 & 1 & 0 \\ x & 0 & 1 \\ 1 & 0 & 0 \end{bmatrix}$$

Using the principle of mathematical induction, we can show that

$$Q^n = \begin{bmatrix} t_{n+1}(x) & t_n(x) & t_{n-1}(x) \\ xt_n(x) + t_{n-1}(x) & xt_{n-1}(x) + t_{n-2}(x) & xt_{n-2}(x) + t_{n-3}(x) \\ t_n(x) & t_{n-1}(x) & t_{n-2}(x) \end{bmatrix}$$

See Exercise 4.

Since $|Q| = 1$, it follows that $|Q^n| = 1$; that is,

$$\begin{vmatrix} t_{n+1}(x) & t_n(x) & t_{n-1}(x) \\ xt_n(x) + t_{n-1}(x) & xt_{n-1}(x) + t_{n-2}(x) & xt_{n-2}(x) + t_{n-3}(x) \\ t_n(x) & t_{n-1}(x) & t_{n-2}(x) \end{vmatrix} = 1$$

Now, multiply row 1 by x^2 and add to row 2; then exchange rows 1 and 2. This yields the *tribonacci polynomial identity*:

$$\begin{vmatrix} t_{n+2}(x) & t_{n+1}(x) & t_n(x) \\ t_{n+1}(x) & t_n(x) & t_{n-1}(x) \\ t_n(x) & t_{n-1}(x) & t_{n-2}(x) \end{vmatrix} = -1$$

In particular, let $x = 1$. We then get the following tribonacci identity:

$$\begin{vmatrix} T_{n+2} & T_{n+1} & T_n \\ T_{n+1} & T_n & T_{n-1} \\ T_n & T_{n-1} & T_{n-2} \end{vmatrix} = -1$$

EXERCISES 47

1. Find $t_{-1}(x)$.
2. Find $t_{11}(x)$ and $t_{12}(x)$.
3. Find Q^2 and Q^3.
4. Establish the formula for Q^n.
5. In 1973, V. E. Hoggatt, Jr., and M. Bicknell defined the nth *quadranacci number* T_n^* by $T_n^* = T_{n-1}^* + T_{n-2}^* + T_{n-3}^* + T_{n-4}^*$, where $n \geq 5$, $T_1^* = 1 = T_2^*$, $T_3^* = 2$ and $T_4^* = 4$. Compute T_5^*, T_6^*, and T_7^*.
6. In 1973, V. E. Hoggatt, Jr., and M. Bicknell also introduced a new family of polynomials t_n^*, called *quadranacci polynomials*. They are defined by $t_n^*(x) = x^3 t_{n-1}^*(x) + x^2 t_{n-2}^*(x) + x t_{n-3}^*(x) + t_{n-4}^*(x)$, where $n \geq 5$, $t_{-2}^*(x) = t_{-1}^*(x) = t_0^*(x) = 0$, and $t_1^* = 1$. Find $t_3^*(x)$, $t_4^*(x)$, $t_5^*(x)$, and $t_6^*(x)$.
7. Find $t_n^*(1)$.

The quadranacci polynomials are generated by the Q-matrix

$$Q = \begin{bmatrix} x^3 & 1 & 0 & 0 \\ x^2 & 0 & 1 & 0 \\ x & 0 & 0 & 1 \\ 1 & 0 & 0 & 0 \end{bmatrix} \quad \text{(Hoggatt and Bicknell, 1973)}$$

8. Find Q^2 and Q^3.
9. Find Q^n (Hoggatt, Jr. and Bicknell, 1973).
10. Find $|Q^n|$ (Hoggatt and Bicknell, 1973).
11. Prove that

$$\begin{vmatrix} T_{n+3}^* & T_{n+2}^* & T_{n+1}^* & T_n^* \\ T_{n+2}^* & T_{n+1}^* & T_n^* & T_{n-1}^* \\ T_{n+1}^* & T_n^* & T_{n-1}^* & T_{n-2}^* \\ T_n^* & T_{n-1}^* & T_{n-2}^* & T_{n-3}^* \end{vmatrix} = (-1)^{n+1} \quad \text{(Hoggatt and Bicknell, 1973)}$$

FUNDAMENTALS

This Appendix presents the fundamental symbols, definitions, and facts needed to pursue the essence of the theory of Fibonacci and Lucas numbers. For convenience, we have omitted examples and all proofs. (For a complete study of these fundamentals, we invite you to refer to the author's book on number theory.)

SEQUENCES

The sequence $s_1, s_2, s_3, \ldots, s_n, \ldots$ is denoted by $\{s_n\}_1^\infty$ or simply $\{s_n\}$. The nth term s_n is the *general term* of the sequence. Sequences can be classified as finite or infinite, as the next definition shows.

Finite and Infinite Sequences

A sequence is *finite* if its domain is finite; otherwise, it is *infinite*.

The Summation Notation

$$\sum_{i=k}^{i=m} a_i = a_k + a_{k+1} + \cdots + a_m$$

The variable i is the *summation index*. The values k and m are the *lower* and *upper limits* of the index i. The "i =" above the Σ is usually omitted; in fact, the indices above and below the Σ are also omitted, when there is no confusion. Thus

$$\sum_{i=k}^{i=m} a_i = \sum_{i=k}^{m} a_i = \sum_{k}^{m} a_i$$

The index i is a dummy variable; you can use any variable as the index without affecting the value of the sum, so

$$\sum_{l}^{m} a_i = \sum_{l}^{m} a_j = \sum_{l}^{m} a_k$$

The following results are extremely useful in evaluating finite sums. They can be proven using the principle of mathematical induction (PMI).

Theorem A.1. Let $n \in \mathbf{N}$ and $c \in \mathbb{R}$. Let $a_1, a_2, \ldots,$ and $b_1, b_2, \ldots,$ be any two number sequences. Then:

$$\sum_{1}^{n} c = nc$$

$$\sum_{1}^{n} (ca_i) = c\left(\sum_{1}^{n} a_i\right)$$

$$\sum_{1}^{n} (a_i + b_i) = \left(\sum_{1}^{n} a_i\right) + \left(\sum_{1}^{n} b_i\right)$$

(These results can be extended for any integral lower limit.) ■

Indexed Summation

The summation notation can be extended to sequences with *index sets* I as their domains. For instance, $\sum_{i \in I} a_i$ denotes the sum of the values a_i, as i runs over the various values in I.

Often we need to evaluate sums of the form $\sum_{P} a_{ij}$, where the subscripts i and j satisfy certain properties P.

Multiple summations often arise in mathematics. They are evaluated in the right-to-left fashion. For example, the double summation $\sum_{i}\sum_{j} a_{ij}$ is evaluated as $\sum_{i}\left(\sum_{j} a_{ij}\right)$ and the triple summation $\sum_{i}\sum_{j}\sum_{k} a_{ijk}$ as $\sum_{i}\left[\sum_{j}\left(\sum_{k} a_{ijk}\right)\right]$.

The Product Notation

The product $a_k a_{k+1} \cdots a_m$ is denoted by $\prod_{i=k}^{i=m} a_i$, where *product symbol* $\prod$ is the Greek capital letter pi. As in the case of the summation notation, the "$i =$" below and above the product symbol can be dropped if doing so leads to no confusion:

$$\prod_{k}^{m} a_i = a_k a_{k+1} \cdots a_m$$

Once again, i is just a dummy variable.

The Factorial Function

Let n be a nonnegative integer. The *factorial function* $f(n) = n!$ (read n factorial) is defined by $n! = n(n-1)\cdots 2.1$, where $0! = 1$. Thus $f(n) = n! = \prod_{1}^{n} i$.

FLOOR AND CEILING FUNCTIONS

The *floor* of a real number x, denoted by $\lfloor x \rfloor$, is the greatest integer $\leq x$. The *ceiling* of x, denoted by $\lceil x \rceil$, is the least integer $\geq x$.* The *floor function* $f(x) = \lfloor x \rfloor$ and the *ceiling function* $g(x) = \lceil x \rceil$ are also known as the *greatest integer* function and the *least integer function*, respectively.

Theorem A.2. Let x be any real number and n any integer. Then:

- $\lfloor n \rfloor = n = \lceil n \rceil$
- $\lceil x \rceil = \lfloor x \rfloor + 1 (x \notin \mathbf{Z})$
- $\lfloor x + n \rfloor = \lfloor x \rfloor + n$
- $\lceil x + n \rceil = \lceil x \rceil + n$
- $\left\lfloor \frac{n}{2} \right\rfloor = \frac{n-1}{2}$ if n is odd
- $\left\lceil \frac{n}{2} \right\rceil = \frac{n+1}{2}$ if n is odd ■

THE WELL-ORDERING PRINCIPLE

The principle of mathematical induction is a powerful proof technique we employ often. It is based on the following axiom.

The Well-Ordering Principle (WOP)

Every nonempty set of positive integers has a least element.

MATHEMATICAL INDUCTION

The principle of mathematical induction† (PMI) is a powerful proof technique we use throughout the text.

The next result is the cornerstone of the principle of induction. Its proof follows by the WOP.

*These two notations and the names, *floor* and *ceiling*, were introduced by Kenneth E. Iverson in the early 1960s. Both notations are variations of the original greatest integer notation $[x]$.

†Although the Venetian scientist Francesco Maurocylus (1491–1575) applied it in proofs in a book he wrote in 1575, the term mathematical induction was coined by the English mathematician, Augustus De Morgan (1806–1871).

Theorem A.3. Let S be a set of positive integers satisfying the following properties:

1. $1 \in S$.
2. If k is an arbitrary positive integer in S, then $k + 1 \in S$.

Then $S = \mathbb{N}$ ■

This result can be generalized, as the next theorem shows.

Theorem A.4. Let n_0 be a fixed integer. Let S be a set of integers satisfying the following conditions:

1. $n_0 \in S$.
2. If k is an arbitrary integer $\geq n_0$ such that $k \in S$, then $k+1 \in S$. Then S contains all integers $n \geq n_0$. ■

Theorem A.5. (The Principle of Mathematical Induction). Let $P(n)$ be a statement satisfying the following conditions, where $n \in \mathbf{Z}$:

1. $P(n_0)$ is true for some integer n_0.
2. If $P(k)$ is true for an arbitrary integer $k \geq n_0$, then $P(k + 1)$ is also true.

Then $P(n)$ is true for every integer $n \geq n_0$. ■

Proving a result by PMI involves two key steps:

1. *Basis Step*. Verify that $P(n_0)$ is true.
2. *Induction Step*. Assume $P(k)$ is true for an arbitrary integer $k \geq n_0$ [*inductive hypothesis* (IH)]. Then verify that $P(k + 1)$ is also true.

Summation Facts

1. $\sum_1^n i = \dfrac{n(n+1)}{2}$
2. $\sum_1^n i^2 = \dfrac{n(n+1)(2n+1)}{6}$
3. $\sum_1^n i^3 = \left[\dfrac{n(n+1)}{2}\right]^2$
4. $\sum_1^n ar^{n-1} = \dfrac{a(r^n - 1)}{r - 1}, \qquad (r \neq 1)$

We now turn to a stronger version of the PMI.

Theorem A.6. (The Second Principle of Mathematical Induction). Let $P(n)$ be a statement satisfying the following conditions, where $n \in \mathbf{Z}$:

1. $P(n_0)$ is true for some integer n_0.
2. If k is an arbitrary integer $\geq n_0$ such that $P(n_0), P(n_0 + 1), \ldots, P(k)$ are each true, then $P(k + 1)$ is also true.

Then $P(n)$ is true for every integer $n \geq n_0$. ■

RECURSION

Recursion is one of the most elegant and powerful problem-solving techniques. It is the backbone of most programming languages.

Suppose you would like to solve a complex problem. The solution may not be obvious. It may turn out, however, that the problem could be defined in terms of a simpler version of itself. Such a definition is called a *recursive definition*. Consequently, the given problem can be solved provided the simpler version can be solved.

Recursive Definition of a Function

The *recursive definition* of a function f consists of three parts, where $a \in \mathbf{W}$:

- *Basis Clause*. A few initial values of the function $f(a), f(a+1), \ldots, f(a+k-1)$ are specified. An equation that specifies such initial values is an *initial condition*.
- *Recursive Clause*. A formula to compute $f(n)$ from the k preceding functional values $f(n-1), f(n-2), \ldots, f(n-k)$ is made. Such a formula is called a *recurrence relation* (or *recursion formula*).
- *Terminal Clause*. Only values thus obtained are valid functional values. (For convenience, we drop this clause from the recursive definition.)

Thus the recursive definition of f consists of one or more (a finite number of) initial conditions and a recurrence relation.

The next theorem confirms that a recursive definition is indeed a valid definition.

Theorem A.7. Let $a \in \mathbf{W}$, $X = \{a, a+1, a+2, \ldots\}$, and $k \in \mathbb{N}$. Let $f : X \to \mathbb{R}$ such that $f(a), f(a+1), \ldots, f(a+k-1)$ are known. Let n be a positive integer $\geq a+k$ such that $f(n)$ is defined in terms of $f(n-1), f(n-2), \ldots$ and $f(n-k)$. Then $f(n)$ is defined for every integer $n \geq a$. ■

By virtue of this theorem, recursive definitions are also known as *inductive definitions*.

THE DIVISION ALGORITHM

The division algorithm is an application of the WOP and is often employed to check the correctness of a division problem.

Suppose an integer a is divided by a positive integer b. Then we get a unique *quotient*, q, and a unique *remainder*, r, where the remainder satisfies the condition $0 \leq r < b$; a is the *dividend* and b the *divisor*. This is formally stated as follows.

Theorem A.8. (The Division Algorithm). Let a be any integer and b a positive integer. Then there exist unique integers q and r such that

$$a = b \cdot q + r$$

Dividend → a; Divisor → b; Quotient → q; Remainder → r

where $0 \leq r < b$. ■

Although this theorem does not present an algorithm for finding q and r, traditionally it has been called the *division algorithm*. Integers q and r can be found using the familiar long-division method.

You can see that the equation $a = bq + r$ can be written as

$$\frac{a}{b} = q + \frac{r}{b}$$

where $0 \leq r/b < 1$. Consequently, $q = \lfloor a/b \rfloor$ and $r = a - bq$.

Div and Mod Operators

Two simple and useful operators, *div* and *mod*, are often used in discrete mathematics and computer science to find quotients and remainders:

$$a \text{ div } b = \text{quotient when } a \text{ is divided by } b.$$

$$a \text{ mod } b = \text{remainder when } a \text{ is divided by } b.$$

It now follows from these definitions that $q = a \text{ div } b = \lfloor a/b \rfloor$ and $r = a \text{ mod } b = a - bq = a - b \cdot \lfloor a/b \rfloor$.

TI-86 provides a built-in function *mod* in the MATH NUM menu.

The Divisibility Relation

Suppose $a = bq + 0 = bq$. We then say that b *divides* a, b is a *factor* of a, a is *divisible* by b, or a is a *multiple* of b, and write $b|a$. If b is not a factor of a, we write $b \nmid a$.

Divisibility Properties

Theorem A.9. Let a and b be positive integers such that $a|b$ and $b|a$. Then $a = b$. ■

Theorem A.10. Let a, b, c, s, and t be any integers. Then:

- If $a|b$ and $b|c$, then $a|c$ (*transitive property*).
- If $a|b$ and $a|c$, then $a|(sb + tc)$.
- If $a|b$, then $a|bc$. ■

The expression $sb + tc$ is called a *linear combination* of b and c. Thus, if a is a factor of b and c, then a is also a factor of any linear combination of b and c. In particular, $a|(b + c)$ and $a|(b - c)$.

The floor function can be used to determine the number of positive integers less than or equal to a positive integer a and divisible by a positive integer b, as the next theorem shows.

Theorem A.11. Let a and b be any positive integers. Then the number of positive integers $\leq a$ and divisible by b is $\lfloor a/b \rfloor$. ■

THE PIGEONHOLE PRINCIPLE

Suppose m pigeons fly into n pigeonholes to roost, where $m > n$. What is your conclusion? Since there are more pigeons than pigeonholes, at least two pigeons must roost in the same pigeonhole; in other words, there must be a pigeonhole containing two or more pigeons.

We now state the simple version of the pigeonhole principle.

Theorem A.12. (The Pigeonhole Principle). If m pigeons are assigned to n pigeonholes, where $m > n$, then at least two pigeons must occupy the same pigeonhole. ■

The pigeonhole principle is also called the *Dirichlet box principle* after the German mathematician, Peter Gustav Lejeune Dirichlet (1805–1859), who used it extensively in his work on number theory.

The pigeonhole principle can be generalized as follows.

Theorem A.13. (The Generalized Pigeonhole Principle). If m pigeons are assigned to n pigeonholes, there must be a pigeonhole containing at least $\lfloor (m - 1)/n \rfloor + 1$ pigeons. ■

THE ADDITION PRINCIPLE

Let A be a finite set and $|A|$ the number of elements in A. (Often we use the vertical bars to denote the absolute value of a number, but here it denotes the number of elements in a set. The meaning of the notation should be clear from the context.)

Union and Intersection

Let A and B be any two sets. Their *union* $A \cup B$ consists of elements belonging to A or B; their *intersection* $A \cap B$ consists of the common elements.

With this understood, we can move on to the inclusion–exclusion principle.

Theorem A.14. (The Inclusion–Exclusion Principle). Let A, B, and C be finite sets. Then $|A \cup B| = |A| + |B| - |A \cap B|$ and $|A \cup B \cup C| = |A| + |B| + |C| - |A \cap B| - |B \cap C| - |C \cap A| + |A \cap B \cap C|$. ■

This theorem can be extended to any finite number of finite sets.

Corollary A.1. (Addition Principle). Let A, B, and C be finite, pairwise disjoint sets. Then $|A \cup B| = |A| + |B| - |A \cap B|$ and $|A \cup B \cup C| = |A| + |B| + |C|$. ■

THE GREATEST COMMON DIVISOR

A positive integer can be a factor of two positive integers, a and b. Such factors are *common divisors*, or *common factors*, of a and b. Often we are in their largest common divisor.

> **The Greatest Common Divisor**: The *greatest common divisor* (GCD) of two positive integers a and b is the largest positive integer that divides both a and b; it is denoted by (a, b).

A Symbolic Definition of gcd

A positive integer d is the gcd of two positive integers a and b:

1. if $d|a$ and $d|b$.
2. if $d'|a$ and $d'|b$, then $d'|d$, where d' is also a positive integer.

Relatively Prime Integers

Two positive integers a and b are *relatively prime* if their gcd is 1; that is, if $(a, b) = 1$.

We now turn to a discussion of some interesting and useful properties of gcds.

Theorem A.15. Let $(a, b) = d$. Then $(a/d, b/d) = 1$ and $(a, a - b) = d$. ■

Linear Combination

A *linear combination* of the integers a and b is a sum of multiples of a and b, that is, a sum of the form $sa + tb$, where s and t are any integers.

Theorem A.16. The gcd of the positive integers a and b is a linear combination of a and b. ■

It follows by this theorem that the gcd (a, b) can always be expressed as a linear combination $sa + tb$. In fact, it is the smallest positive such linear combination.

Theorem A.17. Let a, b, and c be any positive integers. Then $(ac, bc) = c(a, b)$. ■

Theorem A.18. Two positive integers, a and b, are relatively prime if and only if there are integers s and t such that $sa + tb = 1$. ■

Corollary A.2. If $a|c$ and $b|c$, and $(a, b) = 1$, then $ab|c$. ■

Remember that $a|bc$ does *not* mean $a|b$ or $a|c$, although under some conditions it does. The next corollary explains when it is true.

Corollary A.3. (Euclid). If a and b are relatively prime, and if $a|bc$, then $a|c$. ■

The definition of gcd can be extended to three or more positive integers, as the next definition shows.

The gcd of *n* Positive Integers

The gcd of $n(\geq 2)$ positive integers $a_1, a_2, \ldots, a_n$ is the largest positive integer that divides each a_i. It is denoted by $(a_1, a_2, \ldots, a_n)$.

The next theorem shows how nicely recursion can be used to find the gcd of three or more integers.

Theorem A.19. Let $a_1, a_2, \ldots, a_n$ be $n(\geq 2)$ positive integers. Then $(a_1, a_2, \ldots, a_n) = ((a_1, a_2, \ldots, a_{n-1}), a_n)$. ■

The next corollary is an extension of Corollary A.3.

Corollary A.4. If $d|a_1a_2 \cdots a_n$ and $(d, a_i) = 1$ for $1 \leq i \leq n - 1$, then $d|a_n$. ■

THE FUNDAMENTAL THEOREM OF ARITHMETIC

Prime numbers are the building blocks of all integers. Integers are made up of primes, and every integer can be decomposed into primes. This result, called the *fundamental theorem of arithmetic*, is the cornerstone of number theory. It appears in Euclid's *Elements*.

Before we state it formally, we need to lay some groundwork in the form of two lemmas and a corollary. Throughout, assume all letters denote positive integers.

Lemma A.1. (Euclid). If p is a prime and $p|ab$, then $p|a$ or $p|b$. ■

The next lemma extends this result to three or more factors using PMI.

Lemma A.2. Let p be a prime and $p|a_1a_2\cdots a_n$, where $a_1, a_2, \ldots, a_n$ are positive integers, then $p|a_i$ for some i, where $1 \le i \le n$. ■

The next result follows nicely from this lemma.

Corollary A.5. If $p, q_1, q_2, \ldots, q_n$ are primes such that $p|q_1q_2\cdots q_n$, then $p = q_i$ for some i, where $1 \le i \le n$. ■

We can state the fundamental theorem of arithmetic, the most fundamental result in number theory.

Theorem A.20. (The Fundamental Theorem of Arithmetic). Every positive integer $n \ge 2$ is either a prime or can be expressed as a product of primes. The factorization into primes is unique except for the order of the factors. ■

A factorization of a composite number n in terms of primes is a *prime factorization* of n. Using the exponential notation, this product can be rewritten in a compact way. Such a product is the *prime-power decomposition* of n; if the primes occur in increasing order, then it is the *canonical decomposition*.

Canonical Decomposition

The *canonical decomposition* of a positive integer n is of the form $n = p_1^{a_1} p_2^{a_2} \cdots p_k^{a_k}$ where $p_1, p_2, \ldots, p_k$ are distinct primes with $p_1 < p_2 < \cdots < p_r$ and each exponent a_i is a positive integer.

THE LEAST COMMON MULTIPLE

The least common multiple of two positive integers, a and b, is closely related to their gcd. We explore two methods for finding the least common multiple of a and b.

Least Common Multiple: The *least common multiple* (LCM) of two positive integers a and b is the least positive integer divisible by both a and b; it is denoted by $[a, b]$.

Next we rewrite a symbolic definition of LCM.

A Symbolic Definition of LCM

The LCM of two positive integers a and b is the positive integer m such that:

- $a|m$ and $b|m$;
- If $a|m'$ and $b|m'$, then $m \le m'$, where m' is a positive integer.

Next we show a close relationship between the gcd and the LCM of two positive integers.

Theorem A.21. Let a and b be positive integers. Then $[a, b] = ab/(a, b)$. ■

Corollary A.6. Two positive integers a and b are relatively prime if and only if $[a, b] = ab$. ■

Again, as in the case of gcd, recursion can be applied to evaluate the LCM of three or more positive integers, as the next result shows.

Theorem A.22. Let $a_1, a_2, \ldots, a_n$ be $n(\geq 2)$ positive integers. Then $[a_1, a_2, \ldots, a_n] = [[a_1, a_2, \ldots, a_{n-1}], a_n]$. ■

Corollary A.7. If the positive integers $a_1, a_2, \ldots, a_n$ are pairwise relatively prime, then $[a_1, a_2, \ldots, a_n] = a_1 a_2 \cdots a_{n-1} a_n$. ■

The converse of this corollary is also true.

Corollary A.8. Let $m_1, m_2, \ldots, m_k$ and a be positive integers such that $m_i | a$ for $1 \leq i \leq k$. Then $[m_1 m_2 \cdots m_k] | a$. ■

MATRICES AND DETERMINANTS

Matrices contribute significantly to the study of Fibonacci and Lucas numbers. They were discovered jointly by two brilliant English mathematicians, Arthur Cayley (1821–1895) and James Joseph Sylvester (1814–1897). The matrix notation allows data to be summarized in a very compact form, and manipulated in a convenient way.

Matrix

A *matrix* is a rectangular arrangement of numbers enclosed by brackets. A matrix with m rows and n columns is an $m \times n$ (read m by n) matrix, its *size* being $m \times n$. If $m = 1$, it is a *row vector*, and if $n = 1$, then it is a *column vector*. If $m = n$, it is called a *square matrix of order* n. Each number in the arrangement is an *element* of the matrix. Matrices are denoted by uppercase letters.

Let a_{ij} denote the element in row i and column j of A, where $1 \leq i \leq m$ and $1 \leq j \leq n$. Then the matrix is abbreviated as $A = (a_{ij})_{m \times n}$, or simply (a_{ij}) if the size is clear from the context.

Equality of Matrices

Two matrices $A = (a_{ij})$ and $B = (b_{ij})$ are *equal* if they have the same size and $a_{ij} = b_{ij}$ for every i and j.

The next definition presents two special matrices.

Zero and Identity Matrices

If every element of a matrix is zero, then it is a *zero matrix*, denoted by O.

Let $A = (a_{ij})_{n\times n}$. Then the elements $a_{11}, a_{22}, \ldots, a_{nn}$ form the *main diagonal* of the matrix A. Suppose

$$a_{ij} = \begin{cases} 1 & \text{if } i = j \\ 0 & \text{otherwise} \end{cases}$$

Then A is called the *identity matrix* of order n; it is denoted by I_n, or I when there is no ambiguity.

Matrices also can be combined to produce new matrices. The various matrix operations are presented below.

Matrix Addition

The *sum* of the matrices $A = (a_{ij})_{m\times n}$ and $B = (b_{ij})_{m\times n}$ is defined by $A + B = (a_{ij} + b_{ij})_{m\times n}$. (You can add only matrices of the same size.)

Negative of a Matrix

The *negative* (or *additive inverse*) of a matrix $A = (a_{ij})$, denoted by $-A$, is defined by $-A = (-a_{ij})$.

Matrix Subtraction

The *difference* $A - B$ of the matrices $A = (a_{ij})_{m\times n}$ and $B = (b_{ij})_{m\times n}$ is defined by $A - B = (a_{ij} - b_{ij})_{m\times n}$. (You can subtract only matrices of the same size.)

Scalar Multiplication

Let $A = (a_{ij})$ be any matrix and k any real number (called a *scalar*). Then $kA = (ka_{ij})$.

The fundamental properties of the various matrix operations are stated in the followin theorem.

Theorem A.23. Let A, B, and C be any $m \times n$ matrices, O the $m \times n$ zero matrix, and c and d any real numbers. Then:

- $A + B = B + A$
- $A + (B + C) = (A + B) + C$
- $A + O = A = O + A$
- $A + (-A) = O = (-A) + A$
- $(-1)A = -A$
- $c(A + B) = cA + cB$
- $(c + d)A = cA + dA$
- $(cd)A = c(dA)$

■

Next, we define the product of two matrices as follows.

Matrix Multiplication

The *product* AB of the matrices $A = (a_{ij})_{m\times n}$ and $B = (b_{ij})_{n\times p}$ is the matrix $C = (c_{ij})_{m\times p}$, where $c_{ij} = a_{i1}b_{1j} + a_{i2}b_{2j} + \cdots + a_{in}b_{nj}$.

The product $C = AB$ is defined only if the number of columns in A equals the number of rows in B. The size of the product is $m \times p$.

The fundamental properties of matrix multiplication are stated in the next theorem.

Theorem A.24. Let A, B, and C be three matrices. Then:

- $A(BC) = (AB)C$
- $AI = A = IA$
- $A(B + C) = AB + AC$
- $(A + B)C = AC + BC$

provided the indicated sums and products are defined. ∎

Next we briefly discuss the concept of the determinant of a square matrix.

DETERMINANTS

With each square matrix $A = (a_{ij})_{n\times n}$, a unique real number can be associated. This number is called the *determinant* of A, denoted by $|A|$. When A is $n \times n$, it is of *order* n.

The determinant of $A = \begin{bmatrix} a & b \\ c & d \end{bmatrix}$, denoted by $|A| = \begin{vmatrix} a & b \\ c & d \end{vmatrix}$, is defined by $|A| = ad - bc$.

Knowing how to evaluate 2×2 determinants, higher-order determinants can be evaluated, but first, a few definitions.

Minors and Cofactors

The determinant of the matrix $(a_{ij})_{n\times n}$ obtained by deleting row i and column j is called the *minor* of the element a_{ij}; it is denoted by M_{ij}. The *cofactor* A_{ij} of the element a_{ij} is defined by $C_{ij} = (-1)^{i+j}M_{ij}$.

Determinant of a Matrix

The *determinant* of a matrix $A = (a_{ij})_{n\times n}$ is defined as

$$|A| = a_{i1}C_{i1} + a_{i2}C_{i2} + \cdots + a_{in}C_{in}.$$

This sum is called the *Laplace expansion* of $|A|$ by the ith row.

The following result about determinants will come in handy in our discussions.

Theorem A.25. Let A and B be two square matrices of the same size. Then

$$|AB| = |A| \cdot |B|$$

∎

CONGRUENCES

One of the most remarkable relations in number theory is the congruence relation, introduced and developed by the German mathematician Carl Friedrich Gauss (1777– 1855). The congruence relation shares many interesting properties with the equality relation, so it is denoted by the congruence symbol $\equiv$. It facilitates the study of the divisibility theory and has many fascinating applications.

We begin our discussion with the following definition.

Congruence Modulo *m*

Let m be a positive integer. An integer a is *congruent* to an integer b *modulo* m if $m|(a-b)$. In symbols, we then write $a \equiv b \pmod{m}$; m is the *modulus* of the *congruence relation*.

If a is not congruent to b modulo m, then a is *incongruent* to b *modulo* m; we then write $a \not\equiv b \pmod{m}$.

We now present a series of properties of congruence. Throughout we assume that all letters denote integers and all *moduli* (plural of modulus) are positive integers.

Theorem A.26. $a \equiv b \pmod{m}$ if and only if $a = b + km$ for some integer k. ■

A Useful Observation: It follows from the definition (also from Theorem A.26) that $a \equiv 0 \pmod{m}$ if and only if $m|a$; that is, an integer is congruent to 0 if and only if it is divisible by m. Thus $a \equiv 0 \pmod{m}$ and $m|a$ mean exactly the same thing.

Theorem A.27.

- $a \equiv a \pmod{m}$. (*Reflexive property*)
- If $a \equiv b \pmod{m}$, then $b \equiv a \pmod{m}$. (*Symmetric property*)
- If $a \equiv b \pmod{m}$ and $b \equiv c \pmod{m}$, then $a \equiv c \pmod{m}$. (*Transitive property*)

■

The next theorem provides another useful characterization of congruence.

Theorem A.28. $a \equiv b \pmod{m}$ if and only if a and b leave the same remainder when divided by m. ■

The next corollary follows from Theorem A.28.

Corollary A.9. If $a \equiv r \pmod{m}$, where $0 \le r < m$, then r is the remainder when a is divided by m, and if r is the remainder when a is divided by m, then $a \equiv r \pmod{m}$. ■

By this corollary, every integer a is congruent to its remainder r modulo m; r is called the *least residue* of a modulo m. Since r has exactly m choices $0, 1, 2, \ldots, (m-1)$, a is congruent to exactly one of them modulo m. Accordingly, we have the following result.

Corollary A.10. Every integer is congruent to exactly one of the least residues 0, $1, 2, \ldots, (m-1)$ modulo m. ■

Returning to Corollary A.10, we find that it justifies the definition of the *mod* operator. Thus, if $a \equiv r \pmod{m}$ and $0 \leq r < m$, then $a \bmod m = r$; conversely, if $a \bmod m = r$, then $a \equiv r \pmod{m}$ and $0 \leq r < m$.

The next theorem shows that two congruences with the same modulus can be added and multiplied.

Theorem A.29. Let $a \equiv b \pmod{m}$ and $c \equiv d \pmod{m}$. Then $a+c \equiv b+d \pmod{m}$ and $ac \equiv bd \pmod{m}$. ■

It follows from Theorem A.29 that one congruence can be subtracted from another provided they have the same modulus, as the next corollary states.

Corollary A.11. If $a \equiv b \pmod{m}$ and $c \equiv d \pmod{m}$, then $a-c \equiv b-d \pmod{m}$. ■

The next corollary also follows from Theorem A.29. Again, we leave its proof as an exercise.

Corollary A.12. If $a \equiv b \pmod{m}$ and c is any integer, then $a+c \equiv b+c \pmod{m}$, $a-c \equiv b-c \pmod{m}$, $ac \equiv bc \pmod{m}$, and $a^2 \equiv b^2 \pmod{m}$. ■

Part (4) of Corollary A.12 can be generalized to any positive integral exponent n, as the next theorem shows.

Theorem A.30. If $a \equiv b \pmod{m}$, then $a^n \equiv b^n \pmod{m}$ for any positive integer n. ■

The following theorem shows that the cancellation property of multiplication can be extended to congruence under special circumstances.

Theorem A.31. If $ac \equiv bc \pmod{m}$ and $(c, m) = 1$, then $a \equiv b \pmod{m}$. ■

Thus we can cancel the same number c from both sides of a congruence, provided c and m are relatively prime.

Returning to Theorem A.31, it can be generalized as follows.

Theorem A.32. If $ac \equiv bc \pmod{m}$ and $(c, m) = d$, then $a \equiv b \pmod{m/d}$. ■

Congruences of two numbers with different moduli can be combined into a single congruence, as the next theorem shows.

Theorem A.33. If $a \equiv b \pmod{m_1}$, $a \equiv b \pmod{m_2}$, $\ldots$, $a \equiv b \pmod{m_k}$, then $a \equiv b \pmod{[m_1, m_2, \ldots, m_k]}$. ■

The next corollary follows easily from this theorem.

Corollary A.13. If $a \equiv b \pmod{m_1}$, $a \equiv b \pmod{m_2}$, $\ldots$, $a \equiv b \pmod{m_k}$, where the moduli are pairwise relatively prime, then $a \equiv b \pmod{m_1 m_2 \cdots m_k}$. ■

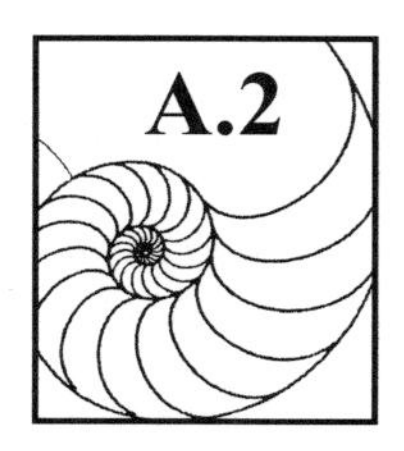

THE FIRST 100 FIBONACCI AND LUCAS NUMBERS

APPENDIX A.2. The First 100 Fibonacci and Lucas Numbers

n	F_n	L_n
1	1	1
2	1	3
3	2	4
4	3	7
5	5	11
6	8	18
7	13	29
8	21	47
9	34	76
10	55	123
11	89	199
12	144	322
13	233	521
14	377	843
15	610	1,364
16	987	2,207
17	1,597	3,571
18	2,584	5,778
19	4,181	9,349
20	6,765	15,127

(*Continued*)

APPENDIX A.2. The First 100 Fibonacci and Lucas Numbers

n	F_n	L_n
21	10,946	24,476
22	17,711	39,603
23	28,657	64,079
24	46,368	103,682
25	75,025	167,761
26	121,393	271,443
27	196,418	439,204
28	317,811	710,647
29	514,229	1,149,851
30	832,040	1,860,498
31	1,346,269	3,010,349
32	2,178,309	4,870,847
33	3,524,578	7,881,196
34	5,702,887	12,752,043
35	9,227,465	20,633,239
36	14,930,352	33,385,282
37	24,157,817	54,018,521
38	39,088,169	87,403,803
39	63,245,986	141,422,324
40	102,334,155	228,826,127
41	165,580,141	370,248,451
42	267,914,296	599,074,578
43	433,494,437	969,323,029
44	701,408,733	1,568,397,607
45	1,134,903,170	2,537,720,636
46	1,836,311,903	4,106,118,243
47	2,971,215,073	6,643,838,879
48	4,807,526,976	10,749,957,122
49	7,778,742,049	17,393,796,001
50	12,586,269,025	28,143,753,123
51	20,365,011,074	45,537,549,124
52	32,951,280,099	73,681,302,247
53	53,316,291,173	119,218,851,371
54	86,267,571,272	192,900,153,618
55	139,583,862,445	312,119,004,989
56	225,851,433,717	505,019,158,607
57	365,435,296,162	817,138,163,596
58	591,286,729,879	1,322,157,322,203
59	956,722,026,041	2,139,295,485,799
60	1,548,008,755,920	3,461,452,808,002

APPENDIX A.2. (*Continued*)

n	F_n	L_n
61	2,504,730,781,961	5,600,748,293,801
62	4,052,739,537,881	9,062,201,101,803
63	6,557,470,319,842	14,662,949,395,604
64	10,610,209,857,723	23,725,150,497,407
65	17,167,680,177,565	38,388,099,893,011
66	27,777,890,035,288	62,113,250,390,418
67	44,945,570,212,853	100,501,350,283,429
68	72,723,460,248,141	162,614,600,673,847
69	117,669,030,460,994	263,115,950,957,276
70	190,392,490,709,135	425,730,551,631,123
71	308,061,521,170,129	688,846,502,588,399
72	498,454,011,879,264	1,114,577,054,219,522
73	806,515,533,049,393	1,803,423,556,807,921
74	1,304,969,544,928,657	2,918,000,611,027,443
75	2,111,485,077,978,050	4,721,424,167,835,364
76	3,416,454,622,906,707	7,639,424,778,862,807
77	5,527,939,700,884,757	12,360,848,946,698,171
78	8,944,394,323,791,464	20,000,273,725,560,978
79	14,472,334,024,676,221	32,361,122,672,259,149
80	23,416,728,348,467,685	52,361,396,397,820,127
81	37,889,062,373,143,906	84,722,519,070,079,276
82	61,305,790,721,611,591	137,083,915,467,899,403
83	99,194,853,094,755,497	221,806,434,537,978,679
84	160,500,643,816,367,088	358,890,350,005,878,082
85	259,695,496,911,122,585	580,696,784,543,856,761
86	420,196,140,727,489,673	939,587,134,549,734,843
87	679,891,637,638,612,258	1,520,283,919,093,591,604
88	1,100,087,778,366,101,931	2,459,871,053,643,326,447
89	1,779,979,416,004,714,189	3,980,154,972,736,918,051
90	2,880,067,194,370,816,120	6,440,026,026,380,244,498
91	4,660,046,610,375,530,309	10,420,180,999,117,162,549
92	7,540,113,804,746,346,429	16,860,207,025,497,407,047
93	12,200,160,415,121,876,738	27,280,388,024,614,569,596
94	19,740,274,219,868,223,167	44,140,595,050,111,976,643
95	31,940,434,634,990,099,905	71,420,983,074,726,546,239
96	51,680,708,854,858,323,072	115,561,578,124,838,522,882
97	83,621,143,489,848,422,977	186,982,561,199,565,069,121
98	135,301,852,344,706,746,049	302,544,139,324,403,592,003
99	218,922,995,834,555,169,026	489,526,700,523,968,661,124
100	354,224,848,179,261,915,075	792,070,839,848,372,253,127

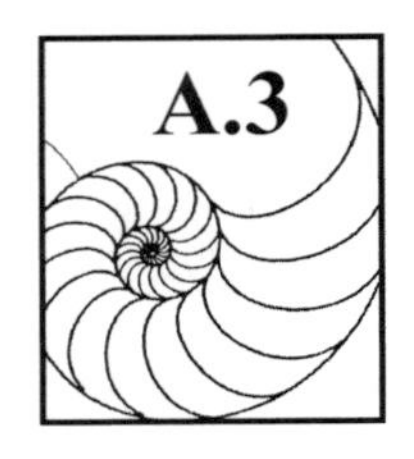

THE FIRST 100 FIBONACCI NUMBERS AND THEIR PRIME FACTORIZATIONS

APPENDIX A.3. The First 100 Fibonacci Numbers and Their Prime Factorizations

n	F_n	Prime Factorization of F_n
1	1	1
2	1	1
3	2	2
4	3	3
5	5	5
6	8	2^3
7	13	13
8	21	$3 \cdot 7$
9	34	$2 \cdot 17$
10	55	$5 \cdot 11$
11	89	89
12	144	$2^4 \cdot 3^2$
13	233	233
14	377	$13 \cdot 29$
15	610	$2 \cdot 5 \cdot 61$
16	987	$3 \cdot 7 \cdot 47$
17	1,597	1597
18	2,584	$2^3 \cdot 17 \cdot 19$
19	4,181	$37 \cdot 113$
20	6,765	$3 \cdot 5 \cdot 11 \cdot 41$

APPENDIX A.3. (*Continued*)

n	F_n	Prime Factorization of F_n
21	10,946	$2 \cdot 13 \cdot 421$
22	17,711	$89 \cdot 199$
23	28,657	28657
24	46,368	$2^5 \cdot 3^2 \cdot 7 \cdot 23$
25	75,025	$5^2 \cdot 3001$
26	121,393	$233 \cdot 521$
27	196,418	$2 \cdot 17 \cdot 53 \cdot 109$
28	317,811	$2 \cdot 13 \cdot 29 \cdot 281$
29	514,229	514229
30	832,040	$2^3 \cdot 5 \cdot 11 \cdot 31 \cdot 61$
31	1,346,269	$577 \cdot 2417$
32	2,178,309	$3 \cdot 7 \cdot 47 \cdot 2207$
33	3,524,578	$2 \cdot 89 \cdot 19801$
34	5,702,887	$1597 \cdot 3571$
35	9,227,465	$5 \cdot 13 \cdot 141961$
36	14,930,352	$2^4 \cdot 3^3 \cdot 17 \cdot 19 \cdot 107$
37	24,157,817	$73 \cdot 149 \cdot 2221$
38	39,088,169	$37 \cdot 113 \cdot 9349$
39	63,245,986	$2 \cdot 233 \cdot 135721$
40	102,334,155	$3 \cdot 5 \cdot 7 \cdot 11 \cdot 41 \cdot 2161$
41	165,580,141	$2789 \cdot 59369$
42	267,914,296	$2^3 \cdot 13 \cdot 29 \cdot 211 \cdot 421$
43	433,494,437	433494437
44	701,408,733	$3 \cdot 43 \cdot 89 \cdot 199 \cdot 307$
45	1,134,903,170	$2 \cdot 5 \cdot 17 \cdot 61 \cdot 109441$
46	1,836,311,903	$139 \cdot 461 \cdot 28657$
47	2,971,215,073	2971215073
48	4,807,526,976	$2^6 \cdot 3^2 \cdot 7 \cdot 23 \cdot 47 \cdot 1103$
49	7,778,742,049	$13 \cdot 97 \cdot 6168709$
50	12,586,269,025	$5^2 \cdot 11 \cdot 101 \cdot 151 \cdot 3001$
51	20,365,011,074	$2 \cdot 1597 \cdot 6376021$
52	32,951,280,099	$3 \cdot 233 \cdot 521 \cdot 90481$
53	53,316,291,173	$953 \cdot 55945741$
54	86,267,571,272	$2^3 \cdot 17 \cdot 19 \cdot 53 \cdot 109 \cdot 5779$
55	139,583,862,445	$5 \cdot 89 \cdot 661 \cdot 474541$
56	225,851,433,717	$3 \cdot 7^2 \cdot 13 \cdot 29 \cdot 281 \cdot 14503$
57	365,435,296,162	$2 \cdot 37 \cdot 113 \cdot 797 \cdot 54833$
58	591,286,729,879	$59 \cdot 19489 \cdot 514229$
59	956,722,026,041	$353 \cdot 2710260697$
60	1,548,008,755,920	$2^4 \cdot 3^2 \cdot 5 \cdot 11 \cdot 31 \cdot 41 \cdot 61 \cdot 2521$

(*Continued*)

APPENDIX A.3. The First 100 Fibonacci Numbers and Their Prime Factorizations

n	F_n	Prime Factorization of F_n
61	2,504,730,781,961	$4513 \cdot 555003497$
62	4,052,739,537,881	$557 \cdot 2417 \cdot 3010349$
63	6,557,470,319,842	$2 \cdot 13 \cdot 17 \cdot 421 \cdot 35239681$
64	10,610,209,857,723	$3 \cdot 7 \cdot 47 \cdot 1087 \cdot 2207 \cdot 4481$
65	17,167,680,177,565	$5 \cdot 233 \cdot 14736206161$
66	27,77,7890,035,288	$2^3 \cdot 89 \cdot 199 \cdot 9901 \cdot 19801$
67	44,915,570,212,853	$269 \cdot 116849 \cdot 1429913$
68	72,723,460,248,141	$3 \cdot 67 \cdot 1597 \cdot 3571 \cdot 63443$
69	117,669,030,460,994	$2 \cdot 137 \cdot 829 \cdot 18077 \cdot 28657$
70	190,392,490,709,135	$5 \cdot 11 \cdot 13 \cdot 29 \cdot 71 \cdot 911 \cdot 141961$
71	308,061,521,170,129	$6673 \cdot 46165371073$
72	498,454,011,879,264	$2^5 \cdot 3^3 \cdot 7 \cdot 17 \cdot 19 \cdot 23 \cdot 107 \cdot 103681$
73	806,515,533,049,393	$9375829 \cdot 86020717$
74	1,304,969,544,928,657	$73 \cdot 149 \cdot 2221 \cdot 54018521$
75	2,111,485,077,978,050	$2 \cdot 5^2 \cdot 61 \cdot 3001 \cdot 230686501$
76	3,416,454,622,906,707	$3 \cdot 37 \cdot 113 \cdot 9349 \cdot 29134601$
77	5,527,939,700,884,757	$13 \cdot 89 \cdot 988681 \cdot 4832521$
78	8,944,394,323,791,464	$2^3 \cdot 79 \cdot 233 \cdot 521 \cdot 859 \cdot 135721$
79	14,472,334,024,676,221	$157 \cdot 92180471494753$
80	23,416,728,348,467,685	$3 \cdot 5 \cdot 7 \cdot 11 \cdot 41 \cdot 47 \cdot 1601 \cdot 2161 \cdot 3041$
81	37,889,062,373,143,906	$2 \cdot 17 \cdot 53 \cdot 109 \cdot 2269 \cdot 4373 \cdot 19441$
82	61,305,790,721,611,591	$2789 \cdot 59369 \cdot 370248451$
83	99,194,853,094,755,497	99194853094755497
84	160,500,643,816,367,088	$2^4 \cdot 3^2 \cdot 13 \cdot 29 \cdot 83 \cdot 211 \cdot 281 \cdot 421 \cdot 1427$
85	259,695,496,911,122,585	$5 \cdot 1597 \cdot 9521 \cdot 3415914041$
86	420,196,140,727,489,673	$6709 \cdot 144481 \cdot 433494437$
87	679,891,637,638,612,258	$2 \cdot 173 \cdot 514229 \cdot 3821263937$
88	1,100,087,778,366,101,931	$3 \cdot 7 \cdot 43 \cdot 89 \cdot 199 \cdot 263 \cdot 307 \cdot 881 \cdot 967$
89	1,779,979,416,004,714,189	$1069 \cdot 1665088321800481$
90	2,880,067,194,370,816,120	$2^3 \cdot 5 \cdot 11 \cdot 17 \cdot 19 \cdot 31 \cdot 61 \cdot 181 \cdot 541 \cdot 109441$
91	4,660,046,610,375,530,309	$13^2 \cdot 233 \cdot 741469 \cdot 159607993$
92	7,540,113,804,746,346,429	$3 \cdot 139 \cdot 461 \cdot 4969 \cdot 28657 \cdot 275449$
93	12,200,160,415,121,876,738	$2 \cdot 557 \cdot 2417 \cdot 4531100550901$
94	19,740,274,219,868,223,167	$2971215073 \cdot 6643838879$
95	31,940,434,634,990,099,905	$5 \cdot 37 \cdot 113 \cdot 761 \cdot 9641 \cdot 67735001$
96	51,680,708,854,858,323,072	$2^7 \cdot 3^2 \cdot 7 \cdot 23 \cdot 47 \cdot 769 \cdot 1103 \cdot 2207 \cdot 3167$
97	83,621,143,489,848,422,977	$193 \cdot 389 \cdot 3084989 \cdot 3610402019$
98	135,301,852,344,706,746,049	$13 \cdot 29 \cdot 97 \cdot 6168709 \cdot 599786069$
99	218,922,995,834,555,169,026	$2 \cdot 17 \cdot 89 \cdot 197 \cdot 19801 \cdot 18546805133$
100	354,224,848,179,261,915,075	$3 \cdot 5^2 \cdot 11 \cdot 41 \cdot 101 \cdot 151 \cdot 401 \cdot 3001 \cdot 570601$

THE FIRST 100 LUCAS NUMBERS AND THEIR PRIME FACTORIZATIONS

APPENDIX A.4. The First 100 Lucas Numbers and Their Prime Factorizations

n	L_n	Prime Factorization of L_n
1	1	1
2	3	3
3	4	2^2
4	7	7
5	11	11
6	18	$2 \cdot 3^2$
7	29	29
8	47	47
9	76	$2^2 \cdot 19$
10	123	$3 \cdot 41$
11	199	199
12	322	$2 \cdot 7 \cdot 23$
13	521	521
14	843	$3 \cdot 281$
15	1,364	$2^2 \cdot 11 \cdot 31$
16	2,207	2207
17	3,571	3571
18	5,778	$2 \cdot 3^3 \cdot 107$
19	9,349	9349
20	15,127	$7 \cdot 2161$

(*Continued*)

APPENDIX A.4. The First 100 Lucas Numbers and Their Prime Factorizations

n	L_n	Prime Factorization of L_n
21	24,476	$2^2 \cdot 29 \cdot 211$
22	39,603	$3 \cdot 43 \cdot 307$
23	64,079	$139 \cdot 461$
24	103,682	$2 \cdot 47 \cdot 1103$
25	167,761	$11 \cdot 101 \cdot 151$
26	271,443	$3 \cdot 90481$
27	439,204	$2^2 \cdot 19 \cdot 5779$
28	710,647	$7^2 \cdot 4503$
29	1,149,851	$59 \cdot 19489$
30	1,860,498	$2 \cdot 3^2 \cdot 41 \cdot 2521$
31	3,010,349	3010349
32	4,870,847	$1087 \cdot 4481$
33	7,881,196	$2^2 \cdot 199 \cdot 9901$
34	12,752,043	$3 \cdot 67 \cdot 63443$
35	20,633,239	$11 \cdot 29 \cdot 71 \cdot 911$
36	33,385,282	$2 \cdot 7 \cdot 23 \cdot 103681$
37	54,018,521	54018521
38	87,403,803	$3 \cdot 29134601$
39	141,422,324	$2^2 \cdot 79 \cdot 521 \cdot 859$
40	228,826,127	$47 \cdot 1601 \cdot 3041$
41	370,248,451	370248451
42	599,074,578	$2 \cdot 3^2 \cdot 83 \cdot 281 \cdot 1427$
43	969,323,029	$6709 \cdot 144481$
44	1,568,397,607	$7 \cdot 263 \cdot 881 \cdot 967$
45	2,537,720,636	$2^2 \cdot 11 \cdot 19 \cdot 31 \cdot 181 \cdot 541$
46	4,106,118,243	$3 \cdot 4969 \cdot 275449$
47	6,643,838,879	6643838879
48	10,749,957,122	$2 \cdot 769 \cdot 2207 \cdot 3167$
49	17,393,796,001	$29 \cdot 599786069$
50	28,143,753,123	$3 \cdot 41 \cdot 401 \cdot 570601$
51	45,537,549,124	$2^2 \cdot 919 \cdot 3469 \cdot 3571$
52	73,681,302,247	$7 \cdot 103 \cdot 102193207$
53	119,218,851,371	119218851371
54	192,900,153,618	$2 \cdot 3^4 \cdot 107 \cdot 1128427$
55	312,119,004,989	$11^2 \cdot 199 \cdot 331 \cdot 39161$
56	505,019,158,607	$47 \cdot 10745088481$
57	817,138,163,596	$2^2 \cdot 229 \cdot 9349 \cdot 95419$
58	1,322,157,322,203	$3 \cdot 347 \cdot 1270083883$
59	2,139,295,485,799	$709 \cdot 8969 \cdot 336419$
60	3,461,452,808,002	$2 \cdot 7 \cdot 23 \cdot 241 \cdot 2161 \cdot 20641$

APPENDIX A.4. (*Continued*)

n	L_n	Prime Factorization of L_n
61	5,600,748,293,801	5600748293801
62	9,062,201,101,803	$3 \cdot 3020733700601$
63	14,662,949,395,604	$2^2 \cdot 19 \cdot 29 \cdot 211 \cdot 1009 \cdot 31249$
64	23,725,150,497,407	$127 \cdot 186812208641$
65	38,388,099,893,011	$11 \cdot 131 \cdot 521 \cdot 2081 \cdot 24571$
66	62,113,250,390,418	$2 \cdot 3^2 \cdot 43 \cdot 307 \cdot 261399601$
67	100,501,350,283,429	$4021 \cdot 24994118449$
68	162,614,600,673,847	$7 \cdot 23230657239121$
69	263,115,950,957,276	$2^2 \cdot 139 \cdot 461 \cdot 691 \cdot 1485571$
70	425,730,551,631,123	$3 \cdot 41 \cdot 281 \cdot 12317523121$
71	688,846,502,588,399	688846502588399
72	1,114,577,054,219,522	$2 \cdot 47 \cdot 1103 \cdot 10749957121$
73	1,803,423,556,807,921	$151549 \cdot 11899937029$
74	2,918,000,644,027,443	$3 \cdot 11987 \cdot 81143477963$
75	4,721,424,167,835,364	$2^2 \cdot 11 \cdot 31 \cdot 101 \cdot 151 \cdot 12301 \cdot 18451$
76	7,639,424,778,862,807	$7 \cdot 1091346396980401$
77	12,360,848,946,698,171	$29 \cdot 199 \cdot 229769 \cdot 9321929$
78	20,000,273,725,560,978	$2 \cdot 3^2 \cdot 90481 \cdot 12280217041$
79	32,361,122,672,259,149	32361122672259149
80	52,361,396,397,820,127	$2207 \cdot 23725145626561$
81	84,722,519,070,079,276	$2^2 \cdot 19 \cdot 3079 \cdot 5779 \cdot 62650261$
82	137,083,915,467,899,403	$3 \cdot 163 \cdot 800483 \cdot 350207569$
83	221,806,434,537,978,679	$35761381 \cdot 6202401259$
84	358,890,350,005,878,082	$2 \cdot 7^2 \cdot 23 \cdot 167 \cdot 14503 \cdot 65740583$
85	580,696,784,543,856,761	$11 \cdot 3571 \cdot 1158551 \cdot 12760031$
86	939,587,134,549,734,843	$3 \cdot 313195711516578281$
87	1,520,283,919,093,591,604	$2^2 \cdot 59 \cdot 349 \cdot 19489 \cdot 947104099$
88	2,459,871,053,643,326,447	$47 \cdot 93058241 \cdot 562418561$
89	3,980,154,972,736,918,051	$179 \cdot 22235502640988369$
90	6,440,026,026,380,244,498	$2 \cdot 3^3 \cdot 41 \cdot 107 \cdot 2521 \cdot 10783342081$
91	10,420,180,999,117,162,549	$29 \cdot 521 \cdot 689667151970161$
92	16,860,207,025,497,407,047	$7 \cdot 253367 \cdot 9506372193863$
93	27,280,388,024,614,569,596	$2^2 \cdot 63799 \cdot 3010349 \cdot 35510749$
94	44,140,595,050,111,976,643	$3 \cdot 563 \cdot 5641 \cdot 4632894751907$
95	71,420,983,074,726,546,239	$11 \cdot 191 \cdot 9349 \cdot 41611 \cdot 87382901$
96	115,561,578,124,838,522,882	$2 \cdot 1087 \cdot 4481 \cdot 11862575248703$
97	186,982,561,199,565,069,121	$3299 \cdot 56678557502141579$
98	302,544,139,324,403,592,003	$3 \cdot 281 \cdot 5881 \cdot 61025309469041$
99	489,526,700,523,968,661,124	$2^2 \cdot 19 \cdot 199 \cdot 991 \cdot 2179 \cdot 9901 \cdot 1513909$
100	792,070,839,848,372,253,127	$7 \cdot 2161 \cdot 9125201 \cdot 5738108801$

REFERENCES

Alexanderson, G. L., 1965, Solution to Problem B-56, *The Fibonacci Quarterly*, 3:2 (April), 159.

Alfred, U., 1963, Problem H-8, *The Fibonacci Quarterly*, 1:1 (Feb.), 48.

Alfred, U., 1963, Problem B-24, *The Fibonacci Quarterly*, 1:4 (Dec.), 73.

Alfred, U., 1964, Problem H-29, *The Fibonacci Quarterly*, 2:1 (Feb.), 49.

Alfred, U., 1964a, "On Square Lucas Numbers," *The Fibonacci Quarterly*, 2:1 (Feb.), 11–12.

Alfred, U., 1964b, "Exploring Fibonacci Magic Squares," *The Fibonacci Quarterly*, 2:3 (Oct.), 216.

Alfred, U., 1966, Problem B-79, *The Fibonacci Quarterly*, 4:3 (Oct.), 287.

Alladi, K., and V. E. Hoggatt, Jr., 1995, "Compositions with Ones and Twos," *The Fibonacci Quarterly*, 13:3 (Oct.), 233–239.

Anaya, R., and J. Crump, 1972, "A Generalized Greatest Integer Function Theorem," *The Fibonacci Quarterly*, 10:2 (Feb.), 207–211.

Anglin, R. H., 1970, Problem B-160, *The Fibonacci Quarterly*, 8:1 (Feb.), 107.

Archibald, R. C., 1918, "Golden Section," *The American Mathematical Monthly*, 25, 232–238.

Avila, J. H., 1964, Solution to Problem B-19, *The Fibonacci Quarterly*, 2:1 (Feb.), 75–76.

Baker A., and H. Davenport, 1969, "The Equations $3x^2 - 2 = y^2$ and $8x^2 - 7 = z^2$," *The Quarterly Journal of Mathematics*, 20 (June), 129–137.

Baravalle, H. V., 1948, "The Geometry of The Pentagon and the Golden Section," *Mathematics Teacher*, 41, 22–31.

Barbeau, E., 1993, Example 2, *The College Mathematics Journal*, 24:1 (Jan.), 65–66.

Barley, W. C., 1973a, Problem B-234, *The Fibonacci Quarterly*, 11:2 (April), 222.

Barley, W. C., 1973b, Problem B-240, *The Fibonacci Quarterly*, 11:3 (Oct.), 335.

Basin, S. L., 1963, "The Fibonacci Sequence as it Appears in Nature," *The Fibonacci Quarterly*, 1:1 (Feb.), 53–56.

Basin, S. L., 1964a, Problem B-23, *The Fibonacci Quarterly*, 2:1 (Feb.), 78–79.

Basin, S. L., 1964b, Problem B-42, *The Fibonacci Quarterly*, 2:4 (Dec.), 329.

Basin, S. L., and V. E. Hoggatt, Jr., 1963a, "A Primer on the Fibonacci Sequence—Part I," *The Fibonacci Quarterly*, 1:1 (Feb.), 65–72.

Basin, S. L., and V. E. Hoggatt, Jr., 1963b, "A Primer on the Fibonacci Sequence—Part II," *The Fibonacci Quarterly*, 1:2 (April), 61–68.

Basin, S. L., and V. Ivanoff, 1963, Problem B-4, *The Fibonacci Quarterly*, 1:1 (Feb.), 74.

Beard, R. S., 1950, "The Golden Section and Fibonacci Numbers," *Scripta Mathematica*, 16, 116–119.

Beiler, A. H., 1966, *Recreations in the Theory of Numbers*, 2nd ed., New York: Dover.

Bell, E. T., 1936, "A Functional Equation in Arithemetic," *Trans. American Mathematical Society*, 39 (May), 341–344.

Bennett, A. A., 1923, "The 'Most Pleasing Rectangle'," *The American Mathematical Monthly*, 30 (Jan.), 27–30.

Beran, L., 1986, "Schemes Generating the Fibonacci Sequence," *The Mathematical Gazette*, 70, 38–40.

Berzsenyi, G., 1977, Problem H-263, *The Fibonacci Quarterly*, 15:3 (Oct.), 373.

Berzsenyi, G., 1979, Problem B-378, *The Fibonacci Quarterly*, 17:2 (April), 185–186.

Bicknell, M., 1970, "A Primer for the Fibonacci Numbers: Part VII," *The Fibonacci Quarterly*, 8:4 (Oct.), 407–420.

Bicknell, M., 1974, Problem B-267, *The Fibonacci Quarterly*, 12:3 (Oct.), 316.

Bicknell, M., and V. E. Hoggatt, Jr., 1963, "Fibonacci Matrices and Lambda Functions," *The Fibonacci Quarterly*, 1:2 (April), 47–52.

Bicknell, M., and V. E. Hoggatt, Jr., 1969, "Golden triangles, Rectangles, and Cuboids," *The Fibonacci Quarterly*, 7:1 (Feb.), 73–91, 98.

Bird, M. T., 1938, Solution to Problem 3802, *The American Mathematical Monthly*, 45 (Nov.), 632–633.

Blank, G., 1956, "Another Fibonacci Curiosity," *Scripta Mathematica*, 21:1 (March), 30.

Blazej, R., 1975, Problem B-298, *The Fibonacci Quarterly*, 13:1 (Feb.), 94.

Blazej, R., 1975, Problem B-294, *The Fibonacci Quarterly*, 13:3 (Oct.), 375.

Bloom, D. M., 1996, Problem 55, *Math Horizons* (Sept.), 32.

Boblett, A. P., 1963, Problem B-29, *The Fibonacci Quarterly*, 1:4 (Dec.), 75.

Brady, W. G., 1971, "The Lambert Function," *The Fibonacci Quarterly*, 9:5 (Dec.), 199–200.

Brady, W. G., 1974, Problem B-253,*The Fibonacci Quarterly*, 12:1 (Feb.), 104–105.

Bridger, C. A., 1967, Problem B-94, *The Fibonacci Quarterly*, 5:2 (April), 203.

Brooke, M., 1962, "Fibonacci Fancy," *Mathematics Magazine*, 35:4 (Sept.), 218.

Brooke, M., 1963, "Fibonacci Formulas," *The Fibonacci Quarterly*, 1:2 (April), 60.

Brooke, M., 1964, "Fibonacci Numbers: Their History Through 1900," *The Fibonacci Quarterly*, 2:2 (April), 149–153.

Brousseau, A., 1967, "A Fibonacci Generalization," *The Fibonacci Quarterly*, 5:2 (April), 171–174.

Brousseau, A., 1968, Problem H-92, *The Fibonacci Quarterly*, 13:1 (Feb.), 94.

Brousseau, A., 1969a, "Summation of Infinite Fibonacci Series," *The Fibonacci Quarterly*, 7:2 (April), 143–168.

Brousseau, A., 1969b, "Fibonacci–Lucas Infinite Series—Research Topic," *The Fibonacci Quarterly*, 7:2 (April), 211–217.

Brousseau, A., 1972, "Ye Olde Fibonacci Curiosity," *The Fibonacci Quarterly*, 10:4 (Oct.), 441–443.

Brown, J. L., Jr., 1964, "A Trigonometric Sum," *The Fibonacci Quarterly*, 2:1 (Feb.), 74–75.

Brown, J. L., Jr., 1965, "Reply to Exploring Magic Squares," *The Fibonacci Quarterly*, 2:3 (April), 146.

Brown, J. L., Jr., 1966, Solution to Problem H-54, *The Fibonacci Quarterly*, 4:4 (Dec.), 334–335.

Brown, J. L., Jr., 1967, Problem H-71, *The Fibonacci Quarterly*, 5:2 (April), 166–167.

Bruckman, P. S., 1973, Solution to Problem B-231, *The Fibonacci Quarterly*, 11:1 (Feb.), 111–112.

Bruckman, P. S., 1973, Problem B-236, *The Fibonacci Quarterly*, 11:2 (April), 223.

Bruckman, P. S., 1973, Solution to Problem B-241, *The Fibonacci Quarterly*, 11:4 (Oct.), 336.

Bruckman, P. S., 1974, Solution to Problem H-201, *The Fibonacci Quarterly*, 12:2 (April), 218–219.

Bruckman, P. S., 1975a, Problem B-277, *The Fibonacci Quarterly*, 13:1 (Feb.), 96.

Bruckman, P. S., 1975b, Problem B-278, *The Fibonacci Quarterly*, 13:1 (Feb.), 96.

Bruckman, P. S., 1979, Problem H-280, *The Fibonacci Quarterly*, 17:4 (Dec.), 377.

Bruckman, P. S., 1980, Problem H-286, *The Fibonacci Quarterly*, 18:3 (Oct.), 281–282.

Burton, D. M., 1989, *Elementary Number Theory*, 3rd ed., Dubuque, IA: W. C. Brown.

Bushman, R. G., 1964, Problem H-18, *The Fibonacci Quarterly*, 2:2 (April), 126–127.

Butchart, J. H., 1968, Problem B-124, *The Fibonacci Quarterly*, 6:4 (Oct.), 289–290.

Byrd, P. F.,"Expansion of Analytic Functions in Polynomials Associated with Fibonacci Numbers," *The Fibonacci Quarterly*, 1:1 (Feb.), 16–29.

Calendar Problems, 1993, *Mathematics Teacher*, 86 (Jan.), 49.

Calendar Problems, 1999, *Mathematics Teacher*, 92 (Oct.), 606–611.

Camfield, W. A., 1965, "Jaun Gris and The Golden Section," *The Art Bulletin*, 47, 128–134.

Candido, G., 1951, "A Relationship Between the Fourth Powers of the Terms of the Fibonacci Series," *Scripta Mathematica*, 17:3–4 (Sept.–Dec.), 230.

Candido, G., 1970, Curiosa 272, *Scripta Mathematica*, 17:3–4 (Sept.–Dec.), 230.

Carlitz, L., 1963, Problem B-19, *The Fibonacci Quarterly*, 1:3 (Oct.), 75.

Carlitz, L., 1964, "A Note on Fibonacci Numbers," *The Fibonacci Quarterly*, 1:2 (Feb.), 15–28.

Carlitz, L., 1965, Solution to Problem H-39, *The Fibonacci Quarterly*, 3:1 (Feb.), 51–52.

Carlitz, L., 1965a, "The Characteristic Polynomial of a Certain Matrix of Binomial Coefficients," *The Fibonacci Quarterly*, 3:2 (April), 81–89.

Carlitz, L., 1965b, "A Telescoping Sum," *The Fibonacci Quarterly*, 3:4 (Dec.), 75–76.

Carlitz, L., 1967a, Problem B-110, *The Fibonacci Quarterly*, 5:5 (Dec.), 469–470.

Carlitz, L., 1967b, Problem B-111, *The Fibonacci Quarterly*, 5:5 (Dec.), 470–471.

Carlitz, L., 1968, Solution to Problem H-92, *The Fibonacci Quarterly*, 6:2 (April), 145.

Carlitz, L., 1969, Problem H-112, *The Fibonacci Quarterly*, 7:1 (Feb.), 61–62.

Carlitz, L., 1970, Problem B-185, *The Fibonacci Quarterly*, 8:3 (April), 325.

Carlitz, L., 1970, Problem B-185, *The Fibonacci Quarterly*, 8:3 (April), 326.

Carlitz, L., 1971a, Problem B-185, *The Fibonacci Quarterly*, 9:1 (Feb.), 109.

Carlitz, L., 1971b, Problem B-186, *The Fibonacci Quarterly*, 9:1 (Feb.), 109–110.

Carlitz, L., 1972, Problem B-213, *The Fibonacci Quarterly*, 10:2 (Feb.), 224.

Carlitz, L., 1972, Solution to Problem B-215, *The Fibonacci Quarterly*, 10:3 (April), 331–332.

Carlitz, L., 1972, "A Conjecture Concerning Lucas Numbers," *The Fibonacci Quarterly*, 10:5 (Nov.), 526, 550.

Carlitz, L., 1977, Problem H-262, *The Fibonacci Quarterly*, 15:4 (Dec.), 372–373.

Carlitz, L., and J. A. H. Hunter, 1969, "Some Powers of Fibonacci and Lucas Numbers," *The Fibonacci Quarterly*, 7:5 (Dec.), 467–473.

Carlitz, L., and R. Scoville, 1971, Problem B-255, *The Fibonacci Quarterly*, 12:1 (Feb.), 106.

Cheves, W., 1970, Problem B-192, *The Fibonacci Quarterly*, 8:4 (Oct.), 443.

Cheves, W., 1975, Problem B-275, *The Fibonacci Quarterly*, 13:1 (Feb.), 95.

Church, C. A., 1964, Problem B-46, *The Fibonacci Quarterly*, 2:3 (Oct.), 231.

Church, C. A., and M. Bicknell, 1973, "Exponential Generating Functions for Fibonacci Identities," *The Fibonacci Quarterly*, 11:3 (Oct.), 275–281.

Cohen, D. I. A., 1978, *Basic Techniques of Combinatorial Theory*, New York: Wiley.

Cohn, J. H. E., 1964, "Square Fibonacci Numbers, Etc.," *The Fibonacci Quarterly*, 2:2 (April), 109–113.

Cook, I., 1990, "The Euclidean Algorithm and Fibonacci," *The Mathematical Gazette*, 74, 47–48.

Coxter, H. S. M., 1953, "The Golden Section, Phyllotaxis, and Wythoff's Game," *Scripta Mathematica*, 19, 135–143.

Coxter, H. S. M., 1969, *Introduction to Geometry*, 2nd ed., New York: Wiley.

Cross, G., and H. Renzi, 1965, "Teachers Discover New Math Theorems," *The Arithmetic Teacher*, 12:8, 625–626.

Cross, T., 1996, "A Fibonacci Fluke," *The Mathematical Gazette*, 80 (July), 398–400.

Cundy, H. M., and A. P. Rollett, 1961, *Mathematical Models*, 2nd ed., New York: Oxford University Press.

"Dali Makes Met," 1965, *Time*, Jan. 24, 72.

Davis, B., 1972, "Fibonacci Numbers in Physics," *The Fibonacci Quarterly*, 10:6 (Dec.), 659–660, 662.

Deily, G. R., 1966, "A Logarithmic Formula for Fibonacci Numbers," *The Fibonacci Quarterly*, 4:1 (Feb.), 89.

DeLeon, M. J., 1982, Solution to Problem B-319, *The Fibonacci Quarterly*, 20:1 (Feb.), 96.

Dence, T. P., 1968, Problem B-129, *The Fibonacci Quarterly*, 6:4 (Oct.), 296.

Deninger, R. A., 1972, "Fibonacci Numbers and Water Pollution Control," *The Fibonacci Quarterly*, 10:3 (April), 299–300, 302.

Desmond, J. E., 1970, Problem B-182, *The Fibonacci Quarterly*, 8:5 (Dec.), 549–550.

DeTemple, D. W., 1974, "A Pentagonal Arch," *The Fibonacci Quarterly*, 12:3 (Oct.), 235–236.

Dickson, L. E., 1952, *History of the Theory of Numbers*, Vol. 1, New York: Chelsea.

Draim, N. A., and M. Bicknell, 1966, "Sums of n-th Powers of Roots of a Given Quadratic Function," *The Fibonacci Quarterly*, 4:2 (April), 170–178.

Drake, R. C., 1970, Problem B-180, *The Fibonacci Quarterly*, 8:5 (Dec.), 547–548.

Duckworth, G. E., 1962, *Structural Patterns and Proportions in Vergil's Aeneid*, Ann Arbor, MI: The University of Michigan Press.

Dudley, U., and B. Tucker, 1971, "Greatest Common Divisors in Altered Fibonacci Sequences," *The Fibonacci Quarterly*, 9:1 (Feb.), 89–91.

Dunn, A., 1980, *Mathematical Bafflers*, New York: Dover.

Dunn, A., 1983, *Second Book of Mathematical Bafflers*, New York; Dover.

Eggar, M. H., 1979, "Applications of Fibonacci Numbers," *The Mathematical Gazette*, 63, 36–39.

Erbacker, J., et al., 1963, Problem H-25, *The Fibonacci Quarterly*, 1:4 (Dec.), 47.

Everman, D., et al., 1960, Problem E1396, *The American Mathematical Monthly*, 67 (Sept.), 694.

Eves, H., 1969, *An Introduction to the History of Mathematics*, 3rd ed., New York: Holt, Rinehart and Winston.

Feinberg, M., 1963a, "Fibonacci–Tribonacci" *The Fibonacci Quarterly*, 1:3 (Oct.), 70–74.

Feinberg, M., 1963b, "A Lucas Triangle," *The Fibonacci Quarterly*, 5:5 (Dec.), 486–490.

Ferns, H. H., 1963, Problem B-106, *The Fibonacci Quarterly*, 5:5 (Dec.), 466–467.

Ferns, H. H., 1964, Problem B-48, *The Fibonacci Quarterly*, 2:3 (Oct.), 232.

Ferns, H. H., 1967, Problem B-106, *The Fibonacci Quarterly*, 5:1 (Feb.), 107.

Ferns, H. H., 1967, Problem B-104, *The Fibonacci Quarterly*, 5:3 (Oct.), 292.

Ferns, H. H., 1968, Problem B-115, *The Fibonacci Quarterly*, 6:1 (Feb.), 92–93.

"The Fibonacci Numbers," 1969, *Time*, April 4.

"Fibonacci or Forgery," 1995, *Scientific American*, 272 (May), 104–105.

Filipponi, P., and D. Singmaster, 1990, "Average Age of Generalized rabbits," *The Fibonacci Quarterly*, 28:3 (Aug.), 281.

Finkelstein, R., 1969, Problem B-143, *The Fibonacci Quarterly*, 7:2 (April), 220–221.

Finkelstein, R., 1973, "On Fibonacci Numbers Which Are One More Than a Square," *J. Reine Angewandte Mathematik*, 262/263, 171–182.

Finkelstein, R., 1975, "On Lucas Numbers Which Are One More Than a Square," *The Fibonacci Quarterly*, 13, 340–342.

Finkelstein, R., 1976, "On Triangular Fibonacci Numbers," *Utilitas Mathematica*, 9, 319–327.

Fisher, G., 1997, "The Singularity of Fibonacci Matrices," *The Mathematical Gazette*, 81 (July), 295–298.

Fisher, K., 1976, "The Fibonacci Sequence Encountered in Nerve Physiology," *The Fibonacci Quarterly*, 14:4 (Nov.), 377–379.

Fisher, R., 1993, *Fibonacci Applications and Strategies for Traders*, New York: Wiley.

Fisk, S., 1963, Problem B-10, *The Fibonacci Quarterly*, 1:2 (April), 85.

"Forbidden Fivefold Symmetry May Indicate Quasicrystal Phase," 1985, *Physics Today* (Feb.), 17–19.

Ford, K., 1993, Solution to Problem 487, *The College Mathematics Journal*, 24:5 (Nov.), 476.

Frankel, E. T., 1970, "Fibonacci Numbers as Paths of a Rook on a Chessboard," *The Fibonacci Quarterly*, 8:5 (Dec.), 538–541.

Freedman, D. L., 1996, "Fibonacci and Lucas Connections," *Mathematics Teacher*, 89 (Nov.), 624–625.

Freeman, G. F., 1967, "On Ratios of Fibonacci and Lucas Numbers," *The Fibonacci Quarterly*, 5:1 (Feb.), 99–106.

Freitag, H. T., 1974a, Problem B-256, *The Fibonacci Quarterly*, 12:2 (April), 221.

Freitag, H. T., 1974b, Problem B-257, *The Fibonacci Quarterly*, 12:2 (April), 222.

Freitag, H. T., 1974c, Problem B-262, *The Fibonacci Quarterly*, 12:3 (Oct.), 314.

Freitag, H. T., 1974d, Problem B-270, *The Fibonacci Quarterly*, 12:4 (Dec.), 405.

Freitag, H. T., 1974e, Problem B-271, *The Fibonacci Quarterly*, 12:4 (Dec.), 405.

Freitag, H. T., 1975a, Problem B-282, *The Fibonacci Quarterly*, 13:2 (April), 192.

Freitag, H. T., 1975b, Problem B-286, *The Fibonacci Quarterly*, 13:3 (Oct.), 286.

Freitag, H. T., 1976, Problem B-314, *The Fibonacci Quarterly*, 14:3 (Oct.), 288.

Freitag, H. T., 1979, Problem B-379, *The Fibonacci Quarterly*, 17:2 (April), 186.

Freitag, H. T., 1982a, Problem B-455, *The Fibonacci Quarterly*, 20:3 (Aug.), 282–283.

Freitag, H. T., 1982b, Problem B-457, *The Fibonacci Quarterly*, 20:4 (Nov.), 367.

Freitag, H. T., 1982c, Problem B-462, *The Fibonacci Quarterly*, 20:4 (Nov.), 369–370.

Freitag, H. T., 1982d, Problem B-463, *The Fibonacci Quarterly*, 20:4 (Nov.), 370.

Ganis, S. E., 1959, "Notes on the Fibonacci Sequence," *The American Mathematical Monthly*, 66 (Feb.), 129–130.

Gardiner, T., 1983, "The Fibonacci Spiral," *The Mathematical Gazette*, 67, 120.

Gardner, M., 1956, *Mathematics Magic and Mystery*, New York: Dover.

Gardner, M., 1959, "Mathematical Games," *Scientific American*, 201 (Aug.), 128–134.

Gardner, M., 1960, "*Mathematical Puzzles and Diversions*," Chicago: The University of Chicago Press.

Gardner, M., 1969, "The Multiple Fascination of the Fibonacci Sequence," *Scientific American*, 211 (March), 116–120.

Gardner, M., 1970a, "Mathematical Games," *Scientific American*, 222 (Feb.), 112–114.

Gardner, M., 1970b, "Mathematical Games," *Scientific American*, 222 (March), 121–125.

Gardner, M., 1994, "The Cult of the Golden Ratio," *Skeptical Inquirer*, 18 (Spring), 243–247.

Garland, T. H., 1987, *Fascinating Fibonaccis*, Palo Alto, CA: Dale Seymour Publications.

Gattei, P., 1990–1991, "The 'Inverse' Differential Equation," *Mathematical Spectrum*, 23:4, 127–131.

Gauthier, N., 1995, "Fibonacci Sums of the Type $\sum r^m F_m$," *The Mathematical Gazette*, 79 (July), 364–367.

Geller, S. P., 1963, "A Computer Investigation of a Property of the Fibonacci Sequence," *The Fibonacci Quarterly*, 1:2 (April), 84.

Gessel, I., 1972, Problem H-187, *The Fibonacci Quarterly*, 10:4 (Oct.), 417–419.

Ghyka, M., 1977, *The Geometry of Art and Life*, New York: Dover.

Ginsburg, J., 1953, "A Relationship Between Cubes of Fibonacci Numbers," *Scripta Mathematica*, 19 (Dec.), 242.

Ginzburg, J., 1948, "Fibonacci Pleasantries," *Scripta Mathematica*, 14, 163–164.

Glaister, P., 1995, "Fibonacci Power Series," *The Mathematical Gazette*, 79 (Nov.), 521–525.

Glaister, P., 1996, "Golden Earrings," *The Mathematical Gazette*, 80 (March), 224–225.

Glaister, P., 1997, "Two Fibonacci Sums — a Variation," *The Mathematical Gazette*, 81 (March), 85–88.

Glaister, P., 1998–1999, "The Golden Ration by Origami," *Mathematical Spectrum*, 3:3, 52–53.

Goggins, J. R., 1973, "Formula for $\pi/4$," *The Mathematical Gazette*, 57 (June), 134.

Good, I. J., "A Reciprocal Series of Fibonacci Numbers," *The Fibonacci Quarterly*, 12:4 (Dec.), 346.

Gordon, J., 1938, *A Step-ladder to Painting*, London: Faber & Faber.

Gould, H. W., 1963a, Problem B-7, *The Fibonacci Quarterly*, 1:3 (Oct.), 80.

Gould, H. W., 1963b, "General Functions for Products of Powers of Fibonacci Numbers," *The Fibonacci Quarterly*, 1:2 (April), 1–16.

Gould, H. W., 1964, "Associativity and the Golden Section," *The Fibonacci Quarterly*, 2:3 (Oct.), 203.

Gould, H. W., 1965, "A Variant of Pascal's Triangle," *The Fibonacci Quarterly*, 3:4 (Dec.), 257–271.

Gould, H. W., 1972, "The Case of the Strange Binomial Identities of Professor Moriarty," *The Fibonacci Quarterly*, 10:4 (Oct.), 381–391, 402.

Gould, S. W., 1957, Problem 4720, *The American Mathematical Monthly*, 64:10 (Dec.), 749–750.

Graesser, R. F., 1964, "The Golden Section," *The Pentagon* (Spring), 7–19.

Graham, R. L., 1963a, Problem H-10, *The Fibonacci Quarterly*, 1:2 (April), 53.

Graham, R. L., 1963b, "Fibonacci Sums," *The Fibonacci Quarterly*, 1:4 (Dec.), 76.

Graham, R. L., 1965, Problem H-45, *The Fibonacci Quarterly*, 3:2 (April), 127–128.

Graham, R. L., et al., 1989, *Concrete Mathematics*, Reading, MA: Addison-Wesley.

Grassl, R. M., 1971a, Problem B-202, *The Fibonacci Quarterly*, 9:5 (Dec.), 546–547.

Grassl, R. M., 1971b, Problem B-203, *The Fibonacci Quarterly*, 9:5 (Dec.), 547–548.

Grassl, R. M., 1973, Problem B-264, *The Fibonacci Quarterly*, 11:3 (Oct.), 333.

Grassl, R. M., 1974, Problem B-279, *The Fibonacci Quarterly*, 12:1 (Feb.), 101.

Grassl, R. M., 1975, Problem B-279, *The Fibonacci Quarterly*, 13:3 (Oct.), 286.

Grimaldi, R., 1996, MAA Minicourse No. 13, Orlando, FL, January.

Gross, G., and H. Renzi, 1965, "Teachers Discover New Math Theorem," *Arithmetic Teacher*, 12:8, 625–626.

Gudder, S., 1964, *A Mathematical Journey*, 2nd ed., New York: McGraw-Hill.

Guillotte, G. A. R., 1971a, Problem B-206, *The Fibonacci Quarterly*, 9:5 (Dec.), 550.

Guillotte, G. A. R., 1971b, Problem B-207, *The Fibonacci Quarterly*, 9:5 (Dec.), 551.

Guillotte, G. A. R., 1972a, Problem B-210, *The Fibonacci Quarterly*, 10:2 (Feb.), 222.

Guillotte, G. A. R., 1972b, Problem B-218, *The Fibonacci Quarterly*, 10:3 (April), 335–336.

Guillotte, G. A. R., 1973, Problem B-241, *The Fibonacci Quarterly*, 11:3 (Oct.), 335–336.

Guy, R. K., 1994, *Unsolved Problems In Number Theory*, 2nd ed., New York: Spring-Verlag.

Halton, J. H., 1965, "On a General Fibonacci Identity," *The Fibonacci Quarterly*, 3:1 (Feb.), 31–43.

Halton, J. H., 1967, "Some Applications Associated with Square Fibonacci Numbers," *The Fibonacci Quarterly*, 5:4 (Nov.), 347–355.

Hansen, R. T., 1972, "Generating Identities for Fibonacci and Lucas Triples," *The Fibonacci Quarterly*, 10:6 (Dec.), 571–578.

Hardy, G. H., and E. M. Wright, 1995, *An Introduction to the Theory of Numbers*, 5th ed., New York: Oxford University Press.

Harris, V. C., 1965, "On Identities Involving Fibonacci Numbers," *The Fibonacci Quarterly*, 3:3 (Oct.), 214–218.

Harvey, M., and P. Woodruff, 1995, "Fibonacci Numbers and Sums of Inverse tangents," *The Mathematical Gazette*, 79 (Nov.), 565–567.

Heaslet, M. A., 1938, Solution to Problem 3801, *The American Mathematical Monthly*, 45 (Nov.), 636–637.

Heath, R. V., 1950, "Another Fibonacci Curiosity," *Scripta Mathematica*, 16:1–2 (March–June), 128.

Higgins, F., 1976, Problem B-305, *The Fibonacci Quarterly*, 14:2 (April), 189.

Hillman, A. P., 1971, Editorial Note, *The Fibonacci Quarterly*, 9:5 (Dec.), 548.

Hoare, G., 1989, "Regenerative Powers of Rabbits," *The Mathematical Gazette*, 73, 336–337.

Hogben, L., 1938, *Science for the Citizen*, London: Allen & Unwin.

Hoggatt, V. E., Jr., 1963, Problem B-2, *The Fibonacci Quarterly*, 1:1 (Feb.), 73.

Hoggatt, V. E., Jr., 1964, Problem H-39, *The Fibonacci Quarterly*, 2:2 (April), 124.

Hoggatt, V. E., Jr., 1964, Problem H-31, *The Fibonacci Quarterly*, 2:4 (Dec.), 306–307.

Hoggatt, V. E., Jr., 1965, Problem B-60, *The Fibonacci Quarterly*, 3:3 (Oct.), 238.

Hoggatt, V. E., Jr., 1967, Problem H-77, *The Fibonacci Quarterly*, 5:3 (Oct.), 256–258.

Hoggatt, V. E., Jr., 1968a, Problem H-82, *The Fibonacci Quarterly*, 6:1 (Feb.), 52–54.

Hoggatt, V. E., Jr., 1968b, Problem H-88, *The Fibonacci Quarterly*, 6:4 (Oct.), 253–254.

Hoggatt, V. E., Jr., 1969a, *Fibonacci and Lucas Numbers*, Santa Clara, CA: The Fibonacci Association, University of Santa Clara.

Hoggatt, V. E., Jr., 1969b, Problem H-131, *The Fibonacci Quarterly*, 7:3 (Oct.), 285–286.

Hoggatt, V. E., Jr., 1969c, Problem B-149, *The Fibonacci Quarterly*, 7:3 (Oct.), 333–334.

Hoggatt, V. E., Jr., 1969d, Problem B-150, *The Fibonacci Quarterly*, 7:3 (Oct.), 334.

Hoggatt, V. E., Jr., 1970, "An Application of the Lucas Triangle," *The Fibonacci Quarterly*, 8:4 (Oct.), 360–364, 427.

Hoggatt, V. E., Jr., 1971, Problem H-183, *The Fibonacci Quarterly*, 9:4 (Oct.), 389, 288–289.

Hoggatt, V. E., Jr., 1971a, "Some Special Fibonacci and Lucas Generating Functions," *The Fibonacci Quarterly*, 9:2 (April), 121–133.

Hoggatt, V. E., Jr., 1971b, Problem B-193, *The Fibonacci Quarterly*, 9:2 (April), 221–223.

Hoggatt, V. E., Jr., 1971c, Problem B-204, *The Fibonacci Quarterly*, 9:5 (Dec.), 548–549.

Hoggatt, V. E., Jr., 1971d, Problem B-205, *The Fibonacci Quarterly*, 9:5 (Dec.), 549–550.

Hoggatt, V. E., Jr., 1972, Problem B-231, *The Fibonacci Quarterly*, 10:2 (Feb.), 219–220.

Hoggatt, V. E., Jr., 1972, "Generalized Fibonacci Numbers in Pascal's Pyramid," *The Fibonacci Quarterly*, 10:3 (April), 271–276, 293.

Hoggatt, V. E., Jr., 1972a, Problem B-208, *The Fibonacci Quarterly*, 10:2 (Feb.), 220–221.

Hoggatt, V. E., Jr., 1972b, Problem B-211, *The Fibonacci Quarterly*, 10:2 (Feb.), 222–223.

Hoggatt, V. E., Jr., 1972c, "Generalized Fibonacci Numbers in Pascal's Pyramid," *The Fibonacci Quarterly*, 10:3 (April), 271–276, 293.

Hoggatt, V. E., Jr., 1972e, Problem B-216, *The Fibonacci Quarterly*, 10:3 (April), 332–333.

Hoggatt, V. E., Jr., 1972f, Problem H-201, *The Fibonacci Quarterly*, 10:6 (Dec.), 630.

Hoggatt, V. E., Jr., 1972g, Problem B-222, *The Fibonacci Quarterly*, 10:6 (Feb.), 665.

Hoggatt, V. E., Jr., 1973a, Problem B-302, *The Fibonacci Quarterly*, 11:2 (April), 95.

Hoggatt, V. E., Jr., 1973b, Problem B-248, *The Fibonacci Quarterly*, 11:5 (Dec.), 553.

Hoggatt, V. E., Jr., 1973c, Problem B-249, *The Fibonacci Quarterly*, 11:5 (Dec.), 553.

Hoggatt, V. E., Jr., 1976, Problem B-298, *The Fibonacci Quarterly*, 14:1 (Feb.), 94.

Hoggatt, V. E., Jr., 1977, "Number Theory: The Fibonacci Sequence," *1977 Yearbook of Science and the Future*, Chicago: Encyclopedia Britannica.

Hoggatt, V. E., Jr., 1977a, Problem H-257, *The Fibonacci Quarterly*, 15:3 (Oct.), 283.

Hoggatt, V. E., Jr., 1977b, Problem B-313, *The Fibonacci Quarterly*, 14:3 (Oct.), 288.

Hoggatt, V. E., Jr., 1979, Problem B-375, *The Fibonacci Quarterly*, 17:1 (Feb.), 93.

Hoggatt, V. E., Jr., 1982, Problem H-319, *The Fibonacci Quarterly*, 20:1 (Feb.), 96.

Hoggatt, V. E., Jr., and K. Alladi, 1975, "Generalized Fibonacci Tiling," *The Fibonacci Quarterly*, 13:2 (April), 137–145.

Hoggatt, V. E., Jr., and G. E. Bergum, 1974, "Divisibility and Congruence Relations," *The Fibonacci Quarterly*, 12:2 (April), 189–195.

Hoggatt, V. E., Jr., and G. E. Bergum, 1977, "A Problem of Fermat and the Fibonacci Sequence," *The Fibonacci Quarterly*, 15:3 (Oct.), 323–330.

Hoggatt, V. E., Jr., and M. Bicknell, 1964, "Some new Fibonacci Identities," *The Fibonacci Quarterly*, 2:1 (Feb.), 29–32.

Hoggatt, V. E., Jr., and M. Bicknell, 1972, "Convolution Triangles," *The Fibonacci Quarterly*, 10:6 (Dec.), 599–608.

Hoggatt, V. E., Jr., and M. Bicknell, 1973a, "Roots of Fibonacci Polynomials," *The Fibonacci Quarterly*, 11:3 (Oct.), 271–274.

Hoggatt, V. E., Jr., and M. Bicknell, 1973b, "Generalized Fibonacci Polynomials," *The Fibonacci Quarterly*, 11:5 (Dec.), 457–465.

Hoggatt, V. E., Jr., and M. Bicknell, 1974, "A Primer for the Fibonacci Numbers: Part XIV," *The Fibonacci Quarterly*, 12:2 (April), 147–156.

Hoggatt, V. E., Jr., and H. Edgar, 1980, "Another Proof that $\varphi(F_n) \equiv 0 \pmod 4$ For all $n > 4$," *The Fibonacci Quarterly*, 18:1 (Feb.), 80–82.

Hoggatt, V. E., Jr., and D. A. Lind, 1967, "The Heights of Fibonacci Polynomials and an Associated Function," *The Fibonacci Quarterly*, 5:2 (April), 141–152.

Hoggatt, V. E., Jr., and D. A. Lind, 1968, "Symbolic Substitutions into Fibonacci Polynomials," *The Fibonacci Quarterly*, 6:5 (Nov.), 55–74.

Hoggatt, V. E., Jr., and C. H. King, 1963, Problem H-20, *The Fibonacci Quarterly*, 1:4 (Dec.), 52.

Hoggatt, V. E., Jr., and I. D. Ruggles, 1964, "A Primer for the Fibonacci Numbers—Part V," *The Fibonacci Quarterly*, 2:1 (Feb.) 59–65.

Hoggatt, V. E., Jr., and I. D. Ruggles, 1964, "A Primer on the Fibonacci Sequence — Part V," *The Fibonacci Quarterly*, 2:1 (Feb.), 59–65.

Hoggatt, V. E., Jr., et al., 1971, "Twenty-four Master Identities," *The Fibonacci Quarterly*, 9:1 (Feb.), 1–17.

Hoggatt, V. E., Jr., et al., 1972, "Fibonacci and Lucas Triangles," *The Fibonacci Quarterly*, 10:5 (Nov.), 555–560.

Holt, M. H., 1964, "The Golden Section," *The Pentagon* (Spring), 80–104.

Holt, M. H., 1965, "Mystery Puzzler and Phi," *The Fibonacci Quarterly*, 3:2 (April), 135–138.

Hope-Jones, W., 1921, "The Bee and the Regular Pentagon," *The Mathematical Gazette*, 10:150. [Reprinted in 55:392 (March 1971), 220.]

Horadam, A. F., 1962, "Fibonacci Sequences and a Geometrical Paradox," *Mathematics Magazine*, 35:1 (Jan.), 1–11.

Horadam, A. F., 1963, "Further Appearance of the Fibonacci Sequence," *The Fibonacci Quarterly*, 1:4 (Dec.), 41–42, 46.

Horadam, A. F., 1971, "Pell Identities," *The Fibonacci Quarterly*, 9:3 (May), 245–252, 263.

Horner, W. W., 1966, "Fibonacci and Euclid," *The Fibonacci Quarterly*, 4:2 (April), 168–169.

Horner, W. W., 1969, Problem B-146, *The Fibonacci Quarterly*, 7:2 (April), 223.

Horner, W. W., 1973, "Fibonacci and Appollonius," *The Fibonacci Quarterly*, 11:5 (April), 541–542.

Hosoya, H., 1973, "Topological Index and Fibonacci Numbers with Relation to Chemistry," *The Fibonacci Quarterly*, 11:3 (Oct.), 255–265.

Hosoya, H., 1976, "Fibonacci Triangle," *The Fibonacci Quarterly*, 14:2 (April), 173–178.

Householder, J. E., 1963, Problem B-3, *The Fibonacci Quarterly*, 1:1 (Feb.), 73.

Hunter, J. A. H., 1963, "Triangle Inscribed in a Rectangle," *The Fibonacci Quarterly*, 1:3 (Oct.), 66.

Hunter, J. A. H., 1964a, "Fibonacci Geometry," *The Fibonacci Quarterly*, 2:2 (April), 104.

Hunter, J. A. H., 1964b, "The Golden Cuboid," *The Fibonacci Quarterly*, 2:3 (Oct.), 184, 240.

Hunter, J. A. H., 1966a, "Fibonacci Yet Again," *The Fibonacci Quarterly*, 4:3 (Oct.), 273.

Hunter, J. A. H., 1966b, Problem H-79, *The Fibonacci Quarterly*, 4:1 (Feb.), 57.

Hunter, J. A. H., 1969, Problem H-124, *The Fibonacci Quarterly*, 7:2 (April), 179.

Hunter, J. A. H., and J. S. Madachy, 1975, *Mathematical Diversions*, New York: Dover.

Huntley, H. E., 1964, "Fibonacci Geometry," *The Fibonacci Quarterly*, 2:2 (April), 104.

Huntley, H. E., 1970, *The Divine Proportion*, New York: Dover.

Huntley, H. E., 1974a, "The Golden Ellipse," *The Fibonacci Quarterly*, 12:1 (Feb.), 38–40.

Huntley, H. E., 1974b, "Phi: Another Hiding Place," *The Fibonacci Quarterly*, 12:1 (Feb.), 65–66.

Ivanoff, V., 1968, Problem H-107, *The Fibonacci Quarterly*, 6:6 (Dec.), 358–359.

Jackson, W. D., 1969, Problem B-142, *The Fibonacci Quarterly*, 7:2 (April), 220.

Jaiswal, D. V., 1969, "Determinants Involving Generalized Fibonacci Numbers," *The Fibonacci Quarterly*, 7:3 (Oct.), 319–330.

Jaiswal, D. V., 1974, "Some Geometrical Properties of the Generalized Fibonacci Sequence," *The Fibonacci Quarterly*, 12:1 (Feb.), 67–70.

Jarden, D., 1967, "A New Important Formula for Lucas Numbers," *The Fibonacci Quarterly*, 5:4 (Nov.), 346.

Jean, R. V., 1984, "The Fibonacci Sequence," *The UMAP Journal*, 5:1, 23–47.

Jordan, J. H., 1965, "Gaussian Fibonacci and Lucas Numbers," *The Fibonacci Quarterly*, 3:4 (Dec.), 315–318.

Kimberling, C., 1976, Problem E 2581, *The American Mathematical Monthly*, 83:3 (March), 197.

King, B. W., 1968, Solution to Problem B-129, *The Fibonacci Quarterly*, 6:5 (Oct.), 296.

King, B. W., 1971, Problem B-184, *The Fibonacci Quarterly*, 9:1 (Feb.), 107–108.

King, C. H., 1960, "Some Properties of the Fibonacci Numbers," Master's Thesis, San Jose State College, San Jose, CA (June).

King, C. H., 1963, "Leonardo Fibonacci," *The Fibonacci Quarterly*, 1:4 (Dec.), 15–19.

Klarner, D. A., 1966, "Determinants Involving kth Powers from Second Order Sequences," *The Fibonacci Quarterly*, 4:2 (April), 179–183.

Koshy, T., 1996, "Fibonacci Partial Sums," *Pi Mu Epsilon*, 10 (Spring), 267–268.

Koshy, T., 1998, "New Fibonacci and Lucas Identities," *The Mathematical Gazette*, 82 (Nov.), 481–484.

Koshy, T., 1999, "The Convergence of a Lucas Series," *The Mathematical Gazette*, 83 (July), 272–274.

Koshy, T., 2000, "Weighted Fibonacci and Lucas Sums," *The Mathematical Gazette*, 85 (March), 93–96.

Koshy, T., 2001, "A Family of Polynomial Functions," *Pi Mu Epsilon Journal*, 12 (Spring), 195–201.

Koshy, T., 2001, *Elementary Number Theory with Applications*, New York: Academic Press.

Kramer, J., and V. E. Hoggatt, Jr., 1972, "Special Cases of Fibonacci Periodicity," *The Fibonacci Quarterly*, 10:5 (Nov.), 519–522.

Kravitz, S., 1965, Problem B-63, *The Fibonacci Quarterly*, 3:3 (Oct.), 239–240.

Krishna, H. V., 1972, Problem B-227, *The Fibonacci Quarterly*, 10:2 (Feb.), 218.

Land, F., 1960, *The Language of Mathematics*, London: Murray.

Lang, L., 1973, Problem B-247, *The Fibonacci Quarterly*, 11:5 (Dec.), 552.

Law, A. G., 1971, Solution to Problem H-157, *The Fibonacci Quarterly*, 9:1 (Feb.), 65–67.

Ledin, G., 1966, Problem H-57, *The Fibonacci Quarterly*, 4:4 (Dec.), 336–338.

Ledin, G., 1967, Problem H-117, *The Fibonacci Quarterly*, 5:2 (April), 162.

Lehmer, D. H., 1936, Problem E 621, *The American Mathematical Monthly*, 43 (May), 307.

Lehmer, D. H., 1936, Problem 3801, *The American Mathematical Monthly*, 43 (Nov.), 580.

Lehmer, D. H., 1936, Problem 3802, *The American Mathematical Monthly*, 43 (Nov.), 580.

Lehmer, D. H., 1938a, Problem 3801, *The American Mathematical Monthly*, 45 (Nov.), 636–637.

Lehmer, D. H., 1938b, Problem 3802, *The American Mathematical Monthly*, 45 (Nov.), 632–633.

LeVeque, W. J., 1962, *Elementary Theory of Numbers*, Reading, MA: Addison-Wesley.

Libis, L., 1997, "A Fibonacci Summation," *Math Horizons* (Feb.), 33–34.

Lind, D. A., 1964, Problem B-31, *The Fibonacci Quarterly*, 2:1 (Feb.), 72.

Lind, D. A., 1965, Problem H-64, *The Fibonacci Quarterly*, 3:1 (Feb.), 44.

Lind, D. A., 1965, Problem H-54, *The Fibonacci Quarterly*, 3:2 (April), 116.

Lind, D. A., 1965, Problem B-70, *The Fibonacci Quarterly*, 3:3 (Oct.), 235.

Lind, D. A., 1966, Solution to Problem H-57, *The Fibonacci Quarterly*, 4:4 (Dec.), 337–338.

Lind, D. A., 1967, "Iterated Fibonacci and Lucas Subscripts," *The Fibonacci Quarterly*, 5:1 (Feb.), 89–90, 80.

Lind, D. A., 1967, Problem B-103, *The Fibonacci Quarterly*, 5:3 (Oct.), 290–292.

Lind, D. A., 1967a, "The Q matrix as a Counterexample in Group Theory," *The Fibonacci Quarterly*, 5:1 (Feb.), 44.

Lind, D. A., 1967b, Problem B-91, *The Fibonacci Quarterly*, 5:1 (Feb.), 110.

Lind, D. A., 1967c, Problem B-98, *The Fibonacci Quarterly*, 5:2 (April), 206.

Lind, D. A., 1967d, Problem B-99, *The Fibonacci Quarterly*, 5:2 (April), 206–207.

Lind, D. A., 1968, Problem B-134, *The Fibonacci Quarterly*, 6:6 (Dec.), 404–405.

Lind, D. A., 1970, Problem B-165, *The Fibonacci Quarterly*, 8:1 (Feb.), 111–112.

Lind, D. A., 1971, "A Determinant Involving Generalized Binomial Coefficients," *The Fibonacci Quarterly*, 9:2 (April), 113–119, 162.

Lindstrom, P., 1977, Problem B-341, *The Fibonacci Quarterly*, 14:3 (Oct.), 376.

Litvack, B., 1964, Problem B-47, *The Fibonacci Quarterly*, 2:3 (Oct.), 232.

Lloyd, E. K., 1985, "Enumeration," *Handbook of Applicable Mathematics*, New York: Wiley, 531–621.

London, H., and R. Finkelstein, 1969, "On Fibonacci and Lucas Numbers which Are Perfect Powers," *The Fibonacci Quarterly*, 7:5 (Dec.), 476–481, 487.

Long, C. T., 1981, "The Decimal Expansion of 1/89 and Related Results," *The Fibonacci Quarterly*, 19:1 (Feb.), 53–55.

Long, C. T., and J. H. Jordan, 1967, "A Limited Arithmetic on Simple Continued Fractions," *The Fibonacci Quarterly*, 5:2 (April), 113–128.

Lord, G., 1973, Solution to Problem B-248, *The Fibonacci Quarterly*, 11:5 (Dec.), 553.

Lord, G., 1975, Solution to Problem B-286, *The Fibonacci Quarterly*, 13:3 (Oct.), 286.

Lord, G., 1976, Solution to Problem B-311, *The Fibonacci Quarterly*, 14:3 (Oct.), 287.

Lord, N., 1995, "Balancing and Golden Rectangles," *The Mathematical Gazette*, 79 (Nov.), 573–574.

Mana, P., 1969, Problem B-152, *The Fibonacci Quarterly*, 7:3 (Oct.), 336.

Mana, P., 1970, Problem B-163, *The Fibonacci Quarterly*, 8:1 (Feb.), 110.

Mana, P., 1972, Problem B-215, *The Fibonacci Quarterly*, 10:3 (April), 331–332.

Mana, P., 1978, Problem B-354, *The Fibonacci Quarterly*, 16:2 (April), 185–186.

Matiyasevich, Y. V., and R. K. Guy, 1986, "A New Formula for π," *The American Mathematical Monthly*, 93 (Oct.), 631–635.

Maxwell, J. A., 1963, Problem B-8, *The Fibonacci Quarterly*, 1:1 (Feb.), 75.

McCown, J. R., and M. A. Sequeira, 1994, *Patterns in Mathematics*, Boston: PWS.

McNabb, M., 1963, "Phyllotaxis," *The Fibonacci Quarterly*, 1:1 (Feb.), 57–60.

The Mathematics Teacher, 1993, Calendar Problems, 86 (Jan.), 48–49.

Mead, D. G., 1965, Problem B-67, *The Fibonacci Quarterly*, 3:4 (Dec.), 326–327.

Michael, G., 1964, "A New Proof of an Old Property," *The Fibonacci Quarterly*, 2:1 (Feb.), 57–58.

Milsom, J. W., 1973, Solution to Problem B-227, *The Fibonacci Quarterly*, 11:1 (Feb.), 107–108.

Ming, L., 1989, "On Triangular Fibonacci Numbers," *The Fibonacci Quarterly*, 27, 98–108.

Moise, E. E., and E. Downs, 1964, Geometry, *Reading*, MA: Addison-Wesley.

Montgomery, P., 1977, Solution to Problem E 2581, *The American Mathematical Monthly*, 84:6 (June–July), 488.

Monzingo, M. G., 1983, "Why Are 8:18 and 10:09 Such Pleasant Times?," *The Fibonacci Quarterly*, 21:2 (May), 107–110.

Moore, R. E. M., 1970, "Mosaic Units: Patterns in Ancient Mosaics," *The Fibonacci Quarterly*, 8:3 (April), 281–310.

Moore, S., 1983, "Fibonacci Matrices," *The Mathematical Gazette*, 67, 56–57.

Morgan-Voyce, A. M., 1959, "Ladder Network Analysis using Fibonacci Numbers," *IRE. Trans. on Circuit Theory*, CT-6 (Sept.), 321–322.

Moser, L., 1963, Problem B-5, *The Fibonacci Quarterly*, 1:1 (Feb.), 74.

Moser, L., and J. L. Brown, 1963, "Some Reflections," *The Fibonacci Quarterly*, 1:1 (Dec.), 75.

Moser, L., and M. Wyman, 1963, Problem B-6, *The Fibonacci Quarterly*, 1:1 (Feb.), 74.

Myers, B. R., 1971, "Number of Spanning Trees in a Wheel," *IEEE Transactions on Circuit Theory*, CT-18 (March), 280–281.

Myers, B. R., 1972, Advanced Problem 5795, *The American Mathematical Monthly*, 79:8 (Oct.), 914–915.

Neumer, G., 1993, Problem 487, *The College Mathematics Journal*, 24:5 (Nov.), 476.

O'Connell, M. K., and D. T. O'Connell, 1951, Letter to the Editor, *The Scientific Monthly*, 78 (Nov.), 333.

Odom, G., 1986, Problem E3007, *The American Mathematical Monthly*, 93 (Aug.–Sept.), 572.

Ogg, F. C., 1951, Letter to the Editor, *The Scientific Monthly*, 78 (Nov.), 333.

Ogilvy, C. S., and J. T. Anderson, 1966, *Excursions in Number Theory*, New York: Dover.

Padilla, G. C., 1967, Solution to Problem B-98, *The Fibonacci Quarterly*, 5:2 (April), 206.

Padilla, G. C., 1970a, Problem B-172, *The Fibonacci Quarterly*, 8:4 (Oct.), 444–445.

Padilla, G. C., 1970b, Problem B-173, *The Fibonacci Quarterly*, 8:4 (Oct.), 445.

Parker, F. D., 1964, Problem H-21, *The Fibonacci Quarterly*, 2:2 (April), 133.

Parks, P. C., 1963, "A New proof of the Hurwitz Stability Criterion by the Second Method of Liapunov with Applications to "Optimum" Transfer Functions," *Proceedings of the Joint Automatic Control Conference*, University of Minnesota, June.

Peck, C. B. A., 1969, Solution to Problem B-146, *The Fibonacci Quarterly*, 7:2 (April), 223.

Peck, C. B. A., 1970, Editorial Note, *The Fibonacci Quarterly*, 8:4 (Oct.), 392.

Peck, C. B. A., 1971, Solution to Problem B-206, *The Fibonacci Quarterly*, 9:5 (Dec.), 550.

Peck, C. B. A., 1974, Solution to Problem B-253, *The Fibonacci Quarterly*, 12:1 (Feb.), 105.

Peck, C. B. A., 1974, Solution to Problem B-270, *The Fibonacci Quarterly*, 12:4 (Dec.), 405.

Perfect, H., 1984, "A Matrix Limit- (1)," *The Mathematical Gazette*, 68 (March), 42–44.

Peters, J. M. H., 1978, "An Approximate Relation between π and the Golden Ratio," *The Mathematical Gazette*, 62, 197–198.

Pettet, M., 1967, Problem B-93, *The Fibonacci Quarterly*, 5:1 (Feb.), 111.

Phillips, R., 1994, Numbers: Facts, Figures, and Fiction, New York: Cambridge University Press.

Pierce, J. C., 1951, "The Fibonacci Series," *The Scientific Monthly*, 78 (Oct.), 224–228.

Pond, J. C., 1967, Solution to Problem B-91, *The Fibonacci Quarterly*, 5:1 (Feb.), 110.

Prechter, R. R., Jr., and A. J. Frost, 1985, *The Elliot Wave Principle*, Gainesville, GA: New Clasics Library.

Prielipp, B., 1982, Solution to Problem B-463, *The Fibonacci Quarterly*, 20:5 (Nov.), 370.

Rabinowitz, S., 1998, "An Arcane Formula for a Curious Matrix," *The Fibonacci Quarterly*, 36:4 (Aug.), 376.

Raine, C. W., 1948, "Pythagorean Triangles from the Fibonacci Series 1, 1, 2, 3, 5, 8," *Scripta Mathematica*, 14, 164.

Ransom, W. R., 1959, *Trigonometric Novelties*, Portland, ME: J. Weston Walch.

Rao, K. S., 1953, "Some Properties of Fibonacci Numbers," *The American Mathematical Monthly*, 60 (May), 680–684.

Raphael, L., 1970, "Some Results in Trigonometry," *The Fibonacci Quarterly*, 8:4 (Oct.), 371, 392.

Ratliff, T., 1996, "A Few of my Favorite Calc III Things," MAA Fall Meeting, Univ. of Massachusetts, Boston (Nov.).

Rebman, R. R., 1975, "The Sequence: 1 5 16 45 121 320 . . . in Combinatorics," *The Fibonacci Quarterly*, 13:1 (Feb.), 51–55.

Recke, K. G., 1969a, Problem B-153, *The Fibonacci Quarterly*, 7:3 (Oct.), 276.

Recke, K. G., 1969b, Problem B-157, *The Fibonacci Quarterly*, 7:5 (Dec.), 548–549.

Rigby, J. F., 1988, "Equilateral Triangles and the Golden Ratio," *The Mathematical Gazette*, 72 (March), 27–30.

Robbins, N. R., 1981, "Fibonacci and Lucas Numbers of the Forms $w^2 - 1$, $w^3 \pm 1$," *The Fibonacci Quarterly*, 19, 369–373.

Roche, J. W., 1999, "Fibonacci and Approximating Roots," *Mathematics Teacher*, 92 (Sept.), 523.

Rosen, K. H., 1993, *Elementary Number Theory and Its Applications*, 3rd ed., Reading, MA: Addison-Wesley.

Ruggles, I. D., 1963a, Problem B-1, *The Fibonacci Quarterly*, 1:1 (Feb.), 73.

Ruggles, I. D., 1963b, "Some Fibonacci Results Using Fibonacci-type Sequences," *The Fibonacci Quarterly*, 1:2 (April), 75–80.

Ruggles, I. D., and V. E. Hoggatt, Jr., 1963a, "A Primer for the Fibonacci Sequence—Part III," *The Fibonacci Quarterly*, 1:3 (Oct.), 61–65.

Ruggles, I. D., and V. E. Hoggatt, Jr., 1963b, "A Primer on the Fibonacci Sequence—Part IV," *The Fibonacci Quarterly*, 1:4 (Dec.), 65–71.

Satterly, J., 1956a, "Meet Mr. Tau," *School Science and Mathematics*, 56, 731–741.

Satterly, J., 1956b, "Meet Mr. Tau Again," *School Science and Mathematics*, 56, 150–151.

Schub, P., 1950, "A Minor Fibonacci Curiosity," *Scripta Mathematica*, 16:3 (Sept.), 214.

Schub, P., 1970, "A Minor Fibonacci Curiosity," *Scripta Mathematica*, 17:3–3 (Sept.–Dec.), 214.

Scott, J. A., 1996, "A Fractal Curve Suggested by the MA Crest," *The Mathematical Gazette*, 80 (March), 236–237.

Seamons, R. S., 1967, Problem B-107, *The Fibonacci Quarterly*, 5:1 (Feb.), 107.

Sedlacek, J., 1970, "On the Skeletons of a Graph or Digraph," *Proceedings of the Calgary International Conference of Combinatorial Structures and Their Applications*, New York: Gordon & Breach, 387–391.

Shallit, J., 1976, Problem B-311, *The Fibonacci Quarterly*, 14:3 (Oct.), 287.

Shallit, J., 1981, Problem B-423, *The Fibonacci Quarterly*, 19:1 (Feb.), 92.

Shannon, A. G., 1979, Problem B-382, *The Fibonacci Quarterly*, 17:3 (Oct.), 282–283.

Sharpe, B., 1965, "On Sums $F_x^2 \pm F_y^2$," *The Fibonacci Quarterly*, 3:1 (Feb.), 63.

Sharpe, W. E., 1978, "Golden Weaves," *The Mathematical Gazette*, 62, 42–44.

Singh, P., 1985, "The So-called Fibonacci Numbers in Ancient and Medieval India," *Historia Mathematica*, 12, 229–244.

Singh, S., 1989, "The Beast 666," *Journal of Recreational Mathematics*, 21:4, 244.

Smith, K. J., 1995, *The Nature of Mathematics*, 7th ed., Pacific Grove, CA: Brooks/Cole.

Sofo, A., 1999, "Generalizations of a Radical Identity," *The Mathematical Gazette*, 83 (July), 274–276.

Somer, L., 1979, Solution to Problem B-382, *The Fibonacci Quarterly*, 17:4 (Oct.), 282–283.

Squire, W., 1968, Problem H-83, *The Fibonacci Quarterly*, 6:1 (Feb.), 54–55.

Stancliff, F. S., 1953, "A Curious Property of F_{11}," *Scripta Mathematica*, 19:2–3 (June–Sept.), 126.

Stanley, T. E., 1971, Solution to Problem B-203, *The Fibonacci Quarterly*, 9:5 (Dec.), 547–548.

Starke, E. P., 1936, Solution to Problem 621, *The American Mathematical Monthly*, 43 (May), 307–308.

Steinhaus, H., 1969, *Mathematical Snapshots*, 3rd ed., New York: Oxford University Press.

Steinhaus, H., 1979, *One Hundred Problems in Elementary Mathematics*, New York: Dover.

Stephen, M., 1956, "The Mysterious Number PHI," *Mathematics Teacher*, 49 (March), 200–204.

Stern, F., 1979, Problem B-374, *The Fibonacci Quarterly*, 17:1 (Feb.), 93.

Struyk, A., 1970, "One Curiosm Leads to Another," *Scripta Mathematica*, 17:3–4 (Sept.–Dec.), 230.

Swamy, M. N. S., 1966a, "Properties of the Polynomials Defined by Morgan-Voyce," *The Fibonacci Quarterly*, 4:1 (Feb.), 73–81.

Swamy, M. N. S., 1966b, Problem B-84, *The Fibonacci Quarterly*, 4:1 (Feb.), 90.

Swamy, M. N. S., 1966c, Problem B-74, *The Fibonacci Quarterly*, 4:1 (Feb.), 94–96.

Swamy, M. N. S., 1966d, "More Fibonacci Identities," *The Fibonacci Quarterly*, 4:4 (Dec.), 369–372.

Swamy, M. N. S., 1966e, Problem B-83, *The Fibonacci Quarterly*, 4:4 (Dec.), 375.

Swamy, M. N. S., 1967, Problem H-69, *The Fibonacci Quarterly*, 5:2 (April), 163–165.

Swamy, M. N. S., 1968a, Problem H-127, *The Fibonacci Quarterly*, 6:1 (Feb.), 51.

Swamy, M. N. S., 1968b, "Further Properties of Morgan-Voyce Polynomials," *The Fibonacci Quarterly*, 6:2 (April), 167–175.

Swamy, M. N. S., 1970, Problem H-150, *The Fibonacci Quarterly*, 8:4 (Oct.), 391–392.

Swamy, M. N. S., 1971a, Problem H-157, *The Fibonacci Quarterly*, 9:1 (Feb.), 65–67.

Swamy, M. N. S., 1971b, Problem H-158, *The Fibonacci Quarterly*, 9:1 (Feb.), 67–69.

Tadlock, S. B., 1965, "Products of Odds," *The Fibonacci Quarterly*, 3:1 (Feb.), 54–56.

Tallman, M. H., 1963, Problem H-23, *The Fibonacci Quarterly*, 1:3 (Oct.), 47.

Taylor, L., 1982a, Problem B-460, *The Fibonacci Quarterly*, 20:4 (Nov.), 368–369.

Taylor, L., 1982b, Problem B-461, *The Fibonacci Quarterly*, 20:4 (Nov.), 369.

Thoro, D., 1963, "The Golden Ratio: Computational Considerations," *The Fibonacci Quarterly*, 1:3 (Oct.), 53–59.

Tompkins, P., 1976, *Mysteries of the Mexican Pyramids*, New York: Harper & Row.

Trigg, C. W., 1973, "Geometric Proof of a Result of Lehmer's," *The Fibonacci Quarterly*, 11:5 (Dec.), 593–540.

Turner, M. R., 1974, "Certain Congruence Properties (modulo 100) of Fibonacci Numbers," *The Fibonacci Quarterly*, 12:1 (Feb.), 87–91.

Umansky, H. L., 1956a, "Pythagorean Triangles from Recurrent Series," *Scripta Mathematica*, 22:1 (March), 88.

Umansky, H. L., 1956b, "Curiosa, Zero Determinants," *Scripta Mathematica*, 22:1 (March), 88.

Umansky, H. L., 1970, Letter to the Editor, *The Fibonacci Quarterly*, 8:1 (Feb.), 89.

Umansky, H. L., 1973, Problem B-233, *The Fibonacci Quarterly*, 11:2 (April), 221–222.

Umansky, H. L., and M. Tallman, 1968, Problem H-101, *The Fibonacci Quarterly*, 6:4 (Oct.), 259–260.

Usiskin, Z., 1974, Problem B-265, *The Fibonacci Quarterly*, 12:3 (Oct.), 315.

Usiskin, Z., 1974, Problem B-266, *The Fibonacci Quarterly*, 12:3 (Oct.), 315–316.

Vajda, S., 1989, *Fibonacci and Lucas Numbers, and the Golden Section*, New York: Wiley.

Vinson, J., 1963, "The Relation of the Period Modulo m to the Rank of Appartition of m in the Fibonnaci Sequence," *The Fibonacci Quarterly*, 1:2 (April), 37–45.

Vorobev, N. N., 1961, Fibonacci Numbers, New York: Pergamon Press.

Walker, R. C., 1975, "An Electrical Network and the Golden Section," *The Mathematical Gazette*, 59, 162–165.

Wall, C. R., 1960, "Fibonacci Series modulo m," *The American Mathematical Monthly*, 67 (June–July), 525–532.

Wall, C. R., 1964a, Problem B-32, *The Fibonacci Quarterly*, 2:1 (Feb.), 72.

Wall, C. R., 1964b, Problem B-43, *The Fibonacci Quarterly*, 2:4 (Dec.), 329–330.

Wall, C. R., 1964c, Problem B-55, *The Fibonacci Quarterly*, 2:4 (Dec.), 324.

Wall, C. R., 1964d, Problem B-56, *The Fibonacci Quarterly*, 2:4 (Dec.), 325.

Wall, C. R., 1965, Problem B-45, *The Fibonacci Quarterly*, 3:1 (Feb.), 76.

Wall, C. R., 1967, Problem B-131, *The Fibonacci Quarterly*, 5:5 (Dec.), 466.

Wall, C. R., 1968, Problem B-127, *The Fibonacci Quarterly*, 6:4 (Oct.), 294–295.

Wall, C. R., 1985, "On Triangular Fibonacci Numbers," *The Fibonacci Quarterly*, 23, 77–79.

Webb, W. A., and E. A. Parberry, 1969, "Divisibility Properties of Fibonacci Polynomials," *The Fibonacci Quarterly*, 7:5 (Dec.), 457–463.

Weinstein, L., 1966, "A Divisibility Property of Fibonacci Numbers," *The Fibonacci Quarterly*, 4:1 (Feb.), 83–84.

Wessner, J., 1968, Solution to Problem B-127, *The Fibonacci Quarterly*, 6:4 (Oct.), 294–295.

Whitney, R. E., 1966a, "Extensions of Recurrence Relations," *The Fibonacci Quarterly*, 4:1 (Feb.), 37–42.

Whitney, R. E., 1966b, "Composition of Recursive Formulae," *The Fibonacci Quarterly*, 4:4 (Dec.), 363–366.

Whitney, R. E., 1972, Problem H-188, *The Fibonacci Quarterly*, 4:4 (Dec.), 631.

William, D., 1985, "A Fibonacci Sum," *The Mathematical Gazette*, 69, 29–31.

Williams, H. C., 1998, *Edouard Lucas and Primality Testing*, New York: Wiley.

Wlodarski, J., 1971, "The Golden ratio in an Electrical Network," *The Fibonacci Quarterly*, 9:2 (April), 188,194.

Wlodarski, J., 1973, "The Balmer Series and the Fibonacci Numbers," *The Fibonacci Quarterly*, 11:5 (Dec.), 526, 540.

Woko, J. E., 1997, "A Pascal-like Triangle for $\alpha^n + \beta^n$," *The Mathematical Gazette*, 81 (March), 75–79.

Wong, C. K., and T. W. Maddocks, 1975, "A Generalized Pascal's Triangle," 13:3 (April), 134–136.

Woolum, J., 1968, Problem B-119, *The Fibonacci Quarterly*, 6:2 (April), 187.

The World Book Encyclopedia, 1982, Vols. 1–22, Chicago.

Wulczyn, G., 1974, Solution to Problem B-256, *The Fibonacci Quarterly*, 12:2 (April), 221.

Wulczyn, G., 1974, Problem H-120, *The Fibonacci Quarterly*, 12:3 (Oct.), 401.

Wulczyn, G., 1977, Problem B-342, *The Fibonacci Quarterly*, 15:3 (Oct.), 376.

Wulczyn, G., 1978, Problem B-355, *The Fibonacci Quarterly*, 16:2 (April), 186.

Wulczyn, G., 1979, Problem B-384, *The Fibonacci Quarterly*, 17:3 (Oct.), 283.

Wunderlich, M., 1963, "On the Non-existence of Fibonacci Squares," *Mathematics of Computation*, 17, 455.

Wunderlich, M., 1965, "Another Proof of the Infinite Primes Theorem," *The American Mathematical Monthly*, 72 (March), 305.

Yodder, M., 1970, Solution to Problem B-165, *The Fibonacci Quarterly*, 8:1 (Feb.), 112.

Zeitlin, D., 1965, "Power Identities for Sequences Defined by $W_{n+2} = dW_{n+1} - cW_n$," *The Fibonacci Quarterly*, 3:4 (Dec.), 241–256.

Zeitlin, D., 1967, Solution to Problem H-64, *The Fibonacci Quarterly*, 5:1 (Feb.), 74–75.

Zeitlin, D., 1975, Solution to Problem B-277, *The Fibonacci Quarterly*, 13:1 (Feb.), 96.

Zeitlin, D., and F. D. Parker, 1963, Problem H-14, *The Fibonacci Quarterly*, 1:2 (April), 54.

Zerger, M. J., 1992, "The Golden State—Illinois," *Journal of Recreational Mathematics*, 24:1, 24–26.

Zerger, M. J., 1996, "Vol. 89 (and 11 to Centennial)," *Mathematics Teacher*, 89:1 (Jan.), 3, 26.

SOLUTIONS TO ODD-NUMBERED EXERCISES

EXERCISES 2 (p. 14)

1. 1, 1, 2, 3, 5, 8, 13, 21, 34, 55, 89, 144, 233, 377, 610, 987, 1597, 2584, 4181, 6765
3. 2
5. $F_{-n} = (-1)^{n+1} F_n$
7. $L_{-n} = (-1)^n L_n$
9. 20 + 19 + 15 + 5 + 1 = 60 = 17 + 13 + 11 + 9 + 7 + 3
11. $\sum_1^5 F_i = 12$
13. $\sum_1^8 F_i = 33$
15. $\sum_1^3 L_i = 8$
17. $\sum_1^7 L_i = 73$
19. $\sum_1^n L_i = L_{n+2} - 3$
21. $\sum_1^5 F_i^2 = 40$

23. $\sum_{1}^{8} F_i^2 = 714$

25. $\sum_{1}^{3} L_i^2 = 26$

27. $\sum_{1}^{7} L_i^2 = 1361$

29. $\sum_{1}^{n} L_i^2 = L_n L_{n+1} - 2$

31. $F_3 + F_5 = 2 + 5 = 7 = L_4$; $F_6 + F_8 = 8 + 21 = 29 = L_7$

33. We have $a_n = a_{n-1} + a_{n-2} + 1$, where $a_1 = 0 = a_2$. Let $b_n = a_n + 1$. Then $b_n - 1 = (b_{n-1} - 1) + (b_{n-2} - 1) + 1 = b_{n-1} + b_{n-2}$, so $b_n = b_{n+1} + b_{n-2}$, where $b_1 = 1 = b_2$. Thus $a_n = b_n - 1 = F_n - 1, n \geq 1$.

35. (J. L. Brown) Suppose $F_h < F_i < F_j < F_k$ are in AP, so $F_i - F_h = F_k - F_j = d$ (say). Then $d = F_i - F_h < F_i$, whereas $d = F_k - F_j \geq F_k - F_{k-1} = F_{k-2} \geq F_i$, a contradiction.

37. (DeLeon) Since $F_n < x < F_{n+1}$ and $F_{n+1} < y < F_{n+2}$, $F_n + F_{n+1} < x + y < F_{n+1} + F_{n+2}$; that is, $F_{n+2} < x + y < F_{n+3}$, a contradiction.

39. $a_n = \dfrac{k^n + k - 2}{(k-1)k^n}, n \geq 0.$

41. $\dfrac{1}{k-1}$

EXERCISES 3 (p. 49)

1. $b_1 = 1, b_2 = 2$
   $b_n = b_{n-1} + b_{n-2}, n \geq 3$
   $\therefore b_n = F_{n+1}, n \geq 1.$
3. Yes
5. No
7. $l_n = F_n$
9. $2F_n - 1$
11. 10, 15
13. 14, 25

EXERCISES 4 (p. 68)

1. 00000, 10000, 01000, 00100, 10100, 00010, 10010, 01010, 00001, 10001, 01001, 00101, 10101
3. $a_1 = 1, a_2 = 2, a_n = a_{n-1} + a_{n-2}, n \geq 3.$

5. $b_1 = 1, b_2 = 2, b_n = b_{n-1} + b_{n-2}, n \geq 3.$
7. 4, 7
9. $a_1 = 2, a_2 = 3, a_n = a_{n-1} + a_{n-2}, n \geq 3.$
11. None; $\{\{1\}, \emptyset\}$; $\{\{1\}, \emptyset\}, \{\{2\}, \emptyset\}, \{\{1, 2\}, \emptyset\}$
13. $S_n = F_{n(n+1)/2+2} - F_{n(n-1)/2+2}$

EXERCISES 5 (p. 96)

1. Since $\sum_1^1 F_{2i-1} = F_1 = F_2$, the result is true when $n = 1$. Assume it is true for an arbitrary positive integer k. Then $\sum_1^{k+1} F_{2i-1} = \sum_1^k F_{2i-1} + F_{2k+1} = F_{2k} + F_{2k+1} = F_{2k+2}$. Thus, by PMI, the formula is true for every $n \geq 1$.
3. $\sum_1^n L_i = \sum_1^n L_{i+2} - \sum_1^n L_{i+1} = (L_3 + L_4 + \cdots + L_{n+2}) - (L_2 + L_3 + \cdots + L_{n+1}) = L_{n+2} - L_2 = L_{n+2} - 3$
5. $\sum_1^n L_{2i} = \sum_1^{2n} L_i - \sum_1^n L_{2i-1} = (L_{2n+2} - 3) - (L_{2n} - 2) = (L_{2n+2} - L_{2n}) - 1 = L_{2n+1} - 1$
7. Since $\sum_1^1 L_i^2 = L_1^2 = 1 = L_1 L_2 - 2$, the result is true when $n = 1$. Assume it is true for an arbitrary positive integer k. Then $\sum_1^{k+1} L_i^2 = \sum_1^k L_i^2 + L_{k+1}^2 = (L_k L_{k+1} - 2) + L_{k+1}^2 = L_{k+1}(L_k + L_{k+1}) - 2 = L_{k+1} L_{k+2} - 2$. Thus, by PMI, the result is true for every $n \geq 1$.
9. $F_{10} = 55 = 5 \cdot 11 = F_5 L_5$
11. $F_7^2 - F_5^2 = 13^2 - 5^2 = 144 = F_{12}$
13. $L_3 L_1 - L_2^2 = 4 \cdot 1 - 9 \neq (-1)^2$
15. $v_1 = \alpha + \beta = 1$; $v_2 = \alpha^2 + \beta^2 = (\alpha + \beta)^2 - 2\alpha\beta = 1^2 - 2(-1) = 3$; and $v_{n-1} + v_{n-2} = (\alpha^{n-1} + \beta^{n-1}) + (\alpha^{n-2} + \beta^{n-2}) = \alpha^{n-2}(1 + \alpha) + \beta^{n-2}(1 + \beta) = \alpha^{n-2}\alpha^2 + \beta^{n-2}\beta^2 = \alpha^n + \beta^n = v_n$.
17. Since $(\alpha - \beta)\sqrt{5} = 1$, the result is true when $n = 1$. Assume it is true for all positive integers $\leq k$. Then $\sqrt{5}(\alpha F_k + \beta F_k) = (\alpha^{k+1} - \alpha\beta^k + \alpha^k\beta - \beta^{k+1})$; that is, $\sqrt{5}(\alpha + \beta)F_k = (\alpha^{k+1} - \beta^{k+1}) + (\alpha\beta)(\alpha^{k-1} - \beta^{k-1})$. So $\sqrt{5}F_k = (\alpha^{k+1} - \beta^{k+1}) - \sqrt{5}F_{k-1}$; that is, $\sqrt{5}(F_k + F_{k-1}) = (\alpha^{k+1} - \beta^{k+1})$. Thus $F_{k+1} = (\alpha^{k+1} - \beta^{k+1})/\sqrt{5}$. So the result holds for all $n \geq 1$.
19. Clearly, $L_1 = F_1$. Conversely, let $L_n = F_n$. Then $\alpha^n + \beta^n = (\alpha^n - \beta^n)/\sqrt{5}$, so $(\sqrt{5} - 1)\alpha^n = -(\sqrt{5} + 1)\beta^n$. $\alpha^n = (\alpha/\beta)\beta^n$; that is, $\alpha^{n-1} - \beta^{n-1} = 0$. $\therefore F_{n-1} = 0$, so $n = 1$.
21. $x^2 - (L_n + 2k)x + kL_n + k^2 + (-1)^n = 0$

23. Let $S = \sum_{1}^{n} F_i$. Since $\alpha^i = \alpha F_i + F_{i-1}$, $\sum_{0}^{n} \alpha^i = \alpha \sum_{0}^{n} F_i + \sum_{0}^{n} F_{i-1}$. That is, $(\alpha^{n+1}-1)/(\alpha-1) = \alpha S+1+0+(S-F_n)$. This yields $(\alpha F_{n+1}+F_n-1)/(\alpha-1) = (\alpha+1)S + 1 - F_n$ and thus $S = F_{n+2} - 1$.

25. $F_{n+5} = F_{n+4} + F_{n+3} = 2F_{n+3} + F_{n+2} = 3F_{n+2} + 2F_{n+1} = 5F_{n+1} + 3F_n$

27. $5(F_{n-1}F_{n+1} - F_n^2) = (\alpha^{n-1} - \beta^{n-1})(\alpha^{n+1} - \beta^{n+1}) - (\alpha^n - \beta^n)^2 = -3(-1)^{n-1} + 2(-1)^n = 5(-1)^n \therefore \quad F_{n-1}F_{n+1} - F_n^2 = (-1)^n$

29. $F_{2n} = (\alpha^{2n} - \beta^{2n})/\sqrt{5} = (\alpha^n - \beta^n)[(\alpha^n + \beta^n)/\sqrt{5}] = F_n L_n$

31. $F_{n+1}^2 - F_{n-1}^2 = (F_{n+1} + F_{n-1})(F_{n+1} - F_{n-1}) = L_n F_n = F_{2n}$

33. $F_{n+2} - F_{n-2} = (F_n + F_{n+1}) - (F_n - F_{n-1}) = F_{n+1} + F_{n-1} = L_n$

35. $F_{n+1}^2 - F_n^2 = (F_{n+1} + F_n)(F_{n+1} - F_n) = F_{n+2}F_{n-1}$

37. $L_n^2 + L_{n+1}^2 = (\alpha^n + \beta^n)^2 + (\alpha^{n+1} + \beta^{n+1})^2 = \alpha^{2n} + \beta^{2n} + \alpha^{2n+2} + \beta^{2n+2} = \alpha^{2n+1}(\alpha - \alpha^{-1}) + \beta^{2n+1}(\beta - \beta^{-1}) = (\alpha - \beta)(\alpha^{2n+1} - \beta^{2n+1}) = 5F_{2n+1}$

39. $L_n^2 - 4(-1)^n = (\alpha^n + \beta^n)^2 - 4(\alpha\beta)^n = (\alpha^n - \beta^n)^2 = 5F_n^2$

41. $L_{2n} + 2(-1)^n = \alpha^{2n} + \beta^{2n} + 2(\alpha\beta)^n = (\alpha^n + \beta^n)^2 = L_n^2$

43.
$$\begin{aligned} L_{n+2} - L_{n-2} &= (\alpha^{n+2} + \beta^{n+2}) - (\alpha^{n-2} + \beta^{n-2}) = \alpha^n(\alpha^2 - \alpha^{-2}) \\ &\quad - \beta^n(\beta^{-2} - \beta^2) = \alpha^n(\alpha^2 - \beta^2) - \beta^n(\alpha^2 - \beta^2) \\ &= (\alpha^n - \beta^n)(\alpha^2 - \beta^2) = (\alpha^n - \beta^n)(\alpha - \beta) = 5F_n \end{aligned}$$

45. Since $F_n^2 - F_{n-2}^2 = F_{2n-2}$ and $F_n^2 + F_{n-1}^2 = F_{2n-1}$, it follows that $F_n^2 > F_{2n-2}$ and $F_n^2 < F_{2n-1}$. Thus $F_{2n-2} < F_n^2 < F_{2n-1}$.

47. $L_{-n} = \alpha^{-n} + \beta^{-n} = (-\beta)^n + (-\alpha)^n = (-1)^n L_n$

49.
$$\begin{aligned} 1 + \alpha^{2n} &= \alpha^n(\alpha^n + \alpha^{-n}) = \alpha^n[\alpha^n + (-\beta)^n] \\ &= \begin{cases} \alpha^n(\alpha^n - \beta^n) & \text{if } n \text{ is odd} \\ \alpha^n(\alpha^n + \beta^n) & \text{otherwise} \end{cases} \\ &= \begin{cases} \sqrt{5}F_n\alpha^n & \text{if } n \text{ is odd} \\ L_n\alpha^n & \text{otherwise} \end{cases} \end{aligned}$$

51.
$$\begin{aligned} \text{LHS} &= (\alpha^{2m+n} + \beta^{2m+n}) - (\alpha\beta)^m(\alpha^n + \beta^n) \\ &= \alpha^{m+n}(\alpha^m - \beta^m) - \beta^{m+n}(\alpha^m - \beta^m) \\ &= (\alpha^m - \beta^m)(\alpha^{m+n} - \beta^{m+n}) \\ &= 5F_m F_{m+n} \end{aligned}$$

53.
$$\begin{aligned}\sqrt{5}\text{LHS} &= (\alpha^{2m+n} - \beta^{2m+n}) + (\alpha\beta)^m(\alpha^n - \beta^n)\\ &= \alpha^{m+n}(\alpha^m + \beta^m) - \beta^{m+n}(\alpha^m + \beta^m)\\ &= (\alpha^m + \beta^m)(\alpha^{m+n} - \beta^{m+n}) = \sqrt{5}L_m F_{m+n}\\ \therefore \text{LHS} &= L_m F_{m+n}.\end{aligned}$$

55.
$$\begin{aligned}L_{3n} &= \alpha^{3n} + \beta^{3n} = (\alpha^n + \beta^n)(\alpha^{2n} - \alpha^n\beta^n + \beta^n)\\ &= L_n[L_{2n} - (-1)n]\end{aligned}$$

57.
$$\begin{aligned}\sqrt{5}\text{LHS} &= (\alpha^{m+n} - \beta^{m+n}) - (\alpha^{m-n} - \beta^{m-n})\\ &= \alpha^m(\alpha^n - \alpha^{-n}) - \beta^m(\beta^n - \beta^{-n})\\ &= \alpha^m[\alpha^n - (-\beta)^n] - \beta^m[\beta^n - (-\alpha)^n]\\ &= \begin{cases}(\alpha^m - \beta^m)(\alpha^n + \beta^n) & \text{if } n \text{ is odd}\\ (\alpha^m + \beta^m)(\alpha^n - \beta^n) & \text{otherwise}\end{cases}\\ &= \begin{cases}F_m L_n & \text{if } n \text{ is odd}\\ L_m F_n & \text{otherwise}\end{cases}\end{aligned}$$

53. (Homer, Jr.)
$$\begin{aligned}F_{m+n+1} &= F_{m+1}F_{n+1} + F_m F_n\\ F_{m+n-1} &= F_m F_n + F_{m-1}F_{n-1}\end{aligned}$$

$$\therefore F_{m+1}F_{n+1} - F_{m-1}F_{n-1} = F_{m+n+1} - F_{m+n-1} = F_{m+n}$$

61.
$$\begin{aligned}5\cdot\text{RHS} &= 2[\alpha^{2n+4} + \beta^{2n+4} - 2(-1)^{n+2}]\\ &\quad + 2[\alpha^{2n+2} + \beta^{2n+2} - 2(-1)^{n+1}] - [\alpha^{2n} + \beta^{2n} - 2(-1)^n]\\ &= 2L_{2n+4} + 2L_{2n+2} - L_{2n} + 2(-1)^n\\ &= 2L_{2n+4} + (L_{2n+2} + L_{2n+1}) + 2(-1)^n\\ &= 2L_{2n+4} + L_{2n+3} + 2(-1)^n\\ &= L_{2n+4} + L_{2n+5} + 2(-1)^n = L_{2n+6} + 2(-1)^n\\ 5\cdot\text{LHS} &= \alpha^{2n+6} + \beta^{2n+6} - 2(-1)^{n+3} = L_{2n+6} + 2(-1)^n.\end{aligned}$$

$$\therefore \text{LHS} = \text{RHS}.$$

63. By Exercise 62, $F_{n+2}^3 + F_{n+1}^3 - F_n^3 = F_{3n+3}$ and $F_n^3 + F_{n-1}^3 - F_{n-2}^3 = F_{3n-3}$. Subtracting $F_{n+2}^3 + F_{n+1}^3 - 2F_n^3 - F_{n-1}^3 + F_{n-2}^3 = F_{3n+3} - F_{3n-3} = 4F_{3n}$, by Exercise 57. $\therefore F_{n+2}^3 - 3F_n^3 - F_{n-2}^3 = 3F_{3n} + (F_{3n} - F_{n+1}^3 - F_n^3 + F_{n-1}^3) = 3F_{3n}$, by Exercise 62.

65. $x = L_{2n}$, $y = F_{2n}$

67. (Carlitz) $x^2 - x - 1 = (x - \alpha)(x - \beta)$
$x^{2n} - L_n x^n + (-1)^n = x^{2n} - (\alpha^n + \beta^n)x^n + (\alpha\beta)^n = (x^n - \alpha^n)(x^n - \beta^n)$
Since $x - \alpha | x^n - \alpha^n$ and $x - \beta | x^n - \beta^n$, the desired result follows.

69. (Wulczyn)
$$\text{LHS} = (\alpha^{2n} + \beta^{2n}) + (\alpha^2 + \beta^2)(\alpha\beta)^{n-1}$$
$$= (\alpha^{n+1} + \beta^{n+1})(\alpha^{n-1} + \beta^{n-1}) = L_{n+1}L_{n-1}$$

71. (Lord, 1995) $2L_{n-1}^3 + L_n^3 + 6L_{n+1}^2 L_{n-1} = 2L_{n-1}^3$
$$+ (L_{n+1} - L_{n-1})^3 + 6L_{n+1}^2 L_{n-1} = (L_{n+1} + L_{n-1})^3 = (5F_n)^3$$

73. (Zeitlin, 1963) $\alpha^n + \beta^n = L_n$ and $\alpha^n - b^n = \sqrt{5}F_n$. Adding, $2\alpha^n = L_n + \sqrt{5}F_n$. $(1 + \sqrt{5})^n = 2^{n-1}L_n + \sqrt{5}(2^{n-1}F_n)$ $\therefore a_n = 2^{n-1}L_n$ and $b_n = 2^{n-1}F_n$. Thus $2^{n-1}|a_n$ and $2^{n-1}|b_n$.

75. (Bruckman) Since $F_{2n+1}F_{2n-1} - F_{2n}^2 = 1$, $2F_{2n+1}F_{2n-1} - 1 = 2F_{2n}^2 + 1 = F_{2n}^2 + F_{2n+1}F_{2n-1}$. So if $2F_{2n+1}F_{2n-1} - 1$ is a prime, then so are $2F_{2n}^2 + 1$ and $F_{2n}^2 + F_{2n+1}F_{2n-1}$.

77. (J. E. Homer) LHS $= \sum_1^n g_{k+2}F_k + \sum_1^n g_{k+1}F_k - \sum_1^n g_k F_k = \sum_3^n (F_{k-2} + F_{k-1} - F_k)g_k + g_{n+2}F_n + g_{n+1}(F_n + F_{n-1}) + g_2F_1 - g_1F_1 - g_2F_2 = g_{n+2}F_n + g_{n+1}F_{n+1} - g_1$

79. (Yodder) The formula is true when $n = 1$ and 2. Assume it is true for all positive integers $\leq n$, where $n \geq 2$. Let n be odd. Then $f[(7 \cdot 2^n + 1)/3] = f[(7 \cdot 2^{n-1} - 1)/3] + f[(7 \cdot 2^{n-1} + 2)/3]$
$$= f[(7 \cdot 2^{n-1} - 1)/3] + f[(7 \cdot 2^{n-2} + 1)/3]$$
$$= L_n + L_{n-1} = L_{n+1}$$
Similarly, if n is even, $f[(7 \cdot 2^n - 1)/3] = L_{n+1}$. Thus, by the strong version of PMI, the result is true for every $n \geq 1$.

81. $S_n = \sum_1^{n(n+1)/2} F_{2i-1} - \sum_1^{n(n-1)/2} F_{2i-1} = F_{n(n+1)} - F_{n(n-1)}$

83. Area $= (F_{n+1} + F_{n-1})\sqrt{3}F_n/4 = \sqrt{3}L_nF_n/4 = \sqrt{3}F_{2n}/4$

EXERCISES 7 (p. 112)

1. $8a + 11b$
3. $b - a$

5. $A_n = 2F_{n-2} + 3F_{n-1} = 2F_n + F_{n-1} = F_n + F_{n+1} = F_{n+2}$

7. $-\sqrt{5}$

9. $\lim_{n\to\infty} \frac{G_n}{L_n} = \lim_{n\to\infty} \frac{c\alpha^n - d\beta^n}{\alpha^n + \beta^n} = c$

11.
$$\begin{aligned}\sum_1^n G_i &= \sum_1^n (G_{i+2} - G_{i+1}) = \sum_1^n G_{i+2} - \sum_1^n G_{i+1}\\ &= G_{n+2} - b\end{aligned}$$

13.
$$\begin{aligned}\sum_1^n G_{2i} &= \sum_1^{2n} G_i - \sum_1^n G_{2i-1} = (G_{2n+2} - b) - (G_{2n} + a - b)\\ &= G_{2n+2} - G_{2n} - a = G_{2n+1} - a\end{aligned}$$

15. We have $\sum_{i=1}^{10} F_{i+j} = 11F_{j+7}$, $j \geq 0$.

$$\begin{aligned}\therefore\ \sum_1^{10} G_{k+i} &= \sum_1^{10} (aF_{k+i-2} + bF_{k+i-1})\\ &= a\sum_1^{10} F_{k+i-2} + b\sum_1^{10} F_{k+i-1}\\ &= a(11F_{k-2+7}) + b(11F_{k-1+7})\\ &= 11(aF_{k+5} + bF_{k+6}) = 11G_{k+7}\end{aligned}$$

17. Let $S_i = \sum_1^i G_j = G_{i+2} - b$

$$\begin{aligned}\sum_1^{n-1} S_i &= \sum_1^{n-1} G_{i+2} - (n-1)b\\ &= S_{n+1} - (G_1 + G_2) - (n-1)b\\ &= S_{n+1} - a - nb = G_{n+3} - a - (n+1)b\end{aligned}$$

$$\begin{aligned}\sum_1^n iG_i &= \sum_1^n G_i + \sum_2^n G_i + \sum_3^n G_i + \cdots + \sum_n^n G_i\\ &= S_n + (S_n - S_1) + (S_n - S_2) + \cdots + (S_n - S_{n-1})\\ &= nS_n - \sum_1^{n-1} S_i\\ &= n(G_{n+2} - b) - [G_{n+3} - a - (n+1)b]\\ &= nG_{n+2} - G_{n+3} + a + b\end{aligned}$$

19. (J. W. Milsom) The formula works when $n = 1$, so assume it is true for an arbitrary positive integer $k \geq 1$:

$\sum_1^k F_i G_{3i} = F_k F_{k+1} G_{2k+1}$

Then

$$\begin{aligned}\sum_1^{k+1} F_i G_{3i} &= \sum_1^{k} F_i G_{3i} + F_{k+1} G_{3k+3} \\ &= F_k F_{k+1} G_{2k+1} + F_{k+1} G_{3k+3} \\ &= F_{k+1}(F_k G_{2k+1} + G_{3k+3})\end{aligned}$$

Since $G_{m+n} = F_{n-1} G_m + F_n G_{m+1}$,
$G_{3k+3} = G_{(2k+2)+(k+1)} = F_k G_{2k+2} + F_{k+1} G_{2k+3}$.

$$\begin{aligned}\therefore \sum_1^{k+1} F_i G_{3i} &= F_{k+1}[F_k(G_{2k+1} + G_{2k+2}) + F_{k+1} G_{2k+3}] \\ &= F_{k+1}[(F_{k+2} - F_{k+1})(G_{2k+1} + G_{2k+2}) + F_{k+1} G_{2k+3}] \\ &= F_{k+1} F_{k+2}(G_{2k+1} + G_{2k+2}) = F_{k+1} F_{k+2} G_{2k+3}\end{aligned}$$

Thus the result is true for all $n \geq 1$.

21. (Peck) We have the trinomial expansion $(x + y + z)^n = \sum_{i,j,k \geq 0} \binom{n}{i,\, j,\, k} x^i y^j z^k$, where $\binom{n}{i,\, j,\, k} = \dfrac{n!}{i!j!k!}$ and $i + j + k = n$. Let $x = 1$, $y = \alpha$, and $z = -\alpha^2$. Then $x + y + z = 0$.

$$\therefore \sum_{i,j,k} \binom{n}{i,\, j,\, k} \alpha^j (-\alpha^2)^k = 0$$

$$\sum_{i,j,k} \binom{n}{i,\, j,\, k} (-1)^k \alpha^{j+2k} = 0. \tag{1}$$

$$\text{similarly, } \sum_{i,j,k} \binom{n}{i,\, j,\, k} (-1)^k \beta^{j+2k} = 0. \tag{2}$$

Now multiply (1) by c, (2) by d, and then subtract. Divide the result by $\alpha - \beta$ to get the desired result.

23.
$$\begin{aligned}\text{LHS} &= (c\alpha^{n+k} - d\beta^{n+k})(c\alpha^{n-k} - d\beta^{n-k}) \\ &= c^2\alpha^{2n} + d^2\beta^{2n} - cd(\alpha\beta)^{n-k}(\alpha^{2k} + \beta^{2k}) \\ &= 5(\alpha^{2n} + \beta^{2n}) - \mu(-1)^{n-k} L_{2k} \\ &= 5L_{2n} - \mu(-1)^{n-k} L_{2k}\end{aligned}$$

25.
$$5 \cdot \text{LHS} = (c\alpha^n - d\beta^n)^2 + (c\alpha^{n-1} - d\beta^{n-1})^2$$
$$= c^2\alpha^{2n-2}(1+\alpha^2) + d^2\beta^{2n-2}(1+\beta^2) \quad (1)$$
$$= \sqrt{5}(c^2\alpha^{2n-1} - d^2\beta^{2n-1}) \quad (2)$$

$$\sqrt{5} \cdot \text{RHS} = (3a - b)(c\alpha^{2n-1} - d\beta^{2n-1})$$
$$- \mu(\alpha^{2n-1} - \beta^{2n-1})$$
$$= (c+d)(c\alpha^{2n-1} - d\beta^{2n-1})$$
$$- \mu(\alpha^{2n-1} - \beta^{2n-1})$$
$$= c^2\alpha^{2n-1} - d^2\beta^{2n-1} \quad (3)$$

The result follows by (1) and (2).

27. Follows by changing n to $m+n$ in Exercise 26.

29. LHS = $G_m[F_{n+1} + (-1)^n F_{n-1}] + G_{m-1}[1 - (-1)^n]$ = RHS

31. Using Binet's formula for G_i,

$$5 \cdot \text{LHS} = (-1)^{n+k-1}\mu L_{m-n+2k} + (-1)^n \mu L_{m-n}. \quad (1)$$

Using Binet's formula for F_i,

$$5 \cdot \text{RHS} = (-1)^{n+k-1}\mu L_{m-n+2k} + (-1)^n \mu L_{m-n}. \quad (2)$$

$\therefore$ LHS = RHS.

33.
$$5 \cdot \text{RHS} = 5L_{2n-6} - 2\mu(-1)^{n-3} + 20L_{2n-3} - 4\mu(-1)^{n-2}$$
$$= 5L_{2n-6} + 20L_{2n-3} - 2\mu(-1)^n$$
$$= 5L_{2n} - 2\mu(-1)^n = 5 \cdot \text{LHS}$$
$$\therefore \text{LHS} = \text{RHS}.$$

35.
$$5 \cdot \text{LHS} = 5L_{2n} + 5L_{2n+6} = 30L_{2n} + 40L_{2n+1}$$
$$5 \cdot \text{RHS} = 10L_{2n+2} + 10L_{2n+4} = 30L_{2n} + 40L_{2n+1}$$
$$\therefore \text{LHS} = \text{RHS}.$$

37. Since $(x-y)^2 + (2xy)^2 = (x+y)^2$, the result follows with $x = G_m$ and $y = G_n$.

39. Follows from Candido's identity $[x^2 + y^2 + (x+y)^2]^2 = 2[x^4 + y^4 + (x+y)^4]$, with $x = G_m$ and $y = G_n$.

41.
$$\begin{aligned}\text{LHS} &= 5(aF_{n+r-2} + bF_{n+r-1})^2 + 5(aF_{n-r-2} + bF_{n-r-1})^2 \\ &= 5a^2(F_{n-2+r}^2 + F_{n-2-r}^2) + b^2(F_{n-1+r}^2 + F_{n-1-r}^2) \\ &\quad + 2ab(F_{n-2+r}F_{n-1+r} + F_{n-2-r}F_{n-1+r}) \\ &= a^2[L_{2n-4}L_{2r} - 4(-1)^{n-2+r}] + b^2[L_{2n-2}L_{2r} - 4(-1)^{n-1+r}] \\ &\quad + 2ab[L_{2n-3}L_{2r} - 2(-1)^{n-2+r}] \\ &= (a^2L_{2n-4} + b^2L_{2n-2} + 2abL_{2n-3})L_{2r} - 4a^2(-1)^{n+r} \\ &\quad + 4b^2(-1)^{n+r} - 4ab(-1)^{n+r} \\ &= (a^2L_{2n-4} + 2abL_{2n-3} + b^2L_{2n-2})L_{2r} - 4\mu(-1)^{n+r} = \text{RHS}\end{aligned}$$

43. Let $x = G_{n+2}$ and $y = G_{n+1}$. Then $x^2 - y^2 = (x+y)(x-y) = G_{n+3}G_n$, $2xy = 2G_{n+1}G_{n+2}$, and $x^2+y^2 = G_{n+2}^2 + G_{n+1}^2$. Since $(x^2-y^2)^2 + 4x^2y^2 = (x^2+y^2)^2$, the result follows.

45. Let $x = G_n$ and $y = G_{n-1}$. Then $x^2 + xy + y^2 = G_n^2 + G_nG_{n-1} + G_{n-1}^2 = G_{n-1}(G_{n-1} + G_n) + G_n^2 = G_{n-1}G_{n+1} + G_n^2 = 2G_n^2 + \mu(-1)^n$. The result now follows from the identity $(x+y)^5 - x^5 - y^5 = 5xy(x+y)(x^2+xy+y^2)$.

47. Let $x = G_{n-1}$ and $y = G_n$. Then $x^2 + xy + y^2 = 2G_n^2 + \mu(-1)^n$. The result now follows from the identity $x^4 + y^4 + (x+y)^4 = 2(x^2+xy+y^2)^2$.

49. Let $x = G_{n-1}$ and $y = G_n$. Then $x^2+xy+y^2 = 2G_n^2 + \mu(-1)^n$ and $x^4 + 3x^3y + 4x^2y^2 + y^4 = G_{n-1}^4 + 3G_{n-1}^3G_n + 4G_{n-1}^2G_n^2 + 3G_{n-1}G_n^3 + G_n^4 = G_{n-1}^4 + G_n^4 + 3G_{n-1}G_n(G_{n-1}^2 + G_n^2) + 4G_{n-1}^2G_n^2 = G_{n-1}^4 + G_n^4 + 3G_{n-1}G_n[(3a-b)G_{2n-1} - \mu F_{2n-1}] + 4G_{n-1}^2G_n^2$. The result now follows from the identity $x^8 + y^8 + (x+y)^8 = 2(x^2+xy+y^2)^4 + 8x^2y^2(x^4+3x^3y+4x^2y^2+3xy^3+y^4)$.

51. (Lind) $\sum_i^n iG_i = nG_{n+2} - G_{n+3} + a + b$ and $\sum_1^n G_i = G_{n+2} - b$.

$$\therefore A_n = \frac{nG_{n+2} - G_{n+3} + a + b}{G_{n+2} - b}$$

$$A_{n+1} - A_n = \frac{(n+1)G_{n+3} - G_{n+4} + a + b}{G_{n+3} - b} - \frac{nG_{n+2} - G_{n+3} + a + b}{G_{n+2} - b}$$

Since $\lim_{n\to\infty} \dfrac{G_{n+k}}{G_n} = \alpha^k$, it follows that $\lim_{n\to\infty} (A_{n+1} - A_n) = \dfrac{(n+1) - \alpha + 0}{1 - 0} - \dfrac{n - \alpha + 0}{1 - 0} = 1$.

53. (Bruckman)

$$\begin{aligned}A_n &= C_{n+2} - C_{n+1} - C_n \\ &= \sum_0^{n+2} H_iK_{n-i+2} - \sum_0^{n+1} H_iK_{n-i+1} - \sum_0^{n} H_iK_{n-i}\end{aligned}$$

$$= H_{n+2}K_0 + H_{n+1}K_1 - H_{n+1}K_0 + \sum_0^n H_i(K_{n-i+2} - K_{n-i+1} - K_{n-i})$$

$$= H_{n+2}K_0 + H_{n+1}K_1 - H_{n+1}K_0$$

$$= (H_{n+1} + H_n)K_0 + H_{n+1}K_1 - H_{n+1}K_0 = H_{n+1}K_1 + H_nK_0 \qquad (1)$$

$$\begin{aligned}\therefore\ A_{n+1}A_{n-1} - A_n^2 &= (H_{n+2}K_1 + H_{n+1}K_0)(H_nK_1 + H_{n-1}K_0)\\ &\quad - (H_{n+1}K_1 + H_nK_0)^2\\ &= (H_{n+2}H_n - H_{n+1}^2)K_1^2\\ &\quad + (H_{n+2}H_{n-1} - H_{n+1}H_n)K_0K_1\\ &\quad + (H_{n+1}H_{n-1} - H_n^2)K_0^2\end{aligned}$$

But $H_{n+2}H_{n-1} - H_{n+1}H_n = (H_{n+1}+H_n)H_{n-1} - (H_n+H_{n-1})H_n = H_{n+1}H_{n-1} - H_n^2 = (-1)^n\mu$ $\therefore$ $A_{n+1}A_{n-1} - A_n^2 = (-1)^{n+1}\mu K_1^2 + (-1)^n\mu K_0K_1 + (-1)^n\mu K_0^2 = (-1)^n\mu(K_0^2 + K_0K_1 - K_1^2) = (-1)^n\mu[K_0(K_0 + K_1) - K_1^2] = (-1)^n\mu(K_0K_2 - K_1^2) = (-1)^n\mu\nu$. So the characteristic of the sequence $\{A_n\}$ is $\mu\nu$.

It remains to show that $\{A_n\}$ is a GFS. From (1), we have $A_{n-1} + A_{n-2} = (H_nK_1 + H_{n-1}K_0) + (H_{n-1}K_1 + H_{n-2}K_0) = (H_n + H_{n-1})K_1 + (H_{n-1} + H_{n-2})K_0 = H_{n+1}K_1 + H_nK_0 = A_n$. Thus $\{A_n\}$ is a GFS.

55. Follows from Exercise 54 with $p = L_n, q = L_{n+1}, r = L_{n+2}$, and $s = L_{n+3}$.

EXERCISES 8 (p. 130)

1. 610
3. 28,657
5. 610
7. 28,657
9. 144
11. 610
13. 47
15. 1364
17. 47
19. 1364
21. 123
23. 521
25. 4181
27. 4181
29. 1364

31. 4181

33. $L_{n+1} = \alpha L_n + 1/2 + \theta, 0 \le \theta < 1.$

$\therefore \lim_{n\to\infty} \frac{L_{n+1}}{L_n} = \alpha$

35. $L_{n+1} = \frac{L_n + 1 + \sqrt{5L_n^2 - 2L_n + 1}}{2} + \theta$, where $0 \le \theta < 1$.

$\therefore \lim_{n\to\infty} \frac{L_{n+1}}{L_n} = \alpha$

37. $\lfloor(17,711 + 1/2)/\alpha\rfloor = 10,946$

39. 24,476

41. $$\lim_{n\to\infty} \frac{U_n}{F_{n+1}} = \lim_{n\to\infty} \sqrt{\left(\frac{F_n}{F_{n+1}}\right)^2 + \left(\frac{F_{n-2}}{F_{n-1}} \cdot \frac{F_{n-1}}{F_n} \cdot \frac{F_n}{F_{n+1}}\right)^2}$$

$$= \sqrt{\frac{1}{\alpha^2} + \frac{1}{\alpha^6}} = \frac{\sqrt{\alpha^4 + 1}}{\alpha^2} = \frac{\sqrt{3}\alpha}{\alpha^2} = -\sqrt{3}\beta$$

43. Follows by the Pythagorean theorem.

45. $$\lim_{n\to\infty} \frac{x_n}{y_n} = \lim_{n\to\infty} \sqrt{\frac{G_n^2 + G_{n-2}^2}{G_{n-1}^2 + G_{n-3}^2}}$$

$$= \lim_{n\to\infty} \frac{G_n}{G_{n-1}} \sqrt{\frac{1 + (G_{n-2}/G_n)^2}{1 + (G_{n-3}/G_{n-1})^2}} = \alpha \sqrt{\frac{1 + 1/\alpha^2}{1 + 1/\alpha^2}} = \alpha$$

EXERCISES 9 (p. 140)

1. 8
3. 4
5. 8
7. 4
9. $8 = 42 \cdot 1024 - 43 \cdot 1000$
11. $4 = (-85) \cdot 2076 + 164 \cdot 1076$
13. $8 = (-71) \cdot 1976 + 79 \cdot 1776$
15. $4 = (-97) \cdot 3076 + 151 \cdot 1976$
17. By the division algorithm, $a = bq + r$, where q is an integer and $0 \le r < b$. Since $d' = (b, r)$, $d'|b$ and $d'|r$. $\therefore d'|a$. Thus $d'|a$ and $d'|b$. So $d'|(a, b)$; that is, $d'|d$.

19. $$\begin{aligned} L_{n+1} &= 1 \cdot L_n + L_{n-1} \\ L_n &= 1 \cdot L_{n-1} + L_{n-2} \\ &\vdots \end{aligned}$$

$$4 = 1 \cdot 3 + \boxed{1}$$
$$3 = 3 \cdot 1 + 0$$

$$L_{n+1}L_n = \sum_2^n L_i^2 + 3 \cdot 1^2 = \sum_1^n L_i^2 + 3 - 1. \text{ Thus } \sum_1^n L_i^2 = L_{n+1}L_n - 2.$$

EXERCISES 10 (p. 146)

1. Yes
3. Yes
5. No
7. Yes
9. $a_n = 2(-1)^n + 2^n, n \geq 0$
11. $a_n = 3(-2)^n + 2 \cdot 3^n, n \geq 0$
13. $a_n = F_{n+2}, n \geq 0$
15. $L_n = \alpha^n + \beta^n, n \geq 0$

EXERCISES 11 (p. 149)

1. $43 = 34 + 8 + 1$
3. $137 = 89 + 34 + 8 + 5 + 1$
5. $43 = 29 + 11 + 3$
7. $137 = 123 + 11 + 3$
9. $43 = 34 + 8 + 1$

1	2	3	5	8	13	21	34
49*	98	147	245	392*	637	1029	1666*

$43 \cdot 49 = 49 + 392 + 1666 = 2107$

11. $111 = 89 + 21 + 1$

1	2	3	5	8	13	21	34	55	89
121*	242	363	605	968	1573	2541*	4114	6655	10,769*

$111 \cdot 121 = 121 + 2541 + 10{,}769 = 13{,}431$

13. $43 = 29 + 11 + 3$

1	3	4	7	11	18	29
49	147*	196	343	539*	882	1421*

$43 \cdot 49 = 147 + 539 + 1421 = 2107$

15. $111 = 76 + 29 + 4 + 2$

2	1	3	4	7	11	18	29	47	76
242*	121	363	484*	847	1331	2178	3509*	5687	9196*

$111 \cdot 121 = 242 + 484 + 3509 + 9196 = 13{,}431$

17. Let n be a positive integer and let L_m be the largest Lucas number $\leq n$. Then $n = L_m + n_1$, where $n_1 \leq L_m$. Let L_{m_1} be the largest Lucas numbers $\leq n_1$. Then $n = L_m + L_{m_1} + n_2$, where $n_2 \leq L$. Continuing like this, we get $n = L_m + L_{m_1} + L_{m_2} + \cdots$, where $n \geq L_m > L_{m_1} > L_{m_2} \cdots$. Since this sequence of decreasing positive integers terminates, the desired result follows.

EXERCISES 12 (p. 162)

1. $F_5 = \sum_0^2 \binom{4-i}{i} = \binom{4}{0} + \binom{3}{1} + \binom{2}{2} = 1 + 3 + 1 = 5$

3. $F_{11} = \sum_0^5 \binom{10-i}{i} = \binom{10}{0} + \binom{9}{1} + \binom{8}{2} + \binom{7}{3} + \binom{6}{4} + \binom{5}{5} =$
$1 + 9 + 28 + 35 + 15 + 1 = 89$

5. $\sum_0^n \binom{n}{i} L_i = \sum_0^n \binom{n}{i} (\alpha^i + \beta^i) = \sum_0^n \binom{n}{i} \alpha^i + \sum_0^n \binom{n}{i} \beta^i = (1+\alpha)^n +$
$(1+\beta)^n = \alpha^{2n} + \beta^{2n} = L_{2n}$

7. $\sum_0^n \binom{n}{i} F_{i+j} = \frac{1}{\alpha - \beta}\left[\sum_0^n \binom{n}{i} \alpha^{i+j} - \sum_0^n \binom{n}{i} \beta^{i+j}\right] = \frac{1}{\alpha-\beta}[\alpha^j(1 +$
$\alpha)^n - \beta^j(1+\beta)^n] = \frac{\alpha^{2n+j} - \beta^{2n+j}}{\alpha - \beta} = F_{2n+j}$

9. $\sum_0^n \binom{n}{i} (-1)^i F_{i+j} = \frac{1}{\alpha-\beta}\left[\sum_0^n \binom{n}{i} (-1)^i \alpha^{i+j} - \sum_0^n \binom{n}{i} (-1)^i \beta^{i+j}\right]$
$= \frac{1}{\alpha-\beta}\left[\alpha^j \sum_0^n \binom{n}{i} (-\alpha)^i - \beta^j \sum_0^n \binom{n}{i} (-\beta)^i\right] = \frac{1}{\alpha-\beta}[\alpha^j(1-\alpha)^n -$
$\beta^j(1-\beta)^n = \frac{1}{\alpha-\beta}(\alpha^j\beta^n - \beta^j\alpha^n) = \frac{-(\alpha\beta)^j(\alpha^{n-j} - \beta^{n-j})}{\alpha-\beta} = (-1)^{j+1}F_{n-j}$

11. When $n = 4$:

$$\text{LHS} = 5\left[\binom{4}{0} F_0^2 + \binom{4}{1} F_1^2 + \binom{4}{2} F_2^2 + \binom{4}{3} F_3^2 + \binom{4}{4} F_4^2\right]$$

$$= 5(4 + 6 + 16 + 9) = 175$$

$$\text{RHS} = \binom{4}{0}L_0 + \binom{4}{1}L_2 + \binom{4}{2}L_4 + \binom{4}{3}L_6 + \binom{4}{4}L_8$$

$$= 2 + 12 + 42 + 72 + 47 = 175 = \text{LHS}$$

Similarly, when $n = 5$ also, LHS = RHS.

13. $L_{2i} = 5F_i^2 + 2(-1)^i \therefore \sum_0^n \binom{n}{i} L_{2i} = 5\sum_0^n \binom{n}{i} F_i^2 + 2\sum_0^n \binom{n}{i}(-1)^i$
$= 5\sum_0^n \binom{n}{i} F_i^2$

15. $L_i^2 = L_{2i} + 2(-1)^i$, by Exercise 41 in Chapter 5;
$\therefore \sum_0^n \binom{n}{i} L_i^2 = \sum_0^n \binom{n}{i} L_{2i} + 2\sum_0^n \binom{n}{i}(-1)^i = \sum_0^n \binom{n}{i} L_{2i}$

17. $2^n L_n = (1 + \sqrt{5})^n + (1 - \sqrt{5})^n = \sum_0^n \binom{n}{i}[(\sqrt{5})^n + (-\sqrt{5})^n] =$
$2\sum_0^{\lfloor n/2 \rfloor} \binom{n}{2i} 5^i; \therefore 2^{n-1}L_n = \sum_0^{\lfloor n/2 \rfloor} \binom{n}{2i} 5^i$

19.
$$G_i = (c\alpha^i - d\beta^i)/(\alpha - \beta)$$

$$\sum_0^n (-2)^i \binom{n}{i} G_i = \frac{1}{\sqrt{5}} \sum_0^n \binom{n}{i} (-2)^i (c\alpha^i - d\beta^i)$$

$$= \frac{1}{\sqrt{5}}[c(1 - 2\alpha)^n - d(1 - 2\beta)^n]$$

$$= \frac{c(-\sqrt{5})^n - d(\sqrt{5})^n}{\sqrt{5}} = 5^{(n-1)/2}[c(-1)^n - d]$$

21. $\sum_0^n \binom{n}{i}(-1)^i F_{2i} = \frac{1}{\alpha - \beta} \sum_0^n \binom{n}{i}(-1)^i(\alpha^{2i} - \beta^{2i})$. But $(-1)^i(\alpha^{2i} - \beta^{2i}) =$
$(-\alpha^2)^i - (-\beta^2)^i$.

$$\therefore \sum_0^n \binom{n}{i}(-1)^i F_{2i} = \frac{1}{\alpha - \beta} \sum_0^n \binom{n}{i}[(-\alpha^2)^i - (-\beta^2)^i]$$

$$= \frac{1}{\alpha - \beta}[(1 - \alpha^2)^n - (1 - \beta^2)^n]$$

$$= \frac{(-\alpha)^n - (-\beta)^n}{\alpha - \beta} = (-1)^n F_n$$

23.
$$\begin{aligned}\text{LHS} &= \sum_0^n \binom{n}{i}(-1)^i\left(\frac{c\alpha^i - d\beta^i}{\alpha-\beta}\right)\\ &= \frac{c}{\alpha-\beta}\sum_0^n(-\alpha)^i - \frac{d}{\alpha-\beta}\sum_0^n(-\beta)^i\\ &= \frac{c}{\alpha-\beta}(1-\alpha)^n - \frac{d}{\alpha-\beta}(1-\beta)^n = \frac{c\beta^n - d\alpha^n}{\alpha-\beta}\\ &= \frac{c(-\alpha)^{-n} - d(-\beta)^{-n}}{\alpha-\beta} = (-1)^n G_{-n}\end{aligned}$$

25.
$$\begin{aligned}\text{LHS} &= \sum_0^n \binom{n}{i}(-1)^i\left(\frac{c\alpha^{i+j} - d\beta^{i+j}}{\alpha-\beta}\right)\\ &= \frac{c\alpha^j}{\alpha-\beta}\sum_0^n\binom{n}{i}(-\alpha)^i - \frac{d\beta^j}{\alpha-\beta}\sum_0^n\binom{n}{i}(-\beta)^i\\ &= \frac{c\alpha^j}{\alpha-\beta}(1-\alpha)^n - \frac{d\beta^j}{\alpha-\beta}(1-\beta)^n = \frac{c\alpha^j\beta^n - d\beta^j\alpha^n}{\alpha-\beta}\\ &= (\alpha\beta)^n\frac{c\alpha^{j-n} - d\beta^{j-n}}{\alpha-\beta} = (-1)^n G_{j-n}\end{aligned}$$

27.
$$\begin{aligned}\text{LH}S &= \sum_0^n \binom{n}{i}\frac{\alpha^{k+2i} - \beta^{k+2i}}{\alpha-\beta} = \frac{\alpha^k(1+\alpha^2)^n - \beta^k(1+\beta^2)^n}{\alpha-\beta}\\ &= \frac{[\alpha^{n+k} - (-1)^n\beta^{n+k}](\sqrt{5})^n}{\sqrt{5}}\\ &= \begin{cases} 5^{(n-1)/2}L_{n+k} & \text{if } n \text{ is odd}\\ 5^{n/2}F_{n+k} & \text{otherwise}\end{cases}\end{aligned}$$

29. $5^n = (2\alpha - 1)^{2n} = \sum_0^{2n}\binom{2n}{i}(2\alpha)^i(-1)^{2n-i} = \sum_0^{2n}\binom{2n}{i}(-2\alpha)^i$. Similarly, $5^n = (1-2\beta)^{2n} = \sum_0^{2n}\binom{2n}{i}(-2\beta)^i$. Adding, $2\cdot 5^n = \sum_0^{2n}\binom{2n}{i}(-2)^iL_i$. $\therefore\ 5^n = \sum_0^{2n}\binom{2n}{i}(-1)^i2^{i-1}L_i$.

31. (Swamy)

$$\begin{aligned}\sqrt{5}\sum_0^n \binom{n}{k} F_{4mk} &= \sum_0^n \binom{n}{k}(\alpha^{4mk}-\beta^{4mk}) = \sum_0^n \binom{n}{k}\alpha^{4mk} \\ &\quad - \sum_0^n \binom{n}{k}\beta^{4mk} \\ &= (1+\alpha^{4m})^n - (1+\beta^{4m})^n = \alpha^{2mn}(\alpha^{2m}+\alpha^{-2m})^n \\ &\quad - \beta^{2mn}(\beta^{2m}+\beta^{-2m})^n \\ &= \alpha^{2mn}(\alpha^{2m}+\beta^{2m})^n - \beta^{2mn}(\alpha^{2m}+\beta^{2m})^n \\ \therefore \quad &\text{LHS} = L_{2m}^n F_{2mn}\end{aligned}$$

EXERCISES 13 (p. 178)

1. $x^2 - L_n x + (-1)^n = 0$
3. $L_8 = \sum_0^4 \frac{8}{8-i}\binom{8-i}{i} = \binom{\frac{8}{8}}{}\binom{8}{0} + \binom{\frac{8}{7}}{}\binom{7}{1} + \binom{\frac{8}{6}}{}\binom{6}{2} + \binom{\frac{8}{5}}{}\binom{5}{3} + \binom{\frac{8}{4}}{}\binom{4}{4} = 1 + 8 + 20 + 16 + 2 = 47$
5. $r^5 + s^5 = \sum_0^2 A(5,i)p^{5-2i}q^i = p^5 + 5p^3q + 5pq^2$
7. Since $r + s = \sum_0^0 A(1,i)p^{0-2i}q^i = p$ and $r^2 + s^2 = \sum_0^1 A(2,i)p^{2-2i}q^i = A(2,0)p^2 + A(2,1)q = p^2 + 2q$, the formula works when $n = 1$ and 2. Now assume it is true for all positive integers $\le k$. First notice that:

$$\begin{aligned}r^{k+1} + s^{k+1} &= (r^{k+1} + s^{k+1} + rs^k + r^k s) - (rs^k + r^k s) \\ &= (r^k + s^k)(r+s) - rs(r^{k-1} + s^{k-1}) \\ &= p(r^k + s^k) + q(r^{k-1} + s^{k-1}) \qquad (1)\end{aligned}$$

Let $n = k + 1$. Using Eq. (1):

$$\begin{aligned}\text{RHS} &= p\sum_0^{\lfloor k/2\rfloor} \frac{k}{k-i}\binom{k-i}{i} p^{k-2i}q^i \\ &\quad + q\sum_0^{\lfloor (k-1)/2\rfloor} \frac{k-1}{k-1-i}\binom{k-1-i}{i} p^{k-1-2i}q^i\end{aligned}$$

$$= \sum_{0}^{\lfloor k/2 \rfloor} \frac{k}{k-i} \binom{k-i}{i} p^{k+1-2i} q^i$$

$$+ \sum_{0}^{\lfloor (k-1)/2 \rfloor} \frac{k-1}{k-1-i} \binom{k-1-i}{i} p^{k-1-2i} q^{i+1}$$

$$= \sum_{0}^{\lfloor k/2 \rfloor} \frac{k}{k-i} \binom{k-i}{i} p^{k+1-2i} q^i$$

$$+ \sum_{0}^{\lfloor (k+1)/2 \rfloor} \frac{k-1}{k-j} \binom{k-j}{j-1} p^{k+1-2j} q^j \tag{2}$$

Let k be even. Then Eq. (2) yields:

$$\text{RHS} = \sum_{0}^{k/2} \frac{k}{k-i} \binom{k-i}{i} p^{k+1-2i} q^i + \sum_{0}^{k/2} \frac{k-1}{k-i} \binom{k-i}{i-1} p^{k+1-2i} q^i$$

$$= \sum_{0}^{k/2} \left[\frac{k}{k-i} \binom{k-i}{i} + \frac{k-1}{k-i} \binom{k-i}{i-1} \right] p^{k+1-2i} q^i$$

$$= \sum_{0}^{k/2} \left[\frac{k(k-i)!}{(k-i)(k-2i)!i!} + \frac{(k-1)(k-i)!}{(k-i)(k+1-2i)!(i-1)!} \right] p^{k+1-2i} q^i$$

$$= \sum_{0}^{k/2} \frac{(k-i)!}{(k-i)(k+1-2i)!i!} [k(k+1-2i) + (k-1)i] p^{k+1-2i} q^i$$

$$= \sum_{0}^{k/2} \frac{(k+1)(k-i)!}{(k+1-2i)!i!} p^{k+1-2i} q^i$$

$$= \sum_{0}^{k/2} \frac{k+1}{k+1-i} \binom{k+1-i}{i} p^{k+1-2i} q^i$$

$$= \sum_{0}^{\lfloor (k+1)/2 \rfloor} \frac{k+1}{k+1-i} \binom{k+1-i}{i} p^{k+1-2i} q^i \tag{3}$$

Similarly, it can be shown that Eq. (2) leads to Eq. (3) when k is odd. Thus, by the strong version of PMI, the given formula works for all $n \geq 1$.

9. $F_{10} = \sum_{0}^{5} \binom{9-i}{i} = \binom{9}{0} + \binom{8}{1} + \binom{7}{2} + \binom{6}{3} + \binom{5}{4} = 1 + 8 + 21 + 20 + 5 = 55$

11. $r + s = p, r - s = \sqrt{p^2 + 4q} = \Delta$
$32(r^5 - s^5) = (p + \Delta)^5 - (p - \Delta)^5 = (10p^4 + 20p^2\Delta^2 + 2\Delta^4)\Delta = 32(p^4 + 3p^3q + q^2)\Delta$
$\therefore r^5 - s^5 = (p^4 + 3p^2q + q^2)\Delta$

13. Since $r - s = \Delta \sum_0^0 \binom{-i}{i} p^{-2i}q^i = \Delta$ and $r^2 - s^2 = \Delta \sum_0^1 \binom{1-i}{i} p^{1-2i}q^i = \Delta p$, the formula works when $n = 1$ and 2.
Now assume it is true for every positive integer $\le k$. First, notice that:

$$\begin{aligned} r^{k+1} - s^{k+1} &= (r^{k+1} - s^{k+1} - rs^k + r^k s) + (rs^k - r^k s) \\ &= (r^k - s^k)(r + s) - rs(r^{k-1} - s^{k-1}) \\ &= p(r^k - s^k) + q(r^{k-1} - s^{k-1}) \end{aligned} \tag{1}$$

Let $n = k + 1$. Using Eq. (1):

$$\begin{aligned} \text{RHS} &= \Delta p \sum_0^{\lfloor k/2 \rfloor} \binom{k-i-1}{i} p^{k-2i}q^i \\ &\quad + \Delta q \sum_0^{\lfloor (k-1)/2 \rfloor} \binom{k-i-2}{i} p^{k-2i-1}q^i \\ &= \Delta \sum_0^{\lfloor k/2 \rfloor} \binom{k-i-1}{i} p^{k+1-2i}q^i \\ &\quad + \Delta \sum_0^{\lfloor (k-1)/2 \rfloor} \binom{k-2-i}{i} p^{k-1-2i}q^{i+1} \\ &= \Delta \sum_0^{\lfloor k/2 \rfloor} \binom{k-i-1}{i} p^{k+1-2i}q^i \\ &\quad + \Delta \sum_0^{\lfloor (k+1)/2 \rfloor} \binom{k-1-j}{j-1} p^{k+1-2j}q^j \end{aligned} \tag{2}$$

Let k be even. Then Eq. (2) yields:

$$\begin{aligned} \text{RHS} &= \Delta \sum_0^{k/2} \binom{k-i-1}{i} p^{k+1-2i}q^i + \Delta \sum_0^{k/2} \binom{k-1-i}{i-1} p^{k+1-2i}q^i \\ &= \Delta \sum_0^{k/2} \left[\binom{k-i-1}{i} + \binom{k-i-1}{i-1} \right] p^{k+1-2i}q^i \end{aligned}$$

$$= \Delta \sum_{0}^{k/2} \binom{k-i}{i} p^{k+1-2i} q^i$$

$$= \Delta \sum_{0}^{\lfloor (k+1)/2 \rfloor} \binom{k-i}{i} p^{k+1-2i} q^i \tag{3}$$

Similarly, it can be shown that Eq. (2) leads to Eq. (3) when k is odd. Thus, by the strong version of PMI, the given formula works for all $n \geq 1$.

15.
$$\begin{aligned}\sum_{0}^{3} C(6-j, j) &= \sum_{0}^{3} \binom{6-j}{j} + \sum_{0}^{3} \binom{5-j}{j-1} \\ &= \left[\binom{6}{0} + \binom{5}{1} + \binom{4}{2} + \binom{3}{3}\right] \\ &\quad + \left[\binom{5}{-1} + \binom{4}{0} + \binom{3}{1} + \binom{2}{2}\right] \\ &= (1+5+6+1) + (0+1+3+1) = 18\end{aligned}$$

17. $\sum\limits_{k=1}^{n} C(k, j) = \sum\limits_{k=1}^{n} \binom{k}{j} + \sum\limits_{k=1}^{n} \binom{k-1}{j-1} = \binom{n+1}{j+1} + \binom{n}{j} =$ $C(n+1, j+1)$

19. $C(n, n-2) = \binom{n}{n-2} + \binom{n-1}{n-3} = \binom{n}{2} + \binom{n-1}{2} = (n-1)^2$

21.
$$\begin{aligned}\sum_{0}^{\lfloor 6/2 \rfloor} B(6-j, j) &= \sum_{0}^{3} \binom{6-j}{j} + \sum_{0}^{3} \binom{5-j}{j} \\ &= \left[\binom{6}{0} + \binom{5}{1} + \binom{4}{2} + \binom{3}{3}\right] \\ &\quad + \left[\binom{5}{0} + \binom{4}{1} + \binom{3}{2}\right] \\ &= (1+5+6+1) + (1+4+3) = 21\end{aligned}$$

23. $D(n, 2) = \binom{n}{2} + \binom{n-1}{2} = \dfrac{n(n-1)}{2} + \dfrac{(n-1)(n-2)}{2} = (n-1)^2$

EXERCISES 14 (p. 185)

1. $A(n, n) = 1$, $A(n, 0) = F_{2n-1}$, $n \geq 0$;
   $A(n, j) = A(n-1, j) + A(n-1, j-1)$, $n > j$.

3. Let S_n denote the nth sum, where $S_0 = 1$. $S_n = 1 + \sum_0^{n-1} F_{2n-2k} = 1 + \sum_1^n F_{2k} = 1 + (F_{2n+1} - 1) = F_{2n+1}$.

5. (P. S. Bruckman) Let D_n denote the sum of the elements on the nth diagonal, where $D_0 = 1 = D_1$:

$$D_n = \begin{cases} \sum_1^{n/2} F_{4k} + 1 & \text{if } n \text{ is even} \\ \sum_1^{(n+1)/2} F_{4k-2} & \text{otherwise} \end{cases}$$

Let $n = 2m$. Then $D_{2m} = \sum_1^m F_{4k} + 1 = F_1 + \sum_1^m (F_{4k+1} - F_{4k-1}) = \sum_0^{2m} (-1)^i F_{2i+1}$. On the other hand, let $n = 2m+1$. Then $D_{2m+1} = \sum_1^{m+1} F_{4k-2} = \sum_1^{m+1} (F_{4k-1} - F_{4k-3}) = \sum_0^{2m+1} (-1)^{i+1} F_{2i+1}$. Combining the two cases, $D_n = \sum_0^n (-1)^{n-i} F_{2i+1} = \sum_0^n (-1)^{n-i} (F_{i+1}^2 + F_i^2) = \sum_0^n [(-1)^{n-i} F_{i+1}^2 - (-1)^{n-i-1} F_i^2] = F_{n+1}^2 - 0 = F_{n+1}^2$.

7. Let $n = 2m+1$. Then the rising diagonal sum is $\sum_0^m F_{4i+3}$. Using PMI, this sum can be shown to be $F_{2m+2}F_{2m+3} = F_{n+1}F_{n+2}$. On the other hand, let $n = 2m$. Then diagonal sum $= \sum_0^m F_{4i+1} = F_{2m+1}F_{2m+2} = F_{n+1}F_{n+2}$.

9. $S_0(a,b) = a$, $S_1(a,b) = a+b$; $S_n(a,b) = S_{n-1}(a,b) + S_{n-2}(a,b)$, $n \ge 2$.

11. Since $S_0(a,b) = a = aF_1 + bF_0$ and $S_1(a,b) = a+b = aF_2 + bF_1$, the result is true when $n = 0, 1$. Assume it is true for all nonnegative integers $\le k$. Then $S_{k+1}(a,b) = S_k(a,b) + S_{k-1}(a,b) = (aF_{k+1} + bF_k) + (aF_k + bF_{k-1}) = a(F_{k+1} + F_k) + b(F_k + F_{k-1}) = aF_{k+2} + bF_{k+1}$. $\therefore$ By the strong version of PMI, the result follows.

13. $T_n(a,b) = T_{n-1}(a,b) + T_{n-2}(a,b)$, where $T_0(a,b) = a$ and $T_1(a,b) = a-b$. The result now follows by the strong version of PMI, as in Exercise 11.

EXERCISES 15 (p. 195)

1. $H(n,1) = H(n-1,1) + H(n-2,1)$, where $H(1,1) = 1 = H(2,1)$. $\therefore H(n,1) = F_n$.

3. $H(n,n-1) = H(n-1,n-2) + H(n-2,n-3) = H(n-1,1) + H(n-2,1) = H(n,1) = F_n$.

5. $L_{n+2} \equiv (-1)^{j-1} L_{n-2j} \pmod 5$. Let $n = 2(m-1)$ and $j = m-1$. Then $L_{2m} \equiv (-1)^{m-2} L_0 \equiv 2(-1)^m \pmod 5$.

7. Let U, V, W, and X be as in the figure. Then
   $U = A + B$, $V = C + D$, $W = D - C$, and $X = B - A$.

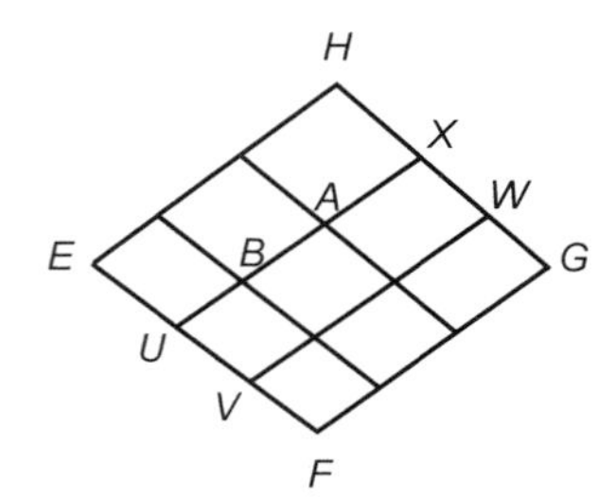

   $\therefore E = V - U = C + D - A - B$, $F = U + V = A + B + C + D$,
   $G = X + W = B + D - A - C$ and $H = W - X = A + D - B - C$.
9. Subtracting Eq. (15.10) from Eq. (15.9),
   $H(n-1, j) + H(n-1, j-1) - H(n-2, j-1) = F_n$.
   That is, $H(n, j) + H(n, j-1) - H(n-1, j-1) = F_{n+1}$.
11. By Eq. (15.10), $H(n, j) + H(n-2, j-1) = F_{n+1}$.
    $\therefore H(n-2, j) + H(n-4, j-1) = F_{n-1}$.
    Subtracting, $H(n, j) - H(n-4, j-2) = F_{n+1} - F_{n-1} = F_n$.

EXERCISES 16 (p. 207)

1. $F_7 = 13$, $F_{21} = 10,946$, and $13|10,946$.
3. $(F_{12}, F_{18}) = (144, 2584) = 8 = F_6 = F_{(12,18)}$
5. $(F_{144}, F_{1925}) = F_{(144,1925)} = F_1 = 1$
7. $18|46,368$, so $L_6|F_{24}$.
9. $(F_{144}, F_{440}) = F_{(144,440)} = F_8 = 21$
11. $\sqrt{5}F_{mn} = \alpha^{mn} - \beta^{mn} = (\alpha^m - \beta^m)[\alpha^{(n-1)m} + \alpha^{(n-2)m}\beta^m - \alpha^{(n-3)m}\beta^{2m} + \cdots + \beta^{(n-1)m}]$. Clearly, $F_m|F_{mn}$.
13. (LeVeque) Suppose $(F_n, F_{n+1}) = d(> 1)$, that is, there is a positive integer n such that F_n and F_{n+1} have a common factor $d > 1$. Then, by the WOP, there is such a least positive integer m. Since $(F_1, F_2) = 1, m > 1$. If $(F_m, F_{m+1}) = d$, then $(F_{m-1}, F_m) = d$. But this contradicts the choice of m.
    $\therefore (F_n, F_{n+1}) = 1$ for all $n \geq 1$.
15. $5|10$, but $L_5 \nmid L_{10}$.
17. $F_4F_8 \nmid F_{32}$
19. $[F_8, F_{12}] = [21, 144] = 1008 \neq F_{24} = F_{[8,12]}$
21. $(F_n, L_n) = 1$ or 2.
23. 5
25. 11

27. 5
29. 11
31. $37|F_{19}$
33. $F_3|F_{3n}$; that is, $2|F_{3n}$.
35. Suppose $3|F_n$; that is, $F_4|F_n$. So $4|n$. Conversely, let $4|n$. Then $F_4|F_n$; that is, $3|F_n$.
37. $5|n$ iff $F_5|F_n$, that is, iff $5|F_n$.
39. $L_{3n} = \alpha^{3n} + \beta^{3n} = (1 + 2\alpha)^n + (1 + 2\beta)^n = \sum_0^n \binom{n}{i} [(2\alpha)^i + (2\beta)^i] = \sum_0^n \binom{n}{i} 2^i L_i$. Since $L_0 = 2$, it follows that $2|L_{3n}$.
41. Let $(F_n, L_n) = 2$. Then $2|F_n$, so $3|n$. Conversely, suppose $3|n$. Let $n = 3m$. Then $2|F_{3m}$ and $2|L_{3m}$. Then $2|(F_{3m}, L_{3m})$. $\therefore$ $(F_n, L_n) = (F_{3m}, L_{3m}) = 2$.
43. Since $L_{i-j} = (-1)^j(F_{i+1}L_j - F_iL_{j+1})$, $L_{(2k-1)n} = L_{2kn-n} = (-1)^n(F_{2kn+1}L_n - F_{2kn}L_{n+1})$. $F_{2n}|F_{2kn}$; that is, $F_nL_n|F_{2kn}$. $\therefore$ $L_n|$RHS; that is, $L_n|L_{(2k-1)n}$.
45. Clearly, $F_m|F_m$. Assume $F_m|F_{mn}$ for all positive integers $n \le k$. Since $F_{m(k+1)} = F_{mk+m} = F_{mk-1}F_m + F_{mk}F_{m+1}$, it follows by the inductive hypothesis that $F_m|F_{m(k+1)}$. $\therefore$ By the strong version of PMI, the result holds for all $n \ge 1$.
47. LHS $= (21 + 34)F_7 = 715$, where $n = 8$. RHS $= [21, 34] + (-1)^8(21, 34) = 714 + 1 = 715 =$ LHS.
49. (Freeman) The identity is true when $n = 2$. Assume it is true for an arbitrary integer $k \ge 2$. Then $F_{2k+1} = F_{2k} + F_{2k-1} = F_kL_k + F_{k+1}L_{k+2} - L_kL_{k+1} = F_kL_k + (F_{k+2} - F_k)L_{k+2} - L_kL_{k+1}$

$$\begin{aligned}
&= F_kL_k + F_{k+2}(L_{k+3} - L_{k+1}) - F_kL_{k+2} - L_kL_{k+1}\\
&= F_{k+2}L_{k+3} - F_{k+2}L_{k+1} - F_kL_{k+1} - (L_{k+2} - L_{k+1})L_{k+1}\\
&= F_{k+2}L_{k+3} - L_{k+1}L_{k+2} + L_{k+1}(L_{k+1} - F_{k+2} - F_k)\\
&= F_{k+2}L_{k+3} - L_{k+1}L_{k+2} + L_{k+1} \cdot 0\\
&= F_{k+2}L_{k+3} - L_{k+1}L_{k+2}.
\end{aligned}$$

Thus, by PMI, the result follows.
51. $n = 5$, so LHS $= (11 + 18)F_6 = 232 = [11, 18] + (11, 18)F_9 =$ RHS.
53. LHS $= (72 + 116)F_7 = 2444 = [72, 116] + (72, 116)F_{11} =$ RHS.
55. Follows since $(F_m, F_n) = F_{(m,n)}$ and $(a, b) = (a, a + b) = (b, a + b)$.
57. (Carlitz) Suppose $F_k|L_n$. Let $n = mk + r$, where $0 \le r < k$. Since $\alpha^n + \beta^n = \alpha^r(\alpha^{mk} - \beta^{mk}) + \beta^{mk}(\alpha^r + \beta^r)$, $F_k|\beta^{mk}L_r$, so $F_k|L_r$. Since $L_r = F_{r-1} + F_{r+1}$, it follows that $L_r < F_{r+2}$ for $r > 2$. Hence we need only consider the case $F_{r+1}|L_r$. Then this implies $F_{r+1}|F_{r-1}$, which is imposiible for $r \ge 2$. Thus $F_k \nmid L_n$ for $k > 4$.
59. $F_{4n} - 1 = F_{4n} - F_2 = F_{(2n+1)+(2n-1)} - F_{(2n+1)-(2n-1)} = F_{2n+1}L_{2n-1}$
61. $F_{4n+2} + 1 = F_{4n+2} + F_2 = F_{(2n+2)+2n} + F_{(2n+2)-2n} = F_{2n+2}L_{2n}$

63. $F_{4n+3} + 1 = F_{4n+3} + F_1 = F_{(2n+2)+(2n+1)} + F_{(2n+2)-(2n+1)} = F_{2n+1}L_{2n+2}$

65. LHS $= (F_{2n}L_{2n+1}, F_{2n+2}L_{2n+1}) = L_{2n+1}(F_{2n}, F_{2n+2}) = L_{2n+1} =$ RHS

67. LHS $= (F_{4n}-1, F_{4n+1}+1) = (F_{2n+1}L_{2n-1}, F_{2n+1}L_{2n}) = F_{2n+1}(L_{2n-1}, L_{2n}) = F_{2n+1} =$ RHS

69. LHS $= (F_{4n+2} - 1, F_{4n+3} + 1) = (F_{2n}L_{2n+2}, F_{2n+1}L_{2n+2}) = L_{2n+2}(F_{2n}, F_{2n+1}) = L_{2n+2} =$ RHS

71. Since $g_0 = 0$, $g_1 = 12$, and $g_{n+2} = 7g_{n+1} - g_n$, the proof follows by the strong version of PMI.

73. (Stanley) $F_{kn+2r} = F_{2r-1}F_{kn} + F_{2r}F_{kn+1}$ and $F_{kn-2r} = F_{2r-1}F_{kn} - F_{2r}F_{kn-1}$.

$$\begin{aligned}\therefore F_{kn-2r} + F_{kn} + F_{kn+2r} &= (2F_{2r-1} + 1)F_{kn} + (F_{kn+1} - F_{kn-1})F_{2r}\\ &= (2F_{2r-1} + 1)F_{kn} + F_{kn}F_{2r}\\ &= (F_{2r} + 2F_{2r-1} + 1)F_{kn} = (L_{2r} + 1)F_{kn}\end{aligned}$$

75. (Lord) When $k = 1, h = 5$ and $5|F5$. $\therefore$ The statement is true when $k = 1$. Assume it is true for k. We have $x^5 - y^5 = (x - y)(x^4 + x^3y + x^2y^2 + xy^3 + y^4)$. Let $x = \alpha^h$ and $y = \beta^h$. Then $F_{5h} = F_h(L_{4h} - L_{2h} + 1)$. But $L_{4h} - L_{2h} + 1 = (5F_{2h}^2 + 2) - (5h^2 - 2) + 1 \equiv 0 \pmod 5$. $\therefore F_{5h} \equiv 0$ (mod 5h). Thus the statement is true for all $k \geq 1$.

EXERCISES 17 (p. 213)

1. $$\begin{aligned}G_{m+n} + G_{m-n} &= G_m[F_{n+1} + (-1)^n F_{n-1}] + G_{m-1}F_n[1 - (-1)^n]\\ &= \begin{cases} G_mL_n & \text{if } n \text{ is even}\\ G_m(F_{n+1} - F_{n-1}) + 2G_{m-1}F_n & \text{otherwise.}\end{cases}\\ &= \begin{cases} G_mL_n & \text{if } n \text{ is even}\\ (G_m + 2G_{m-1})F_n & \text{otherwise}\end{cases}\\ &= \begin{cases} G_mL_n & \text{if } n \text{ is even}\\ (G_{m+1} + G_{m-1})F_n & \text{otherwise}\end{cases}\end{aligned}$$

3. By Exercises 1 and 2,

$$\begin{aligned}G_{m+n}^2 - G_{m-n}^2 &= \begin{cases} (G_{m+1} + G_{m-1})G_mF_{2n} & \text{if } n \text{ is even}\\ (G_{m+1} + G_{m-1})G_mF_{2n} & \text{otherwise}\end{cases}\\ &= (G_{m+1} + G_{m-1})G_mF_{2n}\end{aligned}$$

5. (Swamy)

$$\begin{aligned}\sum_1^n a_{2k-1} &= \sum_1^n (a_{2k-1} + a_{2k}) - \sum_1^n a_{2k} = \sum_1^{2n} a_k - \sum_1^n a_k\\ &= (a_{4n+1} - a) - (a_{2n+1} - a)\\ &= a_{4n+1} - a_{2n+1}\end{aligned}$$

7. $G_{4m+1} + a = G_{4m+1} + G_1 = G_{(2m+1)+2m} + G_{(2m+1)-2m} = G_{2m+1}L_{2m}$, by Exercise 1.
9. $G_{4m+3} + a = G_{4m+3} + G_1 = G_{(2m+2)+(2m+1)} + G_{(2m+2)-(2m+1)}$
$= (G_{2m+3} + G_{2m+1})F_{2m+1}$, by Exercise 1.
11. $G_{4m+1} - a = G_{4m+1} - G_1 = G_{(2m+1)+2m} - G_{(2m+1)-2m}$
$= (G_{2m+2} + G_{2m})F_{2m}$, by Exercise 2.
13. $G_{4m+3} - a = G_{4m+3} - G_1 = G_{(2m+2)+(2m+1)} - G_{(2m+2)-(2m+1)}$
$= G_{2m+2}\, L_{2m+1}$, by Exercise 2.
15. $(G_{4m+1} - a, G_{4m+2} - b) = ((G_{2m+2} + G_{2m})F_{2m}, (G_{2m+3} + G_{2m+1})F_{2m})$
$= F_{2m}$

EXERCISES 18 (p. 225)

1. $\dfrac{2}{x-1} - \dfrac{1}{x+3}$
3. $\dfrac{2}{1+2x} + \dfrac{3}{1-3x}$
5. $\dfrac{1}{2+3x} + \dfrac{2x-1}{x^2+1}$
7. $\dfrac{1-x}{x^2+2} + \dfrac{2x}{x^2+3}$
9. $\dfrac{x-1}{x^2+1} + \dfrac{2x+1}{x^2-x+1}$
11. $a_n = 2^n, n \geq 0$
13. $a_n = 2n - 1, n \geq 1$
15. $a_n = 2^{n+1} - 6n \cdot 2^n, n \geq 0$
17. $a_n = 5 \cdot 2^n - 3^n, n \geq 0$
19. $a_n = F_{n+3}, n \geq 0$
21. $a_n = 3 \cdot 2^n + n \cdot 2^{n+1}, n \geq 0$
23. $a_n = 3(-2)^n + 2^n - 3^n, n \geq 0$
25. $a_n = 2^n + 3n \cdot 2^n - 3^n, n \geq 0$
27. $a_n = n \cdot 2^{n+1} - n^2 2^n, n \geq 0$
29. $a_n = 2(-1)^n - n(-1)^n + 3n^2(-1)^n - 2^{n+1}, n \geq 0$

EXERCISES 19 (p. 237)

1. By Eq. (19.3), $f(x) = \dfrac{e^{\alpha x} - e^{\beta x}}{\sqrt{5}}$. $\therefore f(-x)\dfrac{e^{-\alpha x} - e^{-\beta x}}{\sqrt{5}} = \dfrac{e^{\beta x} - e^{\alpha x}}{\sqrt{5}e^{(\alpha+\beta)x}} = -\dfrac{f(x)}{e^x}$, so $f(x) = -e^x f(-x)$.

3. By Exercise 2, $g(x) = e^{\alpha x} + e^{\beta x}$, so $g(-x) = e^{-\alpha x} + e^{-\beta x} = \dfrac{e^{\alpha x} + e^{\beta x}}{e^x} = \dfrac{g(x)}{e^x}$. $\therefore g(x) = e^x g(-x)$.

5. Let $g(x) = \sum_0^\infty F_{3n}x^n$. Then $4xg(x) = \sum_1^\infty 4F_{3n-3}x^n$ and $x^2 g(x) = \sum_2^\infty F_{3n-6}x^n$, so $(1 - 4x - x^2)g(x) = 2x$, since $F_{3n} = 4F_{3n-3} + F_{3n-6}$. $\therefore g(x) = \dfrac{2x}{1 - 4x - x^2}$.

7. (Hansen) Let $\Delta = 1 - x - x^2$. Then:

$$\begin{aligned}
&\sum_{m=0}^{\infty}(L_m L_n + L_{m-1}L_{n-1})x^m \\
&= L_n \sum_{m=0}^{\infty} L_m x^m + L_{n-1}\sum_{m=0}^{\infty} L_{m-1}x^m \\
&= L_n\left(\frac{2-x}{\Delta}\right) + L_{n-1}\left(\frac{-1+3x}{\Delta}\right) \\
&= \frac{[L_n + (L_n - L_{n-1})] + [2L_{n-1} + (L_{n-1} - L_n)]x}{\Delta} \\
&= \frac{L_n + L_{n-1}x}{\Delta} + \frac{L_{n-2} + L_{n-3}x}{\Delta} \\
&= \sum_{m=0}^{\infty}(L_{m+n} + L_{m+n-2})x^m
\end{aligned}$$

$$\therefore L_m L_n + L_{m-1}L_{n-1} = L_{m+n} + L_{m+n-2} = 5F_{m+n-1}$$

9. Let $A(t) = \dfrac{e^{\alpha t} - e^{\beta t}}{\alpha - \beta} = B(t)$. Then, by Eq. (19.6),

$$\begin{aligned}
\sum_{n=0}^{\infty}\left[\sum_{k=0}^{n}\binom{n}{k}F_k F_{n-k}\right]\frac{t^n}{n!} &= \frac{e^{2\alpha t} + e^{2\beta t} - 2e^{(\alpha+\beta)t}}{(\alpha-\beta)^2} \\
&= \sum_{n=0}^{\infty}\frac{(2^n L_n - 2)}{5}\cdot\frac{t^n}{n!}
\end{aligned}$$

$$\therefore \sum_{k=0}^{\infty}\binom{n}{k}F_k F_{n-k} = \frac{2^n L_n - 2}{5}$$

11. (Padilla) $(1 - x - x^2)A_n(x) = F_n x^{n+2} + F_{n+1}x^{n+1} - x$.
$\therefore A_n(x) = \dfrac{F_n x^{n+2} + F_{n+1}x^{n+1} - x}{1 - x - x^2}$.

13. $(1 - x - x^2)B(x) = -x^2 \sum_0^\infty \dfrac{F_n}{n!}x^n - \sum_0^\infty \dfrac{(n+1)F_{n+1}}{(n+1)!}x^{n+1} + e^x$
$\sqrt{5}(1 - x - x^2)B(x) = \sqrt{5}e^x - x^2(e^{\alpha x} - e^{\beta x}) - x(\alpha e^{\alpha x} - \beta e^{\beta x})$

15. $$\sum_{m=0}^\infty L_{m+n}x^m = \sum_{m=0}^\infty (\alpha^{m+n} + \beta^{m+n})x^m = \alpha^n \sum_0^\infty \alpha^m x^m + \beta^n \sum_0^\infty \beta^m x^m$$
$$= \frac{\alpha^n}{1 - \alpha x} + \frac{\beta^n}{1 - \beta x}$$
$$= \frac{(\alpha^n + \beta^n) + (\alpha^{n-1} + \beta^{n-1})x}{1 - x - x^2} = \frac{L_n + L_{n-1}x}{1 - x - x^2}$$

17. (Carlitz) Let $C(x) = \sum_0^\infty C_n x^n$, where $C_0 = 0$.
$(1 - x - x^2)C(x) = C_1 x + (C_2 - C_1)x^2 + \sum_3^\infty F_n x^n$
$= F_1 x + F_2 x^2 + \sum_3^\infty F_n x^n = \sum_0^\infty F_n x^n = \dfrac{x}{1 - x - x^2}$

$\therefore C(x) = \dfrac{x}{(1 - x - x^2)^2} = x\sum_0^\infty (i+1)(x + x^2)^i = \sum_0^\infty (i+1)x^{i+1}(1+x)^i =$
$\sum_{i=0}^\infty (i+1)x^{i+1} \sum_{j=0}^{i} \binom{i}{j} x^j = \sum_{n=0}^\infty \left[\sum_{i=0}^{n} (i+1)\binom{i}{n-i}\right] x^{n+1}$
$\therefore C_n = \sum_{i=0}^{n} (i+1)\binom{i}{n-i} = \sum_{i=0}^{\lfloor n/2 \rfloor} (n-i+1)\binom{n-i}{i}$

19. Let $A(t) = e^{\alpha t} + e^{\beta t}$ and $B(t) = e^t$. Then $A(t)B(t) = \sum_{n=0}^\infty \left[\sum_{k=0}^{n} \binom{n}{k} L_k\right] \dfrac{t^n}{n!}$.
That is, $\sum_{n=0}^\infty L_{2n}\dfrac{t^n}{n!} = e^{\alpha^2 t} + e^{\beta^2 t} = e^{(\alpha+1)t} + e^{(\beta+1)t} = \sum_{n=0}^\infty \left[\sum_{k=0}^{n} \binom{n}{k} L_k\right] \dfrac{t^n}{n!}$.
$\therefore L_{2n} = \sum_{k=0}^{\infty} \binom{n}{k} L_k$.

21. Let $A(t) = \dfrac{e^{\alpha^2 t} - e^{\beta^2 t}}{\alpha - \beta}$ and $B(t) = e^{-t}$. Then $\dfrac{e^{(\alpha^2-1)t} - e^{(\beta^2-1)t}}{\alpha - \beta} =$
$\sum_{n=0}^\infty \left[\sum_{k=0}^{n} (-1)^{n-k}\binom{n}{k} F_{2k}\right] \dfrac{t^n}{n!}$. That is, $\dfrac{e^{\alpha t} - e^{\beta t}}{\alpha - \beta} =$
$\sum_{n=0}^\infty \left[\sum_{k=0}^{n} (-1)^{n-k}\binom{n}{k} F_{2k}\right] \dfrac{t^n}{n!}$. So $F_n = \sum_{k=0}^{n} (-1)^{n-k}\binom{n}{k} F_{2k}$.

23. (Church and Bicknell) Let $A(t) = \dfrac{e^{\alpha^m t} - e^{\beta^m t}}{\alpha - \beta}$ and $B(t) = e^{\alpha^m t} + e^{\beta^m t}$. This yields $\sum_{n=0}^{\infty} 2^n F_{mn} \frac{t^n}{n!} = \dfrac{e^{2\alpha^m t} - e^{2\beta^m t}}{\alpha - \beta} = \sum_{n=0}^{\infty} \left[\sum_{k=0}^{n} \binom{n}{k} F_{mk} L_{mn-mk} \right] \frac{t^n}{n!}$.
$\therefore \sum_{k=0}^{n} \binom{n}{k} F_{mk} L_{mn-mk} = 2^n F_{mn}$.

25. (Church and Bicknell) Let $A(t) = e^{\alpha^m t} + e^{\beta^m t} = B(t)$. This yields $\left(e^{\alpha^m t} + e^{\beta^m t}\right)^2 = \sum_{n=0}^{\infty} \left[\sum_{k=0}^{n} \binom{n}{k} L_{mk} L_{mn-mk} \right] \frac{t^n}{n!}$.
That is, $\sum_{n=0}^{\infty} (2^n L_{mn} + 2L_m^n) \frac{t^n}{n!} = e^{2\alpha^m t} + e^{2\beta^m t} + 2e^{(\alpha^m + \beta^m)t}$
$= \sum_{n=0}^{\infty} \left[\sum_{k=0}^{n} \binom{n}{k} L_{mk} L_{mn-mk} \right] \frac{t^n}{n!}$
$\therefore \sum_{k=0}^{n} \binom{n}{k} L_{mk} L_{mn-mk} = 2^n L_{mn} + 2L_m^n$

27. (Church and Bicknell)

$$\sum_{n=0}^{\infty} F_{mn} \frac{t^n}{n!} = \frac{e^{\alpha^m t} - e^{\beta^m t}}{\alpha - \beta} = \frac{e^{(\alpha F_m + F_{m-1})t} - e^{(\beta F_m + F_{m-1})t}}{\alpha - \beta}$$

$$= e^{F_{m-1} t} \left(\frac{e^{\alpha F_m t} - e^{\beta F_m t}}{\alpha - \beta} \right) = \sum_{n=0}^{\infty} \left[\sum_{k=0}^{n} \binom{n}{k} F_{m-1}^{n-k} F_m^k F_k \right] \frac{t^n}{n!}$$

The desired result now follows by equating the coefficients of $t^n/n!$.

EXERCISES 20 (p. 246)

1. Yes
3. Yes
5. $\overline{A \quad \alpha \quad C \quad B}$
   $\dfrac{AC}{CB} = \alpha$; that is, $\dfrac{1 - CB}{CB} = \alpha$. $\therefore \dfrac{1}{CB} = \alpha + 1 = \alpha^2$
   Thus, $BC = 1/\alpha^2$ and $AC = \alpha \cdot BC = 1/\alpha$.
7. $t^2 + t - 1 = 0$
9. $-\beta$
11. Let x denote the given sum. Then $x = \sqrt{1 - x}$; that is, $x^2 + x - 1 = 0$. $\therefore x = -\beta$.
13. Let $a/b = c/d = k$. Then $b/a = 1/k = d/c$.
15. Let $a/b = c/d = k$. Then $(a - b)/b = (bk - b)/b = k - 1 = (dk - d)/d = (c - d)/d$.
17. $\dfrac{\text{Sum of the triangular faces}}{\text{Base area}} = \dfrac{4(2b \cdot a)/2}{(2b)^2} = \dfrac{a}{b} = \alpha$, so $a = b\alpha$.
19. $1 + 1/\alpha = (\alpha + 1)/\alpha = \alpha^2/\alpha = \alpha$.

21. $\alpha^2 = \alpha + 1$, so $\alpha^n = \alpha^{n-1} + \alpha^{n-2}$, $n \geq 2$.

23. $\text{LHS} = \dfrac{1/\alpha}{1 - 1/\alpha} = \dfrac{1}{\alpha - 1} = \alpha$

25. $\text{LHS} = \alpha^4 = (\alpha + 1)^2 = 3\alpha + 2$

27. $\alpha\sqrt{3 - \alpha} = \sqrt{3\alpha^2 - \alpha^3} = \sqrt{3(\alpha + 1) - \alpha(\alpha + 1)} = \sqrt{\alpha + 2}$

29. $\alpha + 2 = \dfrac{5 + \sqrt{5}}{2} = \dfrac{10 + 2\sqrt{5}}{4}$ $\therefore \sqrt{\alpha + 2} = \dfrac{\sqrt{10 + 2\sqrt{5}}}{2}$

31. $\cos^2 \pi/10 = \dfrac{1 + \cos \pi/5}{2} = \dfrac{\alpha + 2}{4}$

$$\therefore \cos \pi/10 = \frac{\sqrt{\alpha + 2}}{2}$$

33.
$$\begin{aligned} \nu + \frac{1}{\nu^2} &= \nu + \frac{1}{\nu + 1} = \nu + \frac{\nu - 1}{\nu} = \nu + 1 - \frac{1}{\nu} \\ &= (\nu - 1/\nu) + 1 = \frac{\nu^2 - 1}{\nu} + 1 = 1 + 1 = 2 \end{aligned}$$

35.
$$\begin{aligned} \lim_{n\to\infty} \frac{L_n}{L_{n+1}} &= \lim_{n\to\infty} \frac{\alpha^n + \beta^n}{\alpha^{n+1} + \beta^{n+1}} \\ &= \lim_{n\to\infty} \frac{\alpha^n[1 + (\beta/\alpha)^n]}{\alpha^{n+1}[1 + (\beta/\alpha)^{n+1}]} = 1/\alpha \end{aligned}$$

37.
$$\lim_{n\to\infty} \frac{G_{n+1}}{G_n} = \lim_{n\to\infty} \frac{c\alpha^{n+1} - d\beta^{n+1}}{c\alpha^n - d\beta^n} = \alpha$$

EXERCISES 21 (p. 265)

1. $\pi\alpha$

3. $\dfrac{\pi\alpha^2\sqrt{3 - \alpha}}{24}$

5.
$$\begin{aligned} (x * y) * z &= [a + b(x + y) + cxy] * z \\ &= a + b[a + b(x + y) + cxy + z] + [a + b(x + y) + cxy]cz \\ &= a + b(a + cxy + z + czx + cyz) \\ &\quad + b^2(x + y) + c^2xyz + caz \qquad (1) \end{aligned}$$

Likewise,

$$\begin{aligned} x * (y * z) &= a + b(a + x + cyz + cxy + czx) \\ &\quad + b^2(y + z) + cax + c^2xyz \qquad (2) \end{aligned}$$

Since * is associative, Eqs. (1) and (2) yield $b^2(x-z)+b(z-x)+(z-x)=0$. Since x, y, and z are arbitrary, this implies $b^2-b-1=0$, so $b=\alpha$ or β.

7. (Alexanderson) Let $p(x)=x^n-xF_n-F_{n-1}$, $g(x)=x^2-x-1$, and $h(x)=x^{n-2}+x^{n-3}+2x^{n-4}+\cdots+F_kx^{n-k-1}+\cdots+F_{n-2}x+F_{n-1}$. Then $p(x)=g(x)h(x)$. When $x\geq 0$, $h(x)>0$. When $x>\alpha$, $g(x), h(x)>0$; so $p(x)>0$. When $0\leq x<\alpha$, $g(\alpha)<0$; so $p(\alpha)<0$.

   Suppose $x_k>\alpha$. Since $p(x)>0$, $x_k^n>x_kF_n+F_{n-1}=x_{k+1}^n$, so $x_k>x_{k+1}$. In addition, $x_{k+1}^n=x_kF_n+F_{n-1}>\alpha F_n+F_{n-1}=\alpha^n$. So $x_{k+1}>\alpha$; $\therefore x_k>x_{k+1}>\alpha$. Thus $x_0>\alpha$ implies $x_0>x_1>x_2\cdots>\alpha$. Similarly, if $0\leq x_0<\alpha$, then $0\leq x_0<x_1<x_2<\ldots<\alpha$.

   Thus, in both cases, the sequence $\{x_k\}$ is monotonic and bounded, so x_k converges to a limit l as $k\to\infty$: $l=\sqrt[n]{F_{n-1}+lF_n}$. Since $l^2-lF_n-F_{n-1}=0$, l is the positive zero of $p(x)$. But α is the unique positive zero of $p(x)$, so $l=\lim\limits_{k\to\infty}x_k=\alpha$.

9. Since $|\beta|<1$, Sum $=\dfrac{1}{1-|\beta|}=\dfrac{1}{1+\beta}=\dfrac{1}{\beta^2}=\alpha^2$.

11. $t=\left.\dfrac{x^{t+1}}{t+1}\right|_0^1=\dfrac{1}{t+1}$. So $t^2+t-1=0$; that is, $t=-\alpha,\beta$.

13. (King, 1971) $\dfrac{F_{n+k}}{L_n}=\dfrac{\alpha^{n+k}-\beta^{n+k}}{\sqrt{5}(\alpha^n+\beta^n)}=\dfrac{\alpha_k}{\sqrt{5}}\cdot\dfrac{1-\theta^{n+k}}{1+\theta^n}$, where $|\theta|=\left|\dfrac{\beta}{\alpha}\right|<$, so $\theta^n\to 0$ as $n\to\infty$; $\therefore \lim\limits_{n\to\infty}\dfrac{F_{n+k}}{L_n}=\dfrac{\alpha^k}{\sqrt{5}}$.

15. (Ford) By PMI, $a_n=F_n+k(F_{n+1}-1)$;
   $\therefore \lim\limits_{n\to\infty}\dfrac{a_n}{F_n}=\lim\limits_{n\to\infty}\left[1+k\left(\dfrac{F_{n+1}}{F_n}-\frac{1}{F_n}\right)\right]=1+k\alpha$.

17. By PMI, $b_n=L_n+kF_{n-1}$; $\therefore \lim\limits_{n\to\infty}\dfrac{b_n}{L_n}=\lim\limits_{n\to\infty}\left[1+k\cdot\dfrac{F_{n-1}}{F_n}\cdot\dfrac{F_n}{L_n}\right]=1+k\cdot\dfrac{1}{\alpha}\cdot\dfrac{1}{\sqrt{5}}=1+\dfrac{k}{\sqrt{5}\alpha}$.

19. (Lord) $\sum\limits_0^n\binom{n}{i}\alpha^{3i-2n}=\alpha^{-2n}\sum\limits_0^n\binom{n}{i}(\alpha^3)^i=\alpha^{-2n}(1+\alpha^3)^n=\alpha^{-2n}(2\alpha^2)^n=2^n$.

21. (Ford) Assume $x_n\neq -1$ for every n. By PMI, $x_n=\dfrac{x_0F_{n-1}+F_n}{x_0F_n+F_{n+1}}=\dfrac{F_{n-1}}{F_n}\cdot\dfrac{x_0+F_n/F_{n-1}}{x_0+F_{n+1}/F_n}$; $\therefore x_n$ is defined when $x_0\neq -F_{n+1}/F_n$, $n\geq 1$. When $x_0\neq -F_{n+1}/F_n$, $\lim\limits_{n\to\infty}x_n=\dfrac{1}{\alpha}\cdot\dfrac{x_0+\alpha}{x_0+\alpha}=\dfrac{1}{\alpha}=-\beta$.

23. Clearly, $r\neq 0$. $(r,r^2,\ldots,r^n,\ldots)\in\mathbf{V}$ iff $r^n=r^{n-1}+r^{n-2}$; that is, iff $r^2=r+1$.

25. $a\mathbf{u}+b\mathbf{v}=(a\alpha+b\beta,\ldots,a\alpha^n+b\beta^n,\ldots)=\mathbf{F}$ iff $a\alpha+b\beta=1=a\alpha^2+b\beta^2$. Solving these two equations, we get $a=1/(\alpha-\beta)=-b$.

27. $f^{-1}(x)=(x/A)^{1/n}$ and $f^{(m)}x=An(n-1)\cdots(n-m+1)x^{n-m}$. Then $f^{-1}(x)=f^{(m)}(x)$ implies $(x/A)^{1/n}=An(n-1)\cdots(n-m+1)x^{n-m}$; that is, $x/A=$

$A^{np}[n(n-1)\cdots(n-m+1)]^{np}x^{np(n-m)}$. Then $A^{np+1}[n(n-1)\cdots(n-m+1)]^{np}x^{np(n-m)-1}=0$, so $pn^2-mpn-1=0$. Solving, we get the desired result.

EXERCISES 22 (p. 271)

1. Since $\triangle ABC$ is a golden triangle, $\dfrac{AB}{AC}=\dfrac{BC}{AC}=\alpha$. Since $\triangle ABC\sim\triangle ADC$, $\dfrac{AB}{AC}=\dfrac{AD}{CD}$; $\therefore\ \dfrac{AD}{CD}=\dfrac{AC}{CD}=\alpha$, so $\triangle CAD$ is a golden triangle.

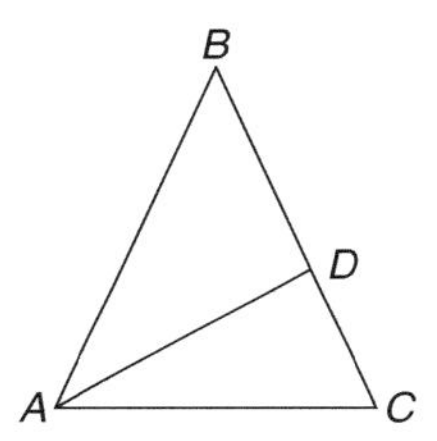

3. Since $\triangle ABC$ is golden, $\dfrac{BC}{AC}=\alpha$ and $\triangle CAD$ is also golden. Since $\triangle ABD$ is isosceles, it follows that $AD=AC=BD$. Let h be the length of the altitude from A to $\overline{BC}$. Then $\dfrac{\text{Area }\triangle ABC}{\text{Area }\triangle BDA}=\dfrac{1/2BC\cdot h}{1/2BD\cdot h}=\dfrac{BC}{BD}=\dfrac{BC}{AC}=\alpha$.
5. $\dfrac{\text{Area }\triangle ABC}{\text{Area }\triangle BDA}=\alpha$; that is, $\dfrac{BC}{BD}=\alpha$; $\therefore\ \dfrac{\text{Area }\triangle ABC}{\text{Area }\triangle CDA}=\dfrac{1/2BC\cdot h}{1/2CD\cdot h}=\dfrac{BC}{CD}=\dfrac{BC}{BC-BD}=\dfrac{1}{1-1/\alpha}=\alpha^2$

EXERCISES 23 (p. 292)

1. $-\beta$
3. $\dfrac{AB}{BC}=\dfrac{BC}{BE}=\dfrac{l}{w}=k$. Let $\dfrac{\text{Area }ABCD}{\text{Area }AEFD}=k$; that is, $\dfrac{AB\cdot BC}{AE\cdot BC}=k$, so $\dfrac{AB}{AE}=k$. Thus $\dfrac{l}{AE}=k$, so $AE=w$. Since $ABCD$ and $BCFE$ are similar, $\dfrac{AB}{BC}=\dfrac{BC}{CF}$; that is, $\dfrac{l}{w}=\dfrac{w}{l-w}$, so $\dfrac{l}{w}=k=\alpha$.

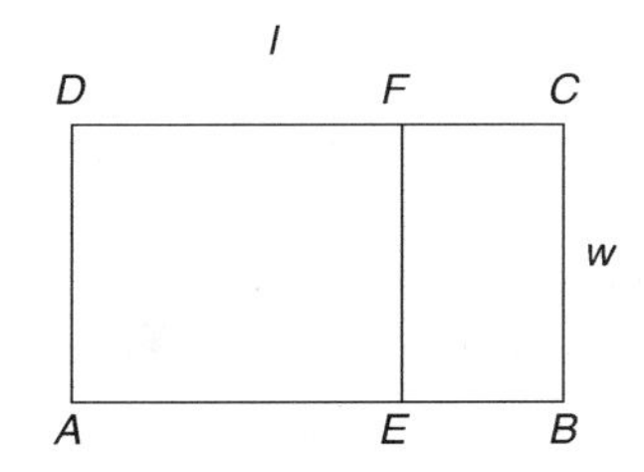

Conversely, let $k = \frac{l}{w} = \alpha$. Then $\frac{AB}{BC} = \frac{BC}{CF} = \alpha$, so $\frac{BC}{AB - AE} = \alpha$. That is, $\frac{1}{\alpha - AE/BC} = \alpha$. So $AE = BC = w$; $\therefore \frac{\text{Area } ABCD}{\text{Area } AEFD} = \frac{AB \cdot BC}{AE \cdot BC} = \frac{AB}{AE} = \frac{AB}{BC} = k = \alpha$

In both cases, $AE = w$; so *AEFD* is a square.

5. Using Figure 23.3, since both are golden rectangles, $\frac{FG}{BG} = \frac{BG}{BC} = \alpha$. Since $\triangle FPG \sim \triangle BPC, \alpha = \frac{FG}{BG} = \frac{FP}{GP}$. Since $\triangle BPG \sim \triangle BPC, \alpha = \frac{BG}{BC} = \frac{BP}{CP}$; $\therefore \frac{FP}{GP} = \frac{BP}{CP} = \alpha$.
7. Since $BGHC$ is a golden rectangle, $\frac{BG}{BC} = \alpha$; that is, $\frac{AC}{BC} = \alpha$. Then $\frac{AB}{BG} = \frac{AC + BC}{BG} = \frac{AC + BC}{AC} = 1 + \frac{BC}{AC} = 1 + 1/\alpha = 1 - \beta = \alpha$; $\therefore ABGF$ is a golden rectangle.
9. $\angle SPQ = 180° - (\angle APS + \angle BPQ) = 180° - (45° + 45°) = 90°$.
   $\therefore$ The parallelogram *PQRS* is a rectangle. Since $\triangle APS \sim \triangle BPQ$, $\frac{AP}{BP} = \frac{PS}{PQ}$. But $\frac{AP}{BP} = \alpha$, so $\frac{PS}{PQ} = \alpha$. Thus *PQRS* is a golden rectangle.

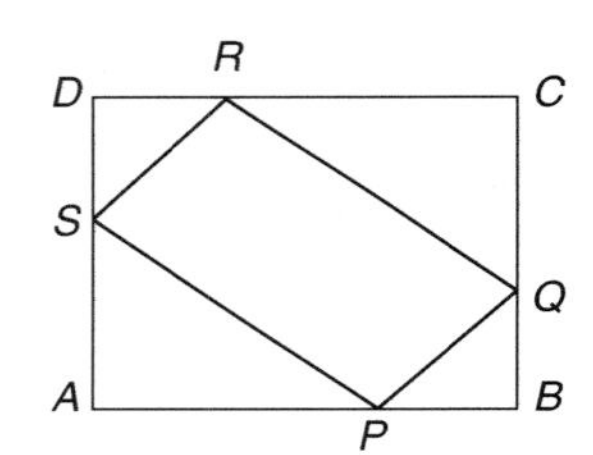

11. Shorter side $= (-1)^{n+1}(bF_n - aF_{n-1})$; longer side $= (-1)^n(bF_{n-1} - aF_{n-2})$.

EXERCISES 24 (p. 307)

1. Follows by Candido's identity $[x^2 + y^2 + (x + y)^2]^2 = 2[x^4 + y^4 + (x + y)^4]$, with $x = F_n$ and $y = F_{n+1}$.
3. Follows by Candido's identity with $x = G_n$ and $y = G_{n+1}$.
5. (Prielipp) $(L_{n-1}L_{n+2})^2 + (2L_nL_{n+1})^2 = [(L_{n+1} - L_n)(L_{n+1} + L_n)]^2 + (2L_nL_{n+1})^2 = (L_{n+1}^2 - L_n^2)^2 + (2L_nL_{n+1})^2 = (L_{n+1}^2 + L_n^2)^2 = (L_{2n} + L_{2n+2})^2$.

EXERCISES 25 (p. 324)

1. $\triangle CQR$ and $\triangle CDR$ are isosceles triangles with $CQ = CR$ and $CR = DR$, respectively. Thus $CQ = CR = DR$. Continuing like this, we get $CQ = CR = DR = DS = ES = ET = AT = AP$.

   Since the point of intersection of any two diagonals originating at adjacent vertices of a regular pentagon divides each diagonal in the Golden Ratio, it follows that $DQ = \alpha DR$, so $RQ = DQ - DR = (\alpha - 1)DR$. Similarly, $ST = (\alpha - 1)SD$; $\therefore QR = ST$. Continuing like this, we get $PQ = QR = RS = ST = TP$. Thus *PQRST* is a regular pentagon.

3. By Heron's formula, Area $= \sqrt{s(s-a)(s-b)(s-c)}$, where $2s = 2AP + TP = 2(a/\alpha^2) + a/\alpha^3 = 2a/\alpha$; $\therefore s = a/\alpha$.

$$\begin{aligned}\text{Area} &= \sqrt{a/\alpha(a/\alpha - a/\alpha^2)(a/\alpha - a/\alpha^2)(a/\alpha - a/\alpha^3)}\\ &= \frac{a^2}{\alpha^4}\sqrt{(\alpha-1)(\alpha-1)(\alpha^2-1)}\\ &= \frac{a^2(\alpha-1)\sqrt{(\alpha^2-1)}}{\alpha^4} = \frac{a^2(\alpha-1)\sqrt{\alpha}}{\alpha^4}\end{aligned}$$

5. Area $\triangle CDS = 1/2 \cdot CS \cdot h = \dfrac{1}{2} \cdot \dfrac{a}{\alpha} \cdot \dfrac{a\sqrt{\alpha+2}}{2\alpha^2} = \dfrac{a^2\sqrt{\alpha+2}}{4\alpha^3}$.

7. Area $SPRD = 2$ (area $\triangle RDS$) $= 2 \cdot \dfrac{1}{2} \cdot \dfrac{a}{\alpha^3} \cdot \dfrac{a\sqrt{\alpha+2}}{2\alpha^2} = \dfrac{a^2\sqrt{\alpha+2}}{2\alpha^5}$.

9. $\alpha^3 : 1$

11. $SR = a/\alpha^3$, $TQ = (a/\alpha^3)\alpha = a/\alpha^2$, and $ZY = SR/\alpha^3 = a/\alpha^6$; $\therefore 2s = 2PV + ZV = 2(SR/\alpha^2) + a/\alpha^6 = 2a/\alpha^5 + a/\alpha^6 = 2a(\alpha+1)/\alpha^6 = 2a/\alpha^4$; so $s = a/\alpha^4$. By Heron's formula,

$$\begin{aligned}\text{Area}\triangle PVZ &= \sqrt{\frac{a}{\alpha^4}\left(\frac{a}{\alpha^4} - \frac{a}{\alpha^5}\right)\left(\frac{a}{\alpha^4} - \frac{a}{\alpha^5}\right)\left(\frac{a}{\alpha^4} - \frac{a}{\alpha^6}\right)}\\ &= \frac{a^2}{\alpha^{10}}\sqrt{(\alpha-1)(\alpha-1)(\alpha^2-1)} = \frac{a^2(\alpha-1)\sqrt{\alpha}}{\alpha^{10}}.\end{aligned}$$

13. Area $\triangle DRS = \dfrac{1}{2} \cdot RS \cdot h = \dfrac{1}{2} \cdot \dfrac{a}{\alpha^3} \cdot \dfrac{a}{\alpha} \cos\pi/5 = \dfrac{a^2\cos\pi/5}{2\alpha^4} = \dfrac{a^2}{4\alpha^3}$

    Desired area = area *PQRST* + 5(area $\triangle$ *DRS*)

$$= \left\{\frac{5a^2}{\alpha\sqrt{3-\alpha}} - \frac{5a^2[\alpha^2 + (\alpha-1)\sqrt{\alpha-1}]}{4\alpha^4}\right\} + \frac{5a^2}{4\alpha^3}$$

$$= \frac{5a^2(\alpha^3 - \sqrt{3-\alpha} - (\alpha-1)\sqrt{3\alpha-4})}{4\alpha^4\sqrt{3-\alpha}}$$

15. By the Pythagorean theorem, $(ar^2)^2 = a^2 + (ar)^2$. Then $r^4 = r^2 + 1$, so $r^2 = \alpha$ and hence $r = \sqrt{\alpha}$.

17. Since $\triangle GOK$ is isosceles, $GK = GO = r$. $\triangle GLD$ and $\triangle GOK$ are similar and golden; $\therefore \dfrac{GD}{LD} = \dfrac{GK}{OK} = \alpha$; that is, $\dfrac{GD}{LD} = \dfrac{r}{OK} = \alpha$; $\therefore GD = \alpha LD$. But $LD = OD = r$, so $GD = r\alpha$.
Since $AD = GD$, $AD = r\alpha$.

19. $x^5 - 1 = (x-1)(x^4 + x^3 + x^2 + x + 1)$
$x^4+x^3+x^2+x+1 = 0$ gives $(x^2+1/x^2)+(x+1/x)+1 = 0$. Let $y = x+1/x$. Then $y^2 + y - 1 = 0$, so $y = -\alpha, -\beta$. When $y = -\alpha$, $x + 1/x = -\alpha$. Then $x^2 + \alpha x + 1 = 0$, so $x = \dfrac{-\alpha \pm \sqrt{\alpha^2 - 4}}{2}$. When $y = -\beta$, $x + 1/x = -\beta$. Then $x^2 + \beta x + 1 = 0$, so $x = \dfrac{-\beta \pm \sqrt{\beta^2 - 4}}{2}$. Thus, the five solutions are $1, \dfrac{-\alpha \pm \sqrt{\alpha^2 - 4}}{2}$, and $\dfrac{-\beta \pm \sqrt{\beta^2 - 4}}{2}$.

21.
$$\begin{aligned} AB^2 &= \left(\frac{\alpha - 1}{2} + \frac{\alpha}{2}\right)^2 + \left(\frac{\alpha\sqrt{3-\alpha}}{2} - \frac{\sqrt{3-\alpha}}{2}\right)^2 \\ &= \frac{(2\alpha - 1)^2 + (3-\alpha)(\alpha - 1)^2}{4} \\ &= \frac{[4(\alpha + 1) - 4\alpha + 1] + [(3-\alpha)(\alpha + 1 - 2\alpha + 1)]}{4} = 3 - \alpha; \end{aligned}$$

$\therefore AB = \sqrt{3-\alpha}$

23. $AE = \sqrt{3-\alpha}$, $BD = \alpha\sqrt{3-\alpha}$, and $h = \dfrac{\alpha}{2} + \dfrac{\alpha - 1}{2} = \dfrac{2\alpha - 1}{2}$.
Area of trapeziod $ABDE = \dfrac{1}{2}(AE + BD)h = \dfrac{1}{2}(\sqrt{3-\alpha} + \alpha\sqrt{3-\alpha}) \cdot \dfrac{2\alpha - 1}{2} = \dfrac{(3\alpha - 1)\sqrt{3-\alpha}}{4}$

25. $PQ^2 = \left(\dfrac{2\beta + 1}{2} + \dfrac{\beta}{2}\right)^2 + \left(\dfrac{\beta\sqrt{3-\alpha}}{2} + \dfrac{\beta\sqrt{3-\alpha}}{2}\right)^2$; $PQ = \dfrac{\sqrt{19\beta + 13}}{2}$

$AP^2 = \left(\dfrac{2\beta + 1}{2} + \dfrac{\alpha}{2}\right)^2 + \left(\dfrac{3\sqrt{3-\alpha}}{2} + \dfrac{\sqrt{3-\alpha}}{2}\right)^2$; $AP = \dfrac{\sqrt{9\beta + 21}}{2}$

$QC^2 = (1 + \beta/2)^2 + \dfrac{\beta^4(3-\alpha)}{4}$; $QC = 3 + 4\beta$

27. $1 : 1/\alpha : 1/\alpha^2 : 1/\alpha^3$

29. $\triangle ACE \sim \triangle AED$; $\therefore \dfrac{AE}{AC} = \dfrac{AD}{AE} = x$ (say). Since $AE = AB$ and $AC = BD$, $\dfrac{AD}{AE} = \dfrac{AB + BD}{AE} = \dfrac{AE + BD}{AE} = 1 + \dfrac{AC}{AE}$; that is $x = 1 + 1/x$, so $x = \alpha$.

31. $F_2 = 2\sum_0^1(-1)^k \cos^{1-k}\pi/5 \sin^k \pi/10$
$= 2[(-1)^0(\alpha/2)(-\beta/2)^0 + (-1)^1(\alpha/2)^0(-\beta/2)] = 1$

$$\begin{aligned} F_4 &= 2^3\sum_0^3(-1)^k \cos^{3-k}\pi/5 \sin^k \pi/10 \\ &= 8[(-1)^0(\alpha/2)^3(-\beta/2)^0 + (-1)(\alpha/2)^2(-\beta/2)] \\ &\quad + [(-1)^2(\alpha/2)(-\beta/2)^2 + (-1)^3(\alpha/2)^0(-\beta/2)^3] \\ &= 8\left[\frac{\alpha^3}{8} + \frac{\alpha^2\beta}{8} + \frac{\alpha\beta^2}{8} + \frac{\beta^3}{8}\right] \\ &= \alpha^3 + \alpha^2\beta + \alpha\beta^2 + \beta^3 = (\alpha+\beta)^3 - 2\alpha\beta(\alpha+\beta) = 1 + 2\cdot 1 = 3 \end{aligned}$$

33. (V. E. Hoggatt) By Eqs. (25.2) and (25.3),
$F_n = \dfrac{2^{n+2}}{5}(\cos^n \pi/5 \sin \pi/5 \sin 3\pi/5 + \cos^n 3\pi/5 \sin 3\pi/5 \sin 9\pi/5)$ and
$(-1)^n F_n = \dfrac{2^{n+2}}{5}(\cos^n 2\pi/5 \sin 2\pi/5 \sin 6\pi/5 + \cos^n 4\pi/5 \sin 4\pi/5 \sin 12\pi/5)$.
Adding, $[1 + (-1)^n]F_n = \dfrac{2^{n+2}}{5}\sum_1^4 \cos^n k\pi/5 \sin k\pi/5 \sin 3k\pi/5$
Thus $\dfrac{2^{n+1}}{5}\sum_1^4 \cos^n k\pi/5 \sin k\pi/5 \sin 3k\pi/5 = \begin{cases} F_n & \text{if } n \text{ is even} \\ 0 & \text{otherwise} \end{cases}$

EXERCISES 26 (p. 331)

1. The asymptotes are $y = \pm\dfrac{x}{\sqrt{\alpha}}$. Solving the equations $y = \pm\dfrac{x}{\sqrt{\alpha}}$ and $y^2 = 4ax$, we get the given points.
3. $PQ = \sqrt{(a\alpha^2 - a\beta^2)^2 + (2a\alpha - 2a\beta)^2} = \sqrt{5a^2(\alpha-\beta)^2} = 5a$
5. $x - \beta y + a\beta^2 = 0, x - \alpha y - 4a + a\alpha = 0.$
7. The slopes of the tangents at P and Q are $1/\alpha$ and $1/\beta$, respectively;
$\therefore \tan\theta = \left|\dfrac{1/\alpha - 1/\beta}{1 + 1/\alpha \cdot 1/\beta}\right| = \frac{\alpha-\beta}{0} = \infty$, so $\theta = \pi/2$.
9. The slopes of the normals at P and Q are $-\alpha$ and $-\beta$, respectively;
$\therefore \tan\theta = \left|\dfrac{-\alpha+\beta}{1+\alpha\beta}\right| = \frac{\alpha-\beta}{0} = \infty$, so $\theta = \pi/2$.
11. $SQ = \sqrt{(a - a\beta^2)^2 + (0 - 2a\beta)^2} = \sqrt{5}a|\beta|$ and
$QR = \sqrt{(a\beta^2 - 0)^2 + (2a\beta + 2a)^2} = \sqrt{5}a\beta^2$; $\therefore SQ : QR = |\beta| : \beta^2 = \alpha : 1$.

EXERCISES 27 (p. 338)

1. $1 + \cfrac{1}{2 + \frac{1}{5+\frac{1}{3}}}$

3. $\frac{52}{23}$

5. $1, \frac{3}{2}, \frac{10}{7}, \frac{43}{30}, \frac{225}{157}$

7. $C_4 = \frac{p_4}{q^4} = \frac{5 \cdot 43 + 10}{5 \cdot 30 + 7} = \frac{225}{157}; C_5 = \frac{p_5}{q^5} = \frac{6 \cdot 225 + 43}{6 \cdot 157 + 30} = \frac{1393}{972}$

9. Since $C_1 = 1 = F_2/F_1$ and $C_2 = 2 = F_3/F_2$, the formula works for $n = 1$ and $n = 2$. Assume it is true for all positive integers $k \le n$, where $n \ge 2 : C_k = p_k/q_k = F_{k+1}/F_k$, so $p_k = F_{k+1}$ and $q_k = F_k$. Then $p_n = 1 \cdot p_{n-1} + p_{n-2} = F_n + F_{n-1} = F_{n+1}$; similarly, $q_n = F_n$; $\therefore C_n = p_n/q_n = F_{n+1}/F_n$. Thus, by PMI, the formula is true for all $n \ge 1$.

11. $C_n = \frac{F_{n+1}}{F_n}$; $\therefore C_n - C_{n-1} = \frac{F_{n+1}}{F_n} - \frac{F_n}{F_{n-1}} = \frac{F_{n+1}F_{n-1} - F_n^2}{F_nF_{n-1}} = \frac{(-1)^n}{F_nF_{n-1}}$; $\therefore \lim_{n\to\infty} (C_n - C_{n-1}) = 0$.

EXERCISES 28 (p. 347)

1. $\sum_1^7 iF_i = 1\cdot1+2\cdot1+3\cdot2+4\cdot3+5\cdot5+6\cdot8+7\cdot13 = 185 = 7\cdot34-55+2 = 7F_9 - F_{10} + 2$.

3. Let $S_j = \sum_i^j F_{2k} = F_{2j+1} - 1$.

$$\begin{aligned}
\text{LHS} &= 2\sum_1^n F_{2k} + 2\sum_2^n F_{2k} + \cdots + 2\sum_n^n F_{2k} \\
&= 2[S_n + (S_n - S_1) + \cdots + (S_n - S_{n-1})] \\
&= 2[nS_n - \sum_1^{n-1} S_j] = 2[n(F_{2n+1} - 1) - (F_{2n} - 1) - n + 1] \\
&= 2(nF_{2n+1} - F_{2n}) = \text{RHS}.
\end{aligned}$$

5. Let $S_j = \sum_1^j L_{2i-1} = L_{2j} - 2$.

$$\text{LHS} = \sum_1^n L_{2i-1} + 2\sum_2^n L_{2i-1} + \cdots + 2\sum_n^n L_{2i-1}$$

$$= S_n + 2[(S_n - S_1) + \cdots + (S_n - S_{n-1})]$$

$$= (2n-1)S_n - 2\sum_1^{n-1} S_j = (2n-1)S_n - 2\sum_1^{n-1}(L_{2j} - 2)$$

$$= (2n-1)(L_{2n} - 2) - 2[(L_{2n-1} - 1) - 2(n-1)]$$

$$= (2n-1)L_{2n} - 2L_{2n-1} = \text{RHS}.$$

7. Let $S_j = \sum_1^j L_{2k} = L_{2j+1} - 1$.

$$\text{LHS} = 2\sum_1^n L_{2i} + 2\sum_2^n L_{2i} + \cdots + 2\sum_n^n L_{2i}$$

$$= 2[S_n + (S_n - S_1) + \cdots + (S_n - S_{n-1})]$$

$$= 2(nS_n - \sum_1^{n-1} S_j) = 2[n(L_{2n+1} - 1) - \sum_1^{n-1}(L_{2j+1} - 1)]$$

$$= 2[n(L_{2n+1} - 1) - (L_{2n} - 3) + (n-1)]$$

$$= 2(nL_{2n+1} - L_{2n} + 2) = \text{RHS}.$$

9. $\text{LHS} = (a-d)\sum_1^n L_i + d\sum_1^n iL_i = (a-d)(L_{n+2} - 3)$

$$+ d(nL_{n+2} - L_{n+3} + 4)$$

$$= (a + nd - d)L_{n+2} - d(L_{n+3} - 7) - 3a = \text{RHS}.$$

11. $\text{LHS} = (a-d)\sum_1^n F_i^2 + d\sum_1^n iF_i^2 = (a-d)F_nF_{n+1}$

$$+ d(nF_nF_{n+1} - F_n^2 + \nu)$$

$$= (a + nd - d)F_nF_{n+1} - d(F_n^2 - \nu) = \text{RHS}.$$

13. Let $S_n = \Sigma iL_i^2$ and $S_n^* = \Sigma(n-i+1)L_i^2$. Then $S_n + S_n^* = (n+1)\Sigma L_i^2 = (n+1)(L_nL_{n+1} - 2)$; $\therefore S_n^* = (n+1)(L_nL_{n+1} - 2) - (nL_nL_{n+1} - L_n^2 + \nu) = L_nL_{n+1} + L_n^2 - 2(n+1) - \nu) = L_nL_{n+2} - 2(n+1) - \nu$.
15. Let $S = \Sigma[a + (i-1)d]L_i^2$ and $S^* = \Sigma[(a + (n-i)d]L_i^2$. Then $S + S^* = (2a + (n-1)d\Sigma L_i^2 = [2a + (n-1)d](L_nL_{n+1} - 2)$; $\therefore S^* = [2a + (n-1)d](L_nL_{n+1} - 2) - (a + nd - d)(L_nL_{n+1} - \nu) + d(L_n^2 - 2n - \nu) = a(L_nL_{n+1} - 2) + d(L_n^2 - 2n - \nu)$.

17. Let $S_j = \Sigma_1^j G_k = G_{j+2} - b$.

$$\begin{aligned}
\sum_1^n iG_i &= \sum_1^n G_i + \sum_2^n G_i + \cdots + \sum_n^n G_i \\
&= S_n + (S_n - S_1) + \cdots + (S_n - S_{n-1}) = nS_n - \sum_1^{n-1} S_i \\
&= n(G_{n+2} - b) - \sum_1^{n-1}(G_{i+2} - b) \\
&= n(G_{n+2} - b) - \sum_1^{n-1} G_{i+2} + (n-1)b \\
&= n(G_{n+2} - b) - (G_{n+3} - b - a - b) + (n-1)b \\
&= nG_{n+2} - G_{n+3} + a + b
\end{aligned}$$

19. $\sum_1^n G_{2i-1} = \sum_1^n G_{2i} - \sum_1^n G_{2i-2} = G_{2n} - G_0 = G_{2n} + a - b$

21. Let $S_j = \sum_1^j G_{2k-1} = G_{2j} + a - b$.

$$\begin{aligned}
\sum_1^n (2i-1)G_{2i-1} &= \sum_1^n G_{2i-1} + 2\sum_2^n G_{2i-1} + \cdots + 2\sum_n^n G_{2i-1} \\
&= S_n + 2(S_n - S_1) + \cdots + 2(S_n - S_{n-1}) \\
&= S_n + 2(n-1)S_n - 2\sum_1^{n-1} S_i = (2n-1)S_n - 2\sum_1^{n-1} S_i \\
&= (2n-1)S_n - 2\sum_1^{n-1}(G_{2i} + a - b) \\
&= (2n-1)S_n - 2\sum_1^{n-1} G_{2i} - 2(n-1)(a-b) \\
&= (2n-1)(G_{2n} + a - b) - 2(G_{2n-1} - a) \\
&\quad - 2(n-1)(a-b) \\
&= (2n-1)G_{2n} - 2G_{2n-1} + 3a - b
\end{aligned}$$

EXERCISES 29 (p. 354)

1. $\sum_{1}^{10} i^2 F_i = 121F_{12} - 23F_{14} + 2F_{16} - 8 = 121 \cdot 144 - 23 \cdot 377 + 2 \cdot 987 - 8$
$= 10{,}719.$

3. $\sum_{1}^{5} i^3 F_i = 216F_7 - 127F_9 + 42F_{11} - 6F_{13} + 50$
$= 216 \cdot 13 - 127 \cdot 34 + 42 \cdot 89 - 6 \cdot 233 + 50 = 880.$

5.
$$\begin{aligned}\sum_{1}^{5} i^4 F_i &= 1296F_7 - 1105F_9 + 590F_{11} - 180F_{13} + 24F_{15} - 416\\ &= 1296 \cdot 13 - 1105 \cdot 34 + 590 \cdot 89 - 180 \cdot 233 + 24 \cdot 610 - 416\\ &= 4072.\end{aligned}$$

7. LHS $= F_1 + 4F_2 + 9F_3 + 16F_4 + 25F_5 + 36F_6 = 1 + 4 \cdot 1 + 9 \cdot 2 + 16 \cdot 3 + 25 \cdot 5 + 36 \cdot 8 = 484.$
RHS $= 49F_8 - 15F_{10} + 2F_{12} - 8 = 49 \cdot 21 - 15 \cdot 55 + 2 \cdot 144 - 8 = 484 =$ LHS.

9. LHS $= F_1 + 8F_2 + 27F_3 + 64F_4 + 125F_5 + 216F_6 = 1 + 8 \cdot 1 + 27 \cdot 2 + 64 \cdot 3 + 125 \cdot 5 + 216 \cdot 8 = 2608.$
RHS $= 343F_8 - 169F_{10} + 48F_{12} - 6F_{14} + 50 = 343 \cdot 21 - 169 \cdot 55 + 48 \cdot 144 - 6 \cdot 377 + 50 = 2608 =$ LHS.

11.
$$\begin{aligned}\text{LHS} &= L_1 + 16L_2 + 81L_3 + 256L_4 + 625L_5 + 1296L_6\\ &= 1 + 16 \cdot 3 + 81 \cdot 4 + 256 \cdot 7 + 625 \cdot 11 + 1296 \cdot 18 = 32{,}368\\ \text{RHS} &= 2401L_8 - 1695L_{10} + 770L_{12} - 204L_{14} + 24L_{16} - 930\\ &= 2401 \cdot 47 - 1695 \cdot 123 + 770 \cdot 322 - 204 \cdot 843 + 24 \cdot 2207 - 930\\ &= 32{,}368 = \text{LHS}.\end{aligned}$$

13. Let $S_j = \sum_{1}^{j} F_t$ and $A_i = \sum_{1}^{i} S_j$. Then $S_j = F_{j+2} - 1$ and $A_i = \sum_{1}^{i}(F_{j+2} - 1) = \sum_{3}^{2+i} F_j - i = \sum_{1}^{2+i} F_j - 2 - i = F_{4+i} - i - 3$, so $A_{n-1} = F_{n+3} - n - 2$.

Using the technique of staggered addition,

$$\begin{aligned}\sum_{1}^{n-1}(2i + 1)S_i &= 3\sum_{1}^{n-1} S_i + 2\sum_{2}^{n-1} S_i + 2\sum_{3}^{n-1} S_i + \cdots + 2\sum_{n-1}^{n-1} S_i \qquad (1)\\ &= 3A_{n-1} + 2(A_{n-1} - A_1) + 2(A_{n-1} - A_2) + \cdots\end{aligned}$$

$$
\begin{aligned}
&+ 2(A_{n-1} - A_{n-2}) \\
&= [3 + 2(n-2)]A_{n-1} - 2\sum_{1}^{n-2} A_i \\
&= (2n-1)A_{n-1} - 2\sum_{1}^{n-2}(F_{4+i} - i - 3) \\
&= (2n-1)A_{n-1} - 2\sum_{1}^{n-2} F_{4+i} + \frac{2(n-2)(n-1)}{2} + 6(n-2) \\
&= (2n-1)A_{n-1} - 2\sum_{5}^{n-2} F_j + (n-2)(n-1) + 6(n-2) \\
&= (2n-1)A_{n-1} - 2\left[\sum_{1}^{n-2} F_j - \sum_{1}^{4} F_j\right] + (n-2)(n+5) \\
&= (2n-1)A_{n-1} - 2[(F_{n+4} - 1) - 7] + (n-2)(n+5) \\
&= (2n-1)(F_{n+3} - n - 2) - 2F_{n+4} \\
&\quad +(n-2)(n+5) + 16 \qquad (2)
\end{aligned}
$$

$$
\begin{aligned}
\sum_{1}^{n} i^2 F_i &= F_1 + 4F_2 + 9F_3 + 16F_4 + \cdots + n^2 F_n \\
&= \sum_{1}^{n} F_i + 3\sum_{2}^{n} F_i + 5\sum_{3}^{n} F_i + \cdots + (2n-1)\sum_{n}^{n} F_i \\
&= S_n + 3(S_n - S_1) + 5(S_n - S_2) + \cdots + (2n-1)(S_n - S_{n-1}) \\
&= \sum_{1}^{n}(2i-1)S_n - [3S_1 + 5S_2 + 7S_3 + \cdots + (2n-1)S_{n-1}] \\
&= \left[\sum_{1}^{n}(2i-1)S_n - \sum_{1}^{n-1}(2i+1)S_i\right] \\
&= n^2 S_n - (2n-1)(F_{n+3} - n - 2) \\
&\quad + 2F_{n+4} - (n-2)(n+5) - 16 \text{ by Eq. (1)} \\
&= n^2(F_{n+2} - 1) - (2n-1)(F_{n+3} - n - 2) + 2F_{n+4} \\
&\quad - (n-2)(n+5) - 16
\end{aligned}
$$

$$= n^2 F_{n+2} - (2n-1)F_{n+3} + 2F_{n+4} - 8$$
$$= (n+1)^2 F_{n+2} - (2n+3)F_{n+4} + 2F_{n+6} - 8$$

EXERCISES 32 (p. 384)

1. $F_n Q + F_{n-1} I = F_n \begin{bmatrix} 1 & 1 \\ 1 & 0 \end{bmatrix} + F_{n-1} \begin{bmatrix} 1 & 0 \\ 0 & 1 \end{bmatrix} = \begin{bmatrix} F_n + F_{n-1} & F_n \\ F_n & F_{n-1} \end{bmatrix} = \begin{bmatrix} F_{n+1} & F_n \\ F_n & F_{n-1} \end{bmatrix} = Q^n.$
3. $\text{LHS} = \sum_1^{n+1} Q^i - \sum_0^n Q^i = Q^{n+1} - Q^0 = Q^{n+1} - I.$
5. Follows by equating the corresponding elements in the last two matrices in the proof of Corollary 32.2.
7. Add Identity (32.10) and Identity (32.23): $2F_{m+n} = F_m(F_{n-1}+F_{n+1}) + F_n(F_{m-1}+F_{m+1}) = F_m L_n + F_n L_m$.
9. Using Exercise 7, $2F_{m-n} = F_m L_{-n} + F_{-n} L_m = (-1)^n F_m L_n + (-1)^{n+1} F_n L_m = (-1)^n (F_m L_n - F_n L_m)$.

11.
$$\begin{aligned} \text{LHS} &= 5(\alpha^m + \beta^m)(\alpha^n + \beta^n) + (\alpha^m - \beta^m)(\alpha^n - \beta^n) \\ &= 6(\alpha^{m+n} + \beta^{m+n}) + 4(\alpha^m \beta^n + \alpha^n \beta^m) \\ &= 6L_{m+n} + 4[\alpha^m(-\alpha)^{-n} + (-\beta)^{-n}\beta^m] \\ &= 6L_{m+n} + 4(-1)^n(\alpha^{m-n} + \beta^{m-n}) \\ &= 6L_{m+n} + 4(-1)^n L_{m-n} \end{aligned}$$

13. Change m to $-m$ in the identity $F_{m+n} = F_{m+1}F_n + F_m F_{n-1}$: $F_{-m+n} = F_{-m+1}F_n + F_{-m}F_{n-1}$; that is, $(-1)^{m-n+1}F_{m-n} = (-1)^m F_{m-1}F_n + (-1)^{m+1} F_m F_{n-1}$. Thus, $F_{m-n} = (-1)^n F_m F_{n-1} - F_{m-1}F_n)$.
15. By Exercises 5 and 13,

$$\begin{aligned} F_{m+n} + F_{m-n} &= F_m F_{n-1}[1 + (-1)^n] + F_n[F_{m+1} - (-1)^n F_{m-1}] \\ &= \begin{cases} L_m F_n & \text{if } n \text{ is odd} \\ F_m L_n & \text{otherwise} \end{cases} \end{aligned}$$

17. By Exercises 6 and 14,

$$\begin{aligned} L_{m+n} + L_{m-n} &= F_{m+1}L_n[1 + (-1)^n] + F_m[L_{n-1} - (-1)^n L_{n+1}] \\ &= \begin{cases} F_m(L_{n-1} + L_{n+1}) & \text{if } n \text{ is odd} \\ 2F_{m+1}L_n + F_m(L_{n-1} - L_{n+1}) & \text{otherwise} \end{cases} \end{aligned}$$

$$= \begin{cases} 5F_m F_n & \text{if } n \text{ is odd} \\ 2F_{m+1}L_n - F_m L_n & \text{otherwise} \end{cases}$$

$$= \begin{cases} 5F_m F_n & \text{if } n \text{ is odd} \\ F_m L_n & \text{otherwise} \end{cases}$$

19. The result is true when $n = 1$. Now assume it is true for an arbitrary positive integer k. Then

$$M^{k+1} = M^k \cdot M = \begin{bmatrix} F_{2k-1} & F_{2k} \\ F_{2k} & F_{2k+1} \end{bmatrix} \begin{bmatrix} 1 & 1 \\ 1 & 2 \end{bmatrix}$$

$$= \begin{bmatrix} F_{2k-1} + F_{2k} & F_{2k-1} + 2F_{2k} \\ F_{2k} + F_{2k+1} & F_{2k} + 2F_{2k+1} \end{bmatrix} = \begin{bmatrix} F_{2k+1} & F_{2k+2} \\ F_{2k+2} & F_{2k+3} \end{bmatrix}$$

Thus, by PMI, the result is true for every $n \geq 1$.

21. (Rabinowitz, 1998) $A_{2n} = \frac{1}{4}\left[\begin{bmatrix} 1 & 3 \\ 3 & 1 \end{bmatrix} A_n^2 - \begin{bmatrix} 2 & 2 \\ 2 & 2 \end{bmatrix} A_n A_{n+1} + \begin{bmatrix} 0 & 2 \\ 2 & 0 \end{bmatrix} A_{n+1}^2\right]$

23. By Cramer's rule, $y = \dfrac{\begin{vmatrix} G_n & G_{n+1} \\ G_{n+1} & G_{n+2} \end{vmatrix}}{\begin{vmatrix} G_n & G_{n-1} \\ G_{n+1} & G_n \end{vmatrix}} = \dfrac{G_n G_{n+2} - G_{n+1}^2}{G_n^2 - G_{n-1}G_{n+1}}$. But $y = 1$. $\therefore G_n G_{n+2} - G_{n+1}^2 = -(G_{n-1}G_{n+1} - G_n^2)$. Let $q_n = G_{n-1}G_{n+1} - G_n^2$, where $q_1 = G_0 G_2 - G_1^2 = (b-a)b - a^2 = -\mu$. Then $q_n = (-1)^{n-1} q_1 = (-1)^n \mu$. $\therefore G_{n-1}G_{n+1} - G_n^2 = (-1)^n \mu$.

25. $\begin{bmatrix} F_{n+1} \\ F_n \end{bmatrix}$

27. $\begin{cases} \begin{bmatrix} F_n \\ -F_{n+1} \end{bmatrix} & \text{if } n \text{ is odd} \\ \begin{bmatrix} -F_n \\ F_{n+1} \end{bmatrix} & \text{otherwise} \end{cases}$

29. $V_m Q^n = (L_{m+1}, L_m) = \begin{bmatrix} F_{n+2} & F_{n+1} \\ F_{n+1} & F_n \end{bmatrix} = \begin{bmatrix} L_{m+1}F_{n+2} + L_m F_{n+1} \\ L_{m+1}F_{n+1} + L_m F_n \end{bmatrix} = (L_{m+n+2}, L_{m+n+1}) = V_{m+n+1}$.

31. $(a+e-b-d)(e+j-h-f) - (b+f-c-e)(d+h-g-e)$

33. $\lambda(P) = \begin{vmatrix} 1 & 1 & 2 \\ 1 & 2 & 3 \\ 2 & 2 & 2 \end{vmatrix} - \begin{vmatrix} 0 & 0 & 1 \\ 0 & 1 & 2 \\ 1 & 1 & 1 \end{vmatrix} = -1$

35. $\lambda(R) = L_{n-1}L_{n+1} - L_n^2 = 5(-1)^{n+1}$

37. $G_{n+k} + G_{n-k} - 2G_n$

39. Notice that $F_{n+1}^2 - F_{n-1}F_n + 2F_nF_{n+1} = F_{n+1}(F_{n+1} + F_n) + F_n(F_{n+1} - F_{n-1}) = F_{n+1}F_{n+2} + F_n^2 = F_{n+2}(F_{n+2} - F_n) + F_n^2 = F_{n+2}^2 - F_n(F_{n+2} - F_n) =$

$F_{n+2}^2 - F_nF_{n+1}$ and $2F_{n+1}^2 + 2F_nF_{n+1} = 2F_{n+1}(F_{n+1} + F_n) = 2F_{n+1}F_{n+2}$. It can be verified that it is true for an arbitrary positive integer n. Then:

$$\begin{aligned}
P^{n+1} &= P^n \cdot P \\
&= \begin{bmatrix} F_{n-1}^2 & F_{n-1}F_n & F_n^2 \\ 2F_{n-1}F_n & F_{n+1}^2 - F_{n-1}F_n & 2F_nF_{n+1} \\ F_n^2 & F_nF_{n+1} & F_{n+1}^2 \end{bmatrix} \\
&= \begin{bmatrix} 0 & 0 & 1 \\ 0 & 1 & 2 \\ 1 & 1 & 1 \end{bmatrix} \\
&= \begin{bmatrix} F_n^2 & F_{n-1}F_n + F_n^2 & F_{n-1}^2 + 2F_{n-1}F_n + F_n^2 \\ 2F_nF_{n+1} & F_{n+1}^2 - F_{n-1}F_n + 2F_nF_{n+1} & 2F_{n+1}^2 + 2F_nF_{n+1} \\ F_{n+1}^2 & F_nF_{n+1} + F_{n+1}^2 & F_n^2 + 2F_nF_{n+1} + F_{n+1}^2 \end{bmatrix} \\
&= \begin{bmatrix} F_n^2 & F_nF_{n+1} & F_{n+1}^2 \\ 2F_nF_{n+1} & F_{n+2}^2 - F_nF_{n+1} & F_{n+1}F_{n+2} \\ F_{n+1}^2 & F_{n+1}F_{n+2} & F_{n+2}^2 \end{bmatrix}
\end{aligned}$$

Thus, by PMI, the formula works for every $n \geq 1$.

41. LHS $= F_n(F_n - F_{n-1}) + F_{n+1}(F_{n+1} - F_n) - 2F_nF_{n-1} = F_nF_{n-2} + F_{n-1}\ (F_{n+1} - F_n) - F_nF_{n-1} = F_nF_{n-2} + F_{n-1}^2 - F_nF_{n-1} = F_nF_{n-2} - F_{n-1}(F_n - F_{n-1}) = F_nF_{n-2} - F_{n-1}F_{n-2} = F_{n-2}(F_n - F_{n-1}) = F_{n-2}^2$

EXERCISES 33 (p. 399)

1. $L_nL_{n+2} - L_{n+1}^2 = 5(-1)^n$
3. (Finkelstein) Let $r = L_{6d}/L_{3d}$ and $s = L_{6d+1} - rL_{3d+1}$. Then, by PMI, $L_{n+6d} = rL_{n+3d} + sL_n$ for all n. In particular, let $n = a$, $a + d$, and $a + 2d$. Hence the rows of the determinant are linearly dependent, so the determinant is zero.
5. Since $G_n = aF_{n-2} + bF_{n-1}$, the determinant is zero by Exercise 3.
7. (Jaiswal) Consider the determinant $D = \begin{vmatrix} G_p & G_{p+m} & G_{p+m+n} \\ G_q & G_{q+m} & G_{q+m+n} \\ G_r & G_{r+m} & G_{r+m+n} \end{vmatrix}$.

   Since $G_{k+m+n} = G_{k+m}F_{n+1} + G_{k+m-1}F_n$, it follows that

$$\begin{aligned}
D &= F_{n+1}\begin{vmatrix} G_p & G_{p+m} & G_{p+m} \\ G_q & G_{q+m} & G_{q+m} \\ G_r & G_{r+m} & G_{r+m} \end{vmatrix} + F_n\begin{vmatrix} G_p & G_{p+m} & G_{p+m-1} \\ G_q & G_{q+m} & G_{q+m-1} \\ G_r & G_{r+m} & G_{r+m-1} \end{vmatrix} \\
&= F_n\begin{vmatrix} G_p & G_{p+m} & G_{p+m-1} \\ G_q & G_{q+m} & G_{q+m-1} \\ G_r & G_{r+m} & G_{r+m-1} \end{vmatrix} = F_n\begin{vmatrix} G_p & G_{p+m-2} & G_{p+m-1} \\ G_q & G_{q+m-2} & G_{q+m-1} \\ G_r & G_{r+m-2} & G_{r+m-1} \end{vmatrix}
\end{aligned}$$

$$= F_n \begin{vmatrix} G_p & G_{p+m-2} & G_{p+m-3} \\ G_q & G_{q+m-2} & G_{q+m-3} \\ G_r & G_{r+m-2} & G_{r+m-3} \end{vmatrix}$$

Thus, by alternately subtracting columns 2 and 3 from one another, the process can be continued to decrease the subscripts. After a certain stage, when m is even, columns 1 and 2 would become identical; and if m is odd, columns 1 and 3 would become identical. In either case, $D = 0$.

The given determinant $\triangle$ can be written as the sum of eight determinants. Using the fact that $D = 0$ and that a determinant vanishes if two columns are identical,

$\triangle = \begin{vmatrix} G_p & G_{p+m} & k \\ G_q & G_{q+m} & k \\ G_r & G_{r+m} & k \end{vmatrix} + \cdots + \cdots = \triangle_1 + \triangle_2 + \triangle_3$ (say). Since $G_{m-1}G_n -$

$G_m G_{n-1} = (-1)^{m-1}\mu F_{n-m}$,

$$\triangle_1 = kF_m \begin{vmatrix} G_p & G_{p-1} & 1 \\ G_q & G_{q-1} & 1 \\ G_r & G_{r-1} & 1 \end{vmatrix} = k\mu F_m[(-1)^{r-1}F_{q-r} + (-1)^{p-1}F_{r-p} + (-1)^{q-1}F_{p-q}]$$

$$\therefore \triangle = k\mu[(-1)^q F_{r-q} - (-1)^p F_{r-p} + (-1)^p F_{q-p}][F_m - F_{m+n} + (-1)^m F_n]$$

9. (Jaiswal) Since $\begin{vmatrix} a & b & c & d \\ b & a & d & c \\ c & d & a & b \\ d & c & b & a \end{vmatrix} = [(a+b)^2-(c+d)^2][(a-b)^2-(c-d)^2]$, the given determinant $\triangle = [(G_{n+3}+G_{n+2})^2 - (G_{n+1}+G_n)^2][(G_{n+3}-G_{n+2})^2 - (G_{n+1}-G_n)^2] = (G_{n+4}^2 - G_{n+2}^2)(G_{n+1}^2 - G_{n-1}^2)$. But $G_{m+1}^2 - G_{m-1}^2 = aG_{2m-2} + bG_{2m-1}$. $\therefore \triangle = (aG_{2n+4} + bG_{2n+5})(aG_{2n-2} + bG_{2n-1})$.

11. Using Theorem 33.7 with $k = 2, m = 1, r = 0$, and $a_n = L_n$,

$$D = (-1)^{n\cdot 2\cdot 3/2}A_2(L_0) = (-1)^{3n}\begin{vmatrix} L_0^2 & L_1^2 & L_2^2 \\ L_1^2 & L_2^2 & L_3^2 \\ L_2^2 & L_3^2 & L_4^2 \end{vmatrix} = (-1)^n \begin{vmatrix} 4 & 1 & 9 \\ 1 & 9 & 16 \\ 9 & 16 & 49 \end{vmatrix}$$

$$= 250(-1)^n$$

13. Using Theorem 33.7 with $k = 3, m = 1, r = 0$, and $a_n = L_n$,

$$D = (-1)^{n\cdot 3\cdot 4/2}A_3(L_0) = A_3(L_0) = \begin{vmatrix} L_0^3 & L_1^3 & L_2^3 & L_3^3 \\ L_1^3 & L_2^3 & L_3^3 & L_4^3 \\ L_2^3 & L_3^3 & L_4^3 & L_5^3 \\ L_3^3 & L_4^3 & L_5^3 & L_6^3 \end{vmatrix}$$

$$= \begin{vmatrix} 4 & 1 & 27 & 64 \\ 1 & 27 & 64 & 343 \\ 27 & 64 & 343 & 1331 \\ 343 & 1331 & 5832 & 24389 \end{vmatrix} = 0$$

15. $g_{n+2}(x) = 2xg_{n+1}(x) + g_n(x), n \geq 0.$

17. (Brown) $Q^n = \begin{bmatrix} F_{n+1} & F_n \\ F_n & F_{n-1} \end{bmatrix}; e^{Q^n} = \begin{bmatrix} \sum_{k=0}^{\infty} \frac{F_{nk+1}}{k!} & \sum_{k=0}^{\infty} \frac{F_{nk}}{k!} \\ \sum_{k=0}^{\infty} \frac{F_{nk}}{k!} & \sum_{k=0}^{\infty} \frac{F_{nk-1}}{k!} \end{bmatrix} = 0$

We have $\sum_{k=0}^{\infty} \frac{F_{nk}}{k!} = \frac{e^{\alpha^n x} - e^{\beta^n x}}{\alpha - \beta}$ *and* $\sum_{k=0}^{\infty} \frac{L_{nk}}{k!} = e^{\alpha^n x} + e^{\beta^n x}$. Since $L_{nk} = F_{nk+1} + F_{nk-1}$,

$$\sum_{k=0}^{\infty} \frac{F_{nk+1} + F_{nk-1}}{k!} = e^{\alpha^n} + e^{\beta^n}; \text{ that is, } \sum_{k=0}^{\infty} \frac{F_{nk+1}}{k!} = (e^{\alpha^n} + e^{\beta^n}) - \sum_{k=0}^{\infty} \frac{F_{nk-1}}{k!} \tag{1}$$

Since $F_{nk+1} = F_{nk} + F_{nk-1}$, we also have

$$\sum_{k=0}^{\infty} \frac{F_{nk+1}}{k!} = \sum_{k=0}^{\infty} \frac{F_{nk}}{k!} + \sum_{k=0}^{\infty} \frac{F_{nk-1}}{k!} = \frac{e^{\alpha^n} - e^{\beta^n}}{\alpha - \beta} + \sum_{k=0}^{\infty} \frac{F_{nk-1}}{k!}. \tag{2}$$

From (1) and (2),

$$\sum_{k=0}^{\infty} \frac{F_{nk+1}}{k!} = \frac{1}{2}\left[(e^{\alpha^n} + e^{\beta^n}) + \frac{e^{\alpha^n} - e^{\beta^n}}{\alpha - \beta}\right] \text{ and}$$

$$\sum_{k=0}^{\infty} \frac{F_{nk-1}}{k!} = \frac{1}{2}\left[(e^{\alpha^n} + e^{\beta^n}) - \frac{e^{\alpha^n} - e^{\beta^n}}{\alpha - \beta}\right].$$

$$\begin{aligned} \therefore \left|e^{Q^n}\right| &= \left(\sum_{k=0}^{\infty} \frac{F_{nk+1}}{k!}\right)\left(\sum_{k=0}^{\infty} \frac{F_{nk-1}}{k!}\right) - \left(\sum_{k=0}^{\infty} \frac{F_{nk}}{k!}\right)^2 \\ &= \frac{1}{4}\left[(e^{\alpha^n} + e^{\beta^n})^2 - \left(\frac{e^{\alpha^n} - e^{\beta^n}}{\alpha - \beta}\right)^2\right] - \left(\frac{e^{\alpha^n} - e^{\beta^n}}{\alpha - \beta}\right)^2 \\ &= e^{\alpha^n + \beta^n} = e^{L_n} \end{aligned}$$

19. (Parker) Let D_n denote the given determinant. Expanding it by the last column, $D_n = a_n D_{n-1} + b_{n-1} D_{n-2}$. Then $g_n = g_{n-1} + g_{n-2}$, where $g_1 = 1$ and $g_2 = 2$; $\therefore g_n = F_{n+1}$.

21. (Parker) Let D_n denote the given determinant. Then $D_1 = a + b$ and $D_2 = a^2 + ab + b^2$. Expanding D_n by row 1, $D_n = (a + b)D_{n-1} - abD_{n-2}$. Solving this second order LHRRWCC, we get

$$D_n = \begin{cases} (n+1)a^n & \text{if } a = b \\ (a^{n+1} - b^{n+1})/(a - b) & \text{otherwise} \end{cases}$$

EXERCISES 34 (p. 413)

1. $F_6 13^5 + F_5 13^6 = 8 \cdot 13^5 + 5 \cdot 13^6 \equiv 134 + 48 \equiv 1 \pmod{181}$; $L_6 13^5 + L_5 13^6 = 18 \cdot 13^5 + 11 \cdot 13^6 \equiv 30 + 178 \equiv 27 \equiv 1 + 2 \cdot 13 \pmod{181}$
3. $F_4 - 11F_5 = 3 - 11 \cdot 5 \equiv 79 \equiv (-1)^5 11^5 \pmod{131}$
5. Let $F_n \equiv 0 \pmod 3$. Then $3|F_n$; that is, $F_4|F_n$. So $n \equiv 0 \pmod 4$. Conversely, let $n \equiv 0 \pmod 4$. Then $4|n$, so $F_4|F_n$; that is, $3|F_n$; $\therefore F_n \equiv 0 \pmod 3$.
7. $F_n \equiv 0 \pmod 5$ iff $5|F_n$, that is, iff $F_5|F_n$. Thus $F_n \equiv 0 \pmod 5$ iff $n \equiv 0 \pmod 5$.
9. Since $5F_n^2 = L_n^2 - 4(-1)^n$, $F_n^2 \equiv L_n^2 \pmod 4$.
11. Since $2L_{m+n} = L_m L_n + 5F_m F_n$, $2L_{m+n} \equiv L_m L_n \pmod 5$.
13. $L_{(2k-1)n} = \alpha^{(2k-1)n} + \beta^{(2k-1)n} = (\alpha^n + \beta^n)[\alpha^{(2k-2)n} - \alpha^{(2k-3)n}\beta^n + \cdots + \beta^{(2k-2)n}] \equiv 0 \pmod{L_n}$.
15. Since $F_{m+n} = F_{m-1}F_n + F_m F_{n+1}$, $F_{n+24} = F_{23}F_n + F_{24}F_{n+1}$, where $F_{23} = 28{,}657 \equiv 1 \pmod 9$ and $F_{24} = 46{,}368 \equiv 0 \pmod 9$; $\therefore F_{n+24} \equiv F_n \pmod 9$.
17. Since $F_3|F_{3n}$, $F_{3n} \equiv 0 \pmod 2$.
19. Since $F_5|F_{5n}$, $F_{5n} \equiv 0 \pmod 5$.
21. Since $2|n$ and $3 \nmid n$, n is of the form $6k + 2$ or $6k + 4$.

    Case 1. Let $n = 6k + 2$. Then $L_n = L_{6k+2} = F_{6k+1}L_2 + F_{6k}L_1$. Since $6|6k$, $8|F_{6k}$; so $F_{6k} \equiv 0 \pmod 4$; $\therefore L_n \equiv F_{6k+1} \cdot 3 + 0 \equiv 3F_{6k+1} \pmod 4$. But $F_{6k+1} \equiv 1 \pmod 4$. Thus $L_n \equiv 3 \pmod 4$.

    Case 2. Let $n = 6k + 4$. Then $L_n = L_{6k+4} = L_{(6k+1)+3} = F_{6k+2}L_3 + F_{6k+1}L_2 \equiv F_{6k+2} \cdot 0 + 1 \cdot 3 \equiv 3 \pmod 4$. Thus, in both cases, $L_n \equiv 3 \pmod 4$.

23. By Exercise 39 in Chapter 16, $2|L_{3n}$. So $L_{3n} \equiv 0 \pmod 2$.
25. Let $(F_n, L_n) = 2$. Then $2|F_n$; that is, $F_3|F_n$; $\therefore n \equiv 0 \pmod 3$. Conversely, let $n = 3m$. Then $2|F_{3m}$, and by Exercise 23, $2|L_{3m}$ also; $\therefore 2|(F_{3m}, L_{3m})$. Suppose $(F_{3m}, L_{3m}) = d > 2$ for all n. Then $F_{3m+3} = F_{3m}F_4 + F_{3m-1}F_3 = 3F_{3m} + 2F_{3m-1}$ and $L_{3m+3} = F_{3m}L_4 + L_{3m-1}L_3 = 7F_{3m} + 4F_{3m-1}$. Then $d|F_{3m+3}$ implies $d|2F_{3m-1}$. This is a contradiction since $d > 2$ and $(F_{3m}, F_{3m-1}) = 1$. Thus $(F_n, L_n) = (F_{3m}, L_{3m}) = 2$.
27. Since $5(F_n^2 + F_{n-2}^2) = 3L_{2n-2} - 4(-1)^n$, by Exercise 36 in Chapter 5, the result follows.
29. (Lind) We have $L_p = (1/2^{p-1}) \sum_{0}^{\lfloor p/2 \rfloor} \binom{p}{2i} 5^i$. Since p is prime, $\binom{p}{j} \equiv 0 \pmod p$ for $0 < j < p$; also $2^{p-1} \equiv 1 \pmod p$, by Fermat's little theorem; $\therefore L_p \equiv 1 \pmod p$.
31. (Wessner) The statement is true for $n = 1$. Assume it is true for every $i \le k$, where $k \ge 1$: $2^i F_i \equiv 2i \pmod 5$, $1 \le i \le k$. Then $2^{k+1}F_k = 2(2^k F_k + 2 \cdot 2^{k-1}F_{k-1}) \equiv 2[2^k + 2 \cdot 2(k-1)] \equiv 2(k-4) \equiv 2(k+1) \pmod 5$; $\therefore$ by the strong version of PMI, the result follows for every $n \ge 1$.

33. (Bruckman) We shall use the identity $F_{5m} = 25F_m^5 + 25(-1)^m F_m^3 + 5F_m$, $m \geq 0$ and PMI. The given result is true when $n = 0$ and $n = 1$. Assume it is true for an arbitrary positive integer k: $F_{5^k} \equiv 5^k \pmod{5^{k+3}}$, so $F_m = m(1 + 125a)$ for some integer a, where $m = 5^k$. By the preceding identity, $F_{5m} = 5^2m^5(1 + 125a)^5 - 5^2m^3(1 + 125a)^3 + 5m(1 + 125a) \equiv 5^2m^5 - 5^2m^3 + 5m \pmod{5^4m}$. Since $k \geq 1$, $5|m$; so $5^2|m^2$. Hence $5^4m|5^2m^3$; $\therefore F_{5m} \equiv 5m \pmod{5^4m}$; that is, $F_{5^{k+1}} \equiv 5^{k+1} \pmod{5^{k+4}}$. Thus, by PMI, the result is true for all $n \geq 0$.

35. (Prielipp) Since $L_n^2 = 5F_n^2 + 4(-1)^n$, $(L_n^2)^2 = (5F_n^2)^2 + 8(-1)^n(5F_n^2) + 4^2$. $\therefore (5F_n^2)^2 + 4^2 \equiv (L_n^2)^2 \pmod{5F_n^2}$.

37. $\sum\limits_{i=1}^{20} F_{n+i} = \sum\limits_{1}^{n+20} F_i - \sum\limits_{1}^{n} F_i = (F_{n+22} - 1) - (F_{n+2} - 1) = F_{n+22} - F_{n+2} = (F_nF_{21} + F_{n+1}F_{22}) - (F_n + F_{n+1}) = F_n(F_{21} - 1) + F_{n+1}(F_{22} - 1) \equiv F_n \cdot 0 + F_{n+1} \cdot 0 \equiv 0 (\bmod F_{10})$, since $F_{21} \equiv F_{22} \equiv 1 \pmod{55}$.

39. 5, 89, 11, 199

41. Let $5|n$. The $5|F_n$, so $F_n = 5m$ for some integer m. Then $F_{F_n} = F_{5m}$. Since $5|F_{5m}$, it follows that $F_{F_n} \equiv 0 \pmod 5$. Conversely, let $F_{F_n} \equiv 0 \pmod 5$. Then $5|F_n$, so $n \equiv 0 \pmod 5$.

43. Follows since $2L_{m+n} = L_mL_n + 5F_mF_n$.

EXERCISES 35 (p. 422)

1. 8
3. 7
5. 6
7. $L_{25} = L_{24} + L_{23} \equiv 7 + 2 \equiv 1 \pmod 8$
9. $L_{i-2} = L_i - L_{i-1} \equiv 3 - 4 \equiv 7 \pmod 8$
11. Since $F_1 \equiv 1 \pmod 4$, the result is true when $n = 0$. Assume it is true for an arbitrary positive integer k: $F_{6k+1} \equiv 1 \pmod 4$. Then $F_{6(k+1)+1} = F_{6k+7} = F_{6k+5} + F_{6k+6} = F_{6k+4} + 2F_{6k+5} = \cdots = 5F_{6k+1} + 8F_{6k+2} \equiv 5 \cdot 1 + 0 \equiv 1 \pmod 4$; $\therefore$ by PMI, the result follows.
13. $F_{6n-1} = F_{6n+1} - F_{6n} \equiv 1 - 0 \equiv 1 \pmod 4$, since $F_6|F_{6n}$.
15. $L_{6n+2} = F_{6n+1} + F_{6n+3} = F_{6n+1} + (F_{6n+1} + F_{6n+2}) = 2F_{6n+1} + (F_{6n+1} + F_{6n}) = 3F_{6n+1} + F_{6n} \equiv 3 \cdot 1 + 0 \equiv 3 \pmod 4$, since $F_6|F_{6n}$.

17.
$$\begin{aligned}
\text{RHS} &= (\alpha^{4n+3} + \beta^{4n+3})[\alpha^{8n+6} + \beta^{8n+6} + 2(\alpha\beta)^{4n+3} + 3] \\
&= (\alpha^{4n+3} + \beta^{4n+3})(\alpha^{8n+6} + \beta^{8n+6} + 1) \\
&= \alpha^{12n+9} + \beta^{12n+9} + (\alpha\beta)^{4n+3}(\alpha^{4n+3} + \beta^{4n+3}) + L_{4n+3} \\
&= L_{12n+9} - L_{4n+3} + L_{4n+3} = L_{12n+9} \\
&= \text{LHS}
\end{aligned}$$

19. Since $L_{m+n} = F_{m-1}L_n + F_m L_{n-1}$, $L_{4n+2} = F_{4n-1}L_2 + F_{4n}L_1 = 3F_{4n-1} + F_{4n} \equiv 0 \pmod 3$.
21. $L_{4n+1} = L_{4n+2} - L_{4n} \equiv 0 - \pm 1 \equiv \pm 1 \pmod 3$, by Exercises 19 and 20.
23. Follows by Exercise 21 in Chapter 34.
25. $L_{m+2k} \equiv (-1)L_m \equiv (-1)^2 L_{m-2k} \equiv (-1)^3 L_{m-4k} \equiv \cdots \equiv (-1)^{\lfloor m/4 \rfloor} L_{m-2\lfloor m/4 \rfloor k} \pmod{L_k}$. Let $k = 2$ and $m = 4n - 2$. Then $L_{4n+2} = L_{(4n-2)+2\cdot 2} \equiv (-1)^{n-1} L_{4n-2-4\lfloor (4n-2)/4 \rfloor} \equiv (-1)^{n-1} L_2 \equiv 0 \pmod 3$.
27. (Kramer and Hoggatt)

$$\begin{aligned}
\text{LHS} &= \frac{\alpha^{5^{n+1}} - \beta^{5^{n+1}}}{\alpha - \beta} = \frac{\alpha^{5^n \cdot 5} - \beta^{5^n \cdot 5}}{\alpha - \beta} \\
&= \frac{\alpha^{5^n} - \beta^{5^n}}{\alpha - \beta} (\alpha^{5^n \cdot 4} + \alpha^{5^n \cdot 3}\beta^{5^n} + \alpha^{5^n \cdot 2}\beta^{5^n \cdot 2} + \alpha^{5^n}\beta^{5^n \cdot 3} + \beta^{5^n \cdot 4}) \\
&= F_{5^n}[\alpha^{5^n \cdot 4} + \beta^{5^n \cdot 4} + (\alpha\beta)^{5n}(\alpha^{5^n \cdot 2} + \beta^{5^n \cdot 2}) + (\alpha\beta)^{5^n \cdot 2}] \\
&= F_{5^n}(L_{4 \cdot 5^n} - L_{2 \cdot 5^n} + 1) = \text{RHS}
\end{aligned}$$

29. (Turner) We have:

$$\begin{aligned}
2^{n-1}F_n &= \binom{n}{1} + 5\binom{n}{3} + 5^2\binom{n}{5} + \cdots + 5^{\lfloor n/2 \rfloor}\binom{n}{\lfloor n/2 \rfloor} \\
&\equiv n + \frac{5(n-1)(n-2)}{6} \pmod{25}
\end{aligned}$$

$\therefore 2^{60k-1}F_{60k} \equiv 60k + 50k(60k-1)(60k-2) \equiv 10k \pmod{25}$. Since $2^{20} \equiv 1 \pmod{25}$, it follows that $F_{60k} \equiv 20k \pmod{25}$. Since $6|60k$, $F_6|F_{60k}$; that is, $8|F_{60k}$. So $F_{60k} \equiv 0 \equiv 20k \pmod 4$. Combining the two congruences yields the desired result. (Follows from Exercise 31 also.)
31. Since $F_{r+s} = F_r F_{s-1} + F_{r+1}F_s$, $F_{60k+n} = F_{60k}F_{n-1} + F_{60k+1}F_n \equiv 20kF_{n-1} + (60k+1)F_n \pmod{100}$, by Exercises 29 and 30.
33. (Peck) Follows since $F_{(n+2)k} - F_{nk} = L_k F_{(n+1)k}$, where k is odd.
35. (Zeitlin) Since $F_{(n+2)k} - L_k F_{(n+1)k} + (-1)^k F_{nk} = 0$, $F_{(n+2)k} - 2F_{(n+1)k} + (-1)^k F_{nk} = (L_k - 2)F_{(n+1)k}$. So, when k is even, $F_{(n+2)k} + F_{nk} \equiv 2F_{(n+1)k} \pmod{L_k - 2}$.
37. Follows since $L_{(2m+1)(4n+1)} - L_{2m+1} = 5F_{(2m+1)2n}F_{(2m+1)(2n+1)}$.
39. (Prielipp) Since $F_1 + F_3 = 3 \equiv 0 \pmod 3$ and $F_4 + F_4 = 6 \equiv 0 \pmod 3$, the statement is true when $n = 0$ and $n = 1$. Assume it is true for all nonnegative integers $\le k$. So $F_{3k-2} + F_{k+2} \equiv 0 \equiv F_{3k+1} + F_{k+3} \pmod 3$. Then $F_{3k-2} + F_{3k+1} + F_{k+4} \equiv 0 \pmod 3$. But $6F_{3k-1} + 4F_{3k-2} + 3F_{3k+1} = F_{3k+4}$, so $F_{3k-2} + F_{3k+1} \equiv F_{3k+4} \pmod 3$; $\therefore F_{3k+4} + F_{k+4} \equiv 0 \pmod 3$. Thus, by the strong version of PMI, the statement is true for all $n \ge 0$.
41. (Somer) Since $L_4 = 7$, the result is true for $n = 2$. Assume it is true for an arbitrary integer $k \ge 2$. Since $L_m^2 = L_{2m} + 2(-1)^m$, $L_{2^{k+1}} = L_{2^k}^2 - 2(-1)^{2^k} \equiv 7^2 - 2 \equiv 7 \pmod{10}$. Thus, by PMI, the result holds for all $n \ge 2$.

43. (Wulczyn) Since $L_6 = 18 = 2 + 2^4$, the result is true when $n = 1$. Assume it is true for an arbitrary $k \geq 1$: $L_{3\cdot 2^k} \equiv 2 + 2^{2k+2} \pmod{2^{2k+2}}$. Since $L_{2m}^2 = L_{4m} + 2$, $L_{3\cdot 2^{k+1}} = L_{3\cdot 2^k}^2 - 2 \equiv 2 + 2^{2k+4} + 2^{4k+4} \pmod{2^{4k+5}} \equiv 2 + 2^{2k+4} \pmod{2^{2k+6}}$; $\therefore$ by PMI, the result is true for all $n \geq 1$.

EXERCISES 36 (p. 440)

1. $\sum_0^\infty F_i x^{i-1} = \dfrac{1}{1 - x - x^2}$. Using $x = 1/2$, $\sum_1^\infty \dfrac{F_i}{2^i} = 2$.

3. $u_n = A\alpha^n + B\beta^n$, where $A = \dfrac{b - a\beta}{\alpha - \beta}$ and $B = \dfrac{a\alpha - \beta}{\alpha - \beta}$.

5. By Exercise 4, $\sum_{i=0}^\infty \dfrac{u_i}{k^{i+1}} = \dfrac{a(k-1) + b}{(k - \alpha)(k - \beta)}$. Let $a = 0$ and $b = 1$.

   Desired sum $= \dfrac{1}{(k - \alpha)(k - \beta)} = \dfrac{1}{k^2 - k - 1}$.

7. Since $\beta < 0$, $1 + \beta^2 > 0$ and $\beta^{2k-1} < 0$; $\therefore 0 > \beta^{2k-1}(1 + \beta^2)$, that is, $\alpha^{2k-1} - \alpha^{2k-1} > \beta^{2k-1} + \beta^{2k+1}$. Then $\alpha^{2k-1} - \beta^{2k-1} > -\beta(\alpha^{2k} - \beta^{2k})$; thus $F_{2k-1}/F_{2k} > -\beta$, so $-m/n \notin (\beta, -\beta)$.
9. Since $F_{2k}/F_{2k-1} \geq 1 > -\beta$, $F_{2k}/F_{2k-1} \notin (\beta, -\beta)$
11. Yes, by Exercise 10.
13. Solving the characteristic equation $t^2 - at - b = 0$, $t = r$ or s. So the general solution is $U_n = Rr^n + Ss^n$, where R and S are to be determined. The two initial conditions yield the linear system $R + S = c$ and $Rr + Ss = d$. Solving, $R = \dfrac{c}{2} + \dfrac{2d - ca}{2\sqrt{a^2 + 4b}} = P$ and $S = \dfrac{c}{2} - \dfrac{2d - ca}{2\sqrt{a^2 + 4b}} = Q$; $\therefore U_n = Pr^n + Qs^n$, $n \geq 0$.

15. Let $a = b = d = 1$, $B = -10$, and $c = 0$. Then $m = 109$ and $N = -10$; $\therefore \dfrac{-10}{(-10)\cdot 109} = \sum_1^\infty \dfrac{F_{i-1}}{(-10)^i}$; that is, $\sum_1^\infty \dfrac{F_{i-1}}{(-10)^i} = \dfrac{1}{109}$.

17. (Pond) Since $\lim_{n\to\infty} \dfrac{a_{n+1}}{a_n} = \lim_{n\to\infty} \dfrac{1/F_{n+1}}{1/F_n} = \lim_{n\to\infty} \dfrac{F_n}{F_{n+1}} = \dfrac{1}{\alpha} < 1$, the series converges by d'Alembert's test.

19. (Lindstrome) Let $S_n = \sum_1^n \dfrac{1}{F_i}$. Then $240S_{13} = 240 + 240 + 120 + 80 + 48 + 30 + \dfrac{240}{13} + \dfrac{240}{21} + \dfrac{240}{34} + \dfrac{240}{55} + \dfrac{240}{89} + \dfrac{240}{144} + \dfrac{240}{233} > 803$. $\therefore S > S_{13} > 803/240$.

21. (Peck) Since $\alpha^{n+1} = \alpha F_{n+1} + F_n$, sum $= \sum_1^{\infty} \frac{1}{\alpha^{n+1}} = \frac{1}{\alpha^2}\left(\frac{1}{1-1/\alpha}\right) = 1.$

23. (Graham) $\sum_1^{\infty} \frac{(-1)^{n+1}}{F_n F_{n+1} F_{n+2}} = \sum_1^{\infty} \frac{F_n F_{n+2} - F_{n+1}^2}{F_n F_{n+1} F_{n+2}} = \sum_1^{\infty}\left(\frac{1}{F_{n+1}} - \frac{F_{n+1}}{F_n F_{n+2}}\right) = \left(\sum_1^{\infty} \frac{1}{F_{n+1}} - \frac{F_{n+2} - F_n}{F_n F_{n+2}}\right) = \sum_1^{\infty}\left(\frac{1}{F_{n+1}} - \frac{1}{F_n} + \frac{1}{F_{n+2}}\right) = \sum_1^{\infty}\left(\frac{1}{F_{n+1}} - \frac{1}{F_n}\right) + \sum_1^{\infty} \frac{1}{F_{n+2}} = -1 + \left(\sum_1^{\infty} \frac{1}{F_n} - 2\right) = -3 + \sum_1^{\infty} \frac{1}{F_n}$. The desired sum now follows.

25. (Parker) $\frac{F_n}{F_{n-1}F_{n+1}} = \frac{F_{n+1} - F_{n-1}}{F_{n-1}F_{n+1}} = \frac{1}{F_{n-1}} - \frac{1}{F_{n+1}}$

$$
\begin{aligned}
\therefore \text{ LHS} &= \sum_2^{\infty}\left(\frac{1}{F_{n-1}} - \frac{1}{F_{n+1}}\right) \\
&= \left(\frac{1}{1} - \frac{1}{2}\right) + \left(\frac{1}{1} - \frac{1}{3}\right) + \left(\frac{1}{2} - \frac{1}{5}\right) + \left(\frac{1}{3} - \frac{1}{8}\right) + \cdots = \frac{1}{1} + \frac{1}{1} \\
&= 2
\end{aligned}
$$

27. As in Exercise 25, LHS $= \sum_2^{\infty}\left(\frac{1}{G_{n-1}} - \frac{1}{G_{n+1}}\right) = \left(\frac{1}{a} - \frac{1}{a+b}\right) + \left(\frac{1}{b} - \frac{1}{a+2b}\right) + \left(\frac{1}{a+b} - \frac{1}{2a+b}\right) + \cdots = \frac{1}{a} + \frac{1}{b}.$

29. RHS $= 1 + \sum_2^{n} \frac{F_{i+1}F_{i-1} - F_i^2}{F_i F_{i-1}} = 1 + \sum_2^{n}\left(\frac{F_{i+1}}{F_i} - \frac{F_i}{F_{i-1}}\right) = 1 + \left(\frac{F_{n+1}}{F_n} - \frac{F_2}{F_1}\right) = \frac{F_{n+1}}{F_n} =$ LHS

31. (Carlitz) Since $F_{2n+1} = F_{n+1}L_{n+2} - F_{n+2}L_n$, $\sum_1^{m} \frac{F_{2n+1}}{L_n L_{n+1} L_{n+2}} = \sum_1^{m}\left(\frac{F_{n+1}}{L_n L_{n+1}} - \frac{F_{n+2}}{L_{n+1}L_{n+2}}\right) = \frac{F_2}{L_1 L_2} - \frac{F_{m+2}}{L_{m+1}L_{m+2}}$. Since $\lim_{m\to\infty} \frac{F_{m+2}}{L_{m+1}L_{m+2}} = 0$, the result follows.

33. (Carlitz) LHS $= \sqrt{5}\sum_0^{\infty} \frac{(-1)^n}{\alpha^{2(2n+1)} - \beta^{2(2n+1)}} = \sqrt{5}\sum_0^{\infty} \frac{(-1)^n}{\alpha^{2(2n+1)}} \cdot \frac{1}{1 - \alpha^{-4(2n+1)}} = \sqrt{5}\sum_{n=0}^{\infty}(-1)^n \sum_{r=0}^{\infty} \alpha^{-2(2r+1)(2n+1)} = \sqrt{5}\sum_{r=0}^{\infty} \frac{\alpha^{-2(2r+1)}}{1 + \alpha^{-4(2r+1)}} = \sqrt{5}\sum_{r=0}^{\infty} \frac{1}{\alpha^{2(2r+1)} + \beta^{2(2r+1)}} =$ RHS

35. LHS $= \lim_{n\to\infty} \sum_1^{n}\left(\frac{1}{F_k F_{k+1}} - \frac{1}{F_{k+1}F_{k+2}}\right) = \lim_{n\to\infty}\left(\frac{1}{F_1 F_2} - \frac{1}{F_{n+1}F_{n+2}}\right) = 1 - 0 = 1.$

37. $\sum_{1}^{n}\left(\frac{1}{F_kF_{k+1}F_{k+2}} - \frac{1}{F_{k+1}F_{k+2}F_{k+3}}\right) = \sum_{1}^{n}\frac{F_{k+3}-F_k}{F_kF_{k+1}F_{k+2}F_{k+3}}$. That is, $\frac{1}{1\cdot 1\cdot 2} - \frac{1}{F_{n+1}F_{n+2}F_{n+3}} = 2\sum_{1}^{n}\frac{F_{k+1}}{F_kF_{k+1}F_{k+2}F_{k+3}}$. As $n \to \infty$, $2\sum_{1}^{\infty}\frac{1}{F_kF_{k+1}F_{k+3}} \to \frac{1}{2} - 0 = \frac{1}{2}$. The desired result now follows.

39. Let $S_n = \sum_{1}^{n}\frac{F_{k+1}}{F_kF_{k+3}} = \frac{1}{2}\sum_{1}^{n}\left(\frac{1}{F_k} - \frac{1}{F_{k+3}}\right)$

$$= \frac{1}{2}\left(\frac{1}{F_1} + \frac{1}{F_2} + \frac{1}{F_3} - \frac{1}{F_{n+1}} - \frac{1}{F_{n+2}} - \frac{1}{F_{n+3}}\right)$$

$$= \frac{1}{2}\left(\frac{5}{2} - \frac{1}{F_{n+1}} - \frac{1}{F_{n+2}} - \frac{1}{F_{n+3}}\right). \therefore \lim_{n\to\infty} S_n = \frac{5}{4}.$$

41. (Mana) Let $\frac{1-x}{1-3x+x^2} = \sum_{0}^{\infty} C_ix^i$. This series converges for $|x| < r$, where r is the zero of of $1-3x+x^2$ with the least absolute value, namely, β^2; $\therefore 1-x = (1-3x+x^2)\sum_{0}^{\infty} C_ix^i$. Equating the coefficients of like terms, $C_0 = 1, C_1 = 2$, and $C_{n+2} - 3C_{n+1} + C_n = 0$ for $n \geq 2$. This implies $C_n = F_{2n+1}$.

43. Notice that $\beta < F_k/L_k < -\beta$. Using Eqs. (36.15) and (36.16), LHS $= \frac{F_k/L_k}{1 - F_k/L_k - F_k^2/L_k^2} = \frac{F_kL_k}{L_k^2 - L_kF_k - F_k^2} = \frac{F_{2k}}{L_k^2 - L_kF_k - F_k^2}$ = RHS

EXERCISES 37 (p. 456)

1. When $n = 5$, LHS $= x\sum_{1}^{5} f_i(x) = x[1 + x + (x^2+1) + (x^3+2x) + (x^4+3x^2+1)] = x^5 + x^4 + 4x^3 + 3x^2 + 3x = (x^5 + 4x^3 + 3x) + (x^4 + 3x^2 + 1) - 1 = f_6(x) + f_5(x) - 1$ = RHS. When $n = 6$, LHS $= x\sum_{1}^{6} f_i(x) = x[1+x+(x^2+1)+(x^3+2x)+(x^4+3x^2+1)+(x^5+4x^3+3x)] = x^6+x^5+5x^4+4x^3+6x^2+3x = (x^6 + 5x^4 + 6x^2 + 1) + (x^5 + 4x^3 + 3x) - 1 = f_7(x) + f_6(x) - 1$ = RHS.

3. $f_{10}(x) = f_{4+5+1}(x) = f_5(x)f_6(x) + f_4(x)f_5(x) = (x^4 + 3x^2 + 1)(x^5 + 4x^3 + 3x) + (x^3 + 2x)(x^4 + 3x^2 + 1) = x^9 + 8x^7 + 21x^5 + 20x^3 + 5x$

5. $f_{10}(x) = \sum_{j=0}^{4}\binom{9-j}{j}x^{9-2j} = \binom{9}{0}x^9 + \binom{8}{1}x^7 + \binom{7}{2}x^5 + \binom{6}{3}x^3 + \binom{5}{4}x = x^9 + 8x^7 + 21x^5 + 20x^3 + 5x$

7. $f_7'(x) = \sum_{1}^{6} f_i(x)f_{7-i}(x) = 2[f_1(x)f_6(x) + f_2(x)f_5(x) + f_3(x)f_4(x)] = 2[(x^5+4x^3+3x) + x(x^4+3x^2+1) + (x^2+1)(x^3+2x)] = 6x^5 + 20x^3 + 12x$

9. (Swamy) LHS $= [(x^3+2x)f_{n+1}(x)+(x^2+1)f_n(x)][(y^3+2y)f_{n+1}(y)+(y^2+1)f_n(y)]-xy[(x^3+1)f_{n+1}(x)+xf_n(x)][(y^3+1)f_{n+1}(y)+yf_n(y)]-(x^2+y^2+2)[xf_{n+1}(x)+f_n(x)][(yf_{n+1}(y)+yf_n(y)]-xyf_{n+1}(x)f_{n+1}(y)+f_n(x)f_n(y)] = f_{n+1}(x)f_{n+1}(y)[(x^3+2x)(y^3+2y)-xy(x^3+1)(y^3+1)-xy(x^2+y^2+2)-xy]+f_{n+1}(x)f_n(y)[(x^3+2x)(y^3+1)-xy^2(x^2+1)-x(x^2+y^2+2)-f_n(x)f_{n+1}(y)[(x^3+1)(y^3+2y)-x^2y(y^3+1)-y(x^2+y^2+2)+f_n(x)f_n(y)[(x^2+1)(y^2+1)-x^2y^2-(x^3+y^3+2)+1] = 0\cdot f_{n+1}(x)f_{n+1}(y)+0\cdot f_{n+1}(x)f_n(y)+0\cdot f_n(x)f_{n+1}(y)+0\cdot f_n(x)f_n(y) = 0.$

11. Notice that $g_1(x) = \sum_{j=0}^{0}\binom{-j}{j}x^{-2j} = 1$ and $g_2(x) = \sum_{j=0}^{0}\binom{1-j}{j}x^{1-2j} = x.$

Besides, $xg_{n-1}(x) + g_{n-2}(x) = \sum_{j=0}^{\lfloor(n-2)/2\rfloor}\binom{n-j-2}{j}x^{n-2j-1} + \sum_{j=0}^{\lfloor(n-3)/2\rfloor}\binom{n-j-3}{j}x^{n-2j-3}$

When n is even, say, $n = 2m$:

$$
\begin{aligned}
\text{RHS} &= \sum_{j=0}^{m-1}\binom{2m-j-2}{j}x^{2m-2j-1} + \sum_{j=0}^{m-2}\binom{2m-j-3}{j}x^{2m-2j-3}\\
&= \sum_{j=0}^{m-1}\binom{2m-j-2}{j}x^{2m-2j-1} + \sum_{j=1}^{m-1}\binom{2m-j-2}{j-1}x^{2m-2j-1}\\
&= \sum_{j=0}^{m-1}\binom{2m-j-2}{j}x^{2m-2j-1} + \sum_{j=0}^{m-1}\binom{2m-j-2}{j-1}x^{2m-2j-1}\\
&= \sum_{0}^{m-1}\left[\binom{2m-j-2}{j}+\binom{2m-j-2}{j-1}\right]x^{2m-2j-1}\\
&= \sum_{0}^{m-1}\binom{2m-j-1}{j}x^{2m-2j-1} = g_{2m}(x)
\end{aligned}
$$

Similarly, when $n = 2m+1$, RHS $= g_{2m+1}(x)$. Thus, in both cases, $g_n(x)$ satisfies the recurrence relation Eq. (37.1). $\therefore g_n(x) = f_n(x)$.

13. Follows by Exercise 10.

15. Let $g(t) = \sum_{0}^{\infty} f_{2n}(x)t^{2n}$. Since $f_0(x) = 0 = f_{2n}(x) - (x^2+2)f_{2n-2}(x) + f_{2n-4}(x)$, it follows that $[1-(x^2+2)t^2+t^4]g(t) = xt^2$; $\therefore g(t) = \dfrac{xt^2}{1-(x^2+2)t^2+t^4}$.

17. RHS $= x\sum_0^\infty f_{2n+2}(x)t^{2n} + \sum_0^\infty f_{2n+1}(x)t^{2n} = \frac{x}{t^2}\sum_0^\infty f_{2m}(x)t^{2m} + \frac{1}{t}\sum_0^\infty f_{2n+1}(x)t^{2n+1} = \frac{x}{t^2}\cdot\frac{xt^2}{1-(x^2+2)t^2+t^4} + \frac{1}{t}\cdot\frac{t-t^3}{1-(x^2+2)t^2+t^4} = \frac{x^2+1-t^2}{1-(x^2+2)t^2+t^4}$

19. $f_1(x) = \frac{\alpha(x)-\beta(x)}{\alpha(x)-\beta(x)} = 1$; $f_2(x) = \frac{\alpha^2(x)-\beta^2(x)}{\alpha(x)-\beta(x)} = \alpha(x)+\beta(x) = x$; and $xf_k(x) + f_{k-1}(x) = x\frac{\alpha^k-\beta^k}{\alpha-\beta} + \frac{\alpha^{k-1}-\beta^{k-1}}{\alpha-\beta} = \frac{\alpha^{k-1}(x\alpha+1)-\beta^{k-1}(x\beta+1)}{\alpha-\beta} = \frac{\alpha^{k-1}(\alpha^2)-\beta^{k-1}(\beta^2)}{\alpha-\beta} = \frac{\alpha^{k+1}-\beta^{k+1}}{\alpha-\beta}$, where $\alpha = \alpha(x)$ and $\beta = \beta(x)$. So $f_n(x)$ is the Fibonacci polynomial.

21. $g(3/2, n) = \sum_{k=0}^{n} \frac{2^k(n+k)!}{(n-k-1)!(2k+1)!}\cdot\frac{1}{2^k} = \sum_{k=0}^{n-1}\frac{(n+k)!}{(2k+1)!(n-k-1)!} = \sum_{k=0}^{n-1}\binom{n+k}{2k+1} = F_{2n}$

23. Since $H_0(x) = 1 = f_1(x) + f_0(x)$, the statement is true when $n = 0$. Assume it is true for all integers $i \le n$: $H_i(x) = f_{i+1}(x) + f_i(x)$. Then $H_{n+1}(x) = xH_n(x) + H_{n-1}(x) = x[f_{n+1}(x) + f_n(x)] + [f_n(x) + f_{n-1}(x)] = [xf_{n+1}(x) + f_n(x)] + [xf_n(x) + f_{n-1}(x)] = f_{n+2}(x) + f_{n+1}(x)$. Thus, by the strong version of PMI, the result is true for all $n \ge 0$.

25. $2x+2$; $4x^2+4x+1$; $8x^3+8x^2+4x+2$; $16x^4+16x^3+12x^2+8x+1$.

27. $\psi_n(1/2) = L_n$

29. Let $g(t) = \sum_0^\infty \psi_n(x)t^n$. Then $(-1+2xt+t^2)g(t) = -2x\psi_0(x) - \psi_0(x) - \psi_1(x)t = -4xt - 2 - t$; $\therefore g(t) = \frac{4xt-t-2}{1-2xt-t^2}$

31. $y_n(1) = F_n$

33. Let $g(t) = \sum_0^\infty y_n(x)t^n$. Then $(1-xt-t^2)g(t) = y_0 + (y_1 - xy_0)t = t$; $\therefore g(t) = \frac{t}{1-xt-t^2}$

35. $y_n(1) = L_n$

37. Let $g(t) = \sum_0^\infty y_n(x)t^n$. Then $(1-xt-t^2)g(t) = y_0 + (y_1 - xy_0)t = 2 + (1-2x)t$; $\therefore g(t) = \frac{2+(1-2x)t}{1-xt-t^2}$

39. $z_n(1) = F_n$

41. Let $g(t) = \sum_0^\infty z_n(x)t^n$. Then $(1-t-xt^2)g(t) = t$, so $g(t) = \frac{t}{1-t-xt^2}$

43. $z_n(1) = L_n$.

45. Let $g(t) = \sum_0^\infty z_n(x)t^n$. Then $(1 - t - xt^2)g(t) = 2 - t$, so $g(t) = \dfrac{2-t}{1-t-xt^2}$.

47. Expanding $D_n(x)$ with respect to row 1, $D_n(x) = 2xD_{n-1}(x) - iA$, where

$$A = \begin{vmatrix} i & i & \dots & 0 \\ 0 & 2x & \dots & 0 \\ \cdot & \cdot & \dots & 0 \\ 0 & \cdot & \dots & 2x \end{vmatrix} = iD_{n-2}(x). \text{ So } D_n(x) = 2xD_{n-1}(x) + D_{n-2}(x);$$

$\therefore D_n(x) = \varphi_n(x)$.

EXERCISES 38 (p. 467)

1. $x^{11} + 11x^9 + 44x^7 + 77x^5 + 55x^3 + 11x$

3. Since $f_2(x) + f_0(x) = x = l_1(x)$ and $f_3(x) + f_1(x) = x^2 + 2 = l_2(x)$, the statement is true when $n = 1$ and $n = 2$. Assume it is true for all positive integers $\le n$, where $n \ge 2$. Then $f_{n+2}(x) + f_n(x) = [xf_{n+1}(x) + f_n(x)] + [xf_{n-1}(x) + f_{n-2}(x)] = x[f_{n+1}(x) + f_{n-1}(x)] + [f_n(x) + f_{n-2}(x)] = xl_n(x) + l_{n-1}(x) = l_{n+1}(x)$. Thus, by the strong version of PMI, the result follows.

5. $xl_n(x) = x[f_{n+1}(x) + f_{n-1}(x)] = [xf_{n+1}(x) + f_n(x)] - [f_n(x) - xf_{n-1}(x)] = f_{n+2}(x) - f_{n-2}(x)$

7. $B(n, 0) = \begin{cases} 0 & \text{if } n \text{ is odd} \\ 2 & \text{otherwise} \end{cases}$
   $B(n, n-1) = 0$; $B(n, n) = 1$; $B(n, j) = B(n-1, j-1) + B(n-2, j)$, where $1 \le j \le n-3$, $n \ge 4$.

9. Let $\alpha = \alpha(x)$ and $\beta = \beta(x)$. Then $l_1(x) = \alpha + \beta = x$; $l_2(x) = \alpha^2 + \beta^2 = (\alpha+\beta)^2 - 2\alpha\beta = x^2 - 2(-1) = x^2 + 2$; and $xl_{n-1}(x) + l_{n-2}(x) = x(\alpha^{n-1} + \beta^{n-1}) + (\alpha^{n-2} + \beta^{n-2}) = \alpha^{n-2}(x\alpha + 1) + \beta^{n-2}(x\beta + 1) = \alpha^{n-2}\alpha^2 + \beta^{n-2}\beta^2 = \alpha^n + \beta^n = l_n(x)$. Thus $l_n(x)$ is the Lucas polynomial.

11. Let $\alpha = \alpha(x)$ and $\beta = \beta(x)$. Then $(\alpha - \beta)^2\text{LHS} = (\alpha^n - \beta^n)^2 + (\alpha^{n+1} - \beta^{n+1})^2 = \alpha^{2n} + \alpha^{2n+2} + \beta^{2n} + \beta^{2n+2} = \alpha^{2n+1}(\alpha + \alpha^{-1}) + \beta^{2n+1}(\beta + \beta^{-1}) = \alpha^{2n+1}(\alpha - \beta) - \beta^{2n+1}(\alpha - \beta) = (\alpha - \beta)(\alpha^{2n+1} - \beta^{2n+1})$; $\therefore \text{LHS} = f_{2n+1}(x)$.

13. Let $\alpha = \alpha(x)$ and $\beta = \beta(x)$. Then LHS $= (\alpha^{n-1} + \beta^{n-1})(\alpha^{n+1} + \beta^{n+1}) - (\alpha^n + \beta^n)^2 = (\alpha\beta)^{n-1}(\alpha^2 + \beta^2) - 2(\alpha\beta)^n = (-1)^{n-1}(x^2 + 2) + 2(-1)^{n-1} = (-1)^{n-1}(x^2 + 4) = \text{RHS}$.

15. $\alpha^n(x) + \beta^n(x) = l_n(x)$ and $\alpha^n(x) - \beta^n(x) = \sqrt{x^2+4}\,f_n(x)$. Adding the two equations yields the desired result.

17. x; $x^2 - 1$; $x^3 - 2x$; $x^4 - 3x^2 + 1$.

19. Clearly, $g_0(x) = 0$ and $g_1(x) = 1$. Let $\gamma = \gamma(x)$ and $\delta = \delta(x)$. Then $xg_{n-1}(x) - g_{n-2}(x) = \dfrac{x(\gamma^{n-1} - \delta^{n-1})}{\gamma - \delta} - \dfrac{\gamma^{n-2} - \delta^{n-2}}{\gamma - \delta} = \dfrac{\gamma^{n-2}(x\gamma - 1) - \delta^{n-2}(x\delta - 1)}{\gamma - \delta} = \dfrac{\gamma^n - \delta^n}{\gamma - \delta} = g_n(x)$.

21. Let $\gamma = \gamma(x)$ and $\delta = \delta(x)$.

$$\begin{aligned}\text{LHS} &= (\gamma^n + \delta^n)^2 - \frac{(x^2-4)(\gamma^n - \delta^n)2}{x^2-4} = (\gamma^n + \delta^n)^2 - (\gamma^n - \delta^n)^2 \\ &= 4(\gamma\delta)^n = 4 \cdot 1^n = 4 = \text{RHS}\end{aligned}$$

23. Clearly, $g_0(x) = \sum_0^{-1} \binom{-i-1}{i} (-1)^i x^{-2i-1} = 0$ and $g_1(x) = \sum_0^{0} \binom{-i}{i} (-1)^i x^{-2i} = 1$. When $n \geq 2$:

$$\begin{aligned}xg_{n-1}(x) - g_{n-2}(x) = &\sum_{i=0}^{\lfloor (n-2)/2 \rfloor} \binom{n-i-2}{i} (-1)^i x^{n-2i-1} \\ &- \sum_{i=0}^{\lfloor (n-3)/2 \rfloor} \binom{n-i-3}{i} (-1)^i x^{n-2i-3} \qquad (1)\end{aligned}$$

Let $n = 2m$:

$$\begin{aligned}\text{RHS} &= \sum_{i=0}^{m-1} \binom{2m-i-2}{i} (-1)^i x^{2m-2i-1} \\ &\quad - \sum_{i=0}^{m-2} \binom{2m-i-3}{i} (-1)^i x^{2m-2i-3} \\ &= \sum_{i=0}^{m-1} \binom{2m-i-2}{i} (-1)^i x^{2m-2i-1} \\ &\quad + \sum_{j=0}^{m-1} \binom{2m-j-2}{j-1} (-1)^j x^{2m-2j-1} \\ &= \sum_0^{m-1} \left[\binom{2m-i-2}{i} + \binom{2m-i-2}{i-1} \right] (-1)^i x^{2m-2i-1} \\ &= \sum_0^{m-1} \binom{2m-i-1}{i-1} (-1)^i x^{2m-2i-1} \\ &= \sum_0^{\lfloor (n-1)/2 \rfloor} \binom{n-i-1}{i} (-1)^i x^{n-2i-1} = g_{2m}(x) = g_n(x)\end{aligned}$$

Similarly, when $n = 2m + 1$, RHS of (1) yields $g_n(x)$.

25. Since $l_{2n}(x) = f_{2n+1}(x) + f_{2n-1}(x)$, $\sum_0^\infty l_{2n}(x)t^{2n} = \sum_0^\infty f_{2n+1}(x)t^{2n} + \sum_0^\infty f_{2n-1}(x)t^{2n} = \dfrac{1-t^2}{1-(x^2+2)t^2+t^4} + f_{-1}(x) + \dfrac{t^2-t^4}{1-(x^2+2)t^2+t^4} = \dfrac{2-(x^2+2)t^4}{1-(x^2+2)t^2+t^4}$, since $f_{-1}(x) = 1$.

27. (Webb and Parberry) Let $x = 2i\cos\theta$, where $0 \le \theta \le \pi$. Then $\alpha(x) = \dfrac{2i\cos\theta + \sqrt{-4\cos^2\theta + 4}}{2} = i\cos\theta + \sin\theta$ and similarly, $\beta(x) = i\cos\theta - \sin\theta$. Then $f_n(2i\cos\theta) = \dfrac{(i\cos\theta+\sin\theta)^n - (i\cos\theta - \sin\theta)^n}{2\sin\theta} = \dfrac{(-i)^n(e^{-n\theta i} + e^{n\theta i})}{2\sin\theta} = \dfrac{(-i)^{n-1}\sin n\theta}{\sin\theta}$. $\therefore f_n(2i\cos\theta) = 0$ iff $\sin n\theta = 0$ and $\sin\theta \ne 0$, that is, iff $\theta = k\pi/n$, where $1 \le k \le n-1$.

29. $xt_{n-1}(x) + t_{n-1}(x) = x[bxf_{n-2}(x) + af_{n-3}(x)] + [bxf_{n-3}(x) + af_{n-4}(x)] = bx^2f_{n-2}(x) + axf_{n-3}(x) + bxf_{n-2}(x) + af_{n-4}(x) = bx[xf_{n-2}(x) + f_{n-3}(x)] + a[xf_{n-2}(x) + f_{n-4}(x)] = bxf_{n-1}(x) + af_{n-2}(x)] = t_n(x)$

EXERCISES 39 (p. 476)

1. $x^5 + 15x^4 + 35x^3 + 28x^2 + 9x + 1$; $6x^5 + 35x^4 + 56x^3 + 36x^2 + 10x + 1$.

3. Notice that $J_1(x) = \sum_0^0 \binom{0}{0} x^0 = 1$ and $J_2(x) = \sum_0^0 \binom{0}{0} x^0 = 1$. Besides,

$$J_{n-1}(x) + xJ_{n-2}(x) = \sum_{j=0}^{\lfloor (n-2)/\rfloor} \binom{\lfloor (n-1)/2\rfloor + j}{\lfloor (n-2)/2\rfloor - j} x^{\lfloor (n-2)/2\rfloor - j} + x \sum_{j=0}^{\lfloor (n-3)/2\rfloor} \binom{\lfloor (n-2)/2\rfloor + j}{\lfloor (n-3)/2\rfloor - j} x^{\lfloor (n-3)/2\rfloor - j}$$

Let n be odd, say, $n = 2k+1$.

$$\begin{aligned}
\text{RHS} &= \sum_{j=0}^{k-1} \binom{k+j}{k-j-1} x^{k-j-1} + \sum_{j=0}^{k-1} \binom{k+j-1}{k-j-1} x^{k-j} \\
&= \sum_1^k \binom{k+j-1}{k-j} x^{k-j} + \sum_0^{k-1} \binom{k+j-1}{k-j-1} x^{k-j} \\
&= \sum_1^k \left[\binom{k+j-1}{j} + \binom{k+j-1}{k-j-1}\right] x^{k-j} + 1 + x^k
\end{aligned}$$

$$= \sum_{1}^{k} \binom{k+j}{j} x^{k-j} + x^k + 1$$

$$= \sum_{0}^{k} \binom{k+j}{k-j} x^{k-j} = J_{2k+1}(x)$$

Similarly, it can be shown that the formula works for $n = 2k$. Thus it holds for all $n \geq 1$.

5. Let r and s be the zeros of $t^2 - t - x$. Then $1 - t - xt^2 = (1 - rt)(1 - st)$. Let $\sum_{0}^{\infty} a_n t^n = \frac{1}{1 - t - xt^2} = \frac{1}{(1 - rt)(1 - st)} = \frac{A}{1 - rt} + \frac{B}{1 - st}$, where $A = \frac{r}{\sqrt{1 + 4x}}$ and $B = -\frac{s}{\sqrt{1 + 4x}}$. Then RHS $= A \sum_{0}^{\infty} r^n t^n + B \sum_{0}^{\infty} s^n t^n$, so $a_n = Ar^n + Bs^n = \frac{r^{n+1} - s^{n+1}}{\sqrt{1 + 4x}} = J_{n+1}(x)$, by Exercise 4; $\therefore \sum_{0}^{\infty} J_{n+1}(x) t^n = \frac{1}{1 - t - xt^2}$.

7. $k_n(1) = L_n$
9. $k_n(1) = L_n$
11. $Q_n(1) = 1$
13. $\lfloor n/2 \rfloor$
15. The characteristic roots, given by $t^2 - tx - x = 0$, are given by $t = \frac{x \pm \sqrt{x^2 + 4x}}{2}$. Let r denote the positive root and s the negative root. Then $r + s = x$, $rs = -x$, and the general solution is $Q_n = Ar^n + Bs^n$. Since $Q_n(1) = 1$ and $Q_n(2) = x$, $Ar + Bs = 1$ and $Ar^2 + Bs^2 = x$. Solving $A = \frac{x - s}{r\sqrt{x^2 + 4x}}$ and $B = \frac{r - x}{s\sqrt{x^2 + 4x}}$, $\therefore \quad Q_n(x) = \frac{(x - s)r^n}{r\sqrt{x^2 + 4x}} + \frac{(r - x)s^n}{s\sqrt{x^2 + 4x}} = \frac{(x - s)r^{n-1} + (r - x)s^{n-1}}{\sqrt{x^2 + 4x}} = \frac{r \cdot r^{n-1} - s \cdot s^{n-1}}{\sqrt{x^2 + 4x}} = \frac{r^n - s^n}{\sqrt{x^2 + 4x}}$
17. The characteristic equation is $u^2 - ux - x = 0$; that is, $1 - xt - xt^2 = 0$. Let $1 - xt - xt^2 = (1 - rt)(1 - st)$. Let $\sum_{0}^{\infty} a_n t^n = \frac{1}{1 - xt - xt^2} = \frac{1}{(1 - rt)(1 - st)} = \frac{A}{1 - rt} + \frac{B}{1 - st}$, where $A = \frac{r}{\sqrt{x^2 + 4x}}$ and $B = -\frac{s}{\sqrt{x^2 + 4x}}$; $\therefore \sum_{0}^{\infty} a_n t^n = A \sum_{0}^{\infty} r^n t^n + B \sum_{0}^{\infty} s^n t^n$, so $a_n = Ar^n + Bs^n = \frac{r^{n+1} - s^{n+1}}{\sqrt{x^2 + 4x}} = Q_{n+1}(x)$. Thus $\sum_{0}^{\infty} Q_{n+1}(x) t^n = \frac{1}{1 - xt - xt^2}$, so $\sum_{0}^{\infty} Q_n(x) t^n = \frac{t}{1 - xt - xt^2}$.

EXERCISES 41 (p. 494)

1. We have $b_n = xB_{n-1} + b_{n-1}$, and $xB_n = (x^2 + x)B_{n-1} + xb_{n-1}$ from Eq. (41.2). Subtracting, $xB_n - b_n = x^2 B_{n-1} + (x-1)b_{n-1}$; that is, $(b_{n+1} - b_n) - b_n = x(b_n - b_{n-1}) + (x-1)b_{n-1}$, using Eq. (41.5). Thus $b_{n+1} = (x+2)b_n - b_{n-1}$; that is, $b_n = (x+2)b_{n-1} - b_{n-2}$, $n \geq 2$.

3. $b_5(x) = \sum_0^5 \binom{5+i}{5-i} x^i = \binom{5}{5} + \binom{6}{4}x + \binom{7}{3}x^2 + \binom{8}{2}x^3 + \binom{9}{1}x^4 + \binom{10}{0}x^5 = x^5 + 9x^4 + 28x^3 + 35x^2 + 15x + 1$

5. $b_n(x) = B_n(x) - B_{n-1}(x) = \sum_0^n \left[\binom{n+i+1}{n-i} - \binom{n+i}{n-i-1}\right] x^i = \sum_0^n \binom{n+i}{n-i} x^i$

7.
$$\begin{aligned}
xB_{n-1}(x) + b_{n-1}(x) &= x\sum_0^{n-1} \binom{n+i}{n-i-1} x^i + \sum_0^{n-1} \binom{n+i-1}{n-i-1} x^i \\
&= \sum_0^n \binom{n+i-1}{n-i} x^i + \sum_0^n \binom{n+i-1}{n-i-1} x^i \\
&= \sum_0^n \left[\binom{n+i-1}{n-i} + \binom{n+i-1}{n-i-1}\right] x^i \\
&= \sum_0^n \binom{n+i}{n-i} x^i = b_n(x)
\end{aligned}$$

9. Since $b_0(1) = 1 = F_1$, the result is true when $n = 0$. Assume it is true for every nonnegative integer $i \leq n$, where $n \geq 0$. Then $b_{n+1}(1) = 3b_n(1) - b_{n-1}(1) = 3F_{2n+1} - F_{2n-1} = 2F_{2n+1} + F_{2n} = F_{2n+1} + F_{2n+2} = F_{2n+3}$. Thus, by the strong version of PMI, the result follows.

11. Since $b_n(x) = B_n(x) - B_{n-1}(x)$, $b_n(1) = B_n(1) - B_{n-1}(1) = F_{2n+2} - F_{2n} = F_{2n+1}$.

13. Using Identity (41.9), $B_{m+n}(1) = B_m(1)B_n(1) - B_{m-1}(1)B_{n-1}(1)$. But $B_k(1) = F_{2k+2}$; $\therefore F_{2m++2n} = F_{2m+2}F_{2n} - F_{2m}F_{2n-2}$; thus $F_{m+2}F_n - F_m F_{n-2} = F_{m+n}$.

15. Using Identity (41.12), $B_{2n-1}(1) = [B_n(1) - B_{n-2}(1)]B_{n-1}(1)$; that is, $F_{4n} = (F_{2n+2} - F_{2n-2})F_{2n}$. Thus $F_{2n} = (F_{n+2} - F_{n-2})F_n$.

17. (Swamy) Using Identity (41.7), $b_{n+1} - b_n = (x+1)b_n - b_{n-1}$; that is, $xB_n = (x+1)b_n - b_{n-1}$, by Identity (41.5).

19. By Identity (41.5), $xB_n = b_{n+1} - b_n$, and $xB_{n-1} = b_n - b_{n-1}$. Adding, $x(B_n + B_{n-1}) = b_{n+1} - b_{n-1}$.

21. Follows from Eq. (41.21).

23. (Swamy) From Eq. (41.9), $(x+2)b_{2n+1} = b_{2n+2} + b_{2n} = (b_{n+1}B_{n+1} - b_nB_n) + (b_nB_n - b_{n-1}B_{n-1}) = b_{n+1}B_{n+1} - b_{n-1}B_{n-1}$.

25. (Swamy) $b_{2n} = B_nb_n - B_{n-1}b_{n-1}$, by Exercise 21, and $b_{2n-1} = b_nB_{n-1} - b_{n-1}B_{n-2}$, by Exercise 22; $\therefore b_{2n} - b_{2n-1} = b_n(B_n - B_{n-1}) - b_{n-1}(B_{n-1} - B_{n-2}) = b_n^2 - b_{n-1}^2$, by Identity (41.6).

27. (Swamy) $B_{2i} = B_i^2 - B_{i-1}^2$, by Identity (41.11); $\therefore \sum_0^n B_{2i} = B_n^2 - B_{-1}^2 = B_n^2 - 0 = B_n^2$.

29. (Swamy) By Identity (41.22), $b_{2i} = b_iB_i - b_{i-1}B_{i-1}$; $\therefore \sum_0^n b_{2i} = b_nB_n - b_{-1}B_{-1} = b_nB_n - 0 = b_nB_n$.

31. (Swamy) By Exercise 25, $b_{2i} - b_{2i-1} = b_i^2 - b_{i-1}^2$; $\therefore \sum_0^n (b_{2i} - b_{2i-1}) = b_n^2 - b_{-1}^2$; that is, $\sum_0^{2n} (-1)^i b_i - b_{-1}^2 = b_n^2 - b_{-1}^2$. Thus $\sum_0^{2n} (-1)^i b_i = b_n^2$.

33. Clearly, $g_1(x) = b_0(x^2) = 1$ and $g_2(x) = xB_0(x^2) = x \cdot 1 = x$. Let $n = 2k$. Then $xg_{n+1} + g_n = xg_{2k+1} + g_{2k} = xb_k(x^2) + xB_{k-1}(x^2) = x[x^2B_{k-1}(x^2) + b_{k-1}(x^2)] + xB_{k-1}(x^2) = x[(x^2+1)B_{k-1}(x^2) + b_{k-1}(x^2)] = xB_k(x^2) = g_{2k+2} = g_{n+2}$. On the other hand, let $n = 2k+1$. Then $xg_{n+1} + g_n = xg_{2k+2} + g_{2k+1} = x[xB_k(x^2)] + b_k(x^2) = x^2B_k(x^2) + b_k(x^2) = g_{2k+3} = g_{n+2}$. Thus $g_n(x) = f_n(x)$.

35. $$b_n(x) = B_n(x) - B_{n-1}(x) = \frac{\sin(n+1)\theta}{\sin\theta} - \frac{\sin n\theta}{\sin\theta} = \frac{\sin(n+1)\theta - \sin n\theta}{\sin\theta}$$
$$= \frac{2\cos(2n+1)\theta/2 \sin\theta/2}{2\sin\theta/2\cos\theta/2} = \frac{\cos(2n+1)\theta/2}{\cos\theta/2}$$

37. Using Exercise 35, $b_n(x) = 0$ iff $\cos(2n+1)\theta/2 = 0$, that is, iff $(2n+1)\theta/2 = (2k+1)\pi/2$, where $0 \le k < n$. Then $\theta = \dfrac{(2k+1)\pi}{2n+1}$, so $x + 2 = 2\cos\theta = 2\cos\dfrac{(2k+1)\pi}{2n+1}$. Thus $x = 2\left[\cos\dfrac{(2k+1)\pi}{2n+1} - 1\right] = -4\sin^2\frac{(2k+1)\pi}{4n+2}$, $0 \le k < n$.

39. (Law) (a) First, we shall show that $y_n = (b_n + b_{n-1})/2 = C_n$:
    (1) $y_0 = (b_0 + b_{-1})/2 = (1+1)/2 = 1$
    (2) $y_1 = (b_1 + b_0)/2 = [(x+1)+1]/2 = (x+2)/2$
    (3) $(x+2)y_{n-1} - y_{n-2} = (x+2)(b_{n-1} + b_{n-2})/2 - (b_{n-2} + b_{n-3})/2 = [(x+2)b_{n-1} - b_{n-2}]/2 + [(x+2)b_{n-2} - b_{n-3}]/2 = (b_n + b_{n-1})/2 = y_n$. So $y_n = C_n$.

    (b) Next we shall show that $z_n = (B_n - B_{n-2})/2 = C_n$:
    (1) $z_0 = (B_0 - B_{-2})/2 = [1 - (-1)]/2 = 1$
    (2) $z_1 = (B_1 - B_{-1})/2 = [(x+2) - 0]/2 = (x+2)/2$

(3) $(x+2)z_{n-1} - z_{n-2} = (x+2)(B_{n-1} - B_{n-3})/2 - (B_{n-2} - B_{n-4})/2 = [(x+2)B_{n-1} - B_{n-2}]/2 - [(x+2)B_{n-3} - B_{n-4}]/2 = (B_n - B_{n-2})/2 = z_n$. So $z_n = C_n$.

Thus $2C_n(x) = b_n(x) + b_{n-1}(x) = B_n(x) - B_{n-2}(x)$.

41. Using Exercise 40, $-C_{n+1}(x)C_{n-1}(x) + C_n^2(x) = |S^{n+1} - S^{n-1}|/2 = |S^{n-1}| \cdot |S^2 - I|/2 = \frac{1}{2} \cdot 1 \cdot \begin{vmatrix} x^2+4x+2 & -(x+2) \\ x+2 & -2 \end{vmatrix} = \frac{-x(x+4)}{4}$. Thus $C_{n+1}(x)C_{n-1}(x) - C_n^2(x) = x(x+4)/4$.

43. $(x^2+4x+2)/2$; $(x+2)(x^2+4x+1)/2$; and $(x^4+8x^3+23x^2+16x+2)/2$

EXERCISES 42 (p. 510)

1. Let $G_n = L_n$. Then $\mu = -5$ and $G_{n+1}(G_n + G_{n+2}) = L_{n+1}(L_n + L_{n+2}) = L_{2n+1} + L_{2n+3}$. Thus Eq. (42.3) follows from Eq. (42.1).

3. Let θ_n = RHS. Then $\tan\theta_n = \dfrac{1/L_{2n} + 1/L_{2n+2}}{1 - 1/L_{2n} \cdot 1/L_{2n+2}} = \dfrac{L_{2n} + L_{2n+2}}{L_{2n}L_{2n+2} - 1} = \dfrac{5F_{2n+1}}{5F_{2n+1}^2} = \dfrac{1}{F_{2n+1}}$; $\therefore \tan^{-1}\dfrac{1}{F_{2n+1}}$ = RHS.

5. Let $2\theta = \tan^{-1} 2$. Then $\tan 2\theta = 2$; that is, $\dfrac{2\tan\theta}{1 - \tan^2\theta} = 2$. Solving, $\tan\theta = \frac{-1\pm\sqrt{5}}{2}$. Since $\tan\theta \ge 0$, $\tan\theta = \frac{-1+\sqrt{5}}{2} = -\beta$; $\therefore \theta = \tan^{-1}\beta = \frac{1}{2}\tan^{-1} 2$.

7. (Peck) We have $\tan^{-1}\dfrac{1}{F_{2n}} = \tan^{-1}\dfrac{1}{F_{2n+2}} + \tan^{-1}\dfrac{1}{F_{2n+1}}$. Since $\tan^{-1} x + \tan^{-1} 1/x = \pi/2$, this implies $\tan^{-1}\frac{1}{F_{2n}} = (\pi/2 - \tan^{-1} F_{2n+2}) + (\pi/2 - \tan^{-1} F_{2n+1}) = \pi - \tan^{-1} F_{2n+2} - \tan^{-1} F_{2n+1}$. This gives the desired result.

EXERCISES 43 (p. 516)

1. $U_5 = F_5 = 5$, $U_6 = F_8 = 21$, $X_4 = L_3 = 4$, and $X_7 = L_{13} = 521$. LHS $= 5U_5U_6 = 525 = X_7 - (-1)^{F_5}X_4$ = RHS

3. $X_5X_6 = L_5L_8 = 11 \cdot 47 = 517 = 521 - 4 = X_7 + (-1)^{F_5}X_4$.

5. LHS $= 5V_5V_6 = 5F_{11}F_{18} = 5 \cdot 89 \cdot 2584 = 1{,}149{,}880 = 1149851 + 29 = L_{29} + L_7 = W_7 - (-1)^{L_5}W_4$ = RHS

7. Since $L_n = \alpha^n + \beta^n$, $W_n = \alpha^{L_n} + \beta^{L_n}$ and $W_{n+1} = \alpha^{L_{n+1}} + \beta^{L_{n+1}}$; $\therefore W_nW_{n+1} = (\alpha^{L_{n+2}} + \beta^{L_{n+2}}) + (\alpha^{L_n}\beta^{L_{n+1}} + \alpha^{L_{n+1}}\beta^{L_n}) = W_{n+2} + [\alpha^{L_{n+1}}(-\alpha^{-L_n}) + \beta^{L_{n+1}}(-\beta^{-L_n})] = W_{n+2} + (-1)^{L_n}(\alpha^{L_{n+1}-L_n} + \beta^{L_{n+1}-L_n}) = W_{n+2} + (-1)^{L_n}(\alpha^{L_{n-1}} + \beta^{L_{n-1}}) = W_{n+2} + (-1)^{L_n}W_{n-1}$.

9. LHS $= 5U_4U_6 = 5F_3F_8 = 5 \cdot 2 \cdot 21 = 210 = 199 + 11 = L_{11} + L_5 = W_5 - (-1)^3X_5$ = RHS

11. Since $\sqrt{5}F_n = \alpha^n - \beta^n, \sqrt{5}U_n = \alpha^{F_n} - \beta^{F_n}$. Then $5U_{n-1}U_{n+1} = (\alpha^{F_{n-1}+F_{n+1}} + \beta^{F_{n-1}+F_{n+1}}) - (\alpha^{F_{n-1}}\beta^{F_{n+1}} + \alpha^{F_{n+1}}\beta^{F_{n-1}}) = (\alpha^{L_n} + \beta^{L_n}) - [\alpha^{F_{n+1}}(-\alpha^{-F_{n-1}}) + \beta^{F_{n+1}}(-\beta^{-F_{n-1}})] = W_n - (-1)^{F_{n-1}}(\alpha^{F_{n+1}-F_{n-1}} + \beta^{F_{n+1}-F_{n-1}}) = Wn - (-1)^{F_{n-1}}(\alpha^{F_n} + \beta^{F_n}) = W_n - (-1)^{F_{n-1}}X_n$.
13. $2L_{13} = \sqrt{5L_{12}^2 - 20(-1)^{12}} + L_{12} = \sqrt{5 \cdot 322^2 - 20} + 322 = 1042$; $\therefore L_{13} = 521$.
15. $2L_{n+1} = 5F_n + L_n = \sqrt{5L_n^2 - 20(-1)^n} + L_n$; $\therefore L_{n+1} = [\sqrt{5L_n^2 - 20(-1)^n} + L_n]/2$.
17. $L_{3n} = \alpha^{3n} + \beta^{3n} = (1 + 2\alpha)^n + (1 + 2\beta)^n = \sum_0^n \binom{n}{i}[(2\alpha)^i + (2\beta)^i] = \sum_0^n \binom{n}{i} 2^i L_i$. Since $L_0 = 2$, it follows that $2|L_{3n}$. Conversely, let $2|L_n$. Since $F_{2n} = F_n L_n$, this implies $2|F_{2n}$; that is, $F_3|F_{2n}$. $\therefore 3|n$, since $3 \not| 2$.
19. 233
21. 24,476

EXERCISES 44 (p. 521)

1. Let $z = a + bi$. Then $\|z\| = a^2 + b^2 \geq 0$.
3. Let $w = a + bi$ and $z = c + di$. Since $wz = (ac - bd) + (ad + bc)i$, $\|wz\| = (ac - bd)^2 + (ad + bc)^2 = (a^2 + b^2)(c^2 + d^2) = \|w\| \cdot \|z\|$.
5. Since $f_1 = 1 = 1 + i \cdot 0 = F_1 + iF_0$ and $f_2 = 1 + i = F_2 + iF_1$, the result is true when $n = 1$ and $n = 2$. Assume it is true for every positive integer $k \leq n$. Then $f_{n+1} = f_n + f_{n-1} = (F_n + iF_{n-1}) + (F_{n-1} + iF_{n-2}) = (F_n + F_{n-1}) + i(F_{n-1} + F_{n-2}) = F_{n+1} + iF_n$. Thus, by the strong version of PMI, the result follows.
7. Let $z = a + bi$. Then $\bar{z} = a - bi$, so $\|\bar{z}\| = a^2 + b^2 = \|z\|$.
9. $f_{-10} = F_{-10} + iF_{-11} = (-1)^{11}F_{10} + i(-1)^{12}F_{11} = -55 + 89i$ and $l_{-10} = L_{-10} + iL_{-11} = (-1)^{10}L_{10} + i(-1)^{11}L_{11} = 123 - 199i$
11. $2m - 1 = 7$ and $2n - 1 = 21$, so $2m - 1|2n - 1$. $F_7 = 13$, $F_3F_{11} - F_4F_{10} = 2 \cdot 89 - 3 \cdot 55 = 13$, so $F_7|F_3F_{11} - F_4F_{10}$.
13. LHS $= (L_{n-1} + iL_{n-2})(L_{n+1} + iL_n) - (L_n + iL_{n-1})^2 = (L_{n-1}L_{n+1} - L_n^2) - (L_{n-2}L_n - L_{n-1}^2) + i(L_{n+1}L_{n-2} - L_nL_{n-1}) = -5(-1)^n + 5(-1)^{n-1} - 5(-1)^{n-1}i = 10(-1)^{n+1} - 5(-1)^{n+1}i = 5(2 - i)(-1)^{n+1} =$ RHS
15. LHS $= (F_n + iF_{n-1})^2 + (F_{n+1} + iF_n)^2 = (F_n^2 + F_{n+1}^2) - (F_{n-1}^2 + F_n^2) + 2i(F_nF_{n-1} + F_{n+1}F_n) = F_{2n+1} - F_{2n-1} + 2i(F_{2n}) = F_{2n} + 2iF_{2n} = (1 + 2i)F_{2n} =$ RHS
17. $f_n l_n = (F_n + iF_{n-1})(L_n + iL_{n-1}) = (F_nL_n - F_{n-1}L_{n-1}) + i(F_nL_{n-1} + F_{n-1}L_n) = F_{2n} - F_{2n-2} + i(2F_{2n-1}) = F_{2n-1} + 2iF_{2n-1} = (1+2i)F_{2n-1} =$ RHS

19. LHS $= (L_n + iL_{n-1})^2 - 5(F_n + iF_{n-1})^2 = (L_n^2 - 5F_n^2) - (L_{n-1}^2 - 5F_{n-1}^2) + 2i(L_nL_{n-1} - 5F_nF_{n-1}) = 4(-1)^n - 4(-1)^{n-1} + 4i(-1)^{n-1} = 4(2-i)(-1)^n =$ RHS
21. LHS $= \sum_1^n (F_{2k-1} + iF_{2k-2}) = \sum_1^n F_{2k-1} + i\sum_1^n F_{2k-2} = F_{2n} + i(F_{2n-1} - 1) = f_{2n} - i =$ RHS
23. LHS $= \sum_1^{2n} (-1)^k(F_k + iF_{k-1}) = \sum_1^{2n}(-1)^kF_k + i\sum_1^{2n}(-1)^kF_{k-1} = \sum_1^n F_{2k} - \sum_1^n F_{2k-1} + i\left[\sum_1^n F_{2k-1} - \sum_1^{n-1} F_{2k}\right] = (F_{2n+1} - 1) - F_{2n} + i[F_{2n} - (F_{2n-1} - 1)] = (F_{2n-1} - 1) + i(F_{2n-2} + 1) = (F_{2n-1} + iF_{2n-2}) = i - 1 = f_{2n-1} + i - 1 =$ RHS
25. $C_n\bar{C}_n = (F_n + iF_{n+1})(F_n - iF_{n+1}) = F_n^2 + F_{n+1}^2 = F_{2n+1}$.

EXERCISES 45 (p. 526)

1. $l(0) = l(w)$ implies $2 = \alpha^w + \beta^w$. $l(z + w) = l(z)$ implies $\alpha^{w+z} + \beta^{w+z} = \alpha^z + \beta^z$; that is, $\alpha^{w+z} + \beta^z(2 - \alpha^w) = \alpha^z + \beta^z$. Then $\alpha^w(\alpha^z - \beta^z) + 2\beta^z = \alpha^z + \beta^z$, so $\alpha^w(\alpha^z - \beta^z) = \alpha^z - \beta^z$. Thus $\alpha^w = 1$, so $\text{Re}(w) = 0$. Let $w = 0 + yi$. Then $\alpha^{yi} = 0$, which is possible only if $y = 0$. Thus $w = 0$, a contradiction.
3. $5\cdot\text{RHS} = (\alpha^{z+1} - \beta^{z+1}) + (\alpha^z - \beta^z) = \alpha^z(\alpha+1) - \beta^z(\beta+1) = \alpha^z\cdot\alpha^2 - \beta^z\cdot\beta^2 = \alpha^{z+2} - \beta^{z+2} = 5f(z+2)$; $\therefore$ RHS $= f(z+2) =$ LHS.
5. LHS $= (\alpha^z + \beta^z)^2 - (\alpha^z - \beta^z)^2 = 4(\alpha\beta)^z = 4(-1)^z = 4e^{\pi zi} =$ RHS
7. $l(-z) = \alpha^{-z} + \beta^{-z} = \dfrac{\alpha^z + \beta^z}{(\alpha\beta)^z} = \dfrac{l(z)}{(-1)^z} = l(z)$.
9. $5 \cdot \text{RHS} = (\alpha^z - \beta^z)(\alpha^{w+1} - \beta^{w+1}) + (\alpha^{z-1} - \beta^{z-1})(\alpha^w - \beta^w) = \alpha^{z+w}(\alpha + \alpha^{-1}) + \beta^{z+w}(\beta + \beta^{-1}) - \alpha^{z-1}\beta^w(\alpha\beta + 1) - \alpha^w\beta^{z-1}(\alpha\beta + 1) = \alpha^{z+w}(\alpha - \beta) - \beta^{z+w}(\alpha - \beta) = \sqrt{5}(\alpha^{z+w} - \beta^{z+w})$; $\therefore$ RHS $= f(z+w) =$ LHS
11. Let $w = 0$ and $z = n$ in Eq. (45.1). Then $F_n = \dfrac{1}{\sqrt{5}}\sum_{k=0}^{\infty}\dfrac{(\ln^k\alpha - \ln^k\beta)n^k}{k!}$.

EXERCISES 46 (p. 532)

1. $B(n, 0) = 1 = B(n, n), n \geq 0$;
   $B(n, j) = B(n-2, j-1) + B(n-1, j-1) + B(n-1, j), n \geq 2, j \geq 1$.

EXERCISES 47 (p. 536)

1. 0
3. $Q^2 = \begin{bmatrix} x^2 & 1 & 0 \\ x & 0 & 1 \\ 1 & 0 & 0 \end{bmatrix}\begin{bmatrix} x^2 & 1 & 0 \\ x & 0 & 1 \\ 1 & 0 & 0 \end{bmatrix} = \begin{bmatrix} x^4 + x & x^2 & 1 \\ x^3 + 1 & x & 0 \\ x^2 & 1 & 0 \end{bmatrix}$

$$Q^3 = \begin{bmatrix} x^2 & 1 & 0 \\ x & 0 & 1 \\ 1 & 0 & 0 \end{bmatrix} \begin{bmatrix} x^4+x & x^2 & 1 \\ x^3+1 & x & 1 \\ x^2 & 1 & 0 \end{bmatrix} = \begin{bmatrix} x^6+2x^3+1 & x^4+x & x^2 \\ x^5+2x^2 & x^3+1 & x \\ x^2 & 1 & 0 \end{bmatrix}$$

5. 8, 15, 29
7. T_n^*

9. $$Q^n = \begin{bmatrix} t_{n+1}^* & t_n^* & t_{n-1}^* & t_{n-2}^* \\ x^2t_n^* + xt_{n-1}^* + t_{n-2}^* & x^2t_{n-1}^* + xt_{n-2}^* + t_{n-3}^* & x^2t_{n-2}^* + xt_{n-3}^* + t_{n-4}^* & x^2t_{n-3}^* + xt_{n-4}^* + t_{n-5}^* \\ xt_n^* + t_{n-1}^* & xt_{n-1}^* + t_{n-2}^* & xt_{n-2}^* + t_{n-3}^* & xt_{n-3}^* + t_{n-4}^* \\ t_n^* & t_{n-1}^* & t_{n-2}^* & t_{n-3}^* \end{bmatrix}$$

11. By Exercise 10, $|Q^n| = (-1)^{n+1}$. Since $t_n^*(1) = T_n^*$, it follows that the given determinant equals $(-1)^{n+1}$.

INDEX

C

D

E

H

I

J

K

L

M

N

O

P

Q

R

S

T

X

Y

Z

PURE AND APPLIED MATHEMATICS

A Wiley-Interscience Series of Texts, Monographs, and Tracts

ADÁMEK, HERRLICH, and STRECKER—Abstract and Concrete Catetories
ADAMOWICZ and ZBIERSKI—Logic of Mathematics
AINSWORTH and ODEN—A Posteriori Error Estimation in Finite Element Analysis
AKIVIS and GOLDBERG—Conformal Differential Geometry and Its Generalizations
ALLEN and ISAACSON—Numerical Analysis for Applied Science
*ARTIN—Geometric Algebra
AUBIN—Applied Functional Analysis, Second Edition
AZIZOV and IOKHVIDOV—Linear Operators in Spaces with an Indefinite Metric
BERG—The Fourier-Analytic Proof of Quadratic Reciprocity
BERMAN, NEUMANN, and STERN—Nonnegative Matrices in Dynamic Systems
BOYARINTSEV—Methods of Solving Singular Systems of Ordinary Differential Equations
BURK—Lebesgue Measure and Integration: An Introduction
*CARTER—Finite Groups of Lie Type
CASTILLO, COBO, JUBETE and PRUNEDA—Orthogonal Sets and Polar Methods in Linear Algebra: Applications to Matrix Calculations, Systems of Equations, Inequalities, and Linear Programming
CHATELIN—Eigenvalues of Matrices
CLARK—Mathematical Bioeconomics: The Optimal Management of Renewable Resources, Second Edition
COX—Primes of the Form $x^2 + ny^2$: Fermat, Class Field Theory, and Complex Multiplication
*CURTIS and REINER—Representation Theory of Finite Groups and Associative Algebras
*CURTIS and REINER—Methods of Representation Theory: With Applications to Finite Groups and Orders, Volume I
CURTIS and REINER—Methods of Representation Theory: With Applications to Finite Groups and Orders, Volume II
DINCULEANU—Vector Integration and Stochastic Integration in Banach Spaces
*DUNFORD and SCHWARTZ—Linear Operators
Part 1—General Theory
Part 2—Spectral Theory, Self Adjoint Operators in Hilbert Space
Part 3—Spectral Operators
FARINA and RINALDI—Positive Linear Systems: Theory and Applications
FOLLAND—Real Analysis: Modern Techniques and Their Applications
FRÖLICHER and KRIEGL—Linear Spaces and Differentiation Theory
GARDINER—Teichmüller Theory and Quadratic Differentials
GREENE and KRANTZ—Function Theory of One Complex Variable
*GRIFFITHS and HARRIS—Principles of Algebraic Geometry
GRILLET—Algebra
GROVE—Groups and Characters
GUSTAFSSON, KREISS and OLIGER—Time Dependent Problems and Difference Methods

HANNA and ROWLAND—Fourier Series, Transforms, and Boundary Value Problems, Second Edition
*HENRICI—Applied and Computational Complex Analysis
Volume 1, Power Series—Integration—Conformal Mapping—Location of Zeros
Volume 2, Special Functions—Integral Transforms—Asymptotics—Continued Fractions
Volume 3, Discrete Fourier Analysis, Cauchy Integrals, Construction of Conformal Maps, Univalent Functions
*HILTON and WU—A Course in Modern Algebra
*HOCHSTADT—Integral Equations
JOST—Two-Dimensional Geometric Variational Procedures
KHAMSI and KIRK—An Introduction to Metric Spaces and Fixed Point Theory
*KOBAYASHI and NOMIZU—Foundations of Differential Geometry, Volume I
*KOBAYASHI and NOMIZU—Foundations of Differential Geometry, Volume II
KOSHY—Fibonacci and Lucas Numbers with Applications
LAX—Linear Algebra
LOGAN—An Introduction to Nonlinear Partial Differential Equations
McCONNELL and ROBSON—Noncommutative Noetherian Rings
MORRISON—Functional Analysis: An Introduction to Banach Space Theory
NAYFEH—Perturbation Methods
NAYFEH and MOOK—Nonlinear Oscillations
PANDEY—The Hilbert Transform of Schwartz Distributions and Applications
PETKOV—Geometry of Reflecting Rays and Inverse Spectral Problems
*PRENTER—Splines and Variational Methods
RAO—Measure Theory and Integration
RASSIAS and SIMSA—Finite Sums Decompositions in Mathematical Analysis
RENELT—Elliptic Systems and Quasiconformal Mappings
RIVLIN—Chebyshev Polynomials: From Approximation Theory to Algebra and Number Theory, Second Edition
ROCKAFELLAR—Network Flows and Monotropic Optimization
ROITMAN—Introduction to Modern Set Theory
*RUDIN—Fourier Analysis on Groups
SENDOV—The Averaged Moduli of Smoothness: Applications in Numerical Methods and Approximations
SENDOV and POPOV—The Averaged Moduli of Smoothness
*SIEGEL—Topics in Complex Function Theory
Volume l—Elliptic Functions and Uniformization Theory
Volume 2—Automorphic Functions and Abelian Integrals
Volume 3—Abelian Functions and Modular Functions of Several Variables
SMITH and ROMANOWSKA—Post-Modern Algebra
STAKGOLD—Green's Functions and Boundary Value Problems, Second Editon
*STOKER—Differential Geometry
*STOKER—Nonlinear Vibrations in Mechanical and Electrical Systems
*STOKER—Water Waves: The Mathematical Theory with Applications
WESSELING—An Introduction to Multigrid Methods
†WHITHAM—Linear and Nonlinear Waves
†ZAUDERER—Partial Differential Equations of Applied Mathematics, Second Edition

*Now available in a lower priced paperback edition in the Wiley Classics Library.
†Now available in paperback.